T0179557

Conversations on Quantum Gravity

The holy grail of theoretical physics is to find the theory of everything that combines all the forces of nature, including gravity. This book addresses the question: How far are we from such discovery? Over the last few decades, multiple roads to finding a quantum theory of gravity have been proposed but no obvious description of nature has emerged in this domain. What is to be made of this situation? This volume probes the state of the art in this daunting quest of theoretical physics by collecting critical interviews with nearly forty leading theorists in this field. These broad-ranging conversations give important insights and candid opinions on the various approaches to quantum gravity, including string theory, loop quantum gravity, causal set theory and asymptotic safety. This unique, readable overview provides a gateway into cutting-edge research for students and others who wish to engage with the open problem of quantum gravity.

JÁCOME (JAY) ARMAS is an assistant professor at the Institute of Physics and coordinator of the Dutch Institute for Emergent Phenomena, University of Amsterdam. He completed his PhD at the Niels Bohr Institute and held postdoctoral positions at the University of Bern and Université Libre de Bruxelles. His research interests span foundational issues in string theory, hydrodynamics and black holes, as well as emergent phenomena in quantum and soft matter. In addition to his research activities, he established the international science outreach platform and event series *Science & Cocktails*. This volume, compiled over more than ten years of mainly in-person interviews, is his gift to the physics community.

Conversations on Quantum Gravity

Edited by

JÁCOME (JAY) ARMAS
University of Amsterdam

CAMBRIDGE
UNIVERSITY PRESS

University Printing House, Cambridge CB2 8BS, United Kingdom

One Liberty Plaza, 20th Floor, New York, NY 10006, USA

477 Williamstown Road, Port Melbourne, VIC 3207, Australia

314–321, 3rd Floor, Plot 3, Splendor Forum, Jasola District Centre, New Delhi – 110025, India

103 Penang Road, #05–06/07, Visioncrest Commercial, Singapore 238467

Cambridge University Press is part of the University of Cambridge.

It furthers the University's mission by disseminating knowledge in the pursuit of education, learning, and research at the highest international levels of excellence.

www.cambridge.org
Information on this title: www.cambridge.org/9781107168879
DOI: 10.1017/9781316717639

First published 2021

Printed in the United Kingdom by TJ Books Limited, Padstow Cornwall

A catalogue record for this publication is available from the British Library.

ISBN 978-1-107-16887-9 Hardback

This book is dedicated to:

Signe, Bertrand and Marguerite
My family
My family in Fabrikken, Christiania
My family-in-law
My closest friends
My closest colleagues

Contents

Preface: Why This Book and Why Did It Take So Long?

In 2010 I was in the second year of my PhD at the Niels Bohr Institute in Copenhagen, spending most of my evenings working at, and co-managing, a cocktail bar and art gallery. In between organising events and washing dishes, I relaxed by reading popular science books on neuroscience, which inspired the book I was writing at the time, *Conjunto Homem* (or *The Set of All Men*), published much later in 2014. In that period of time, I came across *Conversations on Consciousness*, a book by Susan Blackmore that left a deep impression on me. The book consisted of a series of interviews of leading neuroscientists about their views on the still unresolved problem of consciousness. It offered an overview of multiple theories about the mechanisms underlying the phenomenon of consciousness, without making any judgement on whether one theory was more convincing than any other. I was struck by the fact that someone with only a popular science background in neuroscience like me was able to follow these explanations as if looking through an open window to an ongoing scientific endeavour.

I clearly remember thinking, perhaps naively, that a similar book focused on the open problem of finding a theory of quantum gravity could also be made available to anyone trying to make sense of this problem. Not long after, I would embark on an almost 10-year journey to complete *Conversations on Quantum Gravity* – the book you are reading now.

By the middle of 2010 I had closed the bar and moved to Copenhagen's renowned hippie commune Christiania, where I would co-run a cinema for several years. These new personal circumstances provided me with a safe and supporting social environment and an activist attitude that led me – realising that no one in my field would ever write this book, and upholding some idealised sense of duty – to take on the task myself. It was only in the beginning of 2011 that I voiced this idea, discussing it first with Shiraz Minwalla, one of the interviewees and the host of my long-term visit to the Tata Institute for Fundamental Physics in Mumbai. His encouraging words, "Thank you for doing this," together with the support of my PhD supervisor Niels Obers, were sufficient, and so this book began in India.

My original hope was to complete it within two years. This obviously did not happen. Until the very end, I insisted on conducting the interviews in person, leading to a lot of travelling. This was fun but occasionally bewildering, for instance, when I felt as if I was walking on a desert in Texas trying to catch a bus to interview Steven Weinberg.

Additionally, it occurred to me only much later the possibility of a third party transcribing the interviews. Performing such transcription work in the beginning was as time-consuming as demotivating.

But the most important cause for the longevity of this process was, in fact, the attitude of the high-energy physics community towards projects of this kind. As one of the interviewees put it at the end of his interview, "This book may be good for the public and it may be good for the physics community but in the end what matters is the papers that you publish." Indeed, the standard by which one's career in a precarious academic job market is measured, according to the scientific community and the funding bodies, is almost purely based on scientific excellence.

A more wholistic perspective that views the scientist as an agent within the community and in the public sphere is lacking. Additional *duties* or *services* to the community and the public are privileges that one may decide to accept when holding comfortable long-term academic positions. This was not my case and, my Christiania-style non-conformism aside, I did focus on writing papers in order to find the next best job.

There is always space in life for hobbies, and I could have taken this book as one. Instead, I focused on another duty toward society and the scientific community for the past 11 years, namely, that of establishing an international science outreach platform and event series *Science & Cocktails*.[1] The immediate impact of *Science & Cocktails* in people's lives brought much richness to my life, while pushing the book to the sidelines for a time. It was only because of many motivational discussions with friends, family, colleagues and acquaintances, whom I cannot thank enough, that I kept pursuing the book project over the years. In fact, in 2013 I had the pleasure to host Susan Blackmore at *Science & Cocktails*, who was surprised to know that I was inspired by her book and in the process of conducting interviews on quantum gravity. She offered me one of her books with the written dedication "Keep asking those questions!" And I did.

The amount of time that I spent on completing the book over the past years is uncountable. However, though quite painful at times, I would do it all over again, as I have strong faith in its value for the physics community as a whole. Depending on the feedback that I receive from this community, I could be motivated to compile a second volume, since there are at least 37 other scientists whose views and opinions on this subject I would find useful, interesting and unique and which, due to limited time and space, I did not get the chance to record.

[1] http://scienceandcocktails.org.

Introduction

Some books result from a breakthrough. This book results from an impasse: it would have not been conceived if there were overwhelming evidence – or agreement within the broader physics community – pinpointing a given theory, framework or model of quantum gravity. No obviously correct description of nature in this domain has emerged. What is to be made of this situation?

I emphasise the involvement of the broader physics community, in addition to those specialists dedicating their lives to the problem of quantum gravity, in order to keep the larger picture in view. After all, the signatures of gravitational effects in simple quantum systems have to be observed in laboratories, and the signatures of quantum effects in the large-scale structure of the universe have to be observed in the sky. Indeed, whatever theory of quantum gravity one devises, gravity must combine with all the other forces of nature in a way that is consistent with all known observations made at any scale between the Planck length and the entire observable universe, a total difference of roughly 61 orders of magnitude. The whole physical picture must hang together.

Ultimately, a theory of quantum gravity must be able to deliver such an all-encompassing framework that also incorporates the interaction between gravity and matter. Not surprisingly, the problem of quantum gravity is considered by many to be intrinsically intertwined with the problem of unification of all forces.

Any examination into this territory must recognise that even the basic concepts are up for debate. Different people have different notions of what is meant by a *theory of quantum gravity*. Perhaps it is not yet clear whether there should be a theory of quantum gravity at all.

Take the notion of *theory* itself. Is a theory a mathematical structure with an associated *set of fundamental principles* capable of specifying dynamical equations that describe the universe as well as predict the values of all constants of nature? Or do we understand it as a *framework* within which we can construct multiple *models*, one of which – say, the standard model of particle physics – will describe the universe? That is, are there many (billions of?) consistent cosmological histories, which we must narrow down (at least) to those compatible with human existence? Or, finally, do we understand a theory as a *method* that we can use, for instance, to quantise a given classical theory of gravity?

Whether we view a theory as a set of fundamental principles, framework, model, method or more generally as an *approach*,[1] there is no table-top experiment or single cosmological observation that decisively confirms that gravity is a quantum field. The problem of finding a theory of quantum gravity partly rests on the lack of experimental set-ups able to probe deviations from standard (established) physics, as well as on the lack of precision within those set-ups.

Could gravity just be an effectively classical phenomenon? An entropic force caused by some underlying unknown microscopic constituents? Is spacetime itself an emergent structure at long distances and low energies? Questions such as these represent part of the debate around the quest for a theory of quantum gravity. The purpose of this introduction is to try to outline this debate – the set of open questions, related problems and different points of view – as well as to lay out and clarify the justified and unjustified criticisms directed towards those points of view.

The method of the remainder of the book is to present interviews with a select group of theoretical physicists and mathematicians who work on quantum gravity (and related subjects) about the quantum gravity debate, the state of the field, the impact of their own discoveries and the discoveries of others, their motivations to pursue these questions, and the benefit such research can bring to the scientific community and society as a whole.

Originally, I intended to provide a faithful and representative overview of the approaches in this area. This method would have led to an overwhelming majority of interviews within string theory, however. (A rough analysis of papers appearing between 2018 and 2020 on arXiv.org, either related to a given approach to quantum gravity or inspired by it, suggests that 85–90% of this research output is due to string theory.) Although my background is in string theory and I spend a considerable amount of time working in this area – as well as applying some of its lessons and tools to the contexts of hydrodynamics, astrophysical black holes, quantum matter, biomembranes and soft matter – I see value in exploring other approaches. Thus, I have chosen to give more room to a diversity of ideas at the expense of going into much depth in certain areas of string theory.

Nevertheless, I want to express clearly and respectfully that I do not think that the opinions presented here on this subject are necessarily more valuable than those of many of the theorists I did not interview, including the many string theorists who were excluded.

Moreover, I emphasise that the interviewees' positions and attitudes are not always widely shared across communities and do not always constitute the opinion of the majority. For instance, many think that the theories that they work on are not final theories but rather ways of providing clues about, and insights into, more fundamental formulations – an opinion not always shared by others in their field.

In my selection of interviewees, I focused mostly on senior theorists who were responsible for much foundational work. I also opted for the diversity of ideas and opinions within each of the approaches, including several theorists who may not brand themselves

[1] I use the word *approach* to refer broadly to any notion of theory (set of principles, framework, model, method or any other possibility).

as proponents of any (theoretical) approach but who, I believe, have unique and interesting points of view on aspects of quantum gravity. Some approaches to quantum gravity are not directly covered in this book due to lack of space and time, but those approaches constitute smaller, or equally small, research programmes compared with the approaches that are directly covered here. I believe the editorial choices made here faithfully represent the debate and open up an array of useful questions.

Given the inherent nature of the subject matter, most of the discussions in the interviews are meant for specialists or experts, although some are suitable for readers with a general physics education and some are appropriate for those with only a popular science background.

The remainder of this introduction is concerned with summarising key aspects of the debate about quantum gravity and the issues motivating my choice of questions.

Since the advent of quantum mechanics and general relativity in the early 1900s, theorists have wondered about the quantum nature of gravity and the effect of gravity on quantum mechanics. In fact, how these two pillars of physics (quantum mechanics and general relativity) work together in theory and in practice has been the source of heated debate from their introduction, in particular between Niels Bohr and Albert Einstein. Already at that time, it became clear that the successful combination of the two theories, even if just by means of thought experiments, was crucial for the development of each theory, including understanding them as well-defined and consistent parts of physics. In the decades to follow, many theorists sought to combine quantum mechanics with special relativity, culminating in the framework of quantum field theory. This framework led to a stream of striking accomplishments, starting with the understanding of the electromagnetic and weak interactions and ending with the strong interactions, ultimately leading to the standard model of particle physics.

At different stages in this process, several theorists such as Richard Feynman and, in particular, Bryce DeWitt in the 1970s took the newly developed toolkit of quantum field theory (for instance, perturbation theory, renormalisation techniques and the gauge principle) and applied it to gravitational interactions. It became clear at that point that gravity, treating the spacetime metric as a fundamental field, even at the perturbative level, is not akin to all other interactions for which such methods and techniques were ideal. In particular, it turned out that gravity (without any other additional source of matter) is perturbatively non-renormalisable. This implies that as we take further and further perturbative corrections to graviton interaction processes, an infinite number of counterterms with an associated infinite number of free parameters need to be considered in order to obtain finite results. Gravity at increasingly high energies thus becomes, inevitably, non-predictive, and physics, as we know it, is lost.

Despite the lack of experimental confirmation, the opinion of the overwhelming majority of researchers, substantiated by multiple arguments that appeal to consistency, such as causality or puzzles related to the existence of black holes (for instance, singularities, information loss and thermodynamic properties), is that gravity must be inherently quantum, at least in a perturbative regime. This well-founded expectation has led to several ideas and suggestions about how to overcome the problem of non-renormalisability, all of which have been pursued to a certain (variable) extent, for instance:

- **New symmetries.** Gravity, as it stands, can be useful below a given energy scale but something else must complete gravity in the ultraviolet (high-energy) regime that renders it finite. This can entail that another mathematical language, framework or approach substitutes (or completes) gravitational interactions in the ultraviolet. Examples often involve new symmetry principles (for instance, conformal symmetry in various shades [worldsheet, spacetime, local, global, etc.], supersymmetry, E_{10} symmetry, non-commutative geometry, spacetime anisotropy and many others) or some topological completion.
- **Additional matter fields.** Even if gravity by itself is not predictive, it may become predictive when coupled to specific matter content (for instance, axions, scalars, supersymmetric particles and conformal matter, among others), which can contain a finite or infinite reservoir of particles or quantum states at high energies.
- **Other fundamental variables.** Taking the metric field as the fundamental field may not be the correct modus operandi. There are other formulations of gravity (for instance, Plebanski, Regge and higher-gauge formulations) and other potential fundamental variables (for instance, vielbeins and Ashtekar variables) or more general notions of geometry (for instance, non-commutative geometry) which, when appropriately quantised, may not lead to predictive physics in one way or another.
- **Modified infrared dynamics.** Although general relativity appears to be the correct description in the infrared (low-energy) regime, there exist various modifications of general relativity at low energies that are within current bounds set by observations (for instance, versions of higher-derivative theories and versions of modified gravity theories). Thus, perhaps the starting point of the quantisation procedure should be one of these infrared modifications and not conventional general relativity.
- **Non-perturbative renormalisability.** Gravity is perturbatively non-renormalisable, but it may be non-perturbatively renormalisable. Therefore, the formulation of the theory should be non-perturbative from the very beginning. Concrete ideas include the "magical" cancellation of divergences of all physical processes when summing over all perturbative orders; the theory suggests a minimal length scale in the configuration space which the path integral is integrating over; or there is an ultraviolet non-Gaussian fixed point characterised by a finite number of coefficients that determines all other coefficients arising from the process of renormalisation.
- **Spacetime discreteness.** It is a mistake to assume that the gravitational metric, and hence spacetime, is a continuum. Instead, spacetime is inherently discrete and thus a cut-off (of the order of the Planck length) is introduced in the theory by hand (for instance, discrete Regge calculus). A continuum spacetime should emerge at long distances. The discrete nature of spacetime may cure the ultraviolet and short-distance divergences.
- **Modifications of quantum mechanics.** The rules of quantum mechanics and the rules of general relativity may not be compatible with each other. A new mathematical framework is needed to modify the principles of quantum mechanics so as to accommodate

both. Examples include modifications of quantum mechanics (for instance, deterministic versions) and different interpretations (for instance, hidden variable theories and Roger Penrose's gravitational objective reduction).

- **New formulations of quantum field theory.** The same type of logic as in the previous point applies to quantum field theory itself. Perhaps there are new formulations of quantum field theory (QFT) or gauge theory (for instance, axiomatic QFT, algebraic QFT and non-geometric formulations) that will more easily accommodate dynamical, and arbitrary, background geometries.
- **Dualities.** Gravity should not be directly quantised. Instead, it should emerge from some other (finite) fundamental theory. This is the case, for instance, when there are (holographic) dualities between theories without gravity and theories with gravity, implying that the very issue of non-renormalisability can be circumvented (for instance, the holographic ideas of Gerard 't Hooft and Leonard Susskind).

It should be said, first of all, that this list does not exhaust all possibilities for how to deal with some of the problems of quantum gravity and, second, that each of these ideas is not necessarily independent. Concrete proposals approaching the problem may in fact involve a combination of such mechanisms and ideas. I find that a rough classification of approaches is more useful in terms of what aspects of the above list they incorporate (or aim at incorporating) than standard classifications in terms of the type of quantisation procedure (for instance, canonical, covariant and path integral) and whether they are discrete, lattice-type approaches, or continuum. The reason for this preference is that there are several approaches where it is not clear exactly what they contain and some which can include several of these properties (for example, multiple quantisation procedures or lattice and continuum methods).

Based on limited knowledge, it is still useful to attempt to characterise the approaches directly covered within this book with the goal of aiding the reader in understanding some of the interviews. These approaches are as follows.

String theory. Introduces new symmetries such as supersymmetry; an infinite tower of massive particles which in standard versions amounts to the existence of extra dimensions; non-perturbative formulations; new geometric structures and variables; new QFT formulations and dualities such as the anti–de Sitter/conformal field theory (AdS/CFT) correspondence. As a matter of fact, it may include specific realisations of all the ideas in the above list in particular regimes, though not all these realisations are known at this point. Corners of string theory are formulated via a path integral (or covariant) quantisation of a two-dimensional worldsheet theory. It is incorrect, however, to claim that this is the general definition of the theory. Gravity is not quantised directly but emerges from the theory via interactions of closed strings at low energies. Usually included in string theory are other ideas such as E_{10} symmetry and twistor theory, though some think these ideas stand on their own. The former relates to maximal supersymmetric and supergravity theories but the exact theory that exhibits such symmetry is not yet known; it includes matter fields and may provide the required ultraviolet completion.

Loop quantum gravity (LQG). Introduces new variables for dealing with gravity directly; is based on a non-perturbative formulation that suggests a minimal length scale in configuration space; provides examples of new QFTs, in particular topological QFTs with defects. The standard formulation is based on canonical quantisation of general relativity theory in four spacetime dimensions, which has two versions (one which requires the addition of matter) directly formulated in the continuum. Spin foam models are usually considered to be part of LQG, though the relationship between the two in four dimensions has not been fully established. Many view spin foam models as a path integral lattice approach within which a continuum limit (critical point) needs to be found after sending the lattice spacing to zero. In that limit, it is expected that some form of canonical LQG will be found. Others, however, view spin foam models as an inherently discrete approach.

Asymptotic safety. Also known as quantum Einstein gravity (QEG), requires the existence of an ultraviolet fixed point; uses renormalisation group techniques and can generate infrared modifications of gravity. In its most explored formulation, it takes the metric as the carrier field and a class of effective action functionals (truncations) that depends on diffeomorphism-invariant operators. It employs a continuum description. In principle, it can consider other action functionals, other carrier fields, other symmetries and matter fields but does not seem to require it.

Causal dynamical triangulations. A lattice approach which, in its usual formulation, takes general relativity with an additional positive cosmological constant as the bare action in the path integral; introduces new variables (piecewise flat manifolds) and a set of gluing rules that constrain configuration space in the path integral; its ultraviolet finiteness also rests on the possibility of finding an ultraviolet critical point and a continuum limit; it may generate infrared modifications.

Causal sets. Introduces a discrete length scale and takes causality (understood as time order) as a fundamental principle. Classically, it is based on a path integral whose bare action reduces to classical general relativity in some appropriate limit. It is not yet clear what the quantum theory is. It has suggested the study of some nonlocal QFTs and the need for a different interpretation of quantum mechanics.

Non-commutative geometry approach. Generalises the notion of Riemannian geometry to non-commutative geometry in terms of a non-commutative algebra, a Dirac operator and a Hilbert space; gravity, as well as matter, appears to emerge from it using (mostly) the algebra as an input. The theory is at first quantised level, and it is not yet clear how to perform second quantisation and address issues of renormalisability.

Hořava-Lifshitz gravity. Introduces spacetime anisotropy as a symmetry (that is, a Lifshitz-type symmetry in the ultraviolet that reduces to Lorentz symmetry in the infrared); there are several versions of the theory, some of which include additional matter fields;

it can provide infrared modifications of gravity depending on how the coupling constants flow toward the infrared. There are some indications that the theory is renormalisable. It is not yet known exactly how to quantise the theory.

Other approaches. Other ideas include emergent gravity, which takes the holographic principle seriously and suggests different interpretations of infrared physics. In contrast, conformal gravity considers canonical gravity with conformal constraints while crystalline gravity introduces matter (defects) with a finite number of degrees of freedom that can be glued as to form spacetime. These latter three approaches remain, up to date, at the classical level.

Two- and three-dimensional models. Several theorists have worked on two- and three-dimensional toy models of quantum gravity within which it is possible to apply many quantisation procedures and study their effects. These models do not contain propagating degrees of freedom and so cannot describe the real world. However, many think that these models can give useful insights for tackling the four-dimensional (or higher) case. As such, these models are discussed throughout the interviews.

What do we expect from a theory of quantum gravity? A crude list of demands and expectations, which different theorists value in different degrees, includes:

- Ultraviolet and infrared finiteness
- A controlled classical limit
- Internal consistency and compatibility with standard physics
- Unification of all interactions
- A perturbative expansion around given backgrounds
- A non-perturbative definition of the theory
- A first-principle derivation of the dynamics
- A background-independent formulation (potentially with suitable boundary conditions)
- A solution space that includes our universe
- An understanding of the underlying fundamental degrees of freedom
- Well-defined physical observables
- The ability to perform computations, say, of the S-matrix
- A finite number of external parameters as input
- Suggestions of phenomenology[2] and dialogue with experiment
- Potential experimental tests and compatibility with observations
- Insights into neighbouring fields
- A microscopic understanding of black hole entropy
- A solution to the black hole information loss paradox
- A resolution of the big bang singularity and black hole singularities
- The theory incorporates the holographic principle
- The theory is outside the swampland.

[2] The word *phenomenology* refers to predictions, features or mechanisms that can potentially be used to interpret real world data.

The failure to meet these expectations leads to potential criticisms. I turned these demands and expectations into questions directed at the interviewees:

- **Effective versus fundamental.** Is the theory an effective theory or a fundamental theory in certain cases or regimes? Is the theory pointing towards structures and principles that should then be used to construct the underlying theory?
- **Perturbative definition and the existence of our universe.** Is the theory perturbatively defined around any given background? Does the theory contain some background that describes our world? How many viable solutions are there?
- **Ability to compute.** Regardless of the properties the theory has, what can we actually compute? Scattering processes? The size of the universe? The shape of a star? The spectrum of some operator?
- **Ability to predict.** How many parameters need to be fixed by external input for the theory to be predictive? A few? Many? An infinite number?
- **Phenomenology.** Does the theory provide suggestive or concrete examples of phenomenology (for instance, particle physics models and cosmological models)?

These types of questions can be asked of any of the approaches to quantum gravity, and I did ask them in all cases in which the theory was sufficiently developed to give a proper answer.

There are also more specific criticisms of each of the approaches as well as specific questions that can highlight both their shortcomings and their advantages. Below are a few examples of such specific questions that I posed to the interviewees:

- **Questions about string theory.** What is string theory, that is, how do you define it? Is there any hope of finding a controlled cosmological solution that describes our universe? Is there hope of finding a non-perturbative definition of string theory in cosmological backgrounds? Is our universe governed by one of the supersymmetric string theories in ten dimensions or has the string theory that describes our universe not yet been found? Is it reasonable to think that anthropic principles should fix the vacuum we live in? What evidence is there for swampland conjectures? Is there any experiment that will verify string theory? Is there actually an ongoing dialogue between string theory and experiment? How well has string theory reproduced the Bekenstein-Hawking entropy of black holes from a microscopic calculation? Has it been proven that string theory is ultraviolet and infrared finite? If we do live in ten dimensions (measured abstractly in terms of field content), why are four of them large and six of them small? Are there holographic dualities for flat spacetime and cosmological spacetime? If the world is supersymmetric but we do not see superparticles at current accelerator scales, how and why did supersymmetry break?
- **Questions about the AdS/CFT correspondence.** Is AdS/CFT a non-perturbative and background-independent formulation? Isn't it just a conjecture or is there substantial proof? How well defined is the dictionary between bulk and boundary theories? Can we recover all bulk physics from the boundary theory? Is it a case of holography or

photography? What is the quantum geometry in the bulk? Can you solve the black hole information paradox? What does an observer who falls into the black hole experience? What does physics look like at the Planck scale in the bulk? What happens if you scatter particles in the bulk at Planckian energies? Has it led to any predictions about strongly coupled systems? Are there concrete observables? Can you extract flat spacetime physics from it? Does it provide an example of emergent space and emergent time?

- **Questions about LQG.** Does LQG have a classical limit? Is the dynamics of LQG derived from first principles? LQG in the Hamiltonian formalism has two different versions (one which does not require matter and another that does), so are they equivalent? How can you formulate the theory directly in the continuum? Since LQG has a very large (infinite?) number of free parameters due to quantisation ambiguities, how can this be fixed in order for the theory to be predictive? Has the Hamiltonian constraint been solved? Is the standard Hamiltonian constraint correct? Has it been shown that LQG is generally covariant? How does LQG resolve the ultraviolet and infrared divergences? Is there any test of LQG? What are the LQG models for black hole entropy and why are the logarithmic corrections to the entropy in disagreement with semiclassical analyses? How faithful are the results and predictions of loop quantum cosmology? How does it resolve the big bang singularity?
- **Questions about spin foam models.** Spin foam models are portrayed as being part of LQG, but has this connection been formally established? Are spin foam models lattice approaches? Has a continuum limit been shown to exist? If there is a continuum limit, will it lead to the same theory as that of standard canonical LQG? What can you compute with it? How do you resolve quantisation ambiguities? Does it rely on asymptotic safety?
- **Questions about QEG.** Is there evidence that QEG has ultraviolet fixed points? What is the dimensionality of the ultraviolet critical surface? What would happen if you added more higher-curvature corrections? Is it possible to add any kind of matter and still find ultraviolet fixed points? Is there any suggestive phenomenology for particle physics and cosmological models? What is the infrared theory? And what is the picture that we find in the ultraviolet, that is, what is the theory at the Planck scale? Can we approach the renormalisation group flow trajectory near the Planck scale? Could you take different symmetries to begin with and still find fixed points? What is the relation between QEG and LQG and CDT?
- **Questions about CDT.** Is CDT generally covariant? Will the preferred foliation introduced in CDT lead to symmetry violations in the continuum limit? Has the existence of a continuum limit been demonstrated? How reliable are the numerical simulations near the fixed points? Is the continuum limit closer to general relativity or to Hořava-Lifshitz gravity? How big are the universes that can be simulated in a computer? Why is the starting action just the Einstein-Hilbert action with a positive cosmological constant? How does CDT justify the topological restrictions that are imposed in the path integral? Can CDT include matter, black holes, etc.? What can be computed with CDT? What is the theory at the Planck scale?

- **Questions about causal sets.** How are causal sets defined? Does every spacetime have an associated causal set? How to determine which spacetime corresponds to a given causal set? What is the classical causal set theory? What evidence do we have for spacetime discreteness? How to grow a causal set? Is there any hope of finding the quantum theory? Have causal sets been used to predict the smallness of the cosmological constant? How can it make such predictions without being a quantum theory? Are there suggestions for phenomenology?
- **Questions about a non-commutative geometry approach.** How can we intuitively understand non-commutative geometry? Why should this be a better starting point for physics than Riemann geometry? What kind of input was needed in order to get the standard model coupled to gravity? Can you predict the magnitude of all masses and the strength of all interactions? How far are we from a quantum theory? What mathematical structures are needed in order to quantise the theory?
- **Questions about Hořava-Lifshitz gravity.** Has it been proven that Hořava-Lifshitz gravity is a renormalisable theory? How should one quantise the theory? Does it predict infrared modifications of general relativity compatible with observations? What is the concept of a black hole within Hořava-Lifshitz gravity? Should we understand it as being embedded in a larger theory such as string theory? What kind of phenomenology does it predict? What are the consequences of coupling the theory to matter?

The interviews also touch upon many other subjects, including interpretations of quantum mechanics, multiple cosmological models, particle physics models, the role of the QCD axion, the role of anyons, renormalisation, conformal symmetry breaking, black holes, two- and three-dimensional quantum gravity toy models, possible quantum gravity experiments, the large hadron collider (LHC), crystalline gravity, emergent gravity, E_{10} theories, conformal gravity, twistor theory, fluid dynamics, skyrmions, gauge theories and thermalisation processes. Indeed, there is no shortage of vital questions in physics.

By presenting these interviews I intend to address the concerns, doubts, misrepresentations and the misinformation in this critical area of study with the hope that it will lead to clarifications, dialogue and progress in the future. Overall my sense is that the many theoretical developments relating to quantum gravity discussed within these pages are as fascinating as they are in need of experimental testing.

The Interviews

The interviews that follow took place between 2011 and 2020 and are sorted in alphabetical order with regard to the interviewees' last name. The time that each interviewee committed to the interview as well as the length of the interviewee's replies determined the duration and size of each interview. Of these interviews, seven were conducted online, one via email and the remaining in person.

The interviews were transcribed by a third party but I later edited them myself, keeping the same writing style throughout. The interviewees had the opportunity to review and modify the edited transcripts as they pleased which, in many cases, contained additional questions that I added a posteriori over the years and to which the interviewees provided answers. Some of the interviewees spent a considerable amount of time reviewing and editing their transcripts, while some performed only minimal edits or none.

Given the long time span between the date of some of the interviews and the publishing date of this book, some interviewees decided to update part of their answers to better reflect their current understanding of the topics discussed, while others decided to keep their original answers. Overall, most interviewees felt that their answers did not be updated, as their views had not significantly changed over time.

To many of the statements made by the interviewees and to my own questions, I added multiple references. The purpose of these references is to provide extra details that substantiate such statements or to motivate the choice of specific questions. Again, some interviewees reviewed these references carefully, while others did not. The choice of references focuses on the work of each interviewee and does not have the aim of summarising entire research fields. I emphasise that this book is not meant to be a review but instead an overview of opinions by hand-picked theoreticians about their views on different aspects of quantum gravity.

I am the author of all footnotes included in the interviews. These footnotes contain links to relevant material mentioned during the interviews or minor clarifications to some of the statements.

1

Jan Ambjørn

Professor at the Niels Bohr Institute, Copenhagen, Denmark, and at the Institute of Mathematics, Astrophysics and Particle Physics (IMAPP), Radboud University

Date: 13 April 2011. Location: Copenhagen. Last edit: 26 May 2014

In your opinion, what are the main puzzles in high-energy physics at the moment?

The main puzzle is to understand the ultraviolet sector of the world and that includes the ultraviolet sector of quantum field theory and also the ultraviolet sector of gravity, since gravity is non-renormalisable.

Does this mean that you need to find a theory that describes quantum gravity?

Not necessarily. According to our current understanding, only non-Abelian theories with fermions seem to have a well-defined ultraviolet completion. The scalar sector of quantum field theories has a Landau pole in general, which is also the case in the standard model. Assuming the theory to be supersymmetric can facilitate finding a completion but we don't know if supersymmetry is a symmetry of nature or not.

Your recent research has been focused on quantum gravity. Why is it so hard to establish a theory of quantum gravity?

Well ... because gravity is non-renormalisable and we don't have a firmly established framework to think about such theories as quantum field theories. It is possible to tackle gravity as a non-perturbative quantum field theory but we don't have many examples of non-perturbative quantum field theories to guide us, which makes it difficult.

Do you think that string theory solves the problem of quantum gravity?

String theory is an attempt to describe a theory of everything but the only problem with it is that, so far, it has made very little contact with the real world. From this point of view, it is hard to call it a *theory of quantum gravity*. The best understood part of string theory is the case in which you take a string moving in 10-dimensional flat spacetime and quantise it. In this situation, you observe that certain vibrations of the string correspond to spin-two particles that you refer to as gravitons. But how do you get from that description to our world? That is unclear, hence you cannot claim that string theory explains why we seemingly live in a four-dimensional world in which we have a theory of gravity (laughs).

I have nothing against string theory but it is completely unclear whether it has anything to do with the real world or not.

Do you think that, even though it might not have anything to do with real world, it can be used to solve other problems of physics?

It's a very well-known strategy to use a given theory to study other problems than those the theory was meant for. For instance, the theory of epicycles was used to understand the kinematic motion of the planets around the sun besides the motion of anything around the earth, which was its original purpose, and, in fact, the theory of epicycles gives you a perfect valid kinematical description of such systems. From this point of view, string theory can also work as a tool and I have nothing against that.

(laughs) Ok, so you don't believe that it's the right theory, if I can put it this way?

But ... what do you mean by "It's the right theory" (laughs)?

You could have the feeling that string theory, even though it is not clear right now how to make contact with experiment, is the right theory ...

I see little experimental evidence that it is the right theory, in fact, I see none and that is the problem. This is the reason why someone has to try to think in alternative ways. However, I'm not saying that string theory couldn't provide, in the end, some explanation of the world but I must say that it seems that no one is interested in pursuing this direction within string theory anymore. Perhaps only a few string theorists work on trying to use string theory to describe the world, as in starting with some action and showing that the world is a consequence of string theory. Most string theorists have branched off and started using string theory like a theory of epicycles, that is, as a tool to describe the kinematics of something else.

I assume that you don't like too much the anthropic principle, which can be used to pick the right vacua of string theory [1]?

Anthropic reasoning doesn't have anything to do with string theory. It was around long before string theory and it's well documented. In principle, you can never falsify it so can you call it *science*? It's not my cup of tea. Clearly, some people cannot figure out how the world is made up and, since they believe that they're so clever, it must be that the world cannot be explained by ordinary science. I'm more modest and I still believe that we have a chance.

One of the criticisms that people often make to string theory is that it is not background-independent while, for example, loop quantum gravity [2, 3] tries to construct a background-independent theory. Do you think that loop quantum gravity is a good candidate for a theory of quantum gravity?

As far as I know, loop quantum gravity still hasn't made contact with the macroscopic world. From this point of view, it is a little bit of a strange theory, whose proponents claim to know everything at the Planck scale but nothing at the macroscopic scale (laughs). It's up to those pursuing loop quantum gravity to provide a proof that it is a theory of gravity. I haven't fully seen it yet but maybe I'm not too well informed.

Do you think string theory can be formulated in a background-independent way?

The way that string theory is usually formulated assumes the existence of a background metric. From a field theory point of view, you could try to formulate a fully fledged string field theory, which would be background-independent. However, very few people work on that at the moment so it might take some time before it's formulated in that way. However, once that is accomplished, it should be background-independent. Any field theory theory is in principle background-independent since you integrate over all fields contained within the theory.

But why is it important to have a background-independent theory?

I don't understand why you ask this question. When you talk about a *background* you mean the *metric*, right? But string theory is supposed to describe the geometry, which is supposed to evolve dynamically. I mean, look at our real world out here! If you start with string theory in 10-dimensional space, how do you end up in this world right now? Clearly, there has to be a dynamical process in which spin-two 10-dimensional objects condense and have expectation values that describe a completely different universe than the one you originally started with. It doesn't make too much sense to have a theory that depends on the very structures that you originally wanted to describe. Thus, it doesn't make sense to me to have a background-dependent theory. Such a theory would be so ugly that I couldn't imagine it would be the right theory.

Do you think that it is actually possible to have a unified theory of everything or that you have to use different theories to describe different aspects of the world?

The latter possibility makes no sense to me. Of course, it's acceptable that something is best described in one language than in another language but the overall description should be unified in only one theory. You can maybe use different mathematical representations of the same theory to approach different problems but that's a different story.

Do you think that we will be able to see any quantum gravity effects in the near future, for instance, at the LHC?

No. I think that there have been ridiculous claims made by certain people that this could be possible.

So if you cannot really test such theories, what are they useful for?

Well ... it's not really ruled out that they can't be tested but I think that any test has to come from cosmological data in one way or another. It is hard to test it at any other scale, in particular by scattering particles. Such cumulative processes could in the end add up to something significant but it would require some very detailed calculations to show that it would be indeed the case. Such scenarios have been supported by some people in a not so far past but I think that it is still preliminary. At the moment, it seems that the best hope of constructing and testing a consistent theory is to apply the theory to our current universe. As such, I do not see such theory as providing us with multiple predictions for the future in such a way that we can live long enough to verify them (laughs).

Since it is hard to test these theories, would it be reasonable for someone to stand up and say, "This is not really science but religion" because it's a matter of belief in a specific theory of quantum gravity?

Clearly, science might come to a limit in which this will be the case because in order to construct an accurate theory of the universe you need to conduct experiments. However, you don't really have the possibility of making many experiments since you just have this one universe. From this point of view, you might have to establish different kinds of standards for what should be called *science* and *scientific predictions*. At the moment, I don't consider this to be a problem. On the other hand, I must say that you will probably always be able to wave your hands in the air and use something like the anthropic principle to explain the workings of the universe. I tend not to refer to that approach as *science*.

Can you tell me what is the story behind the approach to quantum gravity that you pursue? How did it begin?

The approach that I pursue was developed by myself and collaborators, as well as other groups, in the mid-1980s in order to provide a non-perturbative regularisation of Polyakov's string theory [4–11] and it has been extremely successful in describing two-dimensional quantum gravity and non-critical string theory. In the beginning of the 1990s, it became clear that string theory was not going anywhere and it also became unclear whether it would be able to make any predictions. At that point, one had to think whether or not there could be other descriptions of quantum gravity. The generalisation of our approach to higher dimensions, which is related to what is referred to as *asymptotic safety* [12], was possible and could potentially lead to an alternative theory. Thus, we began pursuing it.

What is asymptotic safety?

Asymptotic safety is a kind of generalisation of Wilsonian standard renormalisation group philosophy according to which a quantum theory needs not to be renormalisable in the ordinary sense [12]. By a theory being *renormalisable* one usually means that it has a Gaussian fixed point and hence a small coupling constant but, in principle, one could have a non-trivial fixed point for which this coupling constant is not small. The Wilsonian approach was exactly developed to deal with such situations, for example, the three-dimensional Wilson-Fischer fixed point governs all behaviour of second-order phase transitions in the real world. However, this is an infrared fixed point, contrary to what is needed in gravity, which is an ultraviolet fixed point. Nevertheless, it is possible that it could also apply to ultraviolet fixed points, although it is not as natural as infrared fixed points from the Wilsonian point of view. Taking this as the starting point, it is then natural to have the right lattice formulation to study these fixed points and that's what we have developed.

Okay, but what kind of theory is this? A theory of the universe, a theory of particles, matter, gravity and interactions?

Well . . . it's a theory of quantum geometry and it's a theory of the universe, for sure.

You called this theory *causal dynamical triangulations* (CDT) [13, 14] . . .

CDT is the end result of various attempts of finding a theory that seems to have an interesting phase structure and in which you can properly define a continuum limit. The word *causal* in CDT is due to the fact that we have incorporated a kind of global time in the approach. In more detail, the word *causal* originates from the fact that we triangulate the universe assuming the existence of a proper time which, when cut into small pieces, gives rise to a connected space.

But what are these triangulations or building blocks (if they can be called that) that you assemble together?

They are just what makes up the lattice and have no physical meaning. It's like in any quantum mechanics course you have had when you were an undergraduate. When you perform the path integral you're chopping up your time interval into ϵ pieces. Here it is exactly the same but you have to be a little bit smarter because you are now dealing with geometries and not only with time. However, in the end, you're just chopping up these geometries.

But I thought that these triangulations, which you refer to as *four-simplices*, need to obey certain rules when you assemble them together . . .

Suppose that you have some manifold with a certain topology which you keep fixed, say $\mathbb{R} \times \Sigma$. As in any attempt to quantise gravity canonically, you assume that you have the real line $\mathbb{R}$, which is the topology of the time part of the manifold, and you have a spatial manifold Σ, which is a space with a certain topology. When you glue the building blocks together to create the corresponding geometry you have to do it in such a way that respects this topology that you keep fixed. This requirement implies certain rules for how to assemble the four-simplices.

You have said a few times that the manifold has to be fixed but shouldn't it be dynamical in a full theory of quantum gravity?

Not necessarily. Our approach consists in assuming vacuum canonical quantum gravity where the manifold structure is fixed. Of course, you can have any geometry on that manifold. This is exactly like when you solve Einstein equations for which, usually, the manifold is fixed and you try to find the geometry.

But the fact that you have this time foliation in $\mathbb{R} \times \Sigma$, doesn't it make a background-dependent theory? Clearly, you don't have time and space on an equal footing ...

But time and space are not on an equal footing in any theory of relativity or in any Lorentzian formulation. You always assume to have a certain signature. Even in string theory a certain metric signature is assumed, which should actually be determined dynamically but this is never discussed within string theory. Why is time being inserted in the theory? Similarly, the starting point of our approach also introduces time.

Shouldn't time emerge? Shouldn't the Lorentzian signature emerge from a fundamental theory?

Well ... now you are demanding more than what you demand from string theory. This is not how we developed our approach. We assume that we have a Lorentzian signature as the starting point.

Is CDT diffeomorphism-invariant?

It is, of course, diffeomorphism-invariant. Diffeomorphism invariance has nothing to do with this discussion. At any point in the process, we are always dealing with geometries. I'm not sure if you understand this difference so let me explain. Our approach only deals with geometries, that is, we never use coordinates to describe the system. Diffeomorphism invariance is something that you really don't want to deal with. It is forced upon you in the usual description in which you use coordinates but in the end you want the description to be independent of the coordinates. The diffeomorphism class of all coordinate transformations yields the geometry-invariant so if you work with the geometry from the get-go, everything is diffeomorphism-invariant. The path integral that defines the theory only sums over geometries, which means that there is no need for gauge fixing of any sort.

You are summing over all geometries but at some point you evolve your system and you get a single de Sitter universe [15]. Why do you get just a single geometry in the end?

If you look at our world and assume that it is described by a quantum theory then you are summing over all geometries and nevertheless you have only one geometry here right now. So what is this geometry? It is the expectation value of the sum over all geometries and around that value there are small quantum fluctuations. This is what we mean when we claim that we have a de Sitter universe. When we look at the theory non-perturbatively, the expectation value of the sum over all of these geometries is, seemingly, a de Sitter universe [15].

And this universe also turns out to be four-dimensional? Is the spacetime dimension a parameter that you put in?

We put in four-dimensional building blocks.

So could it happen that you would get more or less dimensions than four in the end?

That depends on what you mean by *dimension*. Look, you're putting nothing in except the requirement that the sum is over all possible four geometries. Of course, these are just configurations in a path integral, which is how you implement the theory, and none of these configurations are physical, in the same way that the path of a particle is not physical in quantum mechanics. It doesn't make any sense in a quantum theory to talk about the path in the same way that it does not make sense to talk about the geometry, as it is not a quantum object. In other words, you have to define physical observables and those are the ones that you can talk about. In the case of quantum mechanics, one of the observables is the expectation value of the sum over all paths satisfying some boundary conditions and, analogously, the sum over all geometries in the case of the universe. After you sum over all geometries you can ask, "Is there a sensible average geometry around which there are well defined quantum fluctuations?" and "Are these quantum fluctuations determined in terms of the coupling constants in the theory?" That leads you to a Euclidean de Sitter universe [15], since we actually have to rotate everything to Euclidean space in order to perform computer simulations.

Can you then rotate back?

In principle you can rotate back but in practice *rotating back* is somewhat of an unclear statement. We can rotate back in a certain way but that requires a bit of a technical explanation.

But if you could not express your results in Lorentzian signature, would you still claim that the theory is sensible?

It's still sensible. It's exactly the same situation that you have in ordinary quantum field theory when you perform computer simulations. All these computer simulations are done in Euclidean quantum field theory. We then need to do some work to understand what these results correspond to in Lorentzian quantum field theory. For example, there are general axioms stating that correlation functions in the Euclidean sector can be rotated back to certain correlations functions in Minkowski space. In our setup, the issue has to do with what can be rotated back in the case of quantum gravity. That's a complicated discussion but in principle observables can be rotated back to the Lorentzian context.

So do you take as a prediction of the theory that the world we live in is a de Sitter universe?

We would like to make that claim. We have a fixed cosmological constant and ...

Does the theory have anything to say about the value of the cosmological constant?

No, that's a free parameter in the theory.

So couldn't there be a more fundamental theory that fixes the value of the cosmological constant?

Okay, so now you're asking if there is a more fundamental theory that fixes the value of the electric charge and so on, right? That's not our theory. Our theory is exactly like standard quantum field theory so it doesn't fix any of the renormalised coupling constants that enter it. Ordinary quantum field theory doesn't fix the mass, the charge, etc.

So would you say that CDT is a kind of formalism in which you come up with a model and experimentally adjust the constants?

That's what it is all about.

Does the starting point contain any matter?

In principle, you could introduce any kind of matter in it but you first have to define some observables. Since the analysis of CDT is performed numerically by means of Monte Carlo simulations, you have to first understand what are the well-defined physical observables that the simulations should measure. In this case, we are looking at a universe and if you put it in into a computer it corresponds to a 10^4 lattice. So, clearly, we are talking about something which is slightly larger than the Planck scale. In fact, we can calculate the Planck scale in terms of lattice units which yields a result of the order of 20 Planck lengths.

That's the total size of the universe?

That's the size of the universe we can fit into a computer.

Is that big or small?

What's your own verdict?

I think it's quite small (laughs) ...

Yeah ... so before you are carried away and start studying how matter propagates in CDT, you should think about what it means to have matter coupled to a universe which is of the size of 20 Planck lengths.

Would you like to simulate a universe with a much bigger size?

It would be nice to simulate a much bigger universe but there's a limit to what you can do with this numerical approach because it has to fit into a computer and the simulation has to run in a finite amount of time.

Can you do anything besides numerical simulations?

Not in four dimensions at the moment.

Okay, so the claim is that according to the numerical simulations, a universe of such a small size is de Sitter?

Even though it is so small, the end result of the simulation is surprisingly well described by the assumptions of isotropy and homogeneity. What you actually observe is a universe which fits these symmetry assumptions that you started with, namely, isotropy and homogeneity, which is basically de Sitter spacetime [13, 14].

Would I get a de Sitter space if the universe was just one Planck length in size?

There is an interplay between the coupling constants and the size of the universe in the following way. For a universe of a given size there are quantum fluctuations around it and we encounter exactly the same phenomenon as in quantum mechanics, that is, if you are looking at sizes which are comparable to the coupling constants in units of $\hbar$, then quantum fluctuations dominate, making it senseless to claim anything at all. However, if the size is large compared to the coupling constants, then it makes sense to talk about expectation values around which there are small quantum fluctuations. In this latter case, what we observe in the simulations are such expectation values as a function of the volume of the universe until we run out of computer memory and time.

But now going back to the issue of the dimensionality of the universe ... I have read in one of your papers that the universe is four-dimensional at large distances and two-dimensional at small distances [16]. How can that be?

In a quantum universe you do not have a standard quantity to which you can call *dimension*. You have to define what you mean by *dimension*. If you choose a nice four-dimensional metric and expand around it, then of course your space is four-dimensional. But if you have a quantum theory where you are summing over all geometries, even if these geometries are kind of four-dimensional, the result will not be necessarily four-dimensional.

Let's imagine the following thought experiment. Suppose that you did not have an action, such that every geometry is equally represented in the path integral. Then, if you sum over all these geometries, will you obtain something which is four-dimensional? The answer depends entirely on the entropy (number of geometries of the same type) of a certain type of geometry. Let's assume that there are many geometries which look like a very thin string and let's assume that when you make the string longer and thinner the number of such geometries increases. If that's the case, in the end, the effective one-dimensional structures will dominate the path integral. Even if all the geometries you started out with were kind of four-dimensional, it might be that if you introduce a length cutoff the geometry that dominates the path integral will be small one-dimensional strings. In this situation, the other higher dimensional structures are *behind the cutoff*.

In other words, I'm reiterating what string theory likes us to believe about its six extra dimensions, that is, that such extra dimensions are small. In principle, in any theory where you sum over geometries, this kind of situation may well happen and it does happen [17]. If you start out with four-dimensional building blocks and glue them together, in the end, when you take the scaling limit where the size of the individual blocks shrinks to zero, there is no a priori reason to expect that the end product has anything to do with the four-dimensional world.

But you did get a four-dimensional world at large scales, right?

Yes . . . but that's an indication that something is right about the approach that we are taking. Describing a four-dimensional world is a major challenge for a background-independent theory.

But what about the claim that the universe is two-dimensional at small scales [16]?

The structure at small and large distances of a typical geometry, which is generated by some action, does not necessarily agree. At large distances it may effectively look like a four-dimensional world and at small distances it could look completely different, with some fractal structure and so on. At small distances you can encounter ultraviolet divergences which would lead to wild quantum fluctuations so you cannot guide yourself using classical intuition. The way to proceed is to define *dimensionality* in a specific way. For instance, you can define the spectral dimension, measured by studying a non-dynamical diffusion process in these geometries. In this context, you can study the dimensionality of the geometry as a function of diffusion time. This allows you to verify that at large distances you have an effective four-dimensional geometry and at small distances you have a lower dimensionality. I do not claim that it is exactly two-dimensional at small distances because it is a purely numerical study but it is around two dimensions, plus or minus something. This indicates that there is a gradual change in the dimensionality of the geometry from large to small distances.

This kind of phenomenon one could also observe in Hořava-Lifshitz gravity [18, 19], right?

Yes, but in any theory where you have an ultraviolet fixed point you will observe such an effect. What does it mean to have an ultraviolet fixed point? It means that the dimensionless coupling constant approaches a constant value but since in four dimensions the gravitational coupling constant has dimension two, this implies that as you approach the fixed point two dimensions will sort of disappear in such a way that you effectively start with four dimensions and end with two dimensions. From this point of view, it is a good sign that one does observe it. It is not a proof that this ultraviolet fixed point exists but it is not in contradiction with it. In fact, people who are studying renormalisation group flows with such an ultraviolet fixed point do observe the same effect [20]. So, it's not only in Petr Hořava's gravity (Hořava-Lifshitz gravity) that you observe it [18, 19], as it is a generic effect.

Do these two approaches – CDT and Hořava-Lifshitz gravity – have more in common than just this flow from two to four dimensions?

I think they could be quite related.

How could they be related?

Since we treat time and space differently in CDT, our regularised theory could potentially be creating an asymmetry between space and time. When approaching these non-perturbative fixed points, we don't know exactly what to expect. We have explicitly put in an asymmetry between space and time, in the sense that we have a time foliation, so it is possible that the underlying theory will develop dynamically such an asymmetry. At first, we did not think that this could be the case since there are very few fixed points and we expected to find isotropy. This is what usually happens in an ordinary lattice theory, where you have no problem in recovering the continuum limit in the end, which is rotationally invariant. However, this is not necessarily the case.

From this point of view, I think that there can be similarities between the two approaches. In fact, in a certain sense, some of our motivations are quite similar to those that led to Hořava-Lifshitz gravity. In particular, we were interested in a lattice formulation that is as close as possible to canonical quantisation. If you do have a canonically quantised theory, then you expect the theory to be a unitary theory. Within the context of CDT, there is a type of transfer matrix [21], and the existence of a transfer matrix usually implies the existence of unitary time evolution. To devise a unitary theory, which is also ultraviolet finite, was one of the main motivations of Hořava. He concluded that one should add some higher derivative terms in the spatial directions but in the time direction one keeps a second-order time evolution in such a way as to have a unitary theory.

You mentioned earlier that certain geometries have different entropies associated with it . . .

That's a trivial statement. Suppose you have an action and this action only depends on the geometry, for example, Einstein's action depends on the integral of the curvature. There are many geometries which have the same on-shell value of the action. In the path integral, what enters is the number of geometries for which the action has a specific value. So in this way, the entropy – the number of such geometries – of a specific class of geometries enters in the path integral. This entropy of configurations for a given value of the action can play an important role in determining what is the dominant configuration in the path integral.

In one of your papers you state that CDT is an entropic theory of quantum gravity [22]. Is it in any way related to Erik Verlinde's idea of gravity as an entropic force [23]?

No, I would say it is anti-related to it. First of all, this is a quantum theory or, to be more precise, an attempt at formulating a quantum theory since we don't know if the continuum theory exists, while Verlinde's theory is some kind of classical theory. When we discuss the notion of *entropy* in our formalism it is really genuine *entropy*, in the sense of the number of configurations. In fact, in our formalism – and I think this is quite beautiful – this actually singles out, in some way, gravity compared to all other forces.

The partition function of quantum gravity is basically the generating function of the number of geometries. This can be made very precise [13, 14]. In this way, the theory of quantum gravity is entirely given by the counting of geometries. In fact, you can use these piecewise linear manifolds (or four simplicies) to decompose Einstein's action in such a way that you make apparent that you are just counting different four-dimensional triangulations. Therefore, if you could count them, exactly you would have solved quantum gravity completely. However, it is difficult to count them, unfortunately, except in two dimensions. This is the reason why non-critical string theory could be solved analytically, that is, in this context you can count the number of geometries explicitly and analytically.

Do you expect that you will be able to use CDT for other things like computing scattering amplitudes at high energies?

No. That's completely uninteresting I think. That's not what CDT is made for.

Why do you say it is not interesting?

I don't know what you precisely mean by *computing scattering amplitudes*. Usually we use quantum field theory to compute scattering amplitudes at high energies where you have an S-matrix and so on. What exactly do you have in mind? I don't know why you would like to include gravitons in this setup. I still have to remind you that the size of the numerically simulated universe is 20 Planck lengths and, in addition, in order to have an S-matrix you need to have flat spacetime, asymptotic states, etc. In summary, CDT is not the right framework for this.

At some point you write down the wave function of the universe [13, 14]. What does it mean to have a wave function of the universe? Is there a multiverse at play in the context of CDT?

That's a good question. There is no multiverse in our approach. We are not doing anything else than in any kind of mini superspace attempt to quantise gravity. You can always use the Wheeler-DeWitt equations, or the Hamiltonian if you have one, to state that the ground state is the wave function of the universe. The way people define the wave function and apply it to a macroscopic system like the universe is, of course, unclear and we do not improve such issues.

The wave function of the universe is defined in Euclidean space by fixing the geometry at the boundary. Then, you use whatever quantum theory you have to sum over all possible geometries and this sum is what you call the *wave function of the universe*. This, we can clearly calculate and that's what our formalism does. The information that the wave function has is the following. If you ask some question in this universe then you can use the wave function to say that, with a certain probability, this or that observable will take a certain expectation value. The wave function contains this information because, in some sense, it accounts for the probability associated to a given spatial surface compared to another spatial surface.

The true meaning of the wave function of the universe is complicated and no one fully understands it. Many people have tried to shed some light on this matter. In fact, there's a huge literature about mini superspace (which are basically just quantum mechanical models) calculations involving this wave function.

I read in one of your papers that the final picture that emerged from your simulation is that of a universe created by tunnelling from nothing [24]. What does *tunnelling from nothing* mean?

This is a standard process in quantum cosmology. For example, people like Stephen Hawking and Vilenkin have discussed it in detail [25, 26]. Given the wave function of the universe, which you usually calculate in Euclidean signature, you can rotate it back to Lorentzian signature and obtain a wave function in Minkowski space. This Lorentzian wave function typically looks like a potential that starts out with zero four-dimensional volume but also contains a barrier which, once you cross it, will roll down and the four volume will expand. In this sense, there's a kind of tunnelling from zero four volume to a real and finite four volume – this is what *tunnelling from nothing* means.

What the precise interpretation of this is, is difficult since one is dealing with the universe. In particular, Vilenkin has been arguing that the interpretation is that in the beginning the universe has zero volume and quantum mechanically it can tunnel to a situation with a finite macroscopic volume [26]. In simple models you can even calculate the amplitude for such process. Our model is not in disagreement with this.

Is CDT at a point of delivering experimental predictions?

Of course not. CDT is a theory of a quantum universe for which not much is yet known and, additionally, without incorporating the effect of matter in detail it is hard to make any predictions. The problem of dealing with matter, from the point of view of cosmology, is that it is difficult to understand the real time evolution of the system, given that we primarily work with Euclidean time. This is the usual issue that one encounters in ordinary quantum field theory, namely, issues of rotating from Euclidean to Minkowski signature and issues of obtaining the real time evolution. These are of increased difficulty when working with only numerical data. So, unfortunately, you have to think carefully about these problems if you want to make intelligent statements about the coupling to matter. At the moment, I don't have too many intelligent things to say about it.

Do you expect that at some point you will be able to study black hole formation for example?

Black hole formation is usually associated with matter and you have to remember that, in a universe that has the size of 20 Planck lengths, black holes are not the most relevant objects to discuss. However, in principle it should be possible to study them. In this case, we do indeed face the problem of rotation from Euclidean to Lorentzian signature. Black holes hardly exist in Euclidean signature and they are usually associated with topology change. At the moment, we have kept the topology fixed but of course you could have chosen a topology that corresponds to a black hole. However, if you want to study a dynamical process, it should be understood how to do the rotation properly and our main emphasis at the moment is to understand whether there is a continuous theory or not. CDT is a lattice theory and before worrying about all kinds of subtleties we want to understand if, at all, there is a continuum theory.

Why is it difficult to understand whether there is such a limit when taking the lattice spacing to zero?

Maybe the best analogy is that of a spin system on a lattice, in which the spins can be up or down. In this case you ask the analogous question "Is there an underlying continuum field theory associated with such a spin system?" The answer is not obvious at all. Clearly, you have to find a region where there are long-range correlations between these spins otherwise there is no notion of continuum. Additionally, you want to remove the dependence on the lattice spacing because you introduced the lattice to regularise the theory. As such, some of the parameters in the theory need to be large compared to the lattice spacing because otherwise you do not recover continuum physics. This usually implies that you need to find a sort of phase transition where you observe long-range correlations. Once you found that phase transition, you have to study the limit corresponding to a continuum field theory. In summary, there are a number of steps to be taken in order to find this limit, even in an ordinary kind of system, and it is not easy to establish its existence, specially if you do not know in advance that there must be one.

How do you plan on trying to establish this limit?

In order to find the limit one has to identify transition lines in the phase diagram, where the geometry undergoes a transition. However, first of all, you have to understand how to think about these phase transitions in a coordinate-independent way, which is something that people don't usually think about because they are not forced to think about it.

Suppose that you have a scalar field $\phi(x)$. Then you can talk about the correlation between $\phi(x)$ and $\phi(y)$. If you have a theory involving quantum gravity, it makes no sense to talk about the correlator between $\phi(x)$ and $\phi(y)$ because x and y are just coordinates and the theory sums over all metrics. In this context, there is no fixed notion of distance between x and y and hence the correlation length as a function of $x - y$ does not have meaning. This implies that one needs to define invariants in a different way.

Conceptually, the study of invariants is very interesting but there isn't such a great deal of literature on how to formulate these invariants in a non-perturbative quantum gravity setting. In fact, it is interesting to know that Albert Einstein was kind of disgusted with the name *theory of relativity*, which was given to the theory that he developed. He thought it was a completely misleading denomination and he actually wanted to call it *theory of invariants* because what defines the theory are quantities which are invariant under coordinate transformations. Thus, in a sense, it's stupid to refer to Einstein's theory as *theory of relativity* because it's not about what is relative but about what's invariant. In the context of CDT, you are forced to think about not only how to define objects which are invariant under coordinate transformations but also, since you are integrating over all geometries, how to define objects which are invariant under geometry.

In order to summarise this conversation, what do you think are the lessons about quantum gravity that came out from CDT and what do you think were the main advances in the search for quantum gravity in the past, say, 30 years? Do you think we've reached some understanding of it?

No, I don't think so. I think that there are no lessons in the sense that we still don't have a theory that we could call a *theory of quantum gravity*. CDT is a formalism that is trying to define quantum gravity non-perturbatively but we haven't proven that there is a continuum limit yet. As with all other theories of quantum gravity, little has been achieved because we still don't know what is the right theory. In my opinion, we still don't have a natural quantisation procedure. We know, from quantum field theory, that this is not going to be easy to show because it cannot be an ordinary Gaussian fixed point.

However, I would like to say that while working with CDT, it became clear that statements like *geometry is emergent*, or *geometry has to go away at the Planck scale* and so on are naturally encoded in any attempt of defining the theory non-perturbatively via pure path integral formulations. These desirable properties are there because, first of all, from the very beginning you don't have anything. You are summing over all geometries but these by themselves are not observables in the theory like the path of a particle is not an observable, even though you still sum over it in the path integral of the particle. In the same way, geometry is not an observable in quantum gravity so the best you can hope for is the emergence of a kind of classical limit where the quantum fluctuations are small. If you approach smaller and smaller distances then you expect that quantum fluctuations are so large that it makes no sense to talk about geometry. In this sense, geometry kind of resolves itself, which naturally happens at the Planck scale where the quantum fluctuations are large no matter what kind of theory you imagine to be dealing with.

All these words/statements have been used within the context of string theory and I think that they have some validity no matter what framework you are using. What is nice about trying to follow our approach is that there are a lot of concrete questions such as "What are the observables?" These are the sort of problems that are not dealt with too much within string theory. When you look at the details of string theory, you can feel uneasy because it only makes sense on-shell and you have an S-matrix which you don't know exactly where it lives, etc. Usually string theory replies to difficult questions with the answer "The beauty of the theory will clarify this for us." Well that might be (laughs) but at least CDT provides a very concrete framework in which you can address these questions.

I don't think that we have a theory of quantum gravity and I don't think that we have learnt too much about it. Honestly, in a way it's difficult to understand if finding the fixed points of the theory will give you the final answer. In a certain way, I hope not (laughs) but I think we have an obligation to look at it and since so many people in the world are working on something else, I think that some of us should work on this direction, which also has its own beauty.

One often hears that string theory *eats* everything like quantum field theory, fluid mechanics, etc. Do you think that at some point string theory will incorporate CDT?

No, I don't think that string theory *eats* everything. I think that this idea that string theory *eats* everything started with some drawings by Dijkgraaf at one of the lattice conferences and in a way I think it is not really what has happened. Instead, what has happened is the following. People who pursued theoretical physics in the 1970s, in a theoretical avant-garde sense like those working on QCD, were running out of things to explore. There was a lack of experimental evidence for guiding how to proceed in the field theory direction and icons like Alexander Polyakov, and later Edward Witten, told us that we should think about string theory. As such, in the 1980s string theory was a very promising candidate for the theory of everything, so it was an extremely exciting research direction that motivated many people to work on it and be part of a big discovery. However, today we still have QCD and we still have grand unified theories but we don't know how to progress from a field theoretical basis. Hence, string theory hasn't *eaten* anything because the theoretical development in field theory basically stopped there.

Loop quantum gravity is also a counter-example to the statement that string theory has *eaten* everything. In fact, loop quantum gravity has become a larger community, regardless of that being a good or a bad thing. In the context of quantum field theory, there is also a pretty stable community in terms of size that studies lattice gauge theories. This community hasn't been *eaten* by string theory either.

What can I say about the future of string theory? I don't know. I have no strong verdict about it but it is true that there is now a huge mass of people working on it. This fact was also one of the motivations behind the book of Lee Smolin [27], though I think that the book also expressed Lee's own frustrations, which I can kind of understand to some extent. It is a real danger if string theory becomes a self-sustained community decoupled from real physics, so to speak. In a way, this problem is inherent not only to string theory but to theoretical high-energy physics as whole. If theoretical physics doesn't produce anything interesting in the future, it will be cut away because there are many other fields which will want to grab the funding away.

Do you think that the lack of contact with the real world is one of the reasons why the number of jobs in theoretical high-energy physics is decreasing?

I think that it will be difficult to continue to argue that so many people should be working on something that is not necessarily leading anywhere. It might be interesting to pursue these directions for the sake of discovering new mathematical structures or exploring weird physical theories but that would lead to a much harder fight for funding and, in the end, would make it difficult for us to defend our position.

Historically, there have always been many more experimental physicists than theoreticians. In particular, there have almost always been more solid state physicists than high-energy physicists. High-energy physicists have had an amazing success in the twentieth century in creating and developing general relativity and quantum mechanics as well as formulating the standard model. However, after these developments, the success of high-energy physicists has stopped. In the last 30 years there have been no new important discoveries, nothing in fact, and if you wait 30 more years and there are still no new discoveries related to the real world then I think that our community will shrink in a significant way. Perhaps if that happens it is only fair (laughs) since if we have nothing to offer then our existence is useless from that point of view.

There will still be room for theoretically minded people but perhaps only a small number, which I don't think will stop progress. If there is a possibility for making progress some bright people will still be able to do it. Most progress, anyway, is made by a few bright people and not necessarily the ones we think about as being the brightest. In a large community, as high-energy physics is now – and here I'm getting a little bit in line with Lee Smolin though I don't agree with many other things that he claims – there is a heated environment of global discussion and fashionable topics, which might actually not be beneficial for what you might call *deep thinking* about which direction to pursue and how to make progress. Whether *deep thinking* leads to any progress is a complicated discussion because you can find examples where progress is made by people developing random ideas and trying out everything they can as well as examples where it is made by *deep thinkers*. Nevertheless, it is true that the milestones in our field were set by only a few people and some of them have basically existed in a kind of vacuum bubble (laughs). Albert Einstein and Paul Dirac, for example, were almost in their own world, so to speak, in order to make progress. Other people interacted a lot with the physics community. Luckily there are no fixed rules for how progress takes place.

The LHC could perhaps change all of this but it is not clear at the present time. People working in our field, at the moment, will be able to defend their motivations for one more generation but afterwards society will be feeling fatigued, due to our inability to deliver, and other fields will take the funding (laughs). Nuclear physics, for example, dominated the funding in the 1950s, 1960s and even in the 1970s it was quite important theoretically but then it died out. At that time, you could still claim that the study of very special elements enriched with neutrons was important for understanding certain experiments but this line of research became more esoteric as time passed, placing nuclear physicists in a position that could no longer be defended.

Did this happen because nuclear physicists solved all the relevant problems they were meant to solve?

To some extent yes but also what they were claiming to be new problems didn't really *click* with the rest of the world. In the same way, it might not *click* with the rest of the world to discuss different aspects of Calabi-Yau compactifications and so on (laughs). There are many people now working on Calabi-Yau compactifications and they can defend its importance for some time but in the end I think that they will run into similar problems as the nuclear physicists did.

However, if supersymmetry is found LHC it will change everything. If that would be the case, I would basically give up studying my own theory of quantum gravity because extremely strong methods that deal with supersymmetry are naturally incorporated in string theory. It is possible to study supersymmetric quantum field theories just by themselves but it all originated from string theory. It's natural to implement supersymmetry in string theory, though one can discuss how natural it is (laughs), and in a way, it is needed to make sense of string theory while it is not, strictly speaking, needed to make sense of other theories.

I am a bit skeptical about supersymmetry being discovered at the LHC but it is a possibility which, at least for some years, one can still wait with excitement to see if it actually happens. Personally, I hope that it will happen since it would be a fantastic prediction of theoretical physics, as one is led to it by various arguments. It's surely a mathematical possibility though not all mathematical possibilities are realised in nature unless you are an extreme believer in the anthropic principle (laughs). We can only hope that the experimental physicists know what they are doing (laughs).

You mentioned earlier that no important discovery was made in the last 30 years. Is there something you consider a breakthrough in theoretical physics in the past 30 years?

I haven't seen any.

As simple as that (laughs) ?

Yeah, I don't think there has been any.

So you think that after the standard model was formulated not much has happened (laughs)?

Well ... if you think about contributions related to the real world there has been nothing. Come on! It's obvious.

That's a bit pessimistic, no (laughs)?

I don't think you can call string theory a massive scientific breakthrough of any kind because it's not necessarily related to the world. So in terms of physics ... nothing. Come on! It's completely trivial.

Why did you choose to do physics? Why not fishing?

I don't like fishing. Physics and astronomy have always been fascinating to me. I think that many of us entered this field because we were interested in fundamental questions about the universe. I'm very privileged that I am just sitting here and studying the universe using a computer. When I was 13 or 14 years old, I was reading popular books about Einstein's relativity and I was deeply frustrated with them because I couldn't really understand what they were trying to convey. The books were trying to convey important concepts (like curved space, time travelling, etc.) without explaining precisely what they meant (laughs). All this was, of course, deeply fascinating and I think many people have been dragged into theoretical physics due to the excitement provided by the science fiction version of it. I'm definitely one of them (laughs).

What do you think is the role of the theoretical physicist in modern society?

I'm obviously biased but I think that theoretical physics is here to stay as a fundamental pillar in our society to what concerns the understanding of nature. No matter whether or not we make any progress beyond what has been accomplished so far, theoretical physics is here to stay. It is an exact way of describing the world and the way you teach physics to others is via theoretical physics. It is the fastest way to pass this information to others. I personally view it as one of the *unbelievable* achievements of mankind, in particular, the fact that we are able to describe nature in abstract mathematical terms and in a very precise way. Take quantum mechanics as an example. I view quantum mechanics as mankind's greatest intellectual achievement. It is so far from any intuition and still it is unbelievably precise, in fact, so precise that we have not seen any glimpse of its failure, though it might fail at some scale. Any experiment performed so far has confirmed quantum mechanics. In summary, I think this kind of knowledge will always have to pass from generation to generation even if we don't make further progress.

References

[1] L. Susskind, "The anthropic landscape of string theory," arXiv:hep-th/0302219.

[2] H. Nicolai, K. Peeters and M. Zamaklar, "Loop quantum gravity: an outside view," *Class. Quant. Grav.* **22** (2005) R193, arXiv:hep-th/0501114.

[3] T. Thiemann, "Loop quantum gravity: an inside view," *Lect. Notes Phys.* **721** (2007) 185–263, arXiv:hep-th/0608210.

[4] J. Ambjorn, B. Durhuus and J. Frohlich, "Diseases of triangulated random surface models, and possible cures," *Nucl. Phys. B* **257** (1985) 433–449.

[5] J. Ambjorn, B. Durhuus, J. Frohlich and P. Orland, "The appearance of critical dimensions in regulated string theories," *Nucl. Phys. B* **270** (1986) 457–482.

[6] J. Ambjorn, B. Durhuus, and J. Frohlich, "The appearance of critical dimensions in regulated string theories. 2.," *Nucl. Phys. B* **275** (1986) 161–184.

[7] J. Ambjorn and B. Durhuus, "Regularized bosonic strings need extrinsic curvature," *Phys. Lett. B* **188** (1987) 253–257.

[8] J. Ambjorn, B. Durhuus and T. Jonsson, "Scaling of the string tension in a new class of regularized string theories," *Phys. Rev. Lett.* **58** (1987) 2619.

[9] J. Ambjorn, P. de Forcrand, F. Koukiou and D. Petritis, "Monte Carlo simulations of regularized bosonic strings," *Phys. Lett. B* **197** (1987) 548–552.

[10] J. Ambjorn, B. Durhuus, J. Frohlich and T. Jonsson, "Regularized strings with extrinsic curvature," *Nucl. Phys. B* **290** (1987) 480–506.

[11] J. Ambjorn, J. Greensite and S. Varsted, "A nonperturbative definition of 2-D quantum gravity by the fifth time action," *Phys. Lett. B* **249** (1990) 411–416.

[12] S. Weinberg, "Ultraviolet divergences in quantum gravity," in General Relativity: An Einstein Centenary Survey, S. W. Hawking and W. Israel, eds., pp. 790–831. Cambridge University Press, 1979.

[13] J. Ambjorn, A. Goerlich, J. Jurkiewicz and R. Loll, "Nonperturbative quantum gravity," *Phys. Rept.* **519** (2012) 127–210, arXiv:1203.3591 [hep-th].

[14] R. Loll, "Quantum gravity from causal dynamical triangulations: a review," *Class. Quant. Grav.* **37** no. 1, (2020) 013002, arXiv:1905.08669 [hep-th].

[15] J. Ambjorn, A. Gorlich, J. Jurkiewicz and R. Loll, "Planckian birth of the quantum de Sitter universe," *Phys. Rev. Lett.* **100** (2008) 091304, arXiv:0712.2485 [hep-th].

[16] J. Ambjorn, J. Jurkiewicz, and R. Loll, "Spectral dimension of the universe," *Phys. Rev. Lett.* **95** (2005) 171301, arXiv:hep-th/0505113.

[17] S. Carlip, "Dimension and dimensional reduction in quantum gravity," *Class. Quant. Grav.* **34** no. 19, (2017) 193001, arXiv:1705.05417 [gr-qc].

[18] P. Hořava, "Quantum gravity at a Lifshitz point," *Phys. Rev. D* **79** (Apr. 2009) 084008. https://link.aps.org/doi/10.1103/PhysRevD.79.084008.

[19] P. Horava, "Spectral dimension of the universe in quantum gravity at a Lifshitz point," *Phys. Rev. Lett.* **102** (2009) 161301, arXiv:0902.3657 [hep-th].

[20] M. Reuter and F. Saueressig, *Quantum Gravity and the Functional Renormalization Group: The Road towards Asymptotic Safety*. Cambridge Monographs on Mathematical Physics. Cambridge University Press, 2019.

[21] J. Ambjorn, J. Gizbert-Studnicki, A. Goerlich, J. Jurkiewicz and R. Loll, "The transfer matrix method in four-dimensional causal dynamical triangulations," *AIP Conf. Proc.* **1514** no. 1, (2013) 67–72, arXiv:1302.2210 [hep-th].

[22] J. Ambjorn, A. Gorlich, J. Jurkiewicz and R. Loll, "CDT – an entropic theory of quantum gravity," in *International Workshop on Continuum and Lattice Approaches to Quantum Gravity*. 7, 2010. arXiv:1007.2560 [hep-th].

[23] E. P. Verlinde, "On the origin of gravity and the laws of Newton," *JHEP* **04** (2011) 029, arXiv:1001.0785 [hep-th].

[24] J. Ambjorn, J. Jurkiewicz and R. Loll, "Semiclassical universe from first principles," *Phys. Lett. B* **607** (2005) 205–213, arXiv:hep-th/0411152.

[25] J. B. Hartle and S. W. Hawking, "Wave function of the universe," *Phys. Rev. D* **28** (Dec. 1983) 2960–2975. https://link.aps.org/doi/10.1103/PhysRevD.28.2960.

[26] A. Vilenkin, "Approaches to quantum cosmology," *Phys. Rev. D* **50** (1994) 2581–2594, arXiv:gr-qc/9403010.

[27] L. Smolin and P. Smolin, *The Trouble with Physics: The Rise of String Theory, the Fall of a Science, and What Comes Next*. A Mariner Book. Houghton Mifflin Company, 2006.

2

Nima Arkani-Hamed

Professor at the School of Natural Sciences, Institute for Advanced Study, Princeton, NJ

Date: 26 August 2011. Location: Copenhagen. Last edit: 20 October 2020

What do you think are the main problems in theoretical physics at the moment?

I think there are two classes of problems. One class is the class which we know contains real problems. This class includes, for instance, situations where we try to bring together current physical theories in a unified framework, and that framework ends up breaking down. The other class contains problems which we are not absolutely sure are real problems. We suspect that they are real problems but they don't have the same character; that is, they do not involve a contradiction of the physical laws as we know them. Such problems have incredibly peculiar properties that require explanation.

What kind of problems does the first class you mentioned include?

The first class of problems contains the most famous ones, for instance, the problem of combining quantum mechanics and gravity in a single theory. There are many aspects to this problem and one of the first considerations to note is that something goes wrong when spacetime curvatures approach the Planck scale. In such situations, quantum mechanical effects become important. In addition, once you try to resolve the physics at distances and times shorter than the Planck scale, you realise that you have to put so much energy into such a small region of space that you collapse that region into a black hole. When that happens, you can't probe the physics beyond that scale.

If we want to know in detail what happens when we collide particles and energies at one millionth of the Planck scale, we can understand it with very good accuracy. However, as the energy approaches the Planck scale something breaks down and we can't make any accurate prediction. These are well-known short-distance aspects of the quantum gravity problem.

You mentioned that quantum effects become important when curvatures approach the Planck scale. What is the role of quantum mechanics in what you just described?

I'm sort of describing the 1980s understanding of what the problem is, which I think we now understand is not really the problem. It's related to the real problem but it's a small aspect of a much deeper and bigger problem. An elucidating thought experiment is the one that I just described. If you're trying to probe increasingly shorter distances and times in a Lorentz-invariant sense, you know from quantum mechanics and its inherent uncertainty principle that you will need to use increasingly higher centre of mass energies. This is not a problem until you put so much energy into such a small region of space that you collapse the region into a black hole. This happens around the Planck length and Planck time, at which point you can't ignore gravity anymore. The black holes thus created shield all the physics that is taking place at short distances and if you keep pumping in more energy, or if you collide particles in a highly energetic accelerator, you will just end up creating even bigger black holes. So, it seems that there is no operational way in which we can talk about distances and times that are small compared to the Planck length and the Planck time.

But should we be able to?

Indeed, put in a different way, what I just said is a fact and the question now is: What do we do with this fact? How should we interpret it? This issue signals some deeper issue, namely, that it appears that beyond a certain scale we can't make accurate predictions. So what happens if you scatter gravitons with centre of mass energy about 1.7 times the Planck energy? In principle, this is possible to do in an experiment and we, as theoretical physicists, must be able to predict the outcome. However, the thought experiment I just described tells us that our understanding is incomplete.

I think that the fact that there's no operational way of measuring distances and times small compared to the Planck length and time is a primitive indication that the idea of spacetime isn't really a fundamental one. You could take the attitude that such distances and times are really there and it just so happens that we can't measure them. However, we have seen that such arguments have been made over and over again in physics and they can't be right. Quantities that we can't even in principle give operational meaning to may be useful concepts but can't be fundamental. Spacetime has to be emergent from something else. This is a very startling conclusion. It is startling that space has to emerge and particularly startling that time has also to emerge. Physics, if nothing else, is about how things evolve in time. You prepare an initial state, you evolve it and observe what happens in the final state. So, it's a very startling conclusion that we need to get rid of space and time and replace it with something else. We need to come to terms with this fact and the issue is how to make that work with all the different aspects of nature that we know are true.

If the world wasn't Lorentz-invariant, life would be a lot easier because you could claim that space is special and so you wouldn't have to make time also emergent. The problem is that if you break Lorentz invariance you are led to all sorts of dumb ideas such as atoms of spacetime. Atoms of spacetime is a naive idea that some people believe. You could think that if spacetime is emergent then if you look at short scales you will find atoms of spacetime in the same way that if you look at short distances in the surface of this table you find that it is made out of atoms of matter. But such a naive and brute force approach to the problem can't be right. If spacetime is made out of spacetime atoms, such atoms must be small but small in what frame? It's not a Lorentz-invariant question so it's incredibly hard to make any sense out of it; in fact, no one has produced a consistent theory with some kind of ultraviolet cutoff or another brute force way of solving ultraviolet problems using this point of view. I find this a very deep clue that the world is Lorentz-invariant, which simultaneously kills these incredibly naive approaches to trying to solve the problem.

Okay, so Lorentz invariance is a property that should be there ...

I think it should be there but this problem is related to something else and what I just described is still the 1980s viewpoint of the problem of quantum gravity. Before we go into detail let me emphasise an aspect of this discussion which is normally not emphasised enough. Popular accounts of the problem usually suggest that there's an incredibly violent clash between quantum mechanics and general relativity. That's just nonsense, okay? We can definitely talk about quantum mechanics and general relativity in the same sentence without a problem. For instance, John Donoghue has calculated the quantum mechanical correction to Newton's law which is approximately $-GMm/r^2 - 127/(30\pi^2)\ell_p^2/r^2$ [1, 2]. You can calculate this accurately, including loop diagrams with gravitons in it, and there's nothing scary about it. The pieces which are calculable are the pieces that have logarithms associated with the imaginary parts of the diagrams and as such are the pieces associated with long-distance physics. Similarly, we can compute the leading corrections to the Schwarzschild metric [3]. These are minuscule corrections that nobody cares about but they're calculable. The true difficulties are at short distances. We simply don't know what's going on at short distances. So it's not that the whole world just doesn't make sense at all. That would be crazy. Every time something sounds totally nuts, it's because it just isn't that way. It would be crazy if we couldn't talk about things falling on the surface of the earth quantum mechanically (laughs). Of course we can; there's no conflict between gravity and quantum mechanics at that level.

The true problem is a more subtle problem. It's the problem of understanding the interplay between quantum mechanics and gravity at short distances, which has dramatic consequences. In particular, spacetime isn't there. However, the deeper aspect of the problem is that the need to understand the emergence of spacetime is not only relevant at short distances but in a sense relevant at all scales. It hints towards a radical, or zeroth order, change in what you even think are the correct observables once you combine quantum mechanics and gravity. Before you discuss issues about theories and formalisms such as background independence or whether you should use the Palatini procedure or whatnot, it's good to know what is the output. At the end of the day, you will have a bunch of equations and you are going to predict some numbers. What will those numbers correspond to? Hopefully they correspond to some experiment that somebody can do, right? And since your theory is, hopefully, a precise mathematical theory, then you will, in principle, spit out numbers to arbitrary precision, right? It would be similar to using QED to calculate the electron magnetic dipole moment with as many decimal places as you would like. So it would be nice to know what observables are there that can make precise sense observationally.

It is a very well-known story that when going from the classical to quantum realm the issues of observables are incredibly important. The big news is that in the transition from classical to quantum there is a vast reduction in the number of things that we can talk about, right? Classically we could talk about about position x and momentum p but when moving to the quantum realm we can only talk about x or p due to the inherent indeterminism in the theory. So, there are big zeroth-order changes just at the level of the kinematic language. There are things that you are not allowed to talk about as having sharply observable properties. These issues have practically no importance when you are just thinking about kicking balls down inclined planes and so on (laughs). However, the fact that now we must talk about x or p and not x and p is a gigantic difference. People often lament this transition from classical to quantum vastly reducing the number of things that we can predict about the world to the extent that classical physicists were so upset about it at first. However, there is a positive flip side to it which is that there are fewer things that can in principle be talked about. In turn, this means that all things in nature must be more closely related to each other than you thought before, precisely because there are so fewer things that you can actually observe (laughs). If you think about it, this is exactly what happened: before quantum mechanics there were waves and particles whereas after quantum mechanics everything is a particle.

So what is really the issue with defining observables?

Let's talk about it. Let's talk about what you can in principle observe in the world first without gravity and then with gravity. Since we're in Copenhagen, it's particularly fitting to talk about this issue because I think the right way to think about it is following the Copenhagen interpretation of quantum mechanics [4]. In fact, I really think that the proponents of this interpretation understood everything perfectly but didn't say as clearly as they could have (laughs). In the 1980s, with the advent of decoherence [5], their point of view became more explicit. In any case, let us remind ourselves about what we can talk about exactly in the quantum mechanical world. In other words, what experiments could we do in principle with arbitrary resources in order to get a perfectly sharp result? Since this is a question of principle, let us forget about whether a perfectly sharp result can be practically obtained. This idealisation invokes two types of infinities. One of those infinities we often discuss, specially when teaching quantum mechanics to students, while the other infinity we talk much less about. However, the second infinity is very important when gravity comes into play so let us go through both.

The first infinity stems from the fact that quantum mechanics is a probabilistic theory and therefore you have to experiment infinitely many times in order to have a perfectly precise prediction. What happens if you only do the experiment n times? In that case there is an error associated with it; in fact, you won't be able to make predictions with an accuracy better than roughly $1/\sqrt{n}$. So you can imagine yourself sitting in your lab and do the experiment once; write down the answer; do the experiment twice; write down the answer; you make a histogram; do it over a million times and get a very sharp histogram with a width of $1/\sqrt{\text{one million}}$; you keep going to one billion times and the width gets narrower; finally you do it infinitely often and it converges to a central value which you call the result of the experiment. This is the first type of infinity and we care a lot about it because it means that you approach this idealised limit very slowly, with $1/\sqrt{n}$.

The second infinity that we talk less about appears because you have to do the experiment with an apparatus of infinite size. Formally, by infinite size I mean that the number of degrees of freedom of the apparatus or the dimensionality of the Hilbert space of the apparatus has to be infinite. Why is this the case? If you want to record the 10^{300}th decimal place of the electron magnetic dipole moment you have to store it somewhere in your apparatus, right? You might say that you don't have to store the information in the apparatus and that you could store it somewhere else but there's a deeper problem that is intrinsically quantum mechanical, which is the fact that the apparatus itself is a quantum mechanical system. The apparatus and the system are co-evolving unitarily by virtue of Schrödinger's equation. For the apparatus to record a perfectly sharp observation you need decoherence to be perfect and for that to be possible the Hilbert space of the apparatus must be infinite and vastly bigger than the system that is being measured.

There is another way of understanding the issue related to this second infinity. If the whole system is evolving unitarily that means that at some point the measurement will be undone. If the apparatus has a finite size then as the system evolves it performs a periodic motion in a finite-dimensional Hilbert space and eventually you will return to any configuration. This would not allow you to perform the experiment infinitely often with the required precision. More prosaically, if you are measuring 10^{300} decimals then by the time you wrote down the result of your $e^{10^{300}}$ experiment you could have fluctuated into an elephant, spontaneously combusted into dust, your brain could have melted, or you could have written down the wrong answer; that is, any number of things could have happened. That means that there's an intrinsic systematic error that you cannot reduce if you do an experiment with a finite-size apparatus. This intrinsic error is of the order of e^{-n} where n is the number of degrees of freedom the apparatus is made of. We don't usually care about this in practice because it's vastly less important than the first error that I described which is of order $1/\sqrt{n}$.

In conclusion, as a matter of principle, if you want to have exactly sharp observables in quantum mechanics you must repeat the experiment infinitely often and with an infinitely large apparatus. This second form of infinity has a very important role to play in the context of gravity. I would like to emphasise again that this point was understood incredibly well, although explained in a characteristically murky way, by those who developed the Copenhagen interpretation and later clarified by the understanding of decoherence in the 1980s. Quantum mechanics really requires, not just as a mere convenience, that you drag around this infinite apparatus with you everywhere you go when measuring some system. Thus, secretly, there's this gigantic apparatus that's going along with you everywhere and because it is gigantic, it doesn't disturb too much the system you are measuring.

How is this related to quantum gravity?

Okay, let's switch on gravity and see what the problem is. What we will see is that, besides the drastic reduction in observables when moving from classical to quantum, there is also a dramatic reduction in the number of observables once gravity is turned on. As a matter of a fact, there are no local observables in any theory of quantum gravity, okay? This is a deeper problem tied up with the emergence of spacetime and related to the issue of short distances compared to the Planck length.

Let's imagine that we're trying to do an experiment right here in this room, which is of fixed size. I have my classical apparatus I'd have to drag along with me and I will have to make it larger and larger to make an incredibly sharp measurement. However, there's going to be a limit because as it gets heavier and heavier, it will start back-reacting on the geometry in this room. Hence, before I can make it infinitely big, it will become so heavy that it will collapse the whole room into a black hole. When that happens, I can't do the experiment infinitely often because we're all going to get crushed into the singularity of the black hole (laughs). This implies that there is a limit on the accuracy of any local observable; in particular, there will be a systematic experimental error that you cannot do anything to reduce. What is that error? As I mentioned before, it's of order e^{-n} with n being the number of degrees of freedom of the apparatus, which is about to turn into a black hole. However, we know that there exists a bound on the number of degrees of freedom that a black hole can have and that bound is constrained by the entropy of the black hole. So, in fact, the error is roughly of order $e^{-A/G}$ where G is Newton's constant and A the area of the black hole, given that the entropy is $S = A/(4G)$.

We can conclude from this thought experiment that there's no precise observable you can associate with any measurement performed in a finite-size room. So when you tell experimentalists that there's no local observables in quantum gravity and they ask you what the hell are they doing in their lab, the answer is that they are doing great and wonderful stuff but that there is no experiment that they could design in their labs that would reduce the systematic error of what they are measuring to zero.

So you wouldn't call observables what they are measuring in the lab?

They are not precise observables. They are approximate observables just as in quantum mechanics x and p are approximate observables. What one needs to understand is what should the fundamental observables/output of the theory be and what this thought experiment shows is that the output cannot be any local observable because such observables have no meaning. So, if someone comes along and tells you that they have a whiz-bang theory of quantum gravity, you should always ask: What are the observables? If they give you an answer that sounds suspiciously like an experiment you could do in a lab somewhere you know that they must be smoking something, right (laughs)? It's a very deep fact about quantum gravity that there are no local observables.

So what observables are there? Let us imagine that we have a special situation where you can make the apparatus bigger and bigger but you can also move things out further and further apart from each other. So, instead of having a fixed-size room, we now have a room which we can extend as much as we would like. In this situation, you can prepare an initial state of photons or gravitons and you can measure how they scatter arbitrarily accurately by making the apparatus bigger and bigger provided that you also take one photon further and further away from the other photon. This is possible to do as long as you push the infinite apparatus out to some infinity (laughs). In other words, you have to push the apparatus to the boundary of the spacetime. Thus, the only sharp observables in a theory of gravity are those associated with the boundary.

Is this a very practical way of defining observables?

It is practical. In fact, in asymptotically flat spacetime, this observable is the S-matrix. You scatter particles in and they go back out. It is not only practical; it is the bread and butter observable that we talk about all the time in particle physics. Now, you could say that when people are doing experiments at the LHC or at any other place where they scatter things into each other, they are definitely not measuring what comes out at infinity. But this only means that such measurements have an irreducible error associated with them. However, such measurements are continuously connected to a quantity that we can in principle make perfectly accurate by sending it out to infinity. Therefore, the exact observable is the S-matrix associated with the boundary and then it has various approximate cousins which we can compute or which we can measure by doing local experiments. If we content ourselves with the experiments at the LHC then there will be some limit on how accurate we can make any such measurement. As a matter of principle, the only sharp observables in a theory of gravity live on the boundary.

This is a startling conclusion and a deeper version of what originally appeared to be a short-distance problem or the problem of emergent spacetime. It means that no matter what your formalism is, if spacetime is continuum or discrete or whatever the heck it is, the output of the theory can't involve anything that knows about the inside of the spacetime. The output of the theory can only be related to things coming from infinity and going back out to infinity. If we would be talking about anti–de Sitter spacetime (AdS), then the observables are boundary correlation functions.

Is this related to holography?

It sounds suspiciously like the holographic principle, right (laughs)? In a sense, it's the only really sharp version of the holographic principle (laughs). In fact, it was realised long before Gerard 't Hooft and Leonard Susskind [6, 7]. People who thought deeply about quantum gravity in the 1960s, such as Roger Penrose and Bryce DeWitt, realised this point. Penrose had the idea that there was no bulk spacetime. He didn't quite phrase it so poetically but he really understood that this was the point, namely, that you shouldn't talk about individual points in spacetime because of quantum fluctuations. He suggested that instead you should look at things that go out to infinity such as light waves, which I think was a big motivation for twistor theory [8, 9]. Bryce DeWitt understood that in asymptotically flat spacetime the only observable is the S-matrix and that's what motivated him to do the first calculations of scattering amplitudes in gravity (laughs). DeWitt was the first to compute the two-to-two gravity scattering amplitude, which is a horrendous calculation [10–12]. However, he did it because it was the first example of an actual observable (laughs).

With this understanding in mind, you know that you have an approximate theory for how to compute these observables, namely, semiclassical gravity. You scatter things in the bulk, you follow things out to infinity and you can determine to good accuracy these boundary observables. This point of view was unappreciated for a long time but resurfaced with AdS/CFT [13]. In fact, this is one of the things that is beautifully explained in Edward Witten's paper on AdS/CFT [14]. For quite some time, people were not thinking along these abstract lines even though string theory has, from the very beginning, been an S-matrix theory. However, this realisation leads you naturally to another one. Specifically, the fact that observables are associated with the boundary, together with the suspicion that there is no spacetime at distances near the Planck length, leads to the radical viewpoint that not only are the observables living on the boundary but that the theory that calculates those observables is also only living on the boundary. In other words, the spacetime inside is sort of an illusion and semiclassical gravity is just an approximate tool for performing the computation. This means that the bulk doesn't exist at all.

These considerations lead to the really big question: What replaces spacetime at short distances? We know that the problem is more than a short-distance problem and associated with the fact that there are no bulk observables at all. The only observables we can think of live on the boundaries of spacetime. The example in which we understand this perfectly is AdS/CFT where the boundary is asymptotically AdS. AdS/CFT has the nice feature that, as far as all local questions are concerned, you know that the curvature of spacetime doesn't really matter. So, even though all the details are not understood, old questions about quantum gravity involving short-scale physics and Planckian scattering are in principle addressed by AdS/CFT. In order to understand what happens when you scatter gravitons at energies of about 100 times the Planck energy you can consider an imaginary universe that is not quite like ours because whether the curvature is negative or positive doesn't make a difference.

You mentioned that we understand these things within AdS/CFT *in principle* and I just wonder how much can we actually answer in detail? For instance, I remember Steven Giddings' papers on extracting the flat space S-matrix [15] and it was not exactly clear that all technicalities were understood.

I don't think it's as difficult as he says it is. He was pointing out what many people, including me, thought were some extremely technical issues that don't have real physical merit. In fact, people such as Katz, Fitzpatrick, Poland and Simmons-Duffin came up with a much sharper proposal for how to extract the S-matrix from the boundary CFT [16]. Instead of setting up the problem in terms of wave packets, they identified how to take the a specific limit of correlation functions that allows you to extract the flat space S-matrix. It's a very sharp and unambiguous proposal.

Okay. Going back to the previous discussion of observables, what is the role of nonlocality in all of this?

I haven't really used the word *nonlocal* yet because the point I am trying to make is deeper than that. When you say *nonlocal* it gives the impression that it's closely local but with some corrections that aren't local, right? However, the point is that the whole notion of local physics is wrong. Fundamentally, there is no local physics: physics only lives on the boundary. In that sense, it's horrendously nonlocal but obviously whatever this physics is, it will have some description such that when it's appropriate to talk about semiclassical geometry, it's approximately local.

This is a different type of nonlocal theory from what is usually considered. There are many dumb nonlocal theories that you could devise by writing down a Lagrangian with arbitrary interactions between distant points in spacetime. This would also be nonlocal but it's stupid because when you take a semiclassical approximation you would find nonlocal physics. On the other hand, the sort of magic of boundary physics is that it doesn't have spacetime built in it but in some approximation the spacetime appears and it is local.

This is infinitely more interesting than the idea of atoms of spacetime and it has the advantage that it can preserve all symmetries. The inside of spacetime can be fluctuating but whether or not the physics is Lorentz-invariant is a statement that can be measured and has a precise meaning in the boundary. In other words, this is the correct kinematic framework for discussing quantum gravity. We are still far from having concrete answers but, conceptually, the old problems of quantum gravity have been solved and, in principle, we can ask these questions in AdS.

Are you referring again to AdS/CFT for answers?

AdS/CFT tells you that any short-scale physics question about gravity can in principle be embedded into some conformal field theory (CFT) and has some answer. Even though we don't know practically how to do it in many cases, we know the answer. For instance: Is black hole formation a unitary process? Yes, it is.

Is that clear?

I think it's completely clear. Incidentally, there is something that is not very well known but it's true. In fact, it's spelled out in one of my least appreciated papers [17] (laughs). People often say that in principle you can imagine that the spacetime curvature is negative and they point out that there is no AdS solution of our real world. However, this is not true. Even our world has an AdS vacuum. So I think you'd be pretty hard-hearted to think that the problem of scattering gravitons at 100 times the Planck energy in $AdS_3 \times S^1$ is very different from our vacuum. That would be crazy, right?

What is the point that you are actually making in this paper [17]? That the standard model itself has many vacua?

The standard model itself has infinitely many vacua but that's a separate polemic point. For this discussion what is important is that the standard model has one vacuum that looks like AdS_3 and so there is some CFT out there which is dual to the standard model on that vacuum. We don't know what that CFT is yet but in that CFT the problem of black hole formation and evaporation is a manifestly unitary problem.

I'm sorry but is this vacuum a real vacuum?

It's a solution of our world (laughs). It's as real as anything else. Let me say this in another way. If you hand in to theorists that $AdS_3 \times S^1$ world without mentioning how the universe actually looked like, what do you think the theorists would find? Well, they would search around until they would find a CFT that matches that world and if they had infinite computing power they would find that such CFT would contain electrons, protons, mesons, W and Z bosons as well as everything else in such a way that it would look identical to the standard model and its spectrum (laughs). Okay, now is that real or not? I mean, if someone handed you tomorrow a 1+1 dimensional CFT exactly solved and with a spectrum that looked identical to the standard model together with everything else that might come at the LHC and beyond, would you consider that to be real or not? I think that's damn real (laughs). So in that world, I would be very happy to understand graviton scattering at 100 times the Planck energy which would be a very good approximation to the answer in our world, right?

Why do you keep referring to AdS/CFT?

As I mentioned earlier, spacetime has to be emergent and the observables are somehow associated with the boundary. AdS/CFT is a sort of gateway drug to the idea of emergent spacetime (laughs). It sort of warms you up to the idea but it's not the final story. It makes it possible to explore and understand these radical shifts in how we think about observables and spacetime without giving everything up at once. In particular, space, gravity, strings, extra dimensions emerge in AdS/CFT but not time. Time on the boundary is the same as time in the bulk of the space. The two theories that are dual to each other are both standard quantum mechanical systems, right? They both have time and evolve in time. You don't have to invent some completely new sort of kinematic framework and language to describe it. But unfortunately, or fortunately, we don't live in AdS spacetime. The predominant difference between AdS and our world is not at short scales but at very long scales where we find an evolving universe. Even without going into any details about what's difficult about cosmology and accelerating universes, the zeroth-order challenge is to understand how time emerges. This is a really tall order which is going to take us out of our comfort zone.

What exactly needs to be understood about our world?

We need to understand what are the correct observables in the context of cosmology. All cosmology starts with some kind of singularity (some sort of big bang somewhere) and, conceivably, can lead to many potential futures. What observables can we precisely define in this context? There's actually one pitiful observable that we could define. Imagine that the universe eventually expands and becomes approximately flat. For instance, you could consider a Friedmann-Robertson-Walker (FRW) universe with vanishing cosmological constant that expanded out and became infinitely big. The observable we can define precisely is not the S-matrix because we can't control the initial condition; that is, we can't control the initial big bang. However, in the far future you can consider an observer who looks at the universe and counts the total number of, say, red stars and blue stars. You can imagine doing this with an infinitely large apparatus and performing the experiment infinitely often. This observable you can in principle measure and might be the output of some putative theory. However, this is a really meagre observable and is usually referred to as the *wave function of the universe*. This wave function bears the same relation to the usual wave function in the context of quantum mechanics. In fact, the reason why we can make sharp predictions in quantum mechanics is because the wave function includes all possible subsystems that the experiment can be performed in, and is thus an eigenstate of the probability operator. The same happens in the cosmological setting: the universe becomes infinitely big and the wave function contains all possible subsystems, becoming, e.g., an eigenstate of the operator of the fraction of red stars.

The reason why the wave function of the universe is a meagre observable is because in the context of an accelerating universe, such as de Sitter spacetime, it is not an observable. In de Sitter spacetime every infinity that we need for defining precise observables is not accessible to us. In this setting, we're surrounded by a cosmological horizon which means that we can't make an infinitely large apparatus. Additionally, because there is a finite Hawking temperature associated with the cosmological horizon, it means that we are being slowly cooked (laughs). This implies that you can't do the experiment infinitely often as you cannot isolate the experiment. I think this is the deepest conceptual problem in physics today as it combines everything we know, from gravity to quantum mechanics, and ends up in one big fat contradiction. In other words, we don't know how to define observables precisely.

Can one solve these issues within string theory or is the theory incomplete?

In an asymptotically AdS universe, string theory is wonderful and, by the way, is also a quantum field theory. One of the most important lessons that we've learnt is that string theory is not a radical departure from conventional physics as we originally thought. It turns out that nature is quite clever in the way that it unifies all forces. In the 1980s people had this very cookie-cutter picture of what unification should be. The idea is that we should take the weak force, strong force, electromagnetism and gravity, and put them together in a kind of big jigsaw puzzle of a super-force so that everything is unified. That picture is a nice picture but it is incredibly naive and very likely wrong.

I believe that there isn't a unique vacuum of the universe but a gigantic landscape of solutions, which means that there is unification in a much deeper sense. Nowadays, we know that gauge theory and gravity don't have to be unified; that is, instead of being two different pieces of a jigsaw puzzle, they are actually the same thing. This is a much deeper and shocking statement. Furthermore, this notion that there are many possible vacua for the theory is, in a sense, a notion of unification on steroids (laughs). The existence of many vacua means that the forces are unified not only in this world but also in all the other possible worlds. It's a gigantic space of possible consistent theories that include gravity, gauge interactions and everything else. This is a far deeper and more satisfying picture of unification than this old cookie-cutter 1980s picture even though it makes life more difficult for making contact with experiments at accessible energies.

But is string theory also wonderful in flat and de Sitter spacetime?

In flat spacetime there could be a satisfactory answer but we don't yet have the analogue holographic description. However, there should be one. There is a sharp observable there and, in a sense, you could think about these recent developments in understanding amplitudes in gauge theory as proving some hints about what this dual description looks like for flat spacetime [18–25]. This is a line of research still under development but this description will not be the usual quantum field theory because time needs to emerge. Whatever the theory is in flat spacetime it cannot have time in it. This is why the objects that we are seeing in these remarkable mathematical structures are much less familiar and we have a lot less intuition for. However, we could have expected that the theory would start looking funny. The reason is that whatever the theory is, it contains building blocks out of which we can build local and unitary representations but which are definitely not the standard ones we encounter in quantum mechanical systems. Whatever it is, it will not be a mild modification of quantum field theory.

In the case of de Sitter spacetime, one can become suspicious of quantum mechanics. The suspicion is not that quantum mechanics is wrong but that it might not be applicable at all precisely because there is no precise observable that one can define. In the history of physics many of us have wondered if quantum mechanics is wrong or inadequate, most recently in the context of the black hole information paradox [26], but in all previous cases where this suspicion arises there has always been at least a candidate quantum mechanical answer for the questions that are being posed. For instance, in the case of the black hole information paradox, the quantum mechanical answer is that one needs to calculate the exact S-matrix. The S-matrix gives you the information about the radiation that is expelled if you scatter a shell of matter. If you handed in the S-matrix to Stephen Hawking he would have said, "Okay, I'm sorry I was wrong" (laughs). In this context, the debate is not about whether quantum mechanics is wrong but about whether one can calculate the S-matrix exactly. On the other hand, in cosmology the problem is that we don't know what object should be computed.

The foundations of quantum mechanics force you to have a separation between system and apparatus and one of those has to be infinitely big. When you use the words *quantum mechanics, gravity* and *cosmology* in the same sentence, that possibility is taken away from us. Thus the question is: What shall we replace it with? That is the most important problem that we face in the twenty-first century. This problem is suggesting some kind of generalisation or deformation of quantum mechanics, in fact, some extension that reduces to standard quantum mechanics in the limit of infinitely many degrees of freedom but which is rather different when you restrict to a finite number of degrees of freedom. Perhaps such theory will tell us that there is a precise way of defining fuzzy observables but we are really missing a new language which, in my opinion, requires a radical change in thinking similar to the one that took place when transitioning from classical to quantum physics.

So what kind of new mathematical structures do we need?

I don't know what we need but I am saying that there is reason to suspect that we need to modify quantum mechanics. This is why the whole program of trying to understand scattering amplitudes in flat spacetime is a nice intermediate-level problem which we might make progress on. I think this is the correct path for later solving the hard problem. The basic goal underlying the program is to understand whether standard physics can be described in such a way that the notions of locality and unitarity are not central. If such notions are in fact not fundamental then there must be a way of describing standard physics without using them.

There is a precedent in the history of physics where this kind of strategy was used. This was the case of the process of transitioning from classical to quantum physics. In that transition, determinism was lost, which was a big scary thing. If you imagine being a classical physicist in 1780 and you are visited by the ghost of theorist future who tells you that in 1930 determinism will be gone, what would you do with that information? Surely you are going to do something because it would be a massive alteration of your world view. The first thing you would probably do is to start thinking how determinism can be removed from classical physics but no way in hell will you guess the formulation of quantum mechanics. Most likely, you will begin by adding stochastic terms to Newton's law and all sorts of other idiotic things but you wouldn't guess the right answer.

There are two philosophies for how to do physics: there's conservative radicalism and radical conservatism. The former is bad and the latter is good. Conservative radicalism is the philosophy that we should be radical and relax assumptions of established physics. For instance, you can say that the speed of light isn't constant or that there is no Lorentz invariance. Any idiot could try to do that but this dumb strategy never worked in physics. Nature works in more interesting and subtle ways. Radical conservatism, on the other hand, is the philosophy that you should take what is established as seriously as possible and try to understand it in as many different ways as you can to sort of do the groundwork for the next theory.

Following the philosophy of radical conservatism, if you were a classical physicist and you wanted to find out how to give up determinism you could try to formulate classical physics in a way that doesn't make determinism such an important central notion. Of course determinism must be there somehow, as classical physics is deterministic, but you would like to formulate classical physics in such a way that the concept of determinism is not an input from the beginning. And, in fact, that way of formulating classical physics exists: it's the principle of least action. When people such as Euler, Lagrange and others formulated classical physics in this way, they were very surprised by it. Conceptually, it is a completely different starting point to consider the possibility that particles snip out all the ways of going from A to B. Choosing the path that minimises the action doesn't look deterministic at all but because you constrain the variations of the action in a particular way, minimising the action is equivalent, at the end of the day, to Newton's laws. Nevertheless, it's a very important development because it does not make determinism manifest; instead, determinism is an output. Once you have such a formulation where determinism is not an input, you may start to wonder why such formulation is even possible. The reason why it is possible is because the world is in fact quantum mechanical. If you start with quantum mechanics and take the $\hbar \to 0$ limit you end up with the principle of least action, not with Newton's law.

The strategy of taking established physics seriously and rewriting it in ways that don't make the notions that we are worried about to be fundamental is the best way to make progress when we are suspicious that a really big jump toward a new framework is necessary. This is how we should proceed with the notions of locality and unitarity. The structures that are emerging are the result of understanding scattering amplitudes in gauge theory. If at some point these structures are well understood then we will be in a similar situation to that triggered by developing the principle of least action. This would be a dream come true. After that, we can contemplate how it could be modified or what kind of new framework is necessary, for instance, for describing deformations away from local physics.

So ultimately you think that you can use these techniques and insights for understanding gravity itself without strings?

We still refer to the collection of correct ideas about, and consistent theories of, quantum gravity as string theory for historical reasons. We've known for a long time that there are corners of what we refer to as string theory in which there are no weakly coupled strings. For example, we know that the ultraviolet completion of 11-dimensional supergravity does not have any string-like structures in it; instead, it has membranes as microscopic objects. So we don't necessarily need strings to understand some corners of quantum gravity.

In any case, some corners of the theory have a description as weakly coupled strings which, by the way, is an established result that will stick around in physics. We know that there is a QCD string [27] and we know that a string description appears in the limit of strong 't Hooft coupling of $N = 4$ super Yang–Mills theory [13]. These are the type of string theories that were developed around 1985. And it is the set of correct theories of quantum gravity that led in past 15–25 years to the understanding of what is the biggest conceptual problem about quantum gravity. That problem is that we don't know how to define observables in cosmology.

Do we currently have a complete theory of quantum gravity?

Before even talking about any theory, if we are hoping to understand cosmology, we need to understand what we are going to predict. So it is not just a problem for string theory; it is a problem for physics (laughs). In the context of the landscape, eternal inflation, the multiverse and so on, people have tried to justify certain observables [28–30]. For instance, you can imagine there being regions of the multiverse where you tunnel into flat spacetime.

In those regions, you recover that meagre observable that I mentioned which can count the fraction of red stars, blue stars, etc. This is what Leonard Susskind refers to as *census taker observers* [31–33]. In my opinion, it's a desperate attempt to cling to something that you can predict sharply. Even though a complete geometric description of the collision of bubbles of universes in the context of eternal inflation is yet to be understood, this observable could in principle measure how many bubble collisions take place in different parts of the sky. Perhaps this observable is a perfectly sharp observable from the point of view of a census taker observer, but why should we give a damn about such observer (laughs)? We are not that observer, not even close. This is very different from the case of the S-matrix because we can get really close to extracting the actual S-matrix from LHC experiments. Whatever we need to understand about cosmology will ultimately require some conceptual revolution.

From your answers so far, do I understand correctly that you are not a big fan of approaches that predict some kind of atomic/discrete structure of spacetime, such as loop quantum gravity (LQG)?

Yes, I am not a fan. The answer to this kind of question is often portrayed in partisan terms. In my opinion, LQG has not given any new insights into quantum gravity. I think this is because they ignore these deep lessons about gravity and instead focus in an absolute maddeningly way on questions about formalism rather than questions about physics. For instance, they state that the theory should be background-independent. But who wants that? This might be your private philosophical wish of what the equations should look like but the actual physics is most certainly background-dependent. If the actual observables live on the boundary then they will depend whether the boundary is AdS or something else. I think that loop quantum gravity has not yet appreciated the point that there are no local observables.

Furthermore, this aspect of the problem seems to be extremely confused with issues of general covariance in the LQG and general relativity communities. They seem to think that general covariance is a very deep and important thing. However, in fact, it's exactly the opposite: diffeomorphisms don't even exist. Diffeomorphisms are gauge redundancies and not real symmetries. They are mental conveniences that help us talk about physics. We have learnt time and time again in the development of string theory that diffeomorphism invariance is not a deep thing. If it were, it would have been impossible to discover AdS/CFT, which states that gauge theory is equivalent to gravity. Where is the diffeomorphism invariance in the gauge theory? Where is the gauge symmetry in gravity? When I was studying this as a student I was asking these questions but I realised later that it just didn't make sense.

Is AdS/CFT a background-independent formulation of quantum gravity?

It's background-independent once you pick the asymptotics, which is the only degree of background independence that you can ever have. Anyway, LQG people are really obsessed with diffeomorphism invariance, background independence and these sort of formal mathematical technical issues. However, we have so far spent half an hour talking about the physical fact that there are no local observables which they don't appreciate. What do they claim the observables are? I mentioned earlier that if someone comes and hands you a whiz-bang theory of quantum gravity, you should ask: What does it predict and what are its observables? LQG tells you that "the observables are the eigenvalues of the area operator." Okay, great! Let's say this is true, you work till you're blue in the face and you compute the eigenvalues of the area operator up to 3000 decimal places, right? But the question is: What measurement measures that value? It's nonsense! There's no measurement that measure it because there are no local observables (laughs).

There are additional issues. LQG is so obsessed with background independence but, ironically, it cannot describe the backgrounds we care about, such as flat spacetime. Observables live in the boundary, but how can they describe the boundary if they don't even have a spacetime? Also, I find particularly amusing that there are people in that community that claim that LQG breaks Lorentz invariance and others who claim that it doesn't. So again, they are extremely concerned with background independence and yet some claim that there is a notion of a preferred frame and that you can tell which observers are moving and which are not as opposed to a Lorentz-invariant theory. This does not make sense.

Another thing that is discussed within LQG is the nature of time. Because they are so much in love with diffeomorphism invariance they say: "It's very hard to talk about any observables since there's no time." They are really concerned about time, what it is and whether it is there or not. They argue that time is relational, that one should be talking about clocks and how they change, etc. I think that this entire business of relational observables and diffeomorphism invariance is one of these things that if you use words you can spend way too long in issues that get resolved instantly with a few equations (laughs). This issue is all about gauge redundancy. Since there is gauge redundancy, there's no preferred time variable. Different redundancies can undo time translations and so on. So, how do you solve this problem? You just have to fix a physical gauge. In that gauge, every quantity is by definition gauge-invariant. In that gauge there's a Hamiltonian and everything evolves forward in time. In this case all is fine, there is no problem. Now, if you really want to, you can go back and speak about all these things in a relational way.

This is similar to electromagnetism where we fix a gauge and describe how an electron here repels an electron there. You can also try to describe this in a gauge-invariant way which is a pain in the ass because you have to introduce Wilson lines stretching all the way from one electron to the other. The point of gauge symmetry isn't to wallow in its wonderfulness but to use it to your advantage. Gauge symmetry helps you and the way it helps you most is in fixing the damn gauge that makes the physics manifest that you want to make manifest (laughs). So once you fix a nice physical gauge, you don't find any ghosts, you have unitary evolution and in that gauge everything is gauge-invariant by definition. So all these problems with time, relational observables, etc., only arise because there's too many words and too few equations.

But at which point do the actual problems appear? All the issues that I've discussed so far do not show up at any perturbative order in Newton's constant G. These issues only show up at order $e^{-1/G}$ so it's really not a perturbative effect. It's an extremely subtle effect which tells you that the exact theory can only have observables living at the boundary. Anyway, these are the reasons why I am so unimpressed with LQG. It seems to disregard such subtle facts and instead dive into problems which are sort of trivial or wrong.

Before we continue with the issue of gravity, I remember from one of your papers that you studied quantum field theories that exhibit macroscopic nonlocality, such as the Dvali-Gabadadze-Porrati (DGP) model [34]. Is this the same kind of nonlocality and are these theories consistent?

It is possible to write down Lagrangians that look Lorentz-invariant but that can exhibit macroscopic nonlocality. What happens to be the case is that such theories, though naively looking consistent, cannot be ultraviolet complete and so can only be effective. But what is important to stress is that such kind of nonlocality is very different from the kind of nonlocality that I have been telling you about. The type of nonlocality that is important and relevant for gravity is a subtle effect that is irrelevant unless you're close to the Planck length or in macroscopic situations for which its effect is noticeable. I have not stressed enough this latter case. As I mentioned, these effects are of order $e^{-1/G} \sim e^{-S}$ where S is some entropy and hence is small. However, if there is a question or problem in which the number e^{+S} makes an appearance, then there is a chance that these tiny effects can be amplified and change the conclusion from the naive one. This happens in the case of black hole formation and evaporation problems. In this context, in which you have to understand what radiation comes out of the black hole and whether the evolution is unitary or not, you actually have to measure all of the Hawking quanta that come out. But there is an order S quantum that comes out so the number of measurements you have to make is of order e^{+S}. In these cases there is an amplification of incredibly tiny nonlocal effects that can change the answer by order one.

Let us contrast this with the type of macroscopic nonlocality discussed in DGP-like theories [34]. In this case, there is nothing subtle about that type of nonlocality. You can take lumps of material within this theory, shine light with a flashlight and transmit information to Alpha Centauri as fast as you want. There is nothing subtle that needs to be amplified to be noticeable; instead, it's *in your face as big as possible that you can get nonlocality* (laughs). Basically, you can send instantaneous messages across the universe. This type of theory is in a completely different universality class and this massive amount of nonlocality is by no means necessary nor required, neither there is any evidence for it.

If this amount of nonlocality was true then all string theories would be wrong. Several people tried to realise these setups in string theory and always failed. What we found in this paper is that the reason why they failed is because such theories are just extremely nonlocal [34]. I should say, however, that such theories are nonlocal in a much more interesting and subtle way than the absolute dumbest way that you could construct a nonlocal theory which is by adding nonlocal operators by hand (laughs). It's a fascinating theoretical structure and I would be much happier if I knew what its purpose in life was. It's such a nice theoretical object and yet it doesn't seem like it could possibly exist in nature. It would be interesting if it found a home to describe maybe some condensed matter system or something else.

Do you want more tea?

Yes, please.

So the issue of observables is the most important problem we face today, but are there other problems you consider important?

There is a second class of important problems, as I mentioned earlier, which are fine-tuning problems, such as the hierarchy problem and the cosmological constant problem. Famously, you have to fine tune one part in 10^{120} to get a vacuum energy small enough to explain why we have a macroscopic universe, right? Why is there a macroscopic universe at all when there are violent power law quantum fluctuations at short distances? It's a zeroth-order fact about the world which we shockingly don't have a good explanation, right (laughs)? But despite the fact that we don't have a good explanation, we can predict g^{-2} up to 12 decimal places regardless of fine tuning the vacuum energy to one part in 10^{120} or fine tuning the Higgs mass to one part in 10^{32}. There is no inconsistency in the equations if we perform such fine tuning and we can proceed as if everything were fine. This is unlike the problem of observables in quantum gravity, which is for a fact a problem of consistency. However, we suspect that there's something terribly wrong with what we're doing. It's a suspicion and so it doesn't have the same status as the first class of problems.

These fine-tuning problems such as the hierarchy and the cosmological constant problem look very similar. In fact they're just two different terms in the expansion of the vacuum energy, right? There's a piece that doesn't depend on the Higgs field and a piece that depends on the Higgs field, both being leading terms in the effective potential. Obviously, the cosmological constant problem is the most severe of the two so you should try to solve it first. The ideology of naturalness tells us that when we confront problems such as these, new physics should appear at the appropriate scale to actually remove the fine-tuning issues. It is an ideology but it should be stressed that it has been successful several times in the past. In the last century, it worked at least in three different occasions.

The most famous and earliest example was the problem of infinite classical self-energy of the electron. This is a problem that vexed Abraham, Lorentz and others in the late 1800s and early 1900s [35]. If you naively calculate the energy stored in the electric field around the electron you find that it is infinite. The way people tried to make sense of it was by assuming that at some scale L the energy contained down to a distance L is comparable with the energy mc^2 of the electron. The scale obtained in this way is called the *classical radius of the electron* and at that scale new physics should appear. To understand this new physics people tried to build models of the electron as an object with a given shape and size as well as some sort of shell-like models but none of it worked. In fact, these attempts are the analogue of technicolour with respect to the hierarchy problem [36] in the sense that you try to build some composite state at the right scale (laughs). However, the problem did have a solution and quite remarkably both quantum mechanics and relativity needed to be understood. The solution is due to the existence of positrons; in particular, a cloud of electrons and positrons smear out the point-like electron and remove the leading divergence in the self-energy. In fact, it was understood that new physics, namely the existence of positrons, appeared at a scale which is 137 times bigger than the classical radius of the electron, much earlier than naively expected.

The other example where this ideology was fruitful was in understanding pions. There are two pions, π_0 and π_+ and they don't have the same mass. Part of the reason why they don't have the same mass is attributed to the existence of up and down quarks. However, completely disregarding the existence of up and down quarks would still result in the two pions having different masses. Because π_+ is charged and π_0 is not, you can draw a one-loop diagram with a loop of the photon that corrects the mass of the charged pion. That correction is exactly the same encountered in the analogous diagram in the standard model, namely the Higgs quadratic divergence, in which case you run into the same problem. So π_+ should be much heavier than π_0 and new physics has to appear at some scale to tame the divergence. To get an estimate for the cutoff you need to introduce the observed mass splitting between π_+ and π_0, which leads to something like 1.5 GeV. However, it turns out that the cutoff is determined by the ρ meson and, as in the previous example, it appears much earlier than naively expected, at about 770 MeV.

The third example where this point of view worked well was in the prediction of the charm quark. Glashow, Iliopoulos and Maiani (GIM) realised that there was a quadratic divergent contribution to $q\bar{q}$ mixing and that new physics had to appear at some scale [37, 38]. That new physics was the existence of the charm quark. Later, Gaillard and Lee computed that it should appear before 5 GeV [39].

In summary, this ideology worked a number of times in the past so it's not like whoever follows it must be smoking crack (laughs). It has a successful track record. But the difficulty that we face with applying this to the cosmological constant (CC) problem is the fact that the length scale associated with the CC is around the millimetre. By the same token, this would lead you to suspect that new physics must appear at the millimetre scale. However, anyone can pick up a ruler with a millimetre scale on it and see that there isn't any new physics at that scale. Thus, this logic fails miserably for the case of the CC. You could think that the reason it fails is because now you have gravity, cosmology and it's different and confusing. That could very well be correct and not all problems are solved with this ideology. So let us ignore this problem for a while and move on to the hierarchy problem.

In the case of the hierarchy problem there are candidate solutions. One solution is the exact analogue of the case of the positron in solving the issue of the infinite classical self-energy of the electron, which is the discovery of superparticles. In the case of superparticles, there is an extension of spacetime symmetry to supersymmetry that implies that anti-particles have to exist. This leads to a doubling of the spectrum which solves the old hierarchy problem. If we were to find supersymmetry, textbooks would describe the solution to the hierarchy problem precisely in this way. There is additional circumstantial evidence that supersymmetry is right such as unification of couplings. However, I keep worrying about the CC because supersymmetry certainly does not address it.

Is this why you have tried to find alternatives to supersymmetry?

I love supersymmetry. Supersymmetry has been largely responsible for making me interested in particle physics and I worked on it while I was a graduate student [40–44]. I was not a fan of particle physics since the beginning of my studies because it appeared to me as an incredibly long, expensive and boring way of peeling one layer of the onion past another layer as you go to shorter distances. But the picture of gauge coupling unification seems to suggest a completely different point of view and something spectacular happening that was not put in by hand into the theory. Nevertheless, the reason why I began searching for alternatives, apart from intellectual fun, was the worry that something was wrong.

The CC problem sort of casts a giant shadow over every attempt to solve the hierarchy problem because it's a much bigger problem, right? Thus, I was motivated to think about alternative approaches to solving the ordinary hierarchy problem. One of these alternatives are models with large extra dimensions [45–48]. The most interesting aspect of these models is that you could actually live with these extra dimensions, which are gigantic compared to what anyone had discussed before. One of the things I was excited about, even though the original proposal didn't have this feature, was the idea that you could modify the physics at long distances in such a drastic way and not have known it. This seemed to be the kind of idea that was needed in order to solve the CC problem. Since we could tolerate millimetre-size extra dimensions, maybe we could have an infinite continuum of particles at short distances that in some way would help to smooth the CC problem. Furthermore, it could help solving the CC problem in such a way that avoided all the no-go theorems that Steven Weinberg and others had established [49].

Another type of theory that I worked on are more conservative bottom-up theories that could be thought of as extensions of technicolour, which deal with composite Higgs particles [50, 51]. There are a lot of different possibilities for weak-scale physics but I always felt that there was something wrong in all approaches to the CC problem.

Are all these approaches to weak-scale physics compatible?

Not at all. Working on model building is a quite different psychological activity compared to working on the first class of problems which you know will take 100 years to solve. In the case of model building there is a more direct contact with experiment and I found it extremely useful to, while performing such activity, adopt the philosophy which you might refer to as *serial intellectual monogamy*. What this philosophy entails is that while you work on a specific model you're completely devoted to it and you convince yourself that there is a big chance of it being true. This is a psychologically useful point of view because it makes you work harder and overcome obstacles that you wouldn't overcome otherwise. Then, after you are done working on a specific model you move on to something else. So I have no emotional connection to any of these ideas whatsoever. If they're wrong or if they're right I don't really care.

As a matter of a fact, while I was working on these ideas, evidence was found for a non-zero cosmological constant and an expanding universe. The persistent efforts I made on thinking about the cosmological constant 80% of the time during eight years of my life had hardly anything to offer [52–54]. They all now seem to be desperate attempts. The fact that a non-zero value for the CC is the simplest explanation for an expanding universe and that its value was so damn close to the dangerous value that would rip galaxies apart was unexpected to me. In addition, I ran into Weinberg's anthropic argument for the CC as a grad student and it completely hit me like a ton of bricks [55]. I walked around for a week or two thinking, "If this is right, what the hell do we do? If it is right how should we do physics in the future?" I think Joseph Polchinski described his reaction in a similar way. In fact, because I thought so much about the CC problem as a grad student my experience was more like "Oh shit, Weinberg is right" followed by "Fuck!" (laughs). I was worried that it would be the end of physics. Once you accept this kind of worldview, in particular, that the CC is environmentally selected, that there is a multiverse of populated universes, then at that moment it becomes a reasonable solution to the problem.

So you think that Weinberg's argument is the best solution to the CC problem?

I think it's the only scientific solution that I know that actually works. It could have been falsified but it wasn't falsified. People usually refer to this as an anthropic picture but it has very little to do with people. It is a quite binary difference of whether the world is empty or not. If you buy the starting point, namely that there are a lot of different vacua and that there is some mechanism populating all of them such as eternal inflation [56] then in such picture it is not at all surprising that we find ourselves in a world that looks like the one we are in right now. It's not a tautological argument and it's not putting human beings in the centre of all things. In fact, it's quite the opposite, right? With this point of view we are 10^{500} times less significant than we were before this picture of the world came along (laughs). In the same way, we don't think that it's a big mystery that we live on a rock that occupies a volume of 10^{-60} in Hubble units. Thus, it's not a shock that we find ourselves in these very finely tuned pockets that aren't empty. I think this is a completely reasonable explanation and at the moment it is by far the best solution to the problem that I know of. All my own attempts to find dynamical mechanisms to solve the CC problem failed utterly. They don't work morally nor technically.

This anthropic point of view is criticised by many people ...

There are all sorts of people with very violent reactions against this explanation. I have found out that the people who are most vehemently sure that there's some alternative solution to the CC problem are the people who have not spent any time thinking about it. The people who are more sympathetic to it are the people who've spent a huge amount of time thinking about the CC problem such as Weinberg, Polchinski and myself (though I don't put myself in the same category, I however did spend a gigantic amount of time thinking about it).

Weinberg's solution is a simple solution in a way as it doesn't involve any equation but makes a rough prediction for what the value of the CC must be. This way of thinking has the same sort of flavour of other similar solutions in the past in which you change the character of the question that you are trying to answer. It's like you rotate your head 30 degrees and you don't have to do any work at all. However, you have to buy this totally different picture of what the world looks like at gigantic scales. Instead of finding a clever dynamical mechanism, you have to make sense of this crazy picture of eternal inflation, the multiverse and how to think about quantum mechanics and cosmology. Understanding the smallness of the CC becomes inexorably intertwined with the deep problems of gravity that we need to understand anyway. We need to understand how to think about these inflating cosmologies and what the observables are.

What about the hierarchy problem? Do you think there are good solutions to that?

I think we are in a really fascinating and incredibly tense bifurcatory moment as far as particle physics and the status of the hierarchy problem are concerned. There are many candidate solutions to the hierarchy problem, some of which predict unification of the couplings, dark matter and so on, which I find a beautiful body of work. Large extra dimension models are in the class of natural explanations. They would be particularly exciting because there would be all these different measurements that could be performed. But extra dimension models are not such a deep concept compared with supersymmetry and they cannot explain the unification of couplings, unlike supersymmetry.

However, we wouldn't be talking about the anthropic point of view or the multiverse if there was any shred of evidence for this whole kit and caboodle of natural theories even before the LHC started running. If you had told theorists in 1985 that they had solved the hierarchy problem and asked where they expect to see new particles they would mention LEP, not the LHC. As higher energies became available both direct and indirect limits on the masses of the expected particles became more constrained and higher and higher. In the early 2000s and late 1990s, the first run of LEP was ending and people started talking about the *LEP paradox* which was the fact that the solution of the hierarchy problem was in tension with experiment [57].

This tension is still there today and it's hard to pinpoint what is actually wrong. You start thinking if you are wrong about some small detail or about a whole collection of things. You start doubting whether the theory at the weak scale is natural or not. This thought led to what I personally think was my best idea, namely, the idea of split supersymmetry (SUSY) that I developed with Dimopoulos [58, 59]. If split SUSY turned out to be true I would be very proud, not because it would be true, but because it's a deep idea. Split SUSY gives up on naturalness but preserves all other successes of SUSY and has remarkable experimental predictions associated with it [60]. If we would actually observe split SUSY or something similar, I think that would really convince almost everybody that there's some kind of multiverse picture at work because we have sharp evidence for fine tuning. If we would only see the Higgs particle and nothing else then there'll be some kind of moderate evidence for tuning, maybe one part in 1000. But if you see split SUSY and you see half of the superpartners, it could be one part in one million tuned, which is like winning the lottery. If it turns out to also be one part in 1000 we will have to wait for the LHC to see all superpartners. So, this is my picture for what's going to happen at the weak scale; it won't be large extra dimensions or some natural version of SUSY.

Now I would like to ask you about dark matter. Have particle physicists always been interested in dark matter?

I think it has always been a question that particle physicists were interested in because it is a natural prediction of particle physics theories that solve the hierarchy problem. As I mentioned earlier, you can solve the hierarchy problem by introducing new physics at the weak scale and it happens frequently that this new physics includes a new stable particle. In the context of SUSY, the usual candidate dark matter particle is the lightest superpartner. The reason why this is the case is simple. Whatever the theory is at the weak scale, if it introduces new physics then it better respect baryon and lepton number to quite high accuracy because baryon and lepton number conservation isn't badly broken, right? This means that the new particles must be carrying some non-trivial baryon and lepton number. The only thing you now need to assume is that amongst the new particles there will be one fermion whose baryon and lepton number is zero. Any such particle can't decay. Hence, they are candidates for dark matter particles.

There are other remarkable facts about dark matter that point towards the weak scale. If you have some heavy particle created during the big bang and you ask how much of it is left at low temperatures, you can perform a standard calculation in cosmology in order to find the abundance of dark matter that you need to explain current observations. This calculation tells you that this heavy particle must have a mass and cross-section for annihilating comparable to the weak scale. This is a remarkable coincidence that usually is referred to as the *WIMP miracle* [61]. It could have turned out to be 20 orders of magnitude off from the weak scale in either direction but it did not. Because of these reasons, I think particle physicists always had an interest in dark matter.

Recently, there have been many new developments in particle physics partially stimulated by lots of interesting astrophysical anomalies that people were trying to explain with dark matter. It turned out that you could not explain these anomalies with dark matter using the most common ideas of what dark matter could be. New models and ideas were needed and so people became interested in building particle physics models of the dark sector that contained more than just a single particle in order to explain these anomalies.

So the idea is that instead of dark matter being composed of just one particle, it could be a collection of many particles?

Yes, a collection of different particles with their own gauge interactions and so on, potentially giving dark matter explanations to a lot of astrophysical anomalies. Many of these anomalies are still unresolved and so this field is very active. So, instead of thinking about dark matter as some appendage to the particle physics model that allowed many to claim in their talks "I have dark matter," people started thinking about what else could be in the dark sector beyond just one particle.

There is a joke around this. Suppose you're made out of dark matter and you were trying to do cosmology in your world. You discover that you are made of about 30% of dark matter and that there is this big cosmological constant. Later, very precise measurements show that actually 3% of the world is not composed of dark matter nor the cosmological constant. You then ask: What could it be? So you go on and think very hard about it and conclude that it's a single neutral fermion. Then someone else comes along and says: "No, no, no, it's an $SU(2) \times SU(2) \times U(1)$ gauge theory with three families and all that." It sounds completely nuts, right? But of course that it is the right answer and in hindsight we are completely happy with it (laughs).

So this mindset led to a lot of activity in dark matter phenomenology and model building. One fruitful outcome of this endeavour is the realisation that there is a whole other way of doing frontier physics and a whole other set of experimental questions that you could ask which does not require going to very high energies. There can be other sectors with lots of light particles but which have some weak interactions with the standard model sector. These interactions wouldn't be as weak as gravitational interactions but would be weak compared to the standard model by a factor of 10^{-3}. Amazingly, this possibility is compatible with everything we know and it means that you could have a bunch of GeV particles moving around with couplings to ordinary matter of 10^{-3}, which could play a role in dark matter phenomenology.

There is a whole experimental program that is now motivated by this possibility. I think that the future of particle physics will be to explore such possibilities rather than going to higher energies because you could explore it without building gigantic accelerators. In fact what you need to do is to have very high luminosities so that you can efficiently probe that sector. Furthermore, if nature is supersymmetric then this new sector is supersymmetric too. Thus, you could probe the hell out of supersymmetry in that sector rather than busting your ass to go to higher and higher energies in our sector.

If I remember correctly, the theory of dark matter that you proposed [62] solved all these astrophysical anomalies, right?

Amazingly, it did solve the anomalies. But I must say that in the meantime astrophysicists gave astrophysical explanations for all these anomalies as well. For instance, there are predictions for what the cosmic ray spectrum of the positron should look like. Some time ago there was a burst of excitement because the observations of the PAMELA satellite showed that the positron fraction went down and back up again. This is what got me excited because it seemed a quite dramatic behaviour (laughs). Later, astrophysicists came out of the woodwork and said that all this could be explained by pulsars. Particle physicists learnt that pulsars can explain anything. There's even a joke that I like to make which is if the curve of the positron fraction spelled out "I am dark matter" the astrophysicists would say, "No problem, pulsars can do that" (laughs).

Anyway, not all anomalies have yet been satisfactory understood. The astrophysical explanations are not more compelling than the dark matter explanations. Some of them are less compelling and it's ironic because it's standard physics whereas dark matter is new physics. Standard physics is so poorly understood that they have far more free parameters than in dark matter models. These models make a lot of predictions for what PLANCK should see and, in fact, PLANCK, as well as the LHC in some cases, could rule out all these models.

There is also another set of anomalies in direct detection of dark matter that was explained by the same models. In particular, we gave an explanation for the famous observation that the DAMA collaboration made years ago [62], namely, the observation of the annual modulation which measures the motion of the earth through a wind of WIMPS. Our explanation was based on concrete models and older ideas of Neil Weiner and Dave Tucker-Smith on inelastic dark matter [63]. Nowadays, this particular explanation has been conclusively ruled out by the XENON collaboration. The specific anomaly we discussed does not need an explanation because it turned out that it is not actually there.

This is a thriving field driven by a huge amount of data. The next generation of detectors will have a good chance of seeing some evidence for WIMPS. If they don't, I think we will have to wait at least 10 or 15 more years.

In your opinion, what has been the biggest breakthrough in theoretical physics in the past 30 years?

AdS/CFT. No question about it.

Why have you chosen to do physics?

What really drew me into the physics when I was about 13 years old was the realisation that physics had the power to actually predict things. I remember learning that a space shuttle needed to travel at 11 km/s to get out of the earth's gravitational pull. At that point I started wondering where that number came from. And once you learn Newton's laws you can find that number yourself. This had a deep impression on me, that is, the fact that simple laws enabled people to figure things out all by themselves. Then, after you learn about quantum mechanics and relativity you realise that there are deep laws expressed in simple mathematical form organised in multiple layers removed from everyday experience. As you go into it you end up staring at deep questions and driven towards the right answers. This is what I find most interesting about this activity.

What do you think is the role of the theoretical physicist in modern society?

In a world that is becoming increasingly more crazy, I think that the most important thing we have to offer to the world is a unwavering belief in the idea of truth. As a theoretical physicist you spend a lot of time stumbling around but if you get in the basin of attraction of the correct answer, you just feel the truth sucking you in until you uncover it. The power of the belief in the truth, without blinding yourself, is something that has guided us for 2000 years and is the bedrock of western society for the past 400 years. However, I am worried that we are currently losing track of it. So I think, as a theoretical physicist, we have to keep the beacon of truth alive.

Theoretical physicists are human beings with emotions, jealousies, envies, hatreds, loves as everyone else but I think that, even though it's not always obvious from person-to-person interactions as people can be arrogant, theoretical physicists also have a deep kind of humility. This kind of humility is such that you believe in your smallness relative to the kind of questions that you are asking. This is the idea that there is something greater than yourself and that not only you are devoting your life to it but also that entire groups of people devoted generations of their lives to explore these questions and push the limits of knowledge further. It's as noble and pure a quest as human beings have ever embarked on. So, I think it's really important for us to keep showing people that truth matters, it exists, it's worth pursing and that it's bigger than all of us.

References

[1] J. F. Donoghue, "Leading quantum correction to the Newtonian potential," *Phys. Rev. Lett.* **72** (1994) 2996–2999, arXiv:gr-qc/9310024.

[2] J. F. Donoghue, "General relativity as an effective field theory: the leading quantum corrections," *Phys. Rev. D* **50** (1994) 3874–3888, arXiv:gr-qc/9405057.

[3] N. E. J. Bjerrum-Bohr, J. F. Donoghue and B. R. Holstein, "Quantum corrections to the Schwarzschild and Kerr metrics," *Phys. Rev. D* **68** (2003) 084005, arXiv:hep-th/0211071. [Erratum: *Phys. Rev. D* **71** (2005) 069904].

[4] J. Faye, "Copenhagen interpretation of quantum mechanics," in *The Stanford Encyclopedia of Philosophy*, E. N. Zalta, ed. Metaphysics Research Lab, Stanford University, winter 2019 ed., 2019.

[5] H. Zeh, "On the interpretation of measurement in quantum theory," *Found. Phys.* **1** (1970) 69–76.

[6] G. 't Hooft, "Dimensional reduction in quantum gravity," *Conf. Proc. C* **930308** (1993) 284–296, arXiv:gr-qc/9310026.

[7] L. Susskind, "The world as a hologram," *J. Math. Phys.* **36** (1995) 6377–6396, arXiv:hep-th/9409089.

[8] R. Penrose, "Twistor algebra," *J. Math. Phys.* **8** no. 2 (1967) 345–366, https://doi.org/10.1063/1.1705200.

[9] R. Penrose and M. MacCallum, "Twistor theory: an approach to the quantisation of fields and space-time," *Physics Reports* **6** no. 4 (1973) 241–315. http://www.sciencedirect.com/science/article/pii/0370157373900082.

[10] B. S. DeWitt, "Quantum theory of gravity. 1. The canonical theory," *Phys. Rev.* **160** (1967) 1113–1148.

[11] B. S. DeWitt, "Quantum theory of gravity. 2. The manifestly covariant theory," *Phys. Rev.* **162** (1967) 1195–1239.

[12] B. S. DeWitt, "Quantum theory of gravity. 3. Applications of the covariant theory," *Phys. Rev.* **162** (1967) 1239–1256.

[13] J. M. Maldacena, "The large N limit of superconformal field theories and supergravity," *Int. J. Theor. Phys.* **38** (1999) 1113–1133, arXiv:hep-th/9711200.

[14] E. Witten, "Anti-de Sitter space and holography," *Adv. Theor. Math. Phys.* **2** (1998) 253–291, arXiv:hep-th/9802150.

[15] M. Gary and S. B. Giddings, "The flat space S-matrix from the AdS/CFT correspondence?," *Phys. Rev. D* **80** (2009) 046008, arXiv:0904.3544 [hep-th].

[16] A. L. Fitzpatrick, E. Katz, D. Poland and D. Simmons-Duffin, "Effective conformal theory and the flat-space limit of AdS," *JHEP* **07** (2011) 023, arXiv:1007.2412 [hep-th].

[17] N. Arkani-Hamed, S. Dubovsky, A. Nicolis and G. Villadoro, "Quantum horizons of the standard model landscape," *JHEP* **06** (2007) 078, arXiv:hep-th/0703067.

[18] N. Arkani-Hamed, F. Cachazo, C. Cheung and J. Kaplan, "The S-matrix in twistor space," *JHEP* **03** (2010) 110, arXiv:0903.2110 [hep-th].

[19] N. Arkani-Hamed, J. Bourjaily, F. Cachazo, and J. Trnka, "Local Spacetime Physics from the Grassmannian," *JHEP* **01** (2011) 108, arXiv:0912.3249 [hep-th].

[20] N. Arkani-Hamed, J. L. Bourjaily, F. Cachazo, S. Caron-Huot and J. Trnka, "The all-loop integrand for scattering amplitudes in planar N = 4 SYM," *JHEP* **01** (2011) 041, arXiv:1008.2958 [hep-th].

[21] N. Arkani-Hamed, J. L. Bourjaily, F. Cachazo, A. Hodges and J. Trnka, "A note on polytopes for scattering amplitudes," *JHEP* **04** (2012) 081, arXiv:1012.6030 [hep-th].

[22] N. Arkani-Hamed, J. L. Bourjaily, F. Cachazo and J. Trnka, "Local integrals for planar scattering amplitudes," *JHEP* **06** (2012) 125, arXiv:1012.6032 [hep-th].

[23] N. Arkani-Hamed and J. Trnka, "The amplituhedron," *JHEP* **10** (2014) 030, arXiv:1312.2007 [hep-th].

[24] N. Arkani-Hamed, J. L. Bourjaily, F. Cachazo, A. B. Goncharov, A. Postnikov and J. Trnka, *Grassmannian Geometry of Scattering Amplitudes*. Cambridge University Press, 4, 2016. arXiv:1212.5605 [hep-th].

[25] N. Arkani-Hamed, T. Lam and M. Spradlin, "Non-perturbative geometries for planar $\mathcal{N} = 4$ SYM amplitudes," arXiv:1912.08222 [hep-th].

[26] A. Almheiri, D. Marolf, J. Polchinski and J. Sully, "Black holes: complementarity or Firewalls?," *JHEP* **02** (2013) 062, arXiv:1207.3123 [hep-th].

[27] D. Mateos, "String theory and quantum chromodynamics," *Class. Quant. Grav.* **24** (2007) S713–S740, arXiv:0709.1523 [hep-th].

[28] L. Susskind, "The anthropic landscape of string theory," arXiv:hep-th/0302219.

[29] B. Freivogel, Y. Sekino, L. Susskind and C.-P. Yeh, "A holographic framework for eternal inflation," *Phys. Rev. D* **74** (2006) 086003, arXiv:hep-th/0606204.

[30] D. Harlow, S. H. Shenker, D. Stanford and L. Susskind, "Tree-like structure of eternal inflation: a solvable model," *Phys. Rev. D* **85** (2012) 063516, arXiv:1110.0496 [hep-th].

[31] L. Susskind, "The census taker's hat," arXiv:0710.1129 [hep-th].

[32] Y. Sekino and L. Susskind, "Census taking in the hat: FRW/CFT duality," *Phys. Rev. D* **80** (2009) 083531, arXiv:0908.3844 [hep-th].

[33] Y. Sekino, S. Shenker and L. Susskind, "On the topological phases of eternal inflation," *Phys. Rev. D* **81** (2010) 123515, arXiv:1003.1347 [hep-th].

[34] A. Adams, N. Arkani-Hamed, S. Dubovsky, A. Nicolis and R. Rattazzi, "Causality, analyticity and an IR obstruction to UV completion," *JHEP* **10** (2006) 014, arXiv:hep-th/0602178.

[35] F. Rohrlich, "The self-force and radiation reaction," *American Journal of Physics* **68** no. 12 (2000) 1109–1112, https://doi.org/10.1119/1.1286430. https://doi.org/10.1119/1.1286430.

[36] C. T. Hill and E. H. Simmons, "Strong dynamics and electroweak symmetry breaking," *Phys. Rept.* **381** (2003) 235–402, arXiv:hep-ph/0203079. [Erratum: *Phys. Rept.* 390, 553–554 (2004).]

[37] S. L. Glashow, J. Iliopoulos and L. Maiani, "Weak interactions with lepton-hadron symmetry," *Phys. Rev. D* **2** (Oct. 1970) 1285–1292. https://link.aps.org/doi/10.1103/PhysRevD.2.1285.

[38] B. Bjørken and S. Glashow, "Elementary particles and SU(4)," *Phys. Lett.* **11** no. 3 (1964) 255–257. www.sciencedirect.com/science/article/pii/0031916364904330.

[39] M. K. Gaillard and B. W. Lee, "Rare decay modes of the *K* mesons in gauge theories," *Phys. Rev. D* **10** (Aug. 1974) 897–916. https://link.aps.org/doi/10.1103/PhysRevD.10.897.

[40] N. Arkani-Hamed, H.-C. Cheng and L. Hall, "Flavor mixing signals for realistic supersymmetric unification," *Phys. Rev. D* **53** (1996) 413–436, arXiv:hep-ph/9508288.

[41] N. Arkani-Hamed, H. Cheng and L. Hall, "A new supersymmetric framework for fermion masses," *Nucl. Phys. B* **472** (1996) 95–108, arXiv:hep-ph/9512302.

[42] N. Arkani-Hamed, H.-C. Cheng and L. Hall, "A supersymmetric theory of flavor with radiative fermion masses," *Phys. Rev. D* **54** (1996) 2242–2260, arXiv:hep-ph/9601262.

[43] N. Arkani-Hamed, H.-C. Cheng, J. L. Feng and L. J. Hall, "Probing lepton flavor violation at future colliders," *Phys. Rev. Lett.* **77** (1996) 1937–1940, arXiv:hep-ph/9603431.

[44] N. Arkani-Hamed, J. March-Russell and H. Murayama, "Building models of gauge mediated supersymmetry breaking without a messenger sector," *Nucl. Phys. B* **509** (1998) 3–32, arXiv:hep-ph/9701286.

[45] N. Arkani-Hamed, S. Dimopoulos and G. Dvali, "The hierarchy problem and new dimensions at a millimeter," *Phys. Lett. B* **429** (1998) 263–272, arXiv:hep-ph/9803315.

[46] I. Antoniadis, N. Arkani-Hamed, S. Dimopoulos and G. Dvali, "New dimensions at a millimeter to a Fermi and superstrings at a TeV," *Phys. Lett. B* **436** (1998) 257–263, arXiv:hep-ph/9804398.

[47] N. Arkani-Hamed, S. Dimopoulos, G. Dvali and J. March-Russell, "Neutrino masses from large extra dimensions," *Phys. Rev. D* **65** (2001) 024032, arXiv:hep-ph/9811448.

[48] N. Arkani-Hamed, S. Dimopoulos and G. Dvali, "Phenomenology, astrophysics and cosmology of theories with submillimeter dimensions and TeV scale quantum gravity," *Phys. Rev. D* **59** (1999) 086004, arXiv:hep-ph/9807344.

[49] S. Weinberg, "The cosmological constant problem," *Rev. Mod. Phys.* **61** (Jan. 1989) 1–23. https://link.aps.org/doi/10.1103/RevModPhys.61.1.

[50] N. Arkani-Hamed, A. Cohen, E. Katz and A. Nelson, "The littlest higgs," *JHEP* **07** (2002) 034, arXiv:hep-ph/0206021.

[51] N. Arkani-Hamed, A. Cohen, E. Katz, A. Nelson, T. Gregoire and J. G. Wacker, "The minimal moose for a little higgs," *JHEP* **08** (2002) 021, arXiv:hep-ph/0206020.

[52] N. Arkani-Hamed, S. Dimopoulos, N. Kaloper and R. Sundrum, "A small cosmological constant from a large extra dimension," *Phys. Lett. B* **480** (2000) 193–199, arXiv:hep-th/0001197.

[53] N. Arkani-Hamed, L. J. Hall, C. F. Kolda and H. Murayama, "A new perspective on cosmic coincidence problems," *Phys. Rev. Lett.* **85** (2000) 4434–4437, arXiv:astro-ph/0005111.

[54] N. Arkani-Hamed, S. Dimopoulos, G. Dvali and G. Gabadadze, "Nonlocal modification of gravity and the cosmological constant problem," arXiv:hep-th/0209227.

[55] S. Weinberg, "Anthropic bound on the cosmological constant," *Phys. Rev. Lett.* **59** (Nov. 1987) 2607–2610. https://link.aps.org/doi/10.1103/PhysRevLett.59.2607.

[56] A. Vilenkin, "Birth of inflationary universes," *Phys. Rev. D* **27** (Jun. 1983) 2848–2855. https://link.aps.org/doi/10.1103/PhysRevD.27.2848.

[57] R. Barbieri and A. Strumia, "The 'LEP paradox,'" in *4th Rencontres du Vietnam: Physics at Extreme Energies (Particle Physics and Astrophysics)*. 7, 2000. arXiv:hep-ph/0007265.

[58] N. Arkani-Hamed and S. Dimopoulos, "Supersymmetric unification without low energy supersymmetry and signatures for fine-tuning at the LHC," *JHEP* **06** (2005) 073, arXiv:hep-th/0405159.

[59] N. Arkani-Hamed, S. Dimopoulos, G. Giudice and A. Romanino, "Aspects of split supersymmetry," *Nucl. Phys. B* **709** (2005) 3–46, arXiv:hep-ph/0409232.

[60] N. Arkani-Hamed, S. Dimopoulos and S. Kachru, "Predictive landscapes and new physics at a TeV," arXiv:hep-th/0501082.

[61] G. Jungman, M. Kamionkowski and K. Griest, "Supersymmetric dark matter," *Phys. Rept.* **267** (1996) 195–373, arXiv:hep-ph/9506380.

[62] N. Arkani-Hamed, D. P. Finkbeiner, T. R. Slatyer and N. Weiner, "A theory of dark matter," *Phys. Rev. D* **79** (2009) 015014, arXiv:0810.0713 [hep-ph].

[63] D. Tucker-Smith and N. Weiner, "The status of inelastic dark matter," *Phys. Rev. D* **72** (2005) 063509, arXiv:hep-ph/0402065.

3

Abhay Ashtekar

Evan Pugh Professor, Holder of the Eberly Chair and Director at the Institute for Gravitation and the Cosmos, Penn State University

Date: 16 June 2011. Location: State College, PA. Last edit: 9 November 2016

What do you think are the main puzzles in theoretical physics at the moment?

In my opinion, the biggest puzzle is how to reconcile two existent and completely different views of the world. On one hand we have general relativity and on the other we have quantum field theory. These theories are conceptually very different from each other. In fact, in their foundations, they start with almost opposite assumptions. Moreover, their mathematics and conceptual frameworks are very distinct, and hence they give rise to two different views of the world, which seem hard to put together.

However, at a less fundamental level, there are also other puzzles such as what the nature of dark matter is. As an aside, personally, I do not consider that dark energy constitutes a puzzle since at the moment there is a simple explanation of it in terms of the cosmological constant. I do not mean to say that we cannot hope to understand more. But we do have a description that works better than anything else. On the other hand, the nature of dark matter is more of a puzzle, but the issue is at an entirely different level from quantum gravity. I consider it as being more of "a current unknown" than a deep conceptual problem.

Regarding the first puzzle, why is it so hard to combine these two theories?

The reason why it is hard is that it is a really difficult conceptual problem and has remained so up to this day, in spite of the fact that many extremely talented people have worked on it, including Paul Dirac and Richard Feynman [1, 2]. The conceptual difficulty is related to the fact that, in general relativity, gravity is not just another force. Rather, gravity is encoded in the very geometry of space and time and, therefore, once you try to work out a theory which realises correctly the quantum nature of gravity, you are also led to considering *quantum spacetimes*. However, if spacetime geometry is quantum mechanical, the foundation to which much of modern physics is anchored is taken away because we no longer have access to a fixed background spacetime. In quantum field theory, for example, we work with Minkowski space, which is fixed and not affected by anything else, and we study the causal structure of the quantum field theory and ask questions about time evolution using a framework which has been obvious to us since the days of Isaac Newton and

Albert Einstein's special relativity. However, if we no longer have a canonical spacetime background, we lack the very foundations we are used to building upon. Already in general relativity, we lose a non-dynamical spacetime background. But given a solution of Einstein equations we acquire a specific spacetime and hence we can deal with only one spacetime at a time.

Instead, we now have to work with the probability amplitude for various spacetimes. Then several questions naturally appear: What does causality mean, or what does time evolution mean, or what precisely is the goal of physics? In more technical terms, while studying quantum field theories, we typically focus on n-point correlation functions, which are Poincaré-invariant and we know how to deal with them. However, if we attempt to make them diffeomorphism-invariant they become trivial because, other than the δ-distributions, there isn't really any other type of distribution that you can write down which has the required mathematical properties and which is diffeomorphism-invariant. This is again a reflection of the fact that there is no fixed background metric. Sooner or later we will have to come to grips with the fact that we need to construct a meaningful paradigm in the absence of a background metric. This is why combining these two theories is hard.

Are you pointing at the fact that theory should be background-independent?

At a fundamental level the theory would have to be background-independent but of course many of the practical applications of the theory will be obtained by choosing appropriate states which encapsulate appropriate backgrounds.

What do you think is the best candidate in the market for a theory of quantum gravity?

Science is a competition of ideas and the way that science advances is that we first ground ourselves in things that we know very well and then we tackle questions that we do not know. In this process of creation, one has prejudices about what is fundamental and what is not fundamental. Once a theory is constructed, you can test these prejudices by letting nature decide which of these are right. In the art of theory creation these internal biases are absolutely essential because if you start by assuming that all ideas are on the same footing and have the same importance, you would not know where to begin.

Typically, the theories we have constructed so far were built on the following reasonings. One attitude is that we understand general relativity, which is a well tested theory, in which there is a specific spacetime structure. Quantum gravity has to do with quantum aspects of space and time so let's begin with what we understand in general relativity, take those conceptual lessons very seriously, and go ahead. This is the premise of loop quantum gravity (LQG) and therefore the strength of the theory is that right from the beginning it is geared to hit the toughest questions head-on. Or, at least, that is its goal. Another reasoning starts from a quantum field theory or particle physics perspective. In this case, we know very well how to describe scattering amplitudes and we were extremely successful with the other three forces of nature. So let us not think of gravity as a special force; instead, think of gravity as just one other force. Let us not worry about the fundamental conceptual problems of general relativity and, instead, try to focus on those questions that we normally answer in particle physics. This reasoning leads to a whole family of theories, starting from supergravity, higher-derivative theories of gravity and finally culminating in string theory.

There are these two kinds of approaches but of course there are other ideas that people are pursuing [3] such as dynamical triangulations, followed by Jan Ambjørn and Renate Loll [4, 5]. This is an idea which is more on the quantum field theory side but respects the conceptual lessons of general relativity, perhaps more than string theory does, because one does not begin ab initio with a new theory built on supersymmetry or higher dimensions, from which general relativity is just going to pop out. There are also other related ideas such as those pursued by Martin Reuter and others in which one tries to find a nontrivial fixed point of the renormalisation group flow, which leads to an asymptotic safety scenario as opposed to an asymptotic free scenario [6].

Several ideas are being pursued. So what is the best approach? The answer is necessarily going to be strongly driven by your prejudices on what is the most important problem. Since none of the theories is complete, what is best will depend on what you think are the core problems, and on what will follow once the core problems are solved. Given this long introduction, to me the core problem is that we do not know what the fundamental, quantum nature of spacetime is. From this perspective, I think that the ideas that originate from LQG or, to some extent, from approaches like dynamical triangulations are more fundamental. Personally, I prefer LQG for several reasons, which we can discuss later. But, in broad terms, I believe that it comes to grips with some of the deeper physical problems more squarely than what other approaches have managed so far.

Do you think that all these approaches are part of the same big theory or do you think that these are actually separate theories? For example, starting from LQG will one be able to somehow end up in string theory?

Let met try to rephrase your question in a way that I can understand. First of all, I have no doubt that the community as a whole will be successful and there will be a theory of quantum gravity, which unifies at the appropriate level the fundamental ideas of general relativity and quantum field theory, and that these two theories emerge as limiting cases. I believe strongly that such theory will ultimately appear. I also think that it's very likely that this theory will have important elements from LQG and also important elements from string theory. In this sense, I think that these two theories are complementary and not in direct competition.

Regarding the specific question that you asked, whether string theory could come out of LQG, I personally think that it could, not in the sense of the current detailed formulation, but in the sense that fundamental ideas underlying string theory could originate in LQG. It is important to clarify what I mean by this. There are distinct elementary particles and distinct fields but all these can be unified into various excitations of a fundamental object called string. In string theory, there are two a priori elements and, as usual, since they are a priori we rarely examine them. One of the elements is that there is some background spacetime – this element was stronger in the perturbative string era but I think that even in the context of AdS/CFT [7] this is still there and we can discuss it in more detail later.

The second a priori element is really that there is something called the fundamental string and the fundamental string is not like a classical string; instead, it is genuinely a quantum mechanical object. If there are two kinds of a priori elements, perhaps there is a bit of redundancy and perhaps there is really a single a priori element from which both these things can arise. In a certain sense, if you take LQG seriously then, at a fundamental level, spacetime emerges out of the elementary excitations of quantum geometry. These elementary excitations have one spatial dimension and are mathematically described by, for example, spin networks [8–11]. If these are coherently superposed in an appropriate way and if you coarse-grain this quantum state of geometry then it can look like a smooth continuum.

This is the basic idea and in some sense it is radical and in another it is not. It is radical because we are really throwing away the continuum, starting from fundamental principles. On the other hand, it is not that radical because if I look at a sheet of paper, it looks like a continuum but I know that if I put it under an electron microscope, I will see its atomic structure. Why should not the same thing be true about geometry, if in fact geometry is a physical entity as Einstein has taught us? Geometry should also have an atomic structure and this atomic structure is encoded in the various properties of spin networks. The spacetime geometry, which is one of the a priori elements in perturbative string theory, emerges from the spin network. Now, you can also consider excitations of these spin networks. In fact, a long time ago Carlo Rovelli, Lee Smolin and I wrote a paper called "Gravitons and Loops" [12], in which we looked at the excitations which only capture the spin-two part of the possible fluctuations of these spin networks.

Some unpublished work I have, which is almost equally old, shows very clearly that if we look at the excitations of spin networks – one-dimensional excitations of geometry – then in addition to spin-two excitations corresponding to the graviton, there is also a dilaton and an anti-symmetric tensor excitation which naturally appear, without any further assumptions. In fact, one can show that once one imposes Einstein vacuum equations (and some other technical constraints) on these excitations, they kill these other degrees of freedom and only spin-two excitations remain. This means that these other types of excitations are also built into the fabric of the theory.

Since the LQG focus has been on looking at geometry and general relativity, these other excitations have not been studied in any great detail but they are there. These excitations could correspond to the elementary particles similarly to what happens in perturbative string theory. Therefore, to summarise, there's enough room in the theory for spin networks to incorporate both the a priori elements in perturbative string theory in a unified fashion. First, if you look at coherent states of these spin networks, and if you coarse-grain them so that you are well above the Planck length, then you would see spacetime emerging out of it. Second, there is potential for the basic idea of perturbative string theory, namely that various particles which we normally think of as being excitations of independent fields are in fact excitations of extended objects, to be realised in LQG. However, the extended objects are not material objects propagating in some background but really the *excitations of geometry itself*, which at the same time give meaning to geometry if the excitations are coherently superposed.

So why hasn't this work been published yet?

One can easily write down what I just told you but it should be developed in more detail. One of the strengths of string theory is that one also finds interactions between the different excitations just by looking at the string worldsheet and how it branches and joins and so forth. So what I would like to develop requires not only studying the spectrum of particles but also studying the interaction between these particles. As in string theory, in which strings propagate in a Minkowski background, here it is also necessary to find a semiclassical state – a coherent state that is the *canonical* quantum state representing Minkowski space. For example, in electrodynamics, if you give me a classical solution to the source-free Maxwell equations, then there's a canonical state, a coherent state of photons of the quantum Maxwell field which is such that the expectation value is the classical Maxwell field you gave me and the uncertainties in the electric and magnetic field are minimised. In this case you know how to produce the coherent state and you could, of course, add a few more photons to that state and it would still be an excellent approximation to the classical Maxwell field.

But the availability of this *canonical* semiclassical state has been technically and conceptually extremely useful in quantum optics and that's why so much advance could be made. So what is lacking – and this is mainly why this work was not published – is to find the canonical state corresponding to Minkowski space. Once you have this specific state then you can look at perturbations and then from these perturbations one can construct the interactions between the different excitations.

I kept hoping that one of these days either we, or somebody else, would come up with this canonical state but that has not happened so far. It's not that we do not know any state which corresponds to Minkowski space; the problem is that we know several [13, 14]. However, if you produce one state, and I produce another one, and somebody else produces yet another one, even if the expectation values might all correspond to the Minkowski geometry, the fluctuations would be different and, of course, the quantum properties would be different. So there's something that is missing here, namely that we would like to have either some principle which gives us coherent states or a mathematical technique in order to construct such states.

Just to make it clear . . . do you not know what the correct state is or is it the case that there is no semiclassical limit?

The first. But let me make a few remarks on the semiclassical limit of LQG since there has been some confusion in the string community; in fact, David Gross has asked me exactly the same question many times, most recently in India. Within LQG different avenues are being pursued with the same overall goal [3, 9–11, 15]. If we take one of the most established ones that uses the solid procedure that Dirac gave us, which consists of replacing Poisson brackets with commutators, then one does find general relativity [10]. In this sense the theory has a semiclassical limit.

The important question is a separate one, namely: Among the solutions to the quantum equations of LQG, can you find some states which will actually represent classical spacetimes? In other words, you must find the appropriate solutions to the Hamiltonian constraint that have the correct semiclassical limit. However, as I have mentioned earlier, we do not have a *canonical* state representing Minkowski space, or a given solution to Einstein equations. That is what is missing. It is not a trivial question since there are many possible states which to leading order in their quantum fluctuations tend to agree but disagree at the subleading order.

I should also mention that in loop quantum cosmology (LQC) – a reduced set of LQG equations – one can actually solve the equations completely, which is one of its real advantages. In this context, one *can* actually construct semiclassical states and show that these semiclassical states follow the classical trajectory to an extremely good approximation [16–18]. For example, in classical general relativity, if you take a closed universe or a universe with matter (and possibly a negative cosmology constant [17, 19], which string theorists like very much) then the solution of Einstein equations has a property that, in the cosmological setting, it starts with a big bang, expands, reaches a maximum radius and then undergoes classical recollapse. At this classical recollapse, there is a precise radius of the universe, which is well defined because the universe is closed, and also there is a precise energy density. There is a specific relation between the two that can be used to test *quantitatively* whether solutions of loop quantum cosmology have a good semiclassical limit.

This is a nontrivial test since even though quantum corrections should be small away from the Planck era, the universe had such a long time to evolve that these tiny corrections could well accumulate and yield answers that are completely different from those in general relativity. This concern was raised by Green and Unruh in 2004 [20]. But when the calculation was done correctly, LQC passed all these tests and we know today how well it approximates to the classical solution [21]. More precisely, there exist solutions to the full quantum equations of LQC, that is, wave packets which remain sharply peaked around the classical trajectories and which do not spread too much, otherwise we wouldn't have a classical geometry in this room. If we plug in these equations into a computer and let the wave packets evolve, then at the classical recollapse we find excellent quantitative agreements with general relativity. In fact, if you consider a universe which recollapses at a mega parsec, you find agreement to one part in 10^{220}! These corrections are ridiculously tiny, as one would hope. However, the corrections are huge near the putative singularities. There's a new repulsive force that is born out of quantum geometry in the Planck regime which dominates over the classical gravitational attraction and resolves the singularities of cosmology [18, 22].

How does this resolve the singularities of cosmology, exactly?

This is a challenge that has not been taken up in a serious way in string theory mainly because it has proved hard to study the realistic cosmological sector non-perturbatively. Any quantum gravity theory that studies cosmology faces a two-fold challenge. On the one hand you want quantum gravity effects to be extremely strong in the ultraviolet regime so that the big bang or big crunch singularities are resolved, while on the other hand you want these effects to dilute away very quickly and not accumulate over a long period of time such that you end up with a good infrared behaviour; that is, you end up having agreement with general relativity at large scales.

In LQC several different models have been studied, such as models with a cosmological constant, models with anisotropies and models with an infinite number of degrees of freedom [18]. All these models have passed both the ultraviolet and the infrared tests. How does this work exactly? We start with a classical solution to the Einstein equations and construct a quantum state that is sharply peaked around that particular dynamical trajectory at a specified late time. This is our initial data and we now take this wave function and let it evolve using the LQC equations.

Sorry, just one question ... you assume that the universe at late times is a de Sitter universe instead of recovering it at late times?

That's a good point. So the statement is that you start by assuming that at late times the universe is very well approximated by a classical spacetime. How this classical spacetime emerges is not explained within LQC.

Okay, so you were saying that you start with the wave function?

We start with the wave function and then we move forward in time and backward in time. If you move forward in time, then classically the curvature becomes smaller and smaller and this state remains sharply peaked on the classical trajectory. This is also the case in the Wheeler-DeWitt theory, which was constructed many years ago and does not take into account quantum geometry [17]. If you evolve it backward in time, the classical trajectory approaches a big bang singularity; however, the quantum trajectory – the quantum wave function – follows a classical trajectory until the matter density is about 10^{-3} of the Planck density, or the curvature is about 10^{-4} of the Planck curvature, leading to a big deviation from classical gravity [17, 23]. This deviation can be interpreted in a precise sense, namely, as there being a repulsive force which becomes significant when the curvature or the energy density becomes very large. Such repulsive force has its origin in quantum geometry and makes the wave packet undergo a bounce because this quantum repulsion dominates over the classical attraction.

Does this mean that you always get a cyclic universe?

That depends on what the model is. If the model has positive spatial curvature, then you do get a cyclic universe. But if the model is spatially flat with no cosmological constant, for example, then you don't get a cyclic universe. Instead, there is just one quantum bounce and the universe is expanding to its future and contracting to its past. These two branches of the universe are bridged together by a highly quantum mechanical region where the density is roughly the Planck density and the curvature is the Planck curvature. In this region, the classical theory completely fails and what we have is really a quantum mechanical treatment of cosmology.

In a sense the picture that I have in mind is that of a sheet of cloth, which you can think of as being a continuum which is woven by one-dimensional threads. For all practical purposes, the sheet is a two-dimensional continuum but if I were to use a magnifying glass, I would see the threads. Spacetime geometry works in a similar manner, as it can be thought of being a sheet woven by such threads, which normally stick together and have some dynamics. One may study the dynamics of the sheet itself away from the Planck regime. However, in the Planck regime, the sheet gets torn up – the fabric of spacetime gets torn up – and it can no longer be studied using a continuum picture. Instead, if you have the equations that actually govern the individual threads, then you can study what happens to these individual threads as time evolves. These equations are the equations of LQC.

And what are the threads exactly?

The threads are the elementary excitations of geometry and the continuum arises as a coherent state of these threads.

Is it possible to have a better understanding of what exactly these elementary excitations are? For example, in string theory we know that string excitations are at least a particular class of elementary excitations.

In the context of LQG, these excitations are loops of space. Let me be more precise. Geometry is built out of elementary blocks and these building blocks are one-dimensional according to LQG. In LQG, the basic mathematical variables are not metrics but instead gravitational connections and the Wilson loops of these connections are the elementary excitations [9, 10, 24, 25].

One might ask: Are there observables associated with these excitations that we can we measure? Let us compare this situation with the electric field. Faraday lines are the elementary excitations of the classical electric field, so how do we measure them? We take a surface and measure the flux through it. Similarly, in LQG, the flux of these elementary excitations is measured in terms of area [9, 26].

We perceive this room right now as a continuum but in fact it is criss-crossed by these elementary excitations. We see light as a smooth distribution of intensity over there. Of course, we know that what we have is a bunch of photons but there are so many of them and their superposition yields a coherent state, which is the classical Maxwell field that we perceive. Similarly, geometry is not a continuum but instead it is really criss-crossed by these elementary one-dimensional excitations.

And to these one-dimensional excitations you call loops or . . . ?

Originally they were called loops but they aren't really loops [9, 10]. They are spin networks – graphs, in essence – and do not have to be loops necessarily [25, 27]. Nevertheless, the name stuck due to historical reasons.

I find this funny because people say very similar things about the usage of the word *string* in *string theory*. Some people would also say that the word *string* stuck due to historical reasons.

You're right and that's because there are membrane excitations, for example, and strings are not necessarily better than anything else.

Now going back to LQC . . . since you say that singularities are resolved, what happens to the notion of time in the big bang or big crunch? Is there always an arrow of time?

In the simplest models that we consider what is happening is that although the quantum wave function follows the general relativity trajectory in the low curvature limit, when the curvature becomes comparable to the Planck curvature, the trajectory deviates greatly from general relativity and undergoes a quantum bounce. However, the wave function is still reasonably sharply peaked, meaning that the quantum fluctuations don't suddenly grow to the point that the notion of spacetime is completely meaningless. The wave function is peaked around an effective geometry. But this geometry is not the same as that in classical general relativity, and satisfies a set of equations, which are not the same as the classical Einstein equations. There are corrections to the classical geometry and just by inspection you see that these corrections are huge when the density is about 0.41 of the Planck density [17, 18, 23].

The statement is that there are quantum fluctuations and hence properties such as time orientation will have some uncertainty associated with it. Nonetheless in the simplest model, it is possible to address the question of time orientation even in the quantum regime. Of course, due to the presence of fluctuations, things are not as sharp as in the classical theory. The light cones are fluctuating in this regime and so the notion of causality is also fluctuating. Spacetime, or the metric, therefore, also inherits these uncertainties. Under normal conditions these fluctuations are not relevant; just as there is an uncertainty in the position of this table right here but we don't worry about it. The universe behaves in a more classical way than this table here, in the sense that the relative uncertainties are smaller, except in the quantum regime, at densities of the order of, say, 10^{-3} of the Planck density, or at curvatures of the order of, say, 10^{-4} of Planck curvature, at which point you cannot ignore the fluctuations and have to take them into account.

You mentioned that the notion of causality is fluctuating. Does this mean that causality is not fundamental?

Microcausality, formulated in terms of fixed light cones on a given spacetime, which is the way that causality is usually formulated, is not a fundamental principle in quantum gravity in the Planck regime. However, there is a sense in which there is causality, even in this regime, and this is related to your previous question about the arrow of time, which I think that I did not answer fully.

In cosmology we usually define the proper time or conformal time by using a sharp classical metric, right? For example, using this metric we discuss the proper time measured by observers who are following geodesics, or the proper time measured in the frame of some dust particles whirling around some galaxy. However, if there is no fixed spacetime background but only the probability amplitude of spacetime geometries then all these notions of time that we normally use become operators, which are rather different from the usual conventional time variable for which a clock is ticking for each observer.

So how does time arise? This problem actually goes back to the big controversy between Isaac Newton and Gottfried Leibniz on the very nature of time itself [28–30]. Newton postulated the existence of space as a three-dimensional continuum, and the existence of time as a one-dimensional continuum. But unless you make a theological reference to God, which Newton (and others) did, how can you think of time as a constantly ticking clock? In some sense, you may wonder, what is the nature of physical reality?

Let me just make a small detour, which is conceptually important, and then come back to this issue. Time is really a relational notion, namely, it's defined by something that changes with respect to other things. For example, the earth is going around the sun, the moon is going around the earth and how many times the moon goes around the earth while the earth goes around the sun once defines one year. This notion of relational time is perfectly well defined and it is interesting that historically the notion of money also underwent a radical change at some point. In the beginning, exchanges were done using barter systems. For example, you gave me a sack of apples and I gave you a lamb, or you gave me three lambs and I gave you a cow, and so on.

These were all relational types of exchanges but later the idea emerged that these systems were much too complicated and the invention of an abstract notion of money appeared. Money is abstract, especially nowadays. Bank notes are just promises and you cannot eat them, you cannot run a car with them and, in fact, you cannot do anything "real" with them. Nevertheless, money has the universal property that allows us to count, without having to refer to the plethora of relations between items being traded. So we no longer ask if three lambs cost a cow; instead, we ask how much a lamb costs or how much a cow costs.

The same type of reasoning applies to time. Newton's time is an abstraction as it does not care about relations between two events; for example, it does not care if for each period of the earth around sun, the moon goes around the earth approximately 12 times. Instead, Newton's time refers to absolute periods, for example, we say that the earth has this period, the moon has that period, etc. However, time is fundamentally relational and that was Leibniz' point of view but, of course, calculations would have been enormously difficult if Newton hadn't come up with his notion of time, just like all our everyday transactions would have been enormously difficult to carry out if we had not come up with the abstract notion of money [30].

In the quantum domain, we are forced to adopt Leibniz' point of view and regard time as a relational concept since Newton's notion of absolute time became inappropriate. First, this notion depended on each observer – a lesson from general relativity – and, second, the notion of observers and their personal ticking clocks, which strongly depend on a fixed spacetime geometry, had to be given up once we began working with the probability amplitude for spacetime geometries. So how do we implement time in this setting? In cosmology we use the notion of relational time by choosing a judicious matter field such as a massless scalar field. The momentum of the scalar field is a constant of motion and therefore in the language of general relativity, the scalar field is growing monotonically in time. However, in quantum theory there is no time in the general relativistic sense but we still do have a scalar field. Therefore, we take the scalar field to be the clock itself, meaning that we take part of the system as a clock with respect to which the rest of the system evolves. This has been a well-known problem in quantum gravity and in quantum cosmology and, more generally, in LQG. The usage of the scalar field as a clock implements a well-defined notion of time in LQC; for example, if you give me the quantum state of the universe, which is sharply peaked with respect to energy density, anisotropy, Weyl curvature and so on, at a given value of the scalar field, then the equations of LQC tell you, in a deterministic way (so in that sense there is causality), how this wave function evolves, not with respect to some *tick, tick, tick*, but with respect to the scalar field. It is, therefore, a correlation between the change of the scalar field and the change of other quantities.

But how does the scalar field itself evolve?

This notion of time is relational so the scalar field evolves monotonically with respect to itself. If I decide to make the scalar field a clock then everything else evolves with the scalar field, including the scalar field itself. Moreover, you don't need to think of the scalar field as time, as there is no a priori reason why you should. Thinking of a particular field as time makes it easier to do physics but it is not necessary. What matters are the correlations between the different fields, which is a more fundamental description and sometimes called *timeless framework*. The framework I have described above, in which one of the variables, for example a scalar field, is taken to be the time, is called a *deparameterised framework*. To summarise, there is causality and there is determinism, but it's not the same type of microcausality discussed in field theories.

I must mention that we have studied in great detail such a framework, in which everything is relational, and how quantum geometry evolves in that framework [18, 31, 32]. We start with full LQG and then we perform a drastic reduction, basically via a series of approximations that have been spelled out, by focusing on a sector in which the geometry is given by homogeneous isotropic cosmologies and quantum fields propagate on this quantum geometry. In the more general case what we do know is that ...

Sorry, but since you mentioned this drastic reduction, I was wondering if you could comment on the common criticism of LQC, namely, that LQC is just a reduced sector of LQG, so why should this be relevant for cosmology?

Let me tell you what the story is and what is the current state of affairs. In order to explain my viewpoint, it is better to begin with an analogy [33] and, of course, analogies are good to some extent but if you take them too literally they lose their value. Suppose that you want to study the hydrogen atom and for some reason Dirac had not been born yet but somehow we did understand quantum electrodynamics. For some reason someone had written the mathematics of electrodynamics and at some point someone realised that there is a bound state of an electron and a proton – the hydrogen atom. This is actually a hard thing to do because electrodynamics is a perturbative treatment but let us suppose that it happened. In fact, it is not important if it is an electron-proton bound state or some other bound state like electron-positron; what matters is the ability to describe some properties of that bound state. Later, Dirac came along and said, "Wait a minute, by assuming spherical symmetry around the proton and solving the Dirac equation, I can describe the hydrogen atom." If you were a follower of the full theory of quantum electrodynamics you would have said that assuming spherical symmetry would be a very drastic assumption since it eliminates, by fiat, all photons. So one's first reaction would be: Does not all of electrodynamics go completely down the drain with this assumption?

In retrospect, we realise that this objection is misplaced because the Dirac theory works *very* well and one has to perform very sensitive measurements such as the Lamb shift to see its limitations. This is an analogy, it is not a proof of anything, but it is not impossible that LQC is the hydrogen atom of quantum gravity. It's the simplest system with a lot of symmetries in it. Of course, you want to make sure that the quantisation is not ad hoc. So in LQC we first look at quantum geometry that is given to us by full LQG and then construct the quantum description of this reduced system following the methods of LQG. These methods give you a quantum theory, which is distinct from the Wheeler-DeWitt theory [34], even though the system only has a finite number of degrees of freedom. The only vector that is common to the two Hilbert spaces is the zero vector so ...

Before you continue, what you're saying is that it could actually be that this reduced system describes the real world?

That's right, in the following sense. This reduced system cannot describe everything in the real world because it has only certain degrees of freedom but if you take the whole world and if you trace over all other degrees of freedom, keeping only these few degrees of freedom, such as the density, the anisotropy, and the homogenous degrees of freedom of the Weyl curvature, that are relevant for the large-scale structure, then you would get something that I believe would be well approximated by the current LQC.

People have been skeptical about this point of view, especially in old literature, in which people argued that if you start with a reduced system of equations with axisymmetry, solve the equations with that symmetry and make a further symmetry reduction down to spherical symmetry, then you obtain a different result than the result you would obtain if you had started with the reduced set of equations with spherical symmetry from the beginning. It turns out that if, in this process, you trace out the extra degrees of freedom correctly you will get the right answer. In fact, we looked at both anisotropic and isotropic universes in

LQC and we traced out all the degrees of freedom that were anisotropic. We found that, in fact, we obtain exactly the isotropic case [35] and that was a pleasant surprise. I don't expect this to happen always for any model, as there will be some corrections but the question is if these corrections are small. So there is some evidence that integrating out the degrees of freedom would actually give you this reduced system of equations in cosmology, though there is no rigorous proof that this is always the case.

The second point I would like to make is related to the work published in 2010 by the group in Warsaw and Kristina Giesel [36], who is now in Germany. They have introduced a new type of quantisation in full LQG with a scalar field that takes into account the relational time used in LQC. There are still several open issues in their work but this group is very mathematically careful and so it is possible that in fact one will be able to see whether this model in full quantum gravity, which is tailored to LQC, in the sense that it has the same matter field and so on, yields this reduced system once you integrate out certain degrees of freedom and focus on a very small family of observables. So, one should be able to test this in a not too distant future.

Okay, so let's assume that it is right, that it actually happens to be true. Does LQC make any observable predictions?

There is a huge literature on this, I mean, literally hundreds of papers, and some of them are coming from distinguished cosmologists, who obviously know what they're doing [33, 37]. However, the early literature did not take into account all the effects of LQC; they focus on one effect at a time. This is not unusual; for example, when you study perturbation theory of the hydrogen atom, or any atomic physical system in general, you take into account corrections one by one, such as the L-S coupling, spin-spin coupling and relativistic corrections. These are all perturbations of the Hamiltonian and it is easier to study them one by one. So I think that at some stage one has to put it all together because there is a possibility that all the effects cancel each other.

The second thing is that much of this literature focuses on the inflation era, in particular what happens at the onset of inflation, during inflation and after inflation. From just basic cosmology and the WMAP data, we know that at the onset of inflation the density is about 10^{-11} of the Planck density and the curvature is about 10^{-12} of the Planck curvature. Therefore these corrections are going to be extremely small unless there is an unexpected surprise. Many people claim that these corrections are not that small but so far they have not given a mechanism by which there could be an enhancement of such effects. The community is therefore a bit skeptical about this and it is not clear that the actual values that are being computed can be measurable in any way in the near future.

Recently, there has been a shift of focus. People have stopped worrying about what happens during inflation by taking the standpoint that, whatever happens, general relativity is a good enough approximation to the universe in that epoch for the background space-time geometry. Instead, people started studying what happens to the density and curvature between the bounce and the onset of slow roll inflation. The bounce may have produced certain effects that could potentially leave a trace later on, as the universe evolves. A few years ago, most cosmologists would have immediately said that it would be impossible to observe what happened in the Planck era before the onset of inflation because whatever happened then would have been diluted away during inflation. However, this turns out not to be the case and now leading cosmologists also agree with this possibility [38].

I must say that I find inflation impressive because although there are clearly ad hoc elements in it [39–41], it is still a paradigm that starts with relatively few assumptions. The basic idea is that for some reason there was an accelerated expansion in the early history of the universe and at the onset of this accelerated expansion the universe is very well described by the FRW geometries and first-order perturbation theory around them. The original quantum state is the so-called Bunch-Davies vacuum, so the fluctuations over the classical geometry are just "vacuum fluctuations" that grow and become the seed for observable effects as the universe evolves. Moreover, once you perform calculations assuming that inflation took place in the early universe, you find predictions that are verified in the CMB spectrum. Furthermore, if you make computer simulations from the CMB era to the present time, you are led to a large-scale structure that is realised in our universe. Therefore, the large-scale structure of the universe grew out of vacuum fluctuations at the onset of the slow roll phase of inflation – this is a very profound idea, a deep philosophical idea (laughs).

However, even though it is a profound and powerful idea, one may wonder why the fluctuations were in a "vacuum state" in the early universe in the first place, right? And why, to set the initial conditions, one has to wait till the density is diluted away to 10^{-11} of Planck density. It is very weird that one has to set initial conditions long after the Planck era. So one of the remarks I want to make is related not to my own work but to the work of Ivan Agullo, a postdoc of mine [38], which is now very well accepted. In this work one starts out with a state with some number of particles, which, therefore, is not the Bunch-Davies vacuum.

The cosmologists' first reaction to this modification is always to claim that if there's some initial particle density, it must be small enough otherwise you would see deviations from what you observe in the CMB and if it's small enough then it is going to be diluted away because nothing survives inflation and its 65 e-foldings. Even the hydrogen atom expands to something like 250 light years during inflation! However, it is not true that the effects of this initial particle density are diluted away, the reason being that there is stimulated emission. If you have a bunch of particles, they create more particles, just like in a laser, and you can show that the particle density is going to be low but it is not going to be diluted at all in spite of inflation. This has some observational consequences, for example, non-Gaussianities in the CMB spectrum.

This got us excited because now we have a framework for studying the bounce and doing perturbation theory with the quantum fields and evolving the fields up to the onset of inflation. If you assume some symmetry conditions and a preferred state at the bounce and let it evolve, what will you get? Do you get small corrections to the Bunch-Davies vacuum or just the Bunch-Davies vacuum? This is what is being done at the moment and preliminary calculations indicate that what you obtain is close to the Bunch-Davies vacuum but not exactly it, which could have new observational effects such as power suppression at large angular scales and hemispherical anisotropy.

Since I gave a long answer, let me summarise this in three sentences. There is a huge amount of work in the LQC literature but it's not clear that the signatures that they give us are going to be observable because a lot of it begins already in the inflation era and therefore the corrections will be very small. Recently, a framework has been formulated which allows starting at the bounce and doing quantum perturbation theory in *quantum* spacetimes. Finally, in this framework, you don't have to assume what the state was at the onset of inflation; you can evolve it starting from the bounce and there are indications that there will be observational effects.

Since this interview was taped, the interface between theory and observations has evolved very significantly. First, quantum field theory on *quantum* cosmological spacetimes was developed in detail [31, 32]. Second, this theory was applied to cosmological scalar and tensor perturbations. These are now represented by quantum fields propagating on the quantum homogeneous isotropic geometries. This theory extends the inflationary scenario back in time over some 12 orders of magnitude in curvature, all the way to the quantum bounce in the Planck regime. One can therefore specify initial conditions at the bounce, evolve fields and calculate the power spectrum at the end of inflation [42]. Third, this evolution brought out an unanticipated interplay between the ultraviolet and the infrared. The ultraviolet LQC modifications of Einstein equations resolve the big bang singularity and provide a new LQC scale given by the universal value of curvature at the bounce. Those modes of perturbations that have wavelength larger than the curvature radius at the bounce get excited in the Planck regime and are no longer in the Bunch-Davies vacuum at the onset of inflation. Therefore, the predictions for the longest wavelength modes – i.e., for the largest angular scales in the sky – are different from those in the standard inflationary theories based on classical general relativity. In the fourth advance, this possibility has been used to account for the large angular scale anomalies reported by the PLANCK collaboration (see, e.g., [33, 43–45]).

Furthermore, there are new predictions for future observations of the T-E and E-E power spectra at large angular scales that the PLANCK collaboration will release [33, 43, 46]. (T-E stands for temperature-electric polarisation correlation and E-E for electric polarisation–electric polarisation correlations.) Thus the LQC analysis has provided an avenue to relate Planck-scale physics to observations. Finally, there is also interesting feedback from observations to fundamental theory. The observed power suppression can be used to select a narrow class of initial conditions through a quantum generalisation [46] of Roger Penrose's Weyl curvature hypothesis [47]. Thanks to these advances, quantum gravity is no longer confined to a pristine perch of mathematical physics; one can hope to see ramifications of Planck-scale physics in the sky through the physics of the very early universe.

What about dark matter and dark energy? Are they put in by hand or . . . ?

In the inflationary scenario you don't have to put dark matter by hand because until the end of inflation, it doesn't play that much of a role. There is only one scalar field in the problem that is responsible for inflation, and the matter we see now is supposed to be produced only at the end of inflation during the "reheating" phase. The only thing that is put in by hand is the scalar field and some potential for the scalar field. One usually uses the simplest potential that passes all the PLANCK data tests. The cosmological constant, or dark energy, you can put it in or you can just leave it out because its effect in the early universe is completely negligible.

Do you think you could see both dark matter and dark energy as excitations of geometry?

Dark matter, maybe, because if you remember the general philosophy that I mentioned previously, excitations of geometry give rise to matter, so there is no reason to think that it only gives rise to visible matter. It could give rise to every type of matter that is coupled to gravity, and dark matter is coupled to gravity. Dark energy, by contrast, I don't think so. It's something that is going to be put in by hand in the theory and then you work out the non-trivial consequences of it.

Now, I should mention that there are some interesting ideas that appeared due to a combination of LQG with group field theory and which could provide some insight into this question. Group field theory [48] is a kind of complementary approach to spin foams [11, 15, 49] which is a path integral formulation of LQG. I'll come back to it in a minute. This approach suggests that a certain coupling constant, which one introduces in group field theory, can be identified with the cosmological constant [50, 51]. There is reason to believe that, in group field theory, this coupling constant will run. It starts at zero when the cosmological constant is of the order of the Planck scale and then flows to one. If it went to exactly one, the cosmological constant would be zero but if somehow, under this renormalisation group flow, it approaches one, but does not exactly reach one, then this would give rise to a small cosmological constant. This is not backed up by any hard calculation within full LQG but it serves to show that there is potential conceptual evidence that it could work.

It is interesting to know a bit about the group field theory and path integral approaches to LQG. LQG started out in a Hamiltonian framework which has been most useful in dealing with problems that involve extremely strong gravitational fields, like in LQC, black holes and so on. Later a complementary approach, based on path integrals, was developed – this is the spin foam approach, that Carlo Rovelli will probably tell you more about, and that had significant advances in the last few years [15, 49]. Group field theory is close to spin foams, the main idea being that at the fundamental level there is no space, there is no time, there is just some abstract group manifold [48]. It is like the matrix models that you're probably more familiar with. Matrix models are usually used to describe two-dimensional gravity but then there was a generalisation due to Boulatov to three dimensions [52], for which instead of having matrices, the quantum states in the model are functions of some copies of groups. These groups can be the Lorentz group or whatever groups that you're working with, and then somehow in this formalism spacetime can be identified as an emergent concept. So the hope is that such theories go beyond spacetime geometry. Group field theory is a sort of generalisation to four dimensions and what is explored there is the fact that there is a close similarity between the perturbation expansion that is produced in group field theory and the transition amplitudes that are given by the final path integral in the spin foam models. There is a mathematical relation between the two and so these are kind of complementary ways of looking at the covariant approach or path integral approach to LQG.

Okay, so changing slightly the topic ... can you start with any classical theory and apply the LQG formulation? Could you apply it to supergravity, for example? Does everything go? Is it possible that LQG picks out the right theory, or you don't see it like that?

At this stage, I don't see it like that. There are ideas about how one might be able to incorporate matter couplings, like the ones that I mentioned to you previously which involved the scalar field, but, if you look at what has really been established to date through hard results, then the answer is that LQG does not determine matter couplings. New results have emerged leading to the incorporation of supersymmetry and higher dimensions in LQG. These advances are primarily due to the Erlangen group (see, e.g., [53, 54]) and it is still the case that within LQG nothing *requires* us to use these extensions. So far, the emphasis of LQG has been on constructing a quantum theory of gravity; the emphasis is not on unification.

As with general relativity, which opened up surprisingly new windows on modern physics, such as black holes, big bang and gravitational waves, the hope is that by understanding quantum gravity deeply, new unexpected windows will open up. However, this can take a long time. Even in general relativity mainstream astronomers did not believe in black holes until the 1980s; they thought that black holes were mathematical artefacts and do not exist in nature. So, it is not going to be tomorrow that we will see some clear vindication of quantum gravity ideas realised in the matter sector – i.e., in particle physics. But LQC is producing interesting results such as that there is no big bang singularity; there is no "horizon problem" that cosmologists refer to because spacetime is much larger than we thought so that everything had enough time to interact with everything else; and the Planck-scale dynamics near the bounce may leave observable effects such as power suppression at largest angular scales. The emphasis is, therefore, on exploring the drastic paradigm changes that arise by bringing gravity and quantum theory together.

Given that this is the primary focus of LQG, it is not necessary for the theory to have any in-built principle that will incorporate everything. However, there's one kind of misconception that the string theory paradigm gives rise to. In string theory, the point of departure is higher dimensions, supersymmetry and a negative cosmological constant. General relativity is supposed to arise indirectly, in a special limit. Therefore, string theorists sometimes ask: How do you know that Einstein's theory is the right one to start with? Well, how did we know that Maxwell's theory is the right one to base quantisation on for the electromagnetic interaction? The answer is that it is the correct theory at the classical level. If I take electrodynamics and combine with quantum mechanics, I get an effective action that includes all the quantum corrections. These quantum corrections include not only the classical Maxwell Lagrangian but also many terms that appear because of quantum theory.

So even though you begin with a classical theory and then you go to quantum theory it doesn't mean that the final quantum theory will not be radically different from the classical theory that you started with. The equations governing quantum physics are quite different from the equations governing the classical theory. This situation is similar for the gravitational interaction. We do know that Einstein's theory is the right one in the classical limit.

I see. So you're saying that you take Einstein equations, apply the LQG formulation, take a semiclassical limit and you get quantum corrections which have nothing to do with Einstein equations. Is that it?

Right. And in fact we see explicitly in LQC that the effects you get are completely non-perturbative and you could not have imagined them a priori. If you look at the effects on the equations of motion due to LQC, you find that both the Friedmann equation and the Raychaudhuri equation have changed [17, 18] but these modifications are precisely such that they can resolve the big bang singularity and yet reduce to the classical equations as soon as the density is about 10^{-3} of the Planck density.

These modifications are non-trivial. Classically, the Friedmann and the Raychaudhuri equations are non-trivial components of Einstein equations in the cosmological setting, and they imply the equations of motion for matter. For example, if you start with only a scalar field, these equations imply the Klein-Gordon equation and if you start with a perfect fluid they imply the conservation of the fluid's stress-tensor. So at first, you may worry that these modifications would have undesirable consequences. But I was very pleased in the end because this does not happen. The modified equations imply exactly the equations that you want, namely, the standard Klein-Gordon equation and the standard continuity equation. Definitely, there is some coherence here.

So one of the points that I try to get across to string theorists all the time is that the effective equations of LQG are *not* the equations of motion of general relativity. They only reduce to the classical equations in the weak field limit, when the curvature is very low compared to the Planck curvature. The second thing I try to get across is the fact that it is the quantum corrections to these equations that cure the big bang singularity. The idea that one *must* change already the classical theory if one wants to get something novel is a recent invention. This idea wasn't there when one looked at quantum electrodynamics or quantum chromodynamics. There, you start with a familiar Lagrangian that is well established classically. When you do quantum field theory, you can find non-trivial and important corrections.

Okay, but these corrections that you are getting are in the reduced system, right?

The corrections that we are getting from LQC are in the reduced system. Many groups have worked on bridging LQC to full LQG. Now there are several very interesting partial results in favour of LQC [55]. But it is still an ongoing process.

Can you get these equations from the full LQG or does this go back to the same problem of finding the state corresponding to Minkowski spacetime?

Yes, in some sense it goes back to the same problem of finding a canonical background state. On the other hand, one can find quantum corrections by working with any background. For example, if you work with any of the states that correspond to Minkowski spacetime in the classical limit, do you get, say, the graviton propagator? Carlo Rovelli will probably tell you more about this because his group is the one that worked this out but the statement is that you do get the graviton propagator [3, 13, 14, 49]. At low curvatures you find the Minkowski metric, as the quantum corrections are small, and so you get the graviton propagator but the *sub-leading* terms will depend on what the state is. So this is another point that I want to get across, namely, it's not that we don't have a semiclassical limit. Rather, there is no canonical state corresponding to Minkowski space. However, if you give me one of these states, to leading order, you will find the usual graviton propagator. To obtain unambiguous sub-leading terms we need a new principle or a new insight that is missing at the moment.

Is it clear that general covariance is there at the quantum level?

Very good. This question is best addressed in the spin foam framework because it is really a spacetime formulation – a histories formulation – rather than a Hamiltonian formation. In the Hamiltonian formulation, if I just take for example ordinary Maxwell theory in Minkowski spacetime, then I would like to see that quantum electrodynamics is Poincaré-invariant. Although, of course, it is so, it's not easy to see this invariance in the Hamiltonian framework. It's technically harder there than in the path integral or covariant frameworks. If you are working in the Hamiltonian framework then you need to build the boost generators and their action is complicated because you need to go from one space-like slice to a non-trivially related space-like slice. In LQG, it is very similar, namely, in the Hamiltonian framework it is difficult to see what is actually happening to 4D covariance. So it is better to see what actually happens in the spin foam language and, in that language, the statement is that everything is indeed covariant.

So, if you start with some initial state and look at the transition amplitude as in ordinary quantum field theory, and you perform a diffeomorphism leaving the end points intact, would it be the case that you would obtain the same answer? The answer to this question is, formally, yes [3, 11, 49]. Why do I say "formally"? Because you are taking into account all possible quantum histories; as in ordinary field theories, these arguments tend to be formal. In ordinary quantum field theory calculations, mathematicians would not agree that the integral is performed using a well-defined measure. In the spin foam framework the situation is better because the integrations are well defined. But there is an open issue at the next level. One should keep in mind that the transition amplitudes are obtained via a certain perturbative expansion. So strictly, to claim that there is general covariance, you need to make sure that the series converges. There are highly sophisticated mathematical and technical issues that still have to be sorted out, but there is no a priori problem with general covariance. It's not clear that the series converges. But in standard quantum field theories the perturbation expansion does not converge either. So at this moment the answer is a little bit imprecise but not in a drastic way. I mean, I also don't know that quantum electrodynamics is Poincaré-invariant for the same reason – the perturbation theory does not converge (laughs).

In the end, therefore, besides other technical issues such as what you exactly mean by diffeomorphisms and what classes of diffeomorphisms you allow, the problem is more a problem of convergence. People have been dedicating more time to this issue than to that of diffeomorphism invariance, which I think is the right thing to do at this point.

People sometimes raise an objection to LQG, namely, LQG takes Einstein equations and quantises them, but we know that the theory is not renormalisable and that perturbatively you will encounter divergences. Does somehow the quantisation procedure of LQG make it finite and well behaved?

In perturbation theory around Minkowski space, there are divergences in quantum general relativity. When people refer to these divergences they mean ultraviolet divergences. People don't usually speak of infrared divergences but there can also be infrared issues. For example there are infrared issues in QED but we know how to handle them. So let us discuss the ultraviolet divergences. If you assume that there is no built-in cutoff in the theory, then in the path integral language you are summing over all continuum geometries. If so, there is going to be an ultraviolet divergence because of the fact that there is no smallest scale in space and no largest scale in momentum space. The key ingredient of LQG – if there is one thing that LQG brings out as a lesson – is that one should be using quantum geometry, in fact, a specific quantum geometry that came out of a lot of work in the 1990s [9, 10, 24–27, 56, 57]. This quantum geometry is mathematically precisely defined in the sense that you have a regular measure to perform functional integrals. In this framework one can control everything. This is the main point I want to communicate.

The basic feature of quantum geometry is that the geometrical observables have a discrete spectrum, which comes out of very precise kinematics. So, for example, in spin foams, when one integrates over all geometries, one is not integrating over smooth metrics but instead over quantum geometries. What are these quantum geometries? Well, suppose that you give me a manifold and you give me an initial slice and a final slice and suppose that you want to calculate the transition amplitude between the two slices. What you have to do is to introduce a simplicial decomposition of this region that you gave me. Let us focus on ultraviolet issues and return to the infrared problems later. We take N simplices, where N can be a very large number. Now, for each simplicial decomposition, there is a dual two-complex that you construct. This might sound all fancy but it is very simple. Let me just explain using three dimensions instead of four. If you are in three dimensions like in this room for example, and you've got this two-dimensional initial slice and another two-dimensional final slice, then what you are going to do is to build tetrahedrons between them. For each tetrahedron the dual is going to be a point, a vertex up here. If I give you a face of any one tetrahedron, then you look at a line, or a link, which is transverse to it.

So the dual of a triangle is a line. Thus, you can construct a graph with vertices and links. In the simple setting of a three-sphere you can introduce two tetrahedra, then the statement is that the dual would be one point up here, another point up there, and the lines which are dual to the faces meet because the faces meet, so you get a graph which has vertices and links. Faces of tetrahedra are dual to the links of the graph.

This is also true in four dimensions, namely, I give you the initial three-dimensional manifold and I give you an initial spin network state on it, and a final three-manifold with a final state on it, then you can build between them a two complex, which basically means that each of the links of the spin network evolves in time spanning out a two-dimensional surface. The vertices evolve in time spanning out a line and, furthermore, new vertices could be created during evolution. These two complexes have vertices, links and faces and each of them carries some geometrical information. This information is a set of quantum numbers that specify a quantum geometry. This is the basic idea.

So, when you are summing over all these quantum geometries you're not summing over all classical metrics, which is what gave rise to ultraviolet difficulties and non-renormalisability in perturbative treatments. Instead, you are summing over all these two complexes which carry quantum numbers. Can you shrink these two complexes arbitrarily? Well … the area has a minimal eigenvalue and if you were to go below it the triangle would just disappear. Hence, (a) you are summing over all discrete geometries, or quantum geometries, precisely defined and (b) the geometrical operators have discrete eigenvalues – in particular the area has a minimum eigenvalue. As a result, you simply can't take the limit in which you continuously shrink the area to zero. Since you cannot continuously shrink the two-complex, there is no ultraviolet divergence. Earlier, I told you that the spin foam amplitude is written as a perturbation series. That series is basically related to how many simplices you have in between the initial and the final slice. In the dual picture this is the number of vertices you have in a graph – so you get what is called a "vertex expansion." In order to get the total transition amplitude, you have to allow for an arbitrary number of simplices. At any order in this expansion there are no ultraviolet divergences [3, 11, 49]. So, in LQG the claim from spin foams is that there is ultraviolet finiteness at any order in the vertex expansion.

And is the series also convergent?

Whether the whole series is convergent, that we don't know. It is still an open problem. I personally feel that more effort should be given to that, but there's only a finite number of people working on this. A more important problem is that of infrared divergences since even though there is a minimal area for these quantum geometries, there may not be a bound on how large it can be, which can lead to infrared divergences. That is, the area of the tetrahedra associated with each graph might not be bounded from above. However, recently, in the course of last years, two independent groups have shown that if you introduce a positive cosmology constant then, automatically, there is an infrared regulator [58, 59]. This is pleasing. It is surprising and it is technically hard. But it did not come as a shock because something like this was known to happen in three dimensions. In three dimensions, if you take simplicial quantum geometry, for the same reasons, there are also infrared divergences. But it was shown by the mathematicians Turaev and Viro that if you introduce a positive cosmological constant then this theory can be reinterpreted as being a theory with a quantum deformation [60]. This quantum deformation leads automatically to convergence. It is a mathematically precise result in three dimensions and so we had hoped that something similar would happen in four dimensions. And it did.

But have they shown that if there is no cosmological constant there are no infrared divergences?

No, that is still open. It's just that there are infinitely many terms if the cosmological constant vanishes. With the cosmological constant, the sum terminates at a finite order and hence the result is finite. It is just that, really just that (laughs). For zero cosmological constant you have to allow for an infinite number of terms in the vertex expansion and so you could encounter divergences, but no one has performed a systematic study in order to establish whether this is the case or not. Ultimately, the statement is that, in spite of the classical perturbative problems, there are reasons to believe that by summing over quantum geometries instead of smooth continuous geometries, the theory is both ultraviolet and infrared finite, order-by-order in the vertex expansion.

There's another issue that I have heard several string theorists mention. There is some intuition that the existence of a minimal eigenvalue for the area breaks Lorentz invariance at some level. However, there is a paper by Rovelli and Speziale discussing this [61], in which they pointed out that there's no tension between the minimal area eigenvalues and Lorentz invariance. The simplest example they gave is that of the rotation group in which the angular momentum, just like the area, does not change continuously from zero to $1/2$ and one can say that spin $1/2$ is the minimum non-zero eigenvalue. There is absolutely no contradiction between quantisation of angular momentum and invariance under the rotation group. The reason why people might think that there is a contradiction is because of classical intuition. For, in this situation you cannot draw classical pictures of vectors rotating around an axis in the usual way. The same holds for the Lorentz group, as they spelled out in their paper.

This is a trivial comment but I find it amusing that in string theory all problems, like ultraviolet divergences, are solved because strings have a minimum size, while here all problems seem to be solved because of the minimum size for the area?

(laughs) That's exactly right.

I read in many papers of LQG that the area having a minimum size is considered to be a prediction of LQG. I was wondering, is it possible to make an experimental test of this somehow?

Yes, that's a nice question. In the cosmological context, you can track down the minimal eigenvalue of the area operator to the value of the critical density at the bounce [33]. There's an analytic formula relating the two quantities and the critical density blows up as you shrink the minimum area to zero. In LQC, there is a mathematical result which states that density operators in the physical Hilbert space are bounded from above, hence the density of any state in the theory cannot be larger than this critical density. Some potential cosmological observations such as the "anomalies" at large angular scale in the CMB spectrum that we talked about earlier depend on the critical density. Since these effects depend on the existence of the bounce and the details of the bounce depend on the critical density, this could be an indirect way of observing the minimum area of quantum geometries. I think that at the moment there is no way of testing quantisation of area outside the cosmological setting. There are other ways in which the minimum area has a significant role to play, as in black hole entropy, but it's more a sort of theoretical test rather than an observational one.

There is a criticism towards the area operator, namely, that it is not a physical observable in the usual sense that we use in quantum mechanics. Is this a problem?

Yes, area arises as a "kinematical" observable, that's right. The correct statement is that if you add matter to the theory then there will be a physical prescription for which the physical area will be quantised. So, you have to introduce fields or other structures in the theory, for which the area is not defined as the area of an abstract, mathematical surface, but of a physical surface. If you do this then the area observable will become a physical observable instead of a kinematical observable. This is why we can discuss the properties of black holes, which are described in terms of isolated horizons, since these are now physical objects. If you give me a random surface, I can measure its properties and I can tell you if it is an isolated horizon or not, basically by looking at the light rays coming from it.

What is an isolated horizon, exactly?

Isolated horizons are isolated in the sense that they're left to themselves, meaning that nothing is falling across them [62]. So this means that we consider horizons in equilibrium situations, which is the usual setting for calculating their entropy.

Before we move on to the topic of black holes, I have yet another technical question related to this. In LQG one usually defines the area operator but one naturally wonders, if one is to have a fully covariant formulation of LQG, how to define a volume operator instead? Thomas Thiemann showed that this volume operator has the right semiclassical limit [63] if you take lattices with six-valent vertices. I was just wondering, since this is exactly what the spectral triple approach to quantum gravity does [64, 65], if this could be a new way of reformulating LQG or if you think that choosing only one type of lattices is nonsense?

It is certainly not nonsense. But I haven't thought deeply about this. I have followed a bit those papers but cannot recall the details. In my opinion, and this is a sort of gut feeling, I think that it would be a very strong statement to claim that you can only take one specific type of graph. More than anything else, I kind of feel that such results are typically based on the assumptions about what is meant exactly by semiclassical limit, or how it should be obtained. It may be that something has to be generalised or simplified in what Thomas did. These results are valuable, nevertheless, because they specifically say that if you do this and that then for sure you obtain this particular thing. And that's very good to know. But it's like a no-go theorem; usually you can tweak one of its assumptions and avoid it. There is an example of this in LQG that is used all the time. In quantum mechanics, for systems with finite degrees of freedom, there is the well-known Stone–von Neumann no-go theorem, which gives you the uniqueness of the representation of the Heisenberg algebra, generated by the position operator q and the momentum operator p. If you take the exponential forms, $e^{i\lambda q}$ and $e^{i\mu p}$, of these operators with real parameters λ and μ, and assume that the representations of these operators depend continuously on λ and μ, you obtain the Stone–von Neumann uniqueness theorem which says that the only way to do quantum mechanics is to use the standard Schrödinger representation. However, LQG suggests that while the assumption of continuity in both λ and μ is natural in non-relativistic quantum mechanics, it is too strong in the diffeomorphism-invariant context. One of them has to be dropped. As soon as you drop it, there are new possible representations, e.g., the exponentiated Heisenberg algebra used in LQC. So the no-go theorem is bypassed. Something similar may happen with results which say that you have to restrict yourself only to certain types of graphs for such and such a reason. This is what I meant previously by simplifying or generalising Thomas' work. I think that his results are very interesting but I don't think that they are giving a watertight no-go theorem.

Okay, but more concretely, do you see a problem with not taking the sum over all possible graphs?

I don't see a problem with restricting yourself to some category of graphs, at first. Aesthetically, one may be uneasy but I don't see that there is a conceptual problem there. In any case, it would provide a useful subsector of everything that you can do. When we started investigating quantum geometry, in the early days it was important for us to make sure that all operators, such as the area and volume operators, could be consistently defined for all states – spin network states – based on arbitrary graphs. If you are just going to restrict yourself to one specific type of graphs, like those dual to tetrahedra, very well, but you better have good motivation for it. In the case of tetrahedra, there is good reason, since up to three variant graphs you just can't reproduce anything physically useful [9, 10] and the use of tetrahedra is the simplest possibility that comes next. So if you have good motivations to restrict to one specific type of graphs, other than the fact that you can't do otherwise, I think it can be worthwhile exploring.

Okay, now shifting to the topic of black holes. According to LQG, do black holes have a singularity?

When quantum effects are correctly incorporated, there won't be a singularity. However, in LQG the calculations related to black holes are far from being as detailed and precise as they are in LQC, where cosmological singularities are resolved. This is because there is less symmetry in the case of black holes, even when you restrict yourself to spherical, non-rotating ones. In LQC there is a precise notion of singularity resolution and the equations of motion have been defined and analysed in every possible way. The equations of motion that describe black holes in LQG are not at the same level of careful thinking. It is more of a first attempt or perhaps second attempt; it is not mature enough. However, conclusions could be drawn under simplifying assumptions and these investigations show that there is no singularity inside a black hole.

In general, I think that quantum geometry in the Planck regime near the classical singularity of black holes is going to be much more complicated than that in LQC. There will also be quantum bounces but the bounce structure is much more intricate than that in the simplest cosmological models. Anisotropic and the Gowdy inhomogeneous cosmological models already gave us a hint about what the structure would be like, although even these cases are much simpler than black holes. But already there, we do not have a simple bounce. If I take something like a Friedman model which is spatially homogeneous and isotropic, then the only physical observable is density or scalar curvature. In this situation you do find a bounce which is related to a critical value of density but as soon as you introduce anisotropies, components of the Weyl curvature can also lead to bounces. Basically, anytime an observable quantity enters the Planck regime and dilutes away due to quantum geometry effects, there will be a bounce. In black holes, the Weyl curvature blows up as you approach the classical singularity at $r = 0$. However, in LQG, one expects that, as you approach $r = 0$, the Weyl curvature will increase, enter the Planck scale, and then become dilute again because of the repulsive character of quantum geometry effects. This happens but not necessarily in a homogeneous manner as it does in LQC. In the black hole case, there are various "bounces" in both time and space, and so the bounce structure is more complicated.

There are several things that still need to be understood. It may well be the case that, in contrast to LQC, quantum corrections cannot be absorbed in an effective, local metric near the putative singularity. One may just have to live with the full quantum state in the Planck regime and classical geometry could be a good approximation only in the low curvature region. So, for black holes in LQG, the full spacetime will be much larger than what Einstein told us. More precisely, it doesn't end at the singularity because of the bounces. This is something that several people have studied in detail in simple models but so far these calculations do not incorporate the full gravitational collapse. However, there is a simpler model, called CGHS due to Callan, Giddings, Harvey and Strominger [66], which are two-dimensional black holes with one space and one time dimension. It mimics properties of the Schwarzschild black hole. It's not Schwarzschild, but the differences are not that great. People, like Amos Ori, have looked at this case in detail, in particular, by starting with the wave function and evolving it [67–69]. When it goes through the bounce it becomes very quantum mechanical and then it becomes classical again. Something like this is likely to happen in general and this opens a door to understand the spacetime process that will resolve the "information loss" issue.

In the standard Hawking picture, there is a "future edge" to spacetime, represented by a singularity, whence $\mathcal{J}^+$ is not the full future boundary. Therefore, a part of the incoming state can, so to say, fall into the future singularity and only the remainder appears at $\mathcal{J}^+$. Thus, the future edge of spacetime serves as a sink of information. In LQG, because of the singularity resolution there will be no such future edge. Instead, there will be a kind of a pocket region where the classical singularity was. The fluctuations in geometry would be very large in this region but spacetime could become tame on the other side of this pocket. The quantum state would evolve through the pocket and appear on the other side in the quantum extension of spacetime. Thus, because spacetime is larger than its classical counterpart, there is, so to say, "room for information recovery." The quantum S-matrix will then be perfectly well defined. There is a series of papers that I have written with various colleagues for these two-dimensional black holes in which we present strong arguments that this scenario is actually realised [70–72]. I believe that the same thing happens in realistic Schwarzschild-type black holes.

Just to make it clear, when you make all these calculations you actually ascribe a state to the black hole? I mean, can you write the black hole state for the Schwarzschild black hole in LQG? Or that requires the Minkowski state at spatial infinity?

We cannot write the state for the Schwarzschild black hole because it is indeed related to the Minkowski state at infinity. But in the CGHS black hole, we can do it. In this case, you are dealing with the genuine collapse of a scalar field that forms a classical black hole. What we do is to start with a quantum state and evolve this quantum state. To understand its evolution we make a series of approximations and we learn what happens at future infinity given a state in the distant past. So it is a state of spacetime geometry plus the scalar field.

So, given these considerations, can one say that LQG managed to give an explanation for the information loss paradox, or is it too much to ask for?

There are still important open issues. But certainly LQG provides a detailed scenario for how the information can be recovered [73]. A quite detailed scenario, in fact, but it is based on certain assumptions and some hard calculations should be done in order to test if those assumptions are valid or not. However, I should also emphasise that such assumptions are made explicitly and/or implicitly in every other investigation that I'm aware of, where calculations are carried out in spacetime. If they are carried out in some dual conformal field theory, one does not know what happens in spacetime and in essence one just claims that the S-matrix is unitary and therefore the information paradox is resolved – a tautological claim since if we take the S-matrix to be unitary as an assumption then of course information is not lost. On the other hand, if you want to know what happens to the spacetime interior and precisely how the horizon disappeared, in the CGHS model we have very detailed numerical calculations that have shown very interesting universalities in the black hole evaporation process, which were not known before [71, 72, 74].

The CGHS solution is a two-dimensional black hole and its action is very similar to – but not quite the same as – the reduction of the Einstein-Klein-Gordon action in the presence of spherical symmetry. It's really like a spherically symmetric collapse that produces a spherical black hole. This problem has infinitely many degrees of freedom. In four-dimensional classical relativity, for example, this is the problem that people have solved using numerical relativity both in astrophysics and in the so-called critical phenomena. One doesn't know how to solve this problem analytically. In the CGHS case, there occurs a simplification due to a very clever tweak in the action. The model is classically solvable and one can write down the exact solution that produces the black hole dynamically. In fact, I have used this example in the last few years in my relativity courses because it's very useful for people to see a dynamically formed black hole through collapse, instead of just looking at stationary black holes. So, in this setting we start with an initial state with everything being quantum mechanical and evolve this state. However, in order to study Hawking radiation we had to make the mean field approximation, which means that we ignore quantum fluctuations of geometry but not of the matter fields. This approximation is justified if you have a large number of matter fields; it is essentially the same as a large-N approximation that is often done in other areas of physics. In this approximation, we have many copies of scalar fields that collapse and form a black hole while simultaneously taking into account the quantum nature of the fluctuations of the scalar field and replacing all the geometrical operators by their mean values, i.e., not taking into account their fluctuations. Within this setting, the black hole shrinks as the Hawking emission is radiated away and so you can ask various questions about the nature of black holes [70]. These of course depend on the details of the collapse and we considered cases where the black hole is formed by what is called a prompt collapse, that is, a collapse that is caused by sending in a sharp profile of scalar fields.

For black holes created in this way in the early epoch we find what everybody else has found, namely, that for a long time Hawking radiation is negligible, meaning a total of only one Planck mass per scalar field channel is radiated away. But as soon as the dynamical horizon forms the Hawking flux increases at that time and the net flux at infinity increases [70]. (As an aside, the notion of event horizon doesn't make sense in these quantum geometries; you need a more general notion, namely, that of a dynamical horizon, which is something that Badri Krishnan and I had introduced in the context of numerical relativity [75].) It turns out that now the flux of energy at infinity is greater than what you expected from Hawking's argument, which is a consequence of the fact that here you have included backreaction. The radiation keeps increasing and it is not exactly thermal; it is so only at early terms. At late times, the flux at infinity is not compatible with pure thermal radiation [71, 72].

Okay, but have people seen this before?

We were the first group to carry out this calculation with sufficient accuracy. I provided the conceptual framework and the equations but the numerical calculation was carried out by a Princeton group, which is perhaps the best numerical group in the world. To perform this calculation they really needed an incredible accuracy, which had not been tried before. So this work could be out only due to their incredible skill. However, this is still done using the mean field approximation. Nonetheless, because of backreaction the singularity is weakened, but it is still not full quantum gravity since we have ignored the quantum fluctuations of the geometry. In this approximation, the metric is continuous at the singularity, but it's not differentiable. There is an end point in the singularity, from which the last light ray escapes and reaches null infinity. Semiclassical spacetime ends there and to go beyond it you need full quantum gravity.

Nevertheless, we can see that no matter how much energy and matter you send first, that is, regardless of what the details of the original pulse profile are, there is only a small Bondi mass left (of about 0.8 of the Planck mass per evaporation channel) to be evaporated away in the pure quantum mechanical domain. This was a surprise as it could have very well depended on the details of the profile. Instead, it is an universal result. There is also universality in the dynamics. If you study the system between the time that the dynamical horizon is formed and the time that the dynamical horizon shrinks almost to zero you find that the mass-loss profiles are universal. People had not found these features. To go beyond the semiclassical spacetime, we assume that the leftover Bondi mass will be radiated away in some finite amount of time. Then we could show that there is a precise sense in which the S-matrix is unitary. I think that the assumption we make is plausible but the limitation is that this is in the context of the CGHS black hole. The plan was to start this exercise in four dimensions this summer, but this summer is already passing (laughs).

Okay. As far as I know, it was you who first showed in the context of LQG that black hole entropy is proportional to the area. This has been advertised, not by you, as a prediction of LQG. However, this calculation depends on the Barbero-Immirzi parameter. So what is your point of view on this parameter and can we call this a prediction?

So the prediction is really that entropy is proportional to the area. But in LQG we have the Barbero-Immirzi parameter as a quantisation ambiguity and we do not have independent control of it. The viewpoint is the following. We do this calculation for one type of isolated horizon, let us say the spherically symmetric Schwarzschild black hole, and we find that the entropy is proportional to the area. However, it also depends on the Barbero-Immirzi parameter. So we just fix the Barbero-Immirzi parameter in order to match the Bekenstein-Hawking area law. We take this as a measurement of the Barbero-Immirzi parameter [76]. Doing so, it means that we do not have a prediction that the entropy for the Schwarzschild black hole is exactly one quarter of the area plus higher-order quantum corrections. Nevertheless, I can now take other black holes such as rotating black holes, distorted black holes or cosmological horizons and calculate the entropy using that parameter and then you predict that in these cases you get one quarter of the area. As it stands we need an experiment (or some thought experiment) to fix the Barbero-Immirzi parameter but it would be much nicer if we had an alternative way for fixing this parameter, for example, by trying to find agreement with some quantum field theory in, say, Minkowski spacetime, or something like that.

So your point of view is that the Barbero-Immirzi parameter is a constant that has to be fixed?

Yes, I see it just like the θ parameter of QCD. The θ parameter in QCD appears due to a quantisation ambiguity and classically it has no effect on the theory. Classical theories with different θ parameters are all equivalent because they are related by adding a topological term to the action, or by making a canonical transformation in the phase space. On the other hand, quantum mechanically, these are inequivalent theories. The reasoning is the same for the Barbero-Immirzi parameter. Originally, we did not realise that it was related to a topological term but later the topological meaning of this parameter was clarified by other groups (see, e.g., [77]). In any case, it is really like the θ parameter in QCD; it is a quantisation ambiguity and as in QCD, there is no way to fix it other than via a measurement. Of course, when you fix the parameter, you need to make sure that it leads to desirable predictions in *other* experiments and so it would be nice to have further ways of fixing this parameter in LQG. Unfortunately, at the moment, we don't.

Now, I want to go back to something that you mentioned in the beginning of this conversation. Namely, you mentioned that you thought that the AdS/CFT correspondence [7] was not a completely background-independent formulation of quantum gravity. Could you elaborate on this?

Right. Well ... I have raised several issues about AdS/CFT in discussions with string theorists like Juan Maldacena, Joseph Polchinski and various others. I have not heard any good answers. In fact, I have not heard any answer at all. So I think that my points are still valid (laughs). First of all, let me say what I find beautiful and attractive about AdS/CFT and then I'll tell you what are its limitations.

What I find incredibly attractive about AdS/CFT is that it brings out, at some meta-level, a kind of underlying unity of physics. The fact that you can obtain results for a conformal field theory (CFT) – which might be used in an idealised condensed matter system – by doing a calculation of Green's functions on a black hole background is really impressive. (I said "idealised" because it is not a realistic system that condensed matter physicists typically work with.) These two systems are *very* different, and even though the AdS/CFT correspondence only relates to some idealised condensed matter systems, idealised fluids or idealised quark-gluon plasma, I think it is beautiful that it brings out a meta-unity in physics.

For quantum gravity proper, however, I don't think that AdS/CFT has done very much for the following reasons. First of all, it's always in 10 dimensions. Let us consider the best understood case in which you take 10-dimensional spacetime and compactify five dimensions on the S^5. You are then left with five-dimensional anti–de Sitter (AdS) spacetime, for which its boundary is four-dimensional and that's where the super Yang–Mills (SYM) theory – i.e., CFT – lives. This SYM theory is claimed to have the same content as non-perturbative string theory in this AdS background. One of the criticisms I want to make is that people usually just declare this to be the definition of the desired quantum theory of gravity. However, declaring it is not the same as proving it. Now, one might take the attitude of taking this definition at face value and then exploring interesting questions. But how do you use AdS/CFT to understand what happens during and at the end of the black hole evaporation *in spacetime*? Furthermore, physically we are interested in either a positive or a zero cosmological constant, not a negative one. What happens near the singularities in a realistic cosmological context?

If we go back to the 1970s and 1980s where quantum gravity was being studied seriously by people like John Wheeler and Bryce DeWitt we will notice that understanding graviton-graviton scattering was not high on the priority list because it is a rather boring process from a deep quantum gravity perspective, even though it may be useful in some context. On the other hand, the problem of time, the quantum nature of the spacetime geometry, the fate of singularities in the cosmological context or in the black hole context, and Hawking's information paradox were very high on the list. How are these problems addressed using AdS/CFT? A standard argument, for the last of these problems, is to claim that the S-matrix is unitary and hence no information is lost but can you explain to me on the gravity side how exactly information is recovered? And what about the meaning of black hole horizons? Does this concept even make sense in a proper quantum theory of gravity or is it just an approximate concept? What is a quantum black hole? People tell me that a quantum black hole is a thermal state in the dual CFT but I would like to understand what that means in spacetime terms since this characterisation is definitely not what an astrophysicist would use for a black hole.

Nothing, or very little, is ever done from the gravity perspective. This is my second criticism. But even if by tomorrow there's a revolution and someone tells me, "This is exactly what is happening in the gravity side," I will ask: What gravity side are we talking about? We're talking about gravity with a spacetime that has five large non-compact dimensions and five *large* dimensions which are compactified on S^5. Not many people would be happy with five large spacetime dimensions, and they would be totally unhappy with five large compactified dimensions. The radius of these five compactified dimensions is the cosmological radius and not a Planck-size radius. If these five directions were wound up in a Planck-size sphere, then people might be happy, but this is not the case. We're talking about something which most physicists would not recognise in the gravity they experience. This is not what astrophysicists mean by gravity. But the big string theory community simply declares that this is gravity and, not only gravity, but also quantum gravity. Moreover the cosmological constant is necessarily negative in AdS/CFT. This is not the result of some preliminary calculation or a first approximation.

In the beginning I had no problem with it as a first step. But by now almost two decades have passed and we don't really know what to do in the asymptotically flat case, and we don't *really* know what to do in the positive cosmological constant case. So to me it really does not answer the core questions of quantum gravity in a satisfactory way. It could be at most – and this is something that Stephen Shenker said at a KITP workshop in Santa Barbara – only a tiny corner of quantum gravity in which the cosmological constant is negative, in which the internal dimensions are large, and so on. So to me it does not really say something about the real world and I find it disappointing.

I also find disappointing – specially due to my interest in the gravitational aspects – that for several years now the focus of this research has been on non-gravitational aspects like applications of the AdS/CFT conjecture to idealised condensed matter systems, idealised quark-gluon plasma and so on. As I said previously, it is fantastic that gravitational techniques are being extended to these domains but it is not solving any quantum gravity problems.

But you do agree that AdS/CFT is a non-perturbative, background-independent formulation of a theory of quantum gravity? Maybe not the realistic one, but still ...

I do agree that there's a non-perturbative, background-independent definition of a theory. The reason I would not call it a formulation is that to me the dictionary is not sufficiently developed after all these years – it is not rich enough – to call it a formulation. At the moment, I don't know how to calculate the most interesting things – like the end point of black hole evaporation or fate of cosmological singularities – on the gravity side in terms of what I calculate on the CFT side. Everybody has different expectations, of course, but this is my point of view.

I have a very naive question. If it really is a non-perturbative, background-independent theory, could you somehow use the machinery of LQG to describe this theory?

It's a very good question. Did you ask this question to Laurent Freidel? Because he's probably the one who has thought most about this question.

I intended to but I forgot to ask him.

From my perspective – and I must say that I have not studied this in detail – there are too many differences. Most of the emphasis put by people who work on AdS/CFT is on the CFT side; there is relatively little work on the gravity side. Therefore, it becomes difficult to bridge it with LQG, in which the emphasis is on the gravity side. As I mentioned earlier, I think it is possible to make contact with perturbative string theory by looking at the excitations of geometry. If we had the canonical quantum state corresponding to Minkowski spacetime, we would be able to make this contact in a few months. I think that in order to make a connection with AdS/CFT there are too many intermediate steps to take. So I am slightly sceptical. However, if a point of contact could be made, even if just halfway in between, it would represent a significant advance. I would be delighted.

At the time the first black hole entropy calculations in LQG were carried out, I was in one of these six-month workshops at KITP and there were lots of discussion seminars, bringing very diverse people together. It worked very well. Black hole entropy calculations in string theory make use of theories of gauge fields living on D-branes. In LQG, we were also thinking of theories involving gauge fields, in particular, Chern-Simons theories living on three-dimensional isolated horizons. The pictures that we were drawing looked very similar. So we looked for a systematic relation between the two sets of entropy calculations.

We didn't succeed, nor did our friends who work in string theory. I feel that making a connection with AdS/CFT involves difficulties at a higher level.

But it should be possible to make this connection, right? Or could we find something different if we would answer these tough questions in quantum gravity that you mentioned using AdS/CFT?

Yes, it could be something different. It could be a different formulation. It could be something along the lines of non-commutative geometry, although I personally don't think this would be the case. It could be that the physical idea of quantum geometry that we have in LQG would not be realised in this particular model of the AdS/CFT correspondence. However, it is a tantalising possibility and, if it works out, we would definitely learn a lot from each other's approaches.

What do you think has been the biggest breakthrough in theoretical physics in the past 30 years?

I don't think that there has been a truly significant breakthrough on the theory side in the past 30 years, certainly not as significant as understanding the electroweak interactions, for example. Among the theoretical advances that have taken place, I think that there have been impressive developments in understanding the merger of black holes and their emission of gravitational waves. For 25 years or 30 years people have been trying to solve the two-body problem in general relativity but they just didn't make that much progress. However, in the past 15 years this problem was completely solved using numerical techniques that began being developed by the Princeton group that I mentioned earlier. Frans Pretorius made the seminal breakthrough in this area [78], opening the door for others. So now one knows in great detail what happens as classical black holes collide and merge, forming a single black hole as well as how much radiation is produced and what their properties are. This is really important for gravitational wave observatories. Indeed, the templates created through these binary coalescence simulations play a vital role in interpreting the data collected by gravitational wave observatories. The spectacular event observed by the LIGO collaboration in September 2015 provides an outstanding example of this.

Another important breakthrough is the development of the idea of an inflationary scenario in cosmology, namely, that the early universe was in a state of accelerated expansion and that at the onset of inflation the quantum state of the universe is in the Bunch-Davies vacuum.

Personally, I am very interested in ideas that originated in quantum gravity. I think that LQG and string theory have provided many novel ideas. I think that there has been more concrete progress on the gravity side through LQC. In the final picture the approach may turn out to be wrong; we don't know. It may be that LQC predicts that there should be such and such corrections to the power spectrum on large angular scales, and certain non-Gaussianities, but then future observations may reveal that these traits are not there. Still the fact that there are concrete calculations that have created a bridge between observations and fundamental physics at the Planck scale is in itself impressive. On the mathematical and conceptual side, the development of a framework to study quantum geometries and transition amplitudes without a background spacetime is brand new, and their intellectual potential is deep. But again these ideas may not be realised in nature. Similarly, with AdS/CFT, even though I don't think that it brought novel ideas to quantum gravity – I don't think that calculating graviton-graviton scattering unambiguously is one of its deepest problems – it brought out this meta-structure bridging areas of physics that one would have never dreamed of. As I already said, despite the fact that AdS/CFT deals with idealised systems, I find it a deep idea.

Why have you chosen to do physics?

Why have I chosen ...? Yeah, good question. Early on I was really drawn to very basic questions. The fact that simple laws, even just Newton's laws, have such a tremendous scope seemed kind of magical. The fact that the apple falling down from a tree is really described by the same equation that governs the orbit of the moon around earth is a fantastic concept. It is absolutely great that one can make so much sense of the observable universe – I mean the physical universe, which is still very simple compared to the biological universe, issues of consciousness and so on. Still, it's just astounding that we can understand so much starting with basic principles. I find that tremendously beautiful. Humanity has thought about some of these questions about the nature of space and time, and origin of the universe, at least since 600 BC. It was then that Lao Tzu, Gautama Buddha and others made commentaries on these questions. And we find that these questions have answers that you can arrive at by studying the physical properties of the universe. It's very beautiful and that's what drove me to theoretical physics.

There is another thing that attracted me to physics very much, although I now have qualms about it. If you look at something like art or literature, or poetry, you spend a lot of time doing beautiful things but ultimately its value tends to be determined by a few critics. Some painters are considered as being great masters and sometimes I don't find anything great in their works, while others were undervalued and yet I find their work very beautiful. We see this explicitly all the time, how much influence egos of critics and various other human and political forces have. And there are cultural factors. I was very struck that values that formed the basis of judging literature in my mother tongue, Marathi, while I was a middle school student in India were entirely different from the values used to judge English literature that I learned in high school. Science seemed to be so beautiful because it was, at a fundamental level, detached from human pettiness. It was also detached from local cultural norms. Physics, in particular, is very objective. There is something pure and noble about it. Critics and outside judges don't decide. Cultural trends don't decide. Nature decides. This is what drew me to fundamental physics.

However, nowadays I have second thoughts about this. When science is closer to experiment, everything that I said holds. But when science moves away from experiment, as it is unfortunately the case now since technology has not kept pace with the theoretical progress, it is not exactly like that. For example, I think that in many frontier areas of fundamental physics the competition between ideas is driven not just by raw merit; scientific politics is playing a much greater role. The influence that people have makes matters more complicated. In assigning value to an idea, it seems to matter a lot whether you come from the particle physics community or from the general relativity community. It is not simply "pure" science anymore. Ideas are not solely judged objectively. Instead, there are all kinds of nuances, secondary things and prejudice which have become increasingly important. There's considerably more of this than there was when I first entered physics. The development of this field, the influence of ideas, how many people work on it and so on is greatly determined by considerations that are not purely scientific. As the title of Penrose's book says, there is an increasing role of fashion, faith and fantasy [79]. I don't mean to say that everything is like that but there is more of it than when I entered the field.

What do you think is the reason for the string theory community and the LQG community being very distinct?

The real reason is just the fact that they come from two different backgrounds and therefore their sets of values are different from each other. What seems to be important in string theory is unification while in LQG we give more importance to understanding the quantum nature of geometry. As the fields evolved, they became more differentiated. I still write notes every year about the recent developments in string theory, even though the mathematical techniques have become very specialised. In turn, this has made it more difficult to communicate. It is one thing to discuss at a general level, e.g., of our discussion, but if you want to explain to someone precisely, say, what are the obstacles of doing LQG in 10 dimensions beyond what the Erlangen group [53] has done, that would be difficult to explain and it would take a long time. Nevertheless, I encourage my students to learn string theory. For example, one of my students did a project with Radu Roiban in string theory and I really wanted him to do that because it's important for him to understand what the techniques and the philosophy in string theory are. I think it's important that there is communication between these two areas.

There are of course some string theorists who are just interested in science and nothing more, but that's not the case with many of them. Many of them behave as if everybody else is wasting their time and that intolerance is something that I do not find very appealing. When I see statements like "String theory is the language of God" I am distressed because it sounds like a fundamentalist's slogan. This is too much, right? I don't understand such statements; I mean, logically it doesn't make any sense to me. So that kind of spirit, I think, is not very good and I hope people in LQG don't have that spirit. At least, I try to talk to young people in conferences and various places and I try to make the point that they have to be very open to new ideas.

What do you think is the role of theoretical physicists in modern society?

I think that not only the theoretical physicists but anyone who seeks fundamental knowledge is absolutely critical for human societies to flourish. Even though history shows that fundamental physics can lead to unforeseen, spectacular applications, more and more funding agencies seem to demand research with *direct* applications. At the national level in the US, for example, a few years ago the funding agencies coined the phrase *curiosity-driven science*. This carried a negative undercurrent, suggesting that science that is directly applicable should be considered as better, more worthy of funds, than curiosity-driven science. That is not beneficial in expanding the frontiers of human knowledge. People whose pursuits are fundamental are absolutely essential if our society is to keep progressing. Fundamental questions have to be answered with integrity and with a high level of understanding, without worrying about material gains that one can get immediately. Theoretical physicists are not the only ones, obviously, but theoretical physicists, in particular, study nature and the fundamental laws that govern the universe. I think that this is really their role. I think there should always be people who worry about fundamental problems regardless of whether or not there's going to be any application in the immediate future. If no such people exist then humanity doesn't really have a bright future (laughs).

References

[1] P. Dirac, *Lectures on Quantum Mechanics*. Dover Books on Physics. Dover Publications, 2013.

[2] R. Feynman, "Quantum theory of gravitation," *Acta Phys. Polon.* **24** (1963) 697–722.

[3] A. Ashtekar, M. Reuter and C. Rovelli, "From general relativity to quantum gravity," arXiv:1408.4336 [gr-qc].

[4] J. Ambjorn, A. Goerlich, J. Jurkiewicz and R. Loll, "Nonperturbative quantum gravity," *Phys. Rept.* **519** (2012) 127–210, arXiv:1203.3591 [hep-th].

[5] R. Loll, "Quantum gravity from causal dynamical triangulations: a review," *Class. Quant. Grav.* **37** no. 1 (2020) 013002, arXiv:1905.08669 [hep-th].

[6] M. Reuter and F. Saueressig, *Quantum Gravity and the Functional Renormalization Group: The Road towards Asymptotic Safety*. Cambridge University Press, 1, 2019.

[7] J. M. Maldacena, "The large N limit of superconformal field theories and supergravity," *Int. J. Theor. Phys.* **38** (1999) 1113–1133, arXiv:hep-th/9711200.

[8] C. Rovelli and L. Smolin, "Spin networks and quantum gravity," *Phys. Rev. D* **52** (1995) 5743–5759, arXiv:gr-qc/9505006.

[9] J. C. Baez, "Spin networks in nonperturbative quantum gravity," in *The Interface of Knots and Physics*, Louis H. Kauffman, ed., pp. 167–203. 4, American Mathematical Society, 1995. arXiv:gr-qc/9504036.

[10] T. Thiemann, "Modern canonical quantum general relativity," arXiv:gr-qc/0110034.

[11] J. Barrett, K. Giesel, F. Hellmann, L. Jonke, T. Krajewski, J. Lewandowski, C. Rovelli, H. Sahlmann and H. Steinacker, eds., *Proceedings, 3rd Quantum Geometry and Quantum Gravity School: Zakopane, Poland, February 28–March 13, 2011*, vol. QGQGS2011. 2011.

[12] A. Ashtekar, C. Rovelli and L. Smolin, "Gravitons and loops," *Phys. Rev. D* **44** (1991) 1740–1755, arXiv:hep-th/9202054 [hep th].

[13] E. Bianchi, E. Magliaro and C. Perini, "Coherent spin-networks," *Phys. Rev. D* **82** (2010) 024012, arXiv:0912.4054 [gr-qc].

[14] E. Bianchi, E. Magliaro and C. Perini, "LQG propagator from the new spin foams," *Nucl. Phys. B* **822** (2009) 245–269, arXiv:0905.4082 [gr-qc].

[15] A. Perez, "The spin foam approach to quantum gravity," *Living Rev. Rel.* **16** (2013) 3, arXiv:1205.2019 [gr-qc].

[16] A. Ashtekar, T. Pawlowski and P. Singh, "Quantum nature of the big bang," *Phys. Rev. Lett.* **96** (2006) 141301, arXiv:gr-qc/0602086.

[17] A. Ashtekar, T. Pawlowski and P. Singh, "Quantum nature of the big bang: improved dynamics," *Phys. Rev. D* **74** (2006) 084003, arXiv:gr-qc/0607039.

[18] A. Ashtekar and P. Singh, "Loop quantum cosmology: a status report," *Class. Quant. Grav.* **28** (2011) 213001, arXiv:1108.0893 [gr-qc].

[19] E. Bentivegna and T. Pawlowski, "Anti–de Sitter universe dynamics in LQC," *Phys. Rev. D* **77** (2008) 124025, arXiv:0803.4446 [gr-qc].

[20] D. Green and W. G. Unruh, "Difficulties with closed isotropic loop quantum cosmology," *Phys. Rev. D* **70** (2004) 103502, arXiv:gr-qc/0408074.

[21] A. Ashtekar, T. Pawlowski, P. Singh and K. Vandersloot, "Loop quantum cosmology of k = 1 FRW models," *Phys. Rev. D* **75** (2007) 024035, arXiv:gr-qc/0612104.

[22] M. Bojowald, "Absence of singularity in loop quantum cosmology," *Phys. Rev. Lett.* **86** (2001) 5227–5230, arXiv:gr-qc/0102069.

[23] A. Ashtekar, A. Corichi and P. Singh, "Robustness of key features of loop quantum cosmology," *Phys. Rev. D* **77** (2008) 024046, arXiv:0710.3565 [gr-qc].

[24] A. Ashtekar and J. Lewandowski, "Representation theory of analytic holonomy C* algebras," arXiv:gr-qc/9311010.

[25] A. Ashtekar and J. Lewandowski, "Differential geometry on the space of connections via graphs and projective limits," *J. Geom. Phys.* **17** (1995) 191–230, arXiv:hep-th/9412073 [hep-th].

[26] A. Ashtekar and J. Lewandowski, "Quantum theory of geometry. 1: Area operators," *Class. Quant. Grav.* **14** (1997) A55–A82, arXiv:gr-qc/9602046.

[27] A. Ashtekar and J. Lewandowski, "Projective techniques and functional integration for gauge theories," *J. Math. Phys.* **36** (1995) 2170–2191, arXiv:gr-qc/9411046.

[28] J. K. McDonough, "Leibniz's philosophy of physics: against absolute space and time," in *The Stanford Encyclopedia of Philosophy*, E. N. Zalta, ed. Metaphysics Research Lab, Stanford University, fall 2019 ed., 2019.

[29] A. Ashtekar and J. Stachel, *Conceptual Problems of Quantum Gravity*. Einstein Studies. Birkhäuser Boston, 1991.

[30] A. Ashtekar, "Time in fundamental physics," *Stud. Hist. Phil. Mod. Phys.* **52**, 69–74 (2015). www.sciencedirect.com/science/article/abs/pii/S1355219814000902.

[31] A. Ashtekar, W. Kaminski and J. Lewandowski, "Quantum field theory on a cosmological, quantum space-time," *Phys. Rev. D* **79** (2009) 064030, arXiv:0901.0933 [gr-qc].

[32] I. Agullo, A. Ashtekar and W. Nelson, "Extension of the quantum theory of cosmological perturbations to the Planck era," *Phys. Rev. D* **87** no. 4 (2013) 043507, arXiv:1211.1354 [gr-qc].

[33] A. Ashtekar, "Symmetry reduced loop quantum gravity: a bird's eye view," *Int. J. Mod. Phys. D* **25** no. 8 (2016) 1642010, arXiv:1605.02648 [gr-qc].

[34] B. S. DeWitt, "Quantum theory of gravity. I. The canonical theory," *Phys. Rev.* **160** (Aug. 1967) 1113–1148. https://link.aps.org/doi/10.1103/PhysRev.160.1113.

[35] A. Ashtekar and E. Wilson-Ewing, "Loop quantum cosmology of Bianchi I models," *Phys. Rev. D* **79** (2009) 083535, arXiv:0903.3397 [gr-qc].

[36] M. Domagala, K. Giesel, W. Kaminski and J. Lewandowski, "Gravity quantized: loop quantum gravity with a scalar field," *Phys. Rev. D* **82** (2010) 104038, arXiv:1009.2445 [gr-qc].

[37] A. Barrau, T. Cailleteau, J. Grain and J. Mielczarek, "Observational issues in loop quantum cosmology," *Class. Quant. Grav.* **31** (2014) 053001, arXiv:1309.6896 [gr-qc].

[38] I. Agullo and L. Parker, "Non-Gaussianities and the stimulated creation of quanta in the inflationary universe," *Phys. Rev. D* **83** (2011) 063526, arXiv:1010.5766 [astro-ph.CO].

[39] A. H. Guth and S. H. H. Tye, "Phase transitions and magnetic monopole production in the very early universe," *Phys. Rev. Lett.* **44** (Mar. 1980) 631–635. https://link.aps.org/doi/10.1103/PhysRevLett.44.631.

[40] A. Linde, "A new inflationary universe scenario: a possible solution of the horizon, flatness, homogeneity, isotropy and primordial monopole problems," *Phys. Lett. B* **108** no. 6 (1982) 389–393. www.sciencedirect.com/science/article/pii/0370269382912199.

[41] A. Starobinsky, "Dynamics of phase transition in the new inflationary universe scenario and generation of perturbations," *Phys. Letts. B* **117** no. 3 (1982) 175–178. www.sciencedirect.com/science/article/pii/037026938290541X.

[42] I. Agullo, A. Ashtekar and W. Nelson, "The pre-inflationary dynamics of loop quantum cosmology: confronting quantum gravity with observations," *Class. Quant. Grav.* **30** (2013) 085014, arXiv:1302.0254 [gr-qc].

[43] A. Ashtekar and A. Barrau, "Loop quantum cosmology: from pre-inflationary dynamics to observations," *Class. Quant. Grav.* **32** no. 23 (2015) 234001, arXiv:1504.07559 [gr-qc].

[44] I. Agullo, "Loop quantum cosmology, non-Gaussianity, and CMB power asymmetry," *Phys. Rev. D* **92** (2015) 064038, arXiv:1507.04703 [gr-qc].

[45] I. Agullo and N. A. Morris, "Detailed analysis of the predictions of loop quantum cosmology for the primordial power spectra," *Phys. Rev. D* **92** no. 12 (2015) 124040, arXiv:1509.05693 [gr-qc].

[46] A. Ashtekar and B. Gupt, "Quantum gravity in the sky: interplay between fundamental theory and observations," *Class. Quant. Grav.* **34** no. 1 (2017) 014002, arXiv:1608.04228 [gr-qc].

[47] R. Penrose, *The Road to Reality: A Complete Guide to the Laws of the Universe*. Vintage Series. Vintage Books, 2007.

[48] D. Oriti, "Group field theory and loop quantum gravity," 2014. arXiv:1408.7112 [gr-qc]. https://inspirehep.net/record/1312968/files/arXiv:1408.7112.pdf.

[49] C. Rovelli and F. Vidotto, *Covariant Loop Quantum Gravity: An Elementary Introduction to Quantum Gravity and Spinfoam Theory*. Cambridge Monographs on Mathematical Physics. Cambridge University Press, 2015.

[50] A. Ashtekar, M. Campiglia and A. Henderson, "Loop quantum cosmology and spin foams," *Phys. Lett. B* **681** (2009) 347–352, arXiv:0909.4221 [gr-qc].

[51] G. Calcagni, S. Gielen and D. Oriti, "Group field cosmology: a cosmological field theory of quantum geometry," *Class. Quant. Grav.* **29** (2012) 105005, arXiv:1201.4151 [gr-qc].

[52] D. Boulatov, "A model of three-dimensional lattice gravity," *Mod. Phys. Lett. A* **7** (1992) 1629–1646, arXiv:hep-th/9202074.

[53] N. Bodendorfer, T. Thiemann and A. Thurn, "New variables for classical and quantum gravity in all dimensions III. Quantum theory," *Class. Quant. Grav.* **30** (2013) 045003, arXiv:1105.3705 [gr-qc].

[54] N. Bodendorfer, T. Thiemann, and A. Thurn "Towards loop quantum supergravity (LQSG)," *Phys. Lett. B* **711** (2012) 205–211, arXiv:1106.1103 [gr-qc].

[55] J. Engle, C. Fleischhack and E. Alesci, "Panel on symmetry reduction: from LQG to LQC," http://relativity.phys.lsu.edu/ilqgs/.

[56] A. Ashtekar and J. Lewandowski, "Quantum theory of geometry. 2. Volume operators," *Adv. Theor. Math. Phys.* **1** (1998) 388–429, arXiv:gr-qc/9711031.

[57] A. Ashtekar, A. Corichi and J. A. Zapata, "Quantum theory of geometry III: Non-commutativity of Riemannian structures," *Class. Quant. Grav.* **15** (1998) 2955–2972, arXiv:gr-qc/9806041.

[58] M. Han, "4-dimensional spin-foam model with quantum lorentz group," *J. Math. Phys.* **52** (2011) 072501, arXiv:1012.4216 [gr-qc].

[59] W. J. Fairbairn and C. Meusburger, "q-Deformation of Lorentzian spin foam models," *PoS* **QGQGS2011** (2011) 017, arXiv:1112.2511 [gr-qc].

[60] V. Turaev and O. Viro, "State sum invariants of 3 manifolds and quantum 6j symbols," *Topology* **31** (1992) 865–902.

[61] C. Rovelli and S. Speziale, "Lorentz covariance of loop quantum gravity," *Phys. Rev. D* **83** (2011) 104029, arXiv:1012.1739 [gr-qc].

[62] A. Ashtekar and B. Krishnan, "Isolated and dynamical horizons and their applications," *Living Rev. Rel.* **7** (2004) 10, arXiv:gr-qc/0407042 [gr-qc].

[63] K. Giesel and T. Thiemann, "Consistency check on volume and triad operator quantisation in loop quantum gravity. I," *Class. Quant. Grav.* **23** (2006) 5667–5692, arXiv:gr-qc/0507036.

[64] J. Aastrup and J. M. Grimstrup, "Intersecting Connes noncommutative geometry with quantum gravity," *Int. J. Mod. Phys. A* **22** (2007) 1589–1603, arXiv:hep-th/0601127.

[65] J. Aastrup and J. M. Grimstrup, "Quantum holonomy theory," *Fortsch. Phys.* **64** no. 10 (2016) 783–818, arXiv:1504.07100 [gr-qc].

[66] C. G. Callan, S. B. Giddings, J. A. Harvey and A. Strominger, "Evanescent black holes," *Phys. Rev. D* **45** no. 4 (1992) 1005, arXiv:hep-th/9111056.

[67] D. Levanony and A. Ori, "Interior design of a two-dimensional semiclassic black hole," *Phys. Rev. D* **80** (2009) 084008, arXiv:0910.2333 [gr-qc].

[68] D. Levanony and A. Ori, "Interior design of a two-dimensional semiclassical black hole: quantum transition across the singularity," *Phys. Rev. D* **81** (2010) 104036, arXiv:1005.2740 [gr-qc].

[69] A. Ori, "Approximate solution to the CGHS field equations for two-dimensional evaporating black holes," *Phys. Rev. D* **82** (2010) 104009, arXiv:1007.3856 [gr-qc].

[70] A. Ashtekar, V. Taveras and M. Varadarajan, "Information is not lost in the evaporation of 2-dimensional black holes," *Phys. Rev. Lett.* **100** (2008) 211302, arXiv:0801.1811 [gr-qc].

[71] A. Ashtekar, F. Pretorius and F. M. Ramazanoglu, "Evaporation of 2-dimensional black holes," *Phys. Rev. D* **83** (2011) 044040, arXiv:1012.0077 [gr-qc].

[72] A. Ashtekar, F. Pretorius and F. M. Ramazanoglu, "Surprises in the evaporation of 2-dimensional black holes," *Phys. Rev. Lett.* **106** (2011) 161303, arXiv:1011.6442 [gr-qc].

[73] A. Ashtekar, "Issue of information loss: current status," *International Loop Quantum Gravity Seminar* (9 Feb. 2016), http://relativity.phys.lsu.edu/ilqgs/.

[74] A. Ashtekar, "Black hole evaporation: a perspective from loop quantum gravity," *Universe* **6** no. 2 (2020) 21, arXiv:2001.08833 [gr-qc].
[75] A. Ashtekar and B. Krishnan, "Dynamical horizons and their properties," *Phys. Rev. D* **68** (2003) 104030, arXiv:gr-qc/0308033 [gr-qc].
[76] A. Ashtekar, J. Baez, A. Corichi and K. Krasnov, "Quantum geometry and black hole entropy," *Phys. Rev. Lett.* **80** (1998) 904–907, arXiv:gr-qc/9710007.
[77] R. K. Kaul and S. Sengupta, "Topological parameters in gravity," *Phys. Rev. D* **85** (Jan. 2012) 024026. https://link.aps.org/doi/10.1103/PhysRevD.85.024026.
[78] F. Pretorius, "Evolution of binary black hole spacetimes," *Phys. Rev. Lett.* **95** (2005) 121101, arXiv:gr-qc/0507014 [gr-qc].
[79] R. Penrose, *Fashion, Faith, and Fantasy in the New Physics of the Universe*. Princeton University Press, 2017.

4

Jan de Boer

Professor at the Institute of Physics, University of Amsterdam, the Netherlands

Date: 21 October 2020. Via Zoom. Last edit: 1 December 2020

In your opinion, what are the main puzzles in theoretical physics at the moment?

There are many interesting open questions in theoretical physics. For instance, there are many questions related to collective and emergent behaviour of various systems and materials, such as high T_c superconductivity as well as many other systems in the context of soft matter. In addition, there are also many open issues in the domain of quantum gravity and cosmology. Specifically, it is still unclear to what extent the existence of a theory of quantum gravity has actual observable consequences for our universe. Somewhat related are important open questions regarding the nature of dark matter, the physics beyond the standard model and what exactly the world is made of, which have been around for decades.

Other interesting questions deal with the fact that we have a beautiful framework for describing equilibrium physics, namely thermodynamics and its generalisations, but we lack a systematic conceptual framework for non-equilibrium physics that is applicable to many different situations and systems, in particular systems that are far from equilibrium.

Recently, there has been a lot of interesting work on general aspects of dynamical systems. For instance, in classical systems we can distinguish between integrable systems and classically chaotic systems but there is another regime in between where you find quasi-periodic orbits (e.g., as in KAM theory). However, the quantum counterparts of these regimes are not yet clear. There are notions of quantum chaos, though no general agreement about what the definition of it should be, and also there are quantum integrable systems. On the other hand, the "in between" regime, analogous to quasi-periodic orbits in classical systems, is not well understood in the quantum context. It may be related to quantum scars or many-body localisation but it is not yet clear. This type of question about quantum chaotic systems deserves a better conceptual understanding.

You mentioned that one of these open questions is about quantum gravity and to what extent it describes our universe. In your opinion, is there an answer to what is the theory of quantum gravity?

I think string theory is a very good and promising theory of quantum gravity. For instance, within string theory it is possible to compute graviton scattering amplitudes in flat spacetime, including ultraviolet finite loop corrections [1, 2]. Additionally, the AdS/CFT correspondence [3], which is an integral part of string theory, has been extensively applied and tested in many different circumstances. By studying these cases, it has been possible to draw many qualitative general lessons about quantum gravity. However, the detailed technical implementation of string theory in the context of cosmology, as well as the translation of those qualitative lessons to quantitative results in cosmology, is something that has not been fully accomplished.

What kind of qualitative lessons do you have in mind?

One of these lessons is that at sufficiently short distances, shorter than the length scale set by the cosmological constant, the S-matrix that you extract from AdS/CFT agrees with the S-matrix that you find in flat spacetime [4, 5]. As such, there is no reason to believe that this would not be the case in a scenario where you have a positive cosmological constant and at length scales much smaller than the curvature radius of the universe.

There are also lessons arising due to the separation between low- and high-energy degrees of freedom, which in the context of AdS/CFT is usually referred to as the existence of a code subspace in the Hilbert space [6]. In particular, for the purpose of observations you only need to look at the code subspace. Additionally, the high-energy degrees of freedom with which the code subspace interacts can be very complicated, perhaps even nonlocal. The details of this interaction may be model dependent but the fact that many features of this interaction are universal, which in particular is responsible for describing low-energy operators in a theory of gravity via quantum error correction-type phenomena, is an important general lesson. I do not think we have fully explored the importance of this interaction between the visible "code subspace" degrees of freedom and this large number of indistinguishable high-energy degrees of freedom. Gravity is fundamentally different from field theory in that semiclassical gravity has, through black holes, access to some aspects of the high-energy degrees of freedom – a phenomenon which is absent in standard quantum field theory. I have sometimes wondered whether the high-energy degrees of freedom are somehow crucial for things like quantum measurement and decoherence and the emergence of a classical world, but unfortunately do not have anything concrete to say about this.

I should also add to this discussion the recent lessons that we learnt about the importance of non-perturbative semiclassical gravitational configurations, such as islands and wormholes [7–10], which presumably also apply to cosmology.

What are the general lessons in this context of non-perturbative gravitational configurations?

One general lesson is that the island formula roughly relates precise microscopic entropy to semiclassical entropy in a particular system [8]. Semiclassical entropy is relevant for semiclassical observers while the microscopic entropy is relevant for the underlying precise microscopic description. Therefore, the fact that you sometimes need to include apparently disconnected pieces of your universe in computing the semiclassical entropy in order to precisely account for the microscopic entropy is an interesting general insight and there's no reason to believe that it would not apply in a cosmological setting too. I have recently been trying to understand the origin of various wormholes in terms of the aforementioned interaction between low- and high-energy degrees of freedom which I think is a promising direction [11, 12].

You mentioned earlier that string theory is a good candidate for a theory of quantum gravity. What do you think about other approaches like loop quantum gravity (LQG) [13, 14], causal dynamical triangulations (CDT) [15, 16], asymptotic safety [17], non-commutative geometry approaches [18], etc.? Are these approaches that should be pursued?

I think that diversity of ideas is beneficial for physics and since the ultimate test of string theory, which is direct observational confirmation, has not been achieved, strictly speaking, you should be open-minded and also think about other formulations. On the other hand, you have to have honest conversations about the pros and cons of all these other approaches. My impression is that many people live in information bubbles and only listen to people who have similar ideas. As such, they ignore others with different ideas and are not willing to bring meaningful content to intellectual conversations with people with different ideas. This seems to be a general sociological phenomenon nowadays, which is obviously not good for science. I see this phenomenon taking place with all these different approaches. String theory is not innocent in this respect but what strikes me is that many of the papers I've looked at within these other approaches very often only address one particular issue or focus on a particular modification of general relativity. In other words, they do not provide a general computational framework.

Historically, you could argue that one of the important tests of any theory is that in suitable limits it reproduces the results of previous theories. For instance, quantum mechanics when $\hbar \to 0$ reduces to classical mechanics. Thus, any new quantum gravity proposal should definitely in the limit for which $\hbar \to 0$ reduce to general relativity and, for instance, reproduce tree-level graviton scattering amplitudes. If a theory cannot reproduce such limits then it is not yet a useful framework because you cannot be sure that what you have proposed is actually a controlled modification in the appropriate regime of the past theories that you know are correct.

Do you think that all these other approaches are lacking in this respect?

We should not compare apples and oranges. Some of these approaches are not full-fledged developments. That is, they're just an idea by one author written in a handful of papers and in no way comparable to string theory and the detailed technical computations that have been performed within string theory. So the general answer to your question depends on which theory you are actually discussing. I can make a few observations and state some general concerns about some of these approaches which I have, but I should hasten to point out that I do not follow all these approaches on a daily basis.

In the case of asymptotic safety [17], the starting point is general relativity and the aim is to move up in the renormalisation group flow. While following this flow you examine whether or not there are ultraviolet fixed points. So, clearly, at low energies it is compatible with classical general relativity. However, for now it is just a hypothesis and it is not yet clear that such fixed points exist. There is some evidence that if you truncate higher-order operators, and to the extent that you can control the computation, you observe a line in phase space that tends to a point somewhere. However, these are model-specific computations and it appears to be beyond technical control to do a full search of these fixed points. Additionally, if you would find such fixed points it is not clear what the right language is for describing the physics of those fixed points since one is approaching it from the point of view of local quantum field theory. At the fixed point, you will find some kind of scale-invariant theory of manifolds, which means that it will contain geometric fluctuations on all scales. How to describe these fixed points is unknown and so there is a technical gap in this program in order to make a complete theory of quantum gravity. It is of course possible to postulate the existence of these fixed points and extract predictions for particle physics. As such, it is an interesting idea but still far from being a complete theory of quantum gravity.

CDT [15, 16] is also based on the assumption of asymptotic safety because you need to take a continuum limit of some lattice model, which implies that some critical point is required to exist. Usually, in these non-renormalisable models in which you place the theory on a lattice, one has to deal with very difficult, if not impossible, fine-tuning problems in order to approach these critical points. What you typically do is to not include other dangerous operators, restrict to a few couplings and perform a numerical analysis. In principle, since you are making a lattice model of spacetime you could imagine that such model is capable of also describing phases that are not geometrical in the sense that you are dealing with graphs/lattices which have a coarser structure than a smooth manifold. As such, it bears the hope that you could potentially describe some sort of strange fixed point as an abstract graph configuration without a clear geometric interpretation. However, CDT needs to explain why this is the right procedure, why you can ignore all irrelevant operators and why there is no fine-tuning problem. Additionally, causal dynamical triangulations also needs to explain how general covariance is restored as it is explicitly broken in the discrete model since space and time are treated asymmetrically as CDT introduces a time foliation. It is not obvious that violating such a symmetry principle in the beginning will lead to a restoration of that symmetry when taking the continuum limit. In fact, the continuum limit may be more akin to Hořava-Lifshitz gravity [19], as suggested by the analysis of fractal dimensions in the continuum limit [20], than to general relativity. Furthermore, CDT is far from computing a correlator or the four-graviton three-level S-matrix and its loop corrections. In summary, CDT is far from providing a computational framework for these kinds of quantum gravity questions.

LQG [13, 14] is an approach that started with the assumption that you can find an interesting set of classical variables in general relativity that allow you to replace Poisson brackets by commutators. In practice, replacing some brackets by commutators in nonlinear theories is not necessarily straightforward but apparently possible when using Ashtekar variables and leads to a quantum Hilbert space. However, it turns out to be notoriously difficult to impose the Hamiltonian and diffeomorphism constraints in that language. This led to a picture where you have some kind of space of loops in an otherwise topological space, without further structure, in which you need to impose constraints, quantise and ultimately do an actual scattering computation. None of these steps has been properly taken yet. In fact, recent work [21] seems to suggest that the non-renormalisability of quantum gravity may come back when trying to properly impose the constraints. One of the somewhat more phenomenological things you can do is to take a given spin network, as suggested by this theory, and let it evolve. My impression is that there is a big gap between this modern point of view in terms of spin networks and the original Hamiltonian approach to loop quantum gravity that contains all these unsolved problems. Right now, a lot of LQG work is more akin to a phenomenological theory inspired by Ashtekar variables and loop space rather than a theory that has been rigorously and reliably derived from a solid starting point. It is also my impression that there has been progress in understanding the mathematics of LQG but the mathematics used is also very strange because it deals with non-separable Hilbert spaces, which are not Hilbert spaces you typically find in other contexts. For instance, you can treat the harmonic oscillator using a non-separable Hilbert space but that will lead to deviations from the standard harmonic oscillator. It is also possible to use these qualitative lessons and apply them to mini-superspace versions of cosmology leading to loop quantum cosmology (LQC). LQC is also a phenomenological theory because the mini-superspace approximation is not a systematic expansion of full LQG in a small parameter. It has also recently been shown that there is tension between this approach to LQC and general covariance [22]. In summary, LQG has given rise to some phenomenological program, though there are some mathematical physicists who try to fill in the gaps and are critical of these developments. In my opinion, LQG is still very far from a first-principle computation of a four-graviton scattering process.

Do you think LQG has already accomplished a classical limit where they recover general relativity?

No, that has not been done as far as I know. Additionally, there is a reasonably well-known paper by Ashoke Sen [23] that points out that there is a universal model-independent answer to the logarithmic correction to black hole entropy, which you can compute and that any theory of quantum gravity should give. LQG, at least if you use the heuristic models for black hole entropy that have been proposed, does not provide the right answer. The model for a black hole that LQG proposes is basically a spin network from which you remove some links. Then, you interpret the spins that sit on the horizon as the degrees of freedom associated to the horizon and you call that the entropy of the black hole. But why would that even satisfy basic features such as black hole thermodynamics is beyond the control of LQG at this point.

Moving on the Alain Connes' non-commutative approach [18] ... I think that it is still a classical approach. By shifting from a commutative manifold to a non-commutative manifold you replace the commuting set of observables or functions on the manifold by a non-commutative algebra and then you try to follow the same footsteps as in other approaches but now using a non-commutative algebra. However, this non-commutative structure is still a rigid structure and does not quantum-fluctuate; that is, there is no path integral. So perhaps you can use this formalism to make some symmetries more manifest as in the standard model but to perform a proper quantisation, including a quantisation of the geometry, you would have to develop an entirely new calculus. This new calculus would have to allow for a quantisation procedure that respects this whole non-commutative structure. The standard Feynman diagram expansion refers to standard commutative functions and as such will not be so helpful. Additionally, you have to make sure that the theory is renormalisable in whatever new language you develop, which is very complicated.

So you do not think that this is a promising direction?

It's interesting in the sense that any possibility of uncovering a new symmetry principle in physics will lead to progress. If this non-commutative structure would lead to a new symmetry principle and we can quantise the theory while maintaining the symmetry principle somehow, it would be major progress. However, I just don't think that we know how to do that at this point nor is it clear to me that if we knew how to do it, it would solve the standard ultraviolet divergence problems of gravity, that is, whether it will lead to a reasonable path integral.

The generic lesson for curing ultraviolet divergences is that you need simple low-energy operators and a very complicated reservoir of high-energy states. Such a generic structure is needed to address gravity and has been one of the lessons that came out of AdS/CFT. It is not applicable just to string theory. However, all these other approaches do not consider this large bath of high-energy degrees of freedom as only a few low-lying degrees of freedom are quantised. As such, it is hard to reconcile any of these approaches with the lessons that we learnt from string theory.

Doesn't LQG have a mechanism for regularising divergences? Or is it based on asymptotic safety?

As far as I know it is not based on asymptotic safety because it is direct quantisation. It is not clear to me that LQG is ultraviolet finite because you are dealing with non-separable Hilbert spaces and I'm not familiar with new mathematical developments that could potentially allow you to answer this question. Then there still is the question of the proper incorporation of the Hamiltonian constraint.

The usual argument of LQG is that certain operators like the area and volume operators have a discrete spectra [24]. Therefore, there is some minimum length scale that renders the theory ultraviolet finite . . .

Yes but first of all, you can ask whether the area is an observable or not. It's not clear that the area is an observable because only relational observables make sense. Additionally, because we don't know what the Hilbert space is exactly, since the constraints have not been imposed, I am not sure we know the spectrum of these operators very well. Spin foam networks are based on some discretisation of spacetime involving a fundamental length scale, say the Planck scale. Such discretisation process will always cure any ultraviolet divergences because of this cutoff.

Right, and indeed introducing that discretisation is the main point of causal sets [25] in which there is a fundamental discrete length scale. However, in LQG I thought that it is just a regularisation trick and at the end of the day you need to find a continuum limit. In that continuum, the area operator has a discrete spectrum ...

I think it's quite unclear what the precise LQG continuum limit is. I don't know how that limit is obtained and how all the constraints are being imposed, neither what it precisely means to discuss spectra of operators on non-separable Hilbert spaces nor how to set up a graviton scattering problem. Well, I suppose spectral theory of self-adjoint operators on non-separable Hilbert spaces is in principle well understood so maybe the mathematics part of that discussion is fine. LQG contains an interesting set of ideas but one has to be honest and critical about what certain approaches achieve and don't achieve.

Do you also have an opinion about the causal set approach [25]?

No, I'm not so familiar with it but it always appeared to me as a very classical starting point. I'm not sure that *causal structure* is a fundamental concept in quantum gravity. Taking generic lessons from AdS/CFT, it appears that high-energy degrees of freedom are kind of nonlocal and do allow for e^{-S} micro-causality violations, where S is the action or entropy. As I mentioned, the role of islands and wormholes is important in quantum gravity [7–10] and those configurations are at odds with any notion of causal set. In summary, it seems far from a full-fledged quantum theory of gravity.

Since you mentioned it was good to be critical, I wanted to bring some criticism that is usually made toward string theory that has to do with a non-perturbative formulation. For instance, string theory typically starts with the quantisation of a string in a given background, so how can you ever explain the background on which the string is propagating? Is this reasonable criticism?

Yes and no. The question of having a non-perturbative definition of string theory is a reasonable question and you would definitely like to have an answer to it. However, in the context of AdS/CFT we do have an example of a string theory that has a non-perturbative definition. How this works in flat spacetime or in generic cosmological backgrounds is less clear at the moment.

But it's important to find it, right? Is there any hope or indication that there should be such string theories?

There are certainly indications. We do have BFSS theory [26], which is a good description of the discrete light-cone quantisation of M-theory in 11 dimensions. This theory basically reduces to some complicated interacting quantum mechanical system for which there is a non-perturbative definition. Furthermore, matrix string theory in 10 dimensions is another example that describes physics in flat spacetime [27]. The problem of finding a string theory in cosmological backgrounds is that it is hard to find a precise string cosmological background that we have control over.

Is it not possible to make a precise statement about what theory describes flat spacetime holographically?

Not at the same level as AdS/CFT, but as I said we have theories like matrix string theory and BFSS which are flat spacetime theories but in discrete light-cone quantisation. Those are definitely theories that describe quantum gravity in flat spacetime.

Why aren't these good enough?

They might in principle be good enough but it's just very difficult to do interesting computations in these models. They do not involve 't Hooft–type dualities, which means that it is difficult to compare computations in these models to gravitational computations. The reason is that you need to perform such calculations in discrete light-cone quantisation (DLCQ), which implies that the perturbative theories don't really match in a direct way. It is yet another example of a duality between weakly and strongly coupled theories and in that sense it is quite similar to AdS/CFT. However, some computations were carried out. For example, in the 1990s people compared a two-loop scattering computation in BFSS with a suitable scattering process in supergravity with M-theory corrections, which worked fine [28].

Contrary to anti–de Sitter (AdS) spacetime, in flat spacetime we do not have a rigid boundary, which makes it a more difficult problem. This is why in flat spacetime we speak of S-matrix elements, which are murkier than correlators on the boundary of AdS because of infrared divergences, which you have to treat in a suitable way. Perturbative S-matrices are well defined but non-perturbative S-matrices are very complicated objects. Nowadays, people are trying to use *celestial holography* [29, 30] to fix boundary conditions at past and future null infinity. This could be a way of making progress but it's not easy to include massive particles and there are still infrared divergences that must be dealt with. Maybe in some regime you can factor them out as suggested by the recent paper of Andrew Strominger and Nima Arkani-Hamed [31] but I don't know if there is any general statement that can be made at this point.

How would you state the AdS/CFT correspondence?

The statement of the correspondence, roughly speaking, is that non-perturbative string theory in a spacetime with AdS boundary conditions is identical to $\mathcal{N} = 4$ super Yang–Mills (SYM) theory. Well, this is one particular example of AdS/CFT; there are many other examples that involve AdS spaces with different dimensionalities, but this a famous example which involves AdS_5. It is important to emphasise that these do not correspond to string theory in a fixed spacetime, in particular the theory could involve a sum over different spacetimes with different topology. In fact, the only structure which is kept fixed is the AdS boundary conditions. In particular, any correlation function in $\mathcal{N} = 4$ super Yang–Mills theory of gauge-invariant operators, local or nonlocal, is equal to a corresponding string theoretic quantity. As a matter of fact, these types of calculations are the only meaningful calculations that a theory of quantum gravity can deliver in AdS. For instance, you can ask, "What's the partition function of the theory on the four sphere?" However, you cannot ask details about a given operator in the bulk of AdS in an objective fashion because it is strictly speaking not a well-defined entity in the gauge theory. It is possible to try to build a proxy for this operator in the gauge theory à la HKLL [32] but it is only an approximation.

Why do we have a non-perturbative definition of string theory in the context of AdS/CFT?

What makes AdS special is the existence of a well-defined geometry at the boundary. Roughly, it is like putting quantum gravity in a box, which allows you to relate observables in the bulk to well-defined observables at given locations in the boundary. If you would ask me to give you a non-perturbative definition of quantum gravity on a sphere that fluctuates, I would find it very difficult to begin discussing the problem.

But what is the non-perturbative definition of string theory in AdS?

The definition is $\mathcal{N} = 4$ SYM theory.

But AdS/CFT is supposed to provide a duality between string theory in AdS and $\mathcal{N} = 4$ SYM. So what is the non-perturbative definition of string theory in AdS?

On the string theory side what we can do is to use perturbative string theory to check how well it approximates the results obtained from $\mathcal{N} = 4$ SYM. In general, perturbative string theory is a good approximation to the exact answer. Very often, the full non-perturbative result, as obtained from SYM, only differs from perturbative string theory by very small corrections. However, in the case of many observables we actually do have control over non-perturbative string corrections arising from the wrapping of D-branes or NS5 branes [33]. This can be explicitly checked within AdS/CFT as has been done in numerous papers (for a recent state of the art computation, see, for instance, [34]).

Some people understand the duality as $\mathcal{N} = 4$ SYM being dual to classical supergravity with additional quantum corrections. This has, for instance, motivated people to quantise supergravity using LQG methods [35]. Is this reasonable?

That's wrong because we can do computations in $\mathcal{N} = 4$ SYM that correspond to string worldsheet computations in the bulk. If you would just take supergravity and do a one-loop computation, ignoring string theory, you would not get the right answer. An example of such computation is the expectation value of a Wilson loop, which you can evaluate, including quantum corrections, by anchoring a string worldsheet to the boundary of AdS [36, 37]. This has nothing to do with quantum supergravity.

I understand, but coming from a general relativity point of view, what is quantum geometry in the bulk?

Sufficiently complicated quantum geometry questions involve geometric fluctuations that are large. This means that when studying such processes using string theory the calculation becomes more complicated and you have to take into account non-perturbative effects. You could set up a high-energy scattering experiment such that two particles propagate in a classical geometry but when they get close enough the whole notion of classical geometry ceases to exist. To evaluate such process you would need to re-sum all kinds of large metric fluctuations to all orders. This has not been done in string theory because it is very difficult. If you would consider this scattering process in the context of AdS/CFT, it would be related to suitable correlation functions of operators in the gauge theory. But these only probe the S-matrix of the process in which the information about geometric fluctuations is encoded and coarse-grained in a very complicated way. So I think that it is important that we, as low-energy observers, will never detect the individual quantum geometries, and it might not even be a well-posed question to ask, "What is quantum geometry?" As low-energy observers, we in general see statistical averages of these putative quantum geometries, and these statistical averages typically look like semiclassical geometries [38]. In any case, if we want to say anything about the real world we have to first understand it at the perturbative level.

What does spacetime look like at the Planck scale?

I think this is not a good question because it suggests that you can probe spacetime at the Planck scale using a low-energy probe, which is not possible. All you can do is scatter things at the highest energies that we have available and wait long enough to see what's the end product. Based on these measurements you then have to come up with a theory that describes the results. I am doubtful that a geometric language is appropriate once you reach the Planck scale. Since our experiments are, by construction, producing semiclassical output (otherwise we as humans could not process it) we will never directly "see" spacetime at the Planck scale and at best only its imprint on low-energy experiments. So, spacetime at the Planck scale need not "look" like anything.

I am not aware of any compelling evidence that spacetime is discrete. You could argue that it is discrete in the sense that you cannot put an infinite amount of information in a Planck-size cell but that does not mean that you should make a precise lattice model of spacetime, which does not seem like a promising idea to me.There is a general argument by Donald Marolf that states that it's very hard to describe gravity from a continuum limit of a lattice model with a local Hamiltonian because the gravitational Hamiltonian consists only of a boundary term [39].

Some people have criticised AdS/CFT by claiming that it is just a conjecture. Are there striking tests that give strong evidence for the correspondence?

To prove the conjecture you need two different non-perturbative definitions of the theory and to subsequently show that they are identical. At this point we do not have these two independent definitions. However, in a physics sense, AdS/CFT is much more than a conjecture. For example, planar gauge theory turns out to be an integrable system with a complicated spectrum but you can compare that with the perturbative spectrum in the bulk and find agreement [40]. The amount of evidence that you need to use string theory to describe $\mathcal{N} = 4$ SYM is striking.

An interesting state-of-the-art example that I referred to before is of a situation in which you start with gauge theory on a four sphere, deform the shape of the sphere with some parameters and compute the partition function using localisation techniques [34]. This gives you a very complicated expression but from it you can extract the integrated correlation function of four stress tensors on the four sphere, which is a quantity that is not a priori protected by symmetry because it depends on cross-ratios (since it is a four-point function). From that four-point function you extract the flat spacetime S-matrix. This is a pure gauge theory computation as it did not involve gravity at all. Next, you can look at the rather complicated genus one and genus two perturbative higher-curvature corrected string computations, which use indirect arguments and dualities, and find that the two calculations agree. This is pretty striking because it is beyond the planar limit and doesn't rely on integrability. It might still secretly rely on supersymmetry but in a way that is quite mysterious. This is just one instance of the type of evidence there is.

Another type of question that often comes up is related to how well we understand bulk physics from the boundary theory in the context of AdS/CFT. Is there a final answer to this question?

Some of these questions can be answered in principle but some not yet in practice. For instance, at this point we understand in more detail how the black hole information paradox is resolved, thanks to the recent works on islands and in obtaining the Page curve correctly [7–10]. However, we do not yet have a detailed description of the interior of the black hole. Nevertheless, this recent understanding provides a mechanism for obtaining tiny nonlocalities in correlations across the horizon, which were necessary for understanding how information can come out of the black hole. This mechanism is a type of $ER = EPR$ phenomenon [41] which replica wormholes and islands made more precise. The fact that nonlocality was needed was pretty clear because the only other alternative was firewalls [42, 43], which I don't think was a good solution.

But what is the mechanism exactly by which you could have this nonlocal information transfer?

In terms of low-energy effective field theory, the picture is that there are nonlocal couplings between the radiation inside and outside of the horizon of the black hole that scale as e^{-S}, which is just the right amount to get the information out. One could argue that this gives rise to a violation of low-energy effective field theory, but the violations are very small and beyond the normal regime of validity of low-energy effective field theory anyway.

Could you say that the black information problem is now solved?

To some extent yes but in the cosmological context it is less clear. The reason is that in such context, you don't have an infinite reservoir of high-energy states but perhaps only a large but finite reservoir and the question is whether this finiteness is going to affect the calculations.

But if we just stick to the AdS case? Has it been solved?

I think you could say that it has been solved in the sense that you have a unitary dual gauge theory.

Sure, but people would like to know how that works out in the bulk, right?

Yes and we do have part of the mechanism that allows information transfer; namely, the entanglement entropy of the outside radiation follows the Page curve which is needed for unitarity. However, it has not been computed in microscopic detail the quanta present in that radiation. It is also possible to ask whether there are more precise calculations that give you not only the Page curve but also more detailed information. I suspect that in simple correlators you will not see any deviation from the thermal (or Hawking) answer; that is, you need to compute extremely complicated correlation functions to see any deviation from the Hawking radiation results.

As to what concerns the firewall discussion, is there a consensus that there are no firewalls?

I think the majority opinion is that they are not necessary. However, there may still be situations that require firewalls such as cases in which you want to examine the very late time behaviour of black holes (at times e^S) in which you seem to run out of complexity. Leonard Susskind thinks that firewalls may cure that problem [44]. This is not a very practical problem in the sense that a low-energy observer will not be able to perform observations over such long timescales and hence will not be able to see a firewall.

You mentioned that we do not yet have a good description of the interior of a black hole. The discussion on firewalls led to attempts at understanding observers falling into black holes from a boundary point of view in AdS/CFT, for instance the Raju-Papadodimas proposal [45–47]. Does this proposal provide the right understanding or is it not yet clear?

My personal opinion is that there are loose ends in this discussion, due to the frozen vacuum problem. Kyriakos and Suvrat showed that if you assume the black hole interior to be empty, which is a self-consistent assumption, there is a description of observers falling into black holes. However, perhaps you could have assumed something to be inside the black hole, like the famous *pink elephant*, which would also be a self-consistent assumption, and that could have allowed for slightly different equations that would accommodate that assumption. Whenever you have a one-sided black hole you always have to face the frozen vacuum problem, and there is also a version of the problem for two-sided black holes. So I would say that we have a successful answer to the question "What happens to an observer falling into a black hole?" up to the frozen vacuum problem.

AdS/CFT is supposed to provide an example of emergent spacetime. How much do we understand about how to reconstruct spacetime from the boundary theory?

I am not sure that reconstructing spacetime has ever been a major problem. For that purpose there is a simple operational recipe which consists of computing all one-point functions of all supergravity fields in the state that you are interested in, and use those as boundary conditions for the Einstein equations which you integrate inwards from the boundary [48–50]. This defines the initial value problem for solving the required differential equations. This is the starting point, or best guess, for which classical spacetime corresponds to the state you started with but you later need to check whether the resulting spacetime is indeed a classical spacetime; that is, you need to check whether quantum fluctuations in that spacetime are small. In principle, you could prepare a state in the gauge theory that describes the macroscopic superposition of two very different spacetimes, and so you see that generic states do not admit a semiclassical spacetime description. This is analogous to solving the Einstein equations $R_{\mu\nu} - Rg_{\mu\nu}/2 = \langle T_{\mu\nu} \rangle$, where $\langle T_{\mu\nu} \rangle$ is the expectation value of the stress tensor, that is, coupling classical geometry to a quantum state, which is acceptable as long as the quantum fluctuations of $T_{\mu\nu}$ are small compared with its expectation value. That being said, there are a number of methods that can be used to reconstruct the classical geometry, including quantum information theoretic methods based on Ryu-Takayanagi minimal surfaces [51] or the knowledge of singularities of correlation functions [52]. You could call all these methods "tomographic" because you are trying to construct higher-dimensional geometry from lower-dimensional information. Which method is best is just a matter of efficiency.

There is, however, a lot of research on trying to use quantum information theory to understand *bulk reconstruction*, for which the goal is not just reconstructing the spacetime metric but also computing the approximate local low-energy bulk operators. To leading order, this reconstruction is provided by the HKLL prescription [32] (which receives higher-order corrections), but beyond leading order interesting new ingredients come in such as gravitational dressing and quantum error correction [53]. For this type of question, quantum information machinery is very interesting and useful.

Can bulk reconstruction provide a test of the AdS/CFT correspondence?

It is a quite nontrivial test because it is somewhat of a miracle that there is some kind of local low-energy effective field theory in the bulk. It is still a bit of a mystery why on earth a local boundary theory in D spacetime dimensions can be dual to an approximately local $D+1$-dimensional bulk theory. What guarantees bulk locality? Quantum information theory arguments are suitable for understanding bulk locality at AdS scales [53] but it is not yet clear how to understand bulk locality at sub-AdS scales from the boundary theory. I am currently quite interested in trying to use ideas from quantum chaos and algebraic quantum field theory to make progress in this direction [54].

One of the comments you made in the beginning of this conversation was that there doesn't seem to exist any test of string theory. Is there no way of finding evidence for string theory?

A first important test for any theory is internal self-consistency, which is a much more restrictive requirement than one might naively think and which string theory has passed with flying colours. Beyond this, there are some further potential ways of finding evidence for string theory. Gravity is a low-energy effective field theory while string theory provides the complete high-energy behaviour. Nevertheless, gravity alone gives correct black hole thermodynamics, which in AdS/CFT actually controls the high-energy physics of the dual theory. In fact, it is the high-temperature free energy that is computable but the bulk of AdS is just a low-energy effective gravitational theory. In other words, the gravity theory knows about low-energy physics and low-energy scattering processes but also knows some aspects of high-energy physics. This means that gravity does not comply with the standard lore of effective field theory, which usually knows everything below a given cutoff and nothing above it. It is this interplay between the knowledge of high- and low-energy physics, exemplified by the case of black holes, that is sometimes referred to as *UV/IR mixing* and I'm not sure we have fully understood the implications of this fact.

In addition, the swampland programme [55] is aimed at describing the allowed set of low-energy effective field theories, where by *allowed* I mean consistent and ultraviolet-completable. It is quite remarkable that in cases with many supersymmetries the set of allowed low-energy effective field theories agrees precisely with the theories which can be obtained from string theory. It is interesting that the mere existence of a ultraviolet completion imposes restrictions on the coupling constants of the effective theory. For example, the weak gravity conjecture [56] and the trans-Planckian censorship conjecture [57] make certain statements that restrict the allowed theory space. Even in field theory, the sign of a ϕ^4 interaction is fixed by self-consistency arguments.

Another potential source of evidence for string theory, which is difficult but perhaps not impossible, is to prove that the unique consistent completion of the low-energy four-graviton amplitude is the Veneziano amplitude [58]. Bootstrap-type thinking might be helpful here. If this were a true statement, then you would have proven that string theory is the only reasonable ultraviolet extension of gravity. I think that would be very strong evidence that we need string theory.

What about direct experimental implications of string theory?

I think that exploring UV/IR mixing in the context of closed spacetimes such as de Sitter cosmologies is an interesting question that may lead to direct implications. Furthermore,

I think that the prediction that in large classes of strongly coupled systems the viscosity to entropy ratio η/s is $1/(4\pi)$ is a string theory prediction since the viscosity of the horizon is interpreted as a field theory statement [59]. It is striking that experimentally there seems to be a bound on η/s of the same order of magnitude as $1/(4\pi)$ [60]. Gravitational scattering in AdS/CFT also leads to a bound on chaos; namely, there's a bound on the Lyapunov exponent of $2\pi T$, where T is the temperature, which you can argue for in gauge theory using assumptions of large N factorisation [61]. If you had not used AdS/CFT it would be difficult to see any of these bounds.

We find all these bounds within string theory and they are in agreement with the limitations we observe in nature, which is very suggestive. The bound on chaos can actually be measured in some quantum systems, such as a cold atom system, by looking at out-of-time-ordered correlators. These examples are different from observing a new neutrino or a supersymmetric particle but they are strong evidence that string theory is a robust and generic framework that can lead to predictions in strongly coupled theories that have gravity duals and which are similar to those that exist in nature. These predictions are definitely fascinating.

You just mentioned that we could use AdS/CFT to understand generic properties of strongly coupled systems and you also mentioned in the beginning of this discussion that understanding non-equilibrium physics was a major challenge. Have we gained any insight into thermalisation processes using AdS/CFT?

Yes, we have. In AdS/CFT a thermalisation process in the field theory corresponds to black hole formation in AdS which follows certain universal rules. For example, if you throw something into the black hole the information of what was thrown in gets scrambled across the black hole horizon very rapidly. Therefore, in the field theory, a similar statement must apply. This has led to the statement that black holes are the fastest information scramblers in nature, and by AdS/CFT, so are the field theories with a weakly coupled gravitational dual [62]. We have also found some other universal features, for example that thermalisation proceeds from the ultraviolet (UV) to the infrared (IR) at a rate which can be as fast as allowed by causality [63, 64]. There has been a lot of work on this subject and I could give many other examples.

I understand that predictions about strongly coupled theories can be made using string theory but some people criticise that string theory is used like a theory of *epicycles*, meaning that it makes predictions about other systems and not about quantum gravity. Do you agree with this dismissive criticism?

We have learned a lot about quantum gravity from string theory and the reason why we think that the general lessons that we extract from it are valuable is exactly because of the predictions that it makes in all these other situations in which the quantum gravitational aspect of the theory is less pronounced. It gives satisfactory, interesting and meaningful answers that agree with what we see in the real world. In an ideal world we would build a huge accelerator and perform a quantum gravity experiment to probe string theory but that is not possible at the moment. Thus, the best we can hope for is to find indirect evidence for string theory. I think it is very interesting to try to extract these general model-independent predictions from the theory. When I occasionally look at papers that claim to have concrete quantum gravity predictions based on whatever method, these so-called predictions are almost always based on incomplete phenomenological models using various ad hoc assumptions and ad hoc choices of parameters so that pretty much anything goes. Most of that work does not represent genuine progress in our understanding of quantum gravity.

Some people claim that string theory predicts a negative cosmological constant, large extra dimensions and low-energy supersymmetry. Do you think these are string theory predictions?

No. String theory does not predict large extra dimensions and does not predict a negative cosmological constant. Solutions with negative cosmological constant are just a sector of string theory. String theory does predict high-energy supersymmetry because at short distances spacetime is approximately flat and string theory in flat spacetime is supersymmetric and should describe the physics at those scales. Supersymmetry may be broken at some scale, either explicitly or spontaneously, but low-energy supersymmetry is not implied by string theory.

Your answer suggests that the string theory that describes our universe is ultimately one of the consistent critical superstring theories in 10 spacetime dimensions [65].

It's not clear to me. There are many kinds of strongly coupled and weakly coupled solutions and it is not obvious to me that our universe is a weakly coupled solution of any of the known 10- or 11-dimensional supersymmetric theories. Our universe could be a strongly coupled solution of those theories which still has a weakly coupled low-energy gravitational description. At the moment, if we try to create a realistic cosmological solution of string theory, setting all swampland problems aside, you run into the issue that you need to make the fluxes pretty large. This means that the known solutions appear to be on the borderline of invalidating perturbation theory. As such, it is not entirely clear that we have a controlled perturbative vacuum of one of the five supersymmetric theories which describes our universe. However, there are dualities that can help us treat these backgrounds and there are even non-geometric solutions of string theory that could have a perfectly fine bona fide effective low-energy theory of gravity [66–68]. To summarise, at extremely high energies and short distances, we might recover the standard 10- or 11-dimensional theories but I don't think that our world is necessarily a weakly coupled vacuum of any of these theories.

Many people study different Calabi-Yau compactifications in order to find the right vacuum. Do you think that this is something that future generations can still justify doing in the next 10 years, given that so far these searches have not found the right vacuum?

All these searches lead to new knowledge about the theory such as lessons about supersymmetry breaking, how to stabilise the moduli, insights into non-perturbative effects and how D-branes interact as well as interesting mathematics and physics applications. In other words, this activity generates a lot of new related knowledge besides providing a given set of vacua. Additionally, you gain insight into the statistical distribution of string vacua which can have an associated notion of typicality. Perhaps our universe is a very peculiar microstate in a big ensemble but perhaps it falls into a large typical class. These types of questions are very interesting and I think it will take time to answer them.

There's a lot of interesting science coming out of string theory. The vast majority of research is carried out because it contributes to a very big and difficult question, and leads to interesting impact in one way or another. However, we should always remain critical, not only of others but especially of ourselves. For instance, I know people that write long papers on non-boost-invariant fluid dynamics that have a lot of formalism but zero physics applications [69] – just kidding.

Well well ...I wouldn't say zero ...I did have some down-to-earth applications in mind ...

In any case, it's a logical possibility that at some point you will be able to predict the mass of the pion from this fundamental theory and it is also a logical possibility that any consistent and ultraviolet completable theory is part of string theory, which is what the swampland programme is advocating, and the mass of the pion is just what it is. As I said before, it's quite striking to see that in cases in which there is a lot of supersymmetry, the only theories that are consistent are string theories [70].

What do you think has been the biggest breakthrough in theoretical physics in the past 30 years?

I obviously think that AdS/CFT is a great theoretical physics breakthrough. The role of topological phases and topological order in condensed matter physics has also led to a lot of interesting insights, the quantum Hall effect [71] probably being the first example of this type of behaviour.

Why have you chosen to do physics? Why not something else?

I was originally planning to become a mathematician because I like equations but during university I discovered that I liked theoretical physics a bit more because of the conceptual type of questions that were being asked and that I found very interesting and attractive. So, more and more I became a theoretical physicist because I like to work with conceptual ideas but also like mathematics and that's where the two things come together.

What do you think is the role of the theoretical physicist in modern society?

In general, theoretical physicists have exceptionally good analytical thinking skills. Anywhere in society where strong analytic thinking skills are required, theoretical physicists could offer additional value. However, most real world problems also require some other attributes and not all theoretical physicists necessarily have all those other attributes as well. Nevertheless, analytic thinking is required to get an overview of complex problems with many variables as well as many actors and to try to make sense out of them. Besides this, the general methodologies developed in theoretical physics such as for example statistical physics and the theory of critical phenomena have wide-ranging scientific and practical applications. I think these are all valuable contributions to science and society besides performing inspirational research and inspiring the next generation of theoretical physicists who want to think about these deep conceptual problems.

References

[1] J. Scherk and J. H. Schwarz, "Dual models and the geometry of space-time," *Phys. Lett. B* **52** (1974) 347–350.
[2] J. Scherk and J. H. Schwarz, "Dual models for nonhadrons," *Nucl. Phys. B* **81** (1974) 118–144.
[3] J. M. Maldacena, "The large N limit of superconformal field theories and supergravity," *Int. J. Theor. Phys.* **38** (1999) 1113–1133, arXiv:hep-th/9711200.
[4] J. Penedones, "Writing CFT correlation functions as AdS scattering amplitudes," *JHEP* **03** (2011) 025, arXiv:1011.1485 [hep-th].
[5] A. L. Fitzpatrick and J. Kaplan, "Analyticity and the holographic S-matrix," *JHEP* **10** (2012) 127, arXiv:1111.6972 [hep-th].
[6] A. Almheiri, X. Dong and D. Harlow, "Bulk locality and quantum error correction in AdS/CFT," *JHEP* **04** (2015) 163, arXiv:1411.7041 [hep-th].
[7] G. Penington, "Entanglement wedge reconstruction and the information paradox," *JHEP* **09** (2020) 002, arXiv:1905.08255 [hep-th].
[8] A. Almheiri, R. Mahajan, J. Maldacena and Y. Zhao, "The Page curve of Hawking radiation from semiclassical geometry," *JHEP* **03** (2020) 149, arXiv:1908.10996 [hep-th].
[9] A. Almheiri, T. Hartman, J. Maldacena, E. Shaghoulian and A. Tajdini, "Replica wormholes and the entropy of Hawking radiation," *JHEP* **05** (2020) 013, arXiv:1911.12333 [hep-th].
[10] A. Almheiri, T. Hartman, J. Maldacena, E. Shaghoulian and A. Tajdini, "The entropy of Hawking radiation," arXiv:2006.06872 [hep-th].
[11] A. Belin and J. de Boer, "Random statistics of OPE coefficients and Euclidean wormholes," arXiv:2006.05499 [hep-th].
[12] A. Belin, J. De Boer, P. Nayak and J. Sonner, "Charged eigenstate thermalization, Euclidean wormholes and global symmetries in quantum gravity," arXiv:2012.07875 [hep-th].
[13] H. Nicolai, K. Peeters and M. Zamaklar, "Loop quantum gravity: an outside view," *Class. Quant. Grav.* **22** (2005) R193, arXiv:hep-th/0501114.
[14] T. Thiemann, "Loop quantum gravity: an inside view," *Lect. Notes Phys.* **721** (2007) 185–263, arXiv:hep-th/0608210.
[15] R. Loll, "Quantum gravity from causal dynamical triangulations: a review," *Class. Quant. Grav.* **37** no. 1 (2020) 013002, arXiv:1905.08669 [hep-th].
[16] J. Ambjorn, A. Goerlich, J. Jurkiewicz and R. Loll, "Nonperturbative quantum gravity," *Phys. Rept.* **519** (2012) 127–210, arXiv:1203.3591 [hep-th].

[17] M. Reuter and F. Saueressig, *Quantum Gravity and the Functional Renormalization Group: The Road towards Asymptotic Safety*. Cambridge University Press, 1, 2019.

[18] A. Connes, "Gravity coupled with matter and foundation of noncommutative geometry," *Commun. Math. Phys.* **182** (1996) 155–176, arXiv:hep-th/9603053.

[19] P. Horava, "Quantum gravity at a Lifshitz point," *Phys. Rev. D* **79** (2009) 084008, arXiv:0901.3775 [hep-th].

[20] P. Horava, "Spectral dimension of the universe in quantum gravity at a Lifshitz point," *Phys. Rev. Lett.* **102** (2009) 161301, arXiv:0902.3657 [hep-th].

[21] T. Thiemann, "Canonical quantum gravity, constructive QFT and renormalisation," arXiv:2003.13622 [gr-qc].

[22] M. Bojowald, "No-go result for covariance in models of loop quantum gravity," *Phys. Rev. D* **102** no. 4 (2020) 046006, arXiv:2007.16066 [gr-qc].

[23] A. Sen, "Logarithmic corrections to Schwarzschild and other non-extremal black hole entropy in different dimensions," *JHEP* **04** (2013) 156, arXiv:1205.0971 [hep-th].

[24] C. Rovelli and L. Smolin, "Discreteness of area and volume in quantum gravity," *Nucl. Phys. B* **442** (1995) 593–622, arXiv:gr-qc/9411005. [Erratum: *Nucl. Phys. B* **456** (1995) 753–754].

[25] S. Surya, "The causal set approach to quantum gravity," *Living Rev. Rel.* **22** no. 1 (2019) 5, arXiv:1903.11544 [gr-qc].

[26] T. Banks, W. Fischler, S. H. Shenker and L. Susskind, "M theory as a matrix model: a conjecture," *Phys. Rev. D* **55** (1997) 5112–5128, arXiv:hep-th/9610043.

[27] R. Dijkgraaf, E. P. Verlinde and H. L. Verlinde, "Matrix string theory," *Nucl. Phys. B* **500** (1997) 43–61, arXiv:hep-th/9703030.

[28] K. Becker and M. Becker, "A two loop test of M(atrix) theory," *Nucl. Phys. B* **506** (1997) 48–60, arXiv:hep-th/9705091.

[29] J. de Boer and S. N. Solodukhin, "A holographic reduction of Minkowski space-time," *Nucl. Phys. B* **665** (2003) 545–593, arXiv:hep-th/0303006.

[30] S. Pasterski, S.-H. Shao and A. Strominger, "Flat space amplitudes and conformal symmetry of the celestial sphere," *Phys. Rev. D* **96** no. 6 (2017) 065026, arXiv:1701.00049 [hep-th].

[31] N. Arkani-Hamed, M. Pate, A.-M. Raclariu and A. Strominger, "Celestial amplitudes from UV to IR," arXiv:2012.04208 [hep-th].

[32] A. Hamilton, D. N. Kabat, G. Lifschytz and D. A. Lowe, "Local bulk operators in AdS/CFT: a boundary view of horizons and locality," *Phys. Rev. D* **73** (2006) 086003, arXiv:hep-th/0506118.

[33] J. Polchinski, "Tasi lectures on D-branes," in *Theoretical Advanced Study Institute in Elementary Particle Physics (TASI 96): Fields, Strings, and Duality*, C. Efthimiou and B. Greene, eds., pp. 293–356. 11, World Scientific, 1996. arXiv:hep-th/9611050.

[34] S. M. Chester, M. B. Green, S. S. Pufu, Y. Wang and C. Wen, "New modular invariants in $\mathcal{N} = 4$ super-Yang–Mills theory," arXiv:2008.02713 [hep-th].

[35] N. Bodendorfer, T. Thiemann and A. Thurn, "Towards loop quantum supergravity (LQSG) I. Rarita-Schwinger sector," *Class. Quant. Grav.* **30** (2013) 045006, arXiv:1105.3709 [gr-qc].

[36] J. M. Maldacena, "Wilson loops in large N field theories," *Phys. Rev. Lett.* **80** (1998) 4859–4862, arXiv:hep-th/9803002.

[37] S.-J. Rey and J.-T. Yee, "Macroscopic strings as heavy quarks in large N gauge theory and anti–de Sitter supergravity," *Eur. Phys. J. C* **22** (2001) 379–394, arXiv:hep-th/9803001.

[38] V. Balasubramanian, J. de Boer, V. Jejjala and J. Simon, "The library of Babel: on the origin of gravitational thermodynamics," *JHEP* **12** (2005) 006, arXiv:hep-th/0508023.

[39] D. Marolf, "Emergent gravity requires kinematic nonlocality," *Phys. Rev. Lett.* **114** no. 3 (2015) 031104, arXiv:1409.2509 [hep-th].

[40] N. Beisert et al., "Review of AdS/CFT integrability: an overview," *Lett. Math. Phys.* **99** (2012) 3–32, arXiv:1012.3982 [hep-th].

[41] J. Maldacena and L. Susskind, "Cool horizons for entangled black holes," *Fortsch. Phys.* **61** (2013) 781–811, arXiv:1306.0533 [hep-th].

[42] A. Almheiri, D. Marolf, J. Polchinski and J. Sully, "Black holes: complementarity or firewalls?," *JHEP* **02** (2013) 062, arXiv:1207.3123 [hep-th].

[43] A. Almheiri, D. Marolf, J. Polchinski, D. Stanford and J. Sully, "An apologia for firewalls," *JHEP* **09** (2013) 018, arXiv:1304.6483 [hep-th].

[44] L. Susskind, "Black holes at exp-time," arXiv:2006.01280 [hep-th].

[45] K. Papadodimas and S. Raju, "An infalling observer in AdS/CFT," *JHEP* **10** (2013) 212, arXiv:1211.6767 [hep-th].

[46] K. Papadodimas and S. Raju, "State-dependent bulk-boundary maps and black hole complementarity," *Phys. Rev. D* **89** no. 8 (2014) 086010, arXiv:1310.6335 [hep-th].

[47] K. Papadodimas and S. Raju, "Black hole interior in the holographic correspondence and the information paradox," *Phys. Rev. Lett.* **112** no. 5 (2014) 051301, arXiv:1310.6334 [hep-th].

[48] J. de Boer, E. P. Verlinde and H. L. Verlinde, "On the holographic renormalization group," *JHEP* **08** (2000) 003, arXiv:hep-th/9912012.

[49] J. de Boer, "The holographic renormalization group," *Fortsch. Phys.* **49** (2001) 339–358, arXiv:hep-th/0101026.

[50] S. de Haro, S. N. Solodukhin and K. Skenderis, "Holographic reconstruction of spacetime and renormalization in the AdS/CFT correspondence," *Commun. Math. Phys.* **217** (2001) 595–622, arXiv:hep-th/0002230.

[51] V. Balasubramanian, B. D. Chowdhury, B. Czech, J. de Boer and M. P. Heller, "Bulk curves from boundary data in holography," *Phys. Rev. D* **89** no. 8 (2014) 086004, arXiv:1310.4204 [hep-th].

[52] N. Engelhardt and G. T. Horowitz, "Recovering the spacetime metric from a holographic dual," *Adv. Theor. Math. Phys.* **21** (2017) 1635–1653, arXiv:1612.00391 [hep-th].

[53] D. Harlow, "TASI lectures on the emergence of bulk physics in AdS/CFT," *PoS* **TASI2017** (2018) 002, arXiv:1802.01040 [hep-th].

[54] J. De Boer and L. Lamprou, "Holographic order from modular chaos," *JHEP* **06** (2020) 024, arXiv:1912.02810 [hep-th].

[55] C. Vafa, "The string landscape and the swampland," arXiv:hep-th/0509212.

[56] N. Arkani-Hamed, L. Motl, A. Nicolis and C. Vafa, "The string landscape, black holes and gravity as the weakest force," *JHEP* **06** (2007) 060, arXiv:hep-th/0601001.

[57] A. Bedroya and C. Vafa, "Trans-Planckian censorship and the swampland," *JHEP* **09** (2020) 123, arXiv:1909.11063 [hep-th].

[58] G. Veneziano, "Construction of a crossing-symmetric, Regge behaved amplitude for linearly rising trajectories," *Nuovo Cim. A* **57** (1968) 190–197.

[59] P. K. Kovtun, D. T. Son and A. O. Starinets, "Viscosity in strongly interacting quantum field theories from black hole physics," *Phys. Rev. Lett.* **94** (Mar. 2005) 111601. https://link.aps.org/doi/10.1103/PhysRevLett.94.111601.

[60] P. Romatschke and U. Romatschke, *Relativistic Fluid Dynamics In and Out of Equilibrium*. Cambridge Monographs on Mathematical Physics. Cambridge University Press, 5, 2019. arXiv:1712.05815 [nucl-th].

[61] J. Maldacena, S. H. Shenker and D. Stanford, "A bound on chaos," *JHEP* **08** (2016) 106, arXiv:1503.01409 [hep-th].

[62] Y. Sekino and L. Susskind, "Fast scramblers," *JHEP* **10** (2008) 065, arXiv:0808.2096 [hep-th].

[63] V. Balasubramanian, A. Bernamonti, J. de Boer, N. Copland, B. Craps, E. Keski-Vakkuri, B. Muller, A. Schafer, M. Shigemori and W. Staessens, "Thermalization of strongly coupled field theories," *Phys. Rev. Lett.* **106** (2011) 191601, arXiv:1012.4753 [hep-th].

[64] V. Balasubramanian, A. Bernamonti, J. de Boer, N. Copland, B. Craps, E. Keski-Vakkuri, B. Muller, A. Schafer, M. Shigemori and W. Staessens, "Holographic thermalization," *Phys. Rev. D* **84** (2011) 026010, arXiv:1103.2683 [hep-th].

[65] M. Green, M. Green, J. Schwarz and E. Witten, *Superstring Theory, Volume 1: Introduction*. Cambridge Monographs on Mathematical Physics. Cambridge University Press, 1988.

[66] S. Hellerman, J. McGreevy and B. Williams, "Geometric constructions of nongeometric string theories," *JHEP* **01** (2004) 024, arXiv:hep-th/0208174.

[67] C. M. Hull, "A geometry for non-geometric string backgrounds," *JHEP* **10** (2005) 065, arXiv:hep-th/0406102.

[68] J. de Boer and M. Shigemori, "Exotic branes and non-geometric backgrounds," *Phys. Rev. Lett.* **104** (2010) 251603, arXiv:1004.2521 [hep-th].

[69] J. Armas and A. Jain, "Effective field theory for hydrodynamics without boosts," arXiv:2010.15782 [hep-th].

[70] E. Palti, "The swampland: introduction and review," *Fortsch. Phys.* **67** no. 6 (2019) 1900037, arXiv:1903.06239 [hep-th].

[71] F. D. M. Haldane, "Model for a quantum Hall effect without Landau levels: condensed-matter realization of the 'parity anomaly,' " *Phys. Rev. Lett.* **61** (Oct. 1988) 2015–2018. https://link.aps.org/doi/10.1103/PhysRevLett.61.2015.

5

Steven Carlip

Professor at the University of California at Davis

Date: 19 July 2011. Location: Davis, CA. Last edit: 12 March 2019

In your opinion, what are the main puzzles in theoretical physics at the moment?

Well, theoretical physics is a very big area and I can't possibly answer that in general. I have no idea what people in theoretical condensed matter physics or other areas are worrying about. In the areas I work on, the main puzzles include the cosmological constant problem, that is, why the cosmological constant is apparently very nearly zero but not quite zero, and the general problem of how to quantise gravity, which of course is an 80-year-old puzzle by now (laughs) but still one of the main ones. A potentially related puzzle is whether it's possible to unify all fundamental interactions and, if so, how? Another big issue, which in theoretical cosmology goes by the name of the *measure problem* [1], is how to make sense of probabilistic predictions for the universe as a whole. I suspect that in a few years, my answer will be completely different, depending on what comes out of the LHC, because one should at least hope that the big puzzles are determined to some extent by observation and not just by sitting in rooms and philosophising (laughs).

What do we expect to see at the LHC?

I really don't know. I think the most likely thing to show up is something that nobody has thought of yet, which seems to be the historical outcome at big accelerator experiments. Perhaps we will see supersymmetry. However, if we just observe evidence for the minimal supersymmetric standard model, it would boring, but I bet it won't come to that.

Why would that be boring?

Well, because then we have the theoretical framework and we just have to plug a few numbers into it. It wouldn't help us answer any of the other big questions that I just mentioned.

So you think that string theory does not give a good answer to these questions?

I don't think so. I don't want to say that it's impossible that string theory will ever offer an answer, but certainly right now I don't think it does.

I mentioned string theory because no other theory of quantum gravity has incorporated supersymmetry so well besides string theory and you seemed to be sympathetic towards the minimal supersymmetric standard model.

Oh, come on! Some string theorists have been predicting supersymmetry at 5% above the current observed energies for 25 years (laughs). String theory requires worldsheet supersymmetry, but within string theory you can have spacetime supersymmetry or not, as you prefer, and you can break it at whatever scale you want. If you have 10^{500} vacua to choose from [2, 3], you can find plenty where supersymmetry will show up at the LHC and plenty where it won't. So, no, I don't think that finding supersymmetry at the LHC would be evidence for string theory.

Okay. You mentioned earlier the problem of quantising gravity. Why is this problem so difficult?

My best guess is that it's difficult because a basic feature of general relativity, or of any diffeomorphism-invariant theory, is that there are no local observables. Everything we know about quantum field theory involves local observables, and they simply don't exist in quantum gravity. Nowadays there are even good proofs that classical general relativity has no diffeomorphism-invariant local observables [4–6]. So, we're dealing with something that is fundamentally different from any kind of physics we understand. I think this is something that traditional quantum gravity people have understood for a long time and that string theorists are now beginning to run into, in particular, people who work on the AdS/CFT correspondence [7]. For instance, now they understand pretty well that local quantities in spacetime correspond to highly nonlocal, very complicated stuff in the CFT, and that the nice simple local things in the CFT correspond to horribly nonlocal observables in spacetime [8].

Do you think that a theory of quantum gravity has to be nonlocal in general?

In some sense it has to be. The observables in such theory have to be nonlocal. Nonlocal can mean a lot of different things, and the theory doesn't have to be nonlocal in the sense of causality violation or of having effects in a small region instantaneously affecting something in another region. The theory has to be nonlocal in the sense that the observables can't be defined at single points, for a very simple reason: namely, that there's no such thing as a single point in general relativity, because diffeomorphisms move points around.

But is there any evidence for nonlocality in theories of quantum gravity, such as in the AdS/CFT correspondence, for example?

Well, in AdS/CFT it's clear that nice observables in the CFT correspond to nonlocal observables in spacetime [8]. In 2+1-dimensional gravity, it's clear that the observables that you can build in the quantum theory are nonlocal [9]. However, as I mentioned before, nonlocality is already a feature of classical general relativity. For instance, if you pay attention to how the Jet Propulsion Lab[1] discusses the planetary ephemerides, you'll notice that they don't mention the position that planet Mercury is in now. Instead, they talk about the amount of time it would take for a radar pulse emitted at such and such a point on the Earth to return to that point, after reflecting off Mercury, as measured by an atomic clock at that location. That's a nonlocal observable.

In your opinion, what is the best candidate for a theory of quantum gravity right now?

I have no idea. There are a lot of things that one can try, and so far I don't see any particular evidence that any of them is right. The 80-year-old tradition in this subject has been that you find really promising new approaches, you get some distance, you learn things, and then you can't go any further [10, 11]. Maybe loop quantum gravity or string theory or causal dynamical triangulations or causal set theory is going to break this pattern and actually go all the way, but I wouldn't bet a lot of money on any of them.

Which directions should one pursue?

You have to try a lot of things. I think it's much too soon to say that a specific approach is *the* promising approach and that we shouldn't work on other things. We have to try different big programs like string theory or loop quantum gravity, and we have to try all sorts of small programs like just trying to understand quantum black holes or trying to understand lower-dimensional models. I think it's going to take all of those directions to reach a proper understanding of quantum gravity.

Do you think that all of those approaches you mentioned are completely independent and mutually exclusive or do you think that they all kind of describe some part of a big puzzle?

I doubt that they're all part of a single description. I could be wrong, but it seems unlikely to me. In general, there's no reason to believe that a classical theory should lead to a unique quantum theory. Given that nature is fundamentally quantum mechanical, a quantum theory should have a classical limit, but there's no reason to expect that you can't have more than one quantum theory with the same classical limit.

Is quantum gravity necessary?

Well, I've spent my professional life looking for a quantum theory of gravity, so obviously I think it's at least desirable. "Something" is certainly necessary. The source of gravity is the stress-energy tensor of quantum fields, which is a quantum operator that has to somehow couple to the curvature of spacetime. Simply sticking classical general relativity and quantum field theory together makes no sense, mathematically or physically. There are ways to couple classical spacetime to quantum matter, but the only ones I know of look ad

[1] www.jpl.nasa.gov/.

hoc and kludgy. It could also be that we need some change in quantum mechanics as well as gravity, but I'd still probably call that "quantum gravity."

There are some nice arguments that quantum gravity is required for consistency of quantum mechanics [12]. It would be much more satisfying to have genuine experimental evidence, though [13]. There are some interesting recent proposals that may be almost within reach [14]. I'd be shocked if they led to anything really unexpected, but it would certainly make life interesting.

Since you mentioned earlier lower-dimensional models, why is the problem of quantising gravity in 2+1 dimensions much simpler than in 3+1 dimensions?

Because in 2+1 dimensions, at least if there's no matter present, there are no propagating degrees of freedom. So instead of dealing with a quantum field theory, you're dealing with quantum mechanics, with a finite number of degrees of freedom that are inherently global and not local. This means that the technical problems are so much simpler that you can actually carry some of these programs all the way through [15, 16].

You showed that there were many ways of quantising gravity in 2+1 dimensions [15, 16]. What does this mean? That there is no unique quantum gravity theory?

Well, in one sense it's certainly true that there's no unique theory. If you look at some particular approach to quantisation – for example at the covariant canonical quantisation on a spacetime where the space is a torus, which is the simplest non-trivial example – what you find is that there's a set of theories that are labeled by representations of the group of large diffeomorphisms and those theories are different [15, 16]. It's very much analogous to finding quantum field theories with the requirement of Lorentz invariance. In that context, you would find that there are quantum field theories of particles with arbitrary spin and arbitrary mass, which are certainly inequivalent.

Most likely you can't perform experiments in 2+1-dimensional gravity. But suppose for some reason you are after the simplest representation, where wave functions are invariant rather than covariant under large diffeomorphisms. In this case, you can ask whether other approaches to quantisation, like the Wheeler-DeWitt equation, loop quantum gravity approaches or spin foams, are equivalent to, say, covariant canonical quantisation. The answer is that we're not entirely sure [17–19], which is tied together to a fundamental ambiguity present in any quantum theory. This ambiguity is related to the fact that you're interested in some observables that are operators made up from the fundamental operators in your theory [20]. However, there are always ordering ambiguities. If you look back at the history of basic quantum mechanics, for instance back before the Schrödinger equation first appeared, people were working very hard at understanding the hydrogen atom. At some point they decided, based on reasonable arguments, that the right variables to use were action-angle variables. If you write down your basic Heisenberg commutation relations and the Hamiltonian for the hydrogen atom in terms of action-angle variables, you get the wrong answer. The answer you get is "equivalent" to the right answer, but with the wrong operator ordering. To solve this problem, Schrödinger decided to do everything in Cartesian coordinates and to define operator ordering in those coordinates. This procedure gives you the right Hamiltonian.

I don't know, and I don't think anybody knows in any fundamental way, why the correct Hamiltonian of the hydrogen atom is the one that you write in Cartesian coordinates rather than in action-angle variables or using some other canonical transformation. It's just what it is. The same type of problem occurs in 2+1-dimensional gravity. In particular, in order to decide whether all these other quantisation procedures are equivalent or not, you can't just look at the Hamiltonians, but instead you have to ask whether there is some operator ordering that would make them equivalent. There are many cases in which we know that the answer is positive [18, 19], but in the case of the full Wheeler-DeWitt equation we don't know the answer [17].

Are all these quantisation procedures in 2+1-dimensional gravity consistent?

As far as we know, yes.

Okay, so all of them exhibit general covariance, a semiclassical limit, etc.?

All of them have a semiclassical limit. General covariance is tricky to define at this level. Some of these quantisations require choosing a time coordinate, and you can ask whether you have equivalent quantisations if you make different choices of this time coordinate [20]. I don't think the answer is known. There are particular cases where it's known that there are particular changes that you can make that don't affect the quantisation, but for a general choice I think it's not known.

So what about the finiteness of the theory?

It depends on the questions you're asking. There aren't any ultraviolet divergences, because there are no propagating degrees of freedom to run around loose. There are various divergences coming from topology [21, 22], and it's not clear what to make of those. For example, it's not clear whether in a path integral formulation of the theory you should sum over all spacetime topologies or not. If you do, there are sometimes divergences that may have interesting physical significance. If you don't sum over topologies – if you're just looking at a fixed spatial topology – then there aren't any divergences.

All the quantisations are completely different, so how do you know which one is the right one? Is there a way of testing this?

You don't know, and you can't do experiments in 2+1-dimensional gravity.

Do you expect the same ambiguities to be present in 3+1-dimensions and that it's possible to quantise 3+1-dimensional gravity in many different ways?

Some of the ambiguities are certainly going to be there, such as the operator ordering ambiguities, for instance. There's no way to escape those, and I don't know of any way of resolving them without experimental evidence. You might be able to argue that some orderings are much nicer looking than others and hope that that gives you some guidance, but it's not going to tell you for certain. This issue is tricky, too, because if you take some operator that depends on ordering – and hence you have an ambiguity – all that really means is that there's more than one quantum operator. Hence, there is more than one thing that you can observe. So in some sense the ordering ambiguities aren't ambiguities in quantum mechanics, they're ambiguities related to the fact that you don't know which one of these operators corresponds to something that you go out and measure. So this is a question you can only answer by going out and measuring something.

In your papers where you looked at different ways of quantising gravity [15, 16], you did not use string theory techniques, right?

It depends on what you mean by string theory techniques. There's lots of very interesting stuff that you can do in 2+1-dimensional gravity that uses the AdS/CFT correspondence.

Yes, but you didn't make use of that, right?

Right, because what I was working on was spatially compact universes, where there isn't a boundary [16]. However, I've also worked on 2+1-dimensional black holes, which are really the first case of the AdS/CFT correspondence that was ever discovered, even before it had a name [23–26]. In that sense, there are string theory-like techniques that are often relevant.

As far as I understood, forgetting about the AdS/CFT correspondence, your analysis was quite general and not dependent on any background or boundary conditions, right?

Yes, for a compact universe there certainly doesn't seem to be any need for string theory in 2+1-dimensional quantum gravity [16].

So the claim that people normally make that quantising gravity requires unifying all forces of nature is not really true in 2+1 dimensions?

In 2+1 dimensions it's certainly not the right claim. In higher dimensions, I don't know. My personal guess would be that it is not necessary, but we don't know how to quantise gravity, so all I could do is give you a guess.

I understand. One of the techniques you used in 2+1 dimensions was loop quantum gravity and you managed to perform a consistent quantisation, right?

Yeah.

Doesn't that mean that maybe loop quantum gravity is the right direction to go in four dimensions? Or is this statement too far-fetched?

I don't know; 2+1 dimensional gravity is so much simpler that it's just not clear. As I said, I think the basic issue in any quantum theory of gravity is constructing and understanding the appropriate nonlocal observables. In 2+1 dimensions we know exactly what those observables are, and so we can treat them with loop quantum gravity or covariant canonical quantisation or path integrals or any number of other ways. In 3+1 dimensions or higher, we basically don't know what any of the observables are. We can construct analogs of the observables in 2+1 dimensions, but they're not very interesting, as they don't give you information about the local properties of gravity. Loop quantum gravity has a hard time reconstructing a classical limit. In order to do that, you would have to understand what the observables are in loop quantum gravity and how those give you a classical limit. That is something I don't think we know in 3+1 dimensions.

So what lessons can we take from 2+1 dimensions and apply it to 3+1 dimensions? Is there any feature from 2+1 dimensions that we expect in 3+1 dimensions?

There are a few things. For one thing, it's an existence proof. It states that, in some cases at least, it's possible to quantise gravity as a theory in itself. It's also a non-uniqueness proof, and it gives us reason to doubt the claim that a given quantum theory of gravity that someone might come up with must be the right one. We've learned a fair amount about how to reconstruct some local geometry from nonlocal observables. This knowledge isn't at the required level to approach 3+1 dimensions, because in 2+1 dimensions spacetime has a constant curvature: any spacetime in 2+1 dimensions can be built by taking pieces of, say, flat space if there's no cosmological constant, and gluing them together. We've learnt how to use observables in a quantum theory of gravity to describe that gluing, which in 2+1 dimensions yields the complete geometrical information. In 3+1 dimensions it doesn't; instead, it gives just a tiny piece. But this tells us that maybe this is the kind of thing that we have to look for.

If we had added matter to general relativity in 2+1 dimensions, could we have proceed in the same way and used different quantisation procedures?

Nobody has succeed in doing that, except that you can add point particles and construct things that are sort of like local observables localised at the particular locations of the particles. However, nobody knows how to add fields or continuous matter in a way that you can quantise.

Could it be that it's not possible to do it using the same techniques?

It can't be done with the same techniques, because once you add matter you have propagating degrees of freedom. Even though they aren't independent gravitational degrees of freedom that are propagating, the gravitational field gets combined with the matter degrees of freedom, so you can't use these simple global techniques anymore.

An obstruction that people usually discuss when talking about quantum gravity is that gravity is non-renormalisable by itself. So how is it possible that you can quantise it just as it is?

In 2+1 dimensions there are no propagating degrees of freedom, so, as I mentioned, there are no ultraviolet divergences. In 3+1 dimensions I don't know. In this case, you have to be careful about what you mean by renormalisability and how important it is.

There is an interesting proposal by Steven Weinberg of what's called asymptotic safety [27–30]. There are known cases of quantum field theories that exhibit this property, and he suggests that gravity may be one of those. These are field theories that are not renormalisable, but that have a non-Gaussian ultraviolet fixed point. More generally, they have a finite-dimensional ultraviolet (UV) fixed surface (technically, a UV fixed point with a finite number of relevant directions). This means that if you start on this ultraviolet fixed surface, your theory is determined by a finite number of parameters. If you then follow the renormalisation group flow down to low energies, you'll find an infinite number of terms in the effective action, because the theory is non-renormalisable, but the coupling constants in those terms are all determined by a finite number of parameters. Hence, this is a case of a theory that is not renormalisable, but it's almost as good as being renormalisable, in the sense that even if you have an infinite number of coupling constants, they're all determined by a finite number of parameters. There's an active research program by people like Martin Reuter on trying to see whether this really works for quantum gravity [31, 32].

I don't know how one could actually decide if this approach is the correct one. What one can do is to look at various truncations and see if there are ultraviolet fixed points, and ask how many parameters determine the system. But if you're picky, you can always say, "What if you add one more term in your effective action, are you sure that this behaviour is still present?" So it's a hard thing to figure out how to prove, but it's possible that's what gravity is like. On the other hand, it's also possible that gravity is actually finite. It's possible that even if quantum gravity has an infinite number of divergences perturbatively, it may be finite once you re-sum the entire series in perturbation theory.

Is this last possibility the type of argument that loop quantum gravity invokes?

Well, in some sense yes. Loop quantum gravity gives you finite answers because there is a scale, the Planck scale, which is built in from the start. But it's not clear that the sums that compose the path integral in loop quantum gravity are finite, for instance.

There's old work from the early 1960s by Bryce DeWitt [33], and then late 1960s by Isham, Salam and Strathdee [34], who showed that if you take particular infinite sums of Feynman diagrams in quantum gravity, even though they individually diverge, they add up to finite answers. Nobody has gone much further with that, because nobody knows how to take it beyond those particular infinite series where you actually know how to calculate all the terms and how to sum them. But there's certainly no proof that this doesn't happen in general, and there are sort of hand-waving arguments that the divergences in fact do cancel out. Take, for example, the case of a point charge and look at the electromagnetic self energy. Classically, that energy is infinite, but if you include the contribution from the gravitational field, which is negative because it's attractive, you find a binding energy that makes the total energy finite. So, who knows, maybe there's something analogous taking place in quantum gravity.

What is your opinion about emergent theories of gravity such as that pursued by Erik Verlinde [35, 36]?

I actually have a paper on this [37]. The trouble is that the word "emergent" means many different things to different physicists, even more if you add philosophers. In some vague sense, gravity is almost certain to be emergent, in the sense that quantum spacetime, whatever that is, is unlikely to look much like classical spacetime.

But there's a naive sense of "emergent," in which you assume an ordinary spacetime and somehow have gravity emerge in that, which is very hard to get to work. There's a theorem called the Weinberg-Witten theorem [38] that says, basically, that to get gravity right (i.e., to get a pure massless spin 2 graviton), you need something like exact diffeomorphism invariance, and purely nonlocal observables. It's also incredibly hard to get the strong principle of equivalence – universal coupling of gravity to all forms of energy, including its own energy. Among other things, this requires a very particular kind of nonlinearity, which holds in general relativity but seems highly unnatural elsewhere.

I think if you want "emergence," you have to be willing to go all the way, and say that spacetime and gravity emerge together. I also think Visser has ruled out the details of Verlinde's proposal [39], which as far as I know has never been refuted.

Since you mentioned quantum spacetime, how does spacetime look like at the Planck scale? Is it discrete? What do you find in 2+1 dimensions?

I don't know whether spacetime even exists at the Planck scale. I don't have an answer to that. In 2+1 dimensions we find that some observables that you might expect to probe the structure of spacetime at the Planck scale have discrete spectra and others don't. So you have to decide what kind of measurements you're talking about (laughs).

At some point, you argued that at the Planck scale spacetime was effectively two-dimensional [40, 41].

That argument I was making works above the Planck length. I was careful with this precise statement because we don't know how to get to the Planck scale, for which you presumably really need a full quantum theory of gravity. However, there's some evidence that at somewhere around 10 and 100 times the Planck length, you have some sort of effective two-dimensional behaviour. If that's right, then it would give you a new set of approximations that you might be able to take down towards the Planck scale to see what happens there, although I don't know how you can do that yet.

But what kind of arguments do you have for this claim?

There are little pieces of evidence from a bunch of different approaches [40–43]. None of them is in itself very convincing, but taken together they suggest that maybe this is what's going on at small scales. In causal dynamical triangulations, for instance, you can measure the spectral dimension (that is, the dimension as seen by a random walker) and you find that it drops from four at large distances to two at about 15 times the Planck scale [44]. In the asymptotic safety program, you find that fields acquire large anomalous dimensions near an ultraviolet fixed point such that the field theory looks like as if it's a two dimensional field theory [45]. In string theory, there's an old result by Atick and Witten stating that if you look at a gas of strings at very high temperature (above the Hagedorn temperature) the number of degrees of freedom drops and you effectively have a description in terms of a two-dimensional field theory [46]. There are also pieces of evidence originating from other places.

I'm a bit puzzled about your comment regarding string theory. I was under the impression that string theory couldn't say much about spacetime at the Planck scale because strings are placed on a background that is treated classically.

Well, no. There are different things you can do in string theory. If you're just working in perturbative string theory around a fixed background then that's true. But, for instance, Atick and Witten essentially looked at the sum of the perturbation series for high-temperature strings in a flat background [46]. So there are things in string theory that you can do that maybe probe the Planck scale. Though it's not entirely clear that this allows you to look all the way to the Planck scale, it seems to allow you to look at energies slightly below the Planck energy. There is also some debate in the literature whether, at least in some string theories, you can use D0 branes to probe distances much smaller than the Planck scale [47].

Another property that you argued that should be present at the Planck scale is *asymptotic silence* [48]. What does that mean?

This is a wild guess. If you look at a classical spacetime near a generic spacelike singularity, what you find is that light cones become very sharply gravitationally focused. As a result, small regions of space basically can't communicate with their neighbours, because it takes a very long time for light from one small region to reach the next region due to this focusing. In cosmology, what this leads to is what's called BKL behaviour [49], which is a particular kind of chaotic behaviour in which, in some sense, small regions look effectively two-dimensional with a sort of random choice of two dimensions. If you look at the Wheeler-DeWitt equation in the strong coupling limit, which is equivalent to looking at it at or near the Planck scale, you find a similar kind of decoupling of nearby points. So I'm guessing that maybe this effective two-dimensional behaviour is because of this kind of decoupling (the asymptotic silence). It's called *asymptotic silence* because you can't "hear" your neighbour, since you're effectively disconnected from your neighbour.

Mosna, Pitelli and I have looked at the effects of vacuum fluctuations of the stress-energy tensor on light cones, and have found evidence that if you look at slightly above the Planck scale (15 or 20 times the Planck length), the effect of vacuum fluctuations is to give you strong focusing of light [48]. Possibly, that would lead to something like asymptotic silence.

In your work, you also discussed as many ways of describing the microscopics of black holes as there are ways of quantising 2+1-dimensional gravity [24]. What do we learn from all these descriptions of black holes that are supposed to give rise to exactly the same result for the Bekenstein-Hawking entropy?

What I think it means is that there is some underlying symmetry, in particular conformal symmetry, near the horizon that's strong enough to determine the entropy regardless of the details of the microscopic theory [50–52]. There's a nice field theory analogy to this, namely Goldstone's theorem, which states that if you have a broken symmetry then you necessarily have a certain number of massless bosons, with certain properties that you can predict just by the pattern of the symmetry breaking. But the theorem doesn't tell you anything about how these Goldstone bosons are constructed from the fundamental microscopic fields in the theory. It only tells you that the symmetry alone is enough to give you knowledge of some of these patterns. My guess at this point – and I think there is evidence for this, although it is not completely established – is that the same thing is true near a black hole horizon. Essentially, the presence of a horizon, or the imposition within the quantum theory of the existence of a horizon with certain properties, breaks symmetries in a way that determines the basic thermodynamic properties.

So all these different approaches you discuss are relying on the same symmetries in order to achieve the same result?

I think so, but I can't prove it general. There are a couple of cases that I can actually identify the symmetry in a particular approach, but in general I don't know how to do that [25, 50, 51, 53]. However, it seems to be a plausible guess at least.

I noticed that one of the papers you wrote in 1999 discussed the existence of conformal symmetries near the horizon [26]. Is this what the Kerr/CFT proposal is getting at [54]?

The idea that there is a conformal symmetry near the horizon for a generic black hole, I wrote in 1999 but I didn't have all the details right, though it was basically right. I think the Kerr/CFT correspondence is probably a particular case of that. However, it's also true that there are often a number of different dual two-dimensional conformal field theories (CFTs) that you can write for the same black hole. Maybe string theorists would just say that they are just dual to each other, so obviously the same. I don't know how to show that within the context of quantum gravity.

In 2+1 dimensions, could you analyse the quantum properties of black holes and derive their entropy as well?

Sort of. In 2+1 dimensions there are no propagating degrees of freedom, so there's this mystery of how a black hole can have such a large entropy, because there are no obvious degrees of freedom to account for it. There is an answer to that, for which the simplest way to describe is using AdS/CFT language, although not necessarily string theory language. Black holes in 2+1 dimensions exist only if there's a negative cosmological constant, so they're asymptotically AdS. If you look at the metric of these black holes, the statement that there are no propagating degrees of freedom translates into the statement that if you stick in a perturbation to the metric, you can always remove it by a diffeomorphism. However, if you're in an asymptotically anti–de Sitter space, the diffeomorphisms you're allowed at infinity are restricted. As a result, the degrees of freedom that you would normally say are all just equivalent under diffeomorphisms aren't equivalent at infinity, because you can't do arbitrary diffeomorphisms at infinity [55].

It's possible to write down an effective action for these degrees of freedom by just looking at the Einstein-Hilbert action, imposing the boundary conditions at infinity, sticking in all of the boundary terms and pulling out the degrees of freedom that correspond to diffeomorphisms that don't obey the boundary conditions. When you do that, you find a two-dimensional conformal field theory, in particular a Liouville theory. This Liouville theory has the right central charge, that is, the central charge of the BTZ black hole, and you can use the Cardy formula and get the entropy of the black hole [56]. You then have to ask whether those degrees of freedom are really there in the Liouville theory. It turns out that we don't know the answer, because Liouville theory is this peculiar two-dimensional conformal field theory in which there are two different sectors of states. One sector, which is called the normalisable sector, is well understood and doesn't have the right number of degrees of freedom. In the other sector, which is called the non-normalisable sector, the Cardy formula tells you that it does have the right number of degrees of freedom. But nobody really knows how to quantise it, although a former student of mine, Yujun Chen, had some interesting ideas that I think might be in the right direction [57]. Going back to your question, the answer is yes: you can look at the quantum theory of these effective boundary degrees of freedom and it looks as if it probably gives you the right entropy, though I would say that it's not fully settled.

You also looked at the entropy of black holes within the context of loop quantum gravity [58]. Ashoke Sen has also shown recently that loop quantum gravity computations don't agree with Euclidean gravity methods [59]. How should we interpret these results?

What Sen looks at is not the leading order entropy, but the logarithmic corrections. These are very sensitive to exactly what thermodynamic ensemble you're looking at. Even in ordinary thermodynamics of an ideal gas, for instance, the canonical and microcanonical entropies differ by logarithmic terms. Sen is doing a semiclassical computation [59]; for pure gravity, the logarithmic terms he's getting are basically the contribution of a gas of gravitons around the black hole. These are probably not there at all in the loop quantum gravity calculations, which are purely counting horizon states and don't really know how to deal with near-horizon bulk corrections. The loop quantum gravity calculations, on the other hand, have logarithmic terms that are basically combinatoric, coming from the detailed distribution of horizon states [60]. These are probably not there in Sen's partition function calculation; they go beyond a semiclassical path integral. So I don't think there's really anything to be concluded yet – the calculations just aren't commensurate.

In your opinion, what has been the biggest breakthrough in theoretical physics in the past 30 years?

In the past 30 years ... not much (laughs). The past 30 years have been kind of dull when restricting to gravity and high-energy theory. There have been all sorts of interesting things going on in condensed matter theory, for instance, but I just don't know enough about that to claim anything. In high-energy theory, the standard model is just sitting there being standard, and it works too well, right? I would say there has been a lot of work on that, but it's all been filling in details. There are lots of interesting ideas about going beyond the standard model, but it's just a zoo of ideas, and we have no clue which one of them, if any, is right. My guess, if I had to bet on this, is that it will turn out to be something that nobody has thought of.

Lots of interesting stuff in cosmology has happened, but it's mostly been breakthroughs in observational cosmology rather than theory. The fact that observations have confirmed the standard model of cosmology so well is remarkable, and the observation that the universe is accelerating is remarkable, but those aren't really breakthroughs in theory. There have been many interesting ideas in high-energy theoretical physics, none of which we have any strong reason to believe is right. If we look back in 10 years, we might be able to say there was this amazing breakthrough in 1992, where somebody proposed some extension of the standard model and looks like it's correct, but I don't know which one it is, so I can't answer that question.

Why have you chosen to do physics, why not something else?

Because it's so much fun.

What do you think is the role of the theoretical physicist in modern society?

Well, I think the real answer is that it's part of the nature of human beings to want to understand the universe, and that the job of theoretical physicists and experimental physicists is to try to move towards that goal and let the people who aren't theoretical physicists know something about what we've learnt. As a very nice side effect, which is not really the reason that most people do theoretical physics, when you understand more about the universe you have more control over it. So all of modern technology is based on progress in theoretical physicists that was thought to be of purely intellectual interest at the time it was made.

References

[1] G. W. Gibbons and N. Turok, "The measure problem in cosmology," *Phys. Rev. D* **77** (2008) 063516, arXiv:hep-th/0609095 [hep-th].
[2] S. Kachru, R. Kallosh, A. D. Linde and S. P. Trivedi, "De Sitter vacua in string theory," *Phys. Rev. D* **68** (2003) 046005, arXiv:hep-th/0301240.
[3] L. Susskind, "The anthropic landscape of string theory," arXiv:hep-th/0302219.
[4] P. G. Bergmann, "Observables in general relativity," *Rev. Mod. Phys.* **33** (Oct. 1961) 510–514. https://link.aps.org/doi/10.1103/RevModPhys.33.510.

[5] C. G. Torre, "Gravitational observables and local symmetries," *Phys. Rev. D* **48** (Sep. 1993) R2373–R2376. https://link.aps.org/doi/10.1103/PhysRevD.48.R2373.

[6] J. M. Pons, D. C. Salisbury and K. A. Sundermeyer, "Observables in classical canonical gravity: folklore demystified," *J. Phys. Conf. Ser.* **222** (2010) 012018, arXiv:1001.2726 [gr-qc].

[7] J. M. Maldacena, "The large N limit of superconformal field theories and supergravity," *Int. J. Theor. Phys.* **38** (1999) 1113–1133, arXiv:hep-th/9711200.

[8] A. Hamilton, D. N. Kabat, G. Lifschytz and D. A. Lowe, "Holographic representation of local bulk operators," *Phys. Rev. D* **74** (2006) 066009, arXiv:hep-th/0606141 [hep-th].

[9] S. Carlip, "Lectures on (2+1) dimensional gravity," *J. Korean Phys. Soc.* **28** (1995) S447–S467, arXiv:gr-qc/9503024 [gr-qc].

[10] S. Carlip, "Quantum gravity: a progress report," *Rept. Prog. Phys.* **64** (2001) 885, arXiv:gr-qc/0108040 [gr-qc].

[11] S. Carlip, D.-W. Chiou, W.-T. Ni and R. Woodard, "Quantum gravity: a brief history of ideas and some prospects," *Int. J. Mod. Phys. D* **24** no. 11 (2015) 1530028, arXiv:1507.08194 [gr-qc].

[12] A. Belenchia, R. M. Wald, F. Giacomini, E. Castro-Ruiz, C. Brukner and M. Aspelmeyer, "Quantum superposition of massive objects and the quantization of gravity," *Phys. Rev. D* **98** no. 12 (2018) 126009, arXiv:1807.07015 [quant-ph].

[13] S. Carlip, "Is quantum gravity necessary?," *Class. Quant. Grav.* **25** (2008) 154010, arXiv:0803.3456 [gr-qc].

[14] S. Bose, A. Mazumdar, G. W. Morley, H. Ulbricht, M. Toroš, M. Paternostro, A. Geraci, P. Barker, M. Kim and G. Milburn, "Spin entanglement witness for quantum gravity," *Phys. Rev. Lett.* **119** no. 24 (2017) 240401, arXiv:1707.06050 [quant-ph].

[15] S. Carlip, "Six ways to quantize (2+1)-dimensional gravity," in *5th Canadian Conference on General Relativity and Relativistic Astrophysics (5CCGRRA) Waterloo, Canada, May 13–15, 1993*, pp. 215–234. 1993. arXiv:gr-qc/9305020 [gr-qc].

[16] S. Carlip, "Quantum gravity in 2+1 dimensions: the case of a closed universe," *Living Rev. Rel.* **8** (2005) 1, arXiv:gr-qc/0409039 [gr-qc].

[17] S. Carlip, "Notes on the (2+1)-dimensional Wheeler-DeWitt equation," *Class. Quant. Grav.* **11** (1994) 31–40, arXiv:gr-qc/9309002 [gr-qc].

[18] S. Carlip and J. E. Nelson, "Comparative quantizations of (2+1)-dimensional gravity," *Phys. Rev. D* **51** (1995) 5643–5653, arXiv:gr-qc/9411031 [gr-qc].

[19] S. Carlip, "A phase space path integral for (2+1)-dimensional gravity," *Class. Quant. Grav.* **12** (1995) 2201–2208, arXiv:gr-qc/9504033 [gr-qc].

[20] S. Carlip, "The modular group, operator ordering, and time in (2+1)-dimensional gravity," *Phys. Rev. D* **47** (1993) 4520–4524, arXiv:gr-qc/9209011 [gr-qc].

[21] S. Carlip, "The sum over topologies in three-dimensional Euclidean quantum gravity," *Class. Quant. Grav.* **10** (1993) 207–218, arXiv:hep-th/9206103 [hep-th].

[22] S. Carlip and R. Cosgrove, "Topology change in (2+1)-dimensional gravity," *J. Math. Phys.* **35** (1994) 5477–5493, arXiv:gr-qc/9406006 [gr-qc].

[23] A. Strominger, "Black hole entropy from near horizon microstates," *JHEP* **02** (1998) 009, arXiv:hep-th/9712251 [hep-th].

[24] S. Carlip and C. Teitelboim, "Aspects of black hole quantum mechanics and thermodynamics in (2+1)-dimensions," *Phys. Rev. D* **51** (1995) 622–631, arXiv:gr-qc/9405070 [gr-qc].

[25] S. Carlip, "Entropy from conformal field theory at Killing horizons," *Class. Quant. Grav.* **16** (1999) 3327–3348, arXiv:gr-qc/9906126 [gr-qc].
[26] S. Carlip, "Black hole entropy from horizon conformal field theory," *Nucl. Phys. Proc. Suppl.* **88** (2000) 10–16, arXiv:gr-qc/9912118 [gr-qc]. [10(1999)].
[27] S. Weinberg, "Ultraviolet divergences in quantum gravity," in *General Relativity: An Einstein Centenary Survey*, S. W. Hawking and W. Israel, eds., pp. 790–831. Cambridge University Press, 1979.
[28] S. Weinberg, "What is quantum field theory, and what did we think it is?," in *Proceedings of the Conference on Historical Examination and Philosophical Reflections on the Foundations of Quantum Field Theory*, March 1–3, 1996, T. Y. Cao, ed., 241–251. Boston, 1996.
[29] S. Weinberg, "Living with infinities," 2009. arXiv:0903.0568 [hep-th]. https://inspirehep.net/record/814639/files/arXiv:0903.0568.pdf.
[30] S. Weinberg, "Effective field theory, past and future," *PoS* **CD09** (2009) 001, arXiv:0908.1964 [hep-th].
[31] M. Niedermaier and M. Reuter, "The asymptotic safety scenario in quantum gravity," *Living Rev. Rel.* **9** (2006) 5–173.
[32] M. Reuter and F. Saueressig, "Quantum Einstein gravity," *New J. Phys.* **14** (2012) 055022, arXiv:1202.2274 [hep-th].
[33] R. Utiyama and B. S. DeWitt, "Renormalization of a classical gravitational field interacting with quantized matter fields," *J. Math. Phys.* **3** (1962) 608–618.
[34] C. J. Isham, A. Salam and J. A. Strathdee, "Infinity suppression gravity modified quantum electrodynamics," *Phys. Rev. D* **3** (1971) 1805–1817.
[35] E. P. Verlinde, "On the origin of gravity and the laws of Newton," *JHEP* **04** (2011) 029, arXiv:1001.0785 [hep-th].
[36] E. P. Verlinde, "Emergent gravity and the dark universe," *SciPost Phys.* **2** no. 3 (2017) 016, arXiv:1611.02269 [hep-th].
[37] S. Carlip, "Challenges for emergent gravity," *Stud. Hist. Phil. Sci.* **B46** (2014) 200–208, arXiv:1207.2504 [gr-qc].
[38] S. Weinberg and E. Witten, "Limits on massless particles," *Physics Letters B* **96** no. 1 (1980) 59 – 62. www.sciencedirect.com/science/article/pii/0370269380902129.
[39] M. Visser, "Conservative entropic forces," *JHEP* **10** (2011) 140, arXiv:1108.5240 [hep-th].
[40] S. Carlip, "Spontaneous dimensional reduction in short-distance quantum gravity?," *AIP Conf. Proc.* **1196** no. 1 (2009) 72, arXiv:0909.3329 [gr-qc].
[41] S. Carlip, "The small scale structure of spacetime," in *Foundations of Space and Time: Reflections on Quantum Gravity*, G. Ellis, J. Murugan and A. Weltman, eds., pp. 69–84. Cambridge University Press, 2009.
[42] S. Carlip, "Dimensional reduction in causal set gravity," *Class. Quant. Grav.* **32** no. 23 (2015) 232001, arXiv:1506.08775 [gr-qc].
[43] S. Carlip, "Dimension and dimensional reduction in quantum gravity," *Class. Quant. Grav.* **34** no. 19 (2017) 193001, arXiv:1705.05417 [gr-qc].
[44] J. Ambjorn, J. Jurkiewicz and R. Loll, "Spectral dimension of the universe," *Phys. Rev. Lett.* **95** (2005) 171301, arXiv:hep-th/0505113.
[45] O. Lauscher and M. Reuter, "Fractal spacetime structure in asymptotically safe gravity," *JHEP* **10** (2005) 050, arXiv:hep-th/0508202.
[46] J. J. Atick and E. Witten, "The Hagedorn transition and the number of degrees of freedom of string theory," *Nucl. Phys.* **B310** (1988) 291–334.

[47] M. R. Douglas, D. N. Kabat, P. Pouliot and S. H. Shenker, "D-branes and short distances in string theory," *Nucl. Phys. B* **485** (1997) 85–127, arXiv:hep-th/9608024.

[48] S. Carlip, R. A. Mosna and J. P. M. Pitelli, "Vacuum fluctuations and the small scale structure of spacetime," *Phys. Rev. Lett.* **107** (2011) 021303, arXiv:1103.5993 [gr-qc].

[49] V. A. Belinsky, I. M. Khalatnikov and E. M. Lifshitz, "Oscillatory approach to a singular point in the relativistic cosmology," *Adv. Phys.* **19** (1970) 525–573.

[50] S. Carlip, "Symmetries, horizons, and black hole entropy," *Gen. Rel. Grav.* **39** (2007) 1519–1523, arXiv:0705.3024 [gr-qc]. [Int. J. Mod. Phys.D17,659(2008)].

[51] S. Carlip, "Black hole entropy and the problem of universality," in *Quantum Mechanics of Fundamental Systems: The Quest for Beauty and Simplicity: Claudio Bunster Festschrift*, pp. 91–106. 2009. arXiv:0807.4192 [gr-qc].

[52] S. Carlip, "Black hole entropy from Bondi-Metzner-Sachs symmetry at the horizon," *Phys. Rev. Lett.* **120** no. 10 (2018) 101301, arXiv:1702.04439 [gr-qc].

[53] S. Carlip, "Black hole thermodynamics from Euclidean horizon constraints," *Phys. Rev. Lett.* **99** (2007) 021301, arXiv:gr-qc/0702107 [gr-qc].

[54] M. Guica, T. Hartman, W. Song and A. Strominger, "The Kerr/CFT correspondence," *Phys. Rev. D* **80** (2009) 124008, arXiv:0809.4266 [hep-th].

[55] S. Carlip, "Dynamics of asymptotic diffeomorphisms in (2+1)-dimensional gravity," *Class. Quant. Grav.* **22** (2005) 3055–3060, arXiv:gr-qc/0501033 [gr-qc].

[56] S. Carlip, "Conformal field theory, (2+1)-dimensional gravity, and the BTZ black hole," *Class. Quant. Grav.* **22** (2005) R85–R124, arXiv:gr-qc/0503022 [gr-qc].

[57] Y. Chen, "Quantum Liouville theory and BTZ black hole entropy," PhD thesis, UC, Davis, 2004. wwwlib.umi.com/dissertations/fullcit?p3137525.

[58] S. Carlip, "A note on black hole entropy in loop quantum gravity," *Class. Quant. Grav.* **32** no. 15 (2015) 155009, arXiv:1410.5763 [gr-qc].

[59] A. Sen, "Logarithmic corrections to Schwarzschild and other non-extremal black hole entropy in different dimensions," *JHEP* **04** (2013) 156, arXiv:1205.0971 [hep-th].

[60] I. Agullo, J. Fernando Barbero, E. F. Borja, J. Diaz-Polo and E. J. S. Villasenor, "Detailed black hole state counting in loop quantum gravity," *Phys. Rev. D* **82** (2010) 084029, arXiv:1101.3660 [gr-qc].

6
Alain Connes

Professor at the Collège de France, IHES, and a Distinguished Professor at Ohio State University

Date: 18 February 2016. Location: Paris. Last edit: 1 October 2019

From your point of view, what are the main problems in theoretical physics at the moment?

I do not have the pretension of answering this question as if I were a true physicist. I'm a mathematician, not a physicist, but I have been, for all my life, fascinated by some problems in physics. In particular, I have been fascinated by three main problems. The first problem is the problem of understanding what time is and the role of time at the quantum level.

The second problem, which I have spent a considerable amount of time on, is the problem of understanding renormalisation in a mathematical way. Physicists have found and developed renormalisation, which is an amazing technique, and have applied it successfully in many cases. However, the mathematical meaning of this technique has always been rather obscure. It took a lot of time and work but now I think that, thanks to my collaboration with Dirk Kreimer, we finally understand renormalisation in a precise mathematical way.

The third problem that has fascinated me deeply is the subtleness of reality exhibited at the quantum level. standard model discovered that you cannot naively manipulate observables in microscopic systems; instead, you need, for example, to take care about the order of observables in a given product. The consequence of this discovery is the realisation that our mathematical ideas about geometry, space and time in particular, are very naive, in the sense that they are based on sets, and that the quantum reality is much more subtle.

These are three leitmotifs underlying my reflections about parts of human thought that can be relevant for physics but, of course, being a mathematician, I am more interested in trying to understand the mathematical meaning of something rather than the impact that this understanding can have on making actual measurements. The latter, you know, is what physicists actually do (laughs). My motivation is of a rather different nature.

You mentioned that at the moment we understand renormalisation at a mathematical level. What is the mathematical meaning of renormalisation?

This line of work is the one that I have pursued, mostly with Dirk KreimerKreimer, Dirk, starting in the beginning of the year 2000 [1–5]. As far as I am concerned, I consider the result of this work to be the final understanding of the meaning of renormalisation from a mathematical point of view. We have understood that renormalisation is, in fact, the result of the Birkhoff decomposition of a loop with values in a group, which is associated with Feynman graphs [2–4].

This procedure that we have discovered is in fact subsuming all the technically complicated methods that physicists use when they perform renormalisation in the dimensional regularisation scheme, which involves all the very complicated procedures of taking sub-graphs, sub-divergences and all that [6]. Understanding this precisely was a wonderful episode. The fact that the physics community did not give it any attention is totally irrelevant for me. I have been driven just by my need for understanding what renormalisation is and I had spent years and years reading books on renormalisation such as the collected papers on quantum electrodynamics by Julian Schwinger [7]. It was quite clear to me after reading these books that there was no clue at all that this could have any mathematical meaning. No clue whatsoever. So the miracle happened when I met Dirk Kreimer, who had discovered the Hopf algebra of Feynman graphs [8], while at the same time I had been working, together with Henri Moscovici, on Hopf algebras for other reasons [9, 10]. So I was prepared. Together with Dirk Kreimer, it took us more than two years of reflection and understanding in order to see the link with a very deep mathematical theory which is the theory of Birkhoff decompositions [11]. This had been developed by Grothendieck [12] and is related to the Riemann-Hilbert problem.

Though the main breakthrough was done with Dirk Kreimer, I went further, together with Matilde Marcolli [13–15], because we understood the mathematical abstract formulation of the Riemann-Hilbert problem which is related to the Birkhoff decomposition. The iterative procedure which is used in physics when subtracting sub-divergences was understood as being exactly the mathematical procedure which is used in the Birkhoff decomposition when you take values in a nilpotent group to decompose a loop. If the loop is defined on the equator of the Riemann sphere you can write it as a ratio of two loops, one in the upper hemisphere and one in the lower hemisphere, both being holomorphic. The meaning of renormalisation is then the following: when you decompose the loop in dimensions $4 - \epsilon$, you find something which is meaningful for generic values of ϵ but when you put these values on the Riemann sphere you find that $\epsilon = 0$ is a singularity. However, the mathematical way of dealing with this singularity is to perform the Birkhoff decomposition and simply evaluate the holomorphic part at $\epsilon = 0$. This procedure exactly agrees with the very complicated combinatoric method which is used by physicists to extract a finite result.

This indicates that, first of all, there is a conceptual meaning to renormalisation. Second, it gives a deep relation with Galois theory of ambiguity because the Riemann-Hilbert problem is a special case of Galois theory, where the symmetry group one deals with is a nilpotent formal group. It provides a universal, infinite-dimensional formal group acting on physical constants as the proper incarnation of the renormalisation group. These breakthroughs were enough for me to come to terms with my discontent with the method used by physicists. The sociological fact that they have not been understood by physicists nor are they part of their general knowledge is totally unimportant to me.

Does this new understanding of renormalisation give you any clues about why gravity is non-renormalisable in four dimensions?

No. The issues with gravity are related to the problems I mentioned earlier, namely the fine structure of spacetime and that thinking of it as a four-dimensional continuum is not the right thing. The usual picture of spacetime as the ultimate structure is a very naive picture. It appears to me that many people confuse this model of spacetime with reality. Reality is much more interesting but it takes a lot of work and thinking in order to try to go beyond the naive picture.

So you think that this confusion is underlying most theories of quantum gravity?

Yes, sure. It is rather striking that, in the various attempts of quantising gravity that I know of, the important message of quantum physics as Von Neumann formulated, namely, that the real birthplace of everything is Hilbert space, is not being taken into account. A real approach to quantum gravity better start with the fact that geometry should spring from Hilbert space. I find it quite an ill-conceived idea that one could quantise gravity as one quantises all other fields because I don't believe that geometry makes any sense, say, above the Planck energy. This belief is shared by many people. I think that in the same way that the Brout-Englert-Higgs particle emerges from symmetry breaking [16, 17], the actual geometry of spacetime also emerges from symmetry breaking and does not make sense beyond a certain energy level. One of my leitmotifs for developing non-commutative geometry has been to try to understand how geometry can actually emerge from purely Hilbert space considerations.

How can spacetime emerge from Hilbert space?

I have a little bit of an understanding of how spacetime can emerge in this way and it goes back to one of the most striking features of quantum mechanics which parallels the discussion of a very fundamental notion in mathematics, namely the notion of real variables. One of the really new, totally unexpected features of quantum mechanics is the fact that outcomes of microscopic experiments can't be repeated. Even if you consider just a single slit experiment and you shoot electrons or photons through a slim slit of size comparable to the wavelength of the particles, the exact location where the particle will land on a target on the other side of the slit is something that cannot be reproduced. What one can predict is the probability of the particle arriving somewhere but arriving at some fixed spot is something which, from the principles of quantum mechanics, cannot be reproduced. This means that there is some fundamental randomness which is inherent to quantum physics.

On the mathematical side of things, there is a related question that in fact goes back to Newton, which is simply, "What is a real variable?" If you ask a mathematician for their view on this question the most likely answer you will get is that "a real variable is given by a set X and a map from this set X to the real numbers." However, if you think more deeply about it, you will find out that this answer is very unsatisfactory and that is because you can't have coexistence of continuous variables, namely variables which can take a continuous range of possible values, and discrete variables, namely variables which can only affect a countable set of possible values, say, with finite multiplicity for each of them. The reason for this is that if you have a discrete variable then the original set X that you're dealing with will have to be countable, which then implies that it does not allow for continuous variables. The amazing answer that mathematics provides, but which would not have been detected if it were not for the formalism of quantum mechanics by Von Neumann, is that a real variable is just a self-adjoint operator in the Hilbert space. As you know, there is only one Hilbert space which is the Hilbert space with countable basis and Von Neumann tells you that this Hilbert space has variables with discrete spectrum. For instance, take a description of Hilbert space by giving a countable orthogonal basis and take an operator which is diagonal in that basis. This operator has countable spectrum but it coexists with operators which have continuous spectrum. Indeed one could have described that same Hilbert space as being the space of square integrable functions on an interval and of course one will have continuous variables there, given by multiplication operators. The beauty of this formalism is that continuous and discrete variables can coexist. All the properties of real variables are there because if I have a self-adjoint operator it has a spectrum which is composed of the possible values of the variable and it has spectral multiplicity which is the number of times a value can be affected. It fits with reality in the sense that the quantum variables are operators in Hilbert space but the new key fact is that the discrete variables do not commute with the continuous variables. If they would commute, they would be functions defined on the same space X, which is not possible. So they coexist but they do not commute and it is precisely this lack of commutativity which is the new ingredient that makes quantum mechanics exciting and which renders the framework of non-commutative geometry [18–21] appropriate and useful.

Before discussing further non-commutative geometry I would like to return to one of the interesting statements that you made, namely, that you did not believe that gravity could be quantised as all other fields. Does this mean that you don't agree, for example, with the way that loop quantum gravity is quantising gravity?

There are good things in both loop quantum gravity or string theory but in each of them I can see some basic problem. All these approaches carry the stigmas of ordinary geometrical thinking. I want to understand how geometry can spring from Hilbert space and that seems to be missing in these approaches.

What has happened in the process which led to understanding this emergence of geometry from the quantum is quite instructive. When Heisenberg found his commutation relations involving P and Q (momentum and position) there was already quite a hint, a bit of truth, in it, in the sense that when you take the spectrum of either P or Q you find a real line. The other operator is a differentiation operator and it gives you a geometrical structure for that line. However, the way things evolved from these commutation relations is that they were interpreted in terms of Lie group representations. Group representations, applied to the Poincaré group, give a beautiful conceptual notion of particle and the theory unveiled big pieces of landscape, but finite-dimensional Lie groups can hardly lead us to the arbitrary geometries that we observe in gravity and to the variables that we have in gravity. Simple Lie groups are like isolated diamonds or gems and they don't allow for this enormous variability which you have in gravity. So, the development of non-commutative geometry has shown that it is possible to have geometry emerging from purely Hilbert space considerations but in order to do that you apply representation theory to more elaborate forms of Heisenberg's commutation relations involving P and Q. In order to obtain these relations one needed to take a step back and understand how to give more flexibility to both P and Q. The additional flexibility for P was not so hard to find. In fact, it was found by Paul Dirac when he realised how to assemble several momenta together to form a single expression, a single operator, that actually contains in itself all the components of the momenta. This is the Dirac operator which uses gamma matrices. So, the understanding of how to give more flexibility to P stemmed from Dirac's work and one can really explain in very simple terms how far this understanding is grounded in physics and in the understanding of geometry, in particular in the measurement of lengths, as I will do now.

Many formalisms of geometry start with the Riemannian paradigm as a prerequisite, that is, the idea that geometry is given by the measurement of lengths and this measurement of lengths is actually governed locally by providing the square of the line element $ds^2 = g_{\mu\nu}dx^\mu dx^\nu$ and that's it. Now, it turns out that this idea was even questioned by Riemann himself. He wrote in his inaugural lecture on the foundation of geometry that it's questionable whether the texture of space (or spacetime) will obey this "Riemannian paradigm" at any scale, the reason being that the notions of light ray and of solid body on which his intuition was grounded would cease to make sense for very tiny scales.

At the end of the eighteenth century the desire to unify the measurement of distances led to a concrete realisation of a unit of length which was called "mètre étalon" and was conserved near Paris in the form of a platinum bar. Later, the relevance of this choice was put into question in the early 1920s because people found out that the mètre étalon was actually changing length and so of course this was very problematic. They found this out because they measured the mètre étalon by comparing it with the wavelength of a fixed atomic transition of krypton. The outcome of this observation, many years later, was that physicists shifted the definition of the unit of length from a platinum metal bar to the wavelength of a certain transition of krypton and later, eventually, to a hyperfine transition of caesium, which was much more convenient since it gives a microwave of the order of 3 cm.

Once you think more deeply about this, you find that the reason why their classical unit of length needed to be localised was because first it should be quite small, since it is supposed to represent ds, and because it commuted with the coordinates it had to be localised somewhere (as it happened this was near Paris). However, the new unit of length which is given by a wavelength of the hyperfine transition of Caesium – it is called a unit of time using the speed of light as the conversion factor –, is in fact of spectral nature and it's no longer commuting with the coordinates. In fact it involves the Dirac operator or the Dirac propagator, which is the inverse of the Dirac operator. Obviously, because it doesn't commute with the coordinates, it doesn't need to be localised and in particular it's obvious that it should be used as a unit of length if we want for instance to unify the metric system in our galaxy. It's much easier to tell people in nearby stars that our unit of length is a certain transition in the helium or hydrogen spectrum rather than to tell them that they need to come to Paris and compare their unit with the metal bar which is located there (laughs).

It's pretty obvious from the purely physical standpoint that this is a much better definition of the unit of length and mathematically speaking it implies that one actually replaces the ds^2 of Riemannian geometry by a very subtle square root of it, where the square root is taken through the Clifford algebra and where the infinitesimal line element which was formulated in terms of infinitesimal variables by Riemann is replaced by an infinitesimal which is an operator in Hilbert space. As turns out, there is a natural notion of infinitesimal among Hilbert space operators. This is just a parenthesis but this notion of infinitesimal was in fact predicted by Isaac Newton in the sense that he said explicitly that an infinitesimal should not be a number but instead it should be a variable! Moreover Newton gave the definition of an infinitesimal variable and, when you translate it in terms of the understanding of variables using Hilbert space operators, it gives precisely what we call a compact operator. These compact operators have exactly all the properties you would dream of for infinitesimals, they form an ideal among operators, etc.

When you think about the line element for spacetime, you find out that the mathematical formulation consisting of the inverse of the Dirac operator, in the language of physics, is encoded by what is usually referred to as the fermion propagator. This fermion propagator enters Feynman graphs as the internal legs of graphs involving fermions. When you look at quantum field theory textbooks à la Richard Feynman, you find that this propagator is a small tiny line joining two very nearby events. In a sense, it is a qualitative appearance of an infinitesimal line element but there it is actually much more than that and what is striking is the fact that when you do quantum field theory, this line element actually acquires quantum corrections by being dressed. This means that if you take from the beginning the correct version of the line element for geometry, you soon understand that there are quantum corrections to the measurement of lengths, to the geometry, which are given by the dressing of the fermion propagator. Moreover the gauge fields appear as additional terms in the Dirac operator and this embodies exactly the intuition of Bernhard Riemann that in case his paradigm would fail at small distances the geometry should be based on the forces that hold things together!

All these elements that you are describing, such as the Dirac operator, are the basis for non-commutative geometry?

Exactly. In non-commutative geometry, the P of Heisenberg is replaced by the Dirac operator and its inverse plays the role of the line element, which is a purely Hilbert space definition and hence inherently quantum [19]. To what concerns the Q of Heinsenberg, it took us a long time to understand how to give flexibility to it. Eventually, this understanding appeared in papers I wrote together with Ali Chamseddine and Viatcheslav Mukhanov [22–24].

The idea is also very simple and it has to do with the beauty of complex numbers. Complex numbers are actually a good way of understanding pairs of real numbers. When you do planar geometry you will find a lot of nicer proofs by using complex numbers than by using pairs of real numbers. It turns out that when you extend the framework of ordinary geometry by accepting that coordinates might no longer commute, then you are no longer limited to complex numbers, to two-dimensional spaces, and can instead access higher-dimensional spaces while zipping the coordinates into a single operator. An example of these higher-dimensional numbers is quaternions. This idea allows one to understand that in many geometric instances it's easier to comprehend the geometry not by looking at functions with real values or complex values on a space but by looking at matrices of functions which are slightly non-commutative. From the mathematical standpoint, it is easier to present an algebra of functions (consisting of a single operator and a Clifford algebra) instead of having to give generators and relations which are a little bit ad hoc and artificial. What we have found is that understanding the Q of Heisenberg in a more flexible geometrical framework requires giving Q as a single operator as long as you also provide the Clifford algebra. The consequence of this is that instead of obtaining just the commutative algebra of functions in the space that you consider, you obtain a slightly non-commutative algebra of functions, which is equivalent, as far as the naive notion of points is concerned, to the original one. By following this line of thought, what my collaborators and I found is that in order to obtain the four-dimensional continuum as a space, you have to introduce a minimal amount of non-commutativity which is governed by two Clifford algebras of five variables [22–24]. These Clifford algebras are exactly the algebras which, in my work with Chamseddine, had to be put in by hand in order to obtain the Lagrangian of gravity coupled to the standard model [25–31].

So, this was a big revelation because our motivation came from geometry and not from the standard model or anything like that. It turns out that from pure geometrical considerations which are subtle but natural, you recover the Lagrangian of gravity coupled to the standard model [22–24]. Why are these considerations subtle? They are subtle because one had to prove a deep mathematical fact which is that one can reconstruct any four-dimensional compact manifold, fulfilling one condition which I will explain later, out of two maps to the sphere of dimension four. This is a kind of embedding in the product of two 4-spheres. The very intriguing fact was that the problem of finding this embedding, or this immersion, in the product of two spheres was a purely mathematical problem that did not involve a priori the notion of spin. However, it turned out that a necessary condition for this to be true was that the square of the second Stiefel-Whitney class was vanishing. This condition, which is the one I alluded to earlier, is slightly weaker than requiring the manifold to be a spin manifold since the latter is equivalent to the vanishing of the second Stiefel-Whitney class, and this condition is sufficient. What we've discovered is that all compact spin manifolds endowed with a Riemannian metric of integral volume (>4) give solutions of a higher form of Heisenberg's commutation relation which involves, instead of P, the Dirac operator and instead of Q, an operator Y which gives the coordinates on the space. One can then develop the spectral action, which is a very basic principle, by computing Seeley-deWitt coefficients and so forth, and obtain a very specific Lagrangian [22–24].

This Lagrangian has several new features. The first feature is that now the volume is quantised. In other words, the higher form of the Heisenberg commutation relation is related to the index problem and the integer which appears in this index gives you that the volume of the manifold is a multiple of the Planck volume. This volume will be very large as the integer index but it's an integer and this new feature appears for the first time in the spectral action. What was very puzzling to us before this integrality result was that when you compute the spectral action there is a huge cosmological term in the action. This implied that when you compute the equations of motion, performing variations of the action, clumsy huge terms would appear. However, now because the volume is quantised, it's no longer there since, being an integer, you cannot vary it continuously, so it doesn't enter. The first term that actually appears is precisely the variation of the Einstein action with the correct sign. So when you develop this spectral action you find the Einstein action and you also find the Yang–Mills action for the gauge fields and so on. These gauge fields appear because of the slight non-commutativity of the coordinates. What happens is that when you look at the metric in that space, the metric has an overall usual spacetime metric which is determined in terms of the Dirac operator but it also has gauge fields, which account for the inner part of the metric. Exactly as Riemann foresaw, the metric contains in it all the bosonic forces; they are embodied by the gauge fields which are involved in the Dirac operator. In addition, you also get a scalar field, the Higgs field, appearing naturally; it comes from a computation and is not added artificially. So for us, the discovery of the Brout-Englert-Higgs particle was so important because it was the spin-zero field which was actually observed and if our theory has any value it is to place this scalar field on the same footing as the gravitational and gauge fields. All these fields together tell you what is the geometry of the space that you're considering. You don't have to do anything extra, they're just there from the beginning. Another very convincing fact is that the computation of the spectral action spits out subtle mechanisms such as the "see-saw" mechanism which are normally added by hand in the construction of the standard model; this also goes for the $V - A$ of Murray Gell-Mann and Richard Feynman.

I remember that in your older papers on non-commutative geometry applied to the standard model and gravity you input by hand the group structure [25–31]. Is this no longer necessary?

Yes, now there's no input any longer. In the beginning we followed a bottom-up approach in order to obtain the standard model from simple non-commutative spaces. In particular, for a long time we were inputting specific groups and trying to see what was necessary in order to obtain gravity coupled to the standard model field content, including the Higgs field. This led us to understand that it wasn't the group structure which was so important but the algebra behind the group. In more recent papers [22–24] we realised that from purely geometric considerations, without knowing anything about the standard model, you can understand the four-dimensional continuum by introducing a slight amount of non-commutativity via two Clifford algebras and as an unexpected by-product you get the standard model.

But then somehow you input non-commutativity?

The role of the slight amount of non-commutativity is, very much as when you collect four real numbers together into a single quaternion, to express the full coordinate system in a much simpler way. So starting with a single operator and the constant Clifford algebras you find that the algebra they generate is non-commutative, and it corresponds to functions on a four-dimensional continuum. These functions have values in two Clifford algebras which are determined, "spit out" uniquely, by the geometric principles. These two algebras consist of two-by-two matrices over quaternions and of four-by-four matrices over complex numbers. The relation between these algebras and the standard model is quite easy to understand. The two quaternions give you $SU(2)_{\text{left}}$ and $SU(2)_{\text{right}}$ while the four by four matrices give you the Pati-Salam $SU(4)$, which then breaks into the $SU(3)$ for the quarks and the $U(1)$ for the leptons. So things emerge with the right quantum numbers and the right hypercharges. It could be an accident but then it would mean that there is a devil in action.

What about the number of generations?

The number of generations is of course a big question, and is surely related to the yet under-developed notion of fundamental group and universal cover in non-commutative geometry. At the moment we have to input it by hand. What needs to be done in order to get a conceptual reason is to classify the corresponding Hilbert space irreducible representations. There is something special that happens with three generations which physicists have known for a long time. Namely, if you take the eigenstates for lower quarks and move them up to the upper sector using the weak isospin group and compare them with the eigenstates for the upper quarks, you find that there is a mixing matrix and this is the Cabibbo-Kobayashi-Maskawa matrix [32, 33]. It's well known that if you have two generations, this matrix can be written in terms of real numbers and it's only with three generations that you have to deal with unavoidable complex numbers which break the symmetry C. However, even though it is well known, we don't have a clue to the meaning of the need for breaking the symmetry C. If we understood conceptually what was the meaning of this phenomenon, we could be better off but at the moment we don't have this understanding. So at that level of our model, what happens is that the masses and the mixing matrices of the leptons and quarks are still put in by hand. Nevertheless, when you write down the spectral action with this input, you find that there is less freedom than what you have when considering the ordinary coupling of gravity to the standard model. In particular, there is a number of predictions when you assume that the model is valid at the unification scale.

What kind of lack of freedom and predictions?

If you want an action in which, for instance, you have unification of couplings, you consider this action as being valid at the unification scale. Then, you use the renormalisation group to follow the trajectory of the coupling constants and you see what happens. There is an expression in the standard model called $Y2$ which is the sum of the squares of the Yukawa couplings of the leptons and quarks with a factor of three arising in front of the quarks, due to the colour. One prediction of our theory is that $Y2 = 4g^2$, where g is the electroweak coupling constant [29]. When we first obtained this result we thought that it was totally absurd but we took this result at unification scale, ran it down and obtained the proper top quark mass. This means that it gives a Yukawa coupling for the top quark of around one.

Is there some sort of fine-tuning going on here?

No, it's not at all fine-tuning. The only thing you have to fix is the unification scale which is a few orders of magnitude below the Planck scale. Then you run down and you get the coupling of the top quark to be around one, which physicists may think is not too impressive because they can argue that this value can be explained by some form of attractor mechanism. However, something that was far more disconcerting is that the theory also predicted the quartic self-coupling of the Higgs (also called the scattering parameter of the Higgs field) to be of around one at the unification scale [29]. It is well known from the physics literature that this implies that the Higgs mass had to be at least around 170 GeV. This prediction is independent of our model and is related to the fact that if the mass of the Higgs field was lower then the quartic term in the potential wouldn't be positive at the unification scale. This was considered to be the core prediction of this theory but it was of course falsified in 2008 by the Tevatron [34].

I was very disappointed. I immediately admitted failure by writing in the noncommutative geometry blog, and for four years I stopped working along these lines. However, in 2012, my collaborator Ali Chamseddine wrote me, in the spring, a key message that saved the stuff and which at first I didn't believe! He wrote that the problem of the instability of the quartic self-coupling, which I just mentioned, had been addressed (independently of our work and model) by three independent groups of physicists [35–39]. It's a well-known analogy that physicists are like bosons and mathematicians are like fermions (laughs). So physicists tend to clump into groups.

Anyway, these three groups had found a way around this difficulty, which consisted in adjoining to the theory a new scalar field, distinct from the Higgs field, and which essentially does not couple to any other field, except to the Higgs field itself. So if I call this new scalar field ϕ and the Higgs field H, then the coupling is of the form $\phi^2 H^2$ with a certain sign. Additionally there is a ϕ^4 term which is just the self-interaction of this new scalar field. These three groups of physicists computed the renormalisation group flow going down from the unification scale taking into account this new field and they found that it allowed for the Higgs mass to be around 125 GeV. In other words, it allows for the stability of the self-coupling of the Higgs field. What my collaborator pointed out was that this new scalar field was there from the beginning in our paper in 2010 [40]!

In that paper we had found this field to be dictated by our theory from the part of the Dirac operator relating the neutrino sector to the antineutrino sector. However, we ignored this scalar field by wrongly taking it to its vacuum value and performing the renormalisation group calculation without it. We just wanted to obtain the standard model and so we didn't want to bother with it. We also didn't believe that this scalar field could affect the renormalisation group results.

Is this scalar field supposed to be a new particle?

Yes, it's supposed to be a new particle but of a mass of around 10^{11} GeV. So it's a very heavy particle and doesn't really affect the low-energy physics. Something that we had found already in 2006 was that by doing the calculation and taking the finite space to have KO-dimension six, one naturally obtained the see-saw mechanism [29]. So back then we realised that this particle needed to have a huge mass but at that time we didn't treat it as a particle and instead set this field to its vacuum value. However, this huge mass had to be there so that the Yukawa coupling for the neutrinos would be lowered by the see-saw mechanism. What was surprising was that we did not have to introduce the see-saw mechanism by hand; it was in the machine, so to speak. I did not even know this mechanism when doing the calculation! So after 2012, I checked everything again, everything worked out and we wrote a paper with Ali, called "The Resilience of the Spectral Standard Model," in which we actually explained that the wrong prediction was the result of ignoring this new field [41].

So when you run things down, everything works out without adjustable parameters?

Everything works but there is one adjustable parameter which is essentially the mass of this new particle. In any case, there are equally interesting results coming from this theory. Recently, we collaborated with Walter van Suijlekom [42–44], which has been very successful in the sense that we have refined the calculations within the model and we also have been able to do perturbations in certain cases. The model is similar to the Pati-Salam model, where instead of having a $U(1)_{\text{right}}$, it has an $SU(2)_{\text{right}}$, which, as you know, is nice because of asymptotic freedom. So we wondered whether because of this nuance, the problem of unification would be resolved, which is not the case in the standard model. What turns out to be the case is that the coupling constants actually unify in the model that we have [43, 44]. So the model can't be very far off. It's not an E_6 nor E_8 model but it's close to the standard model and it has many advantages. One of the advantages is conceptual because the model corresponds exactly to the two algebras that I mentioned earlier and, moreover, it seems to fit with measurements. Physicists have worked out several models of the Pati-Salam type and whether such models could survive experimental data [45–48]. The model we have is one of them, which is very satisfactory.

On the other hand, one has to admit that the story of this approach has been plagued by periods of discouragement. For instance, the spectral action was put forth in 1996 [25] and in 1998 neutrino mixing was discovered [49, 50] which at the time was not allowed by the theory. It was only in 2006 that we realised that we did not have the correct starting point [29]. The KO-dimension of the finite space was taken to be zero at first but when we took it to be six modulo eight then everything worked and the fermion doubling problem was resolved. Several instances of this type occurred and thus the approach seems to have a certain resilience and can be tested against reality, which is trying to kill it all the time (laughs).

It should also be said that the theory also gives a new notion of quantum geometry which is mathematically very far from trivial. The mental picture you get from this approach is beautiful because what it says is that even though spacetime emerges from the product of two 4-spheres of Planckian size, how can it be that continuum large dimensions can emerge from it? It turns out that the product of two 4-spheres can unfold at exponential rate and that the size of the space is given by the sum of the degrees of the maps to the spheres while you have self-maps of the sphere which multiply a degree by any number. Hence, this kind of exponential unfolding acquires a true and simple meaning out of the quantum. This is just an intuition for now but there's a lot to discover.

So if I ask about the status of the theory, you tell me that it predicts a Pati-Salam type model?

At the moment and at reasonable scales, not at the Planck scale, the type of model that it predicts is very close to the standard model but with several additional nuances, none of which disagrees with anything that we know or have measured. You could ask me what the model is at a quantum level, since you are interested in quantum gravity, and in that regard the theory is not really at the quantum level per se; it is at the first quantised level. Hence, the theory is just at the beginning and I think that there are several ways of progressing. The first very concrete process to go through is not to take gravity into account too much but to study how the other quantum fields actually interfere with the metric. This can be done by dressing the Dirac propagator and devising calculations where the role of the propagator affects concrete measurements of distances at very small scales. The Dirac propagator allows for dressing and also allows for fractal dimensions; hence, it has all the flexibilities one can dream of. However, I think that for it to be a fundamental theory, we would have to find an extension of the electroweak symmetry-breaking mechanism to the gravitational sector. In particular, we have to imagine a more fundamental theory in which the geometry doesn't make any sense beyond the Planck scale but appears out of spontaneous symmetry breaking when the temperature is lowered.

What kind of symmetry is broken?

Symmetry of the unitary group of Hilbert space. Technically it's not quite the unitary group of Hilbert space because there is charge conjugation and quaternions in the background, so it's more like a symplectic unitary group. So my belief is that there is an enormous symmetry which is the symplectic unitary group; this symmetry is spontaneously broken and only then the geometry emerges. This is one possible scenario and if it holds it shows that it is vain to try and quantise gravity in a given geometrical background in which case one is quickly confronted with the lack of renormalisability, since at sufficiently high energies the geometry itself will have totally disappeared.

Are you referring to approaches such as string theory?

String theory is a bit like a chameleon which can drape itself as a precursor of any idea provided the latter would succeed, but here I am referring to whatever theory that would not begin from the basic principle that geometry should emerge from the quantum by a symmetry-breaking mechanism. I'm fortified in this respect by concrete experiments such as those carried out by Anton Zeilinger on quantum teleportation [51]. These experiments show that the primary data is really quantum and the usual notion of locality, namely that all phenomena occur in some location in spacetime, is naive. If you look at examples such as entanglement and the spooky action at a distance you find out that even though there is no transmission of information, there is a kind of global quantum coherence of the geometry of nature. This coherence is not at all understandable in a non-Hilbert space theoretical framework.

So my view is that we're confronted with a challenge which is enormous; namely, our minds have been educated, some people say by natural selection, to understand causality (in terms of light cones as in Einstein theory, etc.) and which we seem to have become dependent on. However, what happens in nature is much more interesting, imaginative and quantum but the human mind craves ordinary logic and ends up creating the picture of a white page, i.e., spacetime, and spends its whole life trying to write a coherent past in it. From the quantum point of view, the past is not even fixed in the sense that you could ask questions about the past that would actually change it. An example of this is the famous Wheeler thought experiment [52]. So what is probably the bigger difficulty that we are facing is that we have to obtain a formalism closer to the quantum in which this page on which the past would be written is only a by-product.

One of the things which I've been thinking about for many years is the idea that somehow it is non-commutativity, the randomness inherent to the quantum, which is at the origin of the variability. Moreover, the usual way of doing physics which consists of writing evolution equations, meaning that the fundamental variability is taken to be the passing of time, does not convince me to be the right approach. Instead, the fundamental variability should be taken to be the quantum variability inherent to the reduction of the wave packet and time should be only an emergent phenomenon.

Of course, one of the big problems is to be able to reconcile these various bits and pieces of the puzzle and I don't claim to have reconciled them in any way. They're all very challenging and I don't claim to have any coherent general philosophical approach to reconciling them but I know that each of these questions that I mentioned from the start is absolutely fundamental. When I read books from the period of the discovery of quantum mechanics in the 1920s or 1930s I really regret that now somehow the basic philosophical discussions are no longer part of the curriculum or part of the life of physicists. Nowadays it seems that the HEPTH world is more based on dogmas and on whether you believe in this dogma or that dogma! For example, I remember an instance where I was in a conference in California and I wanted to ask to a group of string theorists why they begin by looking at the S-matrix, since that already presupposes many things. At that time I was told that this discussion should not be taken (laughs). So I find this absolutely ridiculous because when I read these books on the glorious birth of the quantum theory, there were very long and controversial discussions, for instance, between Albert Einstein and Werner Heisenberg, and one of the discussions related to quantum gravity was the well-known episode between Albert Einstein and Niels Bohr during the Solvay Conference. Do you know it?

No, I don't.

Well, then I cannot resist the temptation to recount this so convincing episode. For me this is certainly a dent which has been made in the problem of confronting the quantum with gravity. At some point in the 1930s, Bohr and Einstein had an encounter at a Solvay Conference where Einstein claimed to have a thought experiment giving a counter-example to the uncertainty principle. Einstein had devised an apparatus composed of a little box suspended by a spring and the box has a mark pointing to its height. A bit like a Swiss cuckoo, the box also has a clock which shows the time when it emits a photon. Einstein argued that with this system there was no way to get the uncertainty principle $\delta E \delta t \geq \hbar/2$ because he argued he could measure time as precisely as he wanted using the clock and could also measure the energy of the photon extremely precisely because he could measure the weight of the clock before and after the emission. The reason why Einstein was absolutely sure to be right was that in the measurement of the energy of the photon the gravitational constant was involved because he was measuring the weight using the variation of the height, and this involved the gravitational constant which thus came in the way and prevented getting the expected naked Planck constant.

At that time Bohr was strongly hit and there is a photograph where one can see Einstein triumphantly exiting the room with Bohr behind him totally discouraged. However, instead of giving up, Bohr didn't sleep during that night and he thought and thought and thought and he found the answer and the answer that he found is absolutely amazing. Bohr defeated Einstein using Einstein's own theory because in general relativity in a first approximation when you consider the new metric, the space metric doesn't change; i.e., $dx^2 + dy^2 + dz^2$ doesn't change but the coefficient in dt^2 does change by an addition of twice the Newtonian potential of the point where you are. Obviously, there is a gravitational constant there but what does this change mean for the cuckoo? It means that when the cuckoo has emitted the photon it slightly goes up because it is lighter but as it slightly goes up, time is now passing differently. This gives a correction which actually cancels out the other occurrence of the gravitational constant. So the gravitational constant is actually not involved and if you compute the uncertainty you find exactly $\hbar/2$.

This was a kind of mind-blowing instance (laughs) and it shows that physics doesn't fool around. The quantum is there, gravitation is there and they speak to each other in many instances such as in the context of the well-known Chandrasekhar limit [53]. So, of course, we need a theory that unifies all these processes. It would be foolish to say that you cannot try to do it. My conclusion is that we have to go on but for that we need to think quietly and not let us be blinded by the propaganda of this or that existing theory.

But do you think that your formulation in terms of non-commutative geometry is enough to puzzle things together?

No, I wouldn't say so. As I said, we're still at the first quantised level, so what we have is a little bit like the situation where you have the Wigner theorem saying that irreducible representations of the Poincaré group represent particles. What we have is "particles of geometry," more precisely quanta of geometry. What we need is to obtain the second quantised version and understand it. My belief is that there is a certain quantum statistical system which will produce this springing of geometry from a symmetry breaking phenomenon at low temperature but which above a certain temperature would be chaotic and devoid of geometric interpretation whatsoever. We have some indications about this system, such as the role of scaling, but we're quite far from knowing what the system is. We know that the formalism has certain potentialities for giving rise to a unifying theory due to the fact that it is a formalism based on Hilbert space and higher analogues of the Heisenberg commutation relations.

What kind of mathematics is involved in the second quantisation of the theory?

We have a good model in mathematics for the first quantised theory because in mathematics this spectral approach to geometry has been very much inspired by the index theory of Atiyah-Singer [54] which gave rise to K-homology [55] which many people have worked on, in particular Kasparov [56]. The second quantised part should be related to a known mathematical theory called Quillen's algebraic K-theory [57] so we have a mathematical model of what needs to be done by developing the dual of Quillen's theory and linking it with quantum field theory. However, implementing it in physics is a huge project which requires a lot of work. As a starter, with Ali Chamseddine and Walter van Suijlekom, we are thinking about an interpretation of the spectral action as entropy using second quantised fermions [58].

Would this second quantisation be something that would hold at the Planck scale?

There is a similar example in spirit in number theory where we have a system of quantum statistical mechanics and where phase transitions give, by analogy, a possible scenario as explained in my book with Matilde Marcolli [59]. The number theory system (called the BC-system) has equilibrium states at various temperatures, and above a certain temperature, there is a unique equilibrium state that is extremely chaotic, but when the temperature lowers there is spontaneous symmetry breaking and the equilibrium states become number theoretically meaningful. In fact, it even happens in variants of the above BC system that at higher temperatures there is no equilibrium state at all. So in the context of physics and the standard model, we expect to have a model in which above a certain temperature (of the order of the Planck size) there would be no equilibrium state, so no geometry at all. Then for lower temperatures there is spontaneous symmetry breaking, which is also what happens in the number theory model. The phases of the system, in the case of physics, would be parameterised by actual geometries. The idea is that the mechanism of electroweak symmetry breaking, which is explained in the physics literature – for instance in the physics reports by Sher using the KMS condition, etc. [60] – has an analogue in the case of emergence of geometry. It remains an open question whether the Yukawa couplings are contextual or fundamental. I have absolutely no idea about that but I think this is the most likely scenario.

In one of your papers, it has been said that the volume is quantised [22]. Is this a prediction?

Yes, the volume is quantised, but the quantum of volume is of the Planck size and is thus so tiny that there is no way to detect the distinction with the continuum. This is rather obvious (laughs). Suppose you look at a holomorphic function and you take its restriction to a circle of radius R. Now if there is no zero of the function on the circle of radius R you get a winding number and this winding number counts how many zeros are inside the circle. This is exactly similar to the winding number that governs our quantisation of the volume because the volume is given by a degree in our case. However, what you find is that when the radius increases, when you cross a radius with a zero, you actually add to the quantum number and the winding number increases by an integer. In a similar way, I believe that you can have springing of geometry which appears from quanta via an exponentiation, or e-folding, of geometry. These are all just bits and pieces of the bigger picture and putting them together is for very far in the future. I have thought mainly about three things, namely, operator algebra which was my first work [61], renormalisation and finally the standard model and geometry but that's all.

In the formulation of non-commutative geometry, there is always a Hilbert space. Is that necessary?

Yeah. What emerges from experiments in the quantum domain is that the basic primary data is spectral. If you speak to people who perform experiments in quantum optics, you will realise that they know Hilbert space very well: they work there, it's their home. This framework is more primary than histories on a white page of space. Our mind is not used to that yet but this is nature and we have to adapt.

But I was just wondering if you don't have to give up on the notion of Hilbert space and unitarity due to the problem of time?

No, it's the exact opposite. Von Neumann not only devised the formalism of quantum mechanics but he also studied the problem of subsystems. So in other words, allowing all operators in Hilbert space corresponds to the full knowledge of a certain quantum system. But Von Neumann tackled the problem of what happens if you only have a subsystem. At first he thought about the Hilbert space factoring into a tensor product of two Hilbert spaces and about considering only operators which are only acting in one of the Hilbert spaces. So, with his collaborator Murray, he looked at algebras of operators which have the same properties as operators acting in one of the two Hilbert spaces and he called them *factors* because in his mind they corresponded to a factorisation of the Hilbert space into a tensor product. From there they devised the theory of factors and to their great surprise they found that there were other factorisations that couldn't arise from the Hilbert space factorisation. These are called *non-type I factors*. In my thesis, in 1972, I discovered that the evolution which was associated to states by Tomita-Takesaki was in fact a property of the factor itself [61]. The reason is that you didn't need to know which states were there as they were all the same, modulo inner automorphisms which are always present in the non-commutative case. This allowed me to compute the invariants that I had defined before and to reduce the classification of type III factors to that of type II (those which have a trace) and their automorphisms. I had divided the type III into subclasses III_λ where λ is a real number between 0 and 1 and my reduction was valid except in the case $\lambda = 1$ which was solved later by Takesaki. So after my discovery of the crucial uniqueness modulo inner automorphisms I had the idea that the fact that you are only dealing with a subsystem actually generates time. This is at the origin of what I was explaining about quantum randomness generating time, for which there is a concrete mathematical basis. At some point I collaborated with Carlo Rovelli on this [62]. The meaning of this is much more fundamental; the meaning of it is that probably the quantum is not only at the origin of variability which is more primitive than the passing of time but, even much stronger than that, I believe that entanglement is in fact a synchronisation of clocks at the level of randomness. So this is an idea which I want to pursue. I believe that putting these pieces together is for the new generations to do but they should feel free. There should be no dogmas. Non-commutative geometry is not a dogma and it shouldn't be a dogma. It's a fact of nature. Nature is like that. We have to face it.

There is also another proposal to quantise non-commutative geometry by Aastrup, Grimstrup and Nest [63–67]. Is it similar to your approach or not at all?

No, it's not completely similar. It's a fact, which I have written in my old book of 1994 [19], that the formalism of spectral triples is actually fitting well with quantum field theory. In other words, the spectral triple can be infinite-dimensional and when you look at some examples of constructed quantum field theories, they are given by spectral triples. So that's a fact. However, I don't think that their approach takes into account all the input from algebraic K-theory and all that. So I think that things are more complicated than what they have considered but it's a good idea anyway to try. I mean, nothing is forbidden. On the contrary, we should try and see what comes out.

What do you think has been the biggest breakthrough in theoretical physics in the past 30 years?

In the past 30 years ... I became really impressed by the discovery of the top quark, the discovery of neutrino mixing, of course the discovery of the Brout-Englert-Higgs particle and the discovery of gravitational waves. I wouldn't like to be these people who found gravitational waves because I mean my God, to measure something which is like the width of a hair in a distance from here to the nearest star and make sure that this is right: wow (laughs). They must have had a lot of sleepless nights. I am very impressed by all these discoveries.

But what about discoveries at the theoretical level?

Let me stretch to the last 50 years. From the theoretical point of view, what I find incredibly impressive is the success of the standard model. I find this success mind-blowing. The fact that the standard model, which was devised by Glashow, Weinberg and Salam and shown to be theoretically consistent at the beginning of the 1970s by 't Hooft (and Veltman), is actually working so well is extremely impressive. All the more now that the Brout-Englert-Higgs particle has been observed. This is a perfect instance of a theoretical work which unveils a working of nature, and so far the standard model resisted all attempts by teams of experimentalists trying to put it in default. No, there is no doubt; I know nothing comparable to this achievement at the physics level, even though asymptotic freedom or the GIM mechanism also fares pretty well.

What do you think is the role of the mathematician in modern society?

There is a very simple answer. The role of the mathematician is to create concepts. Let me take one instance, the notion of truth, in order to illustrate what I mean. We all have in mind that something is either true or false. If we attend a debate on politics or another controversial topic, we are prone to say that this guy is right and that guy is wrong and that's our way of making a judgement. Now, it turns out that probably we should be more advanced at the level of the formalisation of the idea of truth. In fact, there is a mathematical concept which has been created by A. Grothendieck which is the concept of topos and which has, thanks to contributions of F. W. Lawvere, a far more sophisticated notion of truth [68]. Technically the "truth values" form an object of the topos and this object classifies sub-objects exactly like the characteristic function (which takes values in the two point set "True, False") of a subset does in the case of the topos of sets. Thus for this simplest topos something is true or false. But as soon as you take a slightly more involved topos, such as the topos of quivers, you get a much more refined notion of truth values and in the case of quivers it involves "making mistakes, corrections, checking" as fundamental parts of the structure. From this example, you witness that mathematics is a factory of concepts which are extremely rich, which are subtle and which of course are hard to grasp by the public in general due to their mathematical precise and involved formulation. This lack of grasp by the public holds at a certain time in the history of civilisation but I believe that in later years these concepts will become common. This sophisticated notion of truth has, by the way, nothing to do with probabilities. It's a very beautiful and precise notion developed by a great genius of mathematics. So, to me, this is the role of mathematics: fabricate concepts and facilitate the process by which the public acquires them. That's it.

References

[1] A. Connes and D. Kreimer, "Hopf algebras, renormalization and noncommutative geometry," *Commun. Math. Phys.* **199** (1998) 203–242, arXiv:hep-th/9808042 [hep-th].

[2] A. Connes and D. Kreimer, "Renormalization in quantum field theory and the Riemann-Hilbert problem. 1. The Hopf algebra structure of graphs and the main theorem," *Commun. Math. Phys.* **210** (2000) 249–273, arXiv:hep-th/9912092 [hep-th].

[3] A. Connes and D. Kreimer, "Renormalization in quantum field theory and the Riemann-Hilbert problem," *JHEP* **09** (1999) 024, arXiv:hep-th/9909126 [hep-th].

[4] A. Connes and D. Kreimer, "Renormalization in quantum field theory and the Riemann-Hilbert problem. 2. The beta function, diffeomorphisms and the renormalization group," *Commun. Math. Phys.* **216** (2001) 215–241, arXiv:hep-th/0003188 [hep-th].

[5] A. Connes and D. Kreimer, "Insertion and elimination: the doubly infinite Lie algebra of Feynman graphs," *Annales Henri Poincare* **3** (2002) 411–433, arXiv:hep-th/0201157 [hep-th].

[6] S. Weinberg, *The Quantum Theory of Fields*, volume 1. Cambridge University Press, 1995.

[7] J. Schwinger, *Selected Papers on Quantum Electrodynamics*. Dover Books on Engineering and Engineering Physics. Dover Publications, 1958.

[8] D. Kreimer, "On the Hopf algebra structure of perturbative quantum field theories," *Adv. Theor. Math. Phys.* **2** (1998) 303–334, arXiv:q-alg/9707029 [q-alg].

[9] A. Connes and H. Moscovici, "Cyclic cohomology and Hopf algebras," *Letters in Mathematical Physics* **48** (1999) 97–108.

[10] A. Connes and H. Moscovici, "Cyclic cohomology and Hopf symmetry," in *Proceedings, Conference Moshe Flato: Quantization, Deformations, and Symmetries, volumes 1 and 2, Dijon, France, September 5–8, 1999*, G. Dito and D. Sternheimer, eds. *Math. Phys. Stud.* **21–22** (2000), pp. 121–148. 2000. arXiv:math/0002125 [math-oa]. http://monge.u-bourgogne.fr/gdito/cmf1999/procCMF1999/cmf1-12-connes.ps.

[11] G. D. Birkhoff, "Singular points of ordinary linear differential equations," *Trans. Amer. Math. Soc.* **10** (1909) 436–470.

[12] A. Grothendieck, "Sur la classification des fibres holomorphes sur la sphere de Riemann," *American Journal of Mathematics* **79** no. 1 (1957) 121–138.

[13] A. Connes and M. Marcolli, "Renormalization and motivic Galois theory," arXiv:math/0409306 [math-nt].

[14] A. Connes and M. Marcolli, "From physics to number theory via noncommutative geometry. II. Renormalization, the Riemann-Hilbert correspondence, and motivic Galois theory," arXiv:hep-th/0411114 [hep-th].

[15] A. Connes and M. Marcolli, "Quantum fields and motives," *J. Geom. Phys.* **56** (2006) 55–85, arXiv:hep-th/0504085 [hep-th].

[16] P. Higgs, "Broken symmetries, massless particles and gauge fields," *Phys. Letts.* **12** no. 2 (1964) 132–133. www.sciencedirect.com/science/article/pii/0031916364911369.

[17] F. Englert and R. Brout, "Broken symmetry and the mass of gauge vector mesons," *Phys. Rev. Lett.* **13** (Aug. 1964) 321–323. https://link.aps.org/doi/10.1103/PhysRevLett.13.321.

[18] A. Connes and J. Lott, "Particle models and noncommutative geometry (expanded version)," *Nucl. Phys. Proc. Suppl.* **18B** (1991) 29–47.

[19] A. Connes, *Noncommutative Geometry*. Elsevier Science, 1995.

[20] A. Connes, "Noncommutative geometry and reality," *J. Math. Phys.* **36** (1995) 6194–6231.

[21] A. Connes, "Gravity coupled with matter and foundation of noncommutative geometry," *Commun. Math. Phys.* **182** (1996) 155–176, arXiv:hep-th/9603053.

[22] A. H. Chamseddine, A. Connes and V. Mukhanov, "Quanta of geometry: noncommutative aspects," *Phys. Rev. Lett.* **114** no. 9 (2015) 091302, arXiv:1409.2471 [hep-th].

[23] A. H. Chamseddine, A. Connes and V. Mukhanov, "Geometry and the quantum: basics," *JHEP* **12** (2014) 098, arXiv:1411.0977 [hep-th].

[24] A. Connes, "Geometry and the quantum," arXiv:1703.02470 [hep-th].

[25] A. H. Chamseddine and A. Connes, "The spectral action principle," *Commun. Math. Phys.* **186** (1997) 731–750, arXiv:hep-th/9606001 [hep-th].

[26] A. H. Chamseddine and A. Connes, "Universal formula for noncommutative geometry actions: unification of gravity and the standard model," *Phys. Rev. Lett.* **77** (1996) 4868–4871, arXiv:hep-th/9606056 [hep-th].

[27] A. H. Chamseddine and A. Connes, "Scale invariance in the spectral action," *J. Math. Phys.* **47** (2006) 063504, arXiv:hep-th/0512169 [hep-th].

[28] A. Connes and A. H. Chamseddine, "Inner fluctuations of the spectral action," *J. Geom. Phys.* **57** (2006) 1–21, arXiv:hep-th/0605011 [hep-th].

[29] A. H. Chamseddine, A. Connes and M. Marcolli, "Gravity and the standard model with neutrino mixing," *Adv. Theor. Math. Phys.* **11** no. 6 (2007) 991–1089, arXiv:hep-th/0610241 [hep-th].

[30] A. H. Chamseddine and A. Connes, "Why the standard model," *J. Geom. Phys.* **58** (2008) 38–47, arXiv:0706.3688 [hep-th].

[31] A. H. Chamseddine and A. Connes, "Conceptual explanation for the algebra in the noncommutative approach to the standard model," *Phys. Rev. Lett.* **99** (2007) 191601, arXiv:0706.3690 [hep-th].

[32] N. Cabibbo, "Unitary symmetry and leptonic decays," *Phys. Rev. Lett.* **10** (Jun. 1963) 531–533. https://link.aps.org/doi/10.1103/PhysRevLett.10.531.

[33] M. Kobayashi and T. Maskawa, "CP-violation in the renormalizable theory of weak interaction," *Progress of Theoretical Physics* **49** no. 2 (Feb. 1973) 652–657, http://oup.prod.sis.lan/ptp/article-pdf/49/2/652/5257692/49-2-652.pdf. https://doi.org/10.1143/PTP.49.652.

[34] Tevatron New Phenomena Higgs Working Group, CDF, D0 Collaboration, G. Bernardi, V. Buescher, W. Fisher, J.-F. Grivaz, M. Herndon, T. Junk, M. Kruse, S. Mrenna and W.-M. Yao, "Combined CDF and D0 upper limits on standard model Higgs boson production at high mass (155–200 GeV/c^2) with 3 fb^{-1} of data," in *Proceedings, 34th International Conference on High Energy Physics (ICHEP 2008): Philadelphia, Pennsylvania, July 30–August 5, 2008*. 2008. arXiv:0808.0534 [hep-ex]. www.interactions.org/cms/?pid=1026564.

[35] J. Elias-Miro, J. R. Espinosa, G. F. Giudice, H. M. Lee and A. Strumia, "Stabilization of the electroweak vacuum by a scalar threshold effect," *JHEP* **06** (2012) 031, arXiv:1203.0237 [hep-ph].

[36] G. Degrassi, S. Di Vita, J. Elias-Miro, J. R. Espinosa, G. F. Giudice, G. Isidori and A. Strumia, "Higgs mass and vacuum stability in the standard model at NNLO," *JHEP* **08** (2012) 098, arXiv:1205.6497 [hep-ph].

[37] C.-S. Chen and Y. Tang, "Vacuum stability, neutrinos, and dark matter," *JHEP* **04** (2012) 019, arXiv:1202.5717 [hep-ph].

[38] O. Lebedev, "On stability of the electroweak vacuum and the Higgs portal," *Eur. Phys. J. C* **72** (2012) 2058, arXiv:1203.0156 [hep-ph].

[39] F. Bezrukov, M. Yu. Kalmykov, B. A. Kniehl and M. Shaposhnikov, "Higgs boson mass and new physics," *JHEP* **10** (2012) 140, arXiv:1205.2893 [hep-ph]. [275(2012)].

[40] A. H. Chamseddine and A. Connes, "Noncommutative geometry as a framework for unification of all fundamental interactions including gravity. Part I," *Fortsch. Phys.* **58** (2010) 553–600, arXiv:1004.0464 [hep-th].

[41] A. H. Chamseddine and A. Connes, "Resilience of the spectral standard model," *JHEP* **09** (2012) 104, arXiv:1208.1030 [hep-ph].

[42] A. H. Chamseddine, A. Connes and W. D. van Suijlekom, "Inner fluctuations in noncommutative geometry without the first order condition," *J. Geom. Phys.* **73** (2013) 222–234, arXiv:1304.7583 [math-ph].

[43] A. H. Chamseddine, A. Connes and W. D. van Suijlekom, "Beyond the spectral standard model: emergence of Pati-Salam unification," *JHEP* **11** (2013) 132, arXiv:1304.8050 [hep-th].

[44] A. H. Chamseddine, A. Connes and W. D. van Suijlekom, "Grand unification in the spectral Pati-Salam model," *JHEP* **11** (2015) 011, arXiv:1507.08161 [hep-ph].

[45] V. D. Barger, E. Ma and K. Whisnant, "General analysis of a possible second weak neutral current in gauge models," *Phys. Rev. D* **26** (1982) 2378.

[46] D. Chang, R. N. Mohapatra and M. K. Parida, "A new approach to left-right symmetry breaking in unified gauge theories," *Phys. Rev. D* **30** (1984) 1052.

[47] V. Elias, "Coupling constant renormalization in unified gauge theories containing the Pati-Salam model," *Phys. Rev. D* **14** (1976) 1896.

[48] V. Elias, "Gauge coupling constant magnitudes in the Pati-Salam model," *Phys. Rev. D* **16** (1977) 1586.

[49] Super-Kamiokande Collaboration, Y. Fukuda et al., "Evidence for oscillation of atmospheric neutrinos," *Phys. Rev. Lett.* **81** (1998) 1562–1567, arXiv:hep-ex/9807003 [hep-ex].

[50] Super-Kamiokande Collaboration, S. Fukuda et al., "Solar B-8 and hep neutrino measurements from 1258 days of Super-Kamiokande data," *Phys. Rev. Lett.* **86** (2001) 5651–5655, arXiv:hep-ex/0103032 [hep-ex].

[51] D. Bouwmeester, J.-W. Pan, K. Mattle, M. Eibl, H. Weinfurter and A. Zeilinger, "Experimental quantum teleportation," *Nature* **390** no. 6660 (Dec. 1997) 575–579, arXiv:1901.11004 [quant-ph].

[52] J. A. Wheeler, "The past and the delayed-choice double-slit experiment," in *Mathematical Foundations of Quantum Theory*, A. R. Marlow, ed. Academic Press, 1978.

[53] S. Chandrasekhar, "The maximum mass of ideal white dwarfs," *Astrophysics Journal* **74** (July 1931) 81.

[54] M. F. Atiyah and I. M. Singer, "The index of elliptic operators on compact manifolds," *Bull. Amer. Math. Soc.* **69** no. 3 (May 1963) 422–433. https://projecteuclid.org:443/euclid.bams/1183525276.

[55] M. F. Atiyah, "Global theory of elliptic operators," in *Proc. Internat. Conf. on Functional Analysis and Related Topics*, pp. 21–30. University of Tokyo Press, 1970.

[56] G. Kasparov, "Topological invariants of elliptic operators. I: K-homology," *Izv. Akad. Nauk SSSR Ser. Mat.* **39** (1975) 796.

[57] D. Quillen, "Higher algebraic K-theory: I," in *Higher K-Theories*, H. Bass, ed., pp. 85–147. Springer, 1973.

[58] A. H. Chamseddine, A. Connes and W. D. Van Suijlekom, "Entropy and the spectral action," *Commun. Math. Phys.* **373** no. 2, (2019) 457–471, arXiv:1809.02944 [hep-th].

[59] A. Connes and M. Marcolli, *Noncommutative Geometry, Quantum Fields and Motives*. American Mathematical Society, 2007.

[60] M. Sher, "Electroweak Higgs potentials and vacuum stability," *Phys. Rept.* **179** (1989) 273–418.

[61] A. Connes, "Une classification des facteurs de type III," *Ann. Sci. Ecole Normale Superieure* **6** (1973) 133–252.

[62] A. Connes and C. Rovelli, "Von Neumann algebra automorphisms and time thermodynamics relation in general covariant quantum theories," *Class. Quant. Grav.* **11** (1994) 2899–2918, arXiv:gr-qc/9406019 [gr-qc].

[63] J. Aastrup, J. M. Grimstrup and R. Nest, "On spectral triples in quantum gravity I," *Class. Quant. Grav.* **26** (2009) 065011, arXiv:0802.1783 [hep-th].

[64] J. Aastrup, J. M. Grimstrup and R. Nest, "On spectral triples in quantum gravity II," *J. Noncommut. Geom.* **3** (2009) 47–81, arXiv:0802.1784 [hep-th].

[65] J. Aastrup, J. M. Grimstrup and R. Nest, "A new spectral triple over a space of connections," *Commun. Math. Phys.* **290** (2009) 389–398, arXiv:0807.3664 [hep-th].

[66] J. Aastrup, J. M. Grimstrup and R. Nest, "Holonomy loops, spectral triples & quantum gravity," *Class. Quant. Grav.* **26** (2009) 165001, arXiv:0902.4191 [hep-th].

[67] J. Aastrup, J. M. Grimstrup, M. Paschke and R. Nest, "On semi-classical states of quantum gravity and noncommutative geometry," *Commun. Math. Phys.* **302** (2011) 675–696, arXiv:0907.5510 [hep-th].

[68] S. MacLane and I. Moerdijk, *Sheaves in Geometry and Logic: A First Introduction to Topos Theory*. Universitext. Springer, 1994.

7

Robbert Dijkgraaf

Director of the Institute for Advanced Study and Leon Levy Professor, Princeton, N.J.

Date: 24 June 2011. Location: Amsterdam. Last edit: 14 August 2020

From your point of view, what are the main problems in theoretical physics at the moment?

I think this is actually a very difficult question to answer, because it is hard to know what you don't know. Therefore, any answer will reflect one's current state of mind. I think that we still lack a solid foundation of the fundamental laws of nature. We have parts of that description that are very successful, but finding a complete description is still a major outstanding problem. Cosmology is an integral part of this complete story, since you want not only the equations that govern the universe, but also to understand the specific solution that describes our universe.

Which parts of this fundamental description are we missing?

We basically think of the universe as a big dynamical system and we just want to find the right system describing the correct physics. I feel that this point of view might be too conventional and insufficient for truly understanding nature. Somehow, we are still married to the Newtonian model and constantly trying to find some kind of infinitesimal description of how the laws evolve or the physical states evolve. This is the view of science answering the question "What is next?" but we should also answer the "Why" question: Why this universe? Why these laws?

Another fundamental aspect of physics that we are struggling with is that we have a very successful model for many different type of particles – the standard model – but that happens to capture just the particles that we have found so far up to a certain energy level. It would be rather a sign of hubris to assume that the energy level that we have reached with the LHC right now would be enough to adequately describe all matter. Clearly, there must be many new particles and physical phenomena beyond what we know at the moment.

Is string theory this complete description that you have alluded to so far?

In some sense the most positive thing about string theory is that it's incomplete. I think that everyone who is working in string theory has the feeling that, while we are certain that some pieces of the puzzle clearly are here to stay, there are still many parts that are missing or that we don't understand. At the same time, it's impossible to disconnect string theory from the physical world because it's so strongly connected already. Personally, I feel that string theory as it stands now only describes a few parts of the puzzle, but whatever we find at the end of the day will for sure contain string theory. However, that might only be a small part of it, perhaps much smaller than we think, as I could imagine that there are many new ingredients that come into play.

We know that the parts of string theory that we best understand are the parts which are not visible in this universe and so in some sense these are either different universes or they are corners of our own universe that we don't yet see. The description of string theory as we teach in class is valid. But how to interpret that description in our world, in our universe, is not easy. If you make a puzzle you always know that the easy parts are the ones on the edges and the corners. The difficult pieces are in the middle. Within string theory, that's where we started, such as in 10-dimensional strings, all kinds of nice supersymmetric limits, which are the easy pieces. However, I think that our current observable universe is probably right in the middle of the puzzle, far away from the edges and corners. Therefore, to which extent we can in the end find a fundamental description of nature and how direct the path is from that description to the kind of string theory models that we have right now, I don't know. My feeling is that it could be a much longer path than we think it is.

You mentioned that string theory is already strongly connected to the physical world ... which parts are you referring to?

Because of all the different types of dualities that we have found in string theory, we know that basically any gauge theory can be reformulated as a particular background in string theory [1–6]. So we know that, in some sense, the gauge theories describing the weak force or the strong interactions can be described by some exotic state in string theory [7–9]. We also roughly know that all string theories are connected to each other. Thus there is this very complicated map of the world that contains all possible quantum field theories (string theories) and they are basically all connected. It's a very complicated map but I don't think the map splits into continents. The standard model is part of that big map of theories but ...

But has the standard model been exactly derived from string theory?

No, it hasn't, because we do know that you have to take various limits, etc., which from the point of view of string theory means that it is a strongly interacting theory, which makes it a very hard task [10, 11].

Okay, so it's more like you have the intuition that it must be there but ...

Well, I would say yes because of course string theory is a theory with gravity, so you have to take various limits where you remove gravity, etc. But I think certainly at the level of effective field theories, the feeling is that it would be very surprising if not all theories were connected via dualities and transitions.

Why is gravity difficult to quantise?

It's difficult because quantising gravity is not about quantising a particular physical force or field; it's about quantising space and time itself. The whole concept of quantisation works well under the assumption of a fixed spacetime background. Space in itself is not the biggest problem, since we have various models where we know that spatial coordinates can arise as effective descriptions from a more fundamental underlying variable. Such is the case of the simplest possible matrix models where space emerges in some kind of limit procedure taking the rank of the matrix to be very large [12–14]. On the other hand, an emergent time coordinate is much harder to imagine. In quantum mechanics, time appears in a very specific way and is not quantised itself. If you would literally quantise general relativity, you know that the dynamical fields are the spacetime coordinates, which includes the time coordinate and that is a fundamental problem in physics.

This is a conceptually very difficult problem. We are good at quantising things that live in spacetime, such as particles, fields, strings, branes, but actually to quantise the spacetime itself, that is almost paradoxical. However big the revolution of quantum mechanics was, it is still a conservative theory because it follows the Newtonian thinking in which you are given the state of the world at a fixed time and then predict how the world evolves, whether it evolves as a dynamical system or whether it evolves as a quantum mechanical wave function. Of course, it's a very big step to go from a classical phase space to a wave function, but you are still following something through time. Now, requiring time itself to be a quantum variable makes this setup difficult to imagine. This is a problem that we haven't solved in string theory. What we have done in string theory is just to find some kind of perturbative way to treat these small variations in the spacetime background as a nicely controlled quantum system. We found a way to quantise the graviton but quantising the graviton is only a small part of the full structure. We do not have a good framework for quantisation in the sense I've just described and that is one of the fundamental reasons why quantising gravity is difficult.

What is the validity of this approach? Can one probe distances at the Planck scale?

To understand physics at such small distances, we know that we basically have to find the right probes to measure them. You would need to build a microscope that can look at Planck scales but you know that if you make it from strings, you are certainly making it out of fuzzy objects. Thus, it's very difficult to know how things look at those scales. I think that a lesson we learnt from string theory is that looking at different scales requires different objects such as strings and branes [15]. The old-fashioned way of thinking that we can decompose the world into mathematical points and use infinitesimal objects for probing spacetime is not suitable. Quantum gravity tells us that space and time lose their meaning at such small scales. Essentially a black hole will form that destroys spacetime [16].

String theory tells us that we can use many objects to probe spacetime, scatter them and reconstruct what the image is. However, when using strings and branes, we also know that the process becomes more complicated because these objects begin to interact. In fact, even the original question of what spacetime looks like at the Planck scale becomes ill-defined because you first have to tell me through which lens (branes, strings, etc.) I should look at it. You may be inclined to think of the world as a kind of spacetime foam in which the underlying concept of quantum gravity is a space of fluctuating metrics that you must quantise. However, this is clearly not happening in string theory because though we don't really know what our fundamental degrees of freedom are, we definitely know that they are not metrics.

Why do you know that they are not metrics?

Because in string theory there is no model or description that I know of that starts from the path integral of all possible spacetime metrics. The metric appears as some kind of effective object. If you start from the string action, the metric appears as a coupling constant on the worldsheet. If you start from a brane point of view, for instance a non-Abelian gauge theory, the metric itself might be a collective phenomenon. The notion of a spacetime foam entails that we are thinking of spacetime fluctuating at the Planck scale and giving rise to small ripples. However, this is not what is happening in string theory because if you zoom in you no longer have a concept of space and time.

Is this along the lines of what Erik Verlinde has been suggesting [17, 18], namely that spacetime is an emergent phenomenon?

Not really, I mean, this idea of emergent spacetime has been there since the very beginning of string theory. The formulation of string theory never starts with a spacetime metric that is dynamical. In some sense, it is always a derived object. The fundamental variables of string theory do not include the spacetime metric as an ingredient. These fundamental parts may include a large-N gauge field or a specific matrix or some kind of small fluctuating loops, but never a quantised metric. So, the metric is always an effective object. It's something that you measure by putting a certain probe into the system. In that sense spacetime is very similar to thermodynamical quantities and of course that connection has a long history. The relation between thermodynamics and general relativity has been driving part of the field for a very long time, starting with the work of Jacob Bekenstein and Stephen Hawking on black holes [19, 20].

So do you think it's a misconception trying to quantise metrics directly?

I think that is true, yes.

Isn't this what loop quantum gravity tries to do?

It tries to do it but of course also from a different angle in the sense that it does replace the metric field with another field, some associated set of gauge fields. I am a strong believer that the metric is an emergent phenomenon. Although nobody knows how it will work with the time coordinate exactly, space is certainly an effective phenomenon. I was already convinced that this was the case from the very beginning, especially in light of all the dualities between branes, matrices and strings [12–14]. This fits very well with another lesson that string theory also has given us, namely, that there is some kind of fundamental length for spacetime, that is, there is no notion of probing increasingly small regions of spacetime. There is some structure at a given small but finite scale. It is similar to what we learn from the material world, right? A picture is made out of individual pixels, a material is made out of molecules and spacetime is made out of something else.

What could be that something else?

That's the big question and string theory has some limiting cases where we actually know what that is. AdS/CFT is a famous example [5]. In that case, string theory in anti–de Sitter spacetime is described in terms of a large-N gauge theory. In that sense, the large-N gauge fields are the fundamental constituents. Matrix models [12–14] are another good example where we know that a large-N matrix constitutes the fundamental degrees of freedom. You can always ask what is the fundamental description and where it comes from; however, you can see in all these examples that spacetime is something that emerges if you have a quantum system with many degrees of freedom.

Do you think that it's important to try to develop different theories of quantum gravity or one should just stick to string theory?

In some sense I actually feel that all these approaches are intimately connected. I can sit here and say, "Well, quantum gravity is not a theory of fluctuating metrics" but, of course, in some approximation it better be like that. For very small fluctuations around a classical metric, we are able to quantise them and if I do this procedure in a complicated way, I might even recover this effective model. I feel that this is the kind of beauty of this subject, in particular that we're trying to approach it from various directions and I don't think they are necessarily mutually exclusive. One of the amazing things when you learn quantum field theory is that there are all these equivalences, which is really, really bizarre. There seems to be one big complicated theory and if you push it to a certain corner then it suggests that its fundamental constituents are of a certain nature [6]. But if you go to another corner, it suggests other fundamental constituents. What I learnt from all this is that neither of these constituents is fundamental because if you have two different sets of what you call fundamental constituents then you already have a problem (laughs).

This suggests that there can be different ways of thinking about the same problem. Even in our physical world, we have examples of this kind. For instance, we can think about QCD in terms of quarks and gluons and it's a very nice way to think about it, but of course it is only useful at very high energies in which quarks are free. At very low energies, it's a completely useless description. So if people say, "Well, what are the fundamental degrees of freedom?" you can even argue about this because nobody has given a mathematical proof of this theory yet. Hence we don't even know if fully understanding quarks and gluons is enough to make sense of QCD.

We know many examples now of gauge theories more complicated than QCD, where even the notion of the gauge group is ambiguous because you can take two limits of the theory and in one limit one you see that it is an $SU(4)$ gauge theory and in another limit it is an $SU(5)$ [21]. So is it $SU(4)$ or $SU(5)$? Apparently, the answer depends on which question you ask of the system. I think this is the most fundamental problem we have: we try to understand this world of fundamental physics, and what we have is not a single framework, but more like an atlas with various maps with a complicated set of instructions for how to go from one map to the other. In order to do so, you have to completely relabel all your fields. Of course, the big questions is "Is there something like a globe?" Can I take the perspective of the moon where I can see the whole theory at once?

Do you believe it's possible to have this perspective?

We don't know. That's of course the most amazing thing. I think it has been an enormous intellectual jump from, say, the 1970s and 1980s, where we really thought that a gauge theory is a fundamental theory (i.e., we know what it is, it has a gauge field, a gauge group, some matter fields, you have some rules and you quantise it) to nowadays where we say, "Well, you have two theories which are totally equivalent, but one looks like it's made out of one gauge group and the other of another gauge group." Apparently, even the gauge group is not a fundamental physical concept. We cannot really see the gauge group; we can only see the invariant excitations, hadrons or glueballs and we can compute their scattering. Asking what are the fundamental degrees of freedom depends on what you think are the questions you want to ask in a certain regime.

We are developing a different way of thinking about these physical theories which is more democratic. However, it's also a little bit disorienting because you feel that if all these beautiful descriptions are not uniquely fundamental, what is? Is there anything? That's the deep question we are facing and I think we need an enormous conceptual breakthrough to answer it. My gut feeling is that the answer will be completely different. At the moment, I feel we're a little bit like describing the universe before we understood the concept of molecules or atoms or something similar. We see all these different manifestations and we somehow see that we have chemical reactions transforming some type of matter into another type. But how is this possible? We should be able to say that it's just a rearrangement of atoms, if we can take the modern chemists' point of view, but that is precisely what we're lacking.

You mentioned that there is no mathematical proof of QCD?

There's no mathematical proof. One of the seven millennium problems is to find a rigorous mathematical definition of Yang–Mills theory – just pure Yang–Mills with general gauge group. To make it rigorous to the point that a mathematician understands what you're talking about. Everybody feels it should be possible, but there's no proof and I am not sure how many are actually working on this problem. It is very hard.

In which sense is the theory ill-defined?

Well, the expectation is that asymptotically free theories should be well defined because at short distances they become weakly interacting. You can place the theory on a lattice (and on a computer) and it looks all very nice. Okay but then just define the theory in the continuum limit. You can get one million dollars if you can find a good definition. There's a reward, but it's very difficult (laughs).

But what's missing there?

In some sense, a four-dimensional path integral is an ugly object. It's not very clear that the very definition of the theory using the path integral makes mathematical sense. This could be a technical problem, in the sense that you just need to find some smart functional analysis to deal with it, or it could be more fundamental. In a way, I even hope it's a more fundamental issue, because that would actually mean that perhaps even the idea of a gauge theory and a gauge symmetry could be only an effective object, like the metric and spacetime in quantum gravity.

Now a different kind of question ... if you have some PhD student asking you which quantum gravity theory one should work on, what would you say?

Well, as I said they are all connected. So, you can start almost anywhere. Mathematicians like to say, "It doesn't matter where you start digging; if you just dig deep enough, you'll find something interesting." I think, in fact, with quantum gravity we are now at that stage. There are so many different approaches, different models, and I think that if you start digging, take any of them, push it further, further and further, you'll actually come closer to the general answer. It's like excavating a huge buried temple complex. So start anywhere you like, but pursue it to the very end.

Do you think that because so much beautiful mathematics has come out of string theory that it's worthwhile pursing it even though it might not be the theory of quantum gravity?

Well, first of all, as I said, I see string theory including quantum gravity intimately connected to various quantum field theories and therefore to the real world. Of course, there's another story emerging. I feel that there's this enormous zoo of quantum theories which mathematically has been hardly explored. If you go through all the mathematical textbooks, how many of these books dig deep into this quantum world? These theories are so incredibly rich. We took a few elements out and made them mathematically nice. But there's a whole bunch of such elements. So I see it as the biggest reservoir of deep mathematical facts and ideas, which I think will all be extremely useful. We haven't explored this at all because it's so difficult. However, in mathematics I think it's the exploration of quantum physics that has been the most exciting story of the last 30 years by far.

Several people have said in the past that string theory must be right because it's so beautiful. Do you think this aesthetic reasoning is good source of guidance?

Here the point is what do you exactly mean by "aesthetic" and whom do you ask, right? Perhaps it's most fair to ask these questions to people outside the field. Mathematicians, for instance, really think that there are beautiful things coming out of string theory. That's remarkable because actually in general when string theorists say that it's a beautiful theory, they don't necessarily see the contact with deep mathematics. I think that the beauty of string theory is its interconnectedness and the deep underlying concepts. So it's not so much, oh, I get the monster group out, or an automorphic form, or E_8 [22, 23]. Or string theory is beautiful because E_8 is beautiful. Some people think E_8 is beautiful but only in a bizarre sense, that is, in the sense that it is a big exception. Why would the universe be described by the big exception? I think it's actually pretty bizarre, right (laughs)? In fact, I don't specifically like E_8 myself. I think in some sense the much more generic Lie groups $SU(N)$ are just as beautiful.

The remarkable thing of string theory is that it contains both the $SU(N)$s, which are like the bread and butter of gauge groups, and E_8, the exception to the rule. It's all there (laughs). So for me, the true beauty of string theory is the way it connects to all these different subjects, even complementary forms of aesthetics. I could give you a list of 100 mathematic subjects which on the surface appear to be completely disjoint and ask for a mathematics textbook where all these 100 mathematical subjects appear. If it's a mathematician who doesn't know anything about physics, they would say it doesn't exist. There's no such thing that connects automorphic forms, the Langlands program [24], E_8, thermodynamics, differential geometry, all these subjects. But then I say, no, no, no, just start to think about how to quantise wiggling pieces of strings and then you'll connect all of these subjects. That's pretty magical! So I think that's where the real beauty is. It's not a beauty that indicates that we're close to the truth but that reflects the strength and depth of the subject. With every step forwards this beauty evolves and matures. In fact, Feynman described this nicely. He said that if you have a good theory often you find it to be beautiful. But then something like an experiment or a problem or some other addition comes, modifies it and makes it ugly again. However, in the next phase, when a new framework emerges, it starts to become beautiful again, but in a deeper, extended way. Also in science, beauty is an acquired taste.

We have gone through these transitions in physics where we at first instance seem to have lost beauty. There's enormous beauty in a Newtonian universe, in classical mechanics. We have lost parts of that because of the indeterminacy of quantum mechanics, where experiments only allow us to state that we have, say, a 57% chance of this outcome happening. You somehow miss some very essential part of what you thought physical theories were all about: determinism. So, in general I think that "coming closer to the truth" also means that you have to give up on certain aspects of conventional beauty and then, if you're lucky, you get rewarded with something new and exciting in return. In many ways it's a bit like in arts, where you have various phases in art history. I don't think that twentieth-century art is intrinsically more beautiful than nineteenth-century or seventeenth-century art. It's however a different kind of beauty, with more expressive power. In physics the same evolution took place. There were many theories which were very beautiful but which were superseded by something else. In that sense, I think that the fact that "string theory is beautiful and therefore it has to be true" is kind of a silly argument.

One of the big criticisms of string theory is that it has no predictive power. Do you think this is reasonable?

Yes, although often it is not so much a criticism of the theory, as more of a criticism of the theorists who worked on it and the claims that were made in the past. Clearly, in the beginning people were thinking that we were very close to very precise predictions. I can understand that excitement, because at that time people felt there's only a very small set of possible backgrounds, perhaps one or two Calabi-Yau spaces. It is always hard to know what you don't know. Now we are aware of the enormous set of possible backgrounds [11], and the expectation of making concrete predictions has become more modest and realistic. In that sense, I feel that currently we are probing the world of fundamental theories rather than actually zooming in on one particular model that gives strong predictions about the universe. From this point of view, we are in a very different game compared to what has been a more common path in high-energy physics, where people try to predict new particles and new interactions.

In this respect, the lack of predictability is definitely a weakness. But then I also feel that the nature of the problems we are addressing, namely a fundamental description of nature, allows us to learn a lot in that process. We're like explorers who think they have landed on an island and want to find the city of gold. But the island turns out to be a complete continent and now we say, let's first map out that continent because we want to understand the large structures here, what is the basic geography, the ecology, etc. In many ways we are only starting to imagine the magnitude of seeking a fundamental theory of nature, the wide range of questions such a theory should answer, from the structure of black holes and the origin of the big bang to the structure of matter and radiation. This means that we have to manage the expectations of our colleagues and of the world at large. We are not about to discover the secret of the universe or something like that. It's a much longer story.

Regarding these different backgrounds or vacua you can choose, do you think that it's reasonable to use anthropic arguments [25] to pick one of them?

I think it's very early to consider such arguments. There are two things to note about anthropic arguments. First, there have been many instances in the history of physics where we could have tried to apply it, but in retrospect we are very happy we did not. But there are other situations in the history of physics where we did apply it and we're very happy we did. For instance, take the periodic system of elements. Why are there all these specific atoms with specific properties? That is a very good question. It has a beautiful bottom-up explanation in terms of quantum mechanics. But, when trying to understand the solar system, people wondered for a very long time why there was a specific number of planets. It seemed an equally good question. But now we know the only correct answer is that there are all these stellar systems in the universe and we just happen to live in one of them. In this particular example we placed the question in an anthropic context. What this teaches us is that the question we were trying to ask, namely why this specific number of planets, was not a good question to begin with. Nobody is currently interested in this question. We have to try to see the bigger picture and ask more fundamental questions, like the shape of the orbits or the formation of planets. So, I think one thing that we have learned time and time again in physics is that the context of a question is often much bigger and complicated and has more ingredients than we thought it had. And this will undoubtedly happen again.

Do you think that nevertheless it will be possible to find or reformulate string theory in such a way that all fundamental constants will be fixed automatically?

It could be possible. It would certainly be the best outcome, like in quantum chromodynamics, a theory without constants, where all properties of hadrons follow from the internal dynamics. Again, we have to answer bigger questions because we don't only want to know all the fundamental constants of nature, we don't only want to know what is our dynamical system, but we also want to know the specific solution to that dynamical system. Nobody knows of a good example where physics answers both at the same time. Perhaps these two questions could even be connected in some way. So we could still be in a situation that we're doing phenomenology of nuclear physics without knowing there is a theory like QCD underlying it with no variable constants.

One of the issues of string theory that people discuss is the fact that it is not background-independent. Does that mean that we have to reformulate the entirety of string theory as a topological string theory [26–31]?

No, no, no. Background independence is there at the fundamental level clearly. Picking a background should be just like picking a state in the theory. In some sense there should be a description that gives us in a natural way all possible Calabi-Yau manifolds, etc., and as I said we know that for sure this is not a theory that has a fluctuating metric or a fluctuating topology or something. So, no.

What is special about topological string theory is that certain quantities do not depend on the metric but depend on a much smaller set of variables. Roughly, in the physical theory you have a Calabi-Yau metric, you have a Kähler structure and the complex structure, and topological string theory only depends on half of them. It's a kind of an intermediate step. Background dependence is not the same thing as *there is no space of backgrounds*, right? Suppose you have the fundamental description of string theory which is composed of little sorts of atoms, which we don't know what they are, but they can assemble themselves to form spaces and they can do so in very different ways. Then there's a huge phase space of solutions and all these various spaces will be backgrounds and they will depend on various structures, so it will depend on the metric etc. The nice thing about topological string theory is that in some sense the space of possible backgrounds is a little bit simpler and you can describe it in an easier way as it has a nice mathematical definition.

However, string theory by itself is not topological, in the sense that it does depend on all parts of the metric and in addition even topological string theory depends on some parts of the metric. The word *topological* is related to topological sigma models [26], which have to do with the fact that it is topological from the worldsheet point of view, but not from spacetime. In fact, if anything it's a theory that is not quantising the space of metrics but it's quantising the space of complex structures that you can put on an algebraic variety. In this sense, it is a toy model of the full string theory but shares many of the same problems with it.

But then wouldn't you say that the whole string theory had to be topological?

In some sense it is, right? So in fact, I even wrote a paper about this a long time ago. If you look at bosonic string theory you can formulate it like a topological string [32]. It's kind of a bizarre formulation because you have to include the ghost action, etc., but it has the same kind of structure. So if you look at the family of topological string theories, even ordinary string theory can be seen as part of that bigger family.

Where do you find topological string theory?

You find it in a particular corner of string theory by taking some limiting procedure. The effect of some terms in the action of superstrings is captured by topological strings.

How important is the study of topological string theory?

I think there are at least two reasons for studying it. One reason is that it allows for an exact mathematical definition. So at least you know precisely what you're computing and of course since it has a nice mathematical definition, you can also use much stronger techniques to approach it. You find yourself in an area of mathematics that has been pretty well mapped out so that we understand what we can compute, perhaps not everything, but at least we understand most of it. Additionally, another nice thing about topological strings is that, since it's also a part of a bigger story, it can give you certain clues about that bigger story. For instance, if you find a symmetry in this corner it is not guaranteed that it's a symmetry of the whole system but it makes you think. If two of these topological strings are connected then you ask, What's the story of the full thing? Might they be connected in the same way? The third aspect is that in some sense it is a simple system and like the harmonic oscillator, it is exactly solvable. These toy models are crucial to test our thinking as a stepping-stone to a fuller understanding.

But should we understand topological string theory as modelling some kind of physical system?

No, I think in a very fundamental way it is not, because we know that the physical world is much richer. However, it does capture essential elements of physical systems, like a cartoon.

I recall that in one of your papers you proved a connection between topological string theory and the topological sector of loop quantum gravity [31]. Do you think this connection also extends to string theory in general?

Topological string theory is not a model that approximates the standard model or something of that kind, as it misses many ingredients, but it is a good toy model for theoretical physics exactly because it can establish many connections to different areas. How are all these different approaches connected? You can ask this question in the context of topological strings and you're in a much better position to get answers, since you are guided by established mathematics. For instance, we know that three-dimensional Chern-Simons theory with compact gauge groups is a rigorously defined mathematical theory which is dual to a string theory. These Chern-Simons theories are in spirit closely related to some loop quantum gravity formulations. So we see that there is a gauge description which is in some sense a reformulation of gravity fields, in turn directly connected to string theory.

This type of relation is quite powerful and can teach us different lessons. For instance, if you compute some knot invariant or some three-manifold invariant and you ask what's the most fundamental way to define this, I would say that's very difficult to pick because this string theory looks very fundamental, but the gauge theory does too. It's a beautiful model where, on the one hand, we find results in terms of the Gromov-Witten invariants of symplectic geometry, while, on the other hand, we find the same results in terms of knot invariants using quantum groups and gauge theories. Thus, it's hard to pick which of these two deep mathematical areas is more fundamental, especially when both areas have been awarded Fields Medals (laughs). From this kind of analysis, I start wondering if there is even a more fundamental description that unites both. Perhaps there's not; perhaps the end of our story is an atlas composed of many maps: the field theory map, the loop map, the string map, etc. And the fact that we can ask this question in a limited arena (topological string theory) is extremely fascinating and teaches us that we have to keep an open mind and that there can always be another fundamental point of view.

There are two main models of topological string theory, models A and B. What is the difference between them and how are they related?

There are two ways in which you can make a string theory topological in the sense that it does not depend on the topology of the worldsheet [26–31]. Once this is done and you look at the spacetime, you find that this procedure removes from the theory half of the background structures. The A model is the theory that depends on the metric structure (Kähler structure) and the B model is the model that depends on the complex structure. At first sight they look as if they require two different problems to be solved. One problem is that of quantising Kähler structures and the other of quantising complex structures. However, the remarkable thing is that there is a symmetry that relates the two and thus they are equivalent.

Do you think that mirror symmetry has been the biggest mathematical achievement of string theory?

No, no, no, no (laughs). If you would list the main math applications of string theory I would say that mirror symmetry is certainly in the top 10, but there are other important achievements. From my point of view, the way we think about current algebras,

Kac-Moody algebras, modular invariance, algebraic curves and the whole structure of Verlinde algebras [33] and category theory is an even more important mathematical application of string theory. In other words, the understanding of how to quantise conformal field theories in Riemann surfaces is perhaps the number one mathematical application, while mirror symmetry [1, 2] I would perhaps place close to number two. However, it would certainly not be number two if we include other forms of duality, like the geometric Langlands program. There are of course numerous other string theory applications in mathematics. It is a rich and generous field.

Now I would like to shift the focus to M-theory [3]. String theorists believe that M-theory is there and that it connects all the five different string theories. Is there a precise formulation of M-theory?

No, there is not, but we know many aspects of it. For instance, matrix theory provides a description of certain M-theory backgrounds. From another point of view, we can also take type IIA string theory and push it to a strongly coupled regime where spacetime becomes 11-dimensional. In this context, you can understand the 10-dimensional theories as compactifications of the 11-dimensional theory. In that sense, people sometimes refer to M-theory as the overarching model but there isn't more to it than just words because to make such claim we need to have a fundamental 11-dimensional theory that you can use to derive the other theories. So far, there is no such thing. I think of M-theory as just a specific regime of string theory. What we have learnt from it so far is that strings are not universally fundamental, as in a specific regime strongly interacting membranes appear to be the more appropriate description of the theory. It is also possible to think about formulating topological M-theory [31] and I believe this direction is still worth pursuing.

Is the M in M-theory standing for matrix?

I don't know (laughs). Witten referred to this overarching model as M-theory two years before matrix theory was formulated, so I don't think he was thinking about matrices at that time. I would go with membrane.

What do you think has been the biggest breakthrough in theoretical physics in the last 30 years?

The biggest breakthrough for me is the absolutely astonishing fact that we have come to understand that the space of all possible quantum field theories is one big connected landscape. I think that's a completely novel point of view and an extremely deep discovery. It puts everything upside-down. What we thought of as a priori fundamental ingredients, such as choices of gauge groups and particles, appear as specific solutions of equations in an encompassing theory. If I would go back in time 30 years and I would make such a wild claim, people wouldn't believe me and would say, "What are you talking about? How could that be? These theories are so fundamentally different!" I would then have to make a very long and complicated argument with all kind of dualities and leaps of faith.

My colleagues in the past would think that I'm a magician, not a physicist. So, I think this is a huge discovery and we are only slowly realising its full impact.

Why have you chosen to do physics?

I chose it because theoretical physics is the unique field that both uses deep mathematics and at the same time is talking about something very real in the world around us. I like to bring these two things together and certainly in fundamental physics you have the biggest set of mathematical tools and ideas, and the subject is also the biggest subject you can possibly study: the whole universe, from the smallest particles to the largest structures! In many ways this subject is one of the deepest in all of science and I also feel that it is one of the most challenging subjects. It's like climbing the highest mountain. Dangerous but exciting!

What do you think is the role that theoretical physicists have to play in modern society?

What theoretical physicists do well is to take part of reality and bring it into a kind of complete quantitative understanding at a very deep level. I know in many other fields there is some kind of physics envy, because of the kind of problems that we have in physics and the way in which nature cooperates in being captured in equations. In other fields such as climate science or the life sciences one struggles to achieve a similar quantitative, mathematical understanding.

I see, but what would you say to a taxpayer who had to pay tax to the LHC experiment instead of paying tax to improve hospitals?

I am convinced that at a very deep level every human being wonders, "Where am I, where are we going, and where do we come from?" These are the kinds of questions that we have been asking as long as humans are around. Now we are in a very fortunate position that instead of just asking these questions, we can partly answer them. And nature has been very cooperative! It wants to give its answers, and we are making huge progress. The only thing that we as physicists can promise is to fully devote our lives to it and do it as efficiently as possible, for instance by doing it all together. We build only one particle accelerator; we are not competing, we are only collaborating, worldwide. The enormity of the questions we ask and the fact that the entire world asks the same questions makes such large-scale experiments a very symbolic and hopeful act for humanity.

As president of the Royal Netherlands Academy of Arts and Sciences[1] one of your tasks is to advise the government on different scientific issues. Do you think that governments in general listen to the scientific opinion on different matters?

[1] At the time this interview was conducted, Robbert Dijkgraaf was the president of the Royal Netherlands Academy of Arts and Sciences.

Currently I think there is a challenge. There were periods in history when the impacts of science were considered obvious. For instance, just after the Second World War with the discovery of the atomic bomb, the prestige and authority of science was very high among politicians and the public at large. I think we live now in a world where many developments take place simultaneously and for science to be heard in that kind of noisy crowd is much more difficult. Funny enough, people in general are getting to know more about science in such a way that many even become proto-scientists, that is, people start to behave a little bit like scientists themselves and become quite sceptical. In the past people were taking the advice from scientists quite seriously I think because scientists intimidated people with their expertise. If a nuclear physicist would say, "Mr. President, you have to build these rockets" then the rockets would be built. But nowadays they would ask many more questions and claim to have heard different theories about the subject matter. So we are living in a different world. However, I think that the impact and the importance of science is only increasing because if you see all the big problems that the world is confronted with it always requires some solution that relies on science, whether it is about our health, food, technology, pandemics, climate, and anything else. So nowadays we ask a lot from science but funny enough the distance between the scientists and the politicians seems to increase because the scientific knowledge has also become really specialised and complex.

Science has made an enormous amount of discoveries and those discoveries have revolutionised industry, health, etc. over the centuries. However, there is something that doesn't change and that's of course the fundamental way in which science works. This is something you have to explain to the world because I think this is where our authority is based; it's also about our methods, not only about our results. Nowadays this is sometimes challenged and I think that we are not doing a good job when it comes to explaining why we are so convinced of our methods. If my astronomer friend is convinced that "we discovered a neutron star so many light years away" I understand why – many experiments were performed and all fits together. However, the general public and politicians don't really have a good feeling why we are so confident about it. We are convinced because we have this scientific method and I think that it hasn't changed in 400 years. People need to get some kind of feeling for what science is.

Do you think that the actual system for evaluating the quality of scientific work based on the number of citations is a fair system?

No, no it's not (laughs) and in fact I'm quite in a good position to say so, because I can look at various fields and I see sometimes that it's very unfair how the value of science is measured. People are able to manipulate the system and in fact the system is also manipulating people. One even speaks about *les measurables*. Nowadays there are countries and fields where you can double your income if you get a certain publication in a specific journal. I think that is not healthy and it's getting uncomfortable. Science is a living, breathing thing. It has these great qualities and objectivity is a key element of the scientific method. However, it appears that we want to apply in some sense the ideas and methods of our field upon ourselves and therefore we ask, "What's my objective value?" We want to capture it in an equation, or a number, but I think we can only do that up to a limited extent. Note that most of these "metrics" are self-inflicted instruments of torture. One thing I learned from science itself is how subtle nature is and how subtle human beings are. That's why we not only have the physical sciences and mathematics, but also have the humanities and social sciences to capture all this. So it's very strange that we think we can capture the value of a scientist in a single number.

References

[1] P. S. Aspinwall, B. R. Greene and D. R. Morrison, "Calabi-Yau moduli space, mirror manifolds and space-time topology change in string theory," *AMS/IP Stud. Adv. Math.* **1** (1996) 213–279, arXiv:hep-th/9309097.

[2] M. Kontsevich, "Homological algebra of mirror symmetry," arXiv:alg-geom/9411018.

[3] E. Witten, "String theory dynamics in various dimensions," *Nucl. Phys. B* **443** (1995) 85–126, arXiv:hep-th/9503124.

[4] M. Duff, "M theory (the theory formerly known as strings)," *Int. J. Mod. Phys. A* **11** (1996) 5623–5642, arXiv:hep-th/9608117.

[5] J. M. Maldacena, "The large N limit of superconformal field theories and supergravity," *Int. J. Theor. Phys.* **38** (1999) 1113–1133, arXiv:hep-th/9711200.

[6] J. Polchinski, "Dualities of fields and strings," *Stud. Hist. Phil. Sci. B* **59** (2017) 6–20, arXiv:1412.5704 [hep-th].

[7] D. J. Gross, "Two-dimensional QCD as a string theory," *Nucl. Phys. B* **400** (1993) 161–180, arXiv:hep-th/9212149.

[8] J. Erdmenger, "QCD and string theory," in *15th International Workshop on Deep-Inelastic Scattering and Related Subjects*, G. Grindhammer and K. Sachs, eds., pp. 139–150. 4, Deutsches Elektronen-Synchrotron, 2007.

[9] D. Mateos, "String theory and quantum chromodynamics," *Class. Quant. Grav.* **24** (2007) S713–S740, arXiv:0709.1523 [hep-th].

[10] P. Candelas, G. T. Horowitz, A. Strominger and E. Witten, "Vacuum configurations for superstrings," *Nucl. Phys. B* **258** (1985) 46–74.

[11] R. Blumenhagen, B. Kors, D. Lust and S. Stieberger, "Four-dimensional string compactifications with D-branes, orientifolds and fluxes," *Phys. Rept.* **445** (2007) 1–193, arXiv:hep-th/0610327.

[12] N. Ishibashi, H. Kawai, Y. Kitazawa and A. Tsuchiya, "A large N reduced model as superstring," *Nucl. Phys. B* **498** (1997) 467–491, arXiv:hep-th/9612115.

[13] L. Motl, "Proposals on nonperturbative superstring interactions," arXiv:hep-th/9701025.

[14] R. Dijkgraaf, E. P. Verlinde and H. L. Verlinde, "Matrix string theory," *Nucl. Phys. B* **500** (1997) 43–61, arXiv:hep-th/9703030.

[15] M. R. Douglas, D. N. Kabat, P. Pouliot and S. H. Shenker, "D-branes and short distances in string theory," *Nucl. Phys. B* **485** (1997) 85–127, arXiv:hep-th/9608024.

[16] S. B. Giddings, "High-energy black hole production," *AIP Conf. Proc.* **957** no. 1 (2007) 69–78, arXiv:0709.1107 [hep-ph].

[17] E. P. Verlinde, "On the origin of gravity and the laws of Newton," *JHEP* **04** (2011) 029, arXiv:1001.0785 [hep-th].

[18] E. P. Verlinde, “Emergent gravity and the dark universe,” *SciPost Phys.* **2** no. 3 (2017) 016, arXiv:1611.02269 [hep-th].
[19] J. D. Bekenstein, “Black Holes and Entropy,” *Phys. Rev. D* **7** (Apr. 1973) 2333–2346. https://link.aps.org/doi/10.1103/PhysRevD.7.2333.
[20] S. Hawking, “Black hole explosions,” *Nature* **248** (1974) 30–31.
[21] N. Seiberg, “Electric-magnetic duality in supersymmetric nonAbelian gauge theories,” *Nucl. Phys. B* **435** (1995) 129–146, arXiv:hep-th/9411149.
[22] D. J. Gross, J. A. Harvey, E. Martinec and R. Rohm, “Heterotic string,” *Phys. Rev. Lett.* **54** (Feb. 1985) 502–505. https://link.aps.org/doi/10.1103/PhysRevLett.54.502.
[23] M. C. Cheng, J. F. Duncan and J. A. Harvey, “Umbral moonshine,” *Commun. Num. Theor. Phys.* **08** (2014) 101–242, arXiv:1204.2779 [math.RT].
[24] A. Kapustin and E. Witten, “Electric-magnetic duality and the geometric Langlands program,” *Commun. Num. Theor. Phys.* **1** (2007) 1–236, arXiv:hep-th/0604151.
[25] L. Susskind, “The anthropic landscape of string theory,” arXiv:hep-th/0302219.
[26] E. Witten, “Topological sigma models,” *Commun. Math. Phys.* **118** (1988) 411.
[27] R. Dijkgraaf and C. Vafa, “Matrix models, topological strings, and supersymmetric gauge theories,” *Nucl. Phys. B* **644** (2002) 3–20, arXiv:hep-th/0206255.
[28] R. Dijkgraaf and C. Vafa, “On geometry and matrix models,” *Nucl. Phys. B* **644** (2002) 21–39, arXiv:hep-th/0207106.
[29] A. Neitzke and C. Vafa, “Topological strings and their physical applications,” arXiv:hep-th/0410178.
[30] M. Aganagic, R. Dijkgraaf, A. Klemm, M. Marino and C. Vafa, “Topological strings and integrable hierarchies,” *Commun. Math. Phys.* **261** (2006) 451–516, arXiv:hep-th/0312085.
[31] R. Dijkgraaf, S. Gukov, A. Neitzke and C. Vafa, “Topological M-theory as unification of form theories of gravity,” *Adv. Theor. Math. Phys.* **9** no. 4 (2005) 603–665, arXiv:hep-th/0411073.
[32] R. Dijkgraaf, H. L. Verlinde and E. P. Verlinde, “Notes on topological string theory and 2-D quantum gravity,” in *Cargese Study Institute: Random Surfaces, Quantum Gravity and Strings*, pp. 91–156. Trieste Spring School 1990:0091–156.
[33] E. P. Verlinde, “Fusion rules and modular transformations in 2D conformal field theory,” *Nucl. Phys. B* **300** (1988) 360–376.

8

Bianca Dittrich

Faculty member at Perimeter Institute for Theoretical Physics, Waterloo, Canada

Date: 6 October 2020. Via Zoom. Last edit: 20 January 2021

What are the main problems in theoretical physics at the moment?

From my personal point of view, the biggest puzzle is to understand the quantum nature of space and time. In other words, to understand how to quantise gravity.

Why do we need to quantise gravity? Is there some phenomena we can't explain at the moment?

There are attempts that try to avoid a theory of quantum gravity but I think that for the coupling between quantum field theory and gravity to be consistent, we need a theory of quantum gravity. This is a general expectation. Additionally, we expect quantum gravity to be relevant for explaining the nature of black holes and early universe physics. Such expectation would benefit from experimental prospects which quantum gravity is a bit short of, but with recent advances it may be possible to obtain some empirical evidence from black holes and cosmology in the coming years.

Do you think that quantising gravity is a difficult problem or has this problem been solved in many different ways and we just don't know which way is the right way?

I do think it's a difficult problem and I don't think it has been solved in any satisfactory way. We don't really have a proposal that provides us with a complete picture of quantum gravity and lots of physical predictions. And it is fair to say that the history of quantum gravity is already a bit too long, not yet having led to a definite answer. However, it actually inspired a lot of developments which are now an integral part of physics, in particular of quantum field theory. For example, gravity provided the first example of a theory with a gauge symmetry and a lot of the tools developed in the context of gravity were later applied to gauge theory. This is clear from the works of Bryce DeWitt [1] and even those of Richard Feynman [2].

But why is it difficult? There are several factors that make gravity very different from other theories. For instance, if we try to quantise gravity perturbatively we find that it is perturbatively non-renormalisable. Even if you try to quantise it non-perturbatively, which many lattice approaches attempted to do using Monte Carlo simulations, you end up in most cases failing. The reason is that the gravitational action, contrary to actions studied in many other contexts, is unbounded from below. Due to this fact, many lattice Euclidean approaches that were pursued in the past, where Wick rotation is performed to go from Lorentzian to Euclidean geometries, have essentially failed. Wick rotation is not well defined if you do not have a fixed notion of time, which is the case of gravity. If you want to understand gravity in the correct way you need to be able to work with the "*i*" in the path integral; that is, you need to understand how to evaluate Lorentzian path integrals. In fact, of all approaches to quantum gravity, only loop quantum gravity (LQG) and spin foam models work with a Lorentzian path integral. However, the drawback is that we haven't yet developed many calculational techniques able to evaluate such path integrals.

But isn't causal dynamical triangulations (CDT) [3] also a lattice approach that is able to deal with Lorentzian path integrals?

I agree that CDT is an interesting idea and, in fact, I started out my life in physics by studying CDT during [studying for] my diploma. However, the way to deal with CDT is actually to perform Monte Carlo simulations, meaning that, effectively, these simulations do not involve the "*i*" in the path integral. It is, nevertheless, quite an important approach because it solved many of the issues of the lattice Euclidean approaches that I mentioned. The reason why it can do better than these approaches is because CDT imposes restrictions on configuration space. It is, however, an open question as to what extent these restrictions are dependent on the preferred slicing that is used in CDT and whether if, at the end of the day, the theory is just a lattice version of Hořava-Lifshitz gravity [4] or general relativity. Renate Loll, one of the proponents of CDT and a good friend of mine, would probably say that it is not Hořava-Lifshitz gravity but I think that it is not clear that it is not Hořava-Lifshitz gravity. The phase diagram of CDT includes Hořava-Lifshitz gravity [3] and it is unclear whether there is a certain limit of the phase diagram in which you recover full diffeomorphism symmetry or Lorentz symmetry. The preferred slicing allows one to define a notion of Wick rotation, and then one does use Monte Carlo simulations. There is, however, the open question of how to translate back the results of the simulations, which are with Euclidean geometries, to geometries with Lorentzian signature. It is, nevertheless, quite surprising that CDT seems to work since it still works with an unbounded action. This suggests that there is something subtle in the way that the measure of CDT is defined, implying that the dominating configurations in Euclidean lattice methods are eliminated within the CDT configuration space.

In any case, I want to stress that Monte Carlo methods are not always available when dealing with Lorentzian path integrals in the non-CDT fashion. In order to understand gravity we need to be able to deal with diffeomorphism (gauge) symmetry, which we do not know how to do in even simpler contexts. The only theories we know that deal exactly with this symmetry are topological quantum field theories. We do not know how to work with quantum field theory (QFT) in curved spacetime on arbitrary slicings. If you had a full theory of quantum gravity, it would imply that we would have understood how to describe

QFT in these scenarios. Quantum gravity, from this point of view, includes everything and that is one of the major reasons why it is a difficult problem. In addition, we do not have direct measurements that indicate whether or not our current frameworks are on the right track.

From what you just said, it appears that the approaches to quantum gravity you prefer are loop quantum gravity and the spin foam approach. Is that the case?

I'm an open person, so I do think we should check out a number of approaches and test out several principles and assumptions. My personal motivation is to understand quantum spacetime and I think that LQG and spin foams have a more direct approach to this question. Quantising geometry is a central aspect of quantum gravity. I think these approaches have made decent progress over the years but there is a big drawback, namely, that we so far do not have reliable tools to perform large-scale computations. At the end of the day, we have a non-perturbative setup with a very complicated prescription for evaluating transition amplitudes and without being able to have recourse to Monte Carlo simulations. There have been works which develop tools for numerical simulations within spin foam models [5–7] and recently I have proposed a simpler model for spin foam amplitudes that I hope will lead to progress in this direction [8].

Other approaches have an easier task in delivering computations, though they also incorporate more input from standard quantum field theory, such as the asymptotic safety programme [9], or CDT with its Monte Carlo simulations. At this point, LQG and spin foams are lacking some fundamental technology for evaluating path integrals and doing lattice computations in the Lorentzian setting. Developing such tools would be great for physics in general.

What is LQG?

I have a more practical approach to this question. Some proponents of LQG see it as a kind of fundamental theory, which is really not my opinion. Historically, LQG resulted from the idea that in order to quantise gravity you needed to somehow quantise geometry. This philosophy was not successful in any rigorous way for a long time, specially when attempting to quantise the spacetime metric [10]. It was only successful when gravity was reformulated as a gauge theory, with the help of Ashtekar variables. This gauge formulation allowed for a rigorous quantisation of the kinematics of the theory. It does look very similar to a lattice gauge theory, however with an arbitrary choice of lattices. One can then relate the state space of a given lattice to the state space on a finer lattice. This does allow one to define a continuum limit. But this is all on the kinematical level. Constructing such a limit for the dynamics is much more subtle and (in my opinion) does require a coarse-graining process, in which finer degrees of freedom are integrated out. And the result of this coarse-graining procedure should give the dynamics on coarser lattices. That is, instead of postulating a "fundamental" dynamics at once on all lattices, one has to construct a dynamics which is consistent under this coarse-graining and renormalisation process.

The Hilbert space structure of LQG has been criticised because it looks rather different from standard quantum field theory. I do not think it should be a concern because the kinematical structure of LQG is similar to the kinematical structure of topological quantum field theory (TQFT) with defect excitations, which has been developed extensively in the past 15 years, in particular, in the context of condensed matter physics. In a TQFT, defects are defined as objects which cannot probe distances; in other words, defects cannot probe how far away another defect is. This implies that if you move the defects around you end up with an equivalent state to the one you started with. This is exactly analogous to what takes place in LQG, in which the defects have a different interpretation depending on what you choose the vacuum to be. In the traditional formulation of LQG, the defects are string-like excitations of geometry and the kinematical vacuum is a spin network without edges, meaning that in the vacuum no geometry is excited and the state has zero physical volume. If you reformulate the theory as a BF theory with defect excitations, the defects are curvature defects, and the vacuum is a state with zero curvature (or flat connection). In the language of condensed matter physics these two vacua describe two different phases of your systems, like the frozen/unfrozen phases of the Ising model.

I should stress that the kinematical Hilbert space of LQG is the same as that of a TQFT but the dynamics is not. In a TQFT, the dynamics is topological and has some features that makes it easy to study. For instance, it allows you to find distinct formulations of the dynamics that do not depend on particular choices of discretisation (or triangulation). This means that if you compute some quantity in a given triangulation that captures the existing defects, and you compute that same quantity in a finer triangulation, you will obtain the same results. On the other hand, in LQG the dynamics is more complicated and you only expect the details of the triangulation to not have any effect on the results in a kind of refinement limit. This is an expectation and to show it we would need to be able to perform calculations with very fine triangulations which currently are out of reach. I should say that there are people who believe that the dynamics of LQG is also that of a TQFT in which you add many defects, thereby simulating a theory with propagating degrees of freedom. I do think that it requires some additional steps because in order to take a continuum limit you not only need to add many defects but you also expect that these defects will begin interacting with each other. This means that you should expect a phase transition.

What is the exact relation between LQG and BF theory?

BF theory is a topological theory [11] and the $SU(2)$ BF theory has the same kinematical structure as LQG as well as spin foam models. Spin foam models are formulated via the Plebanski action principle for gravity [12], which is a BF theory with additional constraints so that the theory is no longer topological but instead has propagating degrees of freedom in four spacetime dimensions. In three dimensions, this formulation, also known as first-order formulation of gravity, is exactly a BF theory, which is clear when you express it in terms of triad variables and an independent connection. In three dimensions, the theory can also be expressed as a Chern-Simons theory.

Are spin foam models equivalent to LQG?

I would say that spin foams are a part of LQG. Historically, LQG has been formulated via canonical quantisation and there has been a lot of debate whether the canonical approach is equivalent to the covariant (or path integral) approach of spin foams. I think that it is possible to make these two approaches equivalent, if one understands both approaches as lattice discretisations, and thus does acknowledge the need to take a refinement limit on both sides. In this limit it should be possible to derive the Hamiltonian constraints for the canonical approach. I do not think that the Hamiltonian constraints which are used currently are completely correct – I rather think that these should be interpreted as lattice discretisations.

The continuum limit of the path integral approach functions as a projector into a space of states while the Hamiltonian constraint is just a different way of specifying the space of states onto which the projector projects. In three dimensions, the equivalence between LQG and spin foam models has been proven [13, 14]. This three-dimensional case is simpler than the four-dimensional one, as gravity is a TQFT in three dimensions, and it is much simpler to take the continuum limit. However, the Hamiltonian constraint used to prove this is not the usual constraint of the canonical approach; that is, it is not the Thiemann constraint [15, 16].

So do I understand from your answer that you think that the spin foam approach is the right way to formulate LQG and not the canonical approach?

As I said, I take a more practical point of view on LQG. The canonical point of view states that LQG is a continuum approach. Thomas Thiemann derived the so-called Hamiltonian constraints which are mathematically well defined in a complicated way and which you can define with all sorts of matter content [15, 17]. Thiemann showed that there is a notion of regularisation independence which is related to the way that he defined the Hamiltonian constraint. In my opinion, this approach does not square fully with lattice approaches. In particular, he used the freedom of adding new degrees of freedom to the lattice, in particular, new edges, in order to solve the problem of anomalies in the algebra which is necessary to have well-defined Hamiltonian constraints [15]. The LQG programme rests on solving these constraints but so far there has not been much computational progress in solving them in the full theory. But a number of people, including myself, do not believe that these are the correct constraints.

At the time Thiemann formulated the constraints, Lee Smolin pointed out that these new degrees of freedom that Thiemann added to the lattice were ultra-local and do not account for interactions [18]. To date, there is still debate on this matter and some people work on trying to improve these constraints [19]. I think that LQG is a lattice theory and as such you first need to discretise the dynamics onto a lattice and then find a continuum limit by refining your lattice. In other words, one should take the regularisation process seriously. Only then can you actually answer physical questions, which should not depend on particular choices of lattices. This is different than the common formulation of LQG, where one does not interpret the spin networks as defined on a lattice but as continuum objects.

Are these approaches diffeomorphism-invariant?

It is hard to say whether the Thiemann constraints formulated in the canonical approach break diffeomorphisms or not. I think that they do not break diffeomorphism symmetry but at the same time, they are likely not describing the right dynamics.

I actually showed, together with a colleague, that discretising theories with propagating degrees of freedom generically leads to breaking diffeomorphism invariance [20–22]. This does hold already on the classical level and it is well known in the context of numerical relativity. In this setting, people discretise the system leading to violation of diffeomorphism invariance but to remedy that they add additional terms that ensure that such violations are small.

This should not be a surprise. You can formulate the standard path integral for the harmonic oscillator which has time reparametrisation invariance (or one-dimensional diffeomorphisms) but if you introduce a generic discretisation you will always break the reparametrisation invariance. The way to proceed is to take a continuum limit of different quantities, say, transition amplitudes, which do not depend on the discretisation. The propagator between two boundary points is one such example, in which in the continuum you restore the diffeomorphism symmetry.

Spin foam models are based on discretisation so they break the symmetry generically, except in the case of BF theory which is topological. The community does not like this result but for me it's a fact of life. The goal of the programme is to use renormalisation tools, or coarse-graining methods, to get a grip over the refinement/continuum limit of the theory. In this setting, the approach is brought closer to ideas like asymptotic safety but with a stronger non-perturbative character. Spin foams compute amplitudes for different choices of triangulation and so the hope is that as you take finer and finer triangulations, that is, as you take the refinement limit, these amplitudes converge to the same value. We hope that in such a refinement limit one also restores diffeomorphism invariance. We indeed found evidence that triangulation invariance, which is being restored in such a refinement limit, is equivalent to diffeomorphism invariance [23–25].

Have the existence of a refinement limit and triangulation invariance been shown in four dimensions?

No, because it requires computing all these amplitudes, but there are indications from simpler systems and from symmetry reductions of four-dimensional gravity [25]. There is debate in the community about whether diffeomorphism symmetry is present in spin foams or not, and as mentioned earlier, I don't think it is. But even those who think that spin foams are already diffeomorphism-invariant by fiat need to deal with the dependence on the choice of triangulation. Here the suggestion is that one needs to sum over all triangulations – but that would also mean to sum over the most refined triangulations, so these have to be computed in any case. We instead propose to compute the refinement limit itself. If one is interested in a very coarse observable, determining the refinement limit might be computationally fast. For observables which need more refined information, the computational effort will be much higher. Recently, we made some progress on this by formulating what we called *effective spin foam models* which work up to a certain scale and are based on the higher gauge formulation of gravity [26, 27]. We hope that with this model we will more easily be able to compute amplitudes. In fact, we have already been able to show for the first time that we can get the correct equations of motion that you expect in this theory of gravity in a certain regime [8]. This had not been accomplished earlier in any other spin foam model. Of course, at the end, we would like to take the refinement limit but we still need to develop some computational techniques for that.

You mentioned several times that you take a more practical approach to LQG but I still haven't understood exactly what you mean. Could you explain?

Some people take the older viewpoint that you can define the Hamiltonian constraint directly in the continuum and you have a theory which you should now use and compute. Similarly, some people construct some spin foam amplitude and state that these amplitudes are continuum objects (more concretely, appear in a certain expansion of the continuum amplitude). But from the lattice point of view this is just an approximation because you introduced a discretisation. It is only in the refinement limit that the true theory/model emerges. This is similar to the example of the harmonic oscillator that I mentioned. You introduce the discretisation in the path integral but I wouldn't claim anything fundamental by evaluating the path integral for a certain interval with 10 intermediate points. Without taking the continuum/refinement limit of the harmonic oscillator, you will not obtain the correct quantisation. In the spin foam setting, it would really not make sense to interpret results for 10 building blocks or 1,000 building blocks or to sum over all of them. Not taking the refinement limit leaves lots of open questions of how one should interpret the results, which do depend on a choice of lattice or triangulation.

In addition, you will run into further issues once you consider renormalisation or coarse-graining. For instance, if you couple matter to the theory you expect the matter couplings to run. There is in fact a debate wether or not Newton's constant runs but there isn't a debate whether or not matter couplings run. If this is a lattice theory then the matter couplings should depend on the lattice constant but the lattice constant in these theories is one of the variables, meaning that the matter couplings should actually depend on the metric variables. This implies that to define these models with a given discretisation you would need to know the complicated dependence of the coupling constants on the metric to begin with, which you cannot guess. For the current construction of the Hamiltonian constraints such a running of the couplings is ignored. The difficulty lies in that you would have to reconstruct or know how the couplings should behave as a function of all kinds of inhomogeneous lattice constants. So, I understand that not taking the refinement limit is more practical, in the sense of being able to compute, but it's not telling us anything fundamental. Only if you take the refinement limit will you determine the full dynamics.

One of the criticisms of LQG is about the existence of states describing Minkowski space. Has this issue been solved?

This issue is also related to our difficulty in performing calculations for lattices with many building blocks and determining the refinement limit in this way. (Finding this state would be much harder than finding the ground state for QCD on a lattice with many building blocks, which is also an extremely hard, not yet solved, problem.) For instance, in the recent paper I mentioned [8], we were able to obtain the Regge version of Einstein equations for small triangulations, which is still discrete, and one would like to go further and take the continuum limit. It is quite hard because you need to evaluate path integrals with the "*i*" in it so the hope is to find approximations.

In any case, let me explain the issue of Minkowski space in more detail. There is an important distinction between kinematical states and physical states. You can write a coherent semiclassical state whose expectation value is sharply peaked around Minkowski space. This is a kinematical state and semiclassical in all variables. And because it is semiclassical in all variables, we know that it cannot be a physical state. Physical states are states that satisfy constraints (such as the Hamiltonian constraint in the canonical formulation) and they would be the states that you would obtain from projecting with the path integral if you could take the refinement limit. So if you want to discuss Minkowski space, it only makes sense to discuss it as a physical state that approximates Minkowski space in the classical regime. In order to find the physical states you need to solve the dynamics of LQG, that is, to solve the Hamiltonian constraint or the Wheeler-DeWitt equation, or take the refinement limit in spin foam models. Neither of these has been done and, in particular in spin foam models, it requires performing a huge and complicated lattice calculation which is hard to do in practice. This issue was also not solved in earlier works on canonical quantisation, prior to LQG, because there were many discussions about how to define the Hamiltonian constraint and how to deal with many ambiguities in operator ordering.

But don't you need to find such states to have a consistent theory?

Yes, I completely agree. Let me say that I worked during my PhD with my supervisor Thomas Thiemann on standard canonical quantisation of LQG for which there is a *master constraint* that you have to solve [28–32]. However, it is so complicated to solve, even for simple triangulations. It's a similar, but harder, problem to solving a really complicated condensed matter Hamiltonian because in LQG you do not have one Hamiltonian that you need to solve but a family of Hamiltonians. The difficulties that you encounter while trying to solve the Ising model in three dimensions are way easier than solving LQG dynamics. So far there is no complete solution of the three-dimensional Ising model available. In condensed matter, there is still a lot of work trying to improve numerical methods for one- and two-dimensional systems, which are much simpler than LQG because they have one, two or a few degrees of freedom per site, and not an infinite number as in LQG.

We have a non-perturbative formulation but there are unfortunately not many techniques available to do computations in this framework. In condensed matter theory, there has been a lot of progress recently in developing path integral techniques such as tensor network techniques that work well in two and three spacetime dimensions. However, in four spacetime dimensions not much is known. I try to develop also these tensor network techniques with the hope that at some point they can be applied to four-dimensional spin foam models [33]. In general, we need more methods for obtaining approximate results which are better controlled than just taking one or a few building blocks or specific symmetry assumptions. In these cases, it is always debatable whether or not such assumptions are justified. For instance, in the canonical approach people have formulated loop quantum cosmology in which there is a huge symmetry reduction to only a few degrees of freedom [34]. In this setting a lot of results can be obtained but it is an open question whether the results obtained from these symmetry assumptions can also be obtained from the full theory. At this point, it is not clear but it should be possible to compare these results to some kind of mean-field approximation (used for solving the Hamiltonian of statistical systems) within the full theory.

You mentioned ambiguities in earlier approaches to canonical quantisation, so I was wondering, what is the role of the Barbero-Immirzi parameter?

One of the big successes of the spin foam models that appeared in 2008 is that it included the Barbero-Immirzi parameter. Earlier models did not and were not correct. I have particular opinions about the Barbero-Immirzi parameter [35] so let me make a few unconnected comments.

When you work with triangulations you want to assign a metric to these triangulations and the easiest and simple way forward is to work with edge lengths. The path integral of your theory is then defined as a sum over the edge lengths and it turns out that in three spacetime dimensions these edge lengths have a discrete spectrum. This is a surprising result of LQG, which is in stark contrast with quantum versions of Regge calculus. On the other hand, in four spacetime dimensions, if you use edge lengths, the conjugate variables are complicated functions of angles [36] and, as such, it is really hard to understand how to quantise the theory. The way forward is not to work with edge lengths but instead with areas, which turn out to give rise to a more general space of geometries. And it turns out that in LQG and spin foams, in four spacetime dimensions, one also uses the area variables as more fundamental variables, and one has a more general space of geometries, which can be interpreted to include torsion degrees of freedom [37]. For this more general space of geometries the Barbero-Immirzi parameter appears (in the Poisson brackets between the basic variables, which are then converted into commutators in the quantum theory). Interestingly, if you restrict your space of geometries to that obtained by considering edge lengths, that is, forbid the torsion degrees of freedom, the Barbero-Immirzi parameter washes away. So the Barbero-Immirzi parameter is tied to allowing a more general space of (quantum) geometries that does include torsion degrees of freedom. This enlargement of the configuration space arises because there is a puzzling quantum anomaly in the constraints that classically reduce the more general configuration space to the length configuration space. This anomaly in the constraint commutator algebra is parametrised by the

Barbero-Immirzi parameter and it prevents the quantum reduction to a Hilbert space describing the length configuration space [8].

I do not know whether one should ultimately adopt the length variables or area variables as fundamental, and connected to that, whether one should allow these torsion degrees of freedom or not. It boils down to a choice of quantum configuration space. Quantising the configuration space based on length variables is very involved, as there are lots of constraints arising from triangle inequalities. (In the continuum, these can be compared to the condition of having a positive definite metric.) This is a very interesting mathematical physics problem whose difficulties are hard to appreciate. It is much harder than quantising a space like $\mathbb{R}^n$ because this reduced phase space has boundaries and a non-trivial topology. On the other hand, there are good reasons to argue that area variables are the best choice since many entropy calculations are based on area calculations, including the developments in holography such as the Ryu-Takayanagi formula [38]. In any case, I do not know for certain whether areas are more fundamental than edge lengths as fundamental variables. As I come from East Germany, I know that taking a fundamentalist approach can be dangerous.

Another comment I would like to make is that I don't think that the Barbero-Immirzi parameter can be arbitrary [26, 27]. In the effective models that I am currently working on, which I mentioned earlier, obtaining the classical Einstein equations is only possible if the Barbero-Immirzi parameter is within a certain range. Also, since the Barbero-Immirzi parameter appears in the area spectrum of LQG it would be unphysical if it turned out to be huge.

In the past, people argued that the Barbero-Immirzi parameter could be fixed via a black hole entropy calculation [39]. Do you agree with this?

On the face of it, as much as I can rely on these microscopic calculations of black hole entropy in the microcanonical ensemble, the Barbero-Immirzi parameter is expected to show up and, if you want to reproduce one specific value then that would fix the parameter. These calculations are nice but since you cannot solve the theory they base themselves on a mixture of kinematics and dynamics. In addition, you assume that the theory on the boundary/horizon of the black hole, which is related to Chern-Simons theory, is sufficient for performing this counting. So whether you agree with the calculation depends on whether you agree with this assumption. In any case, I think that before attempting to understand black hole entropy we still need to show that LQG can deliver a consistent dynamics, a semiclassical limit and that Minkowski spacetime can be recovered in the continuum limit.

Have people been able to compute graviton scattering amplitudes?

What has been computed so far was to obtain the so-called propagator using kinematical observables on a single building block, which you can better understand as a correlation function. However, I think that a propagator, by definition, is an object that goes across many building blocks. So I would like to obtain the propagator on a lattice and compare that to discretised Regge calculus. The calculation does get harder the more building blocks you try to involve.

In our effective spin foam models this may be within reach. Contrary to other models, it is straightforward to evaluate amplitudes on a given simplex and directly check that it has the right semiclassical behaviour. The crucial problem will be to understand what happens at high momentum, which requires many building blocks, and to check whether the high momentum behaviour implies modifications of standard dispersion relations. We already have calculations where we use six building blocks, but we want to increase this number to obtain a more meaningful result.

People usually speak of the eigenvalues of the area operator as being observables. Do you consider these to be the observables of LQG?

There is a technical definition of observables that is not just applicable to LQG but more generally to general relativity. Namely, observables are diffeomorphism-invariant quantities, which we don't have many of. In this technical sense, the area operator is not diffeomorphism-invariant. The eigenvalues of the area operator give you the area of what, exactly? Together with Thomas Thiemann I wrote a paper showing that the spectrum of a non-gauge-invariant operator does generically not predict the spectra of related gauge-invariant operators, which you may be able to construct from the non-gauge-invariant one [40]. Thomas later wrote a paper where he used a particular kind of matter to localise the areas, which then had a discrete spectrum, but the matter was of a very special kind [41]. But the discreteness of the spectrum of the non-diffeomorphism-invariant area operator might have nevertheless a physical meaning: this discrete spectrum turns the path integral over geometric configurations into a summation. The discreteness is also parametrised by the Barbero-Immirzi parameter, so it brings us back in interesting ways to our previous discussion. I do think it has a role to play, in the sense that the spectrum of such operators are summed over in the path integral, but it is not diffeomorphism-invariant.

There is a huge body of work on what are called relational observables. These observables are hard to define and write explicitly. Also, if you could take the refinement limit of the path integral, you could use the path integral to project observables onto physical space, which you would know how to interpret. For instance, as I mentioned, you could obtain the propagator and perhaps obtain modified dispersion relations which would be observable. In the context of loop quantum cosmology, where you can solve the theory, there are several proposals to compute the spectrum of the WMAP and check whether it is redshifted or blueshifted.

However, finding observables in the technical sense is really complicated and what I found, as is the usual case in quantum gravity, is that to construct them you need to solve the dynamics of the theory. It is an intrinsic property of just classical general relativity that to obtain the diffeomorphism-invariant observables you need to determine the solution of a given Cauchy problem. In general such observables (in the canonical formalism) will be complicated functions of your initial Cauchy data, which makes it quite unpractical.

It is usually the case that before you even try to define a theory, you first understand what the observables of the theory are and then you quantise them. In (quantum) general relativity we understand what the diffeomorphism-invariant observables are but it is to hard to explicitly define them and quantise them. This is not a particular feature of LQG – it is a feature of classical general relativity and appears in some form or another in many quantum gravity approaches.

Given that there isn't a single piece of evidence that allows one to pinpoint what is the right theory of quantum gravity, it has been argued that the value of a given theory can also be measured by how many tools and methods it develops that can be applied to other fields of physics like condensed matter. This is, for instance, something that the string theory community is very proud of. Are there tools and methods developed within LQG being applied to other fields?

I think people are not aware that a big part of LQG is related to TQFTs. While I cannot claim that many of the tools used nowadays in TQFTs originated in LQG, it is fair to say that many people who worked in LQG early on also worked on TQFTs, in particular, higher category theory, which is now widely used in TQFTs and in the theory of topological phases in condensed matter theory as well as studied in string theory and quantum field theory.

A set of mathematical results that arose from LQG is related to coherent states and the quantisation of peculiar mathematical spaces, such as the shape of tetrahedra, which is also relevant for Chern-Simons theory. However, I also think that the proliferation of applications of a given theory is also a question of numbers and it certainly makes a difference whether it is a big or small community. In that sense string theory is highly dynamical while small communities, like [those who study] LQG, are more busy with their own field. Also if someone originates from LQG but contributes to some other field it is often not labeled as "a contribution from LQG."

You have also worked recently in holography [42–45], which is an idea that originated in string theory. What was your motivation to pursue this?

I actually started working on it a bit by accident. I was motivated by the calculation by Barnich, Gonzalez, Maloney and Oblak [46] of the one-loop partition function for three-dimensional gravity in asymptotic flat spacetime. The idea was to try to compute the one-loop partition function in simpler contexts using Regge calculus [47]. As a side note, I don't believe in Regge calculus as a fundamental theory of quantum gravity; I just use it as tool. In any case, we generalised this calculation to the case of finite boundaries, and it was quite a surprise that we did also obtain the same partition function as for asymptotic boundaries. The reason is that this partition function is a character of the asymptotic symmetry group, so it was reasonable to expect that this holds only if one has an asymptotic boundary. Doing the calculation for a finite boundary we automatically obtained a description of gravity in terms of a dual field theory, which everyone expected specially in the context of anti–de Sitter (AdS) spacetime with asymptotic boundary, but it worked also for flat spacetime, with finite boundary. We were surprised we could reproduce the full result because it was much simpler than using heat kernel techniques as in [46].

We repeated these calculations for the case of spin foam models [42–44] and there we found lots of dual field theories depending on choices of boundary states. In three spacetime dimensions you can find all sorts of holographic dualities and in fact you can find them by directly integrating all bulk degrees of freedom, except for the degrees of freedom that happen to describe how your boundary is embedded in your three-dimensional flat solution. In the end, the reason why this works so nicely is because the theory is topological and there are no propagating degrees of freedom. So it's quite easy to integrate out the bulk, even directly in the continuum theory [48], except the variables I mentioned that you can also understand as the geodesic distance from a point x in the bulk to the boundary. And there are nice results that arise from this approach such as the fact that the invariance under the choice of x is a gauge symmetry from the boundary point of view. This connects well with the results by Maloney and Witten using Chern-Simons theory [49].

In four spacetime dimensions it is not as simple. If you enforce the theory to be topological then you can get boundary theories with quite similar structures as in three dimensions. However, once you take into account gravitons as we did in flat spacetime [45], it is not clear how to project all the information about the bulk gravitons onto the boundary. I'm still interested in exploring this further and to understand whether assuming AdS space instead allows you to project this information onto the boundary at infinity. I have not considered 10 spacetime dimensions or AdS_5 as I'm trying to understand this problem from a more naive point of view.

Within the LQG community there are discussions about whether gravity is holographic or not. Some people argue that it is clear that gravity is holographic due to the existence of the Regge-Wheeler equation and a boundary Hamiltonian. I believe these arguments are too general and not helpful, as they don't tell me how the boundary theory encodes the bulk physics. These arguments are made very generally without, for instance, any need for string theory. Recently, people also used the successes of tensor network methods in three dimensions to argue for the holographic nature of gravity. My works on holography aim at understanding whether this is true or not and so I want to look more into this and focus on the case of AdS. Several people within the string theory community were very surprised that we could obtain the same results in three dimensions using these methods, which I take as a good sign.

What do you think has been the biggest breakthrough in theoretical physics in the past 30 years?

I think that there have been major breakthroughs in quantum computing and quantum teleportation that changed theoretical physics. The detection of gravitational waves, which is an experimental endeavour, required many theoretical developments. Also, the WMAP experiments confirmed calculations of linearised quantum fields in a given background, which I find quite astonishing.

Why did you choose to do physics?

I was born in East Germany and after the unification of Germany many people lost their jobs. There was a bit of an expectation that the young generation would just end up jobless.

This gave me the freedom to study physics, which was rumoured to not be helpful for employment, since all choices would make me jobless anyway. It turned out that later the young generation did find jobs. In any case, physics impacted me earlier on as I was fascinated by fundamental questions.

What do you think is the role of the theoretical physicist in modern society?

Part of it is to provide a better understanding of the universe. With time, as in the case of quantum computing, this pursuit leads to revolutionary changes within society. In the past, there were a lot of ethical discussions regarding the role of theoretical physicists, specially given the role they played in making nuclear bombs. Nowadays, these ethical discussions happen mostly in the context of biology. However, I think that theoretical physicists should take more responsibility for society apart from the research they carry on.

References

[1] B. S. DeWitt, "Quantum theory of gravity. I. The canonical theory," *Phys. Rev.* **160** (Aug. 1967) 1113–1148. https://link.aps.org/doi/10.1103/PhysRev.160.1113.

[2] R. Feynman, "Quantum theory of gravitation," *Acta Phys. Polon.* **24** (1963) 697–722.

[3] R. Loll, "Quantum gravity from causal dynamical triangulations: a review," *Class. Quant. Grav.* **37** no. 1 (2020) 013002, arXiv:1905.08669 [hep-th].

[4] P. Hořava, "Quantum gravity at a Lifshitz point," *Phys. Rev. D* **79** (Apr. 2009) 084008. https://link.aps.org/doi/10.1103/PhysRevD.79.084008.

[5] B. Bahr, G. Rabuffo and S. Steinhaus, "Renormalization of symmetry restricted spin foam models with curvature in the asymptotic regime," *Phys. Rev. D* **98** no. 10 (2018) 106026, arXiv:1804.00023 [gr-qc].

[6] P. Dona and G. Sarno, "Numerical methods for EPRL spin foam transition amplitudes and Lorentzian recoupling theory," *Gen. Rel. Grav.* **50** (2018) 127, arXiv:1807.03066 [gr-qc].

[7] P. Donà, M. Fanizza, G. Sarno and S. Speziale, "Numerical study of the Lorentzian Engle-Pereira-Rovelli-Livine spin foam amplitude," *Phys. Rev. D* **100** no. 10 (2019) 106003, arXiv:1903.12624 [gr-qc].

[8] S. K. Asante, B. Dittrich and H. M. Haggard, "Discrete gravity dynamics from effective spin foams," arXiv:2011.14468 [gr-qc].

[9] M. Reuter and F. Saueressig, *Quantum Gravity and the Functional Renormalization Group: The Road towards Asymptotic Safety*. Cambridge University Press, 1, 2019.

[10] A. Ashtekar, "Some surprising implications of background independence in canonical quantum gravity," *Gen. Rel. Grav.* **41** (2009) 1927–1943, arXiv:0904.0184 [gr-qc].

[11] G. T. Horowitz, "Exactly soluble diffeomorphism invariant theories," *Commun. Math. Phys.* **125** (1989) 417.

[12] J. F. Plebanski, "On the separation of Einsteinian substructures," *J. Math. Phys.* **18** (1977) 2511–2520.

[13] K. Noui and A. Perez, "Three-dimensional loop quantum gravity: physical scalar product and spin foam models," *Class. Quant. Grav.* **22** (2005) 1739–1762, arXiv:gr-qc/0402110.

[14] A. Perez, "The spin foam approach to quantum gravity," *Living Rev. Rel.* **16** (2013) 3, arXiv:1205.2019 [gr-qc].

[15] T. Thiemann, "Anomaly-free formulation of nonperturbative, four-dimensional Lorentzian quantum gravity," *Phys. Lett. B* **380** (1996) 257–264, arXiv:gr-qc/9606088.

[16] T. Thiemann, "QSD 4: (2+1) Euclidean quantum gravity as a model to test (3+1) Lorentzian quantum gravity," *Class. Quant. Grav.* **15** (1998) 1249–1280, arXiv:gr-qc/9705018.

[17] T. Thiemann, "Quantum spin dynamics. VIII. The master constraint," *Class. Quant. Grav.* **23** (2006) 2249–2266, arXiv:gr-qc/0510011.

[18] L. Smolin, "The classical limit and the form of the Hamiltonian constraint in nonperturbative quantum general relativity," arXiv:gr-qc/9609034.

[19] M. Varadarajan, "Euclidean LQG dynamics: an electric shift in perspective," arXiv:2101.03115 [gr-qc].

[20] B. Dittrich, "Diffeomorphism symmetry in quantum gravity models," *Adv. Sci. Lett.* **2** (Oct. 2008) 151, arXiv:0810.3594 [gr-qc].

[21] B. Bahr and B. Dittrich, "(Broken) gauge symmetries and constraints in Regge calculus," *Class. Quant. Grav.* **26** (2009) 225011, arXiv:0905.1670 [gr-qc].

[22] B. Dittrich, "How to construct diffeomorphism symmetry on the lattice," *PoS* **QGQGS2011** (2011) 012, arXiv:1201.3840 [gr-qc].

[23] B. Bahr and B. Dittrich, "Improved and perfect actions in discrete gravity," *Phys. Rev. D* **80** (2009) 124030, arXiv:0907.4323 [gr-qc].

[24] B. Bahr, B. Dittrich and S. Steinhaus, "Perfect discretization of reparametrization invariant path integrals," *Phys. Rev. D* **83** (2011) 105026, arXiv:1101.4775 [gr-qc].

[25] B. Bahr and S. Steinhaus, "Numerical evidence for a phase transition in 4D spin foam quantum gravity," *Phys. Rev. Lett.* **117** no. 14 (2016) 141302, arXiv:1605.07649 [gr-qc].

[26] S. K. Asante, B. Dittrich, F. Girelli, A. Riello and P. Tsimiklis, "Quantum geometry from higher gauge theory," *Class. Quant. Grav.* **37** no. 20 (2020) 205001, arXiv:1908.05970 [gr-qc].

[27] S. K. Asante, B. Dittrich and H. M. Haggard, "Effective spin foam models for four-dimensional quantum gravity," arXiv:2004.07013 [gr-qc].

[28] B. Dittrich and T. Thiemann, "Testing the master constraint programme for loop quantum gravity. I. General framework," *Class. Quant. Grav.* **23** (2006) 1025–1066, arXiv:gr-qc/0411138.

[29] B. Dittrich and T. Thiemann, "Testing the master constraint programme for loop quantum gravity. III. SL(2,R) models," *Class. Quant. Grav.* **23** (2006) 1089–1120, arXiv:gr-qc/0411140.

[30] B. Dittrich and T. Thiemann, "Testing the master constraint programme for loop quantum gravity. V. Interacting field theories," *Class. Quant. Grav.* **23** (2006) 1143–1162, arXiv:gr-qc/0411142.

[31] B. Dittrich and T. Thiemann, "Testing the master constraint programme for loop quantum gravity. IV. Free field theories," *Class. Quant. Grav.* **23** (2006) 1121–1142, arXiv:gr-qc/0411141.

[32] B. Dittrich and T. Thiemann, "Testing the master constraint programme for loop quantum gravity. II. Finite dimensional systems," *Class. Quant. Grav.* **23** (2006) 1067–1088, arXiv:gr-qc/0411139.

[33] W. J. Cunningham, B. Dittrich and S. Steinhaus, "Tensor network renormalization with fusion charges: applications to 3D lattice gauge theory," *Universe* **6** no. 7, (2020) 97, arXiv:2002.10472 [hep-th].

[34] I. Agullo and P. Singh, “Loop quantum cosmology,” in *100 Years of General Relativity*, A. Ashtekar and J. Pullin, eds., 183–240. World Scientific, 2017. arXiv:1612.01236 [gr-qc].
[35] B. Dittrich and J. P. Ryan, “On the role of the Barbero-Immirzi parameter in discrete quantum gravity,” *Class. Quant. Grav.* **30** (2013) 095015, arXiv:1209.4892 [gr-qc].
[36] B. Dittrich and P. A. Hohn, “Canonical simplicial gravity,” *Class. Quant. Grav.* **29** (2012) 115009, arXiv:1108.1974 [gr-qc].
[37] B. Dittrich and J. P. Ryan, “Phase space descriptions for simplicial 4D geometries,” *Class. Quant. Grav.* **28** (2011) 065006, arXiv:0807.2806 [gr-qc].
[38] S. Ryu and T. Takayanagi, “Holographic derivation of entanglement entropy from AdS/CFT,” *Phys. Rev. Lett.* **96** (2006) 181602, arXiv:hep-th/0603001.
[39] A. Ashtekar, J. Baez, A. Corichi and K. Krasnov, “Quantum geometry and black hole entropy,” *Phys. Rev. Lett.* **80** (Feb. 1998) 904–907. https://link.aps.org/doi/10.1103/PhysRevLett.80.904.
[40] B. Dittrich and T. Thiemann, “Are the spectra of geometrical operators in loop quantum gravity really discrete?,” *J. Math. Phys.* **50** (2009) 012503, arXiv:0708.1721 [gr-qc].
[41] T. Thiemann, “Solving the problem of time in general relativity and cosmology with phantoms and k-essence,” arXiv:astro-ph/0607380.
[42] B. Dittrich, C. Goeller, E. Livine and A. Riello, “Quasi-local holographic dualities in non-perturbative 3D quantum gravity I: convergence of multiple approaches and examples of Ponzano–Regge statistical duals,” *Nucl. Phys. B* **938** (2019) 807–877, arXiv:1710.04202 [hep-th].
[43] B. Dittrich, C. Goeller, E. R. Livine and A. Riello, “Quasi-local holographic dualities in non-perturbative 3D quantum gravity II: from coherent quantum boundaries to BMS_3 characters,” *Nucl. Phys. B* **938** (2019) 878–934, arXiv:1710.04237 [hep-th].
[44] B. Dittrich, C. Goeller, E. R. Livine and A. Riello, “Quasi-local holographic dualities in non-perturbative 3D quantum gravity,” *Class. Quant. Grav.* **35** no. 13 (2018) 13LT01, arXiv:1803.02759 [hep-th].
[45] S. K. Asante, B. Dittrich and H. M. Haggard, “Holographic description of boundary gravitons in (3+1) dimensions,” *JHEP* **01** (2019) 144, arXiv:1811.11744 [hep-th].
[46] G. Barnich, H. A. Gonzalez, A. Maloney and B. Oblak, “One-loop partition function of three-dimensional flat gravity,” *JHEP* **04** (2015) 178, arXiv:1502.06185 [hep-th].
[47] V. Bonzom and B. Dittrich, “3D holography: from discretum to continuum,” *JHEP* **03** (2016) 208, arXiv:1511.05441 [hep-th].
[48] S. K. Asante, B. Dittrich and F. Hopfmueller, “Holographic formulation of 3D metric gravity with finite boundaries,” *Universe* **5** no. 8 (2019) 181, arXiv:1905.10931 [gr-qc].
[49] A. Maloney and E. Witten, “Quantum gravity partition functions in three dimensions,” *JHEP* **02** (2010) 029, arXiv:0712.0155 [hep-th].

9

Fay Dowker

Professor of Theoretical Physics at the Faculty of Natural Sciences, Department of Physics, Imperial College London

Date: 7 October 2020. Via Zoom. Last edit: 24 November 2020

From your point of view, what are the main problems in theoretical physics at the moment?

Theoretical physics is huge and I can't try to answer that question. In my field, the major task at hand is a task that has been around for a long time but takes a particular form today. That task is to create a unified framework for all of fundamental physics. Such a framework was thought to exist in the mid-nineteenth century, what we might call the mechanistic framework set in Newtonian space and time. In that context, there were bodies and particles with forces between them, and laws of motion. But with the work of Faraday and Maxwell and the discovery of the existence of fields as substances in their own right that mechanistic Newtonian framework turned out to be missing something. And then with the advent of relativity and then quantum mechanics, the sense that we have a comprehensive framework, with only details to fill in, has not been recovered. Today we are in a situation where we don't have a unified framework for fundamental physics and in particular there's a disjunction between our understanding of spacetime and our understanding of matter. If we consider any point in history, the overall goal is to understand everything, without limits and boundaries, and that goal hasn't changed but today one particular aspect of it takes the form of what people call *quantum gravity*.

Why is it difficult to find this unified framework?

It's difficult because the starting points of quantum theory and general relativity are very different. The conceptual basis of quantum theory can't be applied directly to gravity. We have quantum field theory in Minkowski spacetime without gravity but we are already uncomfortable with placing quantum theory on a background that is a solution of general relativity, let alone asking how spacetime itself can be part of the quantum whole. The problem is that quantum theory, as is usually understood, forces us into a position where time is treated as a parameter, while general relativity tells us that spacetime, including time, is a dynamical substance. This tension between the two theories leads to many obstructions which we so far have not been able to resolve. One reason for this is, I believe, because much of the community has been thinking about quantum theory as a canonical theory while I think we should be thinking about it in terms of path integrals.

Are you suggesting that quantum mechanics needs to be modified somehow?

Yes and no (laughs). Once one starts thinking in terms of the path integral, the conceptual basis of quantum theory changes its character. The fundamental ingredients used in a path integral are history and event. There is a relation between them because an event is a set of histories. An example of an event is a particle passing through some region of space, a box, say, at any time between time t_1 and t_2. That event corresponds to a set of histories, namely, all the possible trajectories of the particle such that it is the case that it is inside the box at some time between t_1 and t_2.

The idea that an event is a set of realisations or a set of histories is something that's very familiar to people who work on stochastic processes. In fact, Richard Feynman thought about path integrals in exactly those terms. In his book *QED* [1], which is a series of popular lectures he gave in the 1980s, he speaks about event and history. In particular, to calculate the probability of an event we consider all the histories in that event, add up the amplitudes for all those histories and square the absolute value of the sum, very roughly speaking. This path integral or sum over histories perspective chimes with general relativity as the world in general relativity is not made of three-dimensional space with objects moving in it; instead, it's based on a spacetime, that is, a four-dimensional history of spacetime and events in that spacetime. Thus, thinking about quantum theory via a path integral, the conceptual bases of quantum theory and general relativity appear to be a lot closer.

But is there any modification to quantum mechanics due to formulating it in terms of a path integral?

It depends what you mean by "modify." Basing quantum mechanics on the path integral certainly means a modification in its interpretation, away from an interpretation in terms of a wave function. Usually people think of quantum theory in terms of Hilbert space, state vectors and wave functions. The discussion centres about the wave function and whether it is real or not (laughs). When interpreting this structure, there are debates about the need to add new ingredients, such as a new dynamics for the wave function, or extra particle trajectories to the wave function.

However, from a path integral perspective, the wave function is not a fundamental notion so it doesn't need interpreting (laughs). There is still the problem of interpreting quantum mechanics from a path integral perspective but that's a completely different problem. Now the task is to interpret quantum mechanics in terms of history and event, in what is often called the histories approach to quantum mechanics [2, 3], and not in terms of self-adjoint operators, observables or the split between the observer and the observed.

But what's the problem with understanding the harmonic oscillator from a path integral perspective?

The problem is to understand what's going on inside the box where the harmonic oscillator lives using the path integral formulation. How do we talk about the system, how do we picture it? How do we make specific, comprehensible predictions about the system in terms of its events? We don't know fully yet. However, we have a form of an interpretation, if not a full interpretation: what corresponds to the physical world is a yes-no answer to every question one can ask about the world of the form, "Does event E happen?" [4]. The remaining task, then, is to work out what rules such an "answering map" or "co-event" should satisfy.

This interpretational problem is also intertwined with technical issues such as how to properly define the measure over the space of paths, for which there is no solution to date in the continuum, that I know of. For instance, in the example I gave earlier – the event which is the particle passing through the box at any time between two fixed times – how do we calculate the probability of that event? Feynman gives us a set of conceptual rules. But how do we calculate it in practice if we don't have a measure on path space? This is an open technical question.

There are several approaches to quantum gravity that start with a path integral approach, such as spin foams [5], causal dynamical triangulations (CDT) [6], causal sets [7], etc. Are they all equally good or do you have a preference?

Yes, there are are several approaches to the problem of quantum gravity based on the path integral, or that use path integrals at least. It is important to have a range of approaches. I work on the causal set approach [7], which distinguishes itself from some other path integral approaches in that it is fundamentally discrete. By this I mean that the histories are postulated to be fundamentally discrete at the Planck scale. This is different from the familiar path integral in quantum mechanics. There, discretising the system is a regularisation technique. The propagator in quantum mechanics is defined as the strict mathematical limit of a sequence of skeletonised systems. The causal dynamical triangulations approach uses discreteness to define the path integral for quantum gravity in this way. In CDT the path integral is, conceptually, over continuum spacetimes and the discreteness scale to be taken strictly to zero to obtain the true continuum theory. In causal sets, the discreteness scale is not taken to zero – it is physically fixed to be close to the Planck scale though we don't know the exact value. In causal set theory the full theory is discrete, and we are looking for not a continuum limit, but a continuum *approximation* at large scales. One can look for a continuum limit too but a strict continuum limit of a fundamentally discrete theory like causal sets will inevitably miss some of the physics.

I guess you think that discreteness is something really important at the fundamental level?

Yes, I do. It makes things mathematically well defined for free (laughs) because in a causal set sum over histories, for example, if we sum over causal sets of fixed cardinality of $N = 10^{240}$ elements, that's not just a sum rather than an integral; it is a finite sum. And so we don't need to worry about convergence and existence. This is one of the advantages of the discreteness. Besides this, I think that there are a lot of pointers that discreteness is physically demanded of a theory of quantum gravity. The strongest evidence, I think, is the value of black hole entropy being of the order of the area of the horizon in Planck units. The picture that springs to mind is that of the black hole horizon tiled by Planck-sized areas. In a theory in which spacetime is genuinely continuous, in which physics is genuinely continuous, one is going to face challenges in recovering that result, I believe.

As you probably know, there are many results in string theory where black hole microstates are counted accurately without invoking any kind of fundamental discreteness [8, 9].

Yes, that's interesting. It would be interesting to see whether there really is some fundamental discreteness hidden there (laughs). In the original Strominger-Vafa calculation [10], I think there is a discreteness due to the quantisation of charge. There is no underlying Planck length in that calculation because those black holes are extremal and so the entropy is only a function of the charges. I know that there are black hole microstate counting calculations of black hole entropy away from extremality [11] and those cases make me wonder whether there's a secret discreteness there or not.

Okay, going back to the path integral formulation of causal sets. What exactly is the classical action that you put into the path integral?

We have a family of actions [12–14], one action for each spacetime dimension. The path integral sums over all causal sets, each one being weighted by e^{iS} where S is the action. So we have chosen which dimension action to use, which isn't ideal because we would like the dimension to emerge and not have to put it in. Anyway, say we choose the four-dimensional action, S: we can feed any causal set into it and it will give a number. We conjecture that if the causal set is a special one that has a good approximation as a four-dimensional spacetime, then the causal set action will be related to the Einstein-Hilbert action for that approximating spacetime.

Is this something you can show?

It's a conjecture at the statistical level. We have some evidence for the claim that the average value of this action over many causal sets, each of which has the same spacetime as a good approximation, is close to the Einstein-Hilbert action up to an interesting boundary term [15–17]. We don't know about the fluctuations around this mean value. So it may be that for individual causal sets, the actual action is quite far from the mean. There's a lot to explore with regard to this particular action. It is remarkable that the action is very non-local by definition but in certain cases, say, for example, causal diamonds in Minkowski spacetime, where we expect a vanishing Ricci scalar, the mean action does evaluate to close to zero plus this interesting boundary term [15]. We can indeed prove that in the limit in which the discreteness parameter tends to zero, we get a vanishing value for the bulk contribution to the mean, plus a boundary term.

Why do you refer to the boundary term as an *interesting boundary term*?

By conjecture, the boundary term is equal to the co-dimension two volume of what we call the *joint*. For example, for a causal set that has a causal diamond as a good approximation, the boundary term will be the co-dimension two volume, or area of the sphere that is the intersection between the past cone and the future cone. We don't know yet what the specific role of the boundary term plays but it's very suggestive.

In fact, the action is super interesting. The fundamental degrees of freedom of causal sets are causal relations, which are a type of binary relation. That is, they're relations between pairs of points and not relations between triples. This binary relation is reflected in the action, namely, the action is bi-local. In particular, it is a double sum over the elements of the causal sets. As we know, the Einstein-Hilbert action is a single integration of the Ricci scalar over spacetime. The mean causal set action, however, is a double integral over spacetime because it's bi-local. This has all sorts of interesting consequences. For instance, if we have a portion of spacetime M, we can partition it into two pieces, X and Y, $M = X \cup Y$. We can calculate the action for M and the action for X and for Y separately and see that the action is not additive. The action of M is equal to the action of X plus the action of Y plus a kind of cross term. This cross term is a non-local term depending on the causal relations between the points in X and the points in Y. If we look at the action for a portion of spacetime M to the past of some spacelike hypersurface that intersects a black hole horizon we can ask: What do we get if X and Y are inside and outside, respectively, of the horizon? (laughs)

And the answer is? (laughs)

If our conjectures hold, what we get is something that is proportional to the area of the black hole horizon. This is the case because the boundary term has the specific form it has and in particular it is co-dimension two.

So do you actually have a calculation of black hole entropy from causal sets?

Well, we have a conjecture and simulations that never made the light of day in terms of published papers but it's in my PhD student's thesis [18]. We do not have analytic calculations but the simulations are not inconsistent with the conjecture. For the simulations one needs very big causal sets which come with large fluctuations, making things difficult. There are other proposals for entropy of black holes from causal sets that rely on the idea that the horizon is formed of horizon molecules and the entropy counts them like the entropy of a box of gas counts the molecules of the gas [19, 20].

What's the future of causal sets? Do you need to perform large-scale simulations?

That's one direction that we would love to be able to make progress in. Being able to do big simulations more efficiently would be great. There are people who are very good at doing efficient computing and who are doing causal set simulations. The restriction for us is not really computing speed but memory. One cannot efficiently compress the information of a causal set; one needs to store the whole causal set.

Taking one step back, what are actually the principles of the causal set approach to quantum gravity?

The path integral is the foundation stone of the quantum theory of causal sets and ...

Sorry to interrupt but just to be clear, there is a quantum theory of, and quantum results coming out of, causal sets?

I can't claim that. I can't claim that we have a quantum dynamics for causal sets which we believe will be the physically correct quantum dynamics. So, that's work in progress. To formulate this path integral dynamics for causal sets is a major frontier for us. Coming back to your earlier question, the path integral is the basis for the quantum dynamics of the theory. The other principle is that spacetime is fundamentally discrete and takes the form of a discrete order, a causal set, meaning that of all the structures that a continuum spacetime possesses (metric, differentiable structure, topology, etc.) the most fundamental and physical one is causal order. That causal order is going to survive the revolution of moving from classical to quantum gravity.

By *causal order* you mean causality?

For historical reasons, we're stuck with the term *causal set* because of what the corresponding order relation is called in the continuum theory. So in general relativity, we call this order relation the causal order. In general relativity when we say event X is in the causal future of event Y, we just mean mean that X is after Y. So that order perhaps should be called *temporal order*, simply meaning the notion of before and after. This notion is rather different from the other potential meaning of causal which is connected to *causation*, that of Y being the *cause* of X, whatever that means. This gets one totally embroiled in the long-standing philosophical brouhaha about what it means for one event to cause another event. That is not what we are referring to with causal order and causal set. To avoid confusion, it would be better if we could go back and rewrite the textbooks and refer to the spacetime causal order as temporal order or order of precedence or something else. But we're stuck with the term spacetime causal order and hence the term "causal sets" now.

Okay, so the path integral and temporal order is all you need for building up the theory?

The suggestion is that for spacetime itself, setting matter apart for now, the order and the discreteness of the causal set are all one needs for it to be the foundation of an approximately continuum theory of spacetime. As sparse as it seems, the discreteness together with the causal relation are sufficiently rich to contain all geometric information and to give an approximately continuous spacetime at scales much larger than the discreteness scale. This is a central conjecture of the theory and we have lots of evidence for it but no proof to date [7].

How do you construct a causal set that represents a given spacetime and vice versa?

One way is easy, the other way is hard. If we're given a continuum spacetime that is causally nice, that doesn't have closed causal loops, say, it's globally hyperbolic, then there is a way of constructing a causal set to which that spacetime is a good approximation. The way to build this causal set is by a random sampling of the points of the spacetime. In particular, we take points in this spacetime uniformly distributed at random according to the measure given by the volume form, say, $\sqrt{-g}d^4x$ in four dimensions where g is the determinant of the metric. We fix the density of points such that the average number of points that are chosen from a region of volume V equals V in units of a fundamental volume of the order of the Planck volume. Then, these selected points are endowed with the order relation they inherit from their spacetime causal order. So if the chosen point X is in the past light-cone of the chosen point Y, then X precedes Y in the discrete order that we're constructing. That discrete order is a causal set that has spacetime as a good approximation. This is the process we call *sprinkling*, which is a kind of Poisson process [7].

But if you perform this process twice you'll get two different causal sets, right? So how many causal sets represent the same spacetime?

Yes, that's right, there are many causal sets that have the same continuum spacetime as an approximation.

So if I give you two causal sets how do you know that they are well approximated by the same spacetime?

Now you are getting to the reverse engineering problem. In the specific case you ask, I would say that you should construct the approximate spacetimes to each causal set, and check that those two spacetimes are approximately isometric. Both steps are hard (laughs).

But is there a precise procedure for taking a causal set and constructing the continuum spacetime it approximates to?

There is no algorithm that we can apply in practice, to date. In principle, in concept, there is one but it's not a practical method. Take each continuum Lorentzian spacetime in turn and check if the causal set can be embedded into it in such a way that it is a *faithful embedding*. A faithful embedding means the order relation of the causal set has to match the spacetime causal order, and the number of elements in any spacetime region must be equal to the volume of the region in fundamental units, up to fluctuations. We go through the Lorentzian spacetimes until we've found one into which the causal set is faithfully embeddable.

But isn't this an important question for the causal set programme?

Yes, formally it's an important question because we want to know that the approximating spacetime is essentially unique. If a causal set faithfully embeds into spacetime M_1 and into M_2, then M_1 and M_2 are indistinguishable from each other up to the discreteness scale. They might be different but the difference is at the discreteness scale or smaller and no one cares about that. So, this is the central conjecture – the *Hauptvermutung* – of causal set theory [21–23].

Proving this conjecture as a mathematical theorem would certainly be good. In the meantime we do find evidence for it in other ways. In particular, when we know we have a causal set which does faithfully embed, we want to be able to check that we can read off geometrical information from the causal set, such as the spacetime dimension, and show that it agrees with the continuum spacetime that the causal set embeds into. So we make a causal set C by sprinkling at unit density into a Lorentzian spacetime (M, g), then throw away the spacetime information and ask, Can we extract geometric information from the order relation of C alone? The answer is: Yes, we can do quite a lot of this. For flat spacetime sprinklings we can reconstruct, for example, the dimension in various ways [24], the timelike geodesic distance [25], spatial topology [26] and spacelike distance [27]. For curved space sprinklings we have, for example, a Ricci scalar curvature estimator [12], though we haven't learnt how to tame the fluctuations well enough to claim it is very accurate for a single causal set – probably we need to average it over a region to suppress the fluctuations. There is work to do, but even with what we know already, we can get quite a lot of geometric data especially in the flat case.

Now that I understand what a causal set is, can you tell me what is the classical dynamics of the theory?

So far we have only been talking about the kinematics of causal sets, that is, the properties of causal sets. For the classical dynamics we have stochastic models for growing causal sets. It's a type of process in which the causal set grows element by element ...

You mean that this is the theory describing the evolution of the universe starting with the big bang?

Well, we don't believe that these models are quite the right physical theory because they are not quantum, but in concept, yes, they are cosmological theories of a causal set that grows starting from the empty set (laughs) so there is a beginning of time if you like.

But are these models related in any way to the path integral formulation of the theory?

It's in the same spirit as the path integral, but classical. What kind of theory is the path integral? If one thinks of it as a species of measure theory then it falls into a hierarchy of measure theories which includes classical stochastic theory [3]. The dynamics of stochastic processes is mathematically a measure. A simple example of this that physicists know well is Brownian motion (the theory of the dynamics of a Brownian particle) and its discrete version, the random walk (the dynamical model of a walker walking on some discrete background such as a two-dimensional lattice). Mathematical physicists think of this class of models as measure theories. In the case of Brownian motion, the measure is the Wiener measure on Brownian paths (see, e.g., [28]). Given this measure one can ask questions such as: Does the Brownian particle pass through a given region of space? There is a corresponding probability, or measure, for that event to happen. Typically, individual trajectories have measure zero but one can ask about coarse properties of the trajectories, which are measurable events. Path integral quantum theory is also a measure theory.

I understand, but is there a specific limit of that path integral that gives you exactly the stochastic process that you are using for growing the causal set?

That's a very good question but we can't answer it right now because we don't have the quantum path integral for the causal set. But would we expect it to be the case? I am not sure. We could ask the same question about ordinary particle quantum mechanics. If we take the path integral of a particle, is there some kind of limit that gives us Brownian motion rather than classical Hamiltonian dynamics? Anyway, I think that this characterisation of quantum theory as a species of measure theory makes quantum theory more akin to stochastic processes like Brownian motion than it is to canonical Hamiltonian dynamics. There is a genuine distinction between taking the canonical route to quantisation and the path integral route to quantisation.

Earlier you mentioned that you did not have a quantum path integral. Is it the case that you actually do not have it or that it is just quite difficult to do anything with what you have?

We have a few what I'd call toy models of path integrals for growing causal sets. They are heavily based on the classical growth models that I've mentioned already in the sense that what they do is to take those classical growth models and replace transition probabilities with transition amplitudes [29]. However, such models are rather strange in various ways and we have no strong reason to think that they are going to turn out to be physically interesting. Instead, we see them as a sort of proof in principle that such things can be constructed and a playground for exploring issues such as the extension of the quantum measure. So far we're lacking more examples of these path integral growth models for causal sets.

Going back to the classical stochastic models, are these growth models going to yield a causal set at the end of the day that approximates the continuum spacetime of some solution of Einstein's equations?

At the end, if we can't recover dynamically the prediction that our universe looks like a solution of Einstein's equation, then we don't have a successful theory of quantum gravity (laughs). We have to do that but maybe you're asking how far we are from that?

Not exactly. I do not know all the details of the classical growth process but I imagine that you start adding elements and relations between them and so on. So the question is why at the end of the day should I expect that the causal set will have anything to do with Einstein's equations? Is this built into the growth process?

No, it's not built in at all. Let me give you a bit of context. There are two approaches that people are pursuing right now to finding a quantum theory of causal sets [7]. One we've actually already touched on which is to do with what one might call a *state sum model* where we just take all causal sets of a fixed cardinality and weight them by e^{iS} [30–33]. The number of causal sets of cardinality n grows super exponentially with n and we will not be able to perform that sum because there are too many causal sets so we have to do some kind of simulation using Monte Carlo techniques to sample the space of causal sets and see what a typical causal set will look like. To do that one must wrestle with the issue that the weights are not real and positive but complex. Within this approach, we could worry that we are putting in the dimension of spacetime by hand because we are using the action S which was constructed such that it is related to, say, the four-dimensional Einstein-Hilbert action as we discussed earlier. I believe that this approach won't be truly fundamental but is important to pursue because it may be that in some regime of the theory or in some cosmological epoch, this approach is close to the real physics. Thus it's important for us to investigate these state sum models.

On the other hand, the paradigm of growth models is more likely to be the fundamental dynamics because in that case we're not putting in the spacetime dimension or anything constructed from the Einstein-Hilbert action. In the classical case, growth models were constructed by Rafael Sorkin and David Rideout [34] using three physically motivated rules or principles. One rule is that causal sets grow element by element such that when each element is born, it cannot be born before any already existing element, that is, each element is born either to the future of or spacelike to each existing element.

The second rule is that the labelling of the elements is just a coordinate. Since the model is based on a series of births, each birth comes with an integer label. By convention the first element to be born has the label 0, the next has the label 1, and so on. This process endows the causal set elements with labels but we must ensure only the partial order relation of the elements is physical and not the labels. The rule of *discrete general covariance* states that the probability of growing a particular causal set labelled in some way equals the probability of growing an order-isomorphic causal set with a different labelling.

The third rule is what Rideout and Sorkin call *Bell causality* [34]. Roughly speaking, what this condition means is that if we already have grown a partial causal set and a new element is born, the transition probability of that new element attaching itself to some of the causal set elements over here doesn't depend on the structure of the causal set over there. It's a local causality condition, which is the reason why it's called Bell causality and it states that the structure of the causal set in one region does not affect the growth of the causal set in a region spacelike to it.

So these three rules don't really have anything do with the Einstein-Hilbert action directly (laughs). However, the hope is that we can construct quantum growth models constrained by quantum versions of these principles, which will restrict the types of quantum path integrals to an interesting and small enough class for us to investigate. Ultimately, general relativity should emerge from them.

But this is an open question, right?

Oh, yeah. That's a very open question (laughs).

Okay, so focusing on the classical case, I imagine that people have simulated the growth of some causal sets. Do they get at the end of the day a causal set that approximates Minkowski or de Sitter spacetime?

There's a family of such stochastic models and they're relatively unexplored. One of the models, which is the most explored, is called transitive percolation [34]. It's a class of models known to mathematicians as random graph orders [35]. It's a kind of Lorentzian analogue of the Erdös-Renyi graph [36]. The Erdös-Renyi graph is a really interesting object. One way to construct it is the following. Fix a probability P strictly between zero and one. Take the integers $\{1, 2, 3, \ldots, N\}$ as vertices and add an edge between every pair of vertices – independently – with probability P. Then add a new vertex, $N + 1$, and add an edge between it and every already existing vertex with probability P. Add vertex $N + 2$ in the same way. This provides a process by which a graph grows. The interesting – and rather surprising – thing about the Erdös-Renyi random graph is that it turns out to be a deterministic model when run to infinity. And no matter what P is, when we reach infinitely many vertices we always get the same infinite graph, called the Erdös-Renyi graph or the Rado graph.

Transitive percolation is this same process with a couple of tweaks so that it grows a discrete order or causal set, rather than a simple graph. When an edge is added between i and j with $i < j$ that edge is directed; i.e., $i \prec j$ in the order. And after each stage of the process, the transitive closure is taken to make the relation an order. These tweaks make a huge difference. Now, each time we run the growth process to infinity we will get a different causal set each time for fixed P [35]. And for different P the resulting measure on infinite causal sets is different. However, each of the infinite causal sets grown in any of these models for any P has an interesting, bouncing property. As the causal set grows, the maximal number of elements it has is typically of order $1/P$ where P is the probability parameter that defines the model. But every now and then, at random, the width of the causal set suddenly collapses so that there is only one maximal element. After that it expands again until it reaches this typical width of $1/P$ and fluctuates at about that width for a long time before the width collapses to a single maximal element again, ad infinitum. It's a bouncing cosmology. In fact, with probability one, there'll be infinitely many of these big crunch/big bangs in transitive percolation, for any P, which is nice (laughs).

But does this causal set approximate any continuum spacetime?

Between these big crunch/big bang transitions, which are called posts, one could imagine that the causal set is manifold-like in a way that its continuum approximation is that of a cosmological spacetime. On the other hand, the posts are genuine singularities in the continuum description. This is very suggestive as it would mean that the causal set has a continuum approximation in some places while in others it doesn't.

But do you know if in some places it has a continuum approximation?

No, we don't know that.

But hasn't anyone tried to embed parts of the causal set in de Sitter spacetime or something like that?

Yes, so there's been some work by David Rideout and Maqbool Ahmed [37]. They've studied some properties of the causal set in transitive percolation and they show that there are some measures which match up with de Sitter spacetime. But I think none of us believes that the causal set is really approximated by de Sitter.

Okay. Is your point of view that fundamentally we should be talking about causal sets and spacetime as just some description you may have in certain regimes?

Yes, that's right.

And is it your point of view that this model gives you the correct dynamics for the universe or do you still need to find such model?

This model that we have been talking about is transitive percolation and it is classical, meaning there is no interference. I am strongly of the opinion that spacetime really is quantum mechanical. Transitive percolation is not a quantum model. So, I don't think that it will give us what we want, that is, we're not going to get classical spacetime, general relativity or continuum physics from it. It's a toy model that lets us ask some types of questions and practice obtaining answers that we might expect to get from the quantum theory. The most promising direction that I can see for understanding quantum gravity is to find quantum growth models for causal sets, which between posts would generate causal sets that are manifold-like.

Is the theory as it stands Lorentz-invariant?

Yes, Lorentz invariance comes out of the theory. In fact, if causal sets is the right way to think about quantum gravity then Lorentz invariance emerges from the theory in the following sense. Causal order and discreteness are sufficient in order to reconstruct the approximate continuum geometry. Why do we think this is the case? Because there is a theorem in the continuum which says that if we know the causal relation between all points in spacetime, and the spacetime volume of every region then we know the full metric, topology, spacetime dimension, differentiable structure, that is, everything [38–41]. So, the idea from the causal set perspective is that underlying spacetime there is a causal set. The order relation of the causal set will give us the spacetime causal order, and the number of causal set elements that corresponds to a particular spacetime region gives us the spacetime volume [42]. The volume of a spacetime region is a count of how many causal set elements comprise that region. So, the causal set then has within itself all the information we need.

For this correspondence to work it is crucial that the number of causal elements that corresponds to a particular region is approximately the volume of that region in fundamental units. The only way to achieve that is if those points are distributed randomly. Any other way of distributing – for instance, if we try to do it via some coordinate system – won't work because in a boosted coordinate system the distribution shows itself to be non-uniform. Thus, the only way that the correspondence between volume and number of points in the causal set can work is if the points are distributed randomly. And a random distribution is Lorentz-invariant. This is how Lorentz invariance emerges.

Is this proven or is it an expectation?

Strictly, it's an expectation because it depends on the *Hauptvermutung* for one thing. There are certain things which I claimed in this discussion which lack rigorous mathematical proof.

In the past it was claimed that causal sets predicted a very small cosmological constant [22, 43]. Is this in fact a prediction?

Yes, it was a prediction in the historical meaning of the term *prediction* (laughs). The prediction was made prior to the observations that convinced everyone that the expansion of the universe is accelerating and that the cosmological term Λ is of the order of 10^{-120} (see, e.g., [44]). This prediction was made at a time when no theorist, apart from perhaps arguably Steven Weinberg [45], was making such a claim. Rafael Sorkin predicted that we would measure such a small value for the cosmological constant [22, 43]. In fact, in one of the papers Rafael said that "a cosmological term of this order of magnitude is just on the edge of being observable." It turned out that, as you know, a few years later we were able to measure it.

Right. But how can you predict such a thing if there is no quantum dynamics for causal sets?

It is a heuristic argument. It's not based on a fully fledged, agreed-upon quantum dynamics of causal sets, because, as you say, we don't have one. Instead, it's based upon expectations of what that dynamics would be like if we did have one. The prediction relies on the Heisenberg uncertainty relation between time and energy, which is a heuristic that everyone uses (laughs). We can derive a Heisenberg uncertainty relation between x and p using standard canonical quantum theory but there is no understanding of t as an operator. Anyway, no one doubts that there is such a Heisenberg uncertainty relation (laughs) and it's perfectly legitimate to use it, as when discussing resonances in atomic theory. So in the case of the cosmological constant Λ, the prediction is based on a time/energy uncertainty relation where instead of time we have spacetime volume and instead of energy we have Λ.

And you think that, even though it's heuristic, you would still find such a prediction if you had the right quantum dynamics?

Yes, I hope so. It's our one successful prediction-in-advance from quantum gravity so it better survive (laughs). Rafael once told me that "quantum gravity bats a thousand." It's a baseball statistics term. A batter in baseball has a batting average which for some reason is quoted as a number up to a thousand. If a batter hits the ball and gets to first base every time they bat, then they have a batting average of a thousand. A good batter has an average of roughly 300 which means that 30% of the time they make a hit. So, *to bat a thousand* means that every prediction you've made turns out to be right (laughs). So yeah, quantum gravity bats a thousand because it made one prediction-in-advance which is the value of the cosmological term and it turned out to be right (laughs). In the causal set community this prediction certainly served to strengthen our feeling that this direction of research is very promising.

Are there more predictions coming out of causal sets?

Yes, there are. In this paradigm, Λ is not actually a constant. The cosmological constant is as constant as Hubble's constant (laughs). And not only is it not a constant but it fluctuates, and not only does it fluctuate but it fluctuates randomly between positive and negative values throughout cosmic history [22, 43]. That makes it super different from any other model in which Λ is not constant and it gives possibilities for different phenomenology. It obviously brings challenges as well as opportunities compared to other cosmological paradigms. So we might be able to observe these fluctuations in the cosmological constant.

This model, that Rafael and his collaborators called *everpresent* Λ, is a model in which the expected value of Λ is always of the order of the ambient matter density but it fluctuates between positive and negative values throughout cosmic history [46]. There is recent work by Zwane, Afshordi and Sorkin where they tested the everpresent Λ model against a large subset of cosmological constraints. They found that it can generate cosmologies in which Λ fluctuates and which do as well as Λ-CDM in terms of consistency with data [47], including doing better than Λ-CDM in alleviating the tension between high and low redshift measurements of the Hubble parameter, H_0, [48].

Not all cosmologies generated by the model are consistent with data because the model is random, but Zwane et al. found that the cosmologies that are consistent with data are not atypical in the model. We think that those random fluctuations are actually better treated as quantum mechanical fluctuations but these models are just classically stochastic at the moment.

What is the most important breakthrough in theoretical physics in the past 30 years in your point of view?

Again, I can only answer this question within my sphere of research. I think that what has changed in the past 30 years is that the problem of understanding how general relativity and quantum theory can be reconciled has become mainstream within the high-energy theoretical physics community. Nowadays, many more people are being trained in gravity and black hole physics than when I was young. It's not exactly a breakthrough but it's an important development.

Why did you choose to do physics and not something else?

It kind of chose me. I was only interested in maths when I was young. In the English school system you had to choose three subjects to study between the ages of 16 and 18. I chose maths, maths and physics (laughs). And if I had been able to, I would have chosen maths, maths and maths, I think (laughs). I wasn't really interested in physics in school and I went to study maths at university. Then, I found that I wasn't so good at pure maths, I found it difficult. Although I had liked it a lot before, I started to find it really hard, and simultaneously, I did a first course in general relativity and I thought it was so great and I loved it so much. It attracted me very much and that's when I decided to do physics.

What do you think is the role of the theoretical physicist in modern society?

It's to produce new, accurate and reliable knowledge about the world. That's at the heart of it all. Perhaps something that theoretical physicists could teach people is the idea of unification, which is a pervasive concept and driving heuristic in my field. I don't really mean unification in the sense of a final unified theory, which I'm not sure is a particularly useful idea, but in the sense that the world itself is connected as a unified whole in reality and that when we focus on one particular thing to try to understand it, we always have to ignore the rest of the world and have to make some background assumptions. I think that this kind of awareness can be important for any attempt to understand something, be it a situation, another person or a political choice. It makes sense, in many cases, to make these assumptions and to decide upon and ignore externalities and we need to do it because of how complicated things are. But if we are at least aware of those assumptions, that can be very helpful and gives us flexibility and openness of mind.

References

[1] R. Feynman, *QED: The Strange Theory of Light and Matter*. Universities Press (India) Pvt. Limited, 1985.

[2] J. B. Hartle, "Space-time quantum mechanics and the quantum mechanics of space-time," in *Proceedings of the Les Houches Summer School on Gravitation and Quantizations, Les Houches, France, 6 Jul–1 Aug 1992*, J. Zinn-Justin and B. Julia, eds. North-Holland, 1995. gr-qc/9304006.

[3] R. D. Sorkin, "Quantum mechanics as quantum measure theory," *Mod. Phys. Lett.* **A9** (1994) 3119–3128, gr-qc/9401003.

[4] R. D. Sorkin, "Quantum dynamics without the wave function," *J. Phys.* **A40** (2007) 3207–3222, arXiv:quant-ph/0610204.

[5] A. Perez, "The spin foam approach to quantum gravity," *Living Rev. Rel.* **16** (2013) 3, arXiv:1205.2019 [gr-qc].

[6] R. Loll, "Quantum gravity from causal dynamical triangulations: a review," *Class. Quant. Grav.* **37** no. 1 (2020) 013002, arXiv:1905.08669 [hep-th].

[7] S. Surya, "The causal set approach to quantum gravity," *Living Rev. Rel.* **22** no. 1 (2019) 5, arXiv:1903.11544 [gr-qc].

[8] J. R. David, G. Mandal and S. R. Wadia, "Microscopic formulation of black holes in string theory," *Phys. Rept.* **369** (2002) 549–686, arXiv:hep-th/0203048.

[9] A. Sen, "Microscopic and macroscopic entropy of extremal black holes in string theory," *Gen. Rel. Grav.* **46** (2014) 1711, arXiv:1402.0109 [hep-th].

[10] A. Strominger and C. Vafa, "Microscopic origin of the Bekenstein-Hawking entropy," *Phys. Lett. B* **379** (1996) 99–104, arXiv:hep-th/9601029.

[11] G. T. Horowitz and A. Strominger, "Counting states of near extremal black holes," *Phys. Rev. Lett.* **77** (1996) 2368–2371, arXiv:hep-th/9602051.

[12] D. M. T. Benincasa and F. Dowker, "Scalar curvature of a causal set," *Phys. Rev. Lett.* **104** (May 2010) 181301. https://link.aps.org/doi/10.1103/PhysRevLett.104.181301.

[13] F. Dowker and L. Glaser, "Causal set d'Alembertians for various dimensions," *Class. Quant. Grav.* **30** (2013) 195016, arXiv:1305.2588 [gr-qc].

[14] L. Glaser, "A closed form expression for the causal set d'Alembertian," *Class. Quant. Grav.* **31** (2014) 095007, arXiv:1311.1701 [math-ph].

[15] M. Buck, F. Dowker, I. Jubb and S. Surya, "Boundary terms for causal sets," *Class. Quant. Grav.* **32** no. 20 (2015) 205004, arXiv:1502.05388 [gr-qc].

[16] F. Dowker, "Boundary contributions in the causal set action," *Class. Quant. Grav.* (2020). http://iopscience.iop.org/article/10.1088/1361-6382/abc2fd.

[17] L. Machet and J. Wang, "On the continuum limit of Benincasa-Dowker-Glaser causal set action," *Class Quant. Grav.* **38** (2021) 015010.

[18] D. Benincasa, "The action of a causal set," PhD thesis, Imperial College London, 2013.

[19] D. Dou and R. D. Sorkin, "Black hole entropy as causal links," *Found. Phys.* **33** (2003) 279–296, gr-qc/0302009.

[20] C. Barton, A. Counsell, F. Dowker, D. S. Gould, I. Jubb and G. Taylor, "Horizon molecules in causal set theory," *Phys. Rev. D* **100** no. 12 (2019) 126008, arXiv:1909.08620 [gr-qc].

[21] R. D. Sorkin, "First steps with causal sets," in *Proceedings of the Ninth Italian Conference on General Relativity and Gravitational Physics, Capri, Italy, September 1990*, R. Cianci, R. de Ritis, M. Francaviglia, G. Marmo, C. Rubano and P. Scudellaro, eds., pp. 68–90. World Scientific, Singapore, 1991. www.perimeterinstitute.ca/personal/rsorkin/some.papers/65.capri.pdf.

[22] R. D. Sorkin, "Space-time and causal sets," in *Relativity and Gravitation: Classical and Quantum, Proceedings of the SILARG VII Conference, Cocoyoc, Mexico, December 1990*, J. C. D'Olivo, E. Nahmad-Achar, M. Rosenbaum, M. P. Ryan, L. F. Urrutia and F. Zertuche, eds., pp. 150–173. World Scientific, Singapore, 1991. www.perimeterinstitute.ca/personal/rsorkin/some.papers/66.cocoyoc.pdf.

[23] R. D. Sorkin, "Causal sets: discrete gravity (notes for the Valdivia Summer School)," in *Lectures on Quantum Gravity, Proceedings of the Valdivia Summer School, Valdivia, Chile, January 2002*, A. Gomberoff and D. Marolf, eds. Plenum, 2005. gr-qc/0309009.

[24] D. Meyer, "The dimension of causal sets," PhD thesis, Massachusetts Institute of Technology, 1988. http://hdl.handle.net/1721.1/14328.

[25] G. Brightwell and R. Gregory, "The structure of random discrete space-time," *Phys. Rev. Lett.* **66** (1991) 260–263.

[26] S. Major, D. Rideout and S. Surya, "On recovering continuum topology from a causal set," *J. Math. Phys.* **48** (2007) 032501, arXiv:gr-qc/0604124 [gr-qc].

[27] D. Rideout and P. Wallden, "Spacelike distance from discrete causal order," *Class. Quant. Grav.* **26** (2009) 155013, arXiv:0810.1768 [gr-qc].

[28] H. Kleinert, *Path Integrals in Quantum Mechanics, Statistics, Polymer Physics, and Financial Markets*. EBL-Schweitzer. World Scientific, 2009.

[29] S. Surya and S. Zalel, "A criterion for covariance in complex sequential growth models," 2020.

[30] S. Surya, "Evidence for a phase transition in 2D causal set quantum gravity," *Class. Quant. Grav.* **29** 132001, arXiv:1110.6244 [gr-qc].

[31] L. Glaser, D. O'Connor and S. Surya, "Finite size scaling in 2D causal set quantum gravity," *Class. Quant. Grav.* **35** no. 4 (2018) 045006, arXiv:1706.06432 [gr-qc].

[32] S. Loomis and S. Carlip, "Suppression of non-manifold-like sets in the causal set path integral," *Class. Quant. Grav.* **35** no. 2 (2018) 024002, arXiv:1709.00064 [gr-qc].

[33] W. J. Cunningham and S. Surya, "Dimensionally restricted causal set quantum gravity: examples in two and three dimensions," *Class. Quant. Grav.* **37** no. 5 (2020) 054002, arXiv:1908.11647 [gr-qc].

[34] D. Rideout and R. Sorkin, "A classical sequential growth dynamics for causal sets," *Phys. Rev. D* **61** (2000) 024002, arXiv:gr-qc/9904062.

[35] B. Bollobás and G. Brightwell, "The structure of random graph orders," *SIAM J. Discrete Math.* **10** (1997) 318–335.

[36] P. Erdos and A. Renyi, "On random graphs," *Publicationes Mathematicae* **6** (1959) 290–297.

[37] M. Ahmed and D. Rideout, "Indications of de Sitter spacetime from classical sequential growth dynamics of causal sets," *Phys. Rev. D* **81** (Apr. 2010) 083528. https://link.aps.org/doi/10.1103/PhysRevD.81.083528.

[38] E. Kronheimer and R. Penrose, "On the structure of causal spaces," *Proc. Cambridge Phil. Soc.* **63** (1967) 481–501.

[39] R. Penrose, *Techniques of Differential Topology in Relativity*. SIAM, 1972.

[40] S. Hawking, "Singularities and the geometry of spacetime," *EPJ H* **39** (2014) 413–513.

[41] D. B. Malament, "The class of continuous timelike curves determines the topology of spacetime," *J. Math. Phys.* **18** (1977) 1399–1404.

[42] L. Bombelli, J. Lee, D. Meyer and R. D. Sorkin, "Space-time as a causal set," *Phys. Rev. Lett.* **59** (Aug. 1987) 521–524. https://link.aps.org/doi/10.1103/PhysRevLett.59.521.

[43] R. D. Sorkin, "Forks in the road, on the way to quantum gravity," *Int. J. Theor. Phys.* **36** (1997) 2759–2781, arXiv:gr-qc/9706002.

[44] A. G. Riess et al., "A 2.4% determination of the local value of the Hubble constant," *Astrophys. J.* **826** no. 1 (2016) 56, arXiv:1604.01424 [astro-ph.CO].

[45] S. Weinberg, "Anthropic bound on the cosmological constant," *Phys. Rev. Lett.* **59** (Nov. 1987) 2607–2610. https://link.aps.org/doi/10.1103/PhysRevLett.59.2607.

[46] M. Ahmed, S. Dodelson, P. B. Greene and R. Sorkin, "Everpresent Λ," *Phys. Rev. D* **69** (2004) 103523, arXiv:astro-ph/0209274.

[47] N. Zwane, N. Afshordi and R. D. Sorkin, "Cosmological tests of everpresent Λ," *Class. Quant. Grav.* **35** no. 19 (2018) 194002, arXiv:1703.06265 [gr-qc].

[48] A. G. Riess, "The expansion of the universe is faster than expected," *Nature Rev. Phys.* **2** no. 1 (2019) 10–12, arXiv:2001.03624 [astro-ph.CO].

10

Laurent Freidel

Faculty at Perimeter Institute for Theoretical Physics, Waterloo, Canada

Date: 12 July 2011. Location: Waterloo, Canada. Last edit: 14 October 2020

In your opinion, what are the main problems in theoretical physics at the moment?

There are many. I'm more concerned with theoretical problems such as understanding what is the right theory of quantum gravity. How can we put together quantum mechanics and dynamical spacetime? That is one of the most fundamental questions, along with some questions related to cosmology, such as what really happened near the big bang or what are the constituents of dark matter. However, I think that an experimental breakthrough will be necessary to guide us in the right direction.

Why do we need a theory of quantum gravity? What can't we explain?

We can explain pretty much everything except for 90% of what the universe is made of. So that's quite a lot that we can't explain and we also can't explain how these two theories, that we really know work perfectly well independently, can be put together. These are fundamental limitations in our understanding. There is also a fundamental limitation in our understanding of quantum mechanics.

You mean that quantum mechanics must be modified somehow or do you think it is wrong?

I don't think it's wrong. It might just be true and we are just fundamentally unable to wrap our brains around it. That's one possibility.

But is it a matter of knowing which interpretation of quantum mechanics is the right one or is it a matter of changing some of its principles?

I'm not sure. There are two questions here which are tied up. What is quantum gravity going to teach us about quantum mechanics? Do you have to somehow modify quantum mechanics in order to grasp quantum gravity? I don't really want to bet. It might be that quantum gravity will lead to radical modifications in our understanding of locality and hence maybe in the way we understand quantum mechanics. I think it's good to be open-minded.

In string theory quantum mechanics is not modified in any way . . .

It's perfectly plausible that it does not need to be modified. In fact, loop quantum gravity (LQG) and spin foam theories are based on the usual quantum mechanics. In these cases there is no necessity in changing the framework of quantum mechanics. I think that it's sensible to start first by assuming the correctness of quantum mechanics and then see what the model that you are working on leads you to. They might indicate that one must go beyond quantum mechanics or not at all, even though we still don't really know how to interpret quantum mechanics and don't understand it at the philosophical level.

Why is the problem of combining gravity with quantum mechanics so difficult?

Einstein's theory of gravity tells us not only that the fabric of space and time is something which is relative to the observer but that it also depends on the amount of matter and energy there is. Spacetime is a dynamical quantity, meaning that it's jiggling, it's moving around and changing depending on the objects which are moving in it. It's a dynamical entity and to some extent we understand very well how it works. For instance, if you put lots of matter in the same point you can create black holes (regions of spacetime where the notion of time as seen by an observer comes to a halt near the black hole horizon and from which no information can escape from it). On the other hand, quantum mechanics is a completely different theory that tells us that objects can, for instance, be at two places at the same time or can go through a wall and don't follow continuous paths. So when you try to apply the theory of quantum mechanics to gravity you find many conceptual problems because now you have to imagine that there is some kind of grain of space and grain of time that are jumping around, behaving in this quantum mechanical manner.

In quantum mechanics you can formulate the Heisenberg uncertainty principle that tells you that if you try to probe the position of some object, then you can only probe it to a certain degree of accuracy and that degree of accuracy is limited by the degree of accuracy involved in probing the object's energy. In physics, we look at things by sending light or particles of higher and higher energy. The higher the energy, the smaller the distances that you can see under the microscope. However, at some point, if you want to really look at the smallest distance, you'll have to send such a high amount of energy and localise it somewhere that you are going to create a mini black hole, from which you are not going to be able to extract any information.

So there is this fundamental limitation in physics that tell us that you cannot, given the theory we have, probe the notion of spacetime up to a certain length scale. So people wonder what does really happen at this length scale, what is space and time really made of? At this length scale, or this energy scale, it is going to be very hard to distinguish between spacetime and matter, right? There is no longer a distinction between what is the theatre and who are the actors. We are used to a description of physics where the theory is fixed and the actors are well known but in quantum gravity it is going to be mixed up.

This is one explanation for why it is hard to quantise gravity but there is another more conceptual explanation. There is a clash between these two theories. If you think of them as theories based on principles then, from quantum mechanics, you can extract one principle, namely, the Born reciprocity rule [1], which, by the way, was used by Born to formulate the problem of quantum gravity in 1938 [2]. The Born reciprocity rule expresses the dual nature of objects as particles and waves; that is, you can describe objects in terms of position but you can also describe them in terms of states, which depend on energy. The physics of these two dual descriptions is equivalent; it is a kind of fundamental symmetry behind quantum mechanics making it very different from classical mechanics and the world we experience. In the world we experience, we can distinguish between the position and energy very radically and we have to measure one and the other in order to make sense of the object, whereas in quantum mechanics, these two notions are vaguely unified. On the other hand, generally relativity tells you that there is an imbalance between the notion of energy and the notion of spacetime, thus in some sense breaking this fundamental symmetry. So there are tensions between these two theories, which we know work very well on their own but stand on very different grounds.

What is the best candidate for a theory that combines these two other theories?

It depends on what your requirements are. First of all, there is no theory of quantum gravity which is universally accepted and satisfactory enough to everyone. Therefore, we have different candidates and some are more popular than others. Which one you pick depends a bit on what is your cultural background and what types of questions you want the theory to answer. All candidate theories are incomplete in some sense, so depending on what you think are the most important questions, you are going to think that a given theory is more incomplete than the other one. But if you decide to focus on a different set of questions, it might be the reverse.

There is one big dividing line in the field of quantum gravity. For some people, mainly due to historical reasons, it is believed that the theory of quantum gravity can only exist as a unifying theory of all forces. Finding such theory, therefore, not only combines quantum mechanics with gravity but also solves the problem of unifying all forces of nature in a single framework. A priori, these are two logically different problems and they might overlap or not. So if you believe that the theory of quantum gravity should unify all forces, you are led to string theory. Instead, if you have other questions that you think are more fundamental, related to the fundamental constituents of spacetime and how it looks at the smallest scale, regardless of whether or not it is a unifying theory, you are led to believe that theories of quantum gravity can make sense by themselves as mathematical theories.

Do you believe in that?

I prefer not to be a believer. I think it is a simple first hypothesis and I would like to see why it's not true, if it's not true. So at this stage I don't yet see the necessity of unification for making sense of quantum gravity. There is no logical argument for it. There are certain models that are more natural but it's still a possibility. We have a well-defined mathematical example of a theory of quantum gravity in three dimensions [3–5] so it is perfectly logically possible to have a model of pure quantum gravity that is mathematically consistent.

So in three dimensions you have a theory that makes sense?

Yes, it makes perfect sense. Now, you need matter to probe the theory. It is hard to imagine the notion of observable and observer without having things moving inside the spacetime. If you think more deeply about what is spacetime and try to define it, you will realise that it is the locus of relations between things moving in it. It is not something that you experience directly; instead, what you experience is a web of relations between objects. So, you need some type of matter in order to have some interpretational issues under control but matter coupled to gravity doesn't necessarily imply that you need a unifying theory. No one has demonstrated that it's not possible to develop a theory of quantum gravity in four dimensions without having to recur to unification. If you follow this line of thought you end up with approaches such as LQG or spin foam theories, which are way more advanced.

What questions do you think that string theory is not able to answer?

There are many such questions, such as those that deal with the nature of space and time, in particular with the shortest distance scale or questions regarding the fundamental degrees of freedom and their dynamics. Also, it seems unclear, at the moment, what exactly is the theory behind it and if there is a non-perturbative formulation that you can write down. There are zillions of questions like these, which arise from a quantum gravitational point of view. For example, what happens when you create a black hole and study it quantum mechanically?

After this interview was recorded, there has been a lot of progress, mostly in the context of AdS/CFT, in the description of quantum black holes.

Do you think that this last question cannot be answered using string theory?

No, not yet. Any type of question that has to do with the nature of the fundamental degrees of freedom is not answered by string theory.

Isn't AdS/CFT an exception since string theorists claim that it is a background-independent non-perturbative formulation of quantum gravity with asymptotic anti–de Sitter (AdS) boundary conditions?

Maybe. I think it's an extremely interesting idea but its range of applicability is way beyond string theory. I even worked on a paper expressing my take on it, namely, that any theory of quantum gravity, if there is one, will be an AdS/CFT-type of formulation [6]. However, my main concern with it (and I am not witnessing any developments in this area though I have been pushing people in that direction) is that, since AdS/CFT is a description of the world where in some sense you put some type of theory at infinity (a very special theory that seems to prevent cosmological-type singularities), how do you reconstruct the local bulk physics from this theory at the boundary/infinity? If there is to be some kind of fully embodied holography in this picture, how do you reconstruct local physics?

If there is something happening here in this room either because you or I did something, it won't really affect what happens in some other planet at the same time. Clearly, we can have a description of physics which is local and that's the challenge for a theory of quantum gravity: you have to show that you have a description of physics which is both local and quantum. Somehow, whatever happened in this room, if you wait long enough, will be registered as some form of information at infinity but there is no prescription for taking that information and reconstructing what happened in this room. There's the hope that such is possible and I think this hope is very reasonable but it's a huge challenge. So far, it seems that people have been focusing on the theory at infinity and doing amazing things there, though without knowing how to reconstruct the local physics from those results. This reconstruction will tell you what are the fundamental degrees of freedom. At this point it seems to me that it is a superficial swapping of the problem of quantum gravity with the problem of AdS/CFT and results may sometimes be taken for granted by saying, "Okay, this is AdS/CFT." There's a lot of evidence for it but it does not answer the questions you are looking for, though it gives you a model where you can try to address such questions.

To my knowledge, this correspondence has mostly been used, in a very powerful manner, in the reverse way. People have been placing different theories at the boundary (that is, different conformal field theories (CFTs), or something very similar to CFTs, or some condensed matter system, in fact, a large class of models has been used) and were able to recast several field theory problems in a gravitational language and solve them. However, what you would like to do is exactly the reverse and not only at the classical level but also at the quantum level. If I go and ask one of my colleagues what are the fundamental degrees of freedom, its dynamics, the non-perturbative formulation of the dynamics and how you define locality, there will be no answer.

Okay, you mentioned that every consistent theory of quantum gravity would be formulated in an AdS/CFT-like manner. Why do you say that?

One of the reasons is that the theory of quantum gravity is supposed to solve an equation called the Wheeler-DeWitt equation [7] (the equation that quantum gravitational states are supposed to satisfy) and no one knows how to write this equation in a way that is completely satisfactory. This is what people in LQG have been trying to do and in AdS/CFT no one has been able to write down. What happens, loosely speaking, in the latter context is that you can take a quantum gravitational state and look at how it behaves at infinity. You will find an equation which is way simpler and characterises a CFT. In some sense, this is very generic and if you have any theory of quantum gravity, either string theory or any other theory, and you are able to take this limit you will find some CFT at the boundary. The exact features of this CFT will depend on the matter content and it might be unitary or not. In the case of AdS/CFT, the type of theory at the boundary is almost uniquely fixed by supersymmetry but in general it will be fixed by the type of symmetry you have in the bulk. Generically, I think that whatever quantum gravity theory you start with, you will get some CFT at the boundary. However, whether or not the reverse is true, that is, that you can reconstruct the theory of quantum gravity from the CFT, is what I would like to know and I think more people should work on that question.

Now changing a bit the topic ... can you explain what is the spin foam approach to quantum gravity?

Okay, so I will start first by explaining LQG. LQG is an approach where you take as an assumption that gravity can be quantised independently of other forces of nature. You can couple the theory to matter but that comes only after you have quantised gravity on its own. The formulation of LQG is based on a principle (called the gauge principle) and diffeomorphism invariance. The gauge principle essentially states that excitations of geometry are one-dimensional excitations similar to magnetic field lines. This results in a simplification of the dynamics.

Diffeomorphism invariance, on the other hand, is a symmetry that states that if you choose to label points in your spacetime in a specific way then at the end of the day, the physics must be independent of the labels you use. Physics is all about the relationship between physical objects and so it doesn't really matter how you label them; what really matters is the type of relationship that there is between them. It is an unusual type of symmetry because, in daily life, we are used to labelling things. For example, one label is the time label which is in my clock and I take this time to be real because if I were to change this label, I wouldn't be able to meet you at the right time in the right location. Once you bring the gauge principle and diffeomorphisms together, you can formulate the dynamics of quantum gravity in terms of what we call spin networks (or spin network states). These states are of a different nature as there is some form of nonlocality in them and we draw them on graphs. So, if you see a LQG person doing calculations you'll probably see this person drawing many of these graphs and placing certain labels on them. These graphs represent a type of atom of geometry for which both the area and volume become quantised: geometry is quantised in this framework [8].

It has been established for a long time that there is a preferred basis of states that you should use in order to describe quantum gravity. This basis of states can be seen as a truncation of the theory, which you depict using finite graphs. It is a very different basis of states from the one that people used to like, namely the so-called graviton basis of states. The graviton basis of states is a description of quantum gravity states based on gravitational waves as seen from far away where the gravitational field becomes very weak. In this basis you focus on a truncation of the theory that captures only one such graviton. This truncation, however, usually relies on the existence of a background and people tried to quantise gravity using this graviton basis of states but did not succeed. Therefore, LQG focused on another basis of states, that is, on another way of truncating the dynamics of gravity. Now the main question is how to formulate the theory using this basis of states and its dynamics.

This new basis is slightly more complicated to use than the graviton basis of states because spin network states constitute the spacetime itself. So, it's not like there is a spacetime and on top of that you put excitations (graviton particles moving around as in the graviton basis); instead, it is a more conceptually difficult frame in which there is nothing and you build spacetime using small but complicated spin networks (or graphs). By placing more labels and making bigger graphs you get more complicated geometric entities and these entities, which are fundamental quanta of spacetime, do not resemble the usual classical spacetime. The main task is then to formulate the dynamics of these states and that's what spin foams are really about: they are a quantum mechanical formulation of the dynamics in this basis using techniques which we call covariant techniques or path integral techniques. It's an attempt to formulate the path integral of quantum gravity within this truncation of states at the boundary.

How am I to think of metric in terms of spin networks?

Well ... the metric is some kind of coherent state of spin network states. There's actually a new development that I am participating in that aims at trying to understand if spin networks can be seen as some kind of geometrical object [9]. At some level, they will describe the metric but maybe there's an independent geometrical meaning in general. A preliminary exploration seems to suggest a connection with twistor theory [10].

Am I to understand this approach as being independent of LQG?

Not really, at least not in the way I think about it. However, I do see it as something more general because it has general covariance in-built in the dynamics. Historically, people thought that the dynamics should follow from a Hamiltonian approach but it turns out that there's more flexibility in formulating the dynamics if you use the path integral approach. It is not always clear that these two approaches can be related. In the spin foam formalism you work with a basis of states which is discrete and you extract from it a unitary global time evolution. However, this global time evolution does not have a nice infinitesimal form because not only is space quantised but also the time states are quantised. From the Hamiltonian perspective, you start with a continuous time variable and an infinitesimal time generator from which you can generate a global time evolution. So there's a tension between the two but there are some instances where the connection is established, in particular in models in 2+1 dimensions [11, 12]. In fact, one of the reasons why spin foam models were developed was that it is simply difficult to even write down some type of Hamiltonian dynamics that people would agree on or accept. The spin foam approach is trying to constrain the dynamics using general covariance, thereby bypassing the ambiguity issues in the Hamiltonian approach.

You mentioned that there is an example of a sensible 2+1-dimensional theory of quantum gravity. What do you mean by this exactly? That there is, for example, a semiclassical limit?

A semiclassical limit for sure, but nowadays, due to recent developments, even in four spacetime dimensions there are spin foam models with semiclassical limits [13, 14]. In

2+1 dimensions you have much more than that; namely, you have a consistent dynamics, you know the full basis of states, you know the dynamics exactly and you can prove some version of general covariance at the quantum level [15, 16]. But it's a very simple theory because in 2+1 dimensions there are no local dynamic degrees of freedom, only global degrees of freedom.

But do you actually find Einstein equations in 2+1 dimensions?

Well, you find a quantum version of it.

Okay, but can you take a classical limit and obtain the classical Einstein equations?

Yes, with quantum corrections, of course [11]. Since there is a minimum scale, there is an infinite number of such corrections. Because the dynamics is very simple you can actually re-sum all these corrections. This is related to the fact that there is this minimum scale and you can find an action that is consistent with it and also due to the fact that the series of corrections exhibits the full symmetry of the theory.

Furthermore, there is also an interesting feature in this theory, namely, that it re-establishes the Born reciprocity rule. You can probe the theory, that is, the corresponding gravitational interactions, by introducing matter and seeing the quantum gravity effects. Even though the theory is simple and has almost no dynamics and there are no gravitational waves, the matter will modify the fabric of spacetime in a unique manner [17]. Now, you can wonder what kind of spacetime does the matter that you send in experience and later you realise that quantum gravity curves not only spacetime but also momentum space. In this sense, it is a model that re-establishes the Born reciprocity rule. It is a cute model, for that reason.

Is there any ambiguity in the Hamiltonian in this case?

No. In the spin foam approach there is only one quantisation that people have discussed [3], but no one has proven yet that no other quantisation is possible. There is some work that shows that for a simple geometry (tori), there is one possible parameter ambiguity in the Hamiltonian but the type of model that is solved for all possible genii requires one specific choice of this parameter. So the final answer is no, the symmetry is strong enough in this case to fix completely the dynamics. The model is simple but I think that some of its features could be extracted as generic features even in four dimensions.

You mentioned that there were some spin foam models in 3+1 dimensions for which there is a semiclassical limit. Which model are you referring to?

We have a preferred spin foam model in 3+1 dimensions and we are trying to extrapolate some of the success of 2+1 dimensions to 3+1 dimensions [18, 19]. There has been a sort of convergence in the community towards one spin foam model and one of the reasons people like this model is because you can write the amplitude in terms of the spin network boundary states. You have to add some extra structure and you also have to add some kind of discretisation inside the spacetime but once that is fixed you can take the classic limit. It turns out that in the classical limit you recover a version of gravity which is called Regge dynamics [20]. This type of gravity appears when dealing with excitations which are large compared to the discretisation scale and it is not just general relativity; there are modifications to it.

So you do have a semiclassical limit?

For this model, yes. The semiclassical limit is no longer an issue to me. There are other issues, of course.

Okay. So, if people say that "LQG doesn't have a semiclassical limit" you would disagree?

Yes. If someone says that, it is because that someone has not followed the recent developments in the spin foam formalism.

What are the issues with these models that you were referring to?

One issue is related to obtaining a continuum limit. The main issue, however, is to prove that there is general covariance at the quantum level. That would promote this model to a full theory.

Are the problems of obtaining the continuum limit and that of proving general covariance related?

Yes, they are related. Imagine that you take a regular field theory and you put it on a lattice. You can ask whether the model has some version of the continuum physics or not, that is, if the theory living on the lattice does not change if you refine the lattice (this is called a *perfect action* in renormalisation group language), meaning that you can still capture something exact on the finite lattice, though you have a cutoff. By performing a continuous refinement of the lattice you can look for the continuum limit of the model. Sometimes, if you are lucky enough you can be sitting at a fixed point, which is what happens in 2+1 dimensions [21]. In fact, in this case, one can re-establish exact translation invariance or the exact symmetries of the theory in this discrete model. So the problem of the continuum limit and general covariance are tied up because if you write down models that have some discrete structure and capture the exact symmetry of the theory (like general covariance) then in some sense they capture also features of the continuum limit.

Okay. Just to make it clear ... in 2+1 dimensions besides having a semiclassical limit there is also a continuum limit?

Yes, there is an exact continuum limit [22, 23]. In this case, the symmetry at the quantum level is strong enough to fully constrain the action that governs the dynamics. The theory on the lattice then keeps this symmetry exactly. Then, you can prove that if you refine the lattice nothing happens and hence you've found the perfect action at the discrete level which includes all higher-order corrections. However, we do not expect that level of exactness in 3+1 dimensions because the symmetry is more complicated and the dynamics more involved.

How do you describe gravitons in 3+1 spin foam models?

There's nothing special about that. What are gravitons? Gravitons are just some truncation of the Hilbert space. In the spin foam approach we work with a very different truncation of Hilbert space. There should be an overlap between these bases, namely, the graviton basis and the spin network basis and somehow the graviton basis is a very complicated spin network basis and the spin network is a very complicated graviton combination. Eventually, if it is proven that the spin network basis is a complete basis of states, then there should be a way of describing the graviton.

However, I should say that it is not always clear how to define the graviton. The graviton is a very background dependent entity, so what do you mean by graviton if you only have background dependent states? It will require a bit of gymnastics in order to define it. The graviton is a type of observable that no one really agrees on what it is. There's no clear-cut mathematically exact definition of what a graviton is in a background independent formulation. The graviton is simply a convenient basis and there are people who are trying to emulate graviton calculations using the spin foam approach [24]. In other words, spin network states describe the physics locally while the graviton basis of states describes the physics globally in the very weak field regime. So it's two different bases in two different regimes and there should be an overlap once you have a good definition of what you mean by asymptotic quantum states.

So, what kind of features of quantum gravity do you observe in these models?

As I mentioned earlier, in order to see the features of quantum gravity you need to couple the theory to matter. Once you do that, you begin probing spacetime. In 2+1 dimensions you observe that the momentum space of the particle (matter) that you put in becomes curved in the form of a sphere, whose radius is the Planck length [25, 26]. Furthermore, at the quantum level, the spacetime appears to be non-commutative and hence you start seeing a drastic modification of locality [17, 27]. If I were to propose a quantum gravity experiment then I would focus on departures from locality.

But how would departures from locality be expressed in experiments?

This is going to be the challenge of quantum gravity. For example, going back to the graviton, no one will ever be able to detect a quantum graviton. If we vaguely define the graviton as an excitation of a vacuum state that looks like flat spacetime and if you want to observe one single graviton then you will either need an accelerator of the size of the galaxy (or bigger) or build up a LIGO-type interferometer with mirrors that are going to be so massive that they would sit inside their own Schwarzschild radius (that is, the mirrors would be sitting inside a black hole). So it seems that when people talk about gravitons they forget that they can't really be used to test quantum gravity. We need to find new ways to distinguish between theories of quantum gravity.

The problem of quantum gravity is a two-fold problem. The first problem is that of finding a mathematically consistent theory of quantum gravity that can describe how spacetime looks at the minimum scale and that has a consistent dynamical principle. In my opinion, it does not yet exist. It might be AdS/CFT, it might be a spin foam model but I can't say that it exists at the moment. The second problem concerns how to test these theories. Experiments related to cosmology, for instance, are maybe the way to go. We are definitely not going to detect the graviton so in terms of observation it is not a useful notion at all. So we need to have other insights on how to test quantum gravity.

Is spacetime emergent?

In some sense yes and maybe it's just emerging from quantum spacetime degrees of freedom. The problem with the word "emergent" is that it has different meanings for different people. Sometimes when people ask this question they imagine that there's some kind of non-relativistic system or some fermionic system that emulates gravity degrees of freedom. This type of drastic emergence was pushed by people who work on analogue gravity [28], in which context you find that fluid mechanics can emulate a black hole. For example, you can send a fluid through a throat such that at some point the fluid is so fast that the speed of sound cannot move forward any longer. So if you end up in a region where the fluid is behaving in this manner and the only way that you have to communicate with other regions is to send sound waves, then that region will look like a black hole to you – in this context it is called a dumb hole [29]. So there's a notion of a black hole and a metric which emerges from a theory that is made of molecules. In general, classical spacetime will emerge from a quantum theory; whether that quantum theory is made of some quantum metric or something more radical we do not yet know.

I asked this last question, bearing in mind a string-inspired development by Erik Verlinde [30] in which he explores the idea that the metric is not fundamental but emerges from some microscopic degrees of freedom ...

It's a purely statistical concept. I'm not sure one can attribute this idea to string theory, neither can one say that all string theorists would agree with this idea. This type of idea was around way before Verlinde. Ted Jacobson was the first to promote such an idea, sometimes even in a more comprehensive manner [31]. It's a possibility.

Is this idea compatible with LQG?

It's kind of incompatible. What Verlinde wants to say is that, since gravity emerges then you don't need to quantise gravity. Therefore if someone tries to quantise gravity directly then that somebody will fail. So yes, it is incompatible with LQG. In my opinion, these claims are too vague but I think that it's interesting to pursue. Maybe the motivation to pursue this idea is simply that it's difficult to quantise gravity directly so people might feel pushed toward this direction.

It could turn out that you don't have to quantise gravity. That would be nice (laughs). It's a possibility but whether this is due to desperation or because there is some fundamental reason seems rather unclear at the moment. However, I do think that the idea of Jacobson of interpreting Einstein equations at the classical level as some kind of thermodynamic entity is a piece of the quantum gravity puzzle. There's something different about gravity; that is, the Hamiltonian vanishes. So you are only left with counting (entropy) as you have no local definition of energy. Whether that means that gravity should or should not be quantised I'm not sure one can argue for just from this piece of information.

Now the last topic ... what is the principle of relative locality [32] and where did it come from?

It comes from what we have discussed earlier, namely, from trying to understand what spacetime is really made of. As I mentioned earlier, what you really experience is a web of relationships between objects, and spacetime is just an abstraction. As Albert Einstein taught us, in order to synchronise two clocks you have to make use of a physical process and hence, if you try to define spacetime in terms of a physical process, you realise that spacetime is not something that you directly experience but it's something that you reconstruct.

It turns out that, at low energy and in the absence of quantum gravity, it is very convenient to repackage this reconstruction in terms of the mathematical entity that we call spacetime, which has this notion of absolute locality; that is, we can define exactly points in space at given moments of time independently of the observer. However, in quantum gravity we know that we have to relax this notion of locality but locality is like the big elephant in the room. It is not possible to open a physics book that does not have this assumption of locality, which is not really written anywhere. Therefore, it's even scary to consider putting locality into question and how do you even begin relaxing locality? It seems that our brains are trained to think in a local and causal manner and so it's difficult to make sense of any nonlocal version of the world.

Relative locality is an idea of how we can maybe relax the notion of absolute locality in a controlled manner [32–34]. Relative locality is the notion that spacetime depends on the energy you use to probe it so it is relative to an observer. Different observers see different spacetimes, which is different than saying that different observers use different coordinates to describe events in the same spacetime. This idea states that absolute locality (that every observer sees the same spacetime and that spacetime is not affected by how you probe it) is a low-energy approximation. You can think of absolute locality in a different way. In classical physics, the physics of any object can be described using the phase space where you view exchanges of energy and motion and so on. In this context, there is an underlying hypothesis that you can locally project the physics on this phase space onto spacetime, independently of the energy state of the object. It is this hypothesis that we are questioning. How do we know that this is true or not?

In the same way that we say that Newtonian gravity works very well at low velocities, in Einstein gravity we introduce an energy scale at which this modification of absolute locality becomes visible. This energy scale is most presumably related to the Planck energy scale; that is, it's due to quantum gravity effects. So, what we are doing is focusing our attention in a specific regime of quantum gravity and postulate that locality is modified.

Even just questioning our conception of locality is for me a very important act that can bring new perspectives to the problem of quantum gravity. It also gives new possible experimental windows as it may be a way to test quantum gravity effects. The type of experiment we imagine is a type of experiment that you can even perform in laboratories. The idea is that you take a quantum system and you make this quantum system perform a loop in phase space. The hypothesis of absolute locality says that once you project the dynamics of the phase space, the physics will not depend on the position of the loop in phase space. So you can start designing experiments that test whether this fundamental hypothesis of absolute locality is really true. This is a more phenomenological approach to quantum gravity inspired by the Born reciprocity rule; in fact, one way to have a specific model that we can work with is to give curvature to momentum space [32], which was what we found in 2+1-dimensional spin foam models.

So locality is tied up with this notion of featureless momentum space but it might not be featureless at all. It's a wild idea but after one year of thinking about it, it feels more natural to me. I believed all my entire career that spacetime exists but I no longer do so. I want to see a proof of that. Taking this phenomenological approach is like tackling a conceptual problem of quantum gravity without having to face the full-fledged theory.

Is it the next step to try to take this model with implemented relative locality and quantise it?

I don't think that you should quantise it. It simply gives you a different way of thinking about the problem of quantum gravity in a way that is transversal to all other theories we talked about, like LQG and string theory. It's a different way to address the problem but I think that any theory of quantum gravity should exhibit this principle. In fact, I am working right now, together with some fellow string theorists, on trying to show that this odd idea of momentum space dynamics is in fact present in string theory [35, 36].

Is there any experiment you can make to try to see if nonlocality is there?

Yes, there are several, depending on the feature of momentum space you want to see. It could be curvature, torsion or non-metricity. Each feature leads to different types of effects that are easier to test than others. The most exotic, and less probable, is non-metricity – it's a feature that states that there's a way to move things in momentum space and measure distances in momentum space that does not really match with reality. This can be seen by looking at gamma ray experiments. If this feature is there, it has to be smaller than something which is of the order of the Planck scale.

Is this experiment being done?

It's being done, yes. People look at gamma rays coming from very far away and they measure the time of arrival. It is very easy to use the experiment to put bounds on theory, but the reverse is not easy. With the approach we have developed you can place a bound on detection [37]. I'm also waiting for a group in Vienna that is designing a laboratory experiment in which they take a quantum system and move it around in phase space [38]. The hope is that this can give you a window into quantum gravity, depending on the kind of precision that they get. It's a bizarre idea, but I see it as an opportunity to be able to test any violations of locality.

We have proposed a very specific model that gives geometry to momentum space [32]. Even though the model cannot be considered to be a full-fledged theory, I think that its characteristic features are pretty robust. The other interesting thing about these models is that they were developed based on a simple idea and people can understand it easily without having to belong to a specific school of thought.

What do you think has been the biggest breakthrough in theoretical physics in the past 30 years?

That's a tough one. One of the things that has fascinated me was the WMAP data results [39]. I think it's pretty remarkable that we're able to take these pictures of the CMB background with such precision. It's not exactly theoretical physics but I think this has been one of the main contributions to physics in recent years.

Why have you chosen to do physics?

(laughs) I don't know, I think physics chose me. I didn't choose to do physics, I didn't want to be a physicist; I wanted to maybe be a writer, or a musician or a climber. That would have been my choice, if I'd been able to direct my own path but obviously I was not able to.

What do you think is the role of the theoretical physicist in modern society?

To make people dream and push the boundaries of knowledge. As a collective, theoretical physicists have always been able to make drastic changes in the way we relate to the world and therefore in the society we're in. As an individual, there is some kind of ethos about how one does science and which is important to transmit. I think there are also true benefits from the more collective aspect of theoretical physics, as all things trickle down, for instance from theoretical physics down to engineering and so on. It's maybe better to imagine how the world would be without any theoretical physicists, right? So it's not like you can say that every theoretical physicist is necessary but a world without theoretical physicists would be like a world without music or French wine or sun in the summer; I mean, it would be very boring (laughs).

References

[1] M. Born, "Reciprocity theory of elementary particles," *Rev. Mod. Phys.* **21** (Jul. 1949) 463–473. http://link.aps.org/doi/10.1103/RevModPhys.21.463.

[2] M. Born, "A suggestion for unifying quantum theory and relativity," *Proceedings of the Royal Society of London A: Mathematical, Physical and Engineering Sciences* **165** no. 921 (1938) 291–303, http://rspa.royalsocietypublishing.org/content/165/921/291.full.pdf. http://rspa.royalsocietypublishing.org/content/165/921/291.

[3] G. Ponzano and T. Regge, "Semiclassical limit of Racah coefficients," *Spectroscopic and Group Theoretical Methods in Physics,* F. Bloch, ed., North-Holland, 1968.

[4] V. Turaev and O. Viro, "State sum invariants of 3 manifolds and quantum 6j symbols," *Topology* **31** (1992) 865–902.

[5] C. Rovelli, "The basis of the Ponzano-Regge-Turaev-Viro-Ooguri quantum gravity model in the loop representation basis," *Phys. Rev. D* **48** (1993) 2702–2707, arXiv:hep-th/9304164.

[6] L. Freidel, "Reconstructing AdS/CFT," arXiv:0804.0632 [hep-th].

[7] B. S. DeWitt, "Quantum theory of gravity. I. The canonical theory," *Phys. Rev.* **160** (Aug. 1967) 1113–1148. https://link.aps.org/doi/10.1103/PhysRev.160.1113.

[8] C. Rovelli and L. Smolin, "Discreteness of area and volume in quantum gravity," *Nucl. Phys. B* **442** (1995) 593–622, arXiv:gr-qc/9411005. [Erratum: *Nucl. Phys. B.* **456** (1995) 753–754].

[9] L. Freidel and S. Speziale, "Twisted geometries: a geometric parametrisation of SU(2) phase space," *Phys. Rev. D* **82** (2010) 084040, arXiv:1001.2748 [gr-qc].

[10] L. Freidel and S. Speziale, "From twistors to twisted geometries," *Phys. Rev. D* **82** (2010) 084041, arXiv:1006.0199 [gr-qc].

[11] V. Bonzom and L. Freidel, "The Hamiltonian constraint in 3D Riemannian loop quantum gravity," *Class. Quant. Grav.* **28** (2011) 195006, arXiv:1101.3524 [gr-qc].

[12] S. Alexandrov, "Spin foams and canonical quantization," *Symmetry, Integrability and Geometry: Methods and Applications* (Aug. 2012). http://dx.doi.org/10.3842/SIGMA.2012.055.

[13] F. Conrady and L. Freidel, "On the semiclassical limit of 4D spin foam models," *Phys. Rev. D* **78** (2008) 104023, arXiv:0809.2280 [gr-qc].

[14] J. W. Barrett, R. Dowdall, W. J. Fairbairn, F. Hellmann and R. Pereira, "Lorentzian spin foam amplitudes: graphical calculus and asymptotics," *Class. Quant. Grav.* **27** (2010) 165009, arXiv:0907.2440 [gr-qc].

[15] L. Freidel and D. Louapre, "Diffeomorphisms and spin foam models," *Nucl. Phys. B* **662** (2003) 279–298, arXiv:gr-qc/0212001.

[16] B. Dittrich, "Diffeomorphism symmetry in quantum gravity models," *Adv. Sci. Lett.* **2** no. 2 (Jun. 2009) 151–163.

[17] L. Freidel and E. R. Livine, "3D quantum gravity and effective noncommutative quantum field theory," *Bulg. J. Phys.* **33** no. s1 (2006) 111–127, arXiv:hep-th/0512113.

[18] L. Freidel and K. Krasnov, "A new spin foam model for 4D gravity," *Class. Quant. Grav.* **25** (2008) 125018, arXiv:0708.1595 [gr-qc].

[19] J. Engle, E. Livine, R. Pereira and C. Rovelli, "LQG vertex with finite Immirzi parameter," *Nucl. Phys. B* **799** (2008) 136–149, arXiv:0711.0146 [gr-qc].

[20] T. Regge, "General relativity without coordinates," *Nuovo Cim.* **19** no. 558 (1961).

[21] B. Bahr and B. Dittrich, "Improved and perfect actions in discrete gravity," *Phys. Rev. D* **80** (2009) 124030, arXiv:0907.4323 [gr-qc].

[22] V. Bonzom and M. Smerlak, "Bubble divergences from cellular cohomology," *Lett. Math. Phys.* **93** (2010) 295–305, arXiv:1004.5196 [gr-qc].

[23] J. W. Barrett and I. Naish-Guzman, "The Ponzano-Regge model," *Class. Quant. Grav.* **26** (2009) 155014, arXiv:0803.3319 [gr-qc].

[24] E. Bianchi, L. Modesto, C. Rovelli and S. Speziale, "Graviton propagator in loop quantum gravity," *Class. Quant. Grav.* **23** (2006) 6989–7028, arXiv:gr-qc/0604044.

[25] S. Deser, R. Jackiw and G. 't Hooft, "Three-dimensional Einstein gravity: dynamics of flat space," *Annals Phys.* **152** (1984) 220.

[26] H.-J. Matschull and M. Welling, "Quantum mechanics of a point particle in (2+1)-dimensional gravity," *Class. Quant. Grav.* **15** (1998) 2981–3030, arXiv:gr-qc/9708054.

[27] L. Freidel and S. Majid, "Noncommutative harmonic analysis, sampling theory and the Duflo map in 2+1 quantum gravity," *Class. Quant. Grav.* **25** (2008) 045006, arXiv:hep-th/0601004.

[28] C. Barcelo, S. Liberati and M. Visser, "Analogue gravity," *Living Rev. Rel.* **8** (2005) 12, arXiv:gr-qc/0505065 [gr-qc].

[29] M. Visser, "Survey of analogue spacetimes," *Lect. Notes Phys.* **870** (2013) 31–50, arXiv:1206.2397 [gr-qc].

[30] E. P. Verlinde, "On the origin of gravity and the laws of Newton," *JHEP* **04** (2011) 029, arXiv:1001.0785 [hep-th].

[31] T. Jacobson, "Thermodynamics of space-time: the Einstein equation of state," *Phys. Rev. Lett.* **75** (1995) 1260–1263, arXiv:gr-qc/9504004 [gr-qc].

[32] G. Amelino-Camelia, L. Freidel, J. Kowalski-Glikman and L. Smolin, "The principle of relative locality," *Phys. Rev. D* **84** (2011) 084010, arXiv:1101.0931 [hep-th].

[33] G. Amelino-Camelia, L. Freidel, J. Kowalski-Glikman and L. Smolin, "Relative locality and the soccer ball problem," *Phys. Rev. D* **84** (2011) 087702, arXiv:1104.2019 [hep-th].

[34] G. Amelino-Camelia, L. Freidel, J. Kowalski-Glikman and L. Smolin, "Relative locality: a deepening of the relativity principle," *Gen. Rel. Grav.* **43** (2011) 2547–2553, arXiv:1106.0313 [hep-th].

[35] L. Freidel, R. G. Leigh and D. Minic, "Born reciprocity in string theory and the nature of spacetime," *Phys. Lett. B* **730** (2014) 302–306, arXiv:1307.7080 [hep-th].

[36] L. Freidel, R. G. Leigh and D. Minic, "Metastring theory and modular space-time," *JHEP* **06** (2015) 006, arXiv:1502.08005 [hep-th].

[37] L. Freidel and L. Smolin, "Gamma ray burst delay times probe the geometry of momentum space," arXiv:1103.5626 [hep-th].

[38] I. Pikovski, M. R. Vanner, M. Aspelmeyer, M. Kim and C. Brukner, "Probing Planck-scale physics with quantum optics," *Nature Phys.* **8** (2012) 393–397, arXiv:1111.1979 [quant-ph].

[39] N. Jarosik, C. L. Bennett, J. Dunkley, B. Gold, M. R. Greason, M. Halpern, R. S. Hill, G. Hinshaw, A. Kogut, E. Komatsu, D. Larson, M. Limon, S. S. Meyer, M. R. Nolta, N. Odegard, L. Page, K. M. Smith, D. N. Spergel, G. S. Tucker, J. L. Weiland, E. Wollack and E. L. Wright, "Seven-year Wilkinson Microwave Anisotropy Probe (WMAP) observations: sky maps, systematic errors, and basic results," *The Astrophysical Journal Supplement Series* **192** no. 2 (2011) 14. http://stacks.iop.org/0067-0049/192/i=2/a=14.

11

Steven Giddings

Professor at the Department of Physics, University of California, Santa Barbara

Date: 20 July 2011. Location: Santa Barbara, CA. Last edit: 16 October 2020

What do you think are the main problems in theoretical physics at the moment?

There are a few prominent problems at the moment such as the mystery of quantum gravity, the nature of electroweak symmetry breaking, the smallness and origin of the cosmological constant as well as the nature of dark matter.

Why do you think we need a theory of quantum gravity?

So far no one has managed to describe a consistent theory where you have classical gravity and quantum matter. It's very difficult to reconcile two very different approaches to physics in a single theory. We need a theory of quantum gravity because we need a quantum theory of nature since nature appears to be fundamentally quantum mechanical.

Why is this problem of reconciling the two theories so difficult?

It's hard to match the basic features of classical physics with quantum physics. Quantum physics is probabilistic. For instance, the likelihood of finding a particle in one location or in another location is determined by a probability distribution. The difficult question is how to reconcile that probabilistic account with the classical account of the gravitational field, which doesn't obey quantum principles and isn't governed by probabilities.

What do you think is the most satisfactory theory of quantum gravity at the moment?

I don't think we have a successful theory at the moment. We have two very popular alternatives, loop quantum gravity and string theory. But to me it's not clear that they fully address the most profound puzzles of gravity.

What puzzles do you have in mind?

One of the puzzles that I have thought a lot about concerns what happens when you scatter particles at very high energies. In principle, if you have a complete theory of quantum gravity you should be able to understand the outcome of this gedanken experiment. This is an experiment that pushes the theory to its limit and the theory must provide an answer. Within this context, we encounter deep and profound issues related to black hole information and unitarity. It remains to be seen whether approaches such as loop quantum gravity or string theory can address these problems or if we in fact need new principles.

Has this type of scattering been considered within loop quantum gravity?

People have tried to describe particle scattering in loop quantum gravity [1]. In that setting it is harder to begin with because you first need to find a vacuum, say, Minkowski spacetime, and in that vacuum describe excitations corresponding to two asymptotic particles being scattered. So that first needs to be carefully set up and from there one has to try to calculate the correct amplitudes. String theory, on the other hand, is very well adapted to describe scattering in Minkowski spacetime but it has other limitations in the regime that we are most interested in, namely, the ultra-high-energy regime [2–4].

Before I ask you about these limitations, can you explain why you are concerned in understanding this high-energy regime above the Planck scale?

Being able to address this regime is sort of a litmus test for understanding whether a given theory of quantum gravity is a complete theory. If the theory in question is Lorentz-invariant and has some very weak notion of locality, it is possible to set up the problem of ultra-high-energy collisions. Thus, any such theory better be able to provide some answer to this problem.

In practice, if you have Lorentz invariance in a theory, you can think of a state that corresponds to a given particle and boost that particle, by means of a Lorentz transformation, to arbitrarily high energies. In other words, in a theory respecting the basic properties of special relativity, you can take a particle and accelerate it to very high velocities or just run extremely fast in the opposite direction to the particle motion. These two possibilities correspond to the same physics because of Lorentz invariance (laughs). Thus, you can take two particles and accelerate them independently to arbitrarily high energies each at its own end of the universe. Since there is some approximate notion of locality, this process can be done independently for each particle, as there won't be any interference. You then let the two particles head toward each other, ultimately leading to a collision. In theories of quantum gravity that have these two properties, namely Lorentz invariance and a notion of locality, you must be able to understand the outcome of such collisions.

What kind of limitations to acquiring this understanding were you referring to in the context of string theory?

String theory provides us with an approach to calculating amplitudes in perturbation theory. This is suitable when you are dealing with weak interactions. For instance, when you have two highly energetic particles that miss each other by a long distance when they pass one another. In this situation gravity can still be quite weak even if the energies are far above the Planck scale. In fact, this is an analogous situation to the collision of the Earth and the

Moon. In this case, both the Earth and the Moon are objects with energies and masses that far surpass the Planck scale and, even though they are bound together, they are, in effect, undergoing a collision. However, despite their masses being far above the Planck mass, the gravitational interaction between them is quite weak because of the large separation between the two. Hence, in the context of scattering, if the objects miss each other by a long distance the gravitational interaction can be quite weak. On the other hand, if the objects pass one another close enough we enter a regime where gravity is strong.

Is there a caveat here? I mean, do you take into account how the objects modify the spacetime surrounding them as they travel?

Yes. In fact, even in classical gravity we have a pretty good description of the influence of a single particle on the spacetime, so to speak. To be more precise, we know the gravitational solution for a single arbitrarily boosted massless particle, namely the Aichelburg-Sexl shockwave [5]. One of the features of this solution is that it looks just like flat spacetime ahead of the particle, since if the particle is travelling at the speed of light there is no way that information from the particle could have reached you in the far future. Similarly, by the same reasoning, the solution looks like flat spacetime in the far past. It's a quite simple solution where the entire gravitational field is in the transverse plane that moves along with the particle. Because it is flat spacetime ahead of the particle you can take two of these solutions and glue them together in that flat spacetime region without a problem. Thus, classically you know how to describe the particles before the collision. Additionally, their quantum dynamics is also under control.

Okay, so what are the issues with understanding these collisions in string theory?

The real question is what happens when they actually do collide at sufficiently small distances. What is this relevant distance? When you think about the classical picture of such collision you expect to form a black hole if the separation between the two particles is smaller than the Schwarzschild radius of the black hole which has total mass equal to the centre of mass-energy of the collision. Black hole formation is a very strong gravitational phenomenon; in fact, it's the exact opposite regime of weak interactions. This means that perturbation theory breaks down.

String perturbation theory is better behaved than traditional perturbation theory in gravity but that doesn't help you. It is true that we have spent a long time worrying about the infinity of infinities that appear in gravitational perturbation theory, which goes by the name of the *problem of non-renormalisability*. String perturbation theory cures that problem but another issue arises in the high-energy collision context. In particular, if you sum up all the terms in the perturbation theory you find a badly divergent result which clearly doesn't make sense.

The reason why string perturbation theory breaks down in this context is because each term in the perturbative sum is comparable with each other. It's like summing $1+1+1+\cdots$ and that obviously won't lead to a finite predictive answer. Thus, a possible solution to this issue is to devise an alternative approach to computing scattering amplitudes in string theory. There are certainly some potential approaches but we don't yet seem to have a complete description of how to compute such amplitudes at very high energies [3, 4].

So if I understand correctly, we need a non-perturbative formulation of string theory?

Yes, that's another way of saying it. We need some way of performing non-perturbative calculations of scattering amplitudes. In the past years, we have gained a certain amount of non-perturbative information in string theory from D-branes, dualities and so on [6, 7]. However, we apparently need a fairly complete non-perturbative theory to address these questions.

I've read several times in your papers [3, 8, 9] that people have focused a lot of effort on issues of non-renormalisability and singularities but not on issues related to unitarity. What do you exactly mean by this?

When describing what are the main issues with quantum gravity, it's often said that non-renormalisability, the fact that there is an infinity of infinities in the perturbative expansion, is a central problem, along with the issue of classical black hole singularities. However, I think that the more profound issue that drives these other problems is related to unitarity, specifically in the context of high-energy collisions. The reason why I think this is a more profound issue is because we seem to find that we do not know how to solve it regardless of any possible modification of the theory at short distances. This makes us think whether we actually need a completely different way of looking at gravity.

At short distances the issues appear near the Planck scale and so there must be some kind of modification of the theory at those short distances. Let us assume that you can actually make any modification that you can think about. Has any such modification been able to solve the black hole formation and evaporation problem? No one has been able to come up with a consistent modification that solves this problem. This is additional evidence that there needs to be a modification of how we describe gravity, not only at short distances, but also, though this is not usually stressed enough, at long distances at least in certain circumstances.

There is a simple way of understanding why the long-distance behaviour also needs to be modified. In high-energy collisions, the gravitational interaction between the two particles is very strong at distances of the order of the Schwarzschild radius. In this case, classically you form a black hole and the more you increase the energy of the collision the bigger and bigger the black hole becomes. This is a basic feature of gravity and it tells you that the high-energy behaviour is also a long-distance behaviour since you can form an arbitrary large black hole in an arbitrarily high-energy collision. Thus, ultimately, you need to address such issues at arbitrarily long distances.

I've read many times that we can use high-energy scattering to understand how space-time looks at the Planck scale but if you look at the collision of strings, instead of particles, how can you discuss such small scales given than the string has a finite length?

In string theory there is a notion of shortest length which is the characteristic string scale, as strings are extended objects. If you perform high-energy scattering using strings in order to probe shorter and shorter distances, there is a limit to how short those distances can be and that limit is set by the string scale [10]. This is a feature of string theory but also, apparently, a feature of any general theory of gravity. To be more precise, it's not quite the same phenomenon but it has been argued to be related within string theory. And that phenomenon, as I have mentioned before, is the fact that if you start colliding particles at higher and higher energies to probe shorter and shorter distances, as in particle accelerators in the past 100 years, you will ultimately create a black hole.

But do you see this feature appearing in string theory or is it just a classical expectation?

What you do actually see in string theory is that if you collide strings at very high energies, as you increase the energy you begin creating bigger and bigger strings [10]. In other words, the strings become excited and produce longer strings. It sounds similar to the case of producing black holes in classical gravity, although a precise connection has not yet been made. However, there are conjectures that relate the growth of strings and the growth of black holes [11]. It remains to be seen. But in the true high-energy limit, there is good evidence that you do form black holes from colliding strings [10].

Nevertheless, the central question is whether string theory is able to give a complete description of these processes. Showing a precise relation between the growth of excited strings and the growth of black holes in the appropriate limit would be an exciting result.

But can you answer the question of how spacetime looks at such small scales within string theory? Is it a continuum or discrete spacetime?

I think that our picture of spacetime just isn't a good picture at distances smaller than the Planck length, which is a bit disturbing. The fact that black holes grow in size with increasing energy suggests that there is something wrong with the description of the process in terms of classical spacetime even at long distances. In other words, there is some inconsistency when describing these processes in terms of quantum fields interacting and propagating in the background of a classical black hole geometry.

In summary, the real task is to understand how the basic notions of space and time need to be modified to describe these situations. Obviously, these notions are very useful in all experimental contexts so far (laughs) but they are only approximate notions and must be derived from some fundamental underlying theory.

Should this fundamental theory have Lorentz invariance, obey quantum mechanics or be local?

I think there is a deep conflict between Lorentz invariance (or diffeomorphism invariance), quantum mechanics and locality. People have been trying to understand this tension for over 30 years and haven't found a simple answer. It is likely that one of these properties has to be relaxed but what is the most conservative option to discard?

Lorentz invariance has been extremely well tested experimentally. It's very hard to formulate a consistent theory with Lorentz invariance violation that doesn't lead to conflict with observation once you combine it with quantum mechanics. In particular, you have a hard time dealing with the fact that black holes exist in nature (laughs). It doesn't look particularly attractive to modify Lorentz invariance and it is not clear that it would help you if you did.

Quantum mechanics is also very robust and experimentally verified in multiple situations. There is an enormous history of people trying to modify quantum mechanics in various ways and, typically, those modifications don't lead to anything sensible. It appears to be quite hard to modify it in any way that could help.

Locality can be formulated in a precise way in quantum field theory; in fact, it's one of the basic principles of quantum field theory. However, once you include gravity in a quantum field theory framework you don't understand how to define locality. Locality refers to locations in space and time but the symmetries of gravity involve arbitrary changes in the coordinates describing points in spacetime. Hence, you don't have a rigid way of specifying position in space and time. From the start, locality looks like a soft principle and the more you probe the notion of locality in the context of gravity the more you see that it is a less robust concept.

That being said, however, all natural phenomena that we've seen so far are very well described by local quantum field theory, which has this notion of locality built into it. For certain, locality is a very good approximate notion at the scales we probed so far. Furthermore, naive modifications of locality typically lead to inconsistencies when compared with what we observe. At the same time, a theory of quantum gravity needs to face the black hole information paradox and in that context locality is the cheapest principle to give up but also the most conservative. In addition to appearing necessary in the study of black holes, there are indications that nonlocality is needed at long distances from the study of inflationary cosmology. Thus, there is some tension between these different observations and the real challenge is to understand better this tension.

Why is nonlocality needed for the black hole information paradox and in inflationary cosmology?

The issues arising in inflationary cosmology at very late times parallel the issues arising when studying black holes, so let us focus on the latter. The basic puzzle, which is usually referred to as the *black hole information paradox*, arises from trying to understand aspects of black holes assuming Lorentz invariance, quantum mechanics and locality. It's useful to understand how the paradox is formulated to understand why nonlocality seems necessary to solve it [12–17].

Quantum mechanics always describes evolution from pure states to pure states. A less fancy way of stating this is to say that in quantum mechanics evolution is reversible. You can start out with an initial state and evolve it forward in time, or you start with the final state and evolve it to the initial state. There is a time reversal symmetry in effect. How does this work in the context of black hole formation and evaporation? Consider an initial state of two particles that later undergoes a high-energy collision and forms a black hole. Classically there is a very good description of the geometry of a black hole and if the black hole is large, the curvature near the horizon is small. In this situation, quantum field theory should be valid near the event horizon. In fact, Stephen Hawking was the first to understand the consequences of this assumption [18]. He found that particles are produced in pairs, roughly speaking, near the horizon and one of the particles of the pair often travels out to infinity. With time, the black hole radiates away its mass – a process which is now called *Hawking evaporation*.

In summary, you start with a pure state that leads to a collision, forms a black hole and that black hole evaporates with time. The initial pure state contained some amount of information. If you have trouble imagining why it contained some information you can think of the initial state as being the collision of two very energetic computers with a lot of information in them which ultimately forms a black hole (laughs). Thus, you can ask: What happened to that information after the evaporation process? Once you look closely at the Hawking evaporation process you realise that the information contained in the initial state is not carried away to infinity. The black hole becomes smaller and smaller until it evaporates down to a Planck-size object while the information does not leave the black hole. The reason why the information did not come out is due to locality. For the information inside of the black hole to reach a faraway observer, signals would have to propagate from inside the black hole to the outside, which involves sending information faster than the speed of light. This would only be possible if the theory was nonlocal.

From what I just mentioned, there are two possibilities for the fate of that information. The first possibility is that the black hole completely evaporates and the information is destroyed. However, this would imply a violation of quantum mechanics because if the information is destroyed then there's no way, for instance, to take the final state and evolve it backwards in time to get the initial state. This means that we lost the reversibility property of quantum mechanics. If you go ahead and assume that such violations of quantum mechanics may happen then you will discover violent violations of energy conservation [19]. Following this chain of logic you find that it implies that the temperature anywhere in empty space would be around the Planck temperature. All this is completely crazy and doesn't seem like a good alternative.

The second possibility is that the information is not destroyed but instead is left behind in some kind of small object of Planck size. What would such objects look like? They would be some kind of elementary particles but which could contain an arbitrarily large amount of information, since you could have collided energetic particles in order to form an arbitrarily large back hole. What else can you say about such objects? We know that every kind of particle can pair-produce and there's nothing that forbids their pair-production in different processes [20–22]. You might say that if pair-production takes place it must be extremely small. However, the probability of producing a pair is proportional to the number of possible internal states. You can calculate the probability of producing a pair in a given state and multiply that by the number of internal states to find the probability of producing a pair. However, the number of internal states can be as large as you want it to be. Thus, you quickly conclude that if such objects existed, they would be produced all the time in generic physical processes with sufficient energy. This is obviously crazy as well.

Both these possibilities suggest that there is no simple way out of this paradox. Could you perhaps modify the basic assumptions as to resolve this tension? Could locality violation change the story? Well, if there were a mechanism that would allow relaying information faster than the speed of light so that information could escape the black hole, the paradox would be solved. This is one of the motivations for trying to abandon locality as a fundamental notion [12–17].

Do you understand nonlocality as signals propagating faster than light?

The basic statement of locality is the statement that you can't send signals faster than the speed of light. In a Lorentz-invariant theory, sending signals faster than the speed of light is equivalent to sending signals instantaneously to a faraway observer. Moreover, it's also equivalent to sending signals back in time. In this sense, it is acausal. The fact that such things can happen is one of the reasons why locality, at least in local quantum field theory, is such a sacred principle. Sending signals back in time would lead you to many other paradoxes. So modifying locality must be done in such a way that doesn't led to immediate inconsistencies.

Nevertheless, the game of modifying locality gets you on the dangerous terrain populated by people who think crazy thoughts. You better not be able to send messages back in time so that you can win money at the race track (laughs) or to communicate with people in the next star system instantaneously. I don't think that such processes should be part of the correct description of nature. So the question is: How can we have a theory that doesn't have a precise underlying notion of locality, yet it's consistent? Perhaps only a weaker notion is necessary, such as causality, and not a sharp notion of locality as in quantum field theory. Furthermore, there should only be important violations of locality in very special contexts. Black holes are one of those contexts as well as inflationary cosmology but there may be other situations. Figuring out how exactly one should make such modifications is still one of the most important questions.

Many people claim that AdS/CFT [23] is a non-perturbative formulation of quantum gravity. Because the conformal field theory (CFT) is unitary it is usually said that there cannot be loss of information for black holes in the bulk. Does AdS/CFT somehow introduce nonlocality as to solve the issues that you mentioned?

Many people in the string theory community believe that AdS/CFT will realise a scenario similar to what I've described earlier. In particular, the notion of holography is intrinsically nonlocal; in fact, it's a big departure from local quantum field theory. It's also widely believed that AdS/CFT is a non-perturbative formulation and that the boundary CFT description has unitary evolution and thus there is no information loss. However, there is a serious set of questions about whether the AdS/CFT correspondence holds at a fine-grain level, that is, whether all precise details of phenomena in the bulk anti–de Sitter (AdS) spacetime are captured using the boundary theory, and how.

This is a set of questions that I have worked a fair amount on and had many dialogues with others on how to test AdS/CFT at a fine-grain level [24–29]. This is a way of testing whether AdS/CFT is in fact a fully non-perturbative formulation of quantum gravity. At this point, there are several nontrivial issues that arise and we don't know yet how to solve them. So we cannot exclude the possibility that there is some coarse-graining taking place when translating phenomena in the bulk to the boundary. It is possible that you could be effectively losing degrees of freedom or detailed information about the bulk. This is not unfamiliar in the context of field theory, where one can start with a precise short-distance theory and obtain an effective theory when looking at long-distance behaviour.

A lot of new and fascinating results have been found when applying the AdS/CFT correspondence in different situations, such as in applications to condensed matter systems. But I wonder whether such results that use the boundary theory are not just a consequence of universality and asymptotic symmetries. In summary, I think that if you are going to make statements such as "the AdS/CFT is a fully non-perturbative formulation of quantum gravity" then you better test it in all possible ways.

What are the issues you encountered while testing AdS/CFT?

As you test it, you find that the boundary theory must have certain properties in order to overcome the challenges. At the moment we are not sure whether it has those properties or not. The correspondence relates a theory in the bulk and a theory on the boundary with one less dimension. The ultimate question you would like to answer is: How do you obtain all the detailed dynamics in that extra dimension from the lower-dimensional theory? For instance, locality should be recovered in that bulk AdS space at least in all familiar circumstances, but how do you recover the local physics? Ultimately, the theory must be nonlocal, but as I said, locality should be violated only in special circumstances. Requiring that the boundary CFT can determine the local physics in the bulk theory implies a nontrivial set of constraints. In addition, if you think about the properties of scattering amplitudes which you must recover in small enough regions of the bulk AdS which you can consider approximately flat, you encounter yet another set of nontrivial constraints [26–28].

So do you agree with people who claim that the black hole information paradox has been resolved in AdS/CFT?

The most common statement is the following. We know the boundary theory is quantum mechanical and unitary. We also know that there is a correspondence between the boundary physics and the bulk physics. Therefore we know that the bulk physics is unitary and thus we're done: the paradox is solved. However, the missing piece is whether there is a fully fine-grained relationship between bulk and boundary physics and whether boundary unitarity correctly maps to statements about unitarity in the bulk physics. This is something I have tried to probe and it's not that simple [28, 29].

It is not clear at least to me that the paradox has been resolved. If it is resolved I would like to understand in detail how exactly it is resolved. This brings us to another set of questions. If the paradox is resolved then I would like to give a fairly good local description of the fate of the information that is thrown into the black hole, what happens to Hawking evaporation and what do observers falling into the black hole experience. Naively, nothing unusual happens to an infalling observer when crossing the horizon, only at the singularity violent things may happen. Is this the case within AdS/CFT? This is the type of question which we still don't have a good answer for in terms of approximately local observables.

What do you need exactly to show within AdS/CFT in order to understand these issues?

You need to understand precisely the map between boundary and bulk theories. The commonly discussed map is between correlation functions in the boundary theory and scattering amplitudes in the bulk, which you can define using the S-matrix. Considering the case where the curvature radius of the AdS curvature is very large so that in small enough regions the bulk space is approximately flat, it would be an incredible accomplishment if you could extract from the boundary theory the flat space S-matrix and show that it has all the required properties to describe the unitary formation and evaporation of the black hole. However, there's still another type of question related to approximately local observations, for instance, of infalling observables, which would not involve the S-matrix. Tackling this other type of question requires a deeper understanding of the correspondence.

Are you not satisfied with other proposals for solving the black hole information paradox such as the fuzzball proposal [30] or the Horowitz-Madalcena final state proposal [31]?

Let's consider these two proposals in turn. There is actually a more generic proposal than the fuzzball proposal which I made some time ago [32]. This proposal considers the possibility of having a drastic modification of the black hole all the way to the vicinity of the horizon. I called this realisation a *massive remnant*. The idea is that as the black hole evaporates, far away the object looks just like a normal black hole but there will be strong modifications to the expected black hole geometry near the horizon. Fuzzballs are an example of this kind of tiny massive remnant. But what are the issues with such objects? The main issue is the fact that if you try to make such description coincide with the usual black hole description you encounter the need for some mechanism to relay information faster than the speed of light. The main objection that was raised at the time was that such mechanism was nonlocal. Later on, holography was developed and so people understood the need for some kind of nonlocality. However, this type of nonlocality is not the same as that in massive remnants.

On the other hand, it seems to me that something along the lines of the Horowitz-Maldacena final state proposal could be part of the real story [31]. In this case you fix the final state of the black hole by imposing suitable boundary conditions at the singularity. The result of this assumption is that information can be transmitted out of the black hole by some kind of teleportation. However, a detailed understanding of the dynamics of such mechanism is still yet to be accomplished. Regardless, I think that this idea may at least be a facet of the full answer.

Okay. I would like to you ask you about black hole production. A friend of mine couldn't sleep for four days after the LHC started running because he thought that the entire world would be swallowed by a black hole. Is it possible to produce black holes at the LHC?

In familiar four-dimensional spacetime, that is, three spatial dimensions plus one time dimension, it is not possible [33–36]. The basic reason is because according to quantum mechanics, particle wave functions are spread out with a characteristic wavelength that depends on their momentum. This characteristic length is much bigger than the size of the black hole they would form in a TeV collision. Thus, there is no way in which they could form a black hole. However, there are certain fairly special scenarios, notably those involving extra space dimensions where the strength of the gravitational force could be strong enough so that black holes could be formed at the TeV scale at the LHC [33]. It would be quite interesting if we could see such phenomena.

What kind of signatures would you see of such black hole formation?

Actually, Scott Thomas and I investigated these signatures several years ago [33]. The small black holes that you would create via the collisions would disappear after approximately 10^{-27} seconds due to Hawking evaporation. As the black hole evaporates you would be able to detect Hawking radiation emanating from the black holes, which has a very characteristic and quantifiable signature. Now that the LHC is in operation, experimentalists have been putting bounds on these signatures. So far there's no evidence that black hole production is taking place but we have to see what the LHC will ultimately reveal.

If such black holes would be produced, would they swallow the whole earth?

It's reasonable to ask whether there are risks associated with the potential production of black holes, especially if you are not a physicist working on these topics. And, in fact, there are some very good explanations for why the idea that it would destroy the planet is just crazy. The main explanation is that if such black holes were created at the LHC they would evaporate almost instantaneously [33, 37]. Black hole evaporation is a very robust prediction, which several people, starting with Hawking [18], have shown using basic principles of quantum field theory in a classical black hole geometry. The fact that black holes radiate is a robust result, which does not require a complete theory of quantum gravity to be derived. This is one of the reasons why if such black holes were produced, there wouldn't be an associated risk.

Another reason why one should not be alarmed by it is the fact that if you would be able to produce them at the LHC, then such black holes would be produced constantly due to cosmic ray collisions in the upper atmosphere, whose energy is even higher that what LHC can attain. The same events would take place due to cosmic ray collisions in the sun and other stars [37]. This is yet another safety check.

What do you think has been the biggest breakthrough in theoretical physics in the past 30 years?

I think that on the experimental side, but still relevant to theory, there have been some quite remarkable discoveries and observations such as the smallness of the cosmological constant, whose value is difficult to explain using known physics. Additionally, there is an increasingly robust set of indications that point to the existence of dark matter. On the theoretical side, AdS/CFT is a remarkable development [23], as well as configurations of extra dimensions in string theory or the string landscape [38–42]. However, the question is whether this is the right description of nature. Resolving the existence of black holes with the principles of quantum mechanics is expected to be part of the next great breakthrough!

Why have you chosen to do physics?

Because it's the best way we have to find accurate answers addressing the fundamental questions of nature, the universe and the structure of reality.

What do you think is the role of theoretical physicists in modern society?

I think there are many important roles. One is to give a disciplined and precise way to approach and answer the most basic questions of our existence in this amazingly subtle, mysterious and complex universe. But there are others. To serve as leaders in furthering a scientific and rational approach to life on our planet is one. And to provide models for successful cooperation between very diverse people is another.

References

[1] L. Modesto and C. Rovelli, "Particle scattering in loop quantum gravity," *Phys. Rev. Lett.* **95** (Nov. 2005) 191301. https://link.aps.org/doi/10.1103/PhysRevLett.95.191301.

[2] S. B. Giddings, "Locality in quantum gravity and string theory," *Phys. Rev. D* **74** (2006) 106006, arXiv:hep-th/0604072.

[3] S. B. Giddings, "Is string theory a theory of quantum gravity?," *Found. Phys.* **43** (2013) 115, arXiv:1105.6359 [hep-th].

[4] S. B. Giddings, "The gravitational S-matrix: Erice lectures," *Subnucl. Ser.* **48** (2013) 93–147, arXiv:1105.2036 [hep-th].

[5] P. Aichelburg and R. Sexl, "On the gravitational field of a massless particle," *Gen. Rel. Grav.* **2** (1971) 303–312.

[6] A. Sen, "An introduction to nonperturbative string theory," in *A Newton Institute Euroconference on Duality and Supersymmetric Theories*, D. I. Olive and P. C. West, eds., pp. 297–413. Cambridge University Press, 1999. arXiv:hep-th/9802051.
[7] J. Polchinski, "Tasi lectures on D-branes," in *Theoretical Advanced Study Institute in Elementary Particle Physics (TASI 96): Fields, Strings, and Duality*, C. Efthimiou and B. Greene, eds., pp. 293–356. World Scientific, 1997. arXiv:hep-th/9611050.
[8] S. B. Giddings, M. Schmidt-Sommerfeld and J. R. Andersen, "High energy scattering in gravity and supergravity," *Phys. Rev. D* **82** (2010) 104022, arXiv:1005.5408 [hep-th].
[9] S. B. Giddings, "Black holes, quantum information, and unitary evolution," *Phys. Rev. D* **85** (2012) 124063, arXiv:1201.1037 [hep-th].
[10] S. B. Giddings, D. J. Gross and A. Maharana, "Gravitational effects in ultrahigh-energy string scattering," *Phys. Rev. D* **77** (2008) 046001, arXiv:0705.1816 [hep-th].
[11] G. T. Horowitz and J. Polchinski, "A correspondence principle for black holes and strings," *Phys. Rev. D* **55** (1997) 6189–6197, arXiv:hep-th/9612146.
[12] S. B. Giddings, "Black hole information, unitarity, and nonlocality," *Phys. Rev. D* **74** (2006) 106005, arXiv:hep-th/0605196.
[13] S. B. Giddings, "Nonlocality versus complementarity: a conservative approach to the information problem," *Class. Quant. Grav.* **28** (2011) 025002, arXiv:0911.3395 [hep-th].
[14] S. B. Giddings, "Nonviolent nonlocality," *Phys. Rev. D* **88** (2013) 064023, arXiv:1211.7070 [hep-th].
[15] S. B. Giddings, "Nonviolent information transfer from black holes: a field theory parametrization," *Phys. Rev. D* **88** no. 2 (2013) 024018, arXiv:1302.2613 [hep-th].
[16] S. B. Giddings, "Black holes, quantum information, and the foundations of physics," *Phys. Today* **66** no. 4 (2013) 30–35.
[17] S. B. Giddings, "Black holes in the quantum universe," *Phil. Trans. Roy. Soc. Lond. A* **377** no. 2161 (2019) 20190029, arXiv:1905.08807 [hep-th].
[18] S. Hawking, "Black hole explosions," *Nature* **248** (1974) 30–31.
[19] T. Banks, L. Susskind and M. E. Peskin, "Difficulties for the evolution of pure states into mixed states," *Nucl. Phys. B* **244** (1984) 125–134.
[20] D. Garfinkle, S. B. Giddings and A. Strominger, "Entropy in black hole pair production," *Phys. Rev. D* **49** (1994) 958–965, arXiv:gr-qc/9306023.
[21] S. B. Giddings, "Constraints on black hole remnants," *Phys. Rev. D* **49** (1994) 947–957, arXiv:hep-th/9304027.
[22] S. B. Giddings, "Why aren't black holes infinitely produced?," *Phys. Rev. D* **51** (1995) 6860–6869, arXiv:hep-th/9412159.
[23] J. M. Maldacena, "The large N limit of superconformal field theories and supergravity," *Int. J. Theor. Phys.* **38** (1999) 1113–1133, arXiv:hep-th/9711200.
[24] S. B. Giddings, "The boundary S matrix and the AdS to CFT dictionary," *Phys. Rev. Lett.* **83** (1999) 2707–2710, arXiv:hep-th/9903048.
[25] S. B. Giddings, "Flat space scattering and bulk locality in the AdS/CFT correspondence," *Phys. Rev. D* **61** (2000) 106008, arXiv:hep-th/9907129.
[26] M. Gary, S. B. Giddings and J. Penedones, "Local bulk S-matrix elements and CFT singularities," *Phys. Rev. D* **80** (2009) 085005, arXiv:0903.4437 [hep-th].
[27] M. Gary and S. B. Giddings, "The flat space S-matrix from the AdS/CFT correspondence?," *Phys. Rev. D* **80** (2009) 046008, arXiv:0904.3544 [hep-th].
[28] M. Gary and S. B. Giddings, "Constraints on a fine-grained AdS/CFT correspondence," *Phys. Rev. D* **94** no. 6 (2016) 065017, arXiv:1106.3553 [hep-th].

[29] S. B. Giddings, "Holography and unitarity," arXiv:2004.07843 [hep-th].
[30] S. D. Mathur, "The fuzzball proposal for black holes: an elementary review," *Fortsch. Phys.* **53** (2005) 793–827, arXiv:hep-th/0502050.
[31] G. T. Horowitz and J. M. Maldacena, "The black hole final state," *JHEP* **02** (2004) 008, arXiv:hep-th/0310281.
[32] S. B. Giddings, "Black holes and massive remnants," *Phys. Rev. D* **46** (1992) 1347–1352, arXiv:hep-th/9203059.
[33] S. B. Giddings and S. D. Thomas, "High-energy colliders as black hole factories: the end of short distance physics," *Phys. Rev. D* **65** (2002) 056010, arXiv:hep-ph/0106219.
[34] S. B. Giddings, "Black hole production in TeV scale gravity, and the future of high-energy physics," *eConf* **C010630** (2001) P328, arXiv:hep-ph/0110127.
[35] D. M. Eardley and S. B. Giddings, "Classical black hole production in high-energy collisions," *Phys. Rev. D* **66** (2002) 044011, arXiv:gr-qc/0201034.
[36] S. B. Giddings, "Black holes in the lab?," *Gen. Rel. Grav.* **34** (2002) 1775–1779, arXiv:hep-th/0205205.
[37] S. B. Giddings and M. L. Mangano, "Astrophysical implications of hypothetical stable TeV-scale black holes," *Phys. Rev. D* **78** (2008) 035009, arXiv:0806.3381 [hep-ph].
[38] R. Bousso and J. Polchinski, "Quantization of four form fluxes and dynamical neutralization of the cosmological constant," *JHEP* **06** (2000) 006, arXiv:hep-th/0004134.
[39] S. B. Giddings, S. Kachru and J. Polchinski, "Hierarchies from fluxes in string compactifications," *Phys. Rev. D* **66** (2002) 106006, arXiv:hep-th/0105097.
[40] S. Kachru, R. Kallosh, A. D. Linde and S. P. Trivedi, "De Sitter vacua in string theory," *Phys. Rev. D* **68** (2003) 046005, arXiv:hep-th/0301240.
[41] M. R. Douglas, "The statistics of string/M theory vacua," *JHEP* **05** (2003) 046, arXiv:hep-th/0303194.
[42] L. Susskind, "The anthropic landscape of string theory," arXiv:hep-th/0302219.

12

Rajesh Gopakumar

Senior professor at the Tata Institute of Fundamental Research and centre director of the International Centre for Theoretical Sciences (ICTS-TIFR), Bangalore

Date: 16 February 2011. Location: Allahabad. Last edit: 5 June 2014

In your opinion, what are the main problems in theoretical physics at the moment?

In theoretical physics as a whole?

In high-energy physics would suffice ...

In high-energy physics there is of course the standard list of questions that string theory is aiming to solve including the problem of quantum gravity or ideas of how to extend physics beyond the standard model but also problems like, for instance, understanding QCD non-perturbatively, which is something people have studied for a long time but haven't made so much progress on. In the context of gravity, there are issues regarding singularities, which we still don't understand very well, such as black hole singularities and the big bang singularity. These are obvious issues that suggest that there is a theory beyond Einstein's theory and for which I think that string theory is still very far (as is true of any other framework for quantum gravity) from giving a satisfactory explanation to. Even though string theory has done a good job in accounting for microstates of black holes [1], the singularity problem is still an open problem.

Then, there is of course the black hole information paradox, which still hasn't been properly resolved [2]. In principle there is some kind of resolution to the information paradox, at least in situations where one has a unitary dual description, where you believe that there is a unitary theory and information is preserved, but still no one has really traced through how Stephen Hawking's arguments break down and so on. So, that's definitely an important problem to nail down. Another problem is that the existence of dualities in quantum theories in general seems to suggest that there are some aspects of quantum behaviour of physical systems that we don't understand, in a way, I would say, about quantum mechanics itself.

You mentioned that quantum gravity was one of these main puzzles. Why is it so hard to quantise gravity?

Well ... gravity has do with space and time, making it different from other fields which are not obviously related to the geometry, or if you wish, to the setting of your physical theories. We usually talk about physics as happening on a stage and we imagine a lot of actors on that stage doing various things. Then, you try to understand how these actors are interacting with each other but in the case of gravity it is like the whole stage itself is a participant in the drama and that makes it more complicated because the stage reacts to the actors and so you don't have a fixed ground to put your theory on. When the stage becomes a participant you have to rethink all the usual rules so that you can deal with it. In this sense, quantum gravity is special and of course that shows up at least in two different ways. It shows up in the fact that when you perturbatively try to study small fluctuations of the gravitational field at the quantum level you find infinities which are far worse than those that you get in other theories ...

And we don't want infinities?

Yes, we want to make physical sense of what we are doing, specially when we are talking about small fluctuations, which are like the gravitational waves that you might detect from pulsars for example. What are the leading quantum effects there? That's something for which we should have a well-defined finite answer and for which our conventional techniques are not satisfactory.

The second way in which the particularities of quantum gravity show up is in the context of black holes, which I think encompass some of the real problems of quantum gravity. In this setting, we really see the sharp difference between the nature of gravity and other field theories. For instance, the Bekenstein-Hawking entropy does not scale like it would scale in a regular field theory; that is, it scales with the area rather than with the volume. All this seems to suggest that gravity is on a different footing from any local quantum field theory of the kind that we use to describe other forces. So, gravity is at least special and classically gravity itself gives us a hint that it is incomplete because of the existence of these singularities that I just mentioned.

Not to criticise but you suggested earlier that we don't fully understand quantum mechanics yet, so why should we take it for granted and use it to take one step further and approach gravity?

I didn't meant to imply that the framework of quantum mechanics is incomplete. We have a understanding of it which is practical and it is the understanding that we currently use but all these dualities seem to hint at an alternative understanding or a different way of looking at quantum mechanics. Between 1925 and, say, 1940 or so, people who worked on quantum mechanics worked in the Schrödinger or Heisenberg pictures but then in 1940 the path integral (or functional integral) picture was developed and that gave you a new perspective on quantum mechanics which was completely equivalent to the original formulation of quantum mechanics.

That was, of course, useful and insightful. So, I think that, similarly, there are probably other ways of formulating or looking at quantum mechanics which we haven't yet realised or understood. However, I think that perhaps trying to directly understand or find these new perspectives may be less fruitful than trying to understand some of the questions which are raised both in quantum gravity and in AdS/CFT-type dualities [3]. I don't think that we should stop using quantum theory but I think we will naturally be led to these new formulations by trying to understand these other questions.

Based on what you have said so far it seems that you believe that string theory is a good candidate for a theory of quantum gravity. Do you see it also as a computational tool that can be used to address other problems of physics?

It can be both. Definitely, I think that string theory has largely given, in the case of asymptotically anti–de Sitter (AdS) spaces, a definition of what a quantum theory of gravity is. So, if you loosely categorise all spacetimes in terms of negative curvature, zero curvature and positive curvature, in a sense, we understand one third of the problem. That is a really big step. I don't think that any of the other so-called candidate theories for quantum gravity have achieved that level of completeness in their description because in the case of AdS spaces you have a very detailed dictionary telling you what the observables are and how to compute them in terms of a field theory.

But why focus on AdS spaces? Do they have any realistic physical interest?

They may not be directly the vacua that we observe in nature, since of course the relevant ones are de Sitter or very close to zero curvature, but nevertheless many of the scenarios for realising de Sitter space in string theory start with an AdS vacuum while de Sitter space is some metastable excitation around that vacuum which would eventually decay into AdS [4]. In this sense, we might be able to view de Sitter space, even though we still don't have that clear picture, as some kind of excitation of AdS space. At least in many of these scenarios for realising a positive cosmological constant vacuum in string theory that's sort of the dominant way in which it seems to happen. So, studying AdS space may not be completely irrelevant for realistic scenarios. Understanding this case is important but it's also important to see whether you can understand flat space and de Sitter space in a similar way or whether you have to always go through AdS, which is not completely clear. So, as far as quantum gravity goes, I think that string theory has given us a lot of encouraging signs that we are on the right track.

What do you think of the criticism that string theory is not a falsifiable theory?

If we speak about string theory as a tool, and not as a theory of quantum gravity applicable to our real world, which we are still far from I think, it has made theoretically falsifiable predictions. String theory has developed many scenarios but none to a level which can be called a theory or a proper model of the real world.

Why not?

Because there's a plethora of models and all of them have some features of the real world but not all the features of the real world. I don't think that we already have, literally, a real physical system in which you would see the whole cosmological evolution of the universe.

But do you think that these questions could be answered by string theory alone?

It is not impossible, I would say, but we still haven't done it. I think that string theory has the right theoretical structures in it to try to address these questions but it still hasn't given us a falsifiable theory of the real world which we can shoot down. However, string theory as a tool or framework has made many theoretically falsifiable predictions by means of dualities between theories which we know independently of string theory. For example, it has made predictions that specific quantum field theories show a certain behaviour of their operators (like the anomalous dimension of these operators behaving as a non-trivial function of the coupling constant or correlation functions behaving in a specific way) [5–8]. These are falsifiable statements about quantum field theory.

But could these be experimentally tested?

No. There are theoretically falsifiable statements and there are experimentally falsifiable statements. Suppose that I propose a theory which is a modification of Newton's theory of gravity – let's say a classical theory of gravity. That theory is immediately theoretically falsifiable if it does not reduce to Newton's theory of gravity in the appropriate limit or, more non-trivially, if it makes a theoretical prediction which is inconsistent with the theoretical predictions of some other framework, for instance, with those of electrodynamics. In electrodynamics we know that things don't move faster than the speed of light and we know that photons have two types of polarisation. So, if you make a prediction which is in contradiction with some other framework which we believe we understand, in that case, you can falsify it just at the level of the theory.

In this sense, string theory has made non-trivial predictions about frameworks that we believe in and which we think we understand, namely quantum field theories. These predictions in many instances are completely novel from the point of view of these quantum field theories even if such quantum field theories are not immediately realisable in nature. Nevertheless, someone can simulate these theories on a computer, for instance supersymmetric Yang–Mills theories or other such theories. Thus, if you make a prediction that a specific observable in these field theories behaves in a certain way, you can check whether that is true and so far what we have seen is that it is true. These are definitely theoretically falsifiable statements which would have been quite easy to falsify, in the sense that they are so detailed and not vague general qualitative statements, but haven't been falsified. In that sense, it shows that string theory is theoretically consistent and that it has a solid theoretical structure. Then, of course, experimentally falsifiability might also follow from this line of development rather than from quantum gravity or particle physics if string theory could make predictions about quantum field theories that have been observed in nature. So far this hasn't really happened because we have been mostly studying supersymmetric quantum field theories in the context of the AdS/CFT duality.

So most of the theoretical predictions are for theories . . .

. . . which we don't really realise in nature, but the framework seems to be more general and you could imagine that it will be possible to make predictions about theories that one could realise in nature. What one can aim at is to state that this or that particular system, which is described by some quantum field theory, can also be described by some other rules, namely those of a string theory, from which one can do computations and get the answers which, in fact, one might even not be able to get in the quantum field theory because it is too complicated or too strongly interacting. This would make clear that these string theories are at least good frameworks for describing specific systems. As a result, you can then state that string theory is a falsifiable theory for describing those particular systems.

This line of thought is similar to that which we employ when discussing quantum field theories: you can discuss specific quantum field theories or you can discuss the general class of quantum field theories. Specific quantum field theories are experimentally falsifiable (like QED, QCD, electroweak interaction theories or those that describe the quantum hall effect and so on) but they belong to the general class of quantum field theories which we know to be theoretically falsifiable in general. Similarly, I would imagine to be able to have individual string theories, maybe on some particular backgrounds, which would be experimentally falsifiable. That is, these string theories would make predictions about particular systems, and would also be part of a larger string theory vacua, which might contain theories that have nothing to do with the real world but would still be theoretically falsifiable or, at least, consistent theories in themselves.

Isn't supersymmetry supposed to be observed at high energies? Would that make this study of such theories more relevant?

Well . . . that's maybe what we will learn in the next few years. To date, there isn't any direct experimental evidence for supersymmetry. It is, definitely, the most popular candidate for what we might see beyond the standard model but there is nothing compelling saying that it has to be supersymmetric. Thus, if supersymmetry is indeed found, that would be excellent and there would be real life systems to which many string models or scenarios would apply. If supersymmetry is not found, at least in the colliders, then that still doesn't mean that there may not be supersymmetry at a higher scale. It just means that supersymmetry is not there at low energies and instead broken at some higher scale. So we can't rule out that possibility.

Could you always increase the scale of supersymmetry breaking as much as you wanted to?

Yeah, but there are some constraints. It's very difficult to arrange supersymmetry to be broken at an arbitrary scale because there has to be a mechanism for spontaneous symmetry breaking, which can be implemented perturbatively or non-perturbatively. In the non-perturbative case, you have to deal with large-scale separations while perturbatively you do not. In any case, you cannot rule out supersymmetry at a higher scale but it would definitely make the study of many supersymmetric field theories more of a theoretical exercise rather than really of immediately physical relevance. We would like to know soon which of these scenarios is the correct one.

Is the LHC able to go the energy range where supersymmetry is supposed to be found?

That's what it has been designed for, that is, to go to the energy scale where one believes that one would be able to see this signature. I hope that, even though with all the problems that have showed up (that the LHC won't really run up to the energies that was supposed to), it doesn't exclude this possibility somehow. People often say that there is still good likelihood of knowing it so let's see how that goes.

Now, changing a bit the topic. Supersymmetric theories are usually formulated in 10 dimensions while non-supersymmetric theories are formulated in 26 dimensions. How come there is this jump from 10 to 26? Do string theories have to be formulated either in 10 dimensions or 26 dimensions, or is there a bit more flexibility?

First, the numbers 10 and 26 have to do with string backgrounds that preserve Lorentz invariance but you can have a non-critical string theory which has lower dimensions [9]. For instance, people have studied $c = 1$ theories which are essentially two-dimensional string theories but in those cases the two-dimensional spacetime is not Lorentz-invariant (you have some other fields turned on that break that symmetry) [10–12]. Therefore, the numbers 10 or 26 are not such hard and fast numbers. These numbers are related to the central charge of the underlying conformal field theory, which in the simplest cases, though not necessarily the case, translates into spacetime dimensions as long as they are very large.

When supersymmetry is present in the theory, there is a reduction from 26 to 10 dimensions because, roughly speaking, you sort of double the number of degrees of freedom. However, for the string theories that might describe QCD or some other realistic field theory through AdS/CFT-like correspondences, it may well be the case that the string is not really propagating in 10 or 26 dimensions but instead could be propagating in five dimensions and still have a central charge that is equal to 10 or 26 because of the non-trivial nature of the five-dimensional spacetime. In summary, the numbers 10 and 26 are sort of the simplest scenarios but are not necessarily fixed.

Okay, now moving toward something that you have mentioned several times, the AdS/CFT correspondence. Could you explain what exactly is this conjecture?

The basic idea is that it is a statement about the equivalence of string theories in AdS spacetimes, namely spacetimes which at infinity look like hyperboloids; that is, they're negatively curved. String theories on such AdS spacetimes are supposed to be exactly equivalent to quantum field theories which live on the boundary of this AdS spacetime. In other words, you have an equivalence between two very different-looking quantum theories: one is a quantum theory of just matter fields which lives in a d-dimensional spacetime (need not even be curved; it could be flat spacetime where gravity plays no role) and the other is a quantum theory of gravity in a spacetime of different dimension (at least with one higher dimension) for which the gravitational degrees of freedom are excited; that is, gravity is dynamical.

The claim is that there is a precise dictionary between all the observables of one theory and the other. Moreover, it is not a trivial equivalence in which you change some variables in one theory and rewrite them in terms of the variables of the other one. Instead, it's a sort of quantum equivalence that relates strongly coupled phenomena in one theory to weakly coupled phenomena in the other theory. Thus it is not the case that the classical theories are equivalent in any sense but only the quantum theories: it is only at the quantum mechanical level that you see this connection. In a sense, even though it originally arose in string theory, it actually makes a statement about the equivalence of two different theoretical entities which are independent of string theory because there is a certain limit in which the string theory in AdS space reduces to Einstein's gravity on that same space. In this situation you are making a statement about how certain quantities in Einstein's gravity in AdS space are related to a very strongly coupled limit of the quantum field theory in the lower-dimensional spacetime.

Therefore, this is a statement about two theories that you can independently verify. In principle, you can calculate everything in Einstein's gravity by solving partial differential equations and also in the quantum field theory, though it is very difficult to compute things at strong coupling but in many cases there are theoretical tools like integrability [13] or numerical methods that place the theory on a lattice, which allow you to do so. This is a surprising statement because it means that there is a relation between the two theories, which no one suspected even though Einstein's gravity was known for 80 years as well as quantum field theory for a similar amount of time. This duality turned out to have many applications in other areas as people seem to be excited at least in the particle physics and condensed matter communities.

According to this duality, can quantum field theory provide us with an alternative description of the geometry of spacetime? Can we just discard geometry?

In principle yes, at least the geometry of AdS space, in the sense that the quantum field theory encodes in it all the information about the geometry but maybe not in a useful way. The geometric description of Einstein's theory is sometimes the most useful or at least the most practical way of studying the geometry of spacetime. This is exactly what the whole idea of holography is about: that you can describe a $d+1$-dimensional system in terms of a quantum field theory in d dimensions [14, 15]. All information is there in this d-dimensional hologram of the $d+1$-dimensional theory. So, just like the holograms in your credit cards, which are two-dimensional objects capturing a sort of three-dimensionality, the quantum field theory is sufficient to capture the higher dimensional theory. However, I think that, while this is definitely true, the interesting thing is the interplay between these two dual descriptions, in particular, that there is one description which is more useful in one regime and precisely when you don't have techniques to deal with the other description and vice versa.

Wouldn't you say that finding one of these dualities for non-supersymmetric theories would be a great achievement for string theory?

I think that in a way we have stumbled upon these dualities through other dualities of string theory which are more general. The AdS/CFT duality is in some sense a limiting example of certain dualities in string theory [3]. We have uncovered special cases involving supersymmetry because that is the setting where these more general dualities in string theory operate. Nevertheless, I think that this picture is broader and you probably don't need all the supersymmetry or you don't need it at all. In the non-supersymmetric cases there should be a general connection with the so-called large-N quantum field theories. Though I haven't mentioned it earlier, large N implies that the quantum field theory has a large number of degrees of freedom in which case the dual gravitational description becomes classical. This regime is where most of the work has been done and I believe that even for a system like large-N QCD in four spacetime dimensions there should be a dual classical string theory, which would be very interesting to learn about.

But how hard is this problem of going away from supersymmetric theories?

Well . . . by now there are already some cases where I think we have indications that these dualities hold even without supersymmetry.

Which cases are you referring to?

There are many examples. I have worked with Cumrun Vafa in one of those examples in the early days of these AdS/CFT-type dualities relating Chern-Simons theories which are non-supersymmetric, albeit topological, to a dual string theory, which is also not supersymmetric because it is a topological string theory, although it arises from an underlying supersymmetric string theory [16–19]. In this context, we have a non-supersymmetric duality between two non-supersymmetric systems but it's in a way just a toy model.

Then, there have been more non-trivial examples such as the conjecture by Igor Klebanov and Alexander Polyakov relating the $O(N)$ vector model, which is a 2+1-dimensional quantum field theory that arises a lot in the study of statistical mechanics of spin systems (for instance the $O(3)$ vector model is used for describing real ferromagnetic spin interactions), to a dual description, at least in the large-N limit, in terms of a classical theory in four-dimensional AdS spacetime of a system which involves not just gravity but various other higher spin fields [20]. There have been many non-trivial checks of this conjecture.

More recently, the work I have been doing with Matthias Gaberdiel has been along these lines where we have, similarly, two-dimensional quantum field theories, which are special quantum field theories called conformal field theories (which have an additional scale invariance) and arise from describing critical phenomena in 1+1 dimensions. For example, the two-dimensional Ising model is described by a two-dimensional conformal field theory at its phase transition. Thus, a large class of these two-dimensional non-supersymmetric conformal field theories in the large-N limit, we conjecture, are dual to a theory involving gravity and other fields in 2+1-dimensional AdS spacetime [21–24].

Do you call these theories string theories?

Both the Polyakov and Klebanov conjecture [20] and the case that we are studying [21–24] are sort of halfway between theories of gravity and string theory. They have an infinite number of fields like string theory has but they have a smaller infinity than in the full string theory. It is sort of a truncated version of the full string theory which keeps in it, in a way, a lot of the stringy behaviour. Usually, many of the AdS/CFT-type dualities are played out between theories which contain just gravity or Einstein's gravity. However, in these cases that I am describing we are extending such ideas beyond gravity. We are still not tackling the full string theory but we are taking one step further and tackling a large sub-sector of it. In this particular case, both sides of the duality are non-supersymmetric, both the theory in the AdS spacetime as well as the two-dimensional conformal field theory, and similarly for the $O(3)$ vector models one seems to see the correspondence working non-trivially. This gives you the hope that you can find such correspondences in more general cases and extend this duality to non-supersymmetric examples in general.

This subject is in its infancy and so I think that there are exciting days ahead that will come by pushing it in this direction. In a way, personally, I feel that string theorists are a little addicted to supersymmetry; that is, they somehow feel that without supersymmetry they cannot do very much. This is not really true as it is just a mental block in the sense that there are many other symmetries and many other tools that you can exploit to still make non-trivial connections and even theoretical checks. I think that we shouldn't necessarily allow ourselves to be too limited by working only with supersymmetry. This is what really interests people in other fields, I mean, if you are going to talk to condensed matter physicists and tell them, "Look here, this duality can say wonderful things about strongly interacting systems," they will want to know, "Can it say anything about my strongly interacting system (which is typically a non-supersymmetric system)?" (laughs). I think we should develop the tools and the ideas in order to move in this direction.

There has been an enormous body of work in higher-spin theories. Could you summarise what have been the main implications of this research for both holography and string theory? Why have people become so interested in these setups and what have we learnt from it?

I think the work of the last few years has led people to seeing that higher-spin theories (a) provide very useful, tractable and often non-supersymmetric toy models of holography, (b) can give some intuition and perhaps concrete lessons on stringy modifications of geometry and (c) potentially help in deciphering AdS/CFT in the weakly coupled regime (though this is yet to be concretely realised). These are some of the reasons people in string theory have been looking at this topic [25].

As to the implications, we have seen a lot of evidence for point (a) in terms of interesting higher-spin duals to several classes of interacting field theories in two and three spacetime dimensions. The field theories are often solvable in their own right (and are closer to real life in the sense of being non-supersymmetric and related to well-studied condensed matter systems) and this feature can enable you to learn more about point (b). Indeed, in the case of some of the two-dimensional conformal field theories with AdS_3 duals, there are interesting black holes and other semiclassical solutions which exhibit new features compared to Einstein gravity. We have had to rethink our notions about the definition of a black hole – since the notion of a horizon is not an invariant one [26]. It has led us to replace geometric notions by more abstract invariant concepts like holonomies. The subject is still being intensively explored and so we are still piecing together the whole picture. However, it promises to be an exciting new feature. I also find it personally satisfying that we have been able to let go of our dependence on supersymmetry by exploring other powerful bosonic symmetries.

How do you feel about these dualities being conjectures?

There are conjectures at very different levels I think. For a mathematician probably most of the things that physicists do will always be conjectures because they do not have the right level of rigour. But there are some examples where we have a sort of physicist's proof, or argument, relating these dual systems which gives you confidence that it holds. We would like to understand many of these conjectures in more detail, but I think that the AdS/CFT conjecture in general is one of the cases which is in principle provable, even in the strong sense of the word, because it refers to an equivalence between two different systems which we can independently define and characterise. String theory in AdS spacetime has its own set of rules which when followed lead you to some answer whereas in quantum field theories there are other sets of rules which lead to some answers. It should be possible to show in general why these sets of different-looking rules for different-looking theories always give the same answers.

Proving it for its own sake is not a good enough motivation since there is so much evidence for it; however, the method by which it could be proven would shed light on why there exists a relation between the two theories. Ultimately, it could help us generalising such correspondences to other scenarios. For instance, if we could show in generality that the rules of quantum field theory and the rules of string theory give the same answers then that could enable us to understand what is, in the most optimistic scenario, the dual to large-N QCD. For that reason, it is useful and important to try to understand why this connection exists and to prove it.

What are the missing ingredients for such proof?

The main technical problems have to do with how exactly one can reorganise the rules of quantum field theory such that they look like those of a string theory. I have worked on this for some time and I think I see that a piece of the puzzle has fallen into place: you can show that the Feynman rules of quantum field theory, which define how you compute things perturbatively in quantum field theory, can be organised, at least for all large-N gauge theories (the context for most such dualities), into rules for the propagation of strings [18, 27–30].

Very crudely speaking, a quantum field theory is a theory of point particles and so the Feynman rules essentially assign certain probably amplitudes for the motion of these point particles which can trace all possible paths. Feynman diagrams are pictorial ways of depicting these paths and how particles interact with each other, describing how particles scatter, interact and move. The Feynman rules of quantum field theory, roughly speaking, tell you that you should take into account all such diagrams. On the other hand, the rules of string theory do not describe paths of particles and because the constituent elements are strings, in particular closed strings (like rubber bands), you have to sum over paths of strings which trace out two-dimensional surfaces instead. Thus, the rules of string theory are phrased in the language of a sum over two-dimensional surfaces which take into account all possible shapes and sizes of two-dimensional surfaces. In the case of point particles in quantum field theory, this is analogous to summing over all possible shapes and lengths of one-dimensional paths.

At first sight, this seems very different: How can the sum over one-dimensional paths be the same as the sum over two-dimensional surfaces? But one thing that I think one can argue, and this is something I spent some time trying to understand, is that there is a very natural way within quantum field theory, at least for large-N gauge theories, which doesn't require any supersymmetry, by which the sum over these paths can be organised as a sum over surfaces [30]. Thus, that part of the puzzle, which shows exactly how you can see the sum over surfaces arising from the sum over paths, can be formally established in general. This is in a sense a step beyond Gerard 't Hooft's original work where he organised large-N field theories in terms of the genus of the Feynman diagrams [31]. In fact, at that time he guessed that Feynman diagrams could be organised by the genus of two-dimensional surfaces in large-N theories. However, the work I have done goes one step beyond 't Hooft's work as it states precisely how each inequivalent two-dimensional surface arises when you sum over the Feynman diagrams of the large-N gauge theory.

This is one part of the puzzle that I think has been solved but the second part, which is what would be needed to make this a complete proof, is yet to be solved. One can show that if you start with a point particle propagating according to the rules of quantum field theory you can reorganise its description in such way as if it followed the rules of a two-dimensional surface (string) moving in spacetime. However, what is the spacetime on which this string is moving and what is the weight that you should assign to the motion of that string? This would depend on the spacetime on which the string is propagating and it should be possible to read off from the rules of quantum field theory. In more general terms, given that you have a quantum field theory X with its specific set of Feynman rules, what is the corresponding spacetime Y on which the string is moving? That is the step that remains to be done in order to show that the rules of quantum field theory are equivalent to the rules of string theory. If you could do that explicitly as I just explained, then you would get an idea of how it could be done in general and not just for some specific quantum field theory X. This would tell you what is the correspondence between X and Y and not just for when X is equal to $N = 4$ super Yang–Mills theory and Y is equal to five-dimensional AdS spacetime. Making this connection in general would achieve the goal of describing any general system X by means of its dual description Y. So, there's still this hurdle that will probably only be solved in an incremental way starting with simple classes of theories (maybe supersymmetric theories) then trying to generalise it.

Do you expect that such a proof will come in the near future?

In 10 years, for instance, I wouldn't say that it is unlikely. There's maybe 50% or more chance of that happening but these predictions are always dangerous (laughs).[1]

Do you think that the understanding of the quantum Hall effect [34–36] was the biggest breakthrough in the past 30 years or so for the physics of strongly coupled systems or was it the AdS/CFT correspondence?

Well ...I think that in a way AdS/CFT has proved to be richer. Maybe it might be a bit premature to call it a breakthrough but it is sort of building up into a breakthrough (laughs). It is definitely a qualitative jump in our understanding. The quantum Hall effect was a very important development in understanding strongly interacting systems, learning about new ground states of theories and seeing them experimentally realised. However, I think that what the AdS/CFT and other string dualities seem to suggest is a more vital framework for understanding strongly interacting systems in general and the nature of quantum interactions. I find fascinating the idea that sometimes quantum effects can be, in a way, geometrised. In AdS/CFT this geometrisation takes place but also in the more conventional dualities in string theory like in M-theory [37], where the string coupling constant (which is a measure of quantum effects in string theory) becomes the radius of a circle. Therefore, in a sense you are giving a geometric interpretation to $\hbar$, though we haven't quite fully understood it. So, in the broader setting this might not be a sharp puzzle like the others about quantum gravity I mentioned at the beginning of this conversation, but I think that there are definitely some clues which indicate that our understanding of quantum mechanics is incomplete, and that is something one has to keep an eye out for.

Why do physics?

(laughs) Well, I always enjoyed it and every year I only feel more and more glad that I am into this subject because it is so rich and fulfilling. It is great to have the luxury to spend time thinking about these stimulating questions and specially if you're lucky to make progress, even if it is small. Sometimes it is the moments of revelation that are really what makes it all worthwhile.

So many people criticise theoretical physicists because there aren't many practical applications, at least not in the near future. What do you think is the role of the theoretical physicist in modern society?

First, when people say that, I think they have a very narrow view. Often, whenever I talk to people who don't really know much about physics or its applications I just pull out my mobile phone or camera and say that this wouldn't have been possible if we hadn't understood the laws of quantum mechanics and the photoelectric effect, which was explained by

[1] A concrete realisation of the above program for proving the AdS/CFT correspondence has recently appeared in the case of AdS_3/CFT_2 in [32, 33].

Albert Einstein in 1905. Mobile cameras and the mechanism for storing pictures depend on the photoelectric effect. Then, of course, the transistor is deeply quantum mechanical as well as various other components that are largely exploited, such as computer chips. Everything that you take for granted around you in your modern life wouldn't have been possible if the laws of quantum mechanics hadn't been uncovered.

Slightly more exotic is the example of the Dirac equation, which Paul Dirac wrote down guided by an aesthetic sense without any real experimental reason and turned out to predict a new particle called a positron. Something that he scribbled down on a piece of paper is now actually used to save lives through the PET (positron emission tomography) scan. Hospitals have it and it all came from a few scribbles on a piece of paper. So I think it is short-sighted to view the scientific enterprise through the prism of "Okay, in 10 years time will you have a product coming out from your research?" (laughs).

I think that the progress of human knowledge led to this and I don't see any reason why it should be different. I also often hear people criticising string theory in the following way: "Okay, maybe the Dirac equation and quantum mechanics were indeed successful but that was at the atomic level and now you're talking about the Planck scale which is 20 orders of magnitude even smaller and as such can't conceivably have any application in the next 1000 years or so" (laughs). But then I tell them about something like the AdS/CFT correspondence because in trying to solve some problems of quantum gravity which may be or may not be verifiable immediately you have ended up learning some things about systems which, at first sight, are very far removed from quantum gravity. So I think that the open-endedness of the scientific exploration is very crucial to be able to stumble upon such discoveries. So, who knows? Maybe you will soon be describing the physics of some material using string theory and that might be the material that will be the next silicon or the next transistor. This is the way that scientific progress unfolds and is always unpredictable. Obviously I don't think that everyone should be doing string theory or theoretical physics but I think that people who are passionate about it . . .

At least you should be doing it (laughs), no?

(laughs) Well . . . people who have the right temperament and passion for it. In my opinion, it is healthy for a society to have a small fraction of people who do that and I definitely think it is healthier than the large fractions of people who go to the financial market (laughs) and then destroy the lives of thousands of people through complete speculation or whatever process. So, theoretical physicists have a role to play in society and ideally even in the educational system where they should be integrated into. The theoretical physicist's training is a very unique one because it develops various skills that are useful even if you chose later in life not to be a theoretical physicist. I have met many people in environmental sciences as well as in other scientific disciplines and even sometimes very far removed from science who had a theoretical physics background that helps them to analyse, phrase and understand problems in a particularly useful way. Ideally, a theoretical physicist who is integrated into the educational system can impart that skill to many people who may want to be theoretical physicists but may also want to use those skills in other areas. All these roles are definitely important. Therefore, I don't understand why we should be too defensive about being theoretical physicists. I think that we have had a good track record of being useful.

References

[1] I. Mandal and A. Sen, "Black hole microstate counting and its macroscopic counterpart," *Nucl. Phys. B Proc. Suppl.* **216** (2011) 147–168, arXiv:1008.3801 [hep-th].

[2] J. Polchinski, "The black hole information problem," in *Theoretical Advanced Study Institute in Elementary Particle Physics: New Frontiers in Fields and Strings*, J. Polchinski, P. Vieira and O. DeWolfe, eds., pp. 353–397. World Scientific, 2017. arXiv:1609.04036 [hep-th].

[3] J. M. Maldacena, "The large N limit of superconformal field theories and supergravity," *Int. J. Theor. Phys.* **38** (1999) 1113–1133, arXiv:hep-th/9711200.

[4] S. Kachru, R. Kallosh, A. D. Linde and S. P. Trivedi, "De Sitter vacua in string theory," *Phys. Rev. D* **68** (2003) 046005, arXiv:hep-th/0301240.

[5] N. Gromov and P. Vieira, "The AdS(5) x S**5 superstring quantum spectrum from the algebraic curve," *Nucl. Phys. B* **789** (2008) 175–208, arXiv:hep-th/0703191.

[6] N. Gromov and P. Vieira, "Complete 1-loop test of AdS/CFT," *JHEP* **04** (2008) 046, arXiv:0709.3487 [hep-th].

[7] N. Gromov, V. Kazakov and P. Vieira, "Exact spectrum of anomalous dimensions of planar N = 4 supersymmetric Yang–Mills theory," *Phys. Rev. Lett.* **103** (2009) 131601, arXiv:0901.3753 [hep-th].

[8] N. Gromov, "Introduction to the spectrum of $N = 4$ SYM and the quantum spectral curve," arXiv:1708.03648 [hep-th].

[9] D. Kutasov, "Some properties of (non)critical strings," in *Spring School on String Theory and Quantum Gravity (to Be Followed by Workshop)*, pp. 102–141. Trieste Spring School, 1991. arXiv:hep-th/9110041.

[10] D. J. Gross and A. A. Migdal, "Nonperturbative two-dimensional quantum gravity," *Phys. Rev. Lett.* **64** (Jan. 1990) 127–130. https://link.aps.org/doi/10.1103/PhysRevLett.64.127.

[11] M. R. Douglas and S. H. Shenker, "Strings in less than one-dimension," *Nucl. Phys. B* **335** (1990) 635.

[12] E. Brezin and V. Kazakov, "Exactly solvable field theories of closed strings," *Phys. Lett. B* **236** (1990) 144–150.

[13] N. Beisert et al., "Review of AdS/CFT integrability: an overview," *Lett. Math. Phys.* **99** (2012) 3–32, arXiv:1012.3982 [hep-th].

[14] G. 't Hooft, "Dimensional reduction in quantum gravity," *Conf. Proc. C* **930308** (1993) 284–296, arXiv:gr-qc/9310026.

[15] L. Susskind, "The world as a hologram," *J. Math. Phys.* **36** (1995) 6377–6396, arXiv:hep-th/9409089.

[16] R. Gopakumar and C. Vafa, "Topological gravity as large N topological gauge theory," *Adv. Theor. Math. Phys.* **2** (1998) 413–442, arXiv:hep-th/9802016.

[17] R. Gopakumar and C. Vafa, "M theory and topological strings. 1," arXiv:hep-th/9809187.

[18] R. Gopakumar and C. Vafa, "On the gauge theory/geometry correspondence," *AMS/IP Stud. Adv. Math.* **23** (2001) 45–63, arXiv:hep-th/9811131.

[19] R. Gopakumar and C. Vafa, "M theory and topological strings. 2," arXiv:hep-th/9812127.

[20] I. Klebanov and A. Polyakov, "AdS dual of the critical O(N) vector model," *Phys. Lett. B* **550** (2002) 213–219, arXiv:hep-th/0210114.

[21] M. R. Gaberdiel, R. Gopakumar and A. Saha, "Quantum W-symmetry in AdS_3," *JHEP* **02** (2011) 004, arXiv:1009.6087 [hep-th].

[22] M. R. Gaberdiel and R. Gopakumar, "An AdS_3 dual for minimal model CFTs," *Phys. Rev. D* **83** (2011) 066007, arXiv:1011.2986 [hep-th].

[23] M. R. Gaberdiel and R. Gopakumar, "Triality in minimal model holography," *JHEP* **07** (2012) 127, arXiv:1205.2472 [hep-th].

[24] M. R. Gaberdiel and R. Gopakumar, "Minimal model holography," *J. Phys. A* **46** (2013) 214002, arXiv:1207.6697 [hep-th].

[25] S. Giombi and X. Yin, "The higher spin/vector model duality," *J. Phys. A* **46** (2013) 214003, arXiv:1208.4036 [hep-th].

[26] M. Ammon, M. Gutperle, P. Kraus and E. Perlmutter, "Black holes in three dimensional higher spin gravity: a review," *J. Phys. A* **46** (2013) 214001, arXiv:1208.5182 [hep-th].

[27] R. Gopakumar, "From free fields to AdS," *Phys. Rev. D* **70** (2004) 025009, arXiv:hep-th/0308184.

[28] R. Gopakumar, "Free field theory as a string theory?," *Comptes Rendus Physique* **5** (2004) 1111–1119, arXiv:hep-th/0409233.

[29] O. Aharony, J. R. David, R. Gopakumar, Z. Komargodski and S. S. Razamat, "Comments on worldsheet theories dual to free large N gauge theories," *Phys. Rev. D* **75** (2007) 106006, arXiv:hep-th/0703141.

[30] R. Gopakumar, "What is the simplest gauge-string duality?," arXiv:1104.2386 [hep-th].

[31] G. 't Hooft, "A planar diagram theory for strong interactions," *Nucl. Phys. B* **72** (1974) 461.

[32] L. Eberhardt, M. R. Gaberdiel and R. Gopakumar, "Deriving the AdS_3/CFT_2 correspondence," *JHEP* **02** (2020) 136, arXiv:1911.00378 [hep-th].

[33] M. R. Gaberdiel, R. Gopakumar, B. Knighton and P. Maity, "From symmetric product CFTs to AdS_3," arXiv:2011.10038 [hep-th].

[34] F. D. M. Haldane, "Model for a quantum Hall effect without Landau levels: condensed-matter realization of the 'parity anomaly,' " *Phys. Rev. Lett.* **61** (Oct. 1988) 2015–2018. https://link.aps.org/doi/10.1103/PhysRevLett.61.2015.

[35] R. B. Laughlin, "Quantized Hall conductivity in two dimensions," *Phys. Rev. B* **23** (May 1981) 5632–5633. https://link.aps.org/doi/10.1103/PhysRevB.23.5632.

[36] D. J. Thouless, "Quantization of particle transport," *Phys. Rev. B* **27** (May 1983) 6083–6087. https://link.aps.org/doi/10.1103/PhysRevB.27.6083.

[37] E. Witten, "String theory dynamics in various dimensions," *Nucl. Phys. B* **443** (1995) 85–126, arXiv:hep-th/9503124.

13

David J. Gross

Permanent member and holder of the Chancellor's Chair Professor of Theoretical Physics at the Kavli Institute for Theoretical Physics and Professor of Physics at the Department of Physics at University of California, Santa Barbara

Date: 28 June 2011. Location: Uppsala. Last edit: 12 December 2018

What are the main problems in theoretical physics at the moment?

There are lots and lots of problems. In a sense, the most important problems are observational. Physics is an experimental science and we're trying to understand the real world. There are various directions and speculations we've been making for quite some time and we lack direct evidence for them.

So you think that the problem with theoretical physics is that the experimentalists don't have enough sensitivity to detect certain things (laughs)?

(laughs) Clearly, it's one of the main obstacles. Absolutely.

But what kind of phenomena do we need evidence for?

Almost every single one of the talks here at Strings invokes supersymmetry in one way or the other.[1] Yet we have no direct evidence for it but it could be around the corner. There are a lot of questions that are conceptual and very difficult and we can get to them but there are also a lot of speculations that have been made over the last 35 years.

What kind of speculations?

Well, starting with unification and supersymmetry, not to speak of string theory for which, at best, there are only very indirect clues and since we're not doing mathematics ...

Are we not?

(laughs) Although it can be confusing somehow, we could be wrong and we need evidence so we hope for clues. Clearly, how to probe nature is the main question and that question is not going to be answered by theorists.

[1] This interview took place during the conference Strings 2011.

But if you would get some kind of experimental evidence for some of these things, you would still have a lot of different problems, right?

Oh, absolutely. There are many problems, including all the problems that led to those speculations, which themselves include new constructs, new ideas, new symmetries, new dynamics, etc. But at some point it's important to make sure that we're connected, that we're talking about the real world and not just about mathematical structures (laughs) and we need some help for that.

This problem of probing nature is sort of in the background for us because there's nothing that theory can do about it. A lot of the questions that are addressed nowadays assume that we know that the basic strategy that led to these new structures has been correct. However, it could be wrong and those problems that led to those speculations are still problems because until we have experimental evidence for the solution, all we have are speculative answers.

When I give a public lecture about string theory, I first discuss what has been achieved in the last century, which is magnificent, and then what questions that gave rise to, what problems emerged from the standard model and also from observation. Then, I discuss the attempts to answer those problems, some of which led to string theory, and then I discuss the problems that string theory has given rise to. The first class of problems comes out of theories that have been and are continuously tested and are extraordinarily accurate. We understand these theories better and better as time passes. They have certain conceptual incompleteness issues and also observational issues but our understanding of these issues has also gotten better and better in the last 20 years. However, some of these problems did not seem to fit into our understanding of physics and so the speculations that try to answer these gave rise to new theoretical structures, which, on the one hand, have problems of their own or have suggestive extensions of physics and, on the other hand, become more and more ambitious and tackle new problems, such as cosmological problems, which were sort of thought to be beyond the reach of theory.

Okay, let's focus on one of these then.

Which one?

Quantum gravity. Why is this a difficult problem?

Well, it's a difficult problem ... of course, we probably will only really understand why it's a difficult problem when we are sure what the answer is. Anyway, we didn't have a theory of quantum gravity and ...

Do we need one?

We definitely need one. When quantum mechanics came along, which was not long after general relativity, there weren't many attempts to quantise gravity. Quantum theory was being developed and gravity was kind of irrelevant from the practical point of view. However, people did realise early on that you couldn't just ignore gravity and have a consistent quantum theory. Classical mechanics is simply wrong as it's inconsistent with quantum mechanics, and part of the world that is classical and dynamical can't be consistent with quantum mechanics as it is now. Of course, the fact is that people didn't particularly worry about gravity and those who did ran into severe problems. Gravity, even just as a classical theory, was really only understood probably in the 1950s and 1960s. It took a long time just to get used to the idea of curved space and to be comfortable with the gauge symmetry. Even Albert Einstein didn't really understand his own theory very well and never believed in horizons and singularities. As field theory matured, people tried more and more to quantise gravity and they were unsuccessful. There were two issues, one of which is for most of us no longer an issue, which was the existence of ultraviolet divergences, that is, singularities in the short-distance behaviour and the non-renormalisability of Einstein's theory.

Why is this not an issue anymore?

You hear people talking about whether supergravity is finite or not, in particular $N = 8$ supergravity [1, 2] but, besides motivating people to do hard calculations, it's not that interesting, at least to me.

Because you think that it's finite or you don't care?

No, I don't think it's finite and it doesn't really matter. Even if it was finite, I wouldn't be interested. It would be some indication that there was some really big symmetry embodied in that theory, which would be interesting but I don't think it would be interesting in a deep sense. I also think that it's unlikely to be true.

So, you think that there's no problem with the theory not being finite?

No. String theories are automatically finite. Our understanding of quantum field theory dramatically changed. So, 40 years ago renormalisability and ultraviolet finiteness played an important role in selecting theories that were sensible. Of course, these field theories are the theories that we now understand to be sort of inevitable at microscopic distances or atomic distances. But that point of view has changed. First of all, string theory did solve that problem.

Is that clear? I thought that there was no rigorous proof of that ...

There is no rigorous proof of almost anything in physics (laughs). There is no rigorous proof of thermodynamics. Rigorous proofs are not that interesting, at least to me. But string theory is clearly finite. People have made calculations and shown that there is no source of divergences at short distances [3–6]. There are many ways of putting it but the kinds of singularities that occur in string perturbation theory were understood very early to be just indications of some instability, i.e., of perturbing around the wrong starting point [7]. So they are infrared divergences.

There is, of course, this fantastic thing that happens in string theory, namely, that you can confuse the infrared with the ultraviolet behaviour, so to speak. The infrared behaviour of closed strings is related to the ultraviolet behaviour of open strings and vice versa [7, 8]. But these aren't ultraviolet singularities, so there's no real way they can occur. They just don't happen. If you chose the wrong starting point you can find singularities, which is just an indication that you shouldn't be perturbing around that point.

I see. What about singularities inside the horizon of black holes? Can they occur in string theory?

No. Gravity shows up in string theory only as an approximation in certain situations. Certainly not inside black holes. It's even hard to describe the inside of a black hole but that's of course the region where, by its very nature, gravitational forces are strong and the corrections become big.

Well, you asked me why quantum gravity is difficult, so let us get back to that. Earlier, people followed the approach of just trying to do a semiclassical treatment of quantum gravity, i.e., a semiclassical quantisation, which means that you start with some kind of nice classical situation such as flat space and then you turn on gravity and see what effects it has in the quantum theory. A particular example is the gravitational effects on the energy levels of hydrogen [9–12]. How big are those effects? You actually get infinite results but we understood how to do deal with it perturbatively in quantum field theory. That was the first triumph of quantum field theory applied to the case of quantum electrodynamics and then we understood how to do it in the case of the strong interactions which is even nicer, as it solves the problem in a beautiful way.

This kind of approach, in the case of gravity, simply failed. This is a problem in itself because in modern language we could say that it just teaches us that Einstein's theory of gravity is what we call today an effective theory – a fact which I think Einstein himself realised. Einstein's original paper in 1915, where he nailed the theory down, states the principles behind the theory, such as the principle of equivalence and the principle of coordinate invariance. But then, he also assumes that the equations of motion only involve two-derivative terms. The only term he allowed was an R (Ricci curvature) term but nothing prevented an infinite number of higher-derivative terms. He would have argued that probably unitarity would fix it or that he didn't want to have run-away solutions to his classical equations. We would argue instead that those terms play no role at low energies but that there can be an infinite number of other such terms once you start taking into account quantum corrections. So Einstein's theory is an effective low-energy theory and requires an ultraviolet completion.

By now we are quite accustomed to the fact that any non-renormalisable-looking theory is perfectly well behaved and that we have to work hard to complete them. Closed-string theory can be regarded as an ultraviolet completion of gravity. And certainly, as a perturbative tool, string theory calculates quantum corrections with no new parameters. So it's finite in that sense and that was always one of its attractive features. This feature, together with the fact that an observational clue led fundamental physics towards considering unification near the Planck scale, resulted in the first superstring revolution in the mid-1980s [13].

So, that issue as far as I'm concerned was solved but that doesn't mean that gravity became easy because it only really illuminated some of the other problems with quantum gravity. I think the biggest success of string theory, as far as quantum gravity goes, is not so much stating that string theory is the ultraviolet completion of quantum gravity or is perturbatively finite but rather that it has illuminated some of the non-perturbative issues of quantum gravity, of which there are lots. These have to do not with the fact that the non-linear nature of gravity technically makes it non-normalisable but, instead, with the basic essence of the theory which is that it's a dynamical theory of space and time.

The way you speak about this problem kind of gives me the feeling that you equate quantum gravity with string theory ...

Yeah, that's all right. It's perfectly okay. It's perfectly cultural to do that because, after all, what we know about gravity, which is mostly about weak-field, long-range, classical gravity is contained in string theory. So, at that level there's no difference. The two of them are identical. One is nice from many points of view and one ugly simply because we don't know what to do when discussing many different questions like questions involving black holes where you have to deal with singularities. So, I'm allowed to take the modern point of view of looking at classical gravity as an effective theory. Moreover, if I want to think about gravity in some background then perturbatively the best picture is string theory, even though I do not know what string theory is exactly. Within string theory, we know that there is no violation of unitarity nor violation of quantum mechanics and we can count how many black holes are there and evaluate their entropy.

So you don't think that other approaches like loop quantum gravity have ...

Loop quantum gravity is total BS. I mean, it's really not worth discussing it. Don't put that in the book. But, it really isn't.

This sentence would earn millions ... (laughs)

I've said it before. It really is.

Okay, so there's nothing else in the market?

Not only is there nothing else in the market but there's no difference. String theory contains everything we know solidly about Einstein's theory. When I first started doing string theory back in the late 1960s, it was totally revolutionary [14]. It seemed like something totally divorced from what we knew which is why it was so attractive to those of us who wanted to find a theory of the strong interactions because we were convinced that field theory and all the standard approaches needed something else. String theory appeared to be revolutionary because there were no rules and, as you know, the concept of *revolutionary* by definition implies that there are no rules. There were no rules, there was no theory.

You were just constructing these S-matrix amplitudes following general principles that were sort of in the mainstream philosophy at the time. In string theory, or in the theory that turned out to be string theory, there was only one way to go. So, it's a great revolution, you don't have chaos. Looks like you have no rules but in fact there's only one way to go and from there you get many wonderful results and insights. You go to this conference (Strings) and most people are talking about field theory and their connections to string theory. It's hard to differentiate the two. Our best formulations of string theory, and therefore of gravity, including approximations where you have an ultraviolet completion of Einstein gravity in anti–de Sitter backgrounds, is in terms of field theories [15]. So half of the talks here are about supersymmetric Yang–Mills theory, which is a very close cousin of the standard model. Super Yang–Mills theory is identical to something that we have no other definition of, which is, by the way, called string theory, whatever that is. We are totally convinced of this though there is no rigorous proof.

Can you explain this point more precisely? In your string theory assessments at Strings conferences you always say that the biggest problem of string theory is that we don't understand what string theory is . . .

You're interviewing all these people, you should ask them, "What is string theory?" Dualities exemplify well what I mean by this problem because they imply that there are different mathematical descriptions of the same thing, right? One of them is $\mathcal{N} = 4$ maximally supersymmetric Yang–Mills theory, which is a very well-defined theory; in fact, it's almost solved [16, 17]. Pretty soon it's going to be solved at least for a large number of colours. Eventually it might be solved to all orders, which will be like solving or constructing string theory to all orders in Newton's constant or in a semiclassical expansion around the anti–de Sitter background.

On the field theory side, one can formulate this theory in a traditional sense, that is, one has a Hamiltonian, Feynman rules and one knows how to perform calculations order by order. In principle one could put that theory into a computer and calculate any observable. In fact, people are even beginning to do just that [18–21]. They have ways of recovering the theory by discretising it and performing the path integral. It's a well-defined problem, though not rigorous (laughs). On the other side of the duality, there's no exact definition of the theory. All there is is a perturbative scheme but no Hamiltonian. There are just these rules we've learned about how to construct the perturbative expansion and we know that this expansion doesn't converge as there are non-perturbative corrections. So it's not the best definition we have in the two sides of the duality. In this particular background where a quantum state is being constructed, the best definition is the dual field theory.

From this example we learn that field theory is much richer and part of this bigger structure. In my opinion, the difficult problems that remain are, on the one hand, ones that this understanding has produced and, on the other hand, the problems of quantum gravity that are qualitative and conceptual and that haven't yet been addressed by this new way of thinking that incorporates quantum gravity in string theory. Perhaps more precisely, problems that regard it as something that emerges from field theory. One has to deal with issues of emergent spacetime and emergent gravity as well as with issues of cosmological singularities and of cosmology in general.

From what you have described, it seems that in your opinion string theory can only be defined via dualities and their associated quantum field theories. Is that the case?

In the end string theory is a set of rules, strategies and methods for constructing quantum states. This is the approach we take in physics, at least when we're not discussing the whole universe. However, what we call a theory is more than just a set of rules for constructing a quantum state perturbatively and semiclassically around some background. That is only an approximation to the construction of a quantum state. It's a very powerful handle but we know that there's more to it as it often breaks down. Nevertheless, in quantum field theory we have a dynamics, which is either given to us or we construct it in order to reproduce the real world with some number of parameters that need to be fixed by experiment. So if that perturbation breaks down, we have, in principle, a Hamiltonian and a definition of the answer via path integrals, all of which are, perhaps, mathematically well defined and can eventually be put in a computer. On the other hand, in the case of string theory that cannot be done: there is not such a starting point, not to mention the issue of background dependence. In any case, even when particularising to a given background, there's no definition of the theory.

Have there been any attempts to define the theory and circumvent these problems?

There have been attempts to develop string field theory with the purpose of dealing with the issue of background dependence and even to obtain a non-perturbative formulation of the theory [22–29]. However, they haven't gotten very far.

Why not?

It's probably the wrong way of thinking about it.

Why would that be the case?

Because it follows the approach of ordinary field theory.

But quantum field theory is not background independent, right?

In quantum field theory there is a principle that can be formulated in a background independent way. A standard way of doing calculations is to find some background that is a good enough approximation to the state you're trying to describe and then perturb around that. Those perturbative expansions are approximations but you still have the principle.

If Einstein gravity were a well-defined quantum field theory and we knew how to complete it, or if it wouldn't have all the problems that it does have, one could imagine defining a path integral which is an integral over all spacetimes with given topologies or all metrics with some boundary conditions. Now, in some cases, clearly in the world we see around us, such integrals are dominated by saddle points, like flat spacetime, and then you can calculate small quantum corrections around that point. Of course, that doesn't work for Einstein gravity but there is something similar in string theory which does work, consisting of an ad hoc set of rules for doing that perturbative expansion. However, in a sense, you don't know what you're perturbing around and you don't have that path integral formulation in a useful way.

Quantum field theory, which I just said could be formulated in a background independent way, is not truly background independent because you have to specify boundary conditions. This means that you cannot perform the integral over all metrics with all possible boundary conditions. You have to specify them, for example, by focusing on all metrics that are asymptotically flat. People like Lee Smolin think that there isn't background independence in string theory; however, modulo the issue of boundary conditions, the approach taken in string theory, which is quantum field theory, is background independent. The issue of fixing boundary conditions is a deep conceptual problem of quantum theory as it indicates a lack of background independence, no matter how you look at it. On the other hand, background independence in the sense that you allow for arbitrary metrics in the sum over all possible histories is realised in string theory in the cases where we have a dual description. In fact, it has proved to be more powerful than the background independence of quantum field theory formulations of gravity because it allows for summing over different topologies in the path integral [30–32]. So, in particular in the case of the AdS/CFT correspondence, where the path integral sums over all metrics and topologies with asymptotically anti–de Sitter boundary conditions and we have a field theory dual description, there is background independence in a very strong sense.

In fact, background independence doesn't bother me; what does bother me is that all of this is emergent. The formulation of the theory is in terms of totally different degrees of freedom. This suggests that the theory should be understood as closed strings originating from an open string field theory. This is a point of view that works for asymptotically anti–de Sitter boundary conditions but it does not allow you to explore all the possible string theories in different backgrounds; neither does any approach to quantum gravity. This is a hard problem because in a sense it tells you that you have to specify what is happening at infinity, including infinitely back in the past where there are cosmological singularities, and far in the future where who knows what's going to happen? Maybe the great rip will take place or every observer will be causally separated from everyone else. These are deep conceptual problems.

Do you think that loop quantum gravity has shed some light on these issues?

The loop guys haven't solved a single thing.

Not even background independence in the way you've been talking about?

No, no. There are things so technically shaky and problematic that it's really hard to talk about them or to seriously discuss whatever it is that they're doing. These are problems that I regard as really tough and they have to do with issues that come up in a theory of gravity for the first time, some of them already recognised by Einstein, who soon after writing his theory, started discussing cosmology. At the fundamental level, if you're dealing with a theory of spacetime, both space and time are dynamical and you can't do what particle physicists like to do, which is to put the world inside a box and ignore the rest of the universe. This is usually followed by some experiment which might take 20 years to prepare, though still only a finite amount of time (laughs).

When space and time are unified in a theory, spacetime is a dynamical object and you have to worry about the whole damn thing: the infinite past, the infinite future and the boundaries. These dualities, not just AdS/CFT but even more dramatically matrix string theory [33], which is basically a theory of quantum mechanics of matrices that seems to be dual to M-theory (which we don't understand either, except at the classical level), are examples where we have degrees of freedom with no space and no gravity though still preserving a notion of time. They are very well defined: you can put them in a computer and perform calculations in M-theory using matrix quantum mechanics. But is this still part of the same theory? Or are they different theories? Is there something else that unifies them all? So what is string theory?

These are conceptual problems that haven't originated from any particular experiment. Unless you go back and look at the big bang and try to find out the beginning of the universe you wouldn't have encountered them. These are conceptual problems that came out of the advances in string theory and in the understanding of quantum gravity. These problems aren't going to necessarily be informed by observation (laughs) and are really tough.

In the end, does it all just boil down to finding the right dual description that represents our world?

Yeah, I have no idea. It's remarkable that it's so difficult. Even the dual description of flat space is difficult and, in particular, when attempting to realise the dual description of de Sitter space we find ourselves in serious difficulty.

I have always argued that string theory hasn't really been revolutionary yet. It's less revolutionary than what we used to think. It's not like when quantum mechanics or general relativity appeared, which led to real revolutions. That hasn't really happened in string theory because you have all these dualities that lead you back to field theory.

Don't you think that these dualities that end up describing gravity in a completely different way are revolutionary enough?

I just think the questions that are being posed require a theory that describes the real world. In the language of spacetime, it has to describe the whole spacetime manifold. This is what it would take to describe the beginning of the universe. In my opinion, you cannot avoid worrying about what happened at the beginning. It is true that we are right here and now but to just be able to describe the world right now is not the desired outcome of a theory of quantum gravity. Describing the world right now, without knowledge of its previous history, might be possible by using an effective description of some kind, like the standard model (laughs). To know the answer to the question, "How did the universe begin?" really requires setting up a new framework.

This question is not a question that physics has any experience in answering. It falls into that category of questions for which we do not know how to begin nor what the rules are. This seems to me to be a more likely breeding ground for a conceptual revolution than answering questions such as how to unify all forces, how to construct a finite semiclassical treatment of gravity or how to understand emergent space and emergent time. These latter questions are very hard, require new ideas, in fact, new accidental discoveries but they do not seem to be as profound as understanding the origin of the universe.

You mentioned emergent space and emergent time. As far as I am aware, there is no example of a situation where we understand emergent time, right?

No, and it's hard to even conceptualise what emergent time actually means. All attempts at understanding it always have a notion of time in one way or another. Physical theories usually begin with some degrees of freedom and some dynamics but what would it mean to have some dynamics if there is no time? People have tried to understand these issues using matrix models, which are composed of some dynamical objects (big matrices). The eigenvalues of these matrices turn out to be positions in space and that is how space emerges. Sometimes those matrices don't commute and so space is non-commutative, so it doesn't behave like space in the usual way. On the other hand, time is always there implicitly in the notion of dynamics. Some people have tried to study Euclidean theories for which there are only matrices and no time. A Japanese group has worked on this in the last 15 years but nothing concrete has come out of it [34, 35]. In particular, it is not clear how they would recover the notions of dynamics and causality. In the end, physics is about predicting the future so the current understanding is not satisfactory.

But do you still have the hunch that it should be possible to formulate a theory of emergent time?

I do but it is just so hard to imagine how such theory would look and how to understand it. On the other hand, space and time are unified in a theory of relativity so I cannot see how you could have emergent space without emergent time. Also, I have the feeling that emergent time is a good way to imagine an answer to how the universe began (laughs) since in this case time is a concept that only makes sense for long times compared to the Planck time. This problem is hard and there are other approaches. I particularly like Erik Verlinde's ideas [36], as vague as they may be, because he is trying to take very seriously the idea of emergent space. On the other hand, he also doesn't deal with emergent time.

Are people in the string community addressing these tough problems?

No, mostly not, which is why I made some nice remarks about Erik Verlinde in my talk, because people who work the hardest are looking for jobs, have to prove themselves, discover things, write papers and have fun. It's pretty dangerous to tackle very difficult problems. It might not even be profitable since many things are discovered accidentally. On the other hand, a lot of people are thinking about the hard issues in different ways. As long as there are smart people entering the field and no lack of smart young people, these problems are going to go away. Every once in a while I've said that people should be motivated to try to tackle hard problems. Inevitably, people tend to move towards where everyone else is moving towards. It's the crowd instinct, which has gotten worse with the internet and too much communication. There are very few isolated works nowadays as people need to get a job, recognition, promotion, etc.

That's a good point. I had never thought about the negative aspects of the internet in this respect ...

The internet made it worse because it's almost impossible to have a protected community. In the past, before the fall of the Soviet Union, the Russian physics community was very different and its spirit was really amazing. It was just a different scientific culture. It had this protected habitat that came up with its own ideas and interests, which was quite valuable for science as a whole. Nowadays we know what happens instantaneously everywhere in the world and everybody follows it all the time in real time and speeds up. So a lot of physics gets sped up that way. It increases collaboration, which is good, but doesn't offer protection to small communities so it makes everybody homogeneous. This happens in all aspects of human culture, not just in science, so it's a pity. The real thing you have to have faith in is that there is a coherent and beautiful structure that is out there and it's real, right? And so if it's real we'll find it (laughs).

Do you think that string theory is only useful at very high energies like near the beginning of the universe?

No, no, no. String theory and quantum field theory are so connected that you can use it for all sorts of things. In fact, technically, the things that fascinate me the most about recent developments that I have worked on a bit are the implications of string theory for gauge theories. The ability to get new insights into supersymmetric gauge theories or to learn about other conformal field theories and quantum critical points for the purpose of condensed matter physics are examples of applications of string theory [37, 38]. The methods are extraordinarily powerful for performing actual calculations in gauge theories.

But suppose that the LHC is running and I can tune the energy of the LHC as I please. When will I see anything coming out of string theory?

Well, it depends on what you mean. String theory is a framework, not a particular theory.

Okay, when will I see a modification of the standard model?

I hope next month (laughs). That's the real question that faces us and the thing that for me is most important is supersymmetry. Supersymmetry is a new feature of the world, of our intellectual framework, which came out of string theory. It was discovered within string theory originally and it's an integral part of it. You can't throw away string theory. It's not a revolution, as I mentioned. It's there if you believe in QCD, which I tend to believe in.

Why wouldn't you believe in QCD?

I do believe in it, even the Swedish Academy has affirmed it (laughs). So you're stuck with string theory since it is there in field theory. There is no question that there is a background which describes QCD. I have no doubt about QCD having a dual string description.

You have tried to formulate the string dual to QCD at least in two spacetime dimensions, right? So what is the problem with four spacetime dimensions?

It's harder. We know enough about supersymmetric and non-supersymmetric Yang–Mills theory to understand how its dual in asymptotically anti–de Sitter space occurs. QCD is asymptotically free and is conformal-invariant as you approach the ultraviolet regime [39] so it will be described by some string theory but it won't be as simple as $\mathcal{N} = 4$ super Yang–Mills theory. $\mathcal{N} = 4$ super Yang–Mills theory is almost solved in the large-N limit, which is actually amazing, and its dual string theory description is formulated in a simple anti–de Sitter space times a 5-sphere. The case of QCD is going to be more complicated.

In two dimensions the problem was much easier. The strategy was to take $\mathcal{N} = 2$ super Yang–Mills theory and solve it, which is pretty easy. Once you have solved it, you can reconstruct the string theory dual to it [40, 41]. Now, if we could solve QCD we, first of all, wouldn't need any help from string theory. So the strategy in his case would be to first find the string theory dual to QCD and then to use it to solve QCD. In any case, having both handles on the theory would be very useful. This is another hint that string theory and quantum field theory are the same thing.

Why do you think that string theory and quantum field theory are the same thing?

String theory and quantum field theory are not theories, they are frameworks. Even if you had never heard about string theory you could sit in the world of quantum field theories and, not just by quantum gravity alone but by other reasons, you could be led into a richer structure. So, I don't see the dividing line anymore. However, the issues that come from this larger structure, like emergent spacetime and the issues of quantum gravity, especially cosmology, are the ones that we're now confronted with and start having the tools to approach them. There have been interesting ideas of how to tackle these issues; none of them have worked successfully so far but at least we can start approaching them.

Now, you may wonder if there is anything that will come from experiment that is going to tell us something about these really conceptual ideas about spacetime or seeing strings and extended objects. Probably not, but supersymmetry is absolutely crucial. Furthermore, a lot of our insights, at least for me, have come from the extrapolation of the existence of supersymmetry. For example, it is absolutely crucial for the unification of couplings near the Planck scale. It could have turned out that those couplings unified, say, at a million TeV, that is, a million times higher than the energy at the LHC and way below the Planck scale. If that had happened, we wouldn't be talking about string theory. Of course we would have been led to string theory in the same way because it's there in field theory but we would have not been thinking about it as a basis for unification. There are clues that for me are crucial, which is one reason why I still have faith that we'll see supersymmetry, and if that turns out not to be the case, I will be shaken a bit because some of those clues will all have been proved to be wrong and we can always be wrong you know.

When you say "see supersymmetry," are you thinking of supersymmetry at any energy scale or low-energy supersymmetry?

Low-energy supersymmetry. Of course that supersymmetry in general, I got to believe is there (laughs), driven just by mathematical elegance in the structure of quantum field theory and it could be there at any energy scale. Aside from that, the scale is the Planck scale and at the beginning of the universe we have to deal with Planckian times and densities so we will not be seeing any evidence of that.

Do we have a good understanding of how spacetime looks like at the Planck scale?

No, although there are people who are working on inflationary models and talk about the Planck scale many times (laughs). Then there are a lot of other big problems like the cosmological constant problem.

How do you answer the cosmological constant problem?

I don't. I just believe that it is there.

What about all these string compactifications [42]? Aren't some of them supposed to have a positive cosmological constant at least?

Well, most of them have zero cosmological constant. They are all sick and unstable. I have expressed my opinion on this many times in many places but anyway, I don't like very much a lot of that way of thinking.

You mean you don't like the idea that we have to choose one of the many vacua to describe our world?

They're not vacua, they're metastable-ish states. The reason I object to the word "vacua" is because it assumes that you have some theory with some Hamiltonian with a potential and then you minimise the energy. But you don't have any of this. In fact, it's nothing like that.

There are rules for constructing quantum states which are, vaguely, sort of compactified strings. Even when discussing compactified strings you are not trying to minimise or find a solution to some kind of theory assuming a particular perturbative expansion. So there is no way in which you can actually find a solution but you argue that there is an effective description that looks like you're minimising some thing. Therefore these are not truly solutions of anything and the argument that is used to justify their importance is that these states are metastable and live long so we can use effective field theory. But what is the question? The question is: What is the universe? It's not what the vacuum is, it's not what the stationary state is and these aren't solutions in addition to having singularities.

So this discussion of choosing the right one ...

From the beginning in 1985, when we were very optimistic, it was clear that there were zillions of solutions. I mean, there is actually an infinite number. This factor of 10^{500} that we usually hear about denote solutions of the *type I like in de Sitter space* but there is an infinite number in de Sitter space as well as in anti–de Sitter space and there is an infinite number of solutions in flat space.

What got us excited probably 25 years ago was that there were some solutions that looked like the real world. It's a very important clue but we still don't know what the rules are that pick out the right solution. I find anthropic arguments, namely that we live in the only world in which we could live and therefore that picks out the solution [43], really philosophically disturbing and not science. In addition, I find the arguments for them weak since they are assuming the existence of this meta theory with a potential that you minimise and find a universe that somehow begins, though we don't worry about how it begins, and then inflates but can get pushed into one of these wells or solutions as it reheats.

Isn't M-theory supposed to be that meta theory?

People don't know anything about M-theory (laughs) ... there's not even any perturbative formulation of that theory.

But do you believe it exists?

Sure, we have a matrix model description of the theory [44] but then you have to reconstruct space and time (laughs). It's a very interesting time. What I'm hoping for in the LHC is supersymmetry, which I think is really important, and then clues that will allow us to again extrapolate and learn something about Planckian physics. We can just extrapolate gravity, extrapolate the other forces of nature and maybe we get some hints from cosmology but we're not going to directly do any measurements. In the end, we need some input from nature.

Is it possible to formulate a theory in which all constants of nature will be fixed?

Well, that has always been my hope. That's one of the reasons that I don't like the anthropic principle. I mean physics has been unbelievably good at predicting constants of nature.

But what about the standard model, you have to fix everything there by experiment, right?

When I discuss this in a public lecture, I try to explain that all of atomic physics, all of chemistry, all of biochemistry, all of biology, even politics therefore, are all describable by a theory, a very beautiful theory which in a sense has only one adjustable parameter. That's pretty good.

Nucleons are effectively infinitely massive so their mass plays very little role. If you assume their existence in the periodic table, then you only have the mass of the electron which sets the scale of atoms. It's the dimensionfull parameter in the game and then you have the fine structure constant that sets the strength of the electromagnetic force and that's it. That constant and quantum mechanics gives all of atomic physics, all of molecular physics, all of chemistry, all of biology and human behaviour with one parameter of magnitude $1/137$. So that's a pretty good achievement.

Now, it's true that if you really want to be precise you should understand precisely the masses of the nucleons and if you want to be even more precise you better understand the nuclear structure and the weak force and so on but still, it's just about everything, which is most of other sciences outside of particle physics. In a reductionistic sense, it is all dependent on just one free parameter.

In the standard model we have a lot of parameters but again, for almost everything that isn't produced in high-energy accelerators, like the structure of stars and nuclear forces, we just need the mass of the up and down quarks and even that doesn't play too much of a role. I mean, I can calculate in terms of one parameter with good approximation the masses of all the hadrons. People have done that [45]. We understand the proton mass, sort of, which by the way solves one of the great hierarchy problems of all times, namely, why is the proton mass so light (10^{-19}) compared to the Planck mass? We have all these other mixings and masses but they're not important for daily life. The standard model really does about everything you need for life on earth (without accelerators) with two or three parameters.

So your intuition is that we only need one parameter to describe the whole thing?

No, in the end you don't need any parameters. That was the great thing and it still is the great thing about string theory and one of the really most mysterious factors in the way I describe my view of string theory. What was so appealing about string theory 25 years ago was that we found the heterotic string [46–48], which could explain the real world or it looked like it could. It has compactifications that look like the real world and there are no free parameters. There is no place in string theory where you can adjust a parameter. You can just adjust the solution.

Not even the string coupling?

No, the string coupling is the expectation value of a field, namely, the dilaton. All parameters in string theory are really fields. Those fields might be frozen in some background and hence have some value but they're all dynamical. That's the amazing thing about string theories; that is, you can't adjust things. Now, you can find solutions like the original flat space 10-dimensional solutions in which you can construct states in a way that the dilaton has a moduli space and hence you can vary it. In fact, that's why, although there are 10^{500} so-called vacua, there is an infinite number of supersymmetric solutions which have continuous moduli. Once you break supersymmetry, these fields are usually all fixed. In the case of anti–de Sitter, there is still a free modulus but in the case of these de Sitter vacua, that degeneracy has lifted and there are no free parameters in those solutions. This was the reason why this was so exciting back then because it looked like that if you could figure out some reason to prefer a particular solution then you could well imagine that everything would be fixed because there is no place you can adjust things once you lift the degeneracy that supersymmetry gives.

So string theory is this amazing thing with all these different representations or different starting points, some of which are even just field theory, for constructing solutions and you don't have any choice once you have chosen your starting point as long as it is a good starting point, in particular that it's not unstable. String theory is a framework which is in a way much less arbitrary than quantum field theory because in quantum field theory you appear to have the ability of putting in different couplings and adjusting them. The problem is that you still don't know what picks the right solution. In quantum field theory you know what picks the right solution: it's called the vacuum, the minimum energy. Of course, you have to pick the right Hamiltonian and the right dynamics, or a dynamics you know how to get a solution to. Often you might have to adjust a bit the terms and the value of the parameters that appear in the terms in order to agree with experiment. In string theory you can't adjust anything in the dynamics and you don't know what the Hamiltonian is. Instead, you have all these ways of constructing solutions and you have to pick one. So it's very weird. It's like you thought you had a unique theory and you sort of have a set of rules, which once you know where to get started lead you to a unique solution but where you get started is pretty arbitrary, which comes back to the question: What is string theory? What are the rules that pick the starting point? I don't find it very satisfactory that there shouldn't be any reason for choosing a particular starting point. There are a whole set of people who think that there is no reason to choose one particular starting point and that you could have ended up in a world that originated from any arbitrary starting point. There are a number of solutions you could execute and there are zillions of worlds out there which we will never communicate with that have in fact picked the wrong solution. So these worlds don't have people or they have people who look a bit weird but we can't know about it anyway because we can never go there. Unfortunately, that's where we are now.

But what could be the principle that would pick the right one?

The answer to this question is not that everything is allowed but that there is a reason, a principle, something I don't know what it is and that usual guesses don't work. This was already a problem in 1985 and the original hope was that since many of these different starting points seemed consistent, they would all turn out to be inconsistent except for one. For example, the solutions could have anomalies or something else would go wrong with them but it doesn't appear to be the case so we don't know what goes wrong.

I would still like to believe in this very strong Einsteinean principle that there are no undetermined constants. Einstein made the very strong statement that fundamental physics has to keep working towards the goal of a theory that is so powerful that if you adjust any of these parameters the whole theory collapses. That's one of the reasons that I really got so excited about string theory back then since it doesn't have any things to adjust. However, it has solution space to adjust. So there are two points of view: either there is some principle that picks the right solution or all these solutions out there in the multiverse are allowed and one of them is picked anthropically. Both are logical, I suppose (laughs).

In your opinion what has been the biggest breakthrough in theoretical physics in the past 30 years?

Well ... it's called, I suppose, string theory and all that it contains. That's about everything interesting that happened in the past 30 years (laughs).

That's it?

No question. Please don't ask me to be more specific within string theory. In fundamental physics there is absolutely no question. It is the unbelievable richness which pervades all of modern physics. It's unavoidable and it's really amazing.

Why have you chosen to do physics?

I got interested in physics at a very young age, about 13 or so, just by reading popular science books. I liked mathematics but theoretical physics seemed to me really exciting because it was dealing with the real world. In the beginning of the twentieth century there was relativity and quantum mechanics as well as all their possible applications and that was awfully exciting. So I really just wanted to be a theoretical physicist at a very early age.

What do you think is the role of the theoretical physicists in modern society?

Among a lot of the sciences, theory is barely tolerated. One of the great strengths of physics is that there is this very strong component of people who don't do experiments and yet have a very close relationship with experimental physicists and are respected. This is certainly not true in biology and even in chemistry. In these areas theory is much less tolerated.

Theoretical physicists should probably play a more important role in society than they do because it is a really great education. It has a 400-year-old tradition of successfully learning more and more about the natural world and it has the other great advantage of not having lost its breath. Physics hasn't broken up that much into subfields. Theoretical physicists, if they get a good education, are pretty broad, much more so than people who are so-called theorists in other fields of science. This is one of the reasons why theoretical physicists are often quite successful in moving into other areas. So that's the argument I often make around the world to get more support for theoretical physics (laughs) but I do think it is true.

Theoretical physics is still changing and growing, now to a large extent into biology where there are wonderful questions to address. So I'm very proud of theoretical physicists. They are very arrogant but they have good reasons to be (laughs). Physics in general is incredibly powerful and successful but theory has played an important role and so has experiment. It's been a marvellous example of how a mature science should develop and it is the most mature of all sciences. Astronomy is older but astronomy after Johannes Kepler and Isaac Newton sort of became an observational science until recently. Nowadays, astronomy has a very similar synergy between theory and observation. The other big area of science, which is biology, is an example of a very young science which hasn't yet reached the kind of maturity that physics has achieved. However, it is beginning to, and physics, I think, provides a good example of how that can work successfully.

References

[1] Z. Bern, J. J. Carrasco, L. J. Dixon, H. Johansson and R. Roiban, "The ultraviolet behavior of N = 8 supergravity at four loops," *Phys. Rev. Lett.* **103** (2009) 081301, arXiv:0905.2326 [hep-th].

[2] J. Bjornsson and M. B. Green, "5 loops in 24/5 dimensions," *JHEP* **08** (2010) 132, arXiv:1004.2692 [hep-th].

[3] J. A. Shapiro, "Loop graph in the dual-tub model," *Phys. Rev. D* **5** (Apr. 1972) 1945–1948. https://link.aps.org/doi/10.1103/PhysRevD.5.1945.

[4] C. Lovelace, "Pomeron form-factors and dual Regge cuts," *Phys. Lett. B* **34** (1971) 500–506.

[5] E. Witten, "Superstring perturbation theory revisited," arXiv:1209.5461 [hep-th].

[6] A. Sen, "Ultraviolet and infrared divergences in superstring theory," arXiv:1512.00026 [hep-th].

[7] A. Sen, "Tachyon dynamics in open string theory," *Int. J. Mod. Phys. A* **20** (2005) 5513–5656, arXiv:hep-th/0410103 [hep-th]. [207(2004)].

[8] A. Sen, "Open closed duality: lessons from matrix model," *Mod. Phys. Lett. A* **19** (2004) 841–854, arXiv:hep-th/0308068 [hep-th].

[9] L. Parker, "One-electron atom in curved space-time," *Phys. Rev. Lett.* **44** (Jun. 1980) 1559–1562. https://link.aps.org/doi/10.1103/PhysRevLett.44.1559.

[10] L. Parker, "One-electron atom as a probe of spacetime curvature," *Phys. Rev. D* **22** (Oct. 1980) 1922–1934. https://link.aps.org/doi/10.1103/PhysRevD.22.1922.

[11] L. Parker and L. O. Pimentel, "Gravitational perturbation of the hydrogen spectrum," *Phys. Rev. D* **25** (Jun. 1982) 3180–3190. https://link.aps.org/doi/10.1103/PhysRevD.25.3180.

[12] E. Fischbach, B. S. Freeman and W.-K. Cheng, "General-relativistic effects in hydrogenic systems," *Phys. Rev. D* **23** (May 1981) 2157–2180. https://link.aps.org/doi/10.1103/PhysRevD.23.2157.

[13] M. B. Green and J. H. Schwarz, "Anomaly cancellations in supersymmetric D = 10 gauge theory and superstring theory," *Phys. Lett. B* **149** no. 1 (1984) 117–122. www.sciencedirect.com/science/article/pii/037026938491565X.

[14] D. J. Gross and C. H. Llewellyn Smith, "High-energy neutrino-nucleon scattering, current algebra and partons," *Nucl. Phys. B* **14** (1969) 337–347.

[15] J. M. Maldacena, "The large N limit of superconformal field theories and supergravity," *Int. J. Theor. Phys.* **38** (1999) 1113–1133, arXiv:hep-th/9711200.

[16] N. Gromov, V. Kazakov and P. Vieira, "Exact spectrum of anomalous dimensions of planar N = 4 supersymmetric Yang–Mills theory," *Phys. Rev. Lett.* **103** (2009) 131601, arXiv:0901.3753 [hep-th].

[17] N. Gromov, V. Kazakov, A. Kozak and P. Vieira, "Exact spectrum of anomalous dimensions of planar N = 4 supersymmetric Yang–Mills theory: TBA and excited states," *Lett. Math. Phys.* **91** (2010) 265–287, arXiv:0902.4458 [hep-th].

[18] D. B. Kaplan and M. Unsal, "A Euclidean lattice construction of supersymmetric Yang–Mills theories with sixteen supercharges," *JHEP* **09** (2005) 042, arXiv:hep-lat/0503039 [hep-lat].

[19] S. Catterall, D. B. Kaplan and M. Unsal, "Exact lattice supersymmetry," *Phys. Rept.* **484** (2009) 71–130, arXiv:0903.4881 [hep-lat].

[20] S. Catterall, E. Dzienkowski, J. Giedt, A. Joseph and R. Wells, "Perturbative renormalization of lattice N = 4 super Yang–Mills theory," *JHEP* **04** (2011) 074, arXiv:1102.1725 [hep-th].

[21] D. Schaich, S. Catterall, P. H. Damgaard and J. Giedt, "Latest results from lattice N = 4 supersymmetric Yang–Mills," *PoS* **LATTICE2016** (2016) 221, arXiv:1611.06561 [hep-lat].

[22] T. Banks and E. J. Martinec, "The renormalization group and string field theory," *Nucl. Phys. B* **294** (1987) 733–746.

[23] R. Brustein and S. P. De Alwis, "Renormalization group equation and nonperturbative effects in string field theory," *Nucl. Phys. B* **352** (1991) 451–468.

[24] E. Witten, "On background independent open string field theory," *Phys. Rev. D* **46** (1992) 5467–5473, arXiv:hep-th/9208027 [hep-th].

[25] W. Siegel, "Introduction to string field theory," *Adv. Ser. Math. Phys.* **8** (1988) 1–244, arXiv:hep-th/0107094 [hep-th].

[26] A. Sen, "Universality of the tachyon potential," *JHEP* **12** (1999) 027, arXiv:hep-th/9911116 [hep-th].

[27] E. Witten, "Noncommutative tachyons and string field theory," 2000. arXiv:hep-th/0006071 [hep-th].

[28] D. J. Gross and W. Taylor, "Split string field theory. 1," *JHEP* **08** (2001) 009, arXiv:hep-th/0105059 [hep-th].

[29] D. J. Gross and W. Taylor, "Split string field theory. 2," *JHEP* **08** (2001) 010, arXiv:hep-th/0106036 [hep-th].

[30] S.-J. Rey, "Holographic principle and topology change in string theory," *Class. Quant. Grav.* **16** (1999) L37–L43, arXiv:hep-th/9807241 [hep-th].

[31] A. Chamblin, R. Emparan, C. V. Johnson and R. C. Myers, "Large N phases, gravitational instantons and the nuts and bolts of AdS holography," *Phys. Rev. D* **59** (1999) 064010, arXiv:hep-th/9808177 [hep-th].

[32] J. M. Maldacena and L. Maoz, "Wormholes in AdS," *JHEP* **02** (2004) 053, arXiv:hep-th/0401024 [hep-th].

[33] R. Dijkgraaf, E. P. Verlinde and H. L. Verlinde, "Matrix string theory," *Nucl. Phys. B* **500** (1997) 43–61, arXiv:hep-th/9703030.

[34] N. Ishibashi, H. Kawai, Y. Kitazawa and A. Tsuchiya, "A large N reduced model as superstring," *Nucl. Phys. B* **498** (1997) 467–491, arXiv:hep-th/9612115.

[35] J. Nishimura, "The origin of space-time as seen from matrix model simulations," *PTEP* **2012** (2012) 01A101, arXiv:1205.6870 [hep-lat].

[36] E. P. Verlinde, "On the origin of gravity and the laws of Newton," *JHEP* **04** (2011) 029, arXiv:1001.0785 [hep-th].
[37] D. J. Gross and H. Ooguri, "Aspects of large N gauge theory dynamics as seen by string theory," *Phys. Rev. D* **58** (1998) 106002, arXiv:hep-th/9805129 [hep-th].
[38] S. Sachdev, "Condensed matter and AdS/CFT," *Lect. Notes Phys.* **828** (2011) 273–311. arXiv:1002.2947 [hep-th].
[39] G. Parisi, "Conformal invariance in perturbation theory," *Phys. Lett. B* **39** (1972) 643–645.
[40] D. J. Gross, "Two-dimensional QCD as a string theory," *Nucl. Phys. B* **400** (1993) 161–180, arXiv:hep-th/9212149.
[41] D. J. Gross and W. Taylor, "Two-dimensional QCD is a string theory," *Nucl. Phys. B* **400** (1993) 181–208, arXiv:hep-th/9301068.
[42] S. Kachru, R. Kallosh, A. D. Linde and S. P. Trivedi, "De Sitter vacua in string theory," *Phys. Rev. D* **68** (2003) 046005, arXiv:hep-th/0301240.
[43] L. Susskind, "The anthropic landscape of string theory," arXiv:hep-th/0302219.
[44] T. Banks, W. Fischler, S. H. Shenker and L. Susskind, "M theory as a matrix model: a conjecture," *Phys. Rev. D* **55** (1997) 5112–5128, arXiv:hep-th/9610043.
[45] A. Chodos, R. L. Jaffe, K. Johnson and C. B. Thorn, "Baryon structure in the bag theory," *Phys. Rev. D* **10** (1974) 2599.
[46] D. J. Gross, J. A. Harvey, E. J. Martinec and R. Rohm, "The heterotic string," *Phys. Rev. Lett.* **54** (1985) 502–505.
[47] D. J. Gross, J. A. Harvey, E. J. Martinec and R. Rohm, "Heterotic string theory. 1. The free heterotic string," *Nucl. Phys. B* **256** (1985) 253.
[48] D. J. Gross, J. A. Harvey, E. J. Martinec and R. Rohm, "Heterotic string theory. 2. The interacting heterotic string," *Nucl. Phys. B* **267** (1986) 75–124.

14

Gerard 't Hooft

Emeritus Professor of Theoretical Physics at the University of Utrecht

Date: 23 June 2011. Location: Utrecht. Last edit: 24 November 2018

In your opinion, what are the main problems in theoretical physics at the moment?

There are many and they are of different nature. There are enormous frontiers where our current knowledge basically ends. Those frontiers are in many parts of physics and the questions that they give rise to are usually worthwhile while some others are utterly uninteresting. Many of these questions are extremely technical and complicated like, for example, how to describe the collective behaviour of a bunch of atoms or how to describe the behaviour of matter inside a neutron star. There are many questions of this nature that are difficult to answer because we are extremely far from experimental evidence and the theoretical calculations that we know how to perform are too complex to be reliable. In addition, there are some very fundamental questions that are related to our lack of understanding of how the world really works. This latter set of questions is the one that I am primarily interested in, as many other people are. Among those questions there is of course the question of how to quantise gravity, or rather the way I like to state it, how to make gravity compatible with quantum mechanics. This entails an all-encompassing theory that contains gravity and quantum mechanics in a natural manner and combines them together in a way that looks obviously right. Today we have many so-called candidate theories of quantum gravity, none of which is completely satisfactory.

Does that list include, for example, string theory?

Well, string theory is a very interesting case and we should probably talk about it more because it is certainly one of the most advanced attempts to combine quantum mechanics with gravity. It is mathematically very detailed and also very suggestive. String theory, in that sense, is a very powerful approach, yet it also has its problems.

What kind of problems?

One problem is that it is usually formulated in terms of perturbative expansions. This criticism is often denied but in reality all you have is a perturbative expansion. Perhaps I see a little more clearly than many other people the dangers of a theory which is really not defined beyond its perturbative expansion. The theory has different points around which you can do a perturbative expansion and the claim is that there is one bigger theory that holds for both large and small coupling constants and has a classical expansion of some sort. However, you really only know how to perform perturbative expansions and you don't even have a way to do the non-perturbative calculation or phrase the equations that have to be solved non-perturbatively, which could be solved if, for example, you had infinite computing power. String theory has not been phrased in these terms. As I mentioned, it is possible to perform the perturbative expansion around different starting points, which, nevertheless, yield the same general features. This gives rise to the gut feeling that one is describing the same theory and therefore one extrapolates the existence of one bigger theory. But one has no idea how that theory actually looks beyond perturbative expansions.

Is this different from the standard model?

Yes it is because in the standard model we have, at least in principle, asymptotic freedom of the strong interactions [1]. So whenever the interaction parameter is large the theory is asymptotically free and that means that at very high energies you can do very precise calculations. So you can at least hope to be able to integrate the equations. Therefore the standard model is defined beyond its perturbative expansion as far as strong coupling is concerned. When the couplings are weak, the $U(1)$ part of the standard model is not asymptotically free. That is a weakness of the theory but there we know that the actual value of the perturbative expansion parameter is $1/137$, which is a small number, and so you can do very accurate calculations though not infinitely accurate. Most of us agree that this means that the standard model is not completely well defined but there are ways it could be improved. So the standard model has limited reliability which implies, for example, that the magnetic moment of the electron can only be calculated up to so many decimal places and then we have to stop. Of course, this doesn't mean that nature doesn't have a way to compute the other decimal places but our theories are simply insufficient for that purpose. Thus, in the case of the standard model we know exactly how it lacks precision though it is precise enough for most applications. If string theory could make a similar statement it would be fine but the claim that is usually made is that the theory is actually well defined beyond perturbation theory. If that is the case then string theory has to explain what that exactly means. String theory has very loose foundations; it is like an enormous impressive castle built on mud and so the castle is beautiful but the foundations are awful.

What about the claims that string theory reduces to the minimal supersymmetric standard model in the low-energy limit [2]?

Well, the true story cannot be quite that simple. There are many low-energy limits, while it is actually very difficult to see how any such limit can lead to exactly three generations of quarks and leptons, and a positive cosmological constant. Furthermore, supersymmetry must be broken, and there must also be some effective mechanism for the redundant space-dimensions to compactify; exactly how all this should take place is not well understood.

Now I do realise that string theory is the only theory at present that got this far; other theories do not even present a clue as for what the quark and lepton generations could be, or what could happen to the cosmological constant. All I am saying here is that the complete picture is not yet in sight, not even in string theory. My main point of criticism towards string theory is that it is venerated as a kind of bible: even if we don't understand it completely, it is a god-given story that must be right, even if we read it wrongly. I am seeing this happen in the discussions about quantised black holes; statements are made that cannot be right, since the calculation can be done in a classical limit, where the effects of internal string structures should be irrelevant. Since so many questions are not answered, I suspect that string theory is a highly incomplete theory, at best.

What if someone comes to you and explains that one shouldn't take string theory as a fundamental theory but instead as a formalism like quantum field theory?

I tend to go in that direction, namely, seeing string theory as a very loose mathematical structure [3]. However, even though that could be extremely useful, it doesn't tell us what's really going on. I should also mention another type of criticism, where only few people seem to be able to follow me, and which is related to the foundations of quantum theory. Quantum theory is based on Hilbert spaces and I happen to have abandoned my beliefs that the fundamental laws of nature should be cast in the form of Hilbert spaces. There should be something more fundamental behind it that is more natural. When it comes to phrasing the ultimate theory, we should abandon Hilbert spaces [4]. We are still very far away from the ultimate theory of nature and that is why the concept of Hilbert space is still very important in our present way of talking about nature. Eventually, I think the true theory should be about equations without Hilbert spaces.

What is the problem with the concept of Hilbert spaces?

The problem is that it produces quantum mechanics as we understand today, which gives rise to probabilities and not certainties. The more you think about it, the less acceptable it looks. This is exactly Albert Einstein's original criticism of quantum theory, which is not a theory that tells you what happens but instead is a theory that gives you a series of probabilities. This is acceptable as long as you are talking about a temporary, imperfect way of describing nature. At the moment we don't have the perfect theory; what we do have is something that approximates it very well. Quantum mechanics can't be the ultimate theory because it deals with chances. The world does not actually work like that. If a meteorite falls, it's not by chance that it falls somewhere and it's not because the wave function happens to be centred around that point. It is really happening in that particular way. So a theory that only tells us that we can compute probabilities but not certainties is an imperfect theory, at least to me. Einstein didn't believe that God was throwing dice and I happen to agree with that.

What about all the experiments that corroborate quantum theory?

I am not against those experiments, of course not. The experiments confirm that the present status of our understanding is in terms of probabilities. The probabilities we compute using quantum mechanics are correct and I accept that. So, I've nothing against doing such experiments. I have nothing against the present status of our understanding which is called quantum mechanics. It works fine but it's just not the ultimate truth and some people don't understand that. If you have a theory which is very good, it doesn't mean that it has to be the ultimate truth. The truth is that something happens with certainty. The truth is also that today we are unable to compute outcomes with certainty and we probably will never be. This universe is composed of roughly 10^{80} particles, so it is inconceivable that we would be able to compute how all particles are behaving simultaneously. Even if we had a computer as big as the universe, it would work slower than the universe itself. So it's difficult to imagine that we will not have to deal with probabilities but that does not change my belief that the ultimate laws of nature are based on certainties and not on probabilities.

So you think that there is an underlying theory of quantum mechanics that is deterministic instead of probabilistic?

Yes. Quantum mechanics is our next best guess because we are unable to compute exactly how things work. The laws of nature will be too complicated for us to follow in all detail and, therefore, at some stage we will always have to do some kind of statistical analysis. In an ordinary computer in which it is difficult to understand what every bit and byte does, we end up pushing buttons that will probably do this specific thing or that specific thing. But at some point the complexity is beyond what we can follow in detail, so our need for descriptions in terms of probabilities is absolutely inevitable.

So the underlying theory, whatever that may be, is incredibly complicated?

Yes, but I think that if we are smart enough, we can figure out how it works. Even if this won't enable us to predict the future any better than today, we should be able to figure out why the best approximation to this theory requires introducing Hilbert spaces. Once we figure that out, we should be able to move beyond Hilbert spaces and understand why the world that we're living in is the way it is.

Isn't this type of thinking about quantum mechanics called the hidden variable interpretation [5, 6]?

Yes, it is a kind of hidden variable theory. I sympathise with hidden variable approaches, even though they have acquired a negative connotation over the years.

Wasn't this kind of interpretation excluded by Bell's theorem [7]?

Yeah, but that was a very naive approach. Operators, such as those that give you the position of particles or the spins of particles in various directions, that do not commute are very difficult to grasp by many people. These operators, however, do not describe the spin of particles nor the position of particles. Instead, they give you some sort of statistical description of as yet unknown variables. My view is that ultimately even particles and fields don't really exist as they are all statistical properties of something more basic, presumably something that happens at the Planck scale. At the Planck scale we have pieces of information, bits and bytes of information floating around and they evolve according to the laws of nature, and it is our job to figure out what they are. So far we haven't found out how these bits and bytes evolve, only that some of their properties can be treated statistically and by some miracle physicists have discovered how some of these properties propagate in time. So a particle with some position and momentum is a description of the probabilistic properties of those underlying degrees of freedom. It doesn't mean that a particle exists at one point or exists at another point with some momentum, and neither does it mean that the particle has some spin in the z-direction. These are all statistical properties of something more complicated underlying it.

Whenever you have two of such properties described by non-commuting operators, it means that neither of the two can possibly describe the real world. They describe some statistical feature of the real world. The relation between that feature and the actual world is rather complicated. Even in a deterministic theory such as that describing the planetary system there are issues with certainties. When you think of the planets moving around the sun you might think that nothing can be more deterministic than that. Actually, however, there's a problem with this system. The system is described by real numbers and real numbers have an infinite number of decimal places. It turns out that we can only measure the first couple of them, not all of them, and this can lead to what we call chaos.

However, even leaving chaos aside for a moment, you could say that the planetary system looks very deterministic. If you would know all the masses, all the positions and all the velocities of all planets at an instant of time you could follow those planets with infinite precision. I'm assuming that I'm working with a model of the planetary system that reduces the planets to masses with position and velocity. Real planets have atmospheres and surface irregularities and so on, so they would behave in a more complicated way. Idealised planets moving in idealised orbits are completely deterministic, you might think. However, in that idealised system I can consider an operator that interchanges Mars with Earth, that is, it places Mars where Earth is and Earth where Mars is. It works as a switch operator. In quantum mechanics we work with such operators all the time, like the operator that flips the spin of the electron, and you can diagonalise these operators and measure their eigenvalues. In the planetary system we can do exactly the same thing and measure the eigenvalue of the operator that interchanges Mars with Earth. If you use the operator twice, you get the original configuration back. This means that the operator has eigenvalues ± 1. Can you measure these eigenvalues? In the case of the planets, of course not, since we know that Earth and Mars are not so easy to interchange. This means that the switch operator is always in a superposition state. Now in this case, we can clearly see that we are in a state in which the Earth is right here and Mars is there. This is the truth of the theory underlying it; we are not in an eigenstate of the switch operator. In quantum mechanics, we are unable to identify which are the position operators and which are the switch operators. In fact, all the operators we use are very complicated superpositions of all these possibilities. In certain circumstances you can diagonalise one operator or another operator. So the microscopic truth is somehow ontological when you diagonalise this operator and sometimes ontological when you diagonalise the other operator but never in both circumstances simultaneously. So in this respect, the planetary system is not so different from quantum mechanics – the only difference is that we know which operators we should diagonalise and which we should not diagonalise. In the world of the sub-atomic particles, we simply do not know which of them we have, but that doesn't mean that such operators do not exist, and as soon as you identify those operators, you will realise that they obey Bell's inequalities. What happens to be the case is that we do not know which operators we should use to apply those inequalities and we end up using the wrong ones, leading to violations of those inequalities. Therefore, I am not troubled by the fact that Bell's inequalities are violated.

However, I'm mystified by the fact that our world is well described by quantum mechanics and that for some reason humanity has been unable to identify which operators tell the truth and which operators are interchangers. There is a marvellous mathematical fact that tells you that you can treat these operators all in the same way. For example, you treat the operator that switches the spin of a particle in the same way as you treat the operator that measures the spin of the particle. When you rotate a system, these operators are interchanged. This is a very miraculous property of the world that we live in but it's a mathematical curiosity and not something that tells us that there is something mystical about quantum mechanics. It is very interesting and I would like to know more about it but it doesn't mean that the fact that certain inequalities are not satisfied should imply that you can't have a hidden variable theory underneath. In the case of the planetary system that I described, the planets themselves are the hidden variables.

In one of your papers you stated a theorem that claimed that for any quantum system there exists at least one deterministic model that reproduces all its dynamics after pre-quantisation [8, 9]. Is this a proof that quantum mechanics is deterministic?

Well ... yes and no. The thing I had in mind in that paper was a formal procedure one can carry out. I was thinking about finite quantum mechanical systems, that is, systems with a finite-dimensional Hilbert space. If the Hilbert space has a finite number of degrees of freedom, I could formally turn that into a deterministic theory of the kind that I've been talking about. However, when the system is really complicated, like our universe, then this deterministic theory becomes hopelessly clumsy and you don't really believe that such a theory is the true ontological theory behind it all. This was a mental exercise, slightly more delicate than the standard example of a cellular automaton because it was a model which exhibits information loss. This model contained orbits that formed attractors and repellers. The attracting orbits finally ended up in periodic orbits and those are the quantum states that I had in mind.

Since you mentioned *information loss*, I remembered that in one of your papers you conjectured that information loss takes place in any deterministic system [10–12]. Why would this be true?

Yes, I think that is true. I think that there should be a theory where you don't have a Hilbert space that allows you to perform probability analysis. If you just take a time-reversible system there is a small problem, which I find difficult to address, namely: you can't identify which states are more probable than others. Naively you would expect that in such a system, every single state the system can be in will be equally probable. Now, in this universe not every state is equally probable. Some are much more probable than others. The vacuum is a very probable state. The vacuum is everywhere or everywhere is nearly vacuum, as if the vacuum was a big attractor. This led me to suspect that some states are more probable than others. It means that some states somehow decay and this decay implies that the information of that nature gets lost. To be precise, you have deterministic laws that tell you that two systems that start off originally being in a different state might after a while evolve to become identical systems.

So different systems might evolve to become identical and after a while you get some states which have more progenitors than the others and those states with more different possible states in the past are more probable than a state which has only one state in the distant past or none. This latter state is a very improbable state. So, noticing that some states are more probable than others makes me believe that probably, in the ultimate fundamental theory of nature, there are different states that might evolve to the same state and that gives you more freedom and also a more intuitive understanding as to why some states seem to be more probable than others. No fundamental contradiction arises when stating this, except that you make your life more difficult. It's more difficult to have laws which do have information loss in them than laws which don't. Technically, it's much harder to work with but I also think that it has to do with what we call holography, that is, the idea that information can be found on the surface of a system and not in the bulk.

Before we get into holography, I have another question regarding information loss. Maybe this is very far-fetched reasoning but if all deterministic systems have information loss and given that you're claiming that there is a deterministic system that describes our world, which in some regime reduces to quantum mechanics, then is it completely useless to try and resolve the black hole information paradox [13]?

Of course not. It is quite conceivable that we can resolve that paradox and I'm actually trying very hard to solve it. I don't know whether it will be possible to resolve it soon so maybe many generations of physicists will break their heads with this problem but I don't see why it should not be possible.

I asked this question because you stated that there was information loss in a deterministic system in any case . . .

Yes but you have to watch out because I'm using the expression *information loss* in two different ways. What I just described was information loss at the level of the fundamental theory where two different states can evolve to become identical states. Now, within the theory of quantum mechanics that is not possible because you have unitary evolution and you have interference. As soon as you've embraced quantum mechanics as the way to handle statistical features of our world, then within that scheme you're no longer allowed to have information loss. But I can express my suspicion that there is an underlying theory of states for which different states can evolve and become the same state. If this is the case and you insist on using quantum mechanics, these states that evolved into a single state in the distant future will all be described by a single element in the Hilbert space, which makes it technically difficult. Thus, there will be a stage for which quantum mechanics will be unsuitable to deal with statistical properties and that may be one of the reasons why the standard model scale and the atomic scale are so far away from the Planck scale. As you know, the Planck scale is many orders of magnitude away from the physical scale and the reason may be that there's information loss as one moves closer to the Planck scale, which makes it very difficult to use quantum mechanics and the usual Hilbert space to describe probabilities. In other words, the connection between states in the Hilbert space that we use to describe the world and the ontological state of the underlying theory becomes wilder and wilder as we move closer to the Planck scale. This means that it might get more and more difficult for us to understand Planck-scale physics using Hilbert space technology and that could be a good motivation for dropping Hilbert space.

Have you been able to make progress and find out what this underlying theory could be?

No . . . it's still a big mystery how to do this correctly. In particular, I have great difficulties in dealing with the symmetries of nature. We observe that nature has several symmetries such as Lorentz invariance, including rotation and translational invariance, leading to several conservation laws. We are familiar with this symmetry group from ordinary physics but it is very hard to realise such symmetry in the models that I am thinking about. It is even more difficult to realise general covariance. These beautiful symmetries of nature are difficult to implement in deterministic models and I don't understand why.

I see. Do people claim that maybe it's because nature is not deterministic?

That may well be and of course people often claim that but then I respond with my Einsteinian objection. There is something Einstein didn't yet know at his time, which is that we have a big bang theory of the universe. The universe started at some point; that is, it started with some initial state. If we believe that quantum mechanics describes that initial state then that initial state is fuzzy. But how can the beginning of the universe be fuzzy? People like Stephen Hawking have their answers [14] but I think their answers are a bit mystical and I have difficulties with those answers.

I think that by far the most natural answer is that quantum mechanics is a technical trick to do statistics using probability considerations, as long as we are forced to deal with approximations, while actually there's nothing fuzzy at all about the fundamental laws of nature. The universe just started with a bang! Something switched on at $t = 0$. I think that there was a rigorously defined $t = 0$ where a very simple universe was being switched on. Whether the universe is compact or non-compact nobody knows today but it would be more beautiful if it would turn out to be a compact universe. Once you switch it on, it takes off just like that. Its evolution is very simple in the beginning but later chaos settles in. This is somewhat analogous to what happens in many number systems such as in prime numbers. Prime numbers start with a very simple and clear sequence but then for larger prime numbers the sequence becomes more and more complicated. This feature of many number systems could also be a feature of our universe.

What kind of arguments have people given that you don't like and consider to be fuzzy?

Well, for instance, Stephen Hawking had this idea that the beginning of the universe was basically described by quantum mechanics and that the universe is characterised by a wave function [14]. I don't think that this can be true because we can't measure that wave function and whenever we do a measurement that wave function collapses and that's really something else than what it was before.

Right. So you don't like this particular idea of a multiverse . . .

Our world doesn't seem to be a multiverse in this sense. The only sense that it could be a multiverse is the sense that many string theorists are advocating today, namely, that there are many different solutions of string theory with the same degree of reality [15]. All these solutions are therefore true universes and we happen to just live in one of them. This idea is met with a lot of opposition and even though I and many others don't like it, it is actually very difficult to refute it.

But do you think that anthropic arguments that fix the universe we live in are reasonable?

I'm not happy with them but it is becoming very hard to say why they can't be true. I'm still trying to find out alternative ideas but it may well be that it's difficult. In any case, it's still far too early to make these claims since we don't understand the underlying physics of string theory well enough. In my opinion, both this question and its answer are premature.

Okay. Going back to the holographic principle that you mentioned earlier, what was the reasoning that led you to formulate the holographic principle [16]?

Well, I like to approach a problem from as many different angles as I can and from an early stage, I realised that black holes were going to be a very important ingredient in any quantum theory of gravity. If you haven't been able to understand what a black hole is then the theory is not complete. Black holes are valid solutions of the equations and must be an integral part of any theory of gravity so they must fit beautifully with everything else we know. However, black holes seem to disobey the laws of quantum mechanics and that can't be right. Here, given what I said earlier, I am thinking of quantum mechanics as an effective description as there may be an underlying deterministic theory. Before trying to find the theory of everything, I thought it was important to see if there is a way of naturally incorporating black holes in a quantum theory. So, we know that we have particles, bound states of particles, a vacuum, excited states of the vacuum and we have black holes, but there could be a domain of physics where distinctions between particles and black holes become fuzzy. This is very often the situation in nature, as sometimes there is a continuous transition between one structure, or object, and another object. So perhaps at the Planck scale there is not much of a distinction between some objects that you call elementary particles and others that you call black holes.

I read in your book *In Search of the Ultimate Building Blocks* [17] that you thought that indeed there was no distinction between particles and black holes at the Planck scale. What is the reason for this?

The situation that I think realises at that scale is that the heavier states will naturally behave like black holes and the lighter ones will naturally behave as elementary particles. In between there will be objects which you may or may not call particles or black holes but it won't matter because there will be a gradual transition between them. Of course, calling an electron a black hole is a bit far-fetched and not very useful because it is too light to be a natural solution to the relevant equations. If you want to have a Hilbert space formulation of the theory you begin with the vacuum state which is nothing other than the flat metric without anything in it. Then you can populate it first with the lightest particles, then heavier particles, bound states, particles that are interacting with each other and finally black holes. All this forms your Hilbert space. What I want to understand is how nature does its bookkeeping; that is, I want a good bookkeeping system that can simultaneously describe elementary particles, black holes and everything else.

Now the nice thing about the standard model is that it tells you exactly how to do your bookkeeping. We say that the various fields in the standard model have spin one, spin one half or spin zero. By Fourier-transforming the fields and defining creation and annihilation operators, you can enumerate all the states of the Hilbert space and it becomes a Fock space of all particles. We understand exactly how to do this for the standard model and a precise bookkeeping prescription is one of its virtues. This precise bookkeeping is missing in quantum gravity and I also think that it is missing in string theory. At a given stage, a theory must be able to state what are the possible states of its Hilbert space and what the Hamiltonian is. For instance, in a non-renormalisable model, there is a problem as to how far Lorentz transformations go, since, at energies that are too great, things run out of control. As I said, my own belief is that the language of Hilbert spaces is not the most appropriate one. What is the most appropriate language? We need a sort of universal language that describes what happens at all scales and, in principle, with infinite precision. Now, the hope to reach infinite precision seems to be a crazy one because it looks as if the total amount of information that you can squeeze into a given volume should then be unlimited. Thus, wondering about bookkeeping prescriptions led me to wonder about how much information one really needs in a certain region of spacetime to have infinite precision.

So how much information can I squeeze into a certain volume? The vacuum is simple, there is nothing there, so I can just characterise the vacuum state with one or zero bits of information. In this case, I don't need to waste many bits and bytes to describe it. Then you can start with Fock space and characterise one electron, two electrons and so on and so forth. If you start characterising states in the Hilbert space in that way, how far can you go? What if you squeeze too many bytes in there? The state gets more and more energy and after a while the gravitational force sets in and the situation becomes unstable, leading to the creation of a black hole. So at that point I thought, "Well ... then I have to stop counting particles. I have to start counting how many states a black hole can be in." Due to Hawking's beautiful result [13], we know that there are particles at the black hole's horizon and those particles have a thermal spectrum that allows you to compute the entropy of a black hole, hence the number of states that a black hole can be in. So if there is too much energy, a black hole is created and you know how many states that can be in and so you can continue with your bookkeeping. So, if I look at a finite box of the universe, how many states can the box be in? Well, you end up creating a black hole and then the black hole is going to get bigger and bigger until it doesn't fit in the box anymore. Then you have to stop and that means that there is a finite number of states that the Hilbert space can have. How big is that number? The counting stops when I put the largest black hole that fits in the box. That black hole has its horizon roughly at the boundary so I just have to count how many states I can put on that horizon. This is the holographic principle and it tells you that in this box, no matter what I put in, I can't put more information in it than this number. If you look at what that number is exactly, it is just proportional to the surface area of the box.

This result is somewhat counterintuitive because you might have expected that it would be proportional to the total volume instead of the surface area. However, when you think more about it you notice that according to general relativity, the inside of the box can be a curved spacetime. So the volume of the box can be much bigger or much smaller than that suggested by its boundary and so the interior of the box shouldn't enter as part of the argument for how much information can be put inside the box. It should only depend on the surface, and actually this makes a lot of sense from a general relativity point of view. There's another way to see why the surface is so important. Imagine that we are sitting inside the box. Whatever it is, it is controlled by a Hamiltonian that gives you the evolution of the system inside the box. The Hamiltonian is also the total energy of the system inside. What does this energy give you? It gives you the total gravitational flux that leaves from the box. This is of course the gravitational field that goes through the surface area of the black hole so if I make any change in the interior of the box I change the Hamiltonian and therefore I change the gravitational field that leaves the box, which means that I'm doing something to the surface. Therefore, in principle the surface does contain all the information of what sits inside the box. From this point of view it is a natural thing to believe in but it is nevertheless very counterintuitive. This leads you to the picture that the surface looks like a hologram of what is inside the box.

Is the universe a big hologram?

In a sense yes but then my next remark is, "Isn't this ridiculous? This cannot be true. This is a contradiction." And I believe that the reason for this is because I insisted on using the language of Hilbert spaces to describe what was going on. This is the weakest part of the argument. I think that you shouldn't use this language but instead use classical deterministic systems. You might say that even if you use such deterministic systems it will have to fit inside the box but I don't think that is the case. The rule is that you have to assign the same element in Hilbert space to all states which can evolve into the same state after a long enough time. So then I say, "Okay, if I have a box with a surface and a bulk, in the bulk there's so much information loss that all that you can keep as information is what you have at the surface because that communicates with the outside world." What sits inside the bulk just disappears.

So the classical underlying theories that I am thinking of don't have a little bit of information loss – they have a lot of information loss. Everything that happens in the bulk will disappear after a while and the only information left will be stored on the surface. Overall, the physical dynamics may still be 3+1-dimensional, not 2+1-dimensional. So my theory is a 3+1-dimensional ontological theory but there's so much information loss that in a formal sense you can put all the *lasting* information on a surface. In addition, if you want to use a Hilbert space description, the number of elements of Hilbert space is dictated by the surface and not by the volume. So that is the general rough picture I arrive at, and of course the next question is: "Now fill in the details: what are the actual laws?" and this is extremely difficult.

Do you think that the AdS/CFT conjecture [18] is a precise formulation of this holographic principle?

Well . . . it really brings it to a very different kind of discussion. The CFT part of the story, to me, looks like a mathematical feature of some of these models, but I have to be a bit careful here because to some extent AdS/CFT does reflect the idea of all the information being stored on the surface of a black hole as opposed to the volume. So in some sense this is right but from some other point of view I don't think that these mathematical models are capturing the kind of physics that I'm thinking of. I must admit that I don't quite know what to say about it also. CFT means "conformal field theory" and that is quite a bit different from what I had been thinking about. Although, more recently, as you may have seen, I've been looking with different eyes at CFTs compared to how I have looked at it earlier [19–22]. So there may be some links that I don't quite yet understand.

Some time ago Erik Verlinde published a paper on entropic gravity [23]. Do you think gravity is holographic?

From our discussion up to now it may be seen that I do attach enormous importance to the whole idea of holography. I think it is a key ingredient in our understanding of physics. However, Erik is now extending the argument into another direction by saying that the gravitational force itself could basically be a way to express the chaos that happens in the bulk and that the origin of gravity is entropic. I think that there might be some truth in these suspicions but the big question is, "Okay, you may say these things; now what?" How do we go from here to a good theory? Erik has the same problem as I have when I investigate things. He's entitled to have this hunch and to follow it as best as he can.

I specifically asked this question because the idea I got from reading your papers was that you don't really see spacetime as an emergent concept [19].

No, I don't.

So I was wondering about what you think of this since Erik's paper is claiming that spacetime is an emergent concept.

All right. I think that he is just taking it one step too far. Quite generally speaking, to arrive at a good theory you have to make one step at a time and the smaller your steps are, the better. Emergent spacetime to me seems to be a very big step and very big steps are usually wrong unless you're very lucky. So I think that if you talk of emergent spacetime you are basically denying that the world is essentially 3+1-dimensional. I think that there are some very important ingredients in a good theory, one of which is causality. Causality is very high on my list of demands for theories.

Yes, but is causality broken if spacetime is emergent?

Well ... causality is a concept of space and time, certainly of time. My preferred definition of "time" is whatever coordinate one needs to define the causal order of events correctly. The reason why I say this is that I believe that there has to be an ultimate theory that not only describes what can happen here and there, but also contains a clause stating in which order you have to perform your calculations. The causal order of events is the order in which one must apply the laws of physics to describe events (thus, causality only means something if you are using a precise *model* that completely explains the observed phenomena).

My definition of time, as the causal order of events, excludes closed, timelike loops, for instance. Therefore, it will remove circular arguments in your book of prescriptions; otherwise, you would run into contradictions. Now it so happens that, in today's models, this definition of time coincides with the fourth coordinate in Minkowski space. Thus, I cannot do without that fourth coordinate. Next, the causal ordering experienced in the world we know is only a partial ordering; when two events are spacelike separated, they don't need to be ordered at all. If you want to keep that, you need to define spacelike distances; that is, the other coordinates of Minkowski space also play an important role. At our present state of understanding, I find that throwing the space coordinates away is an enormous step. The odds against that being right also seem to be enormous, in my view. I would operate more carefully: there might be instances, such as the horizon of a black hole, where causal time and Minkowski time might cease to coincide.

Time is one-dimensional and it's ordered. In any universe, no matter how different it is from ours, there must be such a notion of time that gives you the causal order of events. So time cannot just be "emergent" for that reason; it is a basic ingredient of any book that tells you what the laws of nature are.

Space just localises things in such a way that what happens here is independent from what happens there. That always has been understood as a basic feature of our world. In addition, we have light rays that can emanate from here as well as from there and interact. These define the fastest way information can travel, and they determine which regions can be affected by others, and this means that there are spacelike and timelike separations. Thus, if you drop the notion of space, you drop the notion of locality and a nonlocal set of laws of nature is even more complicated than a local one, which gives you little room for analysing things. I am not prepared to abandon either locality or causality, and this makes it difficult for me to regard space and time as possibly being emergent. This would open a Pandora's box of universes where you can't make much sense about what is happening. So I'm very reluctant to drop the elementary notions of space and time, but everyone is entitled to follow their hunches and the people who believe that these notions are emergent can try to go that way.

I also understood from your papers that you think that the theory that describes physics at the Planck scale should be topological [19]. Why should one expect this?

Well that is the idea that the Planck scale is the ultimate scale of nature and there is nothing at distances smaller than the Planck size. Stating that there's nothing means that there's not even spacetime and there are no particles moving there. So the only spacetime we have at that scale is some mathematical, topological extension of space. What happens in between points at the Planck scale is immaterial because there are no degrees of freedom left, so the only thing left is topological and you might or might not decide to introduce a metric in this topological space. Since there is no further structures or features at this scale, it will be unwise to choose anything else but flat spacetime. So then I reached a theory where space and time is flat between points at the Planck scale [19].

Is this the theory in which you impose local conformal invariance [19–21]?

That's another theory. I haven't been able to understand all the connections between the two. As I said, I like to approach problems from different angles and sometimes these angles do not all end at the same point. This means that there may be contradictions in the theories that I'm trying to pursue.

But why would one think of imposing local conformal invariance in the first place?

The reason is related to trying to better understand black holes. I want to have a theory in which black holes obey some sort of fundamental laws of nature. However, there seems to be a contradicting situation. On the one hand, black holes are locally smooth solutions of general relativity so we should be able to use the laws of nature to describe black holes up to the Planck scale. On the other hand, black holes are forms of matter, which seems to contradict the first observation I made about them. Thus, it seems as if there are different ways to look at a black hole. One way to look at it is to have a coordinate frame that describes an infalling observer. It gives you a certain description of the spacetime near the horizon. Another way to look at it is to have a coordinate frame describing a distant observer who sees only Hawking particles coming out of the black hole and other things going in. It is imperative that the theory that I'm searching for allows different descriptions of what's going on without contradictions.

Is this what people refer as the black hole complementarity principle [24]?

Yes. So I end up with this complementary issue, namely, that the description according to an infalling observer should not disagree with whatever is the description according to a distant observer. So there should be a mapping from one to the other. One year ago I made a rather big step [19] and, as I said, big steps might be wrong but I tried anyway. I took this complementarity principle and required that both observers agree about causality; that is, they should agree whether one event is the cause of the other or if two events do not affect each other. This implies that the structure of the light-cones should agree for both observers. If you have a spacetime metric for any two observers who agree on their light-cone structures then each of their metrics is equivalent, modulo a conformal factor. Since the conformal factor does not affect the light-cone structure, I peel it off from the metric and require that all observers agree on the remaining part of the metric. With this in hand, I am left with a local conformal theory and I can find a map from one observer to another, which is nothing else but a conformal mapping. This means that conformal symmetry is now on par with particle physics and gravity. Because of this, I now certainly see a big and important motivation to have some form of conformal invariance in a physical theory.

Okay. I did notice that you tried to use this reasoning to fix all the constants in the standard model [25]. Is that right?

Well ... that came up as a surprise. I took the standard model of elementary particles together with the Einstein-Hilbert action for gravity and required conformal symmetry in the way that I explained earlier. So I took the conformal factor from the metric, which transforms under conformal transformations, and promoted it to an extra real scalar field. It is a rather minimal change because I just proceeded to construct a theory of quantum gravity as one would naively do, that is, by starting with just the Einstein-Hilbert action, the matter action and to state that they interact.

This scalar field looks exactly like the usual scalar field except for one important point: namely, it comes with the opposite sign in the Lagrangian, which is very important and tells me that the energy of the field is negative. Using gauge fixing for the conformal symmetry, this field becomes a dynamical field while all other fields have positive energy. So it just looks like ordinary field theory except that there are anomalies. I'm very keen on anomalies because they express the fact that there are some infinities in the theory and we have to renormalise the theory. This usually leads to new features that one did not expect a priori.

As you might see in my papers, I actually studied the wrong gravitational anomalies, which I couldn't cure [25]. Maybe it is possible to cure them at very high energies or by adding extra fields such as a spin $3/2$ field but this made me realise that I perhaps took a very big step. Instead, I took a similar approach in spirit as the one taken by Weinberg when he wrote down his model for elementary particles, which he called a *model for leptons* [26]. He knew of the existence of hadrons but he didn't understand them so he left them out of the model. This was a very smart thing to do because, indeed, he didn't understand hadrons. Today we know of the existence of charm quarks and at that time he didn't know anything about them since they had not been invented yet. In the context of my theory, I didn't understand the metric modulo its conformal factor but I did understand the conformal factor. So I've concentrated on this conformal factor which also leads to anomalies but which I can cure by requiring the beta function of the theory to vanish.

This procedure leaves you with a discrete subspace of all theories because demanding all beta functions to vanish forces you to be at a fixed point of the theory. If the theory has 20 constants of nature there may be a bunch of fixed points and you have to be at one of those fixed points. I was very excited about this because this gives me 20 equations for 20 unknowns, and this could imply that everything about the particles will be calculable; that is, the parameters of the theory are no longer freely adjustable. Now I've been trying, together with a student, to actually calculate those fixed points. Unfortunately, it seems that all those fixed points are in the complex plane, which means that these theories are not useful. We are working on it but finding these fixed points seems quite more tough than what I originally thought. So we still don't know whether there are any fixed points but if there are, it means that we have a model where the elementary particles interact in a prescribed way. All the interactions are fixed, the cosmological constant is fixed, all the masses are fixed in terms of the Planck mass and the model is completely calculable.

Are the values for the parameters you get those experimentally measured for the standard model?

No. They don't take anything approximately like the values in the standard model. We are still very far away from that but the whole idea that things become calculable I found very exciting. Even if we don't succeed in getting the right model, we do have a theory that tells us that everything is calculable and that is somewhat opposed to the idea of a multiverse where you are forced to use the anthropic principle.

However, the thing that is not calculable in the model that I am describing is that I can choose the algebra in an innumerable number of ways. Eventually, that might lead me again to a multiverse theory so maybe I end up at the same point where string theory has ended. This would just mean that I, for once, would agree with string theory.

Do you believe that you will be able to find a model that would exactly give all the right values for all parameters?

That would be great, except that I should remember that there was a sector of the theory which I haven't yet solved, which was the sector determining the part of the metric that I separated from the conformal factor. Without being able to solve it, I should not expect too much yet from this theory. An attempt to address this problem leads to a deeper problem with a negative metric spin-two particle, an essential consequence of the fact that I started with a non-renormalisable theory of gravity; so no, this solution to all our problems would be too cheap to be credible.

The fact that this theory predicts the existence of magnetic monopoles doesn't disturb you?

No, because those monopoles are probably of the order of the Planck mass. I cannot allow any Abelian group in the gauge group because that would not be asymptotically free. In fact, that would be the exact opposite, namely, infrared free. This means that I cannot allow for electromagnetism to have a $U(1)$ gauge group. Requiring this condition leads to the presence of magnetic monopoles. So, yes, magnetic monopoles will be there but they will probably be too heavy to be produced in any ordinary experiment.

Okay. You have another approach to quantum gravity which you called *crystalline gravity* [27, 28].

That idea realises something that we have discussed earlier, namely, that at small scales compared to the Planck scale there shouldn't be any gravitational field, no structure left. So the only reasonable structure to consider at those scales is flat spacetime. I have had another graduate student who took up that idea [29] and he had the wise idea to think of this situation as locally flat spacetime or piece-wise flat spacetime. These are flat regions of spacetime which are glued together by possible defects. Defects are unavoidable in this procedure and they are all the structure that I need. In addition, they have one nice property: they are only characterised by real numbers. I could even enforce the theory to be invariant under a discrete subgroup of the Poincare group so that the defects would only be characterised by integers. This is the kind of theory that I want: a theory which has integers as its fundamental physical degrees of freedom and for which these integers can somehow evolve.

These defects are strings, right?

They are strings but they are straight strings. Strings without any structure. Only the end points of the string have physical structure. The straight lines themselves have no excitations. In fact, this is just a 3+1-dimensional extrapolation of what happens to be the case in 2+1 dimensions. In 2+1 dimensions, classical gravity can be described using point particles that are gravitating. Those particles are just point defects. In 2+1 dimensions there are no tidal forces and particles just act as defects [27]. The nice thing about this is that it is completely solvable classically. Classically, these defects just follow straight lines and the only thing to watch out for is how they cross each other, either one way or the other. This is all that matters, and it turned out to be a beautiful model that can be put on a computer. In fact, it is a deterministic theory of the universe where you can place particles and follow how they behave. The behaviour turns out to be rather complicated and the universe may expand or sometimes roll back to a big crunch. The model looks completely solvable but the question is whether being solvable classically means that it is also solvable quantum mechanically. Many people say yes but I say no. It's not solvable quantum mechanically because it's not quantisable.

Why not?

At first it looks perfectly quantisable specially because we know many other theories and models that are solvable classically and also solvable once you quantise them. Examples include the harmonic oscillator, the hydrogen atom with its Keplerian orbits, or sequences of atoms which you can solve classically as well as quantum mechanically. We also know of examples in quantum field theory, for example, a free particle theory can be solved in both cases. This intuition leads us to think that in 2+1 dimensions the model can be solved quantum mechanically. However, the usual prescription of replacing Poisson brackets with commutators is quite difficult to implement as the Poisson brackets are in a funny space and have a rather complicated structure. Moreover, in this model, you are using quantum mechanics in a finite compact universe, which means that you can't really do statistics. This is a fundamental difficulty that tells you that you can't really have wave functions in this universe. In fact, my bad experience in 2+1 gravity, for which a quantum theory doesn't exist, has motivated me to consider deterministic systems that do make sense in 2+1 dimensions.

People like Edward Witten [30] and Steven Carlip [31] claim that such quantum mechanical theories should exist but once I ask them about the Hilbert space and what the wave functions are, I understand that they are not speaking about the kind of common quantum mechanical theory that I can make sense of. I would like to know what the spectrum of states is and what the eigenvalues of the Hamiltonian are and they can't tell me. In my opinion, the reason for this is that there is no quantum theory for gravity in 2+1 dimensions. There is only the classical system. This is like the planetary system that I described earlier; that is, you can have interchange operators but you can't quantise the theory. My belief is that the ultimate theory of the universe allows you to play God: you have all the equations exactly, you switch them on and you watch what happens. So in this case, I can play a little game with classical universes but I can't quantise it, so I can't be a quantum God.

The fact that 2+1 dimensions is full of problems led me to crystalline gravity which is formulated in 3+1 dimensions. Can you play God in 3+1 dimensions? The naive answer is no because now you have to deal with defects which are line-like, which is also what string theory has stated.

Was crystalline gravity inspired by string theory?

Well, maybe in some way. There might be a reason for having to consider lines as the starting point in a theory of gravity. There is an important difference between a theory of points and a theory of lines. When two lines meet they always intersect at some point and something happens. When two points meet in 2 + 1 dimensions nothing has to be added to the physics. In the case of lines something has to be added because when two lines cross each other you have to give a prescription of what to do. This was the subject of that paper.

My PhD student [29] worked on prescriptions for what happens when two lines cross. The most immediate guess is that after they cross there is a third line connecting the two but it's not that simple unfortunately. It doesn't work that way. There must be at least two lines connecting the two but even that is often not enough and you enter a rather messy situation. Apart from that, it's very interesting because the starting point seems to be a fairly solid one. So we investigated it further anyway to see how far we could get.

Why did you call this theory *crystalline gravity*?

Originally, the defects in the theory were characterised by real numbers but, in my opinion, a theory that is described by real numbers isn't a finite theory. So I formulated the theory in terms of line defects that are characterised by integer numbers. In this way, I've made the step from continuous spacetime where defects connect continuous pieces of spacetime to discrete spacetime where defects join discrete regions of spacetime. Discrete spacetime is just like a lattice but this particular lattice is not perfect as it has defects in it and it evolves because the defects evolve according to some dynamics. This reminded me of a crystal; within this theory, one can think of space and time as being described by a crystal structure. The idea of describing space and time as a crystal is not new. My friend Hagen KleinertKleinert, Hagen has been studying crystals from a different perspective [32, 33] but also with gravity in mind. His approach is different than mine and we have disagreements about the way he addresses this question. Anyway, this is the picture that we have in this theory.

Can you quantise the theory?

I haven't even approached the problem of quantisation yet. So far, this is a totally classical picture. Once we begin treating these defects statistically, we will start seeing features of quantum mechanics.

Regarding these two different theories that you have on quantum gravity ... they are both background dependent, right?

Yes, this is the same point I made earlier about space and time being fundamental and not emergent so of course that means that I need a spacetime background as a starting point. But the idea underlying one of the theories is that you should only have topology as the background. Once you fill that topology with a given structure, that structure will evolve according to certain laws [19–21].

Yes, but do you have any idea of how to implement this?

I wish I did; that's the big question.

What do you think has been the biggest breakthrough in theoretical physics in the past 30 years?

I think that there really was a breakthrough in the way we began thinking about particles and it was very fruitful. As you very well know, in the last 30 years string theory made big progress, specially to what concerns supersymmetry and supergravity. Cosmology also made a big move forward and I'm very impressed with that because 30 years ago I thought cosmology was just science fiction and I didn't want to take it very seriously. Astronomy also made enormous advances with large telescope arrays as well as space science and their guided probes that can explore other planets.

In my own field in the last 30 years, I feel that the number of new ideas is decreasing. In any case, and despite the community sometimes overselling their results, I think that string theory has made the biggest contributions. It is a good theoretical achievement that string theorists can be very proud of. However, I don't think that they have the ultimate answer, which would in any case be too much to ask for.

What about other theories like loop quantum gravity or causal dynamical triangulations?

I think that these approaches are lagging behind compared to string theory. String theory is more structured and more informative. If I were to bet, I would bet in in favour of string theory, though, as I have mentioned, I don't believe that it is the complete solution. In fact, I think it is far from being the complete solution though it may well be a part of it, which is already a great achievement.

Do you think that there is very little room within the scientific community to come up with new ideas?

I don't think that is true and I don't understand why people say that. There are lots of people who are running around with different ideas and many of those ideas are wrong. People often don't care too much about that. If you have a wrong idea you can still make a living out of it. There are many people who have new ideas, even crazy ones, and I think that this is the way it should be. We shouldn't be too strict about crazy ideas although sometimes you can go way across the edge. Every now and then I get emails from people who really get crazy ideas. However, very often ideas are not obviously crazy. There are people who have thought a lot about their own ideas and found valid reasons to believe in them. I also have ideas that I'm certain not everybody agrees with and that is fine. We should try to think carefully and precisely about our ideas but the problem is that many people run around with ideas that haven't been thought through sufficiently precisely. Anyway, I think society is pretty tolerant towards people with really crazy ideas. We have to admit that science hasn't solved everything yet and as long as that is the case, we can't be too tough with people who think about things differently. Nevertheless, that does not mean that we shouldn't have certain opinions about which ideas are promising, fruitful or nearly possible and which ideas are just outright rubbish. It turns out that the majority of ideas are just rubbish.

Why have you chosen to do physics?

As a kid I thought that physics was the most exciting thing. I remember that I really loved science in general and I wanted to figure out how the world works. When I was a kid I was also interested in tiny animals like bacteria and so on, and everything I could put my hands on. I was also interested in mathematics. My uncle was a theoretical physicist and he had a role to play in increasing my interest in theoretical physics. I grew up in a time when the atomic bomb had just been detonated and the television, computers, etc. had just been invented. These were all exciting applications of physics and were going to change the world.

There were still many unsolved mysteries like the structure of the atomic nucleus and the nature of the gravitational force. Simultaneously, we wanted to go out into space and conquer planets, moons and asteroids. I was dreaming about all these things. Einstein just imagined things, worked out equations and discovered new laws about space and time. I couldn't imagine something more fantastic than discovering new things in this manner, and this was to me a huge motivation to pursue physics.

I think that everyone is glad that you did. What do you think is the role of the theoretical physicist in modern society?

In general, I think that theoretical physics is a doctrine that teaches us how to solve problems. So, we are problem solvers. We understand how to write equations suitable to a given problem and we have a set of tools to find answers to our problems. We also know how to distinguish between problems that can be solved and those that cannot be solved. In order to do so we use physical laws, mathematical rules and so on. The methodology we employ on a daily basis has a much wider range of applicability than physics and it can be useful in any other context where you have a problem, such as in society.

References

[1] D. J. Gross and F. Wilczek, "Ultraviolet behavior of nonabelian gauge theories," *Phys. Rev. Lett.* **30** (1973) 1343–1346.

[2] G. B. Cleaver, A. E. Faraggi and D. V. Nanopoulos, "String derived MSSM and M theory unification," *Phys. Lett. B* **455** (1999) 135–146, arXiv:hep-ph/9811427 [hep-ph].

[3] G. 't Hooft, "On the foundations of superstring theory," *Found. Phys.* **43** no. 1 (2013) 46–53.

[4] G. 't Hooft, "Superstrings and the foundations of quantum mechanics," *Found. Phys.* **44** (2014) 463–471.

[5] A. Einstein, B. Podolsky and N. Rosen, "Can quantum-mechanical description of physical reality be considered complete?," *Phys. Rev.* **47** (May 1935) 777–780. https://link.aps.org/doi/10.1103/PhysRev.47.777.

[6] D. Bohm, "A suggested interpretation of the quantum theory in terms of 'hidden' variables. I," *Phys. Rev.* **85** (Jan. 1952) 166–179. https://link.aps.org/doi/10.1103/PhysRev.85.166.

[7] J. S. Bell, "On the Einstein Podolsky Rosen paradox," *Physics Physique Fizika* **1** (Nov. 1964) 195–200. https://link.aps.org/doi/10.1103/PhysicsPhysiqueFizika.1.195.

[8] G. 't Hooft, "Emergent quantum mechanics and emergent symmetries," *AIP Conf. Proc.* **957** (2007) 154–163, arXiv:0707.4568 [hep-th].

[9] G. 't Hooft, "The cellular automaton interpretation of quantum mechanics. A view on the quantum nature of our universe, compulsory or impossible?," arXiv:1405.1548 [quant-ph].
[10] G. 't Hooft, "Determinism and dissipation in quantum gravity," *Subnucl. Ser.* **37** (2001) 397–430, arXiv:hep-th/0003005 [hep-th].
[11] G. 't Hooft, "Quantum gravity as a dissipative deterministic system," *Class. Quant. Grav.* **16** (1999) 3263–3279, arXiv:gr-qc/9903084 [gr-qc].
[12] G. 't Hooft, "The mathematical basis for deterministic quantum mechanics," *J. Phys. Conf. Ser.* **67** (2007) 012015, arXiv:quant-ph/0604008 [quant-ph].
[13] S. W. Hawking, "Particle creation by black holes," *Commun. Math. Phys.* **43** no. 3 (Aug. 1975) 199–220. https://doi.org/10.1007/BF02345020.
[14] J. B. Hartle and S. W. Hawking, "Wave function of the universe," *Phys. Rev. D* **28** (Dec. 1983) 2960–2975. https://link.aps.org/doi/10.1103/PhysRevD.28.2960.
[15] L. Susskind, "The anthropic landscape of string theory," arXiv:hep-th/0302219.
[16] G. 't Hooft, "On the quantum structure of a black hole," *Nucl. Phys. B* **256** (1985) 727–745.
[17] G. 't Hooft, *In Search of the Ultimate building Blocks*. Cambridge University Press, 1997.
[18] J. M. Maldacena, "The large N limit of superconformal field theories and supergravity," *Int. J. Theor. Phys.* **38** (1999) 1113–1133, arXiv:hep-th/9711200.
[19] G. 't Hooft, "Quantum gravity without space-time singularities or horizons," *Subnucl. Ser.* **47** (2011) 251–265, arXiv:0909.3426 [gr-qc].
[20] G. 't Hooft, "The conformal constraint in canonical quantum gravity," arXiv:1011.0061 [gr-qc].
[21] G. 't Hooft, "Probing the small distance structure of canonical quantum gravity using the conformal group," arXiv:1009.0669 [gr-qc].
[22] G. 't Hooft, "Local conformal symmetry: the missing symmetry component for space and time," arXiv:1410.6675 [gr-qc].
[23] E. P. Verlinde, "On the origin of gravity and the laws of Newton," *JHEP* **04** (2011) 029, arXiv:1001.0785 [hep-th].
[24] L. Susskind, L. Thorlacius and J. Uglum, "The stretched horizon and black hole complementarity," *Phys. Rev. D* **48** (1993) 3743–3761, arXiv:hep-th/9306069 [hep-th].
[25] G. 't Hooft, "A class of elementary particle models without any adjustable real parameters," *Found. Phys.* **41** (2011) 1829–1856, arXiv:1104.4543 [gr-qc].
[26] S. Weinberg, "A model of leptons," *Phys. Rev. Lett.* **19** (1967) 1264–1266.
[27] G. 't Hooft, "A locally finite model for gravity," *Found. Phys.* **38** (2008) 733–757, arXiv:0804.0328 [gr-qc].
[28] G. 't Hooft, "Crystalline gravity," *Int. J. Mod. Phys. A* **24** (2009) 3243–3255.
[29] M. van de Meent, "Piecewise flat gravity in 3+1 dimensions," PhD thesis, Utrecht University, 2011. arXiv:1111.6468 [gr-qc].
[30] E. Witten, "Three-dimensional gravity revisited," arXiv:0706.3359 [hep-th].
[31] S. Carlip, "Lectures on (2+1) dimensional gravity," *J. Korean Phys. Soc.* **28** (1995) S447–S467, arXiv:gr-qc/9503024 [gr-qc].
[32] H. Kleinert, "Emerging gravity from defects in world crystal," *Braz. J. Phys.* **35** (2005) 359–361.
[33] H. Kleinert and J. Zaanen, "World nematic crystal model of gravity explaining the absence of torsion," *Phys. Lett. A* **324** (2004) 361–365, arXiv:gr-qc/0307033 [gr-qc].

15

Petr Hořava

Professor at the Berkeley Center for Theoretical Physics, Berkeley, CA

Date: 30 June 2011. Location: Uppsala. Last edit: 5 February 2021

What do you think are the main problems in theoretical physics at the moment?

There are many interesting questions but I don't know what I would find most attractive if I were entering the field of physics or the field of science at this moment. Since my undergraduate studies I have been fascinated by the problem of quantum gravity so I don't really have any doubts that it's an interesting direction to explore. But there are other interesting questions as well. Over the past 50 years we have learnt that the community of particle physics and gravity (or the so-called fundamental physics community) tends to look down on more applied fields like condensed matter physics. In my opinion, it's not the correct way of looking at the world. It is more sensible to acknowledge that the community of particle physics and gravity doesn't really have the ambition to understand all the complexities of fundamental physics by devising an ultimate theory of everything. Instead, we have to accept the fact that the world is really structured in several layers. It's fascinating to explore any given layer, understand how it connects to neighbouring layers and uncover how the complexity of that layered structure evolves. These kinds of questions are as fundamental, from the point of view of particle physics and gravity, as from the point of view of condensed matter physics where emergent phenomena and complexity play a key role. The same can be said about the emergence of life and biology viewed from the point of view of complex systems.

If I were now entering the field of physics or similar sciences as a freshman in college, I would probably be very tempted to go into a biologically oriented area. In particular, it would be interesting to apply the knowledge about how the world is organised in physical systems to biological systems. I think this is a really fascinating area of research which hasn't even started developing properly yet. Thus, that's a huge opportunity for the next generations who, presumably, will revolutionise biology in the same way that physics was revolutionised in the twentieth century.

Do you see yourself as a string theorist?

Yes, I do. String theory is not really a theory, in the same way that quantum filed theory is not a theory. String theory is a framework, a mathematical language for formulating the behaviour of many-body systems or systems of many particles where collective effects can emerge. Historically, both in high-energy physics and in condensed matter physics, the mathematical language for studying these phenomena is quantum field theory. However, what we've learnt over the past 15 or 20 years is that string theory is not complementary to quantum field theory; instead, it's really a logical completion or extension of quantum field theory. If you take quantum field theory seriously you can for a certain amount of time ignore string theory but then through features like the large-N expansion [1] or the AdS/CFT correspondence [2] you find that string theory is a necessity. Gravity is present in dualities, so, even if you only care about conventional quantum field theories, sooner or later you will find that it's very useful to use string theory for a given dual description that involves gravity. In summary, string theory cannot be eliminated from standard frameworks like quantum field theory and in that sense it's a mathematical framework which we can use to express physical thoughts.

Do you see string theory as the theory of everything that fixes all constants of nature?

Not at this point. For a few decades I have viewed string theory as a formalism/technology rather than a theory. In fact, I think *technology* is the more appropriate word because it really is a powerful machinery that has a self-consistent internal structure which we are slowly uncovering. Whenever you uncover a new corner of this self-consistent mathematical structure you learn something new about physics and often in areas which appeared to be unrelated to the original motivation behind string theory. In fact, the original motivation, as I'm sure you know, was not even quantum gravity. String theory was discovered essentially by accident by people who were motivated to find a theory of strong interactions and only later was it realised that it can actually provide answers to quantum gravity questions [3].

So, do you see string theory as a theory of quantum gravity?

It certainly is a theory of quantum gravity. I don't know if it is the only possible framework in which we can answer questions about quantum gravity but it's certainly the only one that we have right now that has produced concrete results. You can actually ask very interesting questions, be absolutely confused about how nature can solve or give you answers to those questions, and string theory often gives you, at least conceptually, a logical possible answer. We don't know whether that's the answer which is realised in the real world but it's the only mathematically consistent and physically sensible answer that we can offer at the moment. In that sense, I think string theory has been extremely successful.

Which kind of questions/problems are you referring to?

For example, the famous Bekenstein-Hawking entropy of black holes [4, 5] is predicted by a semiclassical argument in general relativity, which does not require detailed knowledge of the precise microscopic structure of quantum gravity – whatever theory that turns out to be. General relativity predicts that black holes carry thermodynamic entropy and so the next question which we immediately start asking is whether this thermodynamic entropy is just like conventional thermodynamic entropy and whether it has a statistical mechanical explanation in terms of the counting of some kind of microstates. This is a very interesting question to ask because as the word *microstates* suggests, it can lead to the understanding of a microscopic structure, in this case, of gravity. Thus, understanding the nature of microstates of black holes can perhaps teach us more about the microscopic underlying degrees of freedom of quantum gravity. String theory, to this day, is the only framework where you actually can, at least for some classes of black holes in certain dimensions, produce answers that make mathematical sense [6]. In this context, counting microstates amounts to counting the degrees of freedom of fluctuating D-branes.

Is it important to search for alternative theories of quantum gravity besides string theory?

I think you can answer this question from several different points of view. What's the best for the community of physicists or for the general society? Or what is the perspective of string theorists themselves? In both cases I think that the answer is of course that any constructive attempts to come up with either alternatives, counterexamples or parallel structures which have some mathematical validity in them is welcome and is something that we should be encouraging. However, when encouraging for alternatives it's also important to require some results. In light of this, I still don't see convincing alternatives at this point.

In practice, and in line with how it originated, string theory provides results to fundamental questions but also creates ideas that enrich other fields of physics. In some sense, it is a laboratory where you generate new ideas by asking mathematically oriented questions or questions oriented about thought experiments at the very abstract level. Often the answers provide new spinoff ideas that are very useful in areas of physics ranging from particle phenomenology to cosmology and condensed matter physics. In phenomenology, for instance, extra-dimension scenarios, AdS/CFT, Randall-Sundrum/Braneworld models [7–9] had clear precursors in string theory before being accepted as interesting ideas for phenomenology. These precursors originated from asking fundamental and conceptual questions rather than practical questions, such as whether the theory fits experimental data.

Nowadays we're starting to learn that the AdS/CFT correspondence might be a very powerful tool for giving new conceptual insights into how to solve strongly coupled systems of interest in condensed matter physics. Of course we're only in the early stages of this endeavour but it might actually revolutionise the way we think about condensed matter physics at strong coupling. In any case, these ideas did not originate by some drive to apply string theory to the real world, either in particle colliders, cosmological observations or in condensed matter systems, but originated by trying to resolve fundamental theoretical puzzles in the structure of the theory. Thus, it seems that such an approach where one tries to resolve internal tensions and fundamental puzzles within string theory is a really good methodology that even allows you to learn about other areas of physics. Additionally, because there are still many fascinating questions to ask, I think that this laboratory of ideas is really far from being exhausted.

Do you see the theory that you proposed, which nowadays people refer to as Hořava-Lifshitz gravity [10], as an alternative to string theory or should it be embedded in it?

Well, even though we haven't really worked out how it might be embedded into string theory, it's my presumption that if this structure really makes sense it will end up being just another corner of string theory. The reason for this expectation is that the theory is based on very natural concepts originating from how we understand quantum field theory (without gravity) at different length scales. This is the essence of the renormalisation group and it's a very powerful way of thinking about how nature organises itself when you deal with complex systems. This way of thinking was there since the origins of string theory, when it was not as mathematically complex as it is now. In connection with Hořava-Lifshitz gravity, I think that once this kind of bottom-up approach to quantum gravity is developed with improved mathematical precision, we will be able to embed it in some corner of string or M-theory.

A priori, this was certainly not an attempt to create an alternative to string theory; instead, it was actually directly inspired by the success of string theory. String theory is the only successful way we know of combining general relativity with quantum mechanics. This was really a theoretical puzzle that was driving much of the progress in fundamental aspects of theoretical physics over the past century. Our ancestors, who were working on this problem, have always been curious about which of the two paradigms of physics is more fundamental. Is it general relativity, described by some classical geometry, or is it quantum mechanics, described by the Schrödinger equation and a probabilistic interpretation of wave functions? Is it general relativity that needs to be modified in order to find a theory of quantum gravity? Is it quantum mechanics that needs to be modified? Or is it both?

String theory gave us a concrete, mathematically consistent answer to this puzzle. It told us that quantum mechanics is not modified one bit but general relativity is modified in a consistent way. String theory tells us that Einstein's equations are only a first approximation and get systematically corrected. However, it's even more fascinating than that because we can often understand string theory in situations that do not involve just weakly curved smooth geometries. We can understand corners of the theory where the spacetime geometry itself becomes modified and in some cases spacetime appears as a derived structure. This teaches us a lesson which I'm taking very seriously because I highly respect the structure of string theory. The lesson is that we should take quantum mechanics as it is, at least for now, and take the geometry of general relativity as an emergent or derived structure. In particular, the spacetime itself might be emergent as we learnt from various solutions in string theory and M-theory. If it is emergent, then I have a very hard time believing that geometric symmetries, such as Lorentz symmetry, should be fundamental symmetries of nature. If spacetime itself is not fundamental, its symmetries can hardly be fundamental. We don't really know exactly how such symmetries emerge with the emergence of spacetime itself but that is what we need to find out.

I always find these arguments of what is fundamental and what is emergent quite difficult to understand given that you have dualities, for instance, between quantum mechanics and gravity.

Both sides are complementary. It's really an isomorphism between two different ways of thinking about the same quantum mechanical system. The quantum mechanical system does not necessarily tell you that you should interpret it as living in a four-dimensional Minkowski spacetime or that you should interpret that same system as living in a 4+1-dimensional curved spacetime geometry in the bulk. It's just a quantum mechanical system which satisfies axioms of quantum mechanics, for which there is a Hamilton that determines the evolution of the system and for which you can calculate correlation functions of observables. You should not think about it as a duality but really as an isomorphism. You are not comparing quantum mechanics to something else; instead, you are comparing two different ways of representing the same quantum mechanical system in different languages. It's like a translation from one language that involves gravity and strings to another language that involves Yang–Mills fields and no gravity. No matter what, it's still the same quantum mechanical system that we are talking about in two different ways.

I understand but I continue to be confused about how you can determine which description is fundamental and which is emergent given that you have a duality/isomorphism between these two languages. How does this work?

Well, that's a good question and maybe none of the descriptions is fundamental. From my point of view, the word "emergence" is more of a catch word that has been used or sometimes misused over the past five or 10 years in this field. What it embodies is the need to question the fundamental nature of the symmetries so I don't care whether we call it "emergent" or something else. Regardless, I am confident that spacetime itself is not a fundamental structure. In fact, it could be very observer-dependent and dependent on how you view the same quantum mechanical system. This is another lesson that originated in string theory, namely, that you can interpret the same solution to string theory in several different languages. For instance, to different observers it can look like different geometries and to some observers there might not be any simple geometric picture.

One of the big criticisms of string theory is that it doesn't have a background-independent formulation. Do you think this is a problem?

I couldn't disagree more. I think it's actually a virtue because I think the notion of background independence is one of the most overrated concepts that people keep mentioning as a demand for a theory of quantum gravity. I don't really believe in the notion of background independence because I have never heard of a useful technical definition of what background independence means. If it means that quantum gravity should somehow be formulated as one unique theory without specifying, say, boundary conditions at asymptotic infinity then I think that such notion is manifestly nonsense.

We know that things don't work that way and again in string theory we have examples of consistent quantum mechanical systems which happen to be describable, say, in the context of the AdS/CFT correspondence by some prescribed asymptotic behaviour of some fluctuating spacetime geometry near infinity. Near infinity the geometry is really frozen and it's only when you freeze the geometry in a prescribed way that you actually obtain a consistent quantum mechanical system. The Hilbert space of states is not well defined until you specify precisely what the asymptotic behaviour should be. We know of several supersymmetric cases in string theory which are self-consistent quantum mechanical systems but which are certainly not isomorphic to each other. They are all distinct, depending on how we pick the boundary conditions. There are many different ways of ending up with a consistent quantum mechanical system and the choice of system depends on specifying the asymptotic behaviour of the background. To make sense of the quantum system you need to specify boundary conditions but you can argue that, locally, an observer will not be sensitive to a particular solution of the Einstein equations. I am happy with this form of background independence, which is what can be realistically expected instead of a unique quantum gravity theory that is independent of all the details of the background. This latter option makes no sense to me.

But isn't M-theory supposed to be background-independent?

I don't know what that means. You cannot give me the Hilbert space or physical states unless you have specified some particular choice of solution, and that choice of solution implicitly means that you have specified boundary conditions. Different choices of boundary conditions will give you different dynamics for the physical states in the Hilbert space. In other words, you define different quantum mechanical systems whenever you make these choices. These individual solutions might still be solutions of one single meta theory, which you can vaguely call M-theory. We think that these are different solutions and not different theories because there are many duality transformations and continuous deformations that allow you to interpolate between one solution and the other. Some of these transformations are discrete and you end up jumping from one representation to another. So we think that all these seemingly different consistent quantum mechanical systems are actually different solutions of one big theory.

Do you think it's possible to formulate M-theory precisely?

I don't know. It's a fascinating open question whether we can come up with a compact definition of what M-theory is but I wouldn't be surprised if it's impossible. It could be similar to trying to describe a non-trivial differentiable manifold, such as a sphere, with just one coordinate system. What one observer sees as fundamental degrees of freedom in physics could be analogous to the notion of choosing a coordinate system in geometry. Sometimes there are differentiable manifolds which cannot be covered by one coordinate system, so you have to use several different ones and prescribe transformations between them. In physics, it could be analogous in the sense that you may have to specify what you think are the fundamental degrees of freedom patch by patch and then specify rules for how to translate from one language to the other. In this setting, there might not necessarily be a uniform definition of what are the microscopic fundamental degrees of freedom of M-theory as a whole. This would be acceptable and requires that we expand slightly our notion of physical theories. It's certainly not against what we've learnt from other fields of science.

If there are no fundamental degrees of freedom, as you suggest, will you be limited in regards to what questions you can ask?

I am not sure this is a problem. We can certainly ask practical questions and get practical answers. For instance, 40 years ago you might have asked how to make sense of the theory of strong interactions as a quantum field theory. If you work hard enough you can calculate many quantities in QCD, make comparisons to experimental data and get beautiful results. However, for that you do not need to know what is the precise mathematical definition of that quantum field theory. Similarly, in string theory much of the work is performed at this level. The Wilsonian paradigm of modern physics and complex systems has taught us that we should think of nature as structured in layers and it is a virtue to admit our level of ignorance beyond a certain layer. We cannot access higher energies than what we can test experimentally. This, however, does not stop us from working in a regime where we understand precisely what happens in the physical system but we must acknowledge that we will not have answers for everything.

But do you think string theory will ever be able to tell us what happened at the beginning of the expansion of the universe?

We will find out sooner or later. It is hard to say at the moment and I don't feel ready to start addressing this question today. I think this is the most common practical attitude in the field; that is, you want to address questions which are within reach at the moment. Many of the beautiful fundamental overarching questions are not within reach. That's fine; it is just telling you that you should proceed step by step and not try to understand everything in one giant leap.

People have been working in string theory since 1968 when the Veneziano amplitude was first derived [11]. We are still learning every year from Strings conferences that there are many surprises and no one really understands the full picture and how it all fits together. On the other hand, it is true that you can rephrase this as a legitimate criticism of the current status of string theory. In fact, this was also one of the motivations that led me to develop aspects of gravity with anisotropic scaling [10, 12–14]. The criticism is that string theory is in some sense very large as it has many different solutions. We understand the behaviour of a lot of them, especially those with a lot of supersymmetry but we also believe that there are a lot of solutions which have little supersymmetry or no supersymmetry at all, although they are more difficult to analyse. So the theory is really huge and involves lots of interconnected beautiful techniques. But if you are interested in answering some profound but narrowly defined physical questions, say, in QCD or in quantum gravity, do you need this huge machinery of string theory? Do you need to know about all these solutions and different techniques for studying them before you can start answering these questions?

We know that in all other areas of high-energy physics, except for gravity, the answer is no, you don't need string theory. It is true that string theory is often very useful and it gives new insights into the questions you are trying to answer but you do not need string theory to define QCD. QCD is a self-contained formally well-defined quantum field theory, at least by physics standards of rigour, and it does not need string theory as an embedding to become consistent. It is already consistent on its own and you can ask questions about QCD within QCD; that is, you don't need to invoke additional stringy degrees of freedom. However, gravity is often believed to be different. In fact, people often quote some kind of a "folk theorem," without any proof, stating that if you want to combine general relativity and quantum mechanics you are forced to use string theory. I have no idea whether that is true or not but some people have speculated along those lines. So is it possible that there is a smaller theory for gravity, just like there is a smaller theory for quarks and gluons, that does not require string theory? If I'm only interested in asking a quantum gravitation question, and not something that involves unification of all interactions, supersymmetry and the LHC physics, is it necessary for me to learn all of string theory first? To answer this question you need new ideas, given that many have tried to find self-consistent theories of quantum gravity and there aren't many successes.

Do you think that loop quantum gravity is one of these examples?

I'm happy to listen to a presentation of results from that field. People have worked on it for many years and I'm still waiting to see results that will be so convincing it will make me want to jump into that field and contribute. Each theory should speak for itself in the end. What is important is not who are the ones who are doing most advertising for their own fields but the amount of concrete results and number of convincing conceptual breakthroughs that you can attach to each field. In this respect, string theory has a huge record. If there is another field which is capable of producing as many spectacular conceptual breakthroughs then I will be more than happy and I think that this is the attitude of most string theorists, if not all of them. String theorists are always open-minded toward new concepts and ideas assuming they carry some value with them and that they actually help solving some theoretical puzzle. So, if loop quantum gravity leads to new techniques, new understandings of things that we've been struggling with, then I think people will be very open-minded.

Why did you formulate theories with anisotropic scaling with respect to space and time?

Because it's one natural idea, for several different reasons, that hasn't been tried. I've mentioned earlier that I am somewhat reluctant to accept that Lorentz invariance should be a fundamental symmetry of nature. I know of course that we've observed it with incredible precision but in a limited range of energies and scales over which we can actually measure. To promote that symmetry to an exact symmetry at arbitrary short distances and arbitrary high energies seems a little far-fetched, specially since we have examples from string theory which suggest that sometimes even those shortest distances don't quite make sense as in classical general relativity. So if general relativity is willing to be modified into a consistent quantum framework in string theory, perhaps it's also willing to be modified as an effective field theory. How do we modify it? You can list the attempts that people have made before and then you can start listing other possibilities that people haven't yet tried. One of those possibilities is to give up Lorentz symmetry as a fundamental symmetry. From the perspective of the renormalisation group it can make the formulation easier because, as you change the isometry group, the critical dimension of the system changes and the critical dimension typically, in the old language of particle physics, is the dimension at which the theory becomes renormalisable and potentially self-consistent.

Does anisotropy imply renormalisability?

It doesn't imply it but it can help you achieve renormalisability. It's a kind a knob that you can turn and change the dimension to that which a given theory becomes renormalisable [10, 12]. In the case of gravity that is essentially what you would like to do because we know that gravity doesn't make sense as an effective theory and predicts its own breakdown at some scale. In fact, another lesson that originated from string theory is that string theory is unique among all theories of relativistic objects such as particles, strings and membranes of various dimensions. Strings are special for a number of very good reasons. Several of these reasons have to do with the renormalisation group behaviour or the scaling behaviour of various important field theories, one of these field theories being gravity itself. If you look at the behaviour of relativistic gravity as you change the spacetime dimension, it naturally picks one time and one space dimension. In two spacetime dimensions, Newton's constant becomes dimensionless. This happens to be precisely the dimension of the string worldsheet. It's not a coincidence that strings make sense as quantum mechanical objects.

In light of this, gravity appears to make more sense as a quantum theory in two spacetime dimensions rather than in four but we are of course interested in understanding how to make sense of gravity precisely in four dimensions. How do you resolve this discrepancy? You need some new knob that you can tweak and one such knob is precisely this change in the scaling between the various dimensions of spacetime. If you allow for some degree of anisotropy, that gives you a shift in the critical dimension of gravity and you no longer need to be confined to $1+1$ dimension as in the relativistic isotropic case. If you include anisotropy, measured by a dynamical critical exponent z, which takes various integer values then you get to shift the critical dimension of gravity by an integer. Using this, in principle, you can have gravity theories which are consistent in $2+1$ dimensions as well as renormalisable by power counting precisely because you allowed a certain degree of anisotropy.

Mathematically, there are many opportunities for using this idea to make sense of a theory of fluctuating geometries as a self-contained quantum system without invoking extra degrees of freedom. Whether you can do this or not is unrelated to whether it will lead to phenomenologically acceptable predictions compatible with general relativity at long distances. However, as I mentioned earlier, it is often successful to explore a fundamental problem and not try to explore applications too early in the process. So I am interested in first understanding the quantum consistency of the model and then comparing it with experimental data.

As far as I understand, when you first proposed this theory there were issues that actually did not allow making it compatible with general relativity at long distances. Are these issues still there?

There are some versions of the theory which seem compatible with the observational data but often this involves some extra gauge symmetry. Together with Charles Melby-Thomson, we extended the gauge symmetry from the diffeomorphisms that preserve the fixed foliation structure of spacetime to something which we called *non-relativistic general covariance*, which is a symmetry as large as the symmetry in general relativity but the group structure is slightly different [13–15]. Because the group is roughly speaking of the same size as the diffeomorphism group in general relativity, and the fields in the theory are, in some sense, in one-to-one correspondence with those in general relativity you have a better chance that this theory flows at long distances to the conventional physics of general relativity as constrained by observations.

The inconvenience of this approach is that you actually need to introduce various auxiliary fields, which is fine, but once you start asking practical questions, say, about the behaviour of test particles around compact solutions such as the ones that should describe the solar system, then in order to compare with experimental data you need to couple test particles in a way consistent with the gauge symmetry [14]. The fact that the theory is formally more complex than the simplest version of gravity and general relativity opens the door for new coupling constants, new ways in which matter can couple to the background. It is true that you can dial the couplings in such away that you don't violate the experimental tests of general relativity in the solar system but, on the other hand, there doesn't seem to be any obvious reason why the couplings should take those particular values. There is a certain degree of tuning which makes the theory less predictive.

Were you expecting that this wouldn't be the case?

As I said, it is not my highest priority to see whether this theory produces general relativity at long distances. Once you understand if the theory is consistent there are a possible number of applications. Understanding the phenomenological implications for gravity is one of them but there are several others that are of interest to condensed matter physics. In fact, we know that there are many quantum field theories that are non-relativistic and can be tested against experimental data at condensed matter laboratories [16]. At the same time, I suspect that we haven't quite appreciated the possibility that perhaps string theory and gravity don't necessarily have to be relativistic. So we should decouple the notion of relativity and gravitation since a priori they are not identical.

When trying to understand better some of the non-relativistic quantum field theories useful for condensed matter we know that one of the most fascinating recent tools is the gauge/gravity of AdS/CFT duality. Is it logical to assume that all gravity duals will be relativistic theories? It seems intuitively more natural if some of these non-relativistic field theories have gravity duals which involve some degree of non-relativistic behaviour in the bulk. It would be interesting to pursue these ideas and I wouldn't be surprised if there is a big corner of string theory and quantum gravity which is non-relativistic. With this in mind, if it turns out that some non-relativistic version of gravity in 4+1 dimensions is useful for understanding some non-relativistic field theory then who cares whether the bulk five-dimensional non-relativistic gravity system satisfies some hypothetical solar system tests? In this sense, you don't even have the burden of having to reproduce general relativity at long distances since nobody even expects you to.

But you did mention before that you could tune some constants to make it exactly like general relativity, right?

I wouldn't say "exactly like general relativity." You tune the couplings such that at long distances and low energies it lies within the experimental bounds of general relativity. However, as you increase the energies and look at shorter distances, you will end up seeing modifications because, unlike general relativity this theory has higher-derivative terms in the action. There is an almost pathetic range of energies in gravity, compared with the tests of the standard model physics, that haven't yet been tested. Therefore we cannot yet see any visible effects of these higher-order terms.

There are few funny things that happen when you break Lorentz invariance; for instance, higher-energy light can travel much faster than lower-energy light, right?

In this particular context yes but it doesn't have to be the case when you break Lorenz invariance in a generic way. In these models that happens to be the case because you have a preferred foliation, in particular a foliation in slices of constant time. This is my choice of foliation and I could have made another choice. I could have said that the spacetime is foliated by two-dimensional hypersurfaces of one time and one spatial dimension, some other co-dimension foliation or some nested foliation where we have the space itself foliated by hypersurfaces of lower dimension. Thus, there're many choices you can make but in the simplest case spacetime is foliated by slices of constant time. If you place a dynamical system in this foliation, which involves the fluctuating metric on the spacetime, then at short distances the dynamics is sensitive to the foliation but that sensitivity is washed away at long distances. Basically, this is the reason why you can start mimicking something that looks like general relativity at long distances.

This fact gives you the possibility of restoring the approximate relativistic symmetries of general relativity at long distances but once you reach shorter distance the presence of the foliation changes the effective behaviour of particles; in particular, the light cones start opening up, due to the higher-order effects, compared to the long-distance behaviour. As the light-cones open up you can travel faster the higher the energy the particle has but you can never travel infinitely fast unless you have infinite energy. There is no causality restriction precisely because the theory is not subjected to fixed light-cones as there's this preferred foliation. So the particles can travel arbitrarily fast as they reach arbitrarily high energies and, as a result, in this minimal version of gravity with anisotropic scaling, the notion of a black hole horizon is also an approximate notion, which is only useful at low energies. At low energies the information about this underlying foliation structure gets washed away and particles think that they have to respect light-cones, which as a result they also respect horizons. So you can define black hole horizons as in general relativity at low energies and the very low-energy particles would behave as if they were not able to leave a certain causal region of spacetime. Such particles would stay behind the horizon and never escape. However, as you start reaching energies which become sensitive to the existence of this fundamental length scale in the system at which all of a sudden the foliation becomes visible in a continuous way, then those particles would be able to freely leave that region and hence there would be no horizon for them.

What about the black hole singularity, is that also an effective notion?

The existence of singularities and the existence of a horizon don't necessarily have much to do with each other. In conventional general relativity, we define the horizon as the boundary of the causal past of future timelike infinity. This definition has nothing to do with whether or not there's a singularity in the spacetime. It just happens that we know from many of the solutions we care about that singularities are hidden behind the event horizon. So maybe we can live with those singularities; at least, that's the idea behind the cosmic censorship conjecture [17]. So we learned to live with the idea that there are singularities in black hole geometries. However, the notion of horizon is decoupled from the notion of singularity. The more fundamental question to ask, and this is one of the lessons of holography, is what thermodynamic properties such horizons have from the point of view of an observer who is outside the horizon. In particular, it doesn't make sense to ask questions about the singularity if you can't directly probe it. In fact, maybe the singularity does not even have a role to play in our quantum mechanical understanding of the system if you are an outside observer.

Perhaps questions about the singularity have very limited practical sense and perhaps the history of an infalling observer, who could potentially see something different, can be related/complementary to the history of an outside observer. In principle, these two points of view of outside and infalling observers need not be simultaneously consistent within some super big Hilbert space of quantum states. It may be that the observations of an infalling observer make some sense but how much sense is not really known because in principle an infalling observer won't live long enough to accumulate enough statistics in order to be able to apply a probabilistic interpretation of quantum mechanical amplitudes. On the other hand, assuming that quantum mechanics holds in this context can be a very useful strategy and has worked well in the hands of many pioneers in this field, such as Leonard Susskind or Gerard 't Hooft – the inventors of holography [18, 19]. What is clear is that an outside observer has the chance to define their quantum mechanics by observing the system from faraway and living for an infinitely long time while using the usual axioms of quantum mechanics. Within that description, it seems that we know at least in principle how the black hole information paradox is solved using AdS/CFT. It seems that it's a perfectly unitary quantum mechanical process where there is no information loss, in principle, although it is very interesting to ask how these things get practically resolved. We still have to wait a bit more to have a satisfactory answer to this.

How would the holographic principle work in Hořava-Lifshitz gravity?

It would again be an emergent property at low energies. It might sound like an unusual claim but I think it's actually a quite conservative picture of what might be the case. This picture is in fact rooted in our understanding of condensed matter systems. In this context, what often happens is that as you lower the characteristic energy scales at which you study the system, for instance lowering the temperature by tweaking your experimental machinery, the system typically undergoes various phase transitions moving towards more ordered phases. Typically, the more order you have, the more rigidity there is in the system and the increase in order brings with itself a reduction in the effective number of degrees of freedom that are accessible at low energies. This represents a hierarchy of symmetry-breaking patterns that causes the decrease of degrees of freedom.

From experience, when you start at very low energies you see very few degrees of freedom because of the rigidity implied by the symmetry-breaking patterns but as you heat up the system or you probe higher and higher energies all of sudden you start freeing more and more degrees of freedom that were not accessible at lower energies. Perhaps quantum gravity could also work analogously to condensed matter systems, in the sense that at very low energies you have, say, the holographic bound on how many degrees of freedom you can have per spacetime volume. As you know, this leads to the fascinating and very challenging problem of understanding the behaviour of dimensional reduction of the effective degrees of freedom. In particular, you know that you cannot have as many degrees of freedom as you would expect from conventional quantum field theory. Over the past 20 years, this puzzle has led people to think about how to make sense of this from a more microscopic point of view. However, this might just be a low-energy artefact of the ordered phase of gravity. As you probe higher energies you might find out that there is a rigidity which enforces this absence of degrees of freedom. The impossibility of having a volume's worth of degrees of freedom in a gravitational system because of black hole horizons is not there as you probe higher and higher energies and free new degrees of freedom. Perhaps they can even behave in a completely non-holographic way.

What are the fundamental degrees of freedom in the theory? Is it the metric that is being quantised?

This theory has a very pragmatic and down-to-earth approach to defining what the degrees of freedom are. It basically assumes that gravity is really just like Yang–Mills theory; that is, it is defined through a conventional path integral as the gauge theory, where the fundamental degrees of freedom are the Yang–Mills gauge field and possibly quark degrees of freedom, depending on whatever matter you couple to it. Similarly, in the case of gravity, we do not assume that it is a mysterious force made out of something more microscopic that we don't understand, which might eventually be true. Instead, we take it as being fluctuations of spacetime just as in Yang–Mills theory we consider fluctuations of the spacetime gauge connection of some internal symmetry group. In the context of gravity, we postulate a gauge symmetry associated with foliation-preserving diffeomorphisms of spacetime. This implies that all of a sudden you can use standard quantum field theory methods for making sense of various physical observables. In essence, you are basically trying to place gravity on the same footing as Yang–Mills theory.

I guess that if this is the case then you aren't simultaneously thinking about how to make spacetime emergent, right?

You have to put some guards on how revolutionary you want to be. If you are a true revolutionary you may end up doing something which seems very exciting and interesting but you are not going anywhere with it. In this theory, I have already relaxed the geometric structure of spacetime compared with what you take for granted in general relativity. So I have to hold on to something and having a smooth field on a smooth spacetime manifold seems like a good starting point. Just as in Yang–Mills theory it could be that there is a more microscopic understanding of where the so-called fundamental Yang–Mills field comes from – it may come from open strings ending on branes or something similar. The point is that we don't need to know that to define the system self-consistently.

So you could say that in some particular regime the metric constitutes the effective degrees of freedom?

Yes, sure, so why invent something more microscopic if it's efficient just to keep the metric? We would be making conceptual progress if we can make sense of the path integral for a fluctuating metric of spacetime. That is one of the motivations here. Perhaps we can think of gravity after all as being similar to Yang–Mills theory precisely by abandoning some of the prejudices that we've learnt from string theory, which should not be taken so seriously in the context of gravity.

Within this setup can you then have a clear picture of what spacetime looks like at the Planck scale?

I can only speculate at this point but if the theory turns out to be consistent as a quantum field theory, then it appears that there is no need for some quantum spacetime foam. It could be that the main thing that happens between long and short distances in gravity is not topology change; in fact, the topology might be frozen and always be trivial. The scaling nature of spacetime is the pre-geometric structure in a sense; it's not something sensitive to the spacetime metric. In this sense it is a topological notion.

Is the theory topological?

I don't think it's topological in the sense of not having degrees of freedom. The term "topological" can have different meanings. Usually, when one refers to a theory as topological one has in mind that it's a theory with no local propagating degrees of freedom. If it had local propagating degrees of freedom it would not be topological because there would have to be some form of inertia or, in the language of field theory, there would have to be a kinetic term for the fields. If no such kinetic term would be present, you would not know what the word "propagate" means. Having a kinetic term is a very non-topological notion as it seems to be implying some background structure, such as the metric, that allows you to actually write down a kinetic term. Thus, if this happens the theory is not topological.

In Hořava-Lifshitz gravity, the theory has propagating graviton polarisations. It is one of the working assumptions that these graviton polarisations could be weakly coupled at short distances at the cost of having anisotropy between space and time. There is really no reason to say that something weird happens with the degrees of freedom at short distances. It just means that they behave as free gravitons within some approximation. It could be as straightforward as having a simple background geometry, such as a flat spacetime, on which you have a bunch of propagating gravitons. This suggests that it could be analogous to QCD in the sense that you don't have to invent anything exotic to make sense of QCD at very short distances. In fact, at shorter distances (i.e., higher energies) the system becomes simpler as it reduces to a bunch of weakly coupled quarks and gluons. So, Hořava-Lifshitz gravity is in principle similar to QCD but to check this precisely requires a lot more work. The theory is still in its infancy.

What do you have to do to check that this is the case?

You have to just work out the consequences of the quantum field theory formalism for this theory. There will be some interesting obstacles along the way because, unlike in the case of Yang–Mills theory, it's not obvious what is the most natural way of formulating this type of theory. Should you keep around this extra scalar polarisation of the graviton which is present in the minimal formulation? Some people would say that they don't like the scalar graviton because it's not phenomenologically viable but I think the more interesting question is whether it actually creates any difficulties in defining the theory at short distances. If it does then we would have to invent a version, like the one that we discussed recently [13–15], where the scalar graviton is eliminated as a gauge artefact of some enhanced gauge symmetry.

So, first you have to sort out all these things. In particular, what is the most promising version of the theory? This is important before you start jumping to conclusions about the precise behaviour of any particular version of the theory because I think the worst thing that could happen is that you commit to some particular version of the theory for no apparent reason and then you, your students and postdocs work very hard, spend a number of years developing it and it turns out that the particular corner of the theory you committed to is meaningless for some technical reason. So it's better to do this step by step and try to find out what's the most interesting version of the theory.

In one of your papers [15] you suggested that this theory was an entropic theory and that it could bear some relation to Erik Verlinde's entropic gravity [20]. Has this been made more precise?

It has not been made precise but you can also say that what Verlinde's entropic gravity hasn't been made more precise either. At this point, it is hard to compare the two. However, while developing this Hořava-Lifshitz gravity system it actually surprised me, already before Verlinde's papers on entropic gravity, that mathematically the system seems isomorphic in a conceptual sense to the way in which Lars Onsager and his collaborator in the 1950s formulated the theory of non-equilibrium thermodynamics [21, 22]. These ideas seem to be in-built in the formalism. The formalism came out naturally from various arguments but from a mathematical viewpoint it's as if you are sort of postulating that the dynamics of the system is driven by some underlying thermodynamic potentials and forces. In particular, the spatial metric plays a role analogous to a thermodynamic potential of sorts and then there would be a well-defined object which is essentially the variation of some scalar function with respect to the metric, which corresponds to the driving force acting on these thermodynamic potentials. In fact, the interpretation of this scalar function in this formalism is the entropy [15].

The way that Onsager proceeded was by postulating the entropy of a system and then seeing how the system responds to being taken away from equilibrium. It turns out that it responds precisely by trying to maximise the entropy or, more precisely, minimise the entropy production. When that is the case, there are certain effective equations which happen to be isomorphic to this Hořava-Lifshitz gravity system in a particular subset of formulations in which you satisfy the so-called detailed balance condition [10, 15]. Originally, this condition was for me just a technical trick for suggesting the most natural and conceptually simple version of the theory where you have as few coupling constants as possible. This reduction of coupling constants takes place because you impose an additional restriction on the form of the action. It wasn't meant to be important or proposed as a phenomenologically interesting condition but it so happens that if you impose the detailed balance condition, you precisely recover this formal analogy with Onsager's theory of near-equilibrium systems, in which case the origin of the dynamics is due to entropic forces acting on thermodynamic potentials [15].

Is this the correct formulation?

No. As I mentioned, it was originally just a trick for nailing down the simplest possible version of the theory with as few independent couplings as possible. If you formulate the most general version of the theory without imposing the detailed balance condition, you will end up with something more realistic in terms of possible phenomenological applications. However, it also makes it more complicated because the theory generically contains a lot of coupling constants and understanding whether the theory becomes nearly free at short distances would be outside anybody's technical ability. If you really want to get a preview into the dynamics of the theory, you would like to restrict the number of couplings by an additional symmetry principle. There might be other symmetries that you could impose but the detailed balance condition was one that I was aware of because of various connections to dynamical systems.

One of the interesting things that arose from this theory of gravity was the connection with causal dynamical triangulations (CDT), right?

I think that there is almost a consensus forming now in the community, both from the continuum approach and from the lattice approach, that the relation between CDT quantum gravity on the lattice and Hořava-Lifshitz gravity in the continuum is similar to the relation between lattice QCD and the continuum approach to QCD [23–25]. In order to define QCD you can of course try to do it analytically using conventional field theory methods but you quickly find out that the theory in various regimes becomes difficult to understand because it's a strongly coupled theory. If you still want to understand the theory in that regime, you can put it on a lattice and run numerical simulations which allow you to probe the behaviour of the theory even though you may not know how to analytically solve the theory in that regime.

Analogously to QCD, you can apply the same reasoning to a theory of gravity. In retrospect, it appears that if you try to come up with a sensible version of Hořava-Lifshitz gravity on a lattice, you end up with a version of CDT. Similarly, if you want to ask what is the correct continuum arena of field theories that can be interpreted as the continuum limit of the CDT approach to quantum gravity, it's natural to presume that it is this theory of gravity with a preferred foliation of spacetime by slices of constant time. In fact, such foliation is exactly the relevant ingredient added to the pre-Hořava-Lifshitz-type gravity models in order to make sense of the continuum limit. The introduction of this foliation of spacetime changed the qualitative behaviour of the lattice system and all of a sudden produced an object which had a microscopic scale in four spacetime dimensions [26]. This is in strong opposition to the earlier cases where they were trying to discretise the system without a preferred foliation, in turn leading to branch polymer phases and crumbled phases, that is, situations where spacetime wouldn't even have an integer dimension [27, 28]. This was a major breakthrough in the lattice gravity side and they worked on this before Hořava-Lifshitz gravity was developed. As a matter of fact, they had actually no idea about what kind of gravity and continuum formulation would be approximated by the lattice results. As it stands now, it seems that what they were doing without knowing was putting Hořava-Lifshitz gravity on the lattice.

Can you state that CDT and Hořava-Lifshitz gravity are the same theory?

It's too early to say but in my opinion, it's a working assumption. We are in fact working with a few students directly on projects which try to flush this out a little bit more as to obtain a more precise match [24]. I think that the group of Jan Ambjørn, Renate Loll and others is trying to do the same from their point of view [25]. There are also results by other groups, for example recent papers by Sotiriou, Visser and Weinfurtner [29, 30] in which they actually improved the calculation that I did of the spectral dimension [12]. I tried to explain the behaviour near the two extremes of the scaling assuming that the system interpolates between two different fixed points, one fixed point at very short distances which has anisotropy in it and another fixed point at long distances which is isotropic. The predicted behaviour is that the effective spectral spacetime dimension changes from one integer value to another integer value in the simplest approximation.

But the topological dimensions remain four?

Correct. What happens is that the scaling makes some of the probes of spacetime think that they live on a different dimensional space. Topologically they still have the same freedom to move in 3+1 dimensions but in some sense the three dimensions are being squeezed at a different rate compared to how time is being squeezed. This makes those particles/probes think that all these three dimensions of space behave as if they were one legitimate dimension compared to time. Of course, there is a mathematically precise definition of spectral dimension which makes use of an auxiliary diffusion process that you put on the manifold and calculate some return probability for a particle to diffuse back to the origin after a certain amount of proper time. Thus, there is a precise way of measuring some invariant of the ambient geometry.

The advantage of this observable is that it does not rely on whether the ambient geometry is smooth or discrete or whatever. It could even be some very complicated fractal structure. In principle, you can always measure this observable and even predict it. In the case of Hořava-Lifshitz gravity you can predict this observable and in the simplest approximation you find that it has the value of $1+d/z$ where d is the number of spatial dimensions and z is the dynamic critical exponent [12]. This formula tells you that at long distances the spacetime is perceived as four-dimensional by the diffusing particles because they can, at long distances, diffuse in all directions at the same rate. But because of the change in the scaling of spacetime at short distances – the spacetime itself does not change the topology or the number of dimensions – the squeezing on these diffusing particles is forcing them to perceive the three dimensions as if they were equivalent to one conventional dimension. This behaviour is smooth so there is some interpolation between those two results and what the group of people I mentioned earlier have been calculating recently is whether or not in the context of Hořava-Lifshitz gravity you can actually get the entire curve interpolating from one result to the other [29, 30]. This is interesting because the same curve can be extracted from CDT numerical simulations. It is complicated to do this calculation analytically as you are dealing with a complicated nonlinear interacting system but within an amazing precision they were able to get a match with the full numerical curve. This seems to be strong evidence that the basic picture I mentioned earlier is true and that CDT is essentially just a way of putting Hořava-Lifshitz gravity on the lattice.

Is it possible to get a mathematical proof of this?

In principle, yes, but the theories are quite rich on both sides. You can play with various structures like various degrees of anisotropy, different foliations and you can add various degrees of freedom. There's probably some minimal matching, for instance by taking a minimal version of the theory on one side which can be put on the lattice in some minimal way and then you can check whether this really works in detail. The structure of the theories is also very rich in the sense that both the analytic approach and the lattice approach predict not just one phase of spacetime but actually a diagram of phase transitions. There are different phases of spacetime geometry which has to do with the fact that the Lifshitz scaling behaviour at short distances is associated with what condensed matter physicists call *multi-critical points*. Multi-critical points typically show up in condensed matter systems where there are several different phases. If you move away from the multi-critical point in different directions, you probe qualitative different phases of matter. Thus, it seems that gravity can also have these different phases and we see them qualitatively in both approaches. CDT sees at least three different phases in 3+1 dimensions and we also have a phase diagram in the continuum picture with different phases. This is another piece of evidence for the equivalence of the two approaches.

Turning things around a bit, can you make predictions that could be tested or did you not spend too much time thinking about this?

There are many people who have written papers on this and they have chosen various versions of the theory and came to various different conclusions [31–37]. Some of the mechanisms proposed in the literature are fascinating, actually. For instance, there have been very interesting pieces of work suggesting that the anisotropic scaling of spacetime can provide a new mechanism for explaining the near scale invariance of fluctuation spectra in cosmology, thus providing an alternative to inflation [33, 38].

There have also been a number of papers proposing that this modification of gravity can suggest new mechanisms for explaining both dark matter or dark energy or some portion of dark matter. There was a very interesting paper by Shinji Mukohyama and his collaborators where they looked at the so-called projectable version of the theory and saw that the effective modification of Einstein's equations in fact gives precisely enough room to mimic the presence of dark matter [35–37]. This modification is not due to adding extra matter terms to the equation but due to the modifications of the equation itself.

Is it a modified Newtonian dynamics (MOND)-like scenario?

Yes. Philosophically it can be seen in that way except that I think that what we usually refer as MOND is based on very limited conceptual understanding of mathematical structures [39]. So this might actually be a framework in which some of the MOND ideas can be usefully revisited and one can see whether such ideas can be put into a more concrete framework of quantum filed theory. This applies to both dark matter and dark energy in different versions of the theory. Robert Brandenberger and collaborators wrote an interesting paper on these issues [33].

Cumrun Vafa proposed the notion of swampland [40]. Is your theory in the swampland?

I think that the swampland is defined sociologically. What I mean by this is that whether or not anything can be embedded consistently into string theory can only be judged with respect to the particular understanding of string theory that we have at a given moment in time. I have a very strong personal suspicion that we don't really understand enough of string theory in order to ascertain whether a given phenomenological model of particle physics is embeddable into some solution of string theory. It might very well be that at some point someone will come up with a generalisation of what we understand as string theory and uncover another huge class of solutions for which all of a sudden there will be a new way of embedding this previously un-embeddable thing into string theory. Thus, I think that the swampland is, at best, a time-dependent concept, deeply rooted in our limited understanding of string theory.

What do you think has been the biggest breakthrough in theoretical physics in the past 30 years?

I think the biggest breakthrough has been the AdS/CFT correspondence. This discovery is connected to the idea of holography, which originated historically in questions about the Bekenstein-Hawking entropy of black holes, but in the end I think the implications are so dramatic and unexpected, going way beyond what string theorists were trying to do at the time. Gauge/gravity duality changed our way of thinking about systems of many particles, on one hand, and gravity, on the other, connecting them together once and for all into one big framework.

Why have you chosen to do physics?

I was growing up under the communist regime in the former Czechoslovakia, now Czech Republic, and I was certainly looking for something which was sufficiently universal that could not be controlled and manipulated easily by the authorities. Of course at the time, for typical citizens, inside the communist regime the options were limited and certainly physics or natural sciences in general were one of those areas where you would say the communists couldn't really interfere that much or control as much the outcome of your way of thinking. If you were looking for some sort of intellectual freedom, following politics, business or philosophy was not a very good idea. Physics was offering more intellectual freedom than art because in art there was some form of control and interference and you could only say so much. In physics you could say what the physical system looks like, how it behaves, what are the laws of nature and the communists couldn't argue with it.

In any case, I should also say that I've always been attracted to natural sciences since early childhood but when I was in high school I thought I would end up as a biologist. Somehow through philosophy, I got closer to more fundamental questions and when they actually taught us the elements of quantum mechanics in the third year in high school I started thinking that physics seemed more important and addressed more fundamental questions than the ones being asked within the context of biology at the time. Maybe I was wrong, maybe in retrospect I was too rooted in the fundamentalist view of the world, which can be traced back to Einstein and others in physics, who seemed to hold the opinion that the systems that they were trying to understand were key to understanding the entire universe. That might be the case but only at certain scales. Understanding the universe at cosmological scales or understanding the universe at the LHC particle scales might be fascinating but to infer consequences from those scales to the everyday scales of human life is an infinitely complicated process which nobody ever dreams of actually performing. The questions that these fundamental theories of physics are addressing are important but the other fields of human enterprise have very important questions which the physicist will not answer by solving puzzles of quantum gravity.

What do you think is the role of the theoretical physicist in modern society?

That's a very tough question but equally tough for any other creative activity. I think that the obvious role is to contribute constructively to the betterment of humanity as a whole. I see physics essentially as a creative enterprise that creates value even from the point of view of a Darwinian understanding of the world. I think that understanding the world will hopefully help civilisation to survive. I think many physicists have the a priori belief that increasing the level and depth of understanding is always good in the long run, even if it can temporarily have all kinds of unpleasant side effects like strange kinds of weapons produced as a result of the work of the theoretical physicist. What do you think is the role of the theoretical physicist?

I like your point of view but I also think that the role of the theoretical physicist is to communicate how science works. If there were a bit more scientific mindset in politics, the world would probably be better off.

I resonate with that idea a lot but I think the theoretical physicist can only do so much. You have to make all kinds of sacrifices to be able to perform this coherent work and actually push the field somehow at least by a tiny step. It would be great if on top of that one could also spend the other 24 hours every day in a 48-hour day communicating the scientific method. However, we don't have those extra 24 hours. Of course, there are some of us who are more talented and oriented towards communicating science and perhaps it is enough that only some of us engage in this role, while the rest of us can spend our time juggling all these mathematically complicated structures.

References

[1] G. 't Hooft, "A planar diagram theory for strong interactions," *Nucl. Phys. B* **72** (1974) 461.

[2] J. M. Maldacena, "The large N limit of superconformal field theories and supergravity," *Int. J. Theor. Phys.* **38** (1999) 1113–1133, arXiv:hep-th/9711200.

[3] D. Rickles, *A Brief History of String Theory: From Dual Models to M-Theory*. The Frontiers Collection. Springer Berlin Heidelberg, 2014.

[4] J. D. Bekenstein, "Black holes and entropy," *Phys. Rev. D* **7** (Apr. 1973) 2333–2346. https://link.aps.org/doi/10.1103/PhysRevD.7.2333.

[5] S. Hawking, "Black hole explosions," *Nature* **248** (1974) 30–31.

[6] A. Strominger and C. Vafa, "Microscopic origin of the Bekenstein-Hawking entropy," *Phys. Lett. B* **379** (1996) 99–104, arXiv:hep-th/9601029.

[7] L. Randall and R. Sundrum, "Large mass hierarchy from a small extra dimension," *Phys. Rev. Lett.* **83** (Oct. 1999) 3370–3373. https://link.aps.org/doi/10.1103/PhysRevLett.83.3370.

[8] L. Randall and R. Sundrum, "An alternative to compactification," *Phys. Rev. Lett.* **83** (Dec. 1999) 4690–4693. https://link.aps.org/doi/10.1103/PhysRevLett.83.4690.

[9] M. Gogberashvili, "Hierarchy problem in the shell universe model," *Int. J. Mod. Phys. D* **11** (2002) 1635–1638, arXiv:hep-ph/9812296.

[10] P. Horava, "Quantum gravity at a Lifshitz point," *Phys. Rev. D* **79** (2009) 084008, arXiv:0901.3775 [hep-th].

[11] G. Veneziano, "Construction of a crossing-symmetric, Regge behaved amplitude for linearly rising trajectories," *Nuovo Cim. A* **57** (1968) 190–197.

[12] P. Horava, "Spectral dimension of the universe in quantum gravity at a Lifshitz point," *Phys. Rev. Lett.* **102** (2009) 161301, arXiv:0902.3657 [hep-th].

[13] P. Horava and C. M. Melby-Thompson, "Anisotropic conformal infinity," *Gen. Rel. Grav.* **43** (2011) 1391–1400, arXiv:0909.3841 [hep-th].

[14] P. Horava and C. M. Melby-Thompson, "General covariance in quantum gravity at a Lifshitz point," *Phys. Rev. D* **82** (2010) 064027, arXiv:1007.2410 [hep-th].

[15] P. Horava, "General covariance in gravity at a Lifshitz point," *Class. Quant. Grav.* **28** (2011) 114012, arXiv:1101.1081 [hep-th].

[16] C. Xu and P. Horava, "Emergent gravity at a Lifshitz point from a Bose liquid on the lattice," *Phys. Rev. D* **81** (2010) 104033, arXiv:1003.0009 [hep-th].

[17] R. Penrose, "Gravitational collapse: the role of general relativity," *Riv. Nuovo Cim.* **1** (1969) 252–276.

[18] G. 't Hooft, "Dimensional reduction in quantum gravity," *Conf. Proc. C* **930308** (1993) 284–296, arXiv:gr-qc/9310026.

[19] L. Susskind, "The world as a hologram," *J. Math. Phys.* **36** (1995) 6377–6396, arXiv:hep-th/9409089.

[20] E. P. Verlinde, "On the origin of gravity and the laws of Newton," *JHEP* **04** (2011) 029, arXiv:1001.0785 [hep-th].

[21] L. Onsager and S. Machlup, "Fluctuations and irreversible processes," *Phys. Rev.* **91** (Sep. 1953) 1505–1512. https://link.aps.org/doi/10.1103/PhysRev.91.1505.

[22] S. Machlup and L. Onsager, "Fluctuations and irreversible process. II. Systems with kinetic energy," *Phys. Rev.* **91** (Sep. 1953) 1512–1515. https://link.aps.org/doi/10.1103/PhysRev.91.1512.

[23] J. Ambjorn, A. Gorlich, S. Jordan, J. Jurkiewicz and R. Loll, "CDT meets Horava-Lifshitz gravity," *Phys. Lett. B* **690** (2010) 413–419, arXiv:1002.3298 [hep-th].

[24] C. Anderson, S. J. Carlip, J. H. Cooperman, P. Horava, R. K. Kommu and P. R. Zulkowski, "Quantizing Horava-Lifshitz gravity via causal dynamical triangulations," *Phys. Rev. D* **85** (2012) 044027, arXiv:1111.6634 [hep-th].

[25] J. Ambjørn, L. Glaser, Y. Sato and Y. Watabiki, "2D CDT is 2D Hořava–Lifshitz quantum gravity," *Phys. Lett. B* **722** (2013) 172–175, arXiv:1302.6359 [hep-th].

[26] J. Ambjorn, J. Jurkiewicz and R. Loll, "Emergence of a 4-D world from causal quantum gravity," *Phys. Rev. Lett.* **93** (2004) 131301, arXiv:hep-th/0404156.

[27] T. Jonsson and J. F. Wheater, "The spectral dimension of the branched polymer phase of two-dimensional quantum gravity," *Nucl. Phys. B* **515** (1998) 549–574, arXiv:hep-lat/9710024.

[28] J. Ambjorn, J. Jurkiewicz and Y. Watabiki, "On the fractal structure of two-dimensional quantum gravity," *Nucl. Phys. B* **454** (1995) 313–342, arXiv:hep-lat/9507014.

[29] T. P. Sotiriou, M. Visser and S. Weinfurtner, "Spectral dimension as a probe of the ultraviolet continuum regime of causal dynamical triangulations," *Phys. Rev. Lett.* **107** (2011) 131303, arXiv:1105.5646 [gr-qc].

[30] T. P. Sotiriou, M. Visser and S. Weinfurtner, "From dispersion relations to spectral dimension – and back again," *Phys. Rev. D* **84** (2011) 104018, arXiv:1105.6098 [hep-th].

[31] G. Calcagni, "Cosmology of the Lifshitz universe," *JHEP* **09** (2009) 112, arXiv:0904.0829 [hep-th].

[32] E. Kiritsis and G. Kofinas, "Horava-Lifshitz cosmology," *Nucl. Phys. B* **821** (2009) 467–480, arXiv:0904.1334 [hep-th].

[33] R. Brandenberger, "Matter bounce in Horava-Lifshitz cosmology," *Phys. Rev. D* **80** (2009) 043516, arXiv:0904.2835 [hep-th].

[34] E. N. Saridakis, "Horava-Lifshitz dark energy," *Eur. Phys. J. C* **67** (2010) 229–235, arXiv:0905.3532 [hep-th].

[35] S. Mukohyama, "Dark matter as integration constant in Horava-Lifshitz gravity," *Phys. Rev. D* **80** (2009) 064005, arXiv:0905.3563 [hep-th].

[36] S. Maeda, S. Mukohyama and T. Shiromizu, "Primordial magnetic field from non-inflationary cosmic expansion in Horava-Lifshitz gravity," *Phys. Rev. D* **80** (2009) 123538, arXiv:0909.2149 [astro-ph.CO].

[37] S. Mukohyama, "Horava-Lifshitz cosmology: a review," *Class. Quant. Grav.* **27** (2010) 223101, arXiv:1007.5199 [hep-th].

[38] B. Chen, S. Pi and J.-Z. Tang, "Scale invariant power spectrum in Horava-Lifshitz cosmology without matter," *JCAP* **08** (2009) 007, arXiv:0905.2300 [hep-th].

[39] B. Famaey and S. McGaugh, "Modified Newtonian dynamics (MOND): observational phenomenology and relativistic extensions," *Living Rev. Rel.* **15** (2012) 10, arXiv:1112.3960 [astro-ph.CO].

[40] C. Vafa, "The String landscape and the swampland," arXiv:hep-th/0509212.

16

Renate Loll

Professor in Theoretical Physics at the Institute for Mathematics, Astrophysics and Particle Physics of the Radboud University, Nijmegen, and head of the Department of High-Energy Physics, the Netherlands

Date: 8 October 2020. Via Zoom. Last edit: 5 December 2020

What are the main problems in theoretical physics at the moment?

In high-energy theoretical physics the main question is to understand quantum gravity. Then there are other problems such as what is the nature of dark energy (or the cosmological constant), which appears to be a fundamental problem, and what is the nature of dark matter, for which there might be a good particle physics explanation, but which could also necessitate a more fundamental understanding.

Weinberg has given an explanation for dark energy [1]. Is it not a good explanation?

I think we are looking for a more fundamental understanding of dark energy that does not rely on anthropic or stochastic arguments. Anthropic reasoning makes me think that I should just pack my bags and go home! This attitude is not the reason why I started working in theoretical physics. Some anthropic arguments are interesting and should be examined closer, but they do not help us to make progress in quantum gravity. I am also not fully convinced of the problem itself. It may be that we are just misinterpreting the cosmological data. In any case, whether or not we need quantum gravity to explain dark energy is not yet settled.

So what do we need quantum gravity for?

There is clearly a reasonable set of questions involving space, time and gravity at very short distances that we would like to have answers to. The standard application of quantum gravity is to early universe physics. If you go back in the history of our universe, you begin accessing higher and higher energies at smaller and smaller distances. Why should physics stop there? That seems unlikely. In fact, just extrapolating from everything we have learned in the past about the nature of physics, quantum gravity should be there. Of course, we could be deceiving ourselves and nothing can be said about physics at 10^{-35} meters.

Is the problem of quantum gravity a hard problem or there are many solutions to it and we just don't know what's the right answer?

It's a hard problem. We don't really have many candidate solutions, right? It's not that everyone has their own complete quantum gravity solution in a drawer and we just have to agree which is the right one. All approaches are highly incomplete, although not all have the same degree of incompleteness. And none of them has produced any suggestion for phenomenology, so what is our reality check? Even before looking at different phenomenologies, which is a hard nut to crack, you can ask whether we can meaningfully compare different theories. The answer is that as long as they are too incomplete we cannot. It does not make sense to merely compare formalisms because they tend to have completely different starting points. It's conceivable that at the end of the day nature is kind and all these candidate theories describe the same thing, but this can only be determined by computing observables. If the same observable is computed in each theory, we can compare them irrespective of the absence of phenomenology. But even finding observables that we all agree upon and can compute is not easy. In any case, claiming to have solved the problem because the theory looks convincing and fits one's beliefs and prejudices is not how theoretical physics should work.

But why do you think it is a hard problem?

There are many ways of looking at the problem and trying to pinpoint the issue, but I think the main reason is that geometry is dynamical. Gravity is encoded in spacetime and that does not fit well with standard quantum field-theoretic quantisation procedures and perturbative methods, which are useful and powerful elsewhere. There are also technical issues. Already classically, gravity is a very complicated and nonlinear theory and we haven't yet understood all of its implications. For instance, we are still trying to describe the detailed dynamics of black hole collisions, throwing our best numerical methods at it to understand what the theory predicts. Another technical issue, which is also encountered in the context of gauge theories, is gauge fixing and extracting gauge-invariant physics. In the case of gravity, where the role of the gauge group is played by the diffeomorphisms, the problem is even worse. It would therefore be rather naive to expect that solving quantum gravity will be given by a one-liner, with everything falling into place. In addition, the absence of guidance from experiment clearly hampers us in finding the correct answer. It means we don't have enough reality checks for our own wild ideas.

Okay, but you obviously have some favourite approach to this problem, I suppose?

Yes, I only have one life! For the first 10 years after my PhD I worked on loop quantum gravity [2–6], which at the time really was *the* new and exciting approach. I eventually got frustrated trying to make sense of the Hamiltonian constraint and also felt a growing need for a reality check for some of the formal manipulations. I became interested in the approach of dynamical triangulations (DT) to quantum gravity [7], where one has immensely powerful numerical lattice tools, which seemed perfect for that purpose. At the very least, by using such methods one could check whether or not a theory is on the right track. This led me to a Lorentzian formulation of DT quantum gravity, now referred to as causal dynamical triangulations (CDT), which is amenable to these computational techniques [8, 9].

But why do you prefer CDT to loop quantum gravity (LQG)?

As I mentioned, there are serious difficulties in understanding quantum gravity in a non-perturbative, Planckian regime. To access this regime directly one needs effective non-perturbative computational methods, which are still lacking in many approaches. My other motivation is that I do not think that we can get much further within a canonical quantisation programme like LQG. People first tried to canonically quantise gravity using the metric as the basic variable and got bogged down in factor-ordering and renormalisation issues. Rephrasing everything in terms of loops and holonomies, as in LQG, allows you to go further but you still have the dynamics and the Hamiltonian constraint to deal with, which to date remain unsolved problems.

But what about covariant approaches to LQG, such as spin foams?

The community working on canonical quantum gravity has to a large extent given up on the old canonical approach and switched to a more covariant path-integral approach. I am sympathetic to this shift but we have at this point already developed the simpler approach of CDT, where we have access to numerical methods and have already gone much further. I will eventually retire, so my goal is to get a grip on the problem within my lifetime. So far, CDT has not encountered any roadblocks.

You mentioned that you were originally interested in DT. As far as I know DT had also encountered roadblocks. How does CDT overcome these roadblocks?

The essential problem with DT is the lack of higher-order phase transitions. The initial claim of the existence of a second-order phase transition in four spacetime dimensions was subsequently checked by several groups. Based on improved numerical data, they concluded that there is most likely only a first-order phase transition between the two highly degenerate phases of the theory, with no evidence of an extended spacetime [10, 11]. The community threw everything but the kitchen sink at this problem, changing the measure, the action and the gluing rules, but nothing seemed to alter this negative conclusion.

CDT, which at the time was called *Lorentzian dynamical triangulations*, was motivated by trying to resolve this issue [8, 12, 13]. Initially, we wanted to understand how to technically perform the path integral over Lorentzian structures. We began by tackling the two-dimensional case. In fact, I remember at the time explaining the CDT idea to a local expert who told me, "Well, look, . . . it will turn out to be either trivial or equivalent to something we already know." I said, "Thank you very much," but kept pursuing the idea together with a handful of people. We constructed a Lorentzian path integral that was generally inequivalent to the Euclidean one, and not trivially related to it by just sticking imaginary factors of "*i*" in various places. We explicitly demonstrated in two dimensions that the critical exponents in both models are different [8, 14–17]. This was done analytically by taking the Lorentzian model, Wick-rotating it, looking at simple observables such as the spectral dimension, and comparing them to the Euclidean DT model (or Liouville gravity). The observables in both models do not agree.

Why is this the case?

It happens because the configuration space of geometries you're taking the path integral over is different in the Lorentzian and Euclidean settings. Their continuum limits lie in different universality classes.

Why is it important to find second-order or higher-order phase transitions as opposed to first-order or none?

This follows the logic of the standard Wilsonian picture of obtaining a continuum quantum field theory by approaching a critical point that is associated with a second- or higher-order phase transition of the underlying lattice system. Such a limit is characterised by a divergent correlation length in terms of lattice units, and the fact that the dependence on arbitrary details of the regularisation procedure drops out. It appears that Wilson's ideas can also be applied in the context of gravity, by taking into account diffeomorphism-invariance and adopting a suitable, geometric interpretation of what is meant by a correlation length [18, 19].

What are the ingredients that you need to introduce in order to make sense of the Lorentzian path integral?

The approach is Lorentzian in nature, which means that the geometry of the simplicial building blocks is flat Minkowskian instead of flat Euclidean. The gluing rules used to assemble the building blocks into a curved manifold are such that one obtains a well-defined light-cone structure. For instance, you don't allow the gluing of two building blocks if their light-cones at the shared face would meet at right angles. The idea of such a rule is to obtain a well-defined notion of a temporal flow. Secondly, you disallow spatial topology changes as a function of time; that is, you impose a piecewise flat analogue of what in the continuum is called *global hyperbolicity*. This implies, for example, that configurations with certain types of baby universes – small regions of spacetime branching off in the time direction – are not allowed in the path integral. As it turns out, in any spacetime dimension we have explored (two, three and four) imposing this set of conditions yields inequivalent quantum gravity models compared to their Euclidean counterparts [9, 20].

I suppose that the rule that ensures that the light-cone structure is preserved is the origin of the word *causal* in *causal dynamical triangulations*?

Yes, that is correct. At the regularised level, all configurations have a well-defined *causal structure*. This should not be confused with the notion of *causality*, which is a property you associate with matter propagating on spacetime. The presence of a causal structure is a prerequisite that enables you to determine whether given physical processes happen causally, that is, in accordance with cause and effect.

Does the fact that you don't allow for topology change have a big impact? And how can you justify such an assumption?

It does have a huge impact, which relatively speaking is biggest in two spacetime dimensions. A priori, there is no way to justify any particular choice of configuration space for the path integral. In particular, classical arguments do not necessarily hold when dealing with quantum configurations. Requiring a causal structure restricts the configuration space of piecewise flat geometries significantly, but does not eliminate local curvature degrees of freedom. If one wants to get interesting non-perturbative physics, one must be careful not to accidentally kill any degrees of freedom. A posteriori our choice to go causal is justified because it is so far the only known way to tame entropic degeneracies and obtain interesting results and a classical limit in four spacetime dimensions [21–25].

Does this condition mean that you won't get spacetimes with closed timelike curves?

In a trivial sense we do have closed timelike curves, because we usually perform numerical simulations by imposing periodic boundary conditions in time, just for technical simplicity. But let me rephrase the question: Can an arbitrary physical or unphysical solution of the Einstein equations appear as the ground state of a non-perturbative formulation of quantum gravity, perhaps with suitable boundary conditions? I don't see why this should be the case. Having said this, you should keep in mind that this is a lattice approach: whatever conditions we impose, at the end of the day we still need to take a non-trivial scaling limit. And we know that new and unexpected properties can appear in this limit.

There are examples in the CDT setup that illustrate this point. For instance, we fix the spacetime topology in our simulations, which at first sight looks like a rather constraining input. However, because the geometry is dynamical, the system can be driven dynamically to a corner of phase space where it effectively has a different topology. This is exactly what happens in one of the phases of the CDT simulations in four spacetime dimensions! We usually work with configurations of topology $S^3 \times S^1$, but as the system evolves dynamically it appears to want to pinch off and change to an S^4, compatible with a Euclidean de Sitter universe [23, 24, 26]. It illustrates that in non-perturbative geometric models there can be strong dynamical effects, sometimes called *entropic effects*, which drive the system to unexpected parts of phase space with properties that were not originally put into the theory. Another example is a generalised version of CDT quantum gravity in three spacetime dimensions based on configurations of topology $S^2 \times [0, 1]$, where in one of the phases the dominant geometries are driven towards developing local wormholes in the continuum limit, which are associated with topology change [27]. On the other hand, it is difficult to make sense computationally of a sum over topologies even in the two-dimensional case, which means that we are limited in our choice of configuration space in the path integral to include only piecewise flat manifolds of a given topology.

You mentioned that CDT was a lattice approach. What does this really mean? Does it mean that you start with the Einstein-Hilbert action plus a cosmological constant and artificially discretise it?

The general philosophy we follow is analogous to lattice QCD. In CDT, we discretise the curvature degrees of freedom, but not spacetime itself. Spacetime is still continuous and composed of small Minkowski-flat building blocks. This is in contrast with the causal set approach where spacetime consists of discrete "atoms of spacetime" [28]. In CDT, we assemble piecewise flat manifolds (or simplicial manifolds) from four-simplices, which locally look just like pieces of standard Minkowski space. The curvature assignments come in terms of discrete bits of deficit angles. As in lattice QCD one introduces an ultraviolet cutoff, given by the edge length of the building blocks. One must then send this cutoff to zero while renormalising all coupling constants and search for interesting scaling limits and critical properties of the system, which a priori are not guaranteed to exist. Our ambition is to obtain a fundamental theory that describes gravitational physics on all scales, including the Planck scale and beyond. Of course, in quantum gravity we would already be happy to have a computational tool for evaluating observables close to the Planck scale. There is good evidence that CDT does provide this, but to claim that it is *the* fundamental theory of quantum gravity that works on all scales requires among other things a more complete understanding of the renormalisation group flows in phase space and how they approach the second-order phase transition lines we have found [9, 18, 19].

The cosmological constant is added by hand to the action?

All models based on dynamical triangulations need a positive bare cosmological constant, independent of the metric signature and the spacetime dimension. This is needed in the path integral to counterbalance the exponential growth of configurations as a function of the spacetime volume. The CDT path integral is finite as long as the number of building blocks is finite, but in the limit where this number is taken to infinity you need a term of the form $e^{-\Lambda V}$, where V is the volume, to make the path integral well defined [20]. It is straightforward to show that not only the bare cosmological constant must be positive but also the renormalised cosmological constant, with its magnitude being a free parameter. The theory does not make any prediction about the magnitude but it does correctly predict the sign of the cosmological constant observed in nature.

You mentioned that the cosmological constant was a fundamental puzzle, but using CDT, it seems that you will never be able to explain it, right?

Yes, it will remain as a free coupling after renormalisation. There could be some mechanism that fixes its magnitude but right now it's not very obvious what that should be.

What justifies you to take the Einstein-Hilbert action plus a cosmological constant in the first place? You could consider corrections to this action, right?

We know that the Einstein-Hilbert action is the right description at low energies. However, starting with the bare Einstein-Hilbert action does not mean that higher-derivative terms are excluded from the dynamics; in fact, they are generated during renormalisation. The non-perturbative path integral receives an obvious contribution from the bare action, but there are of course competing contributions arising from the measure (or the entropy, the number of states for a given value of the action). Higher-derivative terms are therefore included automatically but they are not associated with tunable bare couplings. We follow Occam's razor by starting with a minimal number of coupling constants in the bare action. Then we analyse the resulting phase diagram and ask all the questions we want to ask from our quantum gravity theory. If we hadn't seen anything interesting at this stage, we would have considered other possibilities such as adding a curvature-squared term to the action and exploring the corresponding enlarged phase space. In fact, this latter approach was taken by the DT community in the past, but in their case has not led to the discovery of any second-order phase transitions.

Right, but generating higher-derivative terms during renormalisation from the bare action is still physically different than allowing for extra couplings, right?

That depends on the nature of those couplings. What I am saying is that even without putting such terms into the bare action we would still generate higher-derivative terms and you could see their effect in specific observables. As I said, there has not been much motivation to include explicit higher-curvature terms in four-dimensional CDT, but we would expect a high degree of universality. This has been verified in considerable detail in two spacetime dimensions, where one can check explicitly that taking different building blocks or adding higher-curvature terms does not make a difference [29].

Okay, so in principle I could take any theory of gravity, such as supergravity, with all kinds of matter, introduce the regulator, take a scaling limit and your method could potentially work. Thus only experiment could fix the correct action?

Yes, but these are all just theoretical possibilities. If you look at them closely you will realise that many don't work for one reason or another, for example, since supersymmetry is not easily realised on a lattice. If you start with rich models with tons of parameters you would at the end of the day have to perform many experiments to fix all the parameters. To date, we don't have any kind of phenomenological observable in any quantum gravity theory that has been fixed by some experiment, which suggests that perhaps this is not the best way to go.

How does CDT resolve the issue that gravity is perturbatively non-renormalisable?

It resolves it by looking for non-perturbative renormalisability and the most concrete idea of how this can be accomplished is asymptotic safety [30]. This idea is based on concepts from standard quantum field theory, where one aims at finding renormalisation group trajectories in the ultraviolet that run into a non-trivial fixed point and hence render the theory asymptotically safe.

So is CDT asymptotically safe, or you expect it to be?

There is a continuum asymptotic safety program that has found evidence for the existence of fixed points [31]. In practice, modulo technical issues, CDT should be able to verify whether or not such fixed points in the ultraviolet exist using lattice methods. Setting up renormalisation group flows in a non-perturbative, background independent regime on a lattice is tricky but we have shown that it is possible in principle [18]. In this pilot study we found renormalisation group trajectories, defined as lines of constant physics in phase space. Currently, we do not have enough observables to reliably determine these trajectories and there are technical issues in following them close to the phase transitions. However, we believe that sooner or later we will be able to provide an independent check for the existence of these fixed points using our approach, which is complementary to what the asymptotic safety programme has been pursuing with truncated actions.

So your point of view is that CDT is the lattice version of *the theory* while the asymptotic safety programme is the continuum version?

If there is *the theory* then there will be many ways of studying it. Once you have a better idea of what this theory might be, everyone will look at it using their own methods. Asymptotic safety is one concrete scenario for non-perturbative renormalisability. However, once we are able to understand better the physics near the phase transitions, CDT may suggest a different Planckian description. Perhaps at such scales the theory is topological or low-dimensional, with a different flavour than asymptotic safety.

So you are saying that CDT could suggest a different scenario for non-perturbative renormalisability?

Yes, or perhaps it will lead to a concrete suggestion for how the theory needs to be completed at high energies or what a suitable effective description of the Planckian dynamics is. I think our imagination is a bit limited in this department, so I would leave this option open. We will have a better understanding once we study how different observables behave at and near the Planck scale.

So at this point there isn't a verification that CDT and the asymptotic safety programme are the same?

No, this is work in progress. What we have in CDT is a computational framework that can be used for measuring observables, the spectral dimension and a few others, with some degree of measuring accuracy in a short-distance regime, within a factor five or 10 from the Planck scale. If you have a different formulation of quantum gravity, you can evaluate that same observable within the same short-distance regime and compare. This may already give us important information even before we have a solid grasp of quantum gravity at the Planck scale and beyond.

Some time ago there was a paper showing the equivalence between CDT and Hořava-Lifshitz gravity in two spacetime dimensions [32]. Should one expect this to hold in four spacetime dimensions?

No, it doesn't hold. In two dimensions there are only two universality classes for pure gravity: DT or Liouville gravity and CDT. When we first studied CDT in two dimensions we computed its propagator and found that it is the same as that of a proper-time description of Euclidean two-dimensional gravity [33]. Hořava-Lifshitz gravity in two dimensions is also in that same universality class and not in the Liouville universality class. Is this also the case in higher dimensions? It is a justified question but it doesn't seem to be the case. First of all, the counting of degrees of freedom in CDT and Hořava-Lifshitz gravity is different in higher dimensions than in two dimensions, where neither have any local degrees of freedom. Breaking full diffeomorphism invariance to foliation-preserving diffeomorphism invariance in Hořava-Lifshitz gravity is associated with an additional scalar degree of freedom. This is not the case in CDT quantum gravity, where there is no such breaking of diffeomorphism symmetry [9].

We were quite excited when we learned from Hořava that in his models (which are now called Hořava-Lifshitz gravity) he had found a similar dimensional reduction for the spectral dimension [34, 35] as we had some years earlier [36]. Together with the foliation in his theory, which morally speaking appeared to be close to our lattice formulation, this suggested that perhaps Hořava-Lifshitz gravity was the continuum version of CDT. We of course looked into this, but were not very successful in establishing a more specific connection. As time passed we began to realise that dimensional reduction seems to be very generic, and that many approaches to quantum gravity are able to reproduce a similar effect [37]. Therefore I would say that dimensional reduction is no longer considered a piece of evidence for an equality of the two approaches.

But in CDT you also introduce a foliation, right? So how do you know whether CDT is foliation-independent; that is, how do you know if it is diffeomorphism-invariant?

In CDT, the foliation is not associated with diffeomorphism invariance or otherwise, since the formulation is explicitly coordinate- and labelling-invariant. The primary function of working with a discrete foliation is that it is a simple and convenient way to enforce a piecewise flat analogue of global hyperbolicity, which implies that the topology of spatial slices does not change in time. For Lorentzian metrics in general relativity global hyperbolicity is usually assumed and is a diffeomorphism-invariant property. There is no a priori reason to enforce it on the quantum configurations, but doing it or not basically makes the difference between CDT and DT. One can still ask whether the presence of a foliation may introduce some additional, unwanted bias on top of the desired suppression of topology changes. You can raise similar concerns in lattice QCD about whether a specific choice of lattice geometry influences the final results. Also, if you place QCD on a hypercubic Euclidean lattice it is at odds with global $SO(4)$-covariance. But if you look at observables you find that for larger lattices this covariance is restored quite fast.

Are you saying that something similar happens for CDT?

No, the details are different and one has to look carefully, but it has a similar flavour. All I am saying is that it is too naive to claim that if you have a lattice structure it is necessarily incompatible with diffeomorphism invariance or some other continuum symmetry. One needs to understand whether a given symmetry plays a role at the microscopic level and check whether and how it is recovered in a continuum limit. In background independent quantum gravity, these questions must be phrased and studied in terms of invariant observables. You must check whether any observable provides you with evidence that, say, local Lorentz symmetry is broken at some scale. This also requires some degree of control over the continuum limit, including the Planckian regime. But does one really expect Lorentz transformations to have a meaningful operational implementation at the Planck scale? I doubt it! We don't have little elevator experiments that we can perform at the Planck scale, but if we zoom out to some scale where geometry can be seen as approximately classical, then you can perform such elevator experiments. Will they tell us that CDT is incompatible with Lorentz invariance? I certainly wouldn't hope so, but I don't know. However, we do now have non-trivial evidence from measuring the new, direction-sensitive Ricci curvature observable that in the emergent de Sitter universe the symmetry between time and spatial directions is restored at the level of quasi-local geometry [38], which is promising and is reminiscent of symmetry restoration in lattice gauge theory.

Another thing we have tried to do at the level of the lattice theory is to disentangle the imposition of a well-behaved causal structure from the presence of a foliation. Together with a former PhD student of mine, we worked on a generalised version of CDT, which has more building blocks than the standard version, but allows us to suppress spatial topology changes without having a global foliated structure [39, 40]. It is a more complicated model but we have studied it extensively in three spacetime dimensions and found very gratifying results. In particular, we could reproduce the de Sitter–like phase previously seen in the standard CDT formulation. The fact that this is a slicing-relaxed version of CDT is good evidence that ultimately the foliation doesn't affect the physics much, since you appear to land in the same universality class. Performing the same analysis in four dimensions is out of reach for technical reasons, but I also wouldn't say that this is the most pressing or interesting thing that CDT quantum gravity has to investigate at this stage.

You mentioned several times the necessity of finding well-defined scaling limits. Is it clear that CDT has such scaling limits?

It's clear that we have higher-order phase transitions [41–43], which are prime candidates for scaling limits and which other discrete quantum gravity approaches have not seen. This means there is a clear route forward and one has to examine these candidates closer. As I mentioned earlier, there are technical difficulties that need to be overcome in reliably measuring the geometry close to the phase transitions. These are well-known and expected lattice effects, but they require extra attention in the context of the dynamical lattices we are dealing with. Standard Monte Carlo techniques need to be adapted and optimised, which ultimately requires a good understanding of the nature of the underlying quantum geometry. Identifying new geometrical observables to help us do this is work in progress; in fact, we recently found a promising new ingredient in the form of a generalised notion of Ricci curvature [44, 45].

What do you exactly look for in your simulations?

The way to think about the simulations is as an experimental lab. The data we look at are the expectation values of quantum observables. Due to the nature of quantum gravity, these observables are typically highly nonlocal and characterise the physics of the non-perturbative vacuum (or ground state) of the quantum-fluctuating geometric ensemble. First you have to think about what are meaningful observables and then implement them in the CDT setup.

The question of observables is a question that concerns any theory of quantum gravity, not only CDT. The key issue is how to deal with the diffeomorphism invariance in the quantum theory. Tullio Regge's famous old paper entitled "General Relativity without Coordinates" [46] is, from the point of view of quantum gravity, a vastly underappreciated idea of how to get rid of coordinates and describe curved geometries in an approximate way in terms of the edge lengths of simplicial manifolds. This means, in fact, that in this description diffeomorphisms are nowhere to be found. Distinct piecewise flat manifolds represent distinct curved geometries. Classically, Einstein's equations can be obtained to lowest order in the edge length [46], and some people have explored this idea as a numerical approximation tool for general relativity. However, it doesn't seem to offer great advantages.

The full power of Regge's idea only becomes apparent in the quantum theory, perhaps not so much in the quantum Regge calculus version [7], but in the DT and CDT versions, which do not have any residual gauge redundancy. In CDT quantum gravity, the building blocks are equilateral up to a binary distinction between spacelike and timelike edges and there are no coordinates involved. In the simulations, we do need to refer to individual building blocks by giving them labels, but the Monte Carlo algorithm is set up in a way to ensure label independence throughout. In other words, Regge's idea allows us to describe pure geometry, avoiding any discussion of gauge-fixing, Faddeev-Popov determinants and similar technically challenging questions. Such questions are very difficult to address in continuum path integral approaches, but related issues also appear in canonical quantisation approaches, where you have a Dirac algebra that encodes diffeomorphism symmetry and you must understand whether or not it survives in the quantum theory.

Having said this, the fact that we can't talk about coordinates when doing ensemble averaging in the quantum theory also means that it does not make sense to talk about a local quantity like a curvature scalar at a coordinate point x, say. To obtain a well-defined observable in pure gravity, one instead has to integrate this scalar over the entire manifold and evaluate its expectation value on the ensemble. Typical quantum observables are therefore nonlocal quantities. Spacetime-averaged observables that have been widely studied are fractal dimensions, in particular, the Hausdorff and spectral dimension [9, 26, 36]. The spectral dimension depends on the spectral properties of a Laplace-type operator and can be used to associate an effective dimensionality with the quantum geometry at a given scale. In CDT, we found that on large scales the spectral dimension asymptotes to the classical value of four, but close to the Planck scale it is compatible with two, within measuring accuracy [36]. We termed this phenomenon *dynamical dimensional reduction.*

What does it mean that the spectral dimension is of around two close to the Planck scale?

The way to interpret it is in terms of the effective dimension that a diffusion process experiences in this quantum geometry. It is not easy to grasp intuitively and we don't understand its phenomenological implications, but it is clearly a type of behaviour that deviates from classical expectations. The curve of the spectral dimension as a function of scale includes also non-integer values, which is not something for which there is a good classical explanation. It's a true quantum signature, or at least that is how people think about this result.

But the Einstein-Hilbert action you took is four-dimensional, right?

You are absolutely right in pointing this out. In the old days when people started pursuing DT, no one would have considered that putting together four-dimensional building blocks and using the four-dimensional Einstein-Hilbert action would lead to a geometry with anything but four spacetime dimensions. Surely, quantum effects might change this on microscopic scales but without doubt spacetime should be four-dimensional on macroscopic scales. But we now know that this is the wrong expectation. Why? First, because the non-perturbative regime allows for large quantum fluctuations of the geometry, and second, because one takes a non-trivial continuum limit. This opens the door for strong entropic effects to take over and dominate the behaviour of the ensemble. For example, Euclidean DT quantum gravity exhibits such strong effects, which in half of the phase space drive the geometry towards branched polymers and in the other half produce a highly degenerate crumpling of the geometry [47]. Neither of these phases is associated with a macroscopic Hausdorff or spectral dimension of four, which is a fascinating thing. One observes in simulations in dimension larger than two that for sufficiently large inverse bare gravitational coupling constant, one always ends up in a branched-polymer phase. This is a generic feature and related to the dominance of the conformal mode in the gravitational path integral [48], and therefore is much more general than DT or CDT.

Okay. So you only have two observables that you look at, the integrated Ricci scalar and the spectral dimension?

There are a few more. Observables that have been studied a lot in non-perturbative models of quantum geometry and gravity are the Hausdorff dimension and the spectral dimension. The spectral dimension I have already discussed. The Hausdorff dimension D_H is another scaling exponent that governs the scaling of the volume of geodesic balls as a function of their radius r and that characterises a given geometry. You extract it from the leading behaviour r^{D_H} of these volumes [26]. The interesting aspect of this observable is that it does not rely on any smooth metric $g_{\mu\nu}$ in order to measure it. Instead, you only need a notion of volume, that is, a measure on the geometries, and a notion of distance.

These dimension observables have been studied for a while. They are extremely useful but not very geometrical in a general relativistic sense. Can we define an observable that is more closely related to local geometry? The first thing that comes to mind is curvature, but the problem with curvature is that it's a very divergent quantum operator, even in the continuum. Therefore, one expects that it needs to be renormalised, specifically in a non-perturbative regime. For a long time it was not clear how to do this and how to construct a renormalised version that is meaningful and finite near the Planck scale. A couple of years ago we finally came up with a concrete suggestion for implementing not just a Ricci scalar but a full-blown Ricci curvature [44, 45]. To my knowledge, this is the first time this has been proposed in non-perturbative quantum gravity. There may not be a unique way of defining a renormalised Ricci curvature but we now have a candidate, which looks very promising in four dimensions [38].

Extracting the metric is not possible?

I would say it's the wrong question. Already classically we don't go and measure components of $g_{\mu\nu}(x)$ when investigating the properties of our own universe, because they are gauge-dependent quantities. We use clocks and rods, that is, coordinate-invariant ways of describing spacetime geometry. If you want to find black hole horizons in nature you don't look for spacetime points where the coefficient of dr^2 in the Schwarzschild metric blows up. Or take numerical relativity, where one uses horizon finders that trace the behaviour of light-like geodesics. Similarly, in quantum gravity, and in particular in a non-perturbative regime, whatever statement you want to make about the universe will not involve $g_{\mu\nu}(x)$ by itself but coordinate-invariant geometric quantities.

I understood that de Sitter spacetime is the dominant contribution to the path integral in CDT [23, 24]?

What we say is that we observe some properties that are compatible with those of a classical de Sitter spacetime [23, 24]. We have been careful in making this statement because we do not want to convey the idea that we somehow found a geometry with a local $g_{\mu\nu}(x)$ that matches that of a classical de Sitter space. This is not the case. We look at a particular metric mode that is easily accessible, namely, the global conformal factor, which measures the spatial three-volume as a function of a lattice version of proper time. We then compare the behaviour of this conformal mode with that of a classical de Sitter geometry and find a beautiful match. Not only that, but we have also measured the quantum fluctuations of the three-volume and they match with a continuum computation one can do in a semiclassical mini-superspace analysis around de Sitter space.

So it is not exactly de Sitter but has some features of it?

The global conformal mode is only one mode of the metric, so we cannot claim that the geometry that emerges in the simulations is the same as that of a classical de Sitter space. Our simulations involve up to 1.5 million building blocks and yield a universe with a size of approximately 20 Planck lengths across. Since this is very small, the quantum fluctuations are very large and it is quite striking that at least one of the modes displays a semiclassical behaviour. The likely reason for this robustness is the global nature of this mode; it is just the integrated volume over a whole spatial slice. One should also keep in mind that aiming for definite statements about the local metric at such small scales probably does not make much sense.

Until very recently, we have not been able to make any further statements about the de Sitter nature of our quantum spacetime and whether or not it resembles a constant-curvature space like the classical de Sitter spacetime. However, as I mentioned before, we now have a proposal for a renormalised integrated Ricci scalar that we can measure in four-dimensional CDT [44, 45]. In an initial study we indeed found that this observable is compatible with a de Sitter spacetime [38]. The next thing to look at are quantum fluctuations of this observable, which are not expected to be small, and check to what extent they can match a semiclassical expectation.

Since you introduced by hand a cosmological constant in the action, would you expect anything else but de Sitter spacetime?

It is important to realise that non-perturbative formulations are a completely different ball game, where naive physical reasoning often goes wrong. People played around with these models for many years and their attitude was that clearly nothing but four-dimensional universes could ever come out of simulations of models that use four-dimensional building blocks to start with. In the meantime, we have understood that in 99% of the cases where you take your favourite building blocks and give them some dynamics, you will find some branched-polymer phase that has nothing to do with higher-dimensional physics.

Once you consider non-perturbative quantum dynamics, you allow large fluctuations on short scales and in principle, even when you zoom out, you shouldn't expect to find any signatures of classical geometry. There are very strong universal, entropic effects that drive these systems to branched-polymer or other degenerate phases which causal sets also encounter [28], as well as tensor models [49]. The lesson is that our classical intuition is worth nothing when looking at non-perturbative quantum gravity. What I find particularly rewarding in this story is that in order to get these very non-trivial results – like the emergence of de Sitter space – we do not need to add any exotic ingredients or depart radically from geometric degrees of freedom. Instead, we take the simplest possible approach to the problem and try to see how little we can get away with to see something interesting, without introducing such ingredients.

Is there any picture coming out of CDT for what spacetime looks like at the Planck scale? Is it continuous or discrete?

The short answer is we don't know; we need more observables to probe it. To answer it you need to have an operational way of defining fundamental discreteness in your formalism. This is different from using discrete ingredients to construct your Planckian theory or simply postulating that there are no distances shorter than the Planck length. There is scant evidence for such claims, which are often driven by wishful thinking and based on classical reasoning and intuition. Instead, you should define what you mean by fundamental discreteness and rephrase it in terms of observables. For instance, in kinematical loop quantum gravity, you have a volume operator which you can study in a non-perturbative regime and which has a discrete spectrum [5, 6, 50]. This is a well-defined operator that suggests the existence of a fundamental discreteness, regardless of what kind of phenomenology it may or may not lead to. Some people argue that the world must be discrete because the continuum real line is an abstraction invented by us for convenience, whereas actual meter readings are not of this kind, but I don't find this a convincing argument. This question needs to be formulated in terms of suitable quantum observables.

What do you think has been the biggest breakthrough in theoretical physics in the past 30 years?

If we consider high-energy theoretical breakthroughs, I don't see any. We have found the Higgs particle but that had been theoretically predicted a long time ago. Do you see anything that could be considered a breakthrough?

I'm obviously biased; I think that the AdS/CFT correspondence [51] is a breakthrough, not only because of what it teaches us about quantum gravity (which you might disagree with) but also because it is a widespread tool in theoretical physics (which is hard to disagree with).

I think it's very hard to measure progress, not just in quantum gravity, but in the whole of theoretical high-energy physics. What is progress? If you take some particular approach you can have an objective notion of technical progress within that approach, but that is usually a long shot from understanding more about nature. If I adopt this higher standard, I really think there hasn't been any theoretical breakthrough in the past 30 years.

Why have you chosen to do physics?

I chose physics because it was supposed to be the most challenging subject. When I was 18 years old, physics was one of the many things I was interested in. With so much choice, getting into physics seemed to be the smart thing to do because it was considered the hardest thing you could pursue. As I continued, I found it very interesting that you could push the boundaries of knowledge by tackling very complex and interesting questions, such as quantum gravity. No regrets here.

What do you think is the role of the theoretical physicist in modern society?

We are part of a tradition where we use our limited tools to learn a lot about the physical world and push the boundaries of knowledge on behalf of humanity, despite not knowing in advance whether it will be successful. In this sense, we are just part of the larger human endeavour to understand nature. Additionally, it is also our task to educate people to think analytically and to provide a world view that is evidence-based. This helps people to understand the world and make progress as a society. There is also value in advocating the attitude we take toward science, specially in current times where we seem to live in partisan societies. As a community, we can work across boundaries and are forced to talk to people with different backgrounds, sets of attitudes and prejudices. We have learnt that this is possible and that we can agree on rules for how to communicate and make progress together. The best examples of this are places like CERN, which can be considered a blueprint for how many nations can work together with a common goal in a meaningful way.

References

[1] S. Weinberg, "Anthropic bound on the cosmological constant," *Phys. Rev. Lett.* **59** (Nov. 1987) 2607–2610. https://link.aps.org/doi/10.1103/PhysRevLett.59.2607.

[2] R. Loll, "Loop approaches to gauge field theory," *Theor. Math. Phys.* **93** (1992) 1415–1432.

[3] A. Ashtekar and R. Loll, "New loop representations for (2+1) gravity," *Class. Quant. Grav.* **11** (1994) 2417–2434, arXiv:gr-qc/9405031.

[4] R. Loll, "Nonperturbative solutions for lattice quantum gravity," *Nucl. Phys. B* **444** (1995) 619–640, arXiv:gr-qc/9502006.

[5] R. Loll, "The volume operator in discretized quantum gravity," *Phys. Rev. Lett.* **75** (1995) 3048–3051, arXiv:gr-qc/9506014.

[6] R. Loll, "Spectrum of the volume operator in quantum gravity," *Nucl. Phys. B* **460** (1996) 143–154, arXiv:gr-qc/9511030.

[7] R. Loll, "Discrete approaches to quantum gravity in four-dimensions," *Living Rev. Rel.* **1** (1998) 13, arXiv:gr-qc/9805049.

[8] J. Ambjørn and R. Loll, "Nonperturbative Lorentzian quantum gravity, causality and topology change," *Nucl. Phys. B* **536** (1998) 407–434, arXiv:hep-th/9805108.

[9] R. Loll, "Quantum gravity from causal dynamical triangulations: a review," *Class. Quant. Grav.* **37** no. 1 (2020) 013002, arXiv:1905.08669 [hep-th].

[10] P. Bialas, Z. Burda, A. Krzywicki and B. Petersson, "Focusing on the fixed point of 4-D simplicial gravity," *Nucl. Phys. B* **472** (1996) 293–308, arXiv:hep-lat/9601024.

[11] S. Catterall, R. Renken and J. Kogut, "Singular structure in 4-D simplicial gravity," *Phys. Lett. B* **416** (1998) 274–280, arXiv:hep-lat/9709007.

[12] J. Ambjørn, J. Jurkiewicz and R. Loll, "A nonperturbative Lorentzian path integral for gravity," *Phys. Rev. Lett.* **85** (2000) 924–927, arXiv:hep-th/0002050.

[13] J. Ambjørn, J. Jurkiewicz and R. Loll, "Dynamically triangulating Lorentzian quantum gravity," *Nucl. Phys. B* **610** (2001) 347–382, arXiv:hep-th/0105267.

[14] J. Ambjørn, R. Loll, J. Nielsen and J. Rolf, "Euclidean and Lorentzian quantum gravity: lessons from two-dimensions," *Chaos Solitons Fractals* **10** (1999) 177–195, arXiv:hep-th/9806241.

[15] J. Ambjørn, K. Anagnostopoulos and R. Loll, "A new perspective on matter coupling in 2-D quantum gravity," *Phys. Rev. D* **60** (1999) 104035, arXiv:hep-th/9904012.

[16] J. Ambjørn, J. Jurkiewicz and R. Loll, "Lorentzian and Euclidean quantum gravity – analytical and numerical results," *NATO Sci. Ser. C* **556** (2000) 381–450, arXiv:hep-th/0001124.

[17] J. Ambjørn, J. Correia, C. Kristjansen and R. Loll, "On the relation between Euclidean and Lorentzian 2-D quantum gravity," *Phys. Lett. B* **475** (2000) 24–32, arXiv:hep-th/9912267.

[18] J. Ambjørn, A. Görlich, J. Jurkiewicz, A. Kreienbuehl and R. Loll, "Renormalization group flow in CDT," *Class. Quant. Grav.* **31** (2014) 165003, arXiv:1405.4585 [hep-th].

[19] J. Ambjørn, J. Gizbert-Studnicki, A. Görlich, J. Jurkiewicz and R. Loll, "Renormalization in quantum theories of geometry," *Front. Phys.* **8** (2020) 247, arXiv:2002.01693 [hep-th].

[20] J. Ambjørn, A. Görlich, J. Jurkiewicz and R. Loll, "Nonperturbative quantum gravity," *Phys. Rept.* **519** (2012) 127–210, arXiv:1203.3591 [hep-th].

[21] J. Ambjørn, J. Jurkiewicz and R. Loll, "Emergence of a 4-D world from causal quantum gravity," *Phys. Rev. Lett.* **93** (2004) 131301, arXiv:hep-th/0404156 [hep-th].

[22] J. Ambjørn, J. Jurkiewicz and R. Loll, "Semiclassical universe from first principles," *Phys. Lett. B* **607** (2005) 205–213, arXiv:hep-th/0411152 [hep-th].

[23] J. Ambjorn, A. Gorlich, J. Jurkiewicz and R. Loll, "Planckian birth of the quantum de Sitter universe," *Phys. Rev. Lett.* **100** (2008) 091304, arXiv:0712.2485 [hep-th].

[24] J. Ambjørn, A. Görlich, J. Jurkiewicz and R. Loll, "The nonperturbative quantum de Sitter universe," *Phys. Rev. D* **78** (2008) 063544, arXiv:0807.4481 [hep-th].

[25] J. Ambjørn, A. Görlich, J. Jurkiewicz, R. Loll, J. Gizbert-Studnicki and T. Trzesniewski, "The semiclassical limit of causal dynamical triangulations," *Nucl. Phys. B* **849** (2011) 144–165, arXiv:1102.3929 [hep-th].

[26] J. Ambjørn, J. Jurkiewicz and R. Loll, "Reconstructing the universe," *Phys. Rev. D* **72** (2005) 064014, arXiv:hep-th/0505154 [hep-th].

[27] J. Ambjørn, J. Jurkiewicz, R. Loll and G. Vernizzi, "Lorentzian 3-D gravity with wormholes via matrix models," *JHEP* **09** (2001) 022, arXiv:hep-th/0106082.

[28] S. Surya, "The causal set approach to quantum gravity," *Living Rev. Rel.* **22** no. 1 (2019) 5, arXiv:1903.11544 [gr-qc].

[29] P. Di Francesco, E. Guitter and C. Kristjansen, "Integrable 2-D Lorentzian gravity and random walks," *Nucl. Phys. B* **567** (2000) 515–553, arXiv:hep-th/9907084.

[30] S. Weinberg, "Ultraviolet divergences in quantum gravity," in *General Relativity: An Einstein Centenary Survey*, S. W. Hawking and W. Israel, eds., pp. 790–831. Cambridge University Press, 1979.

[31] M. Reuter and F. Saueressig, *Quantum Gravity and the Functional Renormalization Group: The Road towards Asymptotic Safety*. Cambridge Monographs on Mathematical Physics. Cambridge University Press, 2019. https://books.google.dk/books?id=L2OptQEACAAJ.

[32] J. Ambjørn, L. Glaser, Y. Sato and Y. Watabiki, "2D CDT is 2D Hořava–Lifshitz quantum gravity," *Phys. Lett. B* **722** (2013) 172–175, arXiv:1302.6359 [hep-th].

[33] R. Nakayama, "2-D quantum gravity in the proper time gauge," *Phys. Lett. B* **325** (1994) 347–353, arXiv:hep-th/9312158.

[34] P. Horava, "Quantum gravity at a Lifshitz point," *Phys. Rev. D* **79** (2009) 084008, arXiv:0901.3775 [hep-th].

[35] P. Horava, "Spectral dimension of the universe in quantum gravity at a Lifshitz point," *Phys. Rev. Lett.* **102** (2009) 161301, arXiv:0902.3657 [hep-th].

[36] J. Ambjørn, J. Jurkiewicz and R. Loll, "Spectral dimension of the universe," *Phys. Rev. Lett.* **95** (2005) 171301, arXiv:hep-th/0505113 [hep-th].

[37] S. Carlip, "Dimension and dimensional reduction in quantum gravity," *Class. Quant. Grav.* **34** no. 19 (2017) 193001, arXiv:1705.05417 [gr-qc].

[38] N. Klitgaard and R. Loll, "How round is the quantum de Sitter universe?," *Eur. Phys. J. C* **80** no. 10 (2020) 990, arXiv:2006.06263 [hep-th].

[39] S. Jordan and R. Loll, "Causal dynamical triangulations without preferred foliation," *Phys. Lett. B* **724** (2013) 155–159, arXiv:1305.4582 [hep-th].

[40] S. Jordan and R. Loll, "De Sitter universe from causal dynamical triangulations without preferred foliation," *Phys. Rev. D* **88** (2013) 044055, arXiv:1307.5469 [hep-th].

[41] J. Ambjørn, S. Jordan, J. Jurkiewicz and R. Loll, "A second-order phase transition in CDT," *Phys. Rev. Lett.* **107** (2011) 211303, arXiv:1108.3932 [hep-th].

[42] J. Ambjørn, S. Jordan, J. Jurkiewicz and R. Loll, "Second- and first-order phase transitions in CDT," *Phys. Rev. D* **85** (2012) 124044, arXiv:1205.1229 [hep-th].

[43] D. Coumbe, J. Gizbert-Studnicki and J. Jurkiewicz, "Exploring the new phase transition of CDT," *JHEP* **02** (2016) 144, arXiv:1510.08672 [hep-th].

[44] N. Klitgaard and R. Loll, "Introducing quantum Ricci curvature," *Phys. Rev. D* **97** no. 4 (2018) 046008, arXiv:1712.08847 [hep-th].

[45] N. Klitgaard and R. Loll, "Implementing quantum Ricci curvature," *Phys. Rev. D* **97** no. 10 (2018) 106017, arXiv:1802.10524 [hep-th].

[46] T. Regge, "General relativity without coordinates," *Nuovo Cim.* **19** (1961) 558–571.

[47] J. Ambjørn and J. Jurkiewicz, "Scaling in four-dimensional quantum gravity," *Nucl. Phys. B* **451** (1995) 643–676, arXiv:hep-th/9503006 [hep-th].

[48] A. Dasgupta and R. Loll, "A proper time cure for the conformal sickness in quantum gravity," *Nucl. Phys. B* **606** (2001) 357–379, arXiv:hep-th/0103186 [hep-th].

[49] R. Gurau and J. P. Ryan, "Melons are branched polymers," *Annales Henri Poincare* **15** no. 11 (2014) 2085–2131, arXiv:1302.4386 [math-ph].

[50] C. Rovelli and L. Smolin, "Discreteness of area and volume in quantum gravity," *Nucl. Phys. B* **442** (1995) 593–622, arXiv:gr-qc/9411005. [Erratum: *Nucl. Phys. B* **456** (1995) 753–754.]

[51] J. M. Maldacena, "The large N limit of superconformal field theories and supergravity," *Int. J. Theor. Phys.* **38** (1999) 1113–1133, arXiv:hep-th/9711200.

17

Juan Maldacena

Carl P. Feinberg Professor at the Institute for Advanced Study, Princeton, NJ

Date: 28 June 2011. Location: Uppsala. Last edit: 27 May 2014

In your opinion, what are the main puzzles in theoretical physics at the moment?

I think that there are puzzles in many areas of theoretical physics so I'll only discuss the ones in string theory and quantum gravity. The main puzzle is to understand the beginning of the universe, in particular, how it began and how to describe the initial singularity.

And to this aim do we need a theory of quantum gravity?

I think so; it is definitely necessary because in the beginning the universe was very small, there were large quantum fluctuations and the creation of entire regions of the universe. To describe these aspects you really need to understand quantum gravity.

Is the problem of finding a consistent theory of quantum gravity a difficult problem?

We don't have many consistent theories of quantum gravity; in fact, we really only have one and that is string theory. The other ones are only attempts and they are not at the level of string theory. There could be other theories and I don't think that one should not explore other possibilities. In fact, my talk today[1] was about the exploration of another possibility [1]. However, I think that string theory is a very concrete framework which we don't understand completely. For example, we don't understand how to describe the big bang. On the other hand, there are some other things about string theory which we understand and which give us confidence that it is on the right track.

Like what kind of things?

Situations in which we have black holes for example, or the fact that string theory is obtained from quantum field theory. String theory is not something that we have invented out of the blue; you get it from quantum field theory in a simple way through dualities [2, 3]. Large-N gauge theories naturally give you string theory [3]. One should recall that string theory was motivated by experiment, in particular by seeing strings in colliders.

[1] This interview took place during the conference Strings 2011.

But didn't it go completely in another direction?

It then went in another direction but it is interesting to recall how this happened. Strings were seen in colliders and then people tried to make a theory of weakly interacting fundamental strings [4] and that took them on a detour through 26 dimensions, then 10 dimensions and so on. But now the discovery of dualities brought it back to theories similar to QCD, like $N = 4$ super Yang–Mills. So that's not too different from the initial motivation.

Thus, first of all, string theories are naturally connected to quantum field theories and, second, it is very important to follow consistency. One should try to build consistent theories and complexify the theories while preserving their consistency, rather than making premature assumptions. For example, it would have been wrong to insist that the string theory which describes gauge theories should live in four dimensions because the right answer is that it is at least in one more dimension such as five dimensions or 10 dimensions. This was something that was understood from studying consistency conditions in higher-dimensional string theories. Gauge theories naturally produce extra dimensions when you take the large-N limit and they also produce strings. Therefore, string theory is not completely crazy, but instead it is within the rules of quantum field theory.

But do you think that we are still far from describing QCD?

Yes, if the objective is to describe QCD then we are far from it. With weakly interacting (almost free) strings you could potentially describe large-N QCD. We are also far from this but nevertheless the description of $N = 4$ super Yang–Mills theory has shed some qualitative insights into QCD, like the relation between black holes and the fluid viscosity of super Yang–Mills [5] and techniques for computing scattering amplitudes [6–10]. But one can also explore the universe of quantum field theories and not only focus on a particular theory. QCD at high energies can be tackled perturbatively and many physical processes can be computed numerically. It would be wonderful to solve large-N QCD but string theory also lets us study other theories that are interactive at high energies, which has led to new insights on quantum field theory.

Do you think that it will be possible to find a similar duality, as the AdS/CFT [3], that would describe realistic theories?

It depends on what you mean by "realistic theories." If you imagine a theory which has the matter content of the standard model but lives in four-dimensional anti–de Sitter (AdS_4) spacetime, it may be possible. But theories that have an expanding universe would require some new idea, but perhaps not so terribly new.

You mean that it is not possible to describe it using dualities?

Not using any known duality. There might be some duality that describes it. People have proposed some ideas, like dS/CFT [11] or FRW/CFT [12], which involves looking at boundaries of bubbles that contain in the interior something that approximates Minkowski space; that is, more precisely, these are open Friedmann-Lemaitre-Robertson-Walker

(FRW) solutions. So there are some ideas for how to approach this problem but these are only ideas because there is no concrete proposal for a specific duality between two specific situations. But it might be possible to find one.

One of the big criticisms made toward string theory is that it is very hard to define it in time-dependent backgrounds. Why is this difficult?

I think that this is an inherent problem in quantum gravity, if you don't have some asymptotically simple region, and by simple I mean something like the boundary of AdS for example. In AdS, if you go far away there is a well-defined time direction and an asymptotically flat space. In general, it is hard to define what the observables are, even in classical general relativity. People like Bryce DeWitt who tried to define the quantum theory in pure general relativity, by trying to naively quantise general relativity, ran into this problem [13].

This problem is not a problem that only exists in string theory. Any approach would have this problem and if it doesn't have this problem it is probably because there is something wrong with it. If anyone tells you that one can do local measurements of the volume of spacetime in a theory, it will probably not make sense because you cannot make any precise measurement of a small spacetime region. If you do a measurement within a small spacetime region you are going to be limited in your precision, which is what you expect on general grounds due to information bounds.

Do you expect that this is going to be solvable at some point within this framework?

Well, I hope it will be solved (laughs). There is no guarantee as it's research and if I knew what the solution was I would have already told you (laughs). String theory has solved other problems which were of similar difficulty as this one, like the problem of black hole entropy or the problem of strong coupling. At some point it was realised that there were dualities in string theory between weak and strong coupling and so you could understand how string theories behaved at strong coupling, which was a dream for many years [2]. The solution turned out to be fairly simple. I suspect that once we know the solution to this problem, it might be as simple as the things we understand now (laughs).

What were the steps that led you to formulate AdS/CFT [3]?

The idea arose from trying to understand the black hole information paradox in string theory. This is also one problem that looked fairly unsolvable and then it started appearing more solvable after the discovery of D-branes [14]. D-branes were somehow on the verge of being black holes and they also had a simple description so they were ideally suitable for this kind of study. People started doing calculations with D-branes such as the absorption cross section of quanta by D-branes [15–17] or the entropy of D-branes [18, 19]. With the entropy of D-branes at hand, one could compare it with the entropy of black holes and it was realised that they had similar values, sometimes agreeing with each other and sometimes disagreeing.

These two calculations looked very similar and promising. Trying to understand the reason for this in a deeper way was what led to AdS/CFT. There were some specific calculations which were inspiring, like trying to solve the wave equation of a perturbation that falls into a black hole near the horizon [16]. That wave equation had some special symmetries that only arose in the near horizon region of the black hole and those were the symmetries that we now understand as the symmetries of AdS. The near horizon geometry was in fact AdS. At the time we didn't understand – at least I didn't understand – what those symmetries were. But as it turned out, these symmetries determined the solutions of the wave equation in the same way that the scaling symmetry of the field theory determined solutions in the field theory. Clearly, there was some relation between the field theory living on the D-branes and the gravity solution but it wasn't clear what the precise relationship was. Neither was it clear if the field theory was living behind the horizon or outside the horizon. Isolating the whole near horizon geometry was the idea behind AdS/CFT and the symmetries were the guiding principles.

Do you feel comfortable with the fact that the AdS/CFT is still a conjecture?

I feel that it is quite proven (laughs). With all these results that use integrability and so on [20], I feel that we have essentially proven it. Of course, it's complicated and there are some conjectures which won't satisfy mathematicians, but this line of research constitutes a proof from a physicist's point of view. In practice, you calculate some quantities in the gravity theory and in the field theory, you compare them and they match. In particular, there was this talk today by Marcos Marino[2] who calculated something starting from field theory principles and noticed that it agreed with the gravity calculation [21–23].

But integrability makes some assumptions, right?

The particular calculations that Marino was discussing did not require integrability; it was just based on supersymmetric localisation techniques [21–23]. Of course, you could argue that these are BPS observables. For me, integrability is a more drastic demonstration because you look at non-supersymmetric objects and you still find agreement.

So you don't think that it is really important to prove it with mathematical rigour?

I think this integrability direction is the direction in which it is most interesting to prove it because it leads to a calculational tool. That is, it leads to a way to calculate observables that you couldn't calculate before. It also leads to a view of four-dimensional super Yang–Mills theory as a solvable theory. I think that this is more interesting than some more abstract argument but it might be interesting to find some abstract argument that you could apply to any case.

2 See http://media.medfarm.uu.se/play/video/2131?module=strings2011&video=14.

What about the work by Hirosi Ooguri and Cumrun Vafa on the worldsheet derivation of the duality in the context of topological string theory [24]?

I think that the work by Hirosi Ooguri and Cumrun Vafa was nice. For the topological string it was pretty clear how to do it. Now, if we embed this in the full superstring theory and we understand it in a concrete way, that would also be wonderful. You would like to find something that is better than the original argument and which would allow you to calculate the string theory precisely. Thus, some technique that would give you what the string theory is, not just the supergravity limit, would be nice. I must also mention that I think that in some sense it is impossible to prove it because the string theory side is not non-perturbatively defined. String theory is only an asymptotic series expansion so the best you could hope for would be to prove it perturbatively.

Were you aware of all the possible spinoffs of AdS/CFT?

No, I wasn't (laughs). I was worried that it would be an empty useless statement (laughs), in the sense that I was stating that the two theories were equal and it could be that there wouldn't be any way to check it by comparing the field theory to the gravity side.

This duality has inspired many people to talk about the emergence of spacetime ...

The picture of emergent spacetime has a long history in string theory. I think the first example is already the worldsheet in some sense via T-duality [25], but maybe you can dispute that. Other examples include the old matrix models where you start with zero dimensions and end up with a one-dimensional field theory [26], the BFSS matrix model which was a formulation of 11-dimensional supergravity [27], and then finally AdS/CFT [3]. All these examples were clearly pointing towards this idea of emergence of spacetime, and they are clearly theories of emergence of spacetime.

In AdS it is very clear that you can start with a four-dimensional theory and get a higher dimensional spacetime. In this context, it is mostly emergent space that we understand instead of emergent spacetime. Our failure to understand emergent time is connected to our failure to understand time-dependent backgrounds. If we understood better how time emerges we probably could understand situations in which we start with a theory with no time and then we dynamically find time.

Do you think that all these interesting things that come out of string theory are pointing toward deeper fundamental principles, like holography, of another theory?

You could wonder whether there is another theory which is not string theory and which is not connected to string theory in any simple way but is still consistent and is the most general theory.

Is it your opinion that such a theory is out there?

I don't think so. I think it is only string theory in the following sense. Even if you allow for strongly interacting string theories – the real perturbative strings you only find in situations where the string coupling is small so it is special in that sense – I think that they are probably all connected to each other in some way.

Has string theory come up with models for inflation?

Yes, there are some models of inflation in string theory [28–30]. They are models for the effective field theory of inflation. You can devise a certain configuration of branes which slowly vary with time but they don't address the problem of the initial singularity; they only address the problem of inflation. In other words, these configurations don't tell how you got there or even if they do tell you, then you got there via some bubble nucleation process and so on. Thus, they are not addressing the problem of the initial singularity or what is now called the *measure problem*. They are just models of the early universe, of slow roll inflation.

You did quite a bit of work on inflation [31] (see also [32–36]). Were there any experimental predictions?

I don't think that there are any clear experimental predictions. There have been some suggestions inspired by inflationary models. For example, the brane–anti-brane inflation scenario suggested the creation of cosmic strings and so people suggested that we should look for cosmic strings that have low tension (compared to what you naively expect the string tension to be) [37–41]. If your expectation of the string tension is based on the heterotic string then it would be too high and would have been already ruled out. It is not a unique prediction but it is something suggestive.

There were also other models that predicted that you have inflation with a slow roll potential and with a bunch of wiggles so people are also looking for these wiggles. Some of these inflationary models predict the production of gravity waves, which requires having a potential with a particular slope, which in some sense wasn't very natural from previous points of view. Normally, people consider potentials that go like ϕ^3, ϕ^4, etc., but not with the absolute value of ϕ or ϕ to some fractional power smaller than two [42, 43]. These are not unique predictions; they are just some models that happened to occur to a person and might not be representative of the entire class of models.

The same could be said about the models that occurred to us [31] (laughs). We don't know how representative these models that occurred to any of us are. There's nothing clearly stated about what's not possible to construct. Field displacements of many times the Planck scale are probably impossible but that's not what's required by experiment anyway (laughs).

One of the biggest criticisms of string theory is that it is not a falsifiable theory. Do you think that this criticism is justified?

I think this is wrong. String theory is falsifiable by a table-top experiment. If you have any experiment that violates quantum mechanics then string theory is automatically falsified because string theory is based on quantum mechanics. What is true is that string theory has not made any clear prediction about any phenomena that we did not know before finding string theory.

Now shifting the topic to the string landscape ... is there a dynamical mechanism in string theory which is able to decide which vacua we live in?

Not that we know of. It would be nice to have one but no one has suggested a reasonable mechanism yet.

Is it reasonable to invoke anthropic arguments [44]?

Certainly, anthropic reasoning is the only explanation that we have for the cosmological constant up to today [45]. I don't find that satisfying but it could be the correct explanation. It is the most reasonable explanation (laughs).

Is there no string theory solution to the cosmological constant?

Well, the anthropic solution is the best string theory solution [44]. There might be a better one in the future; I don't know.

Maybe a nice compactification that gives the right value of the cosmological constant?

Even if you found one that has the right value it doesn't explain it; it only fits that value into string theory. The problem is that the method we have for finding these vacua is statistical in nature because you are arguing that you have a certain range for the cosmological constant and by fine-tuning you can make it as small as it is now. But imagine that someone comes up with the possibility of computing the cosmological constant in all the vacua, then it should be that there is a vacuum like ours. If there isn't, string theory would be wrong (laughs). Certainly, if someone found that vacuum they might find correlations between physical parameters but I think that this is not likely to happen with the current techniques. Certainly, not for the cosmological constant unless you think that there are some properties of the standard model that are independent of the cosmological constant and somehow they are constrained by string theory. But even this is somehow unclear because of the difference in scale between the current accelerator scale and the Planck scale.

But do you think that all these vacua exist and that there is a multiverse?

The vacua seem to exist; no one has found any problem with all these vacua.

But do you think that there are other universes with different laws of nature?

That seems reasonable. Maybe some of them do not have consistent histories, or histories that produce observers, or maybe they are less likely to exist than our vacuum which produces observers. There could be things which are very history-dependent and which might be very tough to understand.

I am asking this because what I would really like to know is if you see string theory as a theory which has not been completely formulated but which will be able to completely fix the vacuum or if you see it more like a formalism in which we need to adjust all the constants . . .

In quantum field theory we choose the standard model and adjust the constants but I hope that string theory somehow predicts the constants. What I suspect is that all these different vacua are necessary for making the early universe smooth, that is, for resolving the big bang singularity. That is what I think is the case.

You mentioned previously that you started studying black holes in order to solve the black hole information paradox. You once proposed that you could solve this problem by imposing some boundary condition at the singularity [46], right?

Imposing it at the singularity turned out to be wrong. It could be that you should replace the whole interior of the black hole by some effective boundary condition. That might still be correct. If it is wrong it is because we cannot think about Hilbert spaces in a very local way in the bulk. That could also be the case.

So is this the final resolution of the paradox, or will Samir Mathur's fuzzball proposal [47] play a role?

I think that a proposal which is somewhat vague could be the correct proposal. Mathur's proposal is vague. The question we want to address is, "What does an observer who falls into the black hole see?" If an observer who falls into the black hole never feels that they are falling into the interior then I think it must be wrong because I think that we should be able to trust gravity in the interior of the black hole for the observer who falls in. Mathur would probably say that if you have a fuzzball, whatever it is, whenever you fall into it you feel that you are falling into the horizon and so on and then it's fine. Regardless of what Mathur wants to call it, this is more or less the same as what we would say. The dominant opinion is that the observer sitting at infinity sees some microstates in some way, and that the interior of the black hole contains some dynamics of these microstates such that the whole spacetime in the interior is encoded in the dynamics in some way, which no one yet understands.

Do you think that the proposal you made is consistent with fuzzballs?

The proposal I made could be consistent with fuzzballs; I don't know. It could be that fuzzballs have another description in which you slice the Hilbert space in some way and then you have an effective boundary (laughs).

Has the information paradox problem been solved in the context of string theory?

There are different versions of the paradox. I think it is solved in the sense that we think that the information does come out of the black hole. There is a theory of quantum gravity which is the one given by the dual field theory [3] and in that theory of quantum gravity there is no information loss. However, it is not solved in the sense that we don't understand how to describe the interior of the black hole. The important question that needs to be answered is, "How do you know that an observer falling into the black hole as described by the dual field theory will feel as if they were falling into the black hole?" But you should answer it using not the supergravity description but instead the dual field theory description. We don't know how to answer this question. Of course, you can answer it if you assume the duality but what is not known is a formalism which is explicitly unitary and which has a Hilbert space and so on, for which you can show that an observer who falls into the black hole feels that they are falling through the black hole.

You have done a lot of work on the study of scattering of gluons [6–9]. Why did you spend so much time with this?

I think that it is interesting to develop techniques that allow you to compute scattering amplitudes. It is true that I am not calculating it in a physical theory, I'm computing it in $N = 4$ super Yang–Mills but it may be inspiring for people who then do the actual calculations in $N = 0$ (laughs).

Have people taken inspiration from $N = 4$ and do other things in physical theories?

Yes, in the some sense, some of the techniques developed for $N = 4$ should be useful for QCD. The tree-level amplitudes that have been studied in $N = 4$ are also used in QCD. This is not the work I've done but work that some others have done. Eventually, the goal would be to understand amplitudes via integrability so that you could solve them for $N = 4$ at any value of the coupling, even in perturbation theory. I think that it might be pretty inspiring for people who actually do calculations in QCD. These calculations are done in a context with similar dynamics when compared to the dynamics which is seen in nature and in colliders. That's the biggest motivation for solving planar Yang–Mills.

Because you mentioned the word "collider," I remembered that you had a paper called "Conformal Collider Physics" [48]. Has this any application to the standard model?

There we were studying what the physics is in an exactly scale-invariant theory. I think that we were saying that the calculations we performed could be useful in the context of proposals that assume conformal symmetry as an extension of the standard model (see [49]). It could be that the new physics we discover is conformal or approximately conformal. Conformal field theories appear in nature and in quantum critical points in condensed matter systems so we shouldn't be so focused on QCD or any particular theory. There's many ways in which quantum field theory can arise and understanding time-dependent dynamics in quantum field theory is interesting.

Why did you spend time studying conformal gravity [1]?

I found it strange that conformal gravity encodes the solutions of ordinary gravity and that conformal gravity is a power-counting renormalisable theory.

Could it describe any ultraviolet completion of some known theory?

This $N = 4$ version is supposed to be finite but it has ghosts so it is kind of cheaply finite (laughs). With ghosts it is easy to find finite theories. However, this cheaply finite theory knows about classical gravity in some way that it seems non-trivial to me.

As far as I understood, you have shown that it does know about it ...

Yes, you could formulate the calculation in its entirety in conformal gravity theory and it is not an outlandish strange thing where you have to put some very strange selection of solutions. You introduce very natural boundary conditions and you select the right solutions. In fact, you put a local boundary condition in the future and it selects the right solution. I find that quite striking.

So, what can you do with it?

My suspicion is that the probability of it leading to a quantum theory is very small. But it would be very interesting if it were true. The product of probability times interest is sufficient (laughs).

What has been the biggest breakthrough in theoretical physics in the past 30 years?

I'm tempted to say string theory but if you consider the problems that string theory has tried to solve like the early universe problem and so on, I think the biggest idea was inflation [50–52] because inflation solved the problem. Inflation gave a very nice mechanism for the formation of structures in the universe and how to choose the initial conditions and so on. It is theoretically simple as it doesn't require any fancy mathematics, but it is pretty nice and explains some of the features of the universe which we observe. From a physical observation point of view I think this is a very important thing. I still think that string theory is the most interesting thing but it still has the problem that we haven't observed it (laughs).

Why have you chosen to do physics?

The answer is very simple: because I could (laughs).

What do you think is the role of the theoretical physics in modern society?

To develop interesting physical theories (laughs). I don't think that it is to solve social problems. In society we play our role by doing what we do best and not trying to jump on other things. Einstein was a wonderful physicist and he was also famous for his peace activism but I don't think he made much of an impact in these other areas.

References

[1] J. Maldacena, "Einstein gravity from conformal gravity," arXiv:1105.5632 [hep-th].
[2] E. Witten, "String theory dynamics in various dimensions," *Nucl. Phys. B* **443** (1995) 85–126, arXiv:hep-th/9503124.
[3] J. M. Maldacena, "The large N limit of superconformal field theories and supergravity," *Int. J. Theor. Phys.* **38** (1999) 1113–1133, arXiv:hep-th/9711200.
[4] *The Birth of String Theory*. Cambridge University Press, 2012.
[5] P. Kovtun, D. T. Son and A. O. Starinets, "Viscosity in strongly interacting quantum field theories from black hole physics," *Phys. Rev. Lett.* **94** (2005) 111601, arXiv:hep-th/0405231.
[6] L. F. Alday and J. Maldacena, "Minimal surfaces in AdS and the eight-gluon scattering amplitude at strong coupling," arXiv:0903.4707 [hep-th].
[7] L. F. Alday, J. Maldacena, A. Sever and P. Vieira, "Y-system for scattering amplitudes," *J. Phys. A* **43** (2010) 485401, arXiv:1002.2459 [hep-th].
[8] L. F. Alday, D. Gaiotto, J. Maldacena, A. Sever and P. Vieira, "An operator product expansion for polygonal null Wilson loops," *JHEP* **04** (2011) 088, arXiv:1006.2788 [hep-th].
[9] D. Gaiotto, J. Maldacena, A. Sever and P. Vieira, "Pulling the straps of polygons," *JHEP* **12** (2011) 011, arXiv:1102.0062 [hep-th].
[10] N. Arkani-Hamed, J. L. Bourjaily, F. Cachazo, S. Caron-Huot and J. Trnka, "The all-loop integrand for scattering amplitudes in planar N = 4 SYM," *JHEP* **01** (2011) 041, arXiv:1008.2958 [hep-th].
[11] A. Strominger, "The dS/CFT correspondence," *JHEP* **10** (2001) 034, arXiv:hep-th/0106113.
[12] Y. Sekino and L. Susskind, "Census taking in the hat: FRW/CFT duality," *Phys. Rev. D* **80** (Oct. 2009) 083531. https://link.aps.org/doi/10.1103/PhysRevD.80.083531.
[13] B. S. DeWitt, "Quantum theory of gravity. I. The canonical theory," *Phys. Rev.* **160** (Aug. 1967) 1113–1148. https://link.aps.org/doi/10.1103/PhysRev.160.1113.
[14] J. Polchinski, "Dirichlet branes and Ramond-Ramond charges," *Phys. Rev. Lett.* **75** (1995) 4724–4727, arXiv:hep-th/9510017.
[15] A. Dhar, G. Mandal. and S. R. Wadia, "Absorption versus decay of black holes in string theory and T symmetry," *Phys. Lett. B* **388** (1996) 51–59, arXiv:hep-th/9605234.
[16] J. M. Maldacena and A. Strominger, "Black hole grey body factors and D-brane spectroscopy," *Phys. Rev. D* **55** (1997) 861–870, arXiv:hep-th/9609026.
[17] S. R. Das, G. Gibbons and S. D. Mathur, "Universality of low energy absorption cross sections for black holes," *Phys. Rev. Lett.* **78** (Jan. 1997) 417–419. https://link.aps.org/doi/10.1103/PhysRevLett.78.417.
[18] A. Strominger and C. Vafa, "Microscopic origin of the Bekenstein-Hawking entropy," *Phys. Lett. B* **379** (1996) 99–104, arXiv:hep-th/9601029.
[19] G. T. Horowitz and A. Strominger, "Counting states of near extremal black holes," *Phys. Rev. Lett.* **77** (1996) 2368–2371, arXiv:hep-th/9602051.
[20] N. Beisert et al., "Review of AdS/CFT integrability: an overview," *Lett. Math. Phys.* **99** (2012) 3–32, arXiv:1012.3982 [hep-th].
[21] N. Drukker, M. Marino and P. Putrov, "From weak to strong coupling in ABJM theory," *Commun. Math. Phys.* **306** (2011) 511–563, arXiv:1007.3837 [hep-th].
[22] N. Drukker, M. Marino and P. Putrov, "Nonperturbative aspects of ABJM theory," *JHEP* **11** (2011) 141, arXiv:1103.4844 [hep-th].

[23] M. Marino, "Lectures on localization and matrix models in supersymmetric Chern-Simons-matter theories," *J. Phys. A* **44** (2011) 463001, arXiv:1104.0783 [hep-th].

[24] H. Ooguri and C. Vafa, "World sheet derivation of a large N duality," *Nucl. Phys. B* **641** (2002) 3–34, arXiv:hep-th/0205297.

[25] B. S. p palan, "Duality in statistical mechanics and string theory," *Phys. Rev. Lett.* **58** (Apr. 1987) 1597–1599. https://link.aps.org/doi/10.1103/PhysRevLett.58.1597.

[26] M. R. Douglas, D. N. Kabat, P. Pouliot and S. H. Shenker, "D-branes and short distances in string theory," *Nucl. Phys. B* **485** (1997) 85–127, arXiv:hep-th/9608024.

[27] T. Banks, W. Fischler, S. H. Shenker and L. Susskind, "M theory as a matrix model: a conjecture," *Phys. Rev. D* **55** (1997) 5112–5128, arXiv:hep-th/9610043.

[28] L. McAllister and E. Silverstein, "String cosmology: a review," *Gen. Rel. Grav.* **40** (2008) 565–605, arXiv:0710.2951 [hep-th].

[29] D. Baumann, "Inflation," in *Theoretical Advanced Study Institute in Elementary Particle Physics: Physics of the Large and the Small*, C. Csaki and S. Dodelson, eds., pp. 523–686. World Scientific, 2011. arXiv:0907.5424 [hep-th].

[30] E. Silverstein, "TASI lectures on cosmological observables and string theory," in *Theoretical Advanced Study Institute in Elementary Particle Physics: New Frontiers in Fields and Strings*, J. Polchinski, P. Vieira and O. DeWolfe, eds., pp. 545–606. World Scientific, 2017. arXiv:1606.03640 [hep-th].

[31] J. M. Maldacena and G. L. Pimentel, "On graviton non-Gaussianities during inflation," *JHEP* **09** (2011) 045, arXiv:1104.2846 [hep-th].

[32] K. Abazajian et al., "Inflation physics from the cosmic microwave background and large scale structure," *Astropart. Phys.* **63** (2015) 55–65, arXiv:1309.5381 [astro-ph.CO].

[33] V. Alba and J. Maldacena, "Primordial gravity wave background anisotropies," *JHEP* **03** (2016) 115, arXiv:1512.01531 [hep-th].

[34] N. Arkani-Hamed and J. Maldacena, "Cosmological collider physics," arXiv: 1503.08043 [hep-th].

[35] J. Maldacena, "A model with cosmological Bell inequalities," *Fortsch. Phys.* **64** (2016) 10–23, arXiv:1508.01082 [hep-th].

[36] P. D. Meerburg et al., "Primordial non-Gaussianity," arXiv:1903.04409 [astro-ph.CO].

[37] G. Dvali and S. Tye, "Brane inflation," *Phys. Lett. B* **450** (1999) 72–82, arXiv:hep-ph/9812483.

[38] C. Burgess, M. Majumdar, D. Nolte, F. Quevedo, G. Rajesh and R.-J. Zhang, "The inflationary brane anti-brane universe," *JHEP* **07** (2001) 047, arXiv:hep-th/0105204.

[39] N. T. Jones, H. Stoica and S. Tye, "Brane interaction as the origin of inflation," *JHEP* **07** (2002) 051, arXiv:hep-th/0203163.

[40] S. Sarangi and S. Tye, "Cosmic string production towards the end of brane inflation," *Phys. Lett. B* **536** (2002) 185–192, arXiv:hep-th/0204074.

[41] S. Kachru, R. Kallosh, A. D. Linde, J. M. Maldacena, L. P. McAllister and S. P. Trivedi, "Towards inflation in string theory," *JCAP* **10** (2003) 013, arXiv:hep-th/0308055.

[42] D. Baumann and L. McAllister, *Inflation and String Theory*. Cambridge Monographs on Mathematical Physics. Cambridge University Press, 5, 2015.

[43] E. Silverstein, "The dangerous irrelevance of string theory," arXiv:1706.02790 [hep-th].

[44] L. Susskind, "The anthropic landscape of string theory," arXiv:hep-th/0302219.

[45] S. Weinberg, "The cosmological constant problem," *Rev. Mod. Phys.* **61** (Jan. 1989) 1–23. https://link.aps.org/doi/10.1103/RevModPhys.61.1.

[46] G. T. Horowitz and J. M. Maldacena, "The black hole final state," *JHEP* **02** (2004) 008, arXiv:hep-th/0310281.

[47] S. D. Mathur, "The fuzzball proposal for black holes: an elementary review," *Fortsch. Phys.* **53** (2005) 793–827, arXiv:hep-th/0502050.

[48] D. M. Hofman and J. Maldacena, "Conformal collider physics: energy and charge correlations," *JHEP* **05** (2008) 012, arXiv:0803.1467 [hep-th].

[49] K. A. Meissner and H. Nicolai, "Conformal symmetry and the standard model," *Phys. Lett. B* **648** (2007) 312–317, arXiv:hep-th/0612165.

[50] A. Starobinsky, "Dynamics of phase transition in the new inflationary universe scenario and generation of perturbations," *Phys. Lett. B* **117** no. 3 (1982) 175–178. www.sciencedirect.com/science/article/pii/037026938290541X.

[51] A. Linde, "A new inflationary universe scenario: a possible solution of the horizon, flatness, homogeneity, isotropy and primordial monopole problems," *Phys. Lett. B* **108** no. 6 (1982) 389 – 393. www.sciencedirect.com/science/article/pii/0370269382912199.

[52] A. H. Guth and S. H. H. Tye, "Phase transitions and magnetic monopole production in the very early universe," *Phys. Rev. Lett.* **44** (Mar. 1980) 631–635. https://link.aps.org/doi/10.1103/PhysRevLett.44.631.

18

Shiraz Minwalla

Faculty member at the Tata Institute for Fundamental Physics

Date: 22 February 2011. Location: Mumbai. Last edit: 22 March 2016

In your opinion, what are the main puzzles in high-energy physics at the moment?

There are several outstanding questions in high-energy physics. I will start with those that are less likely to hold conceptual surprises and then move on to those that are more likely to hold conceptual surprises. The first obvious question is, "What is physics like at the TeV scale?" This is the scale that accelerators are just about to reach and very prosaically the most interesting questions are: "Are there new particles, new interactions, or new symmetries at this scale?" These are the type of questions that arise every time you experimentally access a new energy range but there are some indications that the TeV scale is particularly interesting. This might just be numerology but the TeV scale seems to be singled out in two different ways.

First, it is more or less the scale of mass generation and electroweak symmetry breaking in the standard model. Moreover, the naturalness problem with the Higgs – the physical entity responsible for giving masses to the particles is the Higgs field and masses are hard to stabilise against radiative corrections in a quantum field theory – and hence the fact that we have masses of around the TeV scale suggests that there is some mechanism that protects scalar masses against radiative corrections from approximately the TeV scale and higher. This mechanism is not known experimentally but there are some suggestions that it might be a new symmetry that relates bosonic and fermionic fields – supersymmetry – and we know that the masses of fermionic fields can be easily stabilised by means of chiral symmetry. Therefore, this symmetry could be responsible for protecting masses against radiative corrections. This is one suggestion that something particularly interesting is happening at the TeV scale. To emphasise it again, it is not just a suggestion for new particles that can be found; it is also a suggestion for a new mechanism or perhaps a new underlying symmetry of nature unveiling itself at the TeV scale.

Second, estimates made with a lot of assumptions suggest dark matter, which we are now pretty sure is all around us in the universe...

Why do you say that we are pretty sure about the existence of dark matter?

Because there is so much evidence; in particular, the galaxy rotation curves and cosmological data provide independent evidence for the existence of dark matter. It seems likely that there is a lot of dark matter – cold dark matter – all around the universe and these estimates that I mentioned earlier suggest that dark matter particles also have a mass which is of the order of the TeV scale. Thus, all this suggests, in a phenomenological experimental way, that we might be at the cusp of a grand synthesis in physics.

Let me just give you a really optimistic scenario. Suppose that experiments such as CDMS (cold dark matter search) experiments find some particle with mass of the order of the TeV scale and that the LHC also finds a particle with a mass of the same order. Suppose further that we can convincingly identify the two of them. This would result in a grand synthesis of cosmology and particle physics: two mysteries being solved at once. Perhaps it could even have something to do with supersymmetry, which would then connect to, besides adding extra importance to, the searches for new theoretical structures.

Thus, at the practical level these are the urgent questions facing particle physics today. What is the physics at the next level? Is there a mechanism protecting the Higgs mass? What is that mechanism? Are the particles that you unveil at the next level also the particles that are hovering around the universe as dark matter? If we could answer these questions it would be fantastic. But even supposing we do, that would still leave many conceptual puzzles untouched. In my opinion, the biggest of such conceptual puzzles is, "Why is the universe so big?" This is the question of the cosmological constant.

Suppose that you took a theory of gravity and also took quantum mechanics with it. If you were naive you would say that every term that could be generated would at the length scale be associated with the only coupling constant in the theory, namely, Newton's constant. You would naively expect every term, including the cosmological constant, to get generated at the same scale. This would set the scale of the curvature of the universe and set the size of the universe to be of the order of the Planck scale. If that would be the case, nothing like life as we know would happen and it would be a very different kind of universe from the one in which we live today. On the other hand, our universe is a huge universe compared to the Planck scale, right? How is this possible? In my opinion this is the main physical and important puzzle. I should say that it is not a sharp puzzle, in the sense that it is not a direct mathematical contradiction between two different equations, and it is not sharp perhaps because the bigness of the universe can be achieved through fine tuning, leaving room to wiggle around with anthropic reasonings. Nonetheless, anyone who has thought about this must admit that it is quite a puzzle and I think that addressing this question could change the way we think about physics.

So you don't see the quantisation of gravity as one of these big puzzles and related to the puzzle you just mentioned?

Yes, the quantisation of gravity is a very important challenge in theoretical physics and it is probably tied to this question but I didn't mention it because there are formal questions as well as physical questions and I wanted to stick to the physical questions. Why is the universe so big? That is a question of physics. Now, I think that addressing it will most likely require technical control over a quantum theory of gravity.

Okay, but then let's go back to all the problems of particle physics. The predictions of all these new particles and new underlying symmetries like supersymmetry are based on models that in turn are based on string theory. Is this a true statement?

I think it would be an overexaggerated statement to say that all these models are based on string theory. Most of the models of particle physics that have a reasonable chance of being verified at the LHC are models based on field theory – the Randall-Sundrum models being an exception [1–3]. Some of these models incorporate elements of string theory, for instance the idea of supersymmetry, which I believe first made its way into physics through the string worldsheet, or the idea of extra dimensions exploited in the Randall-Sundrum models. When using these ideas, phenomenologists follow the grand tradition of phenomenology, namely, the Occam's razor philosophy: they make the models as simple as possible without bothering with all the hurdles of string theory.

There are also some models, like little Higgs models [4], that have nothing to do with string theory. In my opinion, the first important thing is for the LHC to find something and if it finds interesting stuff, the first layer of reality is most likely going to be some quantum field theory which might give us hints that string theory is the right answer if, for instance, supersymmetry is found. However, it is unlikely for the LHC to directly probe string theory and it is also unlikely to convincingly silence people who want to say, "This is fine, this may be true but string theory isn't."

Do you think that string theory can provide an answer to the cosmological constant puzzle that you mentioned?

I think that's the kind of question that, given the current level of understanding, string theory is well suited for. Let me tell you a fantasy that I have about this puzzle. I sometimes think that the cosmological constant puzzle is only a puzzle because we think about gravity in the wrong way and that string theory, as well as the AdS/CFT correspondence [5], will educate us on how to think in the right way about gravity.

Suppose that you have a large-N conformal field theory with a large separation of scales between the stress tensor and all other operators. Such a theory is necessarily dual to a gravitational theory with a cosmological constant in which the cosmological constant scale is largely separated from the scale of Newton's constant, that is, the Planck scale. $N = 4$ super Yang–Mills theory is such an example but it does not throw any light on the cosmological constant question because in that case the bulk theory is supersymmetric all the way down to the anti–de Sitter (AdS) scale. In this case there is no puzzle.

However, suppose that you can produce an example of AdS/CFT that has the particular feature of a wide separation of scales between the stress tensor and the rest of the operators in addition to being non-supersymmetric. If you have such an example, you can try to understand how the cosmological constant puzzle translates into field theory language. I have discussed this issue at length with my former student Kyriakos Papadodimas without reaching any definite conclusion. Thus, if we could find one example of a field theory like this …

But do you mean an example where you have a dual theory to a de Sitter spacetime, that is, with positive cosmological constant?

No, I mean AdS spacetime. I should emphasise that the way that string theory has helped us make progress so far is not by solving questions which are directly related to the real world but by solving analogous questions. In my opinion, the cosmological constant puzzle is an example where we will most likely make progress. The real world appears to have a small positive cosmological constant but the puzzle would be as great if we had a small negative cosmological constant (laughs). So if you can produce a theory which has a huge separation of scales between the cosmological constant and the Planck scale, whether positive or negative I don't care, I would be very interested in knowing how it manages to do that. A non-supersymmetric field theory with this large separation of scales between a few operators and all other operators has possibly something to do with this. If we can find an example of AdS/CFT with these features it would be really interesting to unravel how the field theory manages to avoid this puzzle.

There might be many loopholes in what I just said but I mentioned it as an example of the kind of way in which string theory can make valuable contributions to this question without having identified the theory of the real world. If you identify what string vacuum, if any, corresponds to the real world then of course there are a lot of things you can do but in the absence of such identification, what kind of progress can you make? Well . . . it's the kind of progress we have been achieving for a long time: you can ask qualitative questions that have counterparts in our world. Questions about black hole physics, which string theory has been spectacular in addressing, and questions about the cosmological constant, which if you can address in any spacetime would be huge progress. So that's the kind of questions that string theory should be shooting at if it can because it has this fantastic formal consistent structure that in the most controlled setting should allow you to address such questions.

So if I understand you correctly, are you saying that we should use string theory as a tool to solve other problems of physics than those of quantum gravity?

Not exactly. Some of the problems that I have mentioned previously are problems about quantum gravity. String theory is a theory with many different vacua and some of the simple vacua we understand very well. Assuming that our world is some vacuum of string theory we might be able to import qualitative lessons from this simple vacuum to our world, even questions about quantum gravity. For instance, the problem of understanding black hole entropy could have a similar solution in different vacua of string theory. Similarly, the problem of the cosmological constant could have a solution that holds in typical vacua. We might be able to answer these questions about quantum gravity in a qualitative fashion without identifying the correct quantum theory of gravity of the real world.

Of course it is also possible to address questions that have nothing to do with quantum gravity using string theory, such as questions about QCD, fluid dynamics, or condensed matter physics. I think that this is another valuable aspect of string theory; that is, you can also use string theory as a formalism, which is probably the most sophisticated mathematical formalism being explored by human beings. Given that it is a very powerful formalism, you should try to make use of it anywhere you can, not only in order to make progress in other fields but also because of the feedback that it will have on string theory.

In my opinion, one of the weaknesses of our field is that it is not being immediately driven by experiments. We lack guidance not just about whether the theory is right or not but also about what questions to ask next. Making contact with other fields can be very valuable because if you make contact with fields which are guided by phenomena then that guidance can change the directions of development of string theory. A small, but not totally insignificant example, is the discovery of charged black branes – hairy black branes – by Gubser and others, motivated by the idea of trying to understand holographic duals of superconductivity/superfluidity in AdS/CFT [6–8]. This work reveals simple gravitational phenomena that should have been understood 20 years ago, in particular that black branes with charged scalar hair are ubiquitous in AdS space. The problem of superconductivity/superfluidity is natural for a condensed matter physicist but in this case it also taught us something about gravity. This is a simple example that shows that making contact with other fields is valuable for the study of string theory because it will guide us to good questions.

Most of the community is working on string theory and only a very small percentage is working on other theories of quantum gravity. Do you think that it is important to develop these other theories or should everyone be working on string theory?

No, I think that anyone who has an idea should try to follow it. I certainly wouldn't say that the whole community should be working on string theory. People should do what appeals to them and let a thousand flowers bloom (laughs). I should say, though, that in some of the other areas the community is isolated and with their own . . .

Couldn't you say the same about string theory?

Oh surely, string theory is a community in itself and many people accuse it as being a very *mafioso* community (laughs). But some other areas like loop quantum gravity are much smaller and also a community in itself. Now the question is, "Should there be people working on that rather than this?". I think that every young person should decide for themselves. If that question were posed to me, and if I were younger than I am now but with a certain physics maturity, the first question I would ask would be, "What intellectual output has come out of the field?" The best kind of intellectual output that could come out of it is the contact with experiments but none of the approaches has that, at least in the sense of quantum gravity. So what other interesting intellectual output has come out of it? My impression from afar is that string theory has generated a lot of intellectual output in connections with mathematics and other areas of physics like condensed matter physics – first through the study of conformal field theories in the 1980s and more recently through the AdS/CFT correspondence – the study of scattering amplitudes in QCD, the study of supersymmetric gauge theories by Nathan Seiberg and Edward Witten [9, 10] and the study of $\mathcal{N} = 4$ super Yang–Mills theory, which turned out to be such an interesting field theory that we are about to solve completely with field theoretical techniques.

People who wouldn't think about touching string theory 10 years ago are now interested in it just because they want to understand better the structure of S-matrices in QCD. My impression is that other communities have a much smaller intellectual output, at least in the sense measured by making contact with other fields outside their own community. It is hard for me to positively judge if they have accomplished various developments within their own formalism but regarding the impact on other fields it seems to be much less. Now, I am not saying that this is a very good measure of the field, as in the end what we want is the field to be measured by contacts with experiment but, in the absence of that, if I were a young person deciding what to do, this would sway me to study string theory rather than other areas. However, of course, if you're the next Einstein and you have your own crazy new idea you should follow it.

In your opinion, what do you think is the reason to try to conceive another theory which is not string theory?

The reason is that we have not made any contact with experiment. I think that is a good enough reason.

Couldn't it be that there is a big problem with string theory that would lead one to develop another theory?

I don't know of any sharp enough problem with string theory. The dissatisfying thing about string theory is that it has been around for a long time and it doesn't seem closer than 30 years ago with regards to making contact with experiment and models of particle physics. It could be a technological problem, which is what all of us are hoping for, but it could be that we are somehow far from the right picture of the world. String theory is such a large and broad framework that I find it hard to believe that it is completely wrong but string theory is a big beast and we are looking at very small corners of it. These small corners could, in any case, be the right place to look even if string theory is far away from experiment.

When you said that it could be far from the right picture, did you mean that it could be the wrong theory of quantum gravity?

Well ... what would be the worst way that it could be wrong? I think it is very unlikely that is wrong as in *theoretically inconsistent* because so much has been demonstrated. For 30 years people have made calculations and checked calculations that could have been false but haven't turned out to be false.

Could you give an example of this?

Yes, so string theory predicts AdS/CFT in some loose sense and you can test it, right? Something I was involved in as a student in one of my first papers [11] was the following.

We looked at chiral operators in $\mathcal{N} = 4$ super Yang–Mills theory and computed their three-point functions using free field perturbation theory. Then, we took type IIB Einstein gravity in $AdS_5 \times S^5$ and computed the three-point functions using gravity. Parts of the answer are fixed by the symmetries in the problem but there are numbers – an infinite set of numbers – which are not predicted by any known symmetry. All these numbers agreed given a certain renormalisation theorem assuming that AdS/CFT holds. I am giving you this example, not only because I was involved in it but also because, since I was involved in it, I can positively vouch for the fact that the two calculations looked like they had nothing to do with each other. They were totally different calculations and the end output was the same. This kind of situation happens again and again in string theory and it is the type of situation that gives you confidence that ...

But this doesn't say anything about quantum gravity, it's only a relation between two different theories, right?

It tells you that the theoretical structure that you are dealing with is mathematically consistent and this theoretical structure definitely has theories that contain quantum mechanics and gravity. The dual theory in $AdS_5 \times S^5$ is purely gravitational when $N = \infty$ but becomes quantum when you move away from $N = \infty$.

This kind of verification implies that you are dealing with a theoretically consistent structure that incorporates gravity and quantum mechanics. In this sense, it would be like a one-in-a-billion chance that string theory would be wrong; however, there is another sense in which it could be wrong. It could be wrong in the sense that the right theory of quantum gravity for our world is not contained within this framework. I don't think we can rule out this possibility, though unlikely simply because string theory is such an inclusive framework and it has eaten up quantum field theory through the AdS/CFT correspondence, yeah? It absorbs what it meets (laughs). For this reason, I think it's unlikely but you can't rule it out. In addition, it would certainly be dogmatic to rule out such a possibility so if someone comes with a good idea, even if it consists of structures far away from the ones that we are used to, we should give it a fair chance.

String theory is criticised for not being verifiable and sometimes even called speculative science. Is this a reasonable criticism?

I don't really know what people mean when they argue this but let me try to reconstruct their argument. The first immediate criticism of string theory, which is fair, is that it started out as an endeavour to explain particle physics and so far it has not come up with a clear model for it. For instance, we don't have a canonical string theory model for what to expect at the LHC, right? However, to say that it is not verifiable or not predicative is to misunderstand the logical structure of string theory, and the string theory community is partly at fault in this.

In the 1980s many of the eminent string theorists, including Edward Witten, used to go around and give talks stating that the hadronic string could one day predict the ratio of the muon to the electron mass. Underlying these ideas was the assumption, which seemed to be very firmly held at the time, that once you introduce all the physical cuts in the theory, string theory will have a small and finite number of vacua and we just have to search that finite number of vacua in order to find the real world [12]. Already in the late 1980s, once it became clear that there were plenty of Calabi-Yaus [13], but certainly by 1995 when we understood non-perturbative string theory better, in particular that there were many consistent string theories for all values of the coupling [14], such as type IIB string theory in flat space, this view of string theory was no longer tangible. Nevertheless, I think that a lot of people continued to hold this view and propagated it well beyond its range of validity.

I think that the right view of string theory, which current evidence seems to suggest, is that string theory is, in many ways of thinking, a formalism rather than a theory. By this I mean the following. There are many vacua in string theory and these vacua can be very different; in fact, in formal quantum field theory language they would be regarded as different theories. For example, if you take type IIB in $AdS_5 \times S^5$ that corresponds to $\mathcal{N} = 4$ super Yang–Mills theory, but if you take IIB in $AdS_5 \times$ Sasaki-Einstein that corresponds to another quantum field theory. It's the same type IIB theory which you place in different vacua but you get very different physics. From the point of view of the dual description, you don't call it different vacua of the same theory; you call it two different theories. So in that sense, even though it's a true, beautiful and very interesting fact that the local physics of string theory is the same in all the different vacua, from the phenomenologists' point of view, string theory is a formalism within which it is possible to construct many different vacua that result in different theories.

Given the amount of research so far, we have already seen that string theory is broad enough to incorporate large numbers of quantum field theories. It's a formalism broader than the formalism of quantum field theory and it's a beautiful formalism, in particular one that allows you to incorporate gravity and quantum mechanics together. You should think of it like the formalism which generalises the formalism of quantum field theory and that allows the incorporation of gravity. Now, nobody seriously criticises quantum field theory by saying that it is not predictive. It is true that it does not predict the standard model; that is, you need to find the right quantum field theory in order to get the standard model. Analogously, this is exactly the way I think of string theory, in particular: it's a huge formalism and to find the real world we need to find the right vacua with the right theory. There's no chance that we could have gotten the standard model without having the formalism of quantum field theory, even though it wasn't sufficient but nevertheless it was a prerequisite. I think it's huge progress to have the formalism. It is an open question if the formalism is the correct one for the real world but it is the only reasonable candidate so far.

Therefore, if people say that string theory is not predictive they will also have to say that quantum field theory is not predictive. It is true that there are one or two theorems that one can prove in quantum field theory, like CPT theorems. However, very few statements are true in all quantum field theories, certainly not statements about particle masses or electron to muon mass ratios. So, if someone would stand up and say, "Quantum field theory is not predictive," people would say "Of course it's not predicative; it's a formalism and formalisms are not predictive."

So hasn't string theory made any prediction at all? Has it made a prediction of some phenomena that now I can go and do an experiment and see that it is right?

I think that there are many different phenomena that are now understood due to the study of string theory. Let me give you an example that I was involved in since I know it well. A small corner of string theory deals with the study of the structure of fluid dynamics via the AdS/CFT correspondence [15]. The structure of relativistic fluid dynamics of fluids that have a conserved $U(1)$ charge in addition to being characterised by a stress tensor, is an old subject spelt out, for instance, in Landau and Lifshitz's textbook [16] and used in various contexts.

Due to the AdS/CFT correspondence, you can use gravity to derive the equations of charged fluid dynamics for $\mathcal{N} = 4$ super Yang–Mills theory. Doing so you find a surprise; namely, you find a term in the equations that is not in Landau and Lifshitz's book [17, 18]. Now, Landau and Lifshitz had good reasons to write down what they did, so you can go and try to track down this mismatch. If you do this carefully you find that the reason for these differences is due to the fact that the charge current of super Yang–Mills theory has a $U(1)^3$ anomaly. Studying fluid dynamics in the presence of this anomaly forces you to change the fluid equations of motion in a way that no one had previously done before. Once you have the result from the string theory calculation you can understand it in a more general way which applies beyond string theory; in fact, the general understanding applies to quantum field theory in general [19]. For example, it can be applied to the study of neutron stars or quark stars (e.g., [20]). I don't know if you can call this a prediction of string theory or not, because once you have a prediction it is a prediction within quantum field theory; however, the causal chain of events indicates that you would not have this prediction without having performed the calculation in string theory in the first place.

This kind of reasoning is leading string theory slowly in the direction of making contact with the real world. For instance, a famous example is that of the viscosity to entropy ratio, which turns out to be $1/4\pi$ in every theory with a gravity dual description [21]. In the RHIC experiment they model the outcome of the collision of two nuclei by a fluid dynamical phase and agreement with experiment depends on what you assume for the viscosity of the fluid. If the viscosity you take is high as suggested by blind extrapolation of the perturbative data in QCD, it just doesn't fit with experiments. However if it's low, of the order of $1/4\pi$, it fits rather well apparently [22, 23]. Notice that $1/4\pi$ is a sharp prediction for $\mathcal{N} = 4$ super Yang–Mills theory and not a sharp prediction for QCD. However, it has made a big difference in the heavy ion collision community because in the absence of the string theory calculation if someone would say, "Look, I get a good fit if I assume the viscosity to entropy ratio to be small," that someone would be laughed at, since perturbation theory suggests that it's a hundred times that number. Now, you have an exact calculation in a strongly coupled theory which yields a very small result so it seems clear that it is not crazy. My understanding is that it had significant psychological impact, if nothing else, in the heavy ion community.

There are many examples like this which will never give you a Nobel Prize. The impact of string theory on experimental physics has been sort of marginal but it just started and so one day it could result in something dramatic. For instance, in the fluid dynamics context and in some situations, there is an isomorphism between the equations of Einstein gravity and the Navier-Stokes equations [15]. There are several puzzles about the Navier-Stokes equations, the most famous being the theory of turbulence which would be a significant intellectual breakthrough if you could develop it; in fact, the sort of thing that would give you a Nobel prize.

In which way would gravity help you with this problem?

Turbulent fluids are almost chaotic and random. Suppose that you measure the velocity of the fluid at one point, then at another point separated by a distance Δx and then you take the expectation value of the product of the first velocity to the power n and the second velocity to the same power n. Experimentally, these numbers in turbulent flows scale as power laws, in particular as Δx to some power that depends on n. Andrey KolmogorovKolmogorov, Andrey, about 60 years ago, developed a mean field theory that predicted what these power laws are [24, 25]. It turns out that Kolmogorov's prediction is correct for $n = 3$, in which case it can be derived more seriously using Ward identities, but it is wrong for any other n, which is analogous to the problem of anomalous dimensions in field theories. Now, there is some disagreement between experts on this, I believe, but at least the most optimistic experimentalists appear to claim that there is evidence that these numbers are universal in the sense that these power laws don't depend on the precise details on how you set up the turbulent flow, neither on which system you're working with. If this is really the case, and I emphasise again that not everyone believes in this, it is a fantastic problem for a theoretical physicist because there is an infinite set of universal numbers and a theorist should be able to calculate universal numbers, right?

Nevertheless, after 60 years of hitting your head on this problem just using the Navier-Stokes equation, it didn't take you anywhere. However, at present we can recast the fluid equations as gravitational equations and it could be that recasting the hard problem of turbulence in the language of gravity, due to the ultraviolet/infrared separations, will suggest new ways of approximating and thinking that will help crack the problem. If this turns out to be the case, which does not seem impossible, it will lead to serious impact on physics.

A quite different example that would lead to significant progress is the case in which you imagine obtaining the string theory dual to QCD and predict glueball masses to 10% accuracy. You can reasonably expect to be able to predict with 10% accuracy in the limit of large N but taking into account $1/N^2$ corrections one can expect, say, $1/10$ accuracy. Predicting glueball masses to 10% accuracy in the real world would have a serious impact on experiment. Currently, we are far from making such prediction as we never identified the string theory dual to QCD, much less solved it. On the other hand, in the case of $\mathcal{N} = 4$ super Yang–Mills theory we are able to make such predictions. To summarise, at the moment the impact of string theory in experimental physics exists but not in a substantial way. The current level of impact does not justify the original purpose of string theory but the potential for that impact to increase is significant.

Very schematically, how would you tackle the problem of turbulence?

I'll tell you what I have thought for a while without having made much progress. Suppose that you pick up any standard book on turbulence, for instance, the one by Uriel Frisch that you can find somewhere here on my bookshelf [26]. In chapter 2, Frisch talks about the notion of turbulence flows as composed of vortices of various sizes. Vortices of a certain size interact with vortices of slightly smaller size and exchange energy between them. A big vortex and a small vortex do not directly interact with each other. From the point of view of fluid dynamics this seems quite puzzling. If you have a big vortex near a small vortex, intuitively you would expect them to interact. There is good evidence that this is not the right way of thinking about turbulence. Certainly the whole way the problem is posed, as a flow of energy between scales, seems very imprecise, at least to my taste.

Suppose that you take the same problem and view it in terms of AdS/CFT. As in many problems in AdS/CFT, objects that are big in the boundary correspond to objects far away into the bulk and objects that are small in the boundary correspond to objects near the boundary. So it does not seem impossible to me that this separation between big and small objects can be made precise using the precise locality of gravity in the radial direction from the boundary. If it is possible to separate the problem into one that you can start thinking scale by scale, using the extra locality of gravity in the fifth dimension – and I emphasise that this line of thought is not being backed up by any equation of any successful program – perhaps you could begin thinking about turbulence in the correct way. At least at the level of words this doesn't sound ridiculous to me. However, words are cheap in physics; what you need are equations and so far I don't have equations to justify this line of reasoning.

You have mentioned several times the AdS/CFT correspondence. Can you explain what it's really about?

Let me introduce the AdS/CFT correspondence in a slightly unusual way. It has been known for a long time, certainly since the late 1970s with the work of Gerard 't Hooft [27], that any integral involving large-N matrices can be effectively evaluated by saddle-point techniques when N is taken to infinity and holding the correct 't Hooft couplings fixed, that is, when the theory becomes classical. Quantum $U(N)$ gauge theories are, in the end, integrals over $N \times N$ matrices living at each point in space. Thus, large-N quantum field theories become effectively classical theories, which we have known for a long time. However, what everyone has always assumed is that effective classical theories would be messy and complicated, mostly because such classical effective theories were not computable except in the case of the simplest theories, such as the theory of a single Hermitian large $N \times N$ matrix (the Wigner model [28]), the Gross-Witten-Vadya model [29, 30] or $N \times N$ quantum mechanics [31]. They're very simple models for which you can identify the classical system but if you complicate it a little more you cannot calculate it from first principles.

We knew for a long time that large-N QCD and large N super Yang–Mills theory had some effective classical description but it seemed that there was no hope of finding it and, certainly, it appeared that if it existed it would be very complicated. In 1997, Juan Maldacena identified the effective classical description of $\mathcal{N} = 4$ super Yang–Mills theory – a highly supersymmetric cousin of QCD – and what he found was that the effective classical description was not some sort of mess but instead the most beautiful equations in theoretical physics, namely, Einstein equations [5]. This was one of these grand unifications of physics: two of the most beautiful equations in physics – the super Yang–Mills Lagrangian and Einstein's action, both unquestionably among the great equations of theoretical physics which enhanced our understanding of the physical world – were related to each other. A priori, they had nothing to do with each other but you find that if you look at the classical description of large $\mathcal{N} = 4$ super Yang–Mills theory it becomes a theory of gravity. It is a fantastic unification in the conceptual sense as two different structures of theoretical physics which at first appear to have nothing to do with each other actually lead you from one to the other. Moreover, it encompasses structures which appeal not only to the mathematicians but also to the real world.

In the strict large-N limit, super Yang–Mills theory becomes classical Einstein gravity but we have also known for a long time that if you relax the large-N limit, that is, if you start doing perturbation theory in $1/N$, this classical theory becomes quantum. Moving a little bit away from the strict large-N limit generates quantum fluctuations of gravity. Thus, gravity gives you the answer to the field theory question, "What is the classical effective theory of a specific large-N Yang–Mills theory?" But it is also true that the field theory gives you an answer to the gravitational question, "How do we make a quantum theory of gravity?" This results from the fact that field theories at large but finite values of N are very well defined and that taking N away from infinity generates quantum fluctuations of gravity. Therefore, the AdS/CFT correspondence is a remarkable discovery.

Let me just say a little bit more about AdS/CFT. If you want to produce a quantum theory of gravity, what do you need? You want to produce a theory with controllable quantum fluctuations which at the classical level reduces to classical gravity. When you wanted to produce a quantum theory of QED what you did was to promote a classical QED theory to a quantum QED theory, roughly, by replacing numbers by operators. It was a great achievement but the two theories were directly related to each other; that is, quantum QED was produced such that in the limit of $\hbar$ going to zero it reduced to classical QED. That's one way of producing a quantum theory with a given classical limit; namely, you take the same action, put it in a path integral and guarantee that when $\hbar$ goes to zero you get the correct classical limit. Now we understand that there is another way of taking classical limits. This other way of taking classical limits does not amount to taking $\hbar$ to zero but instead N to infinity.

All attempts at making a quantum theory of gravity by taking $\hbar$ to zero have failed. If you just try to put Einstein's action in a path integral it doesn't seem to work. Perhaps the reason is that the right classical limit for gravity is not the $\hbar$ to zero kind of limit but the N to infinity type. This is a marvellous realisation as it enlarges the scope of our understanding of how to quantise theories since there's another way of obtaining classical limits. As I said earlier, it is known since the 1970s that there is another way of getting classical limits but

I think that everyone had assumed that the limit you would get would be a god awful mess and that just turns out not to be the case. This new interconnection between different parts of physics and new ways of making quantum theories is a really important result.

We have a pretty good picture of what it means to take $\hbar$ to zero but what does it really mean to take N to infinity? Why would that be classical?

The basic point is this. In gauge theory, the operators you deal with are gauge-invariant operators, meaning that they are traces of products of matrices. The theory becomes classical in terms of these trace variables. I should emphasise that the fluctuations of matrices are not classical, as they are controlled by $\hbar$, but the fluctuations of the values of the traces are classical. Why is that? That's effectively the law of large numbers which tells you that if you average over N different data points, standard deviations scale like $1/\sqrt{N}$, so fluctuations are suppressed upon averaging. Traces of operators are like averages as they are sums over all the operator eigenvalues divided by N. So when you take N to infinity you are greatly increasing the number of degrees of freedom and, roughly speaking, the law of large numbers kicks in, suppressing fluctuations. That's how you make a classical theory using the law of large numbers, namely, by introducing a lot of degrees of freedom so that the average becomes almost classical.

Now going more into your research on fluid dynamics and conformal field theories ...how does the structure of fluid dynamics fit with the AdS/CFT correspondence?

The AdS/CFT correspondence gives you a dual description of field theories at large N but something that I did not emphasise before, which is true at least in the case of $\mathcal{N} = 4$ super Yang–Mills theory, is that you get a gravitational dual description only at strong coupling. Many aspects of AdS/CFT seem at first sight rather mysterious for several reasons, one of which is that cases which you can clearly calculate within the gravity description you cannot clearly calculate within the field theory description and vice versa. It certainly feels mysterious that gravity and field theory are the same and some of this mystery is hiding behind strong coupling physics. How do you understand field theory at strong coupling? I felt there was never a satisfactory understanding of this question and the duality oftentimes feels a bit of a cheat.

Therefore my motivation for this research direction was basically the following. Let's try to find a general field theory phenomenon that applies whether we are at weak or strong coupling but that would be surprising from the point of view of gravity. In other words, I wanted to find a corner of the AdS/CFT correspondence for which we can understand clearly this duality. There are some properties of field theories that exhibit such features but the most *crying out loud* one is the one that Son, Starinets and Kovtun have emphasised, namely, that all field theories in appropriate circumstances, i.e., at long wavelengths compared to an effective mean free path or mean free time, admit an effective description in terms of the equations of fluid dynamics [21]. This is true whether or not the theory is weakly or strongly coupled; in fact, it is more so for strongly coupled theories because the idea of fluid dynamics is that physics locally equilibrates and it equilibrates faster at strong coupling. Thus, the theory of fluid dynamics, which is based on symmetry principles and the notion of local equilibrium, becomes a better and better approximation as the theory becomes increasingly strongly coupled.

In which situations would you experimentally observe such fluids in the context of quantum field theories?

In order to have all features of a quantum field theory you need to look at fluids like the one you find at RHIC [23], where you are dealing with a relativistic field theory. But, as you know, water flows in the end are excitations within a quantum field theory, as are air flows. These properties are very general and they just depend on locality.

I was interested in knowing whether you could use this to make predictions about conformal field theories and quantum field theories in general?

Yes, I think you can make predictions for quantum field theory in general. The prediction is that there is an effective description at long wavelengths in any conformal field theory. You could take a conformal field theory and tune a magnet to the right point, somehow produce a quantum critical system and observe its dynamics. I think you could experimentally verify that the relativistic Navier-Stokes equations are the correct equations. Fluid dynamics is a very important tool for condensed matter physicists because if you want to study phenomena outside equilibrium it is often the effective description of the system.

Now, you should think of fluid dynamics in the sense of an effective field theory, like a Wilsonian limit of a field theory applied at long wavelengths, except that it's a Wilsonian effective field theory for real-time dynamics at finite temperature. Thus, this very general expectation that the dynamics at long wavelengths and at finite temperature is governed by the equations of fluid dynamics resides on the basic physical input that the system under consideration locally equilibrates. Once it equilibrates, you don't need to worry about all the degrees of freedom of the quantum field theory, but only the parameters of equilibration such as the temperature, charge density and velocity, which can vary over space and time. If you make such assumption, symmetries and other simple thermodynamic considerations inevitably lead you to the equations of fluid dynamics.

From the point of view of field theory this sounds very simple but depends on the assumption of local equilibration, which is very hard to prove. I should have also mentioned that fluid dynamics is a classical description for the same reason that large-N theories are classical, that is, due to the law of large numbers. In particular, the flow of water is effectively classical because the number of water molecules inside the small region you're looking at is very large such that the fluctuations in macroscopic quantities are very small. Focusing on super Yang–Mills theory, in the context I am discussing, you find two different classical descriptions of the same system: the first one is gravity, which is supposed to be exact, and the second one is fluid dynamics, which is supposed to work at long wavelengths. If this statement is true then the exact description should reduce to the approximate description in the right limit. Verifying this was the main motivation behind this line of research [15].

It's not difficult to make this identification. You can identify a subspace of slowly varying solutions in gravity and realise that there is a one-to-one correspondence between these solutions and solutions of certain equations of fluid dynamics. These equations are generated self-consistently from gravity as well as the parameters in these equations yielding, for example, the viscosity to entropy ratio to be $1/4\pi$ and many other such relations. The most important aspect of this work is not the fact that we can calculate parameters using this procedure but the fact that we can understand how the equations of gravity reduce to the Navier-Stokes equations in the appropriate limit. I think of this field-theoretical expectation being realised in the context of gravity as a toy version of the AdS/CFT correspondence.

But isn't this just taking AdS/CFT in some limit?

Yes, exactly, but in this specific limit we have proven the correspondence. AdS/CFT is complicated because it relates a strongly coupled theory to classical gravity. However, this is much simpler because you just have to relate classical gravity to classical fluid dynamics. I think that this result is interesting for many reasons. First, it allows you to predict parameters of fluid dynamics and there are surprises; for instance, charged fluid dynamics didn't fit the standard framework, as I mentioned earlier, neither did the theory of superfluidity [32, 33]. By doing calculations using AdS/CFT, we found that the equations of superfluidity as written in books on the subject, such as in the book of Putterman [34], were not correct. Putterman reports that the equations of superfluidity are characterised by 13 different dissipative parameters but our calculations revealed that there are in fact 14 parameters.

Despite being interesting for clearing issues about fluid dynamics, this type of study is also useful for gravity, even at the conceptual level. Let me give you an example. We're all familiar with how fluid dynamics arises from theories that describe, for instance, water flow: you have a lot of water molecules and you average over them resulting in an effective description which is fluid dynamics. However, we know now that gravity is fluid dynamics in a particular regime and as such the gluons are like water molecules. Thus you can ask, "What is the relation between gravity and field theory?" At least in a certain regime, the answer is that this relation is exactly the same as the relation between water molecules and the variables of fluid dynamics. As such, anyone who would try to quantise fluid dynamics rather than quantising the motion of water molecules would probably be, and justifiably, laughed at. Perhaps there could be some correct features of such quantisation procedure but it would certainly not allow you to extract the correct underlying effective description. Therefore, this suggests that quantising gravity directly is silly: you have to quantise the underlying degrees of freedom that give rise to gravity and not the equations of gravity itself. Pursuing this line of thought could lead you to many interesting insights into the basic structure of gravity as a quantum theory.

However, I should emphasise again the comment I made earlier, namely, that in physics words are cheap. It's easy to use words that sound fancy but what is of value are equations. Everything I am claiming should be backed up by equations. For instance, moving away from the $N = \infty$ limit in this correspondence is a very interesting task. There are a couple of papers in this direction but they are still preliminary. Doing it in a way that keeps this physical motivation in mind would be very interesting. I would love to put the right equations to these words but I haven't yet succeeded.

You mentioned the superfluid story, but how do you promote fluids to superfluids in this setting?

This is actually other people's work (laughs). What is a superfluid? A superfluid is a fluid that includes a conserved $U(1)$ charge and where you have a condensate of a field carrying this conserved charge, like in liquid helium. Having a conserved $U(1)$ charge in gravity is easy as you just need to add the Maxwell term to Einstein's Lagrangian. Adding the condensate is more interesting and no one had thought about it until Gubser realised that it was easy for charged scalar fields to condense inside an anti–de Sitter vacuum [35]. Thus, you promote a fluid to a superfluid by looking at fluid flows not of black branes but of black branes with charged scalar condensates.

The way you established this correspondence was by doing a perturbative calculation, right?

Yes, by doing exactly what we did for normal fluids [15].

But does a perturbative calculation constitute a formal proof? What about non-perturbative corrections?

Fluid dynamics is by nature an expansion in the long wavelength limit. What is the logical structure of fluid dynamics itself? Fluid dynamics, like any effective theory, is an expansion in a parameter, the parameter being $1/\lambda$ where λ is the wavelength of the perturbation. What is the structure of this expansion? Is it convergent? Unlikely. Is it Borel summable? Nobody knows. These are the kind of questions you could ask and they are interesting questions that gravity could help you answer. Nevertheless, whether or not an expansion in a parameter is convergent, it is still a useful expansion.

QED, for example, is solved in an expansion in a parameter. We know that the perturbative expansion of QED is not convergent. There are non-perturbative effects. Nonetheless, at weak coupling, two-loop calculations in QED give fabulous agreement with experiment. At the Harvard physics department they have the theoretical and experimental predictions of the anomalous magnetic moment of the electron written down side by side with 50 decimals of precision and they all agree, despite the fact that the QED expansion is not convergent. So we should distinguish between two things: a good effective description when a parameter is small and a convergent expansion. These are two logical separate aspects. QED is a fancy example but take, for example, the harmonic oscillator and add a X^4 potential. Quantum mechanical perturbation theory is not convergent in this case, yet it is very useful. This has the same logical status as the case of fluid dynamics.

What does it mean that it is not convergent?

Okay, let me assume that the expansion is Borel summable. Suppose that this is the case, then you can ask, "How near to the correct answer can perturbation theory take you?" And the answer typically, under certain assumptions, is e^{-g^2}, so if the coupling g is small then you are very near the correct result but not arbitrarily close to it at finite coupling. So this question is the type of question that you can ask whenever you have a perturbative expansion and I would say that gravity can actually help you answer this question.

Your earlier question was, "Don't you have non-perturbative corrections?" You can ask this in gravity and the answer is, "Of course." Consider a black brane. This black brane has many quasi-normal modes; most of them are massive quasi-normal modes, in the sense that as the wave number k goes to zero, the frequency ω goes to a fixed number set by the temperature. However, the black brane has four quasi-normal modes that are massless. Fluid dynamics is effectively the non-linear theory of these four massless quasi-normal modes obtained after integrating out the massive ones. From the point of view of fluid dynamics, are these non-perturbative effects? Of course, the massive quasi-normal modes are associated with the non-perturbative degrees of freedom that you integrated out. Analogously, the W-boson is the particle/mode for fermionic theory that you integrate out. Consequently, the four-fermionic interaction fails when approaching the mass of the W-boson, as at that scale new propagating degrees of freedom need to be taken into account. In fact, the gravity description makes it very clear that the fluid expansion cannot be convergent and by identifying the degrees of freedom that you have integrated out, you could make precise the structure of the fluid dynamics expansion.

Okay, so now that you have this correspondence, can you use it to make any experimental prediction?

As I said before, unfortunately what string theory has been good at so far is answering formal questions. If there is a formal question about quantum field theory that you are interested in, there is a chance that string theory could help you answer it. By formal questions I mean statements of the following type. Suppose that you make some general considerations and suggest that all field theories have a certain structure of equations of fluid dynamics. You could test this using AdS/CFT.

The interesting aspect of this is that several examples have been tested and what was found was that a number of formal statements were wrong. Some examples are the anomaly term in charged fluid dynamics [17, 18] and statements about the structure of higher-derivative corrections to the Navier-Stokes equations in the absence of charge [15], but perhaps the one which will turn out to be the most useful in the end might be the case of superfluid dynamics [32, 33]. Many statements about superfluidity seem wrong and gravity can help you get the correct version. The work by Subir Sachdev et al. on a theory of conductivity which was not known to condensed matter physicists before is another such example [36]. In summary, you use gravity to guide you to get the right general theory that can then be applied to a larger class of field theories. In this sense, it can, and perhaps already is starting to have an experimental impact.

But isn't there one example of a prediction, for instance, in the superfluid context?

We consider superfluids which have a gravitational dual [32, 33] and there's no indication that liquid helium is one of these. In this superfluid context we are performing the general analysis correctly and since it hadn't been done before, it could have an impact. Unfortunately, we don't have a gravity dual for liquid helium so we can't predict the coefficients of that physical system, which would be fantastic.

Yes, but couldn't you make an experiment and try to fit the results to the structure of the equations?

Yes, that you could try to do. If you ask me, "Has it been done before?" the answer is, "Not to my knowledge."

Well . . . but no one knew that there were 14 parameters anyway . . .

Yes (laughs), but it is a separate question of how easy it is for an experimentalist to measure all these coefficients. That is something I don't know anything about. However, it is still important to have the right equations.

Now I want to ask you another question which is unrelated to fluid dynamics but related to AdS/CFT. I watched one of your public talks online entitled "Space and Time beyond Einstein" that took place somewhere around here.[1] What notion of space and time were you referring to?

In that talk, I described Albert Einstein's notion of spacetime and then I described how the AdS/CFT conjecture generalised this notion. Something I find persistently fascinating, without being able to put any satisfactory equations in it, is the following. $\mathcal{N} = 4$ super Yang–Mills theory, which is a theory in four dimensions, has a classical description in 10 dimensions with genuinely local Lorentz invariance in that 10-dimensional space. Now, suppose that you take $\mathcal{N} = 4$ super Yang–Mills at finite N, say, $N = 2$. In that case, I don't think that there will be any 10-dimensional space, while at $N = \infty$ there is. Thus, as we discussed earlier, moving N away from infinity amounts to effectively quantising gravity. Therefore, at least in this example, space and time in a gravitational theory are not just fluctuating notions but are approximate notions; i.e., they are not exact notions. At finite N, there is no way of formulating the theory in terms of space and time. This is what I was speaking about in that talk, namely, how space and time are emergent notions in the AdS/CFT correspondence.

But do you think that this is just in AdS/CFT or is it a fundamental perspective?

I think it is likely to be fundamental.

So, in your opinion, gravity is not a fundamental force?

[1] See www.youtube.com/watch?v=phP3QdqiaeU.

Yes, I think that's likely to be true. When everything settles down with time, I think we'll find that the classical limit of gravity is one of these large-N limits. That is a very general statement but it's my guess. Space and time emerge precisely only at large N and when you move away from large N these notions are no longer appropriate.

So if I look through the window as I move away from large N, what do I see?

If you were an observer in this space, I suppose that if N was large but finite you would be seeing a fluctuating spacetime. But if you went further away from large N you could ask, "What would a conscious observer in $\mathcal{N} = 4$ super Yang–Mills at large N see?" Well ... he would be some guy who lives in some planet in $AdS_5 \times S^5$ (laughs), and when N moves away from infinity, spacetime starts fluctuating more and more. $N = 2$ conscious observers wouldn't be able to see spacetime at all but I'm really going off on the limit here (laughs). The nature of experience is essentially classical so you would just see a mess: fluctuations become so large that there is no sense in which you could think of space and time.

Why are you doing physics? Why are you not catching coconuts outside?

(laughs) I do it because I enjoy it. I suppose I always wanted to do physics, since I was very young. You know that grown-ups sometimes ask young people, "What do you want to be when you grow up?" I remember that when I was very young I would answer, "A scientist and a cricketer." But at some point I dropped the cricketer (laughs).

That's very Indian ... (laughs)

(laughs) It's a great life.

What do you think is the role of this field in modern society? What do theoretical physicists have to offer?

I think we should be honest about this. The likely technological spinoffs of direct research into quantum gravity in the foreseeable future are few. Of course the technological spinoffs with the more AdS/CFT-inspired research could be more immediate. For instance, you may be able to use string theory to understand turbulence or high T_c superconductivity which could then lead to faster trains or faster airplanes.

I should elaborate a bit more on this. It is clear that a condensed matter physicist feeds into the technological chain, so to the extent that string theory fits into that condensed matter community, it also feeds into that chain in a small way. However, I would say that the real answer to this question is not of technological nature; it's part of it but not all of it. The real answer is that human beings ever since they became beings have looked up to the sky and wondered, "Why are we here? How did it begin? and Where are we going?" These questions are a core part of being, that is, the fact that we are curious creatures that do not live by the brain alone.

It is remarkable how much of our universe we understand. We sit on an obscure little planet two thirds between the centre and the rear of an obscure galaxy and we have this mental picture of the universe that stands across billions of light years away, and we can take that picture and extrapolate back in time and so on. We have already achieved a remarkable amount in answering the kind of questions that people have asked for a million years but we haven't reached the end of that. For instance, to the question, "How did it all begin?" we still don't have an answer.

I think that in the end human beings are willing to take a little bit from their wallet to fund a crazy band of people who might help answer these questions (laughs). If you understand how the universe began, it would be a fundamental advance in human knowledge forever. That is the real reason to study string theory, that is, to answer the core basic questions that human beings deep down within them want to know the answers to. I think that society understands this, sometimes better than scientists do. People want to know the nature of reality deep down whether or not it has a technological impact. That is the real reason why I study string theory and I think that the man on the street understands that and is willing to pay for it without getting into faster trains (laughs).

References

[1] M. Gogberashvili, "Hierarchy problem in the shell universe model," *Int. J. Mod. Phys. D* **11** (2002) 1635–1638, arXiv:hep-ph/9812296.

[2] L. Randall and R. Sundrum, "Large mass hierarchy from a small extra dimension," *Phys. Rev. Lett.* **83** (Oct. 1999) 3370–3373. https://link.aps.org/doi/10.1103/PhysRevLett.83.3370.

[3] L. Randall and R. Sundrum, "An alternative to compactification," *Phys. Rev. Lett.* **83** (Dec. 1999) 4690–4693. https://link.aps.org/doi/10.1103/PhysRevLett.83.4690.

[4] M. Schmaltz and D. Tucker-Smith, "Little Higgs review," *Ann. Rev. Nucl. Part. Sci.* **55** (2005) 229–270, arXiv:hep-ph/0502182.

[5] J. M. Maldacena, "The large N limit of superconformal field theories and supergravity," *Int. J. Theor. Phys.* **38** (1999) 1113–1133, arXiv:hep-th/9711200.

[6] S. S. Gubser, C. P. Herzog, S. S. Pufu and T. Tesileanu, "Superconductors from superstrings," *Phys. Rev. Lett.* **103** (2009) 141601, arXiv:0907.3510 [hep-th].

[7] S. S. Gubser and A. Yarom, "Pointlike probes of superstring-theoretic superfluids," *JHEP* **03** (2010) 041, arXiv:0908.1392 [hep-th].

[8] S. S. Gubser and A. Nellore, "Ground states of holographic superconductors," *Phys. Rev. D* **80** (2009) 105007, arXiv:0908.1972 [hep-th].

[9] N. Seiberg and E. Witten, "Electric-magnetic duality, monopole condensation, and confinement in N = 2 supersymmetric Yang–Mills theory," *Nucl. Phys. B* **426** (1994) 19–52, arXiv:hep-th/9407087. [Erratum: *Nucl Phys. B* **430** (1994) 485–486.]

[10] N. Seiberg and E. Witten, "Monopoles, duality and chiral symmetry breaking in N = 2 supersymmetric QCD," *Nucl. Phys. B* **431** (1994) 484–550, arXiv:hep-th/9408099.

[11] S. Lee, S. Minwalla, M. Rangamani and N. Seiberg, "Three point functions of chiral operators in D = 4, N = 4 SYM at large N," *Adv. Theor. Math. Phys.* **2** (1998) 697–718, arXiv:hep-th/9806074.

[12] P. Candelas, G. T. Horowitz, A. Strominger and E. Witten, "Vacuum configurations for superstrings," *Nucl. Phys. B* **258** (1985) 46–74.

[13] B. R. Greene, "String theory on Calabi-Yau manifolds," in *Theoretical Advanced Study Institute in Elementary Particle Physics (TASI 96): Fields, Strings, and Duality*, C. Efthimiou and B. Greene, eds., pp. 543–726. World Scientific, 1997.

[14] E. Witten, "String theory dynamics in various dimensions," *Nucl. Phys. B* **443** (1995) 85–126, arXiv:hep-th/9503124.

[15] S. Bhattacharyya, V. E. Hubeny, S. Minwalla and M. Rangamani, "Nonlinear fluid dynamics from gravity," *JHEP* **02** (2008) 045, arXiv:0712.2456 [hep-th].

[16] L. Landau and E. Lifshitz, *Fluid Mechanics*, vol. 6. Elsevier Science, 2013.

[17] N. Banerjee, J. Bhattacharya, S. Bhattacharyya, S. Dutta, R. Loganayagam and P. Surowka, "Hydrodynamics from charged black branes," *JHEP* **01** (2011) 094, arXiv:0809.2596 [hep-th].

[18] J. Erdmenger, M. Haack, M. Kaminski and A. Yarom, "Fluid dynamics of R-charged black holes," *JHEP* **01** (2009) 055, arXiv:0809.2488 [hep-th].

[19] D. T. Son and P. Surowka, "Hydrodynamics with triangle anomalies," *Phys. Rev. Lett.* **103** (2009) 191601, arXiv:0906.5044 [hep-th].

[20] C. Hoyos, N. Jokela, M. Jarvinen, J. G. Subils, J. Tarrio and A. Vuorinen, "Transport in strongly coupled quark matter," *Phys. Rev. Lett.* **125** (2020) 241601, arXiv:2005.14205 [hep-th].

[21] P. Kovtun, D. T. Son and A. O. Starinets, "Viscosity in strongly interacting quantum field theories from black hole physics," *Phys. Rev. Lett.* **94** (2005) 111601, arXiv:hep-th/0405231.

[22] P. Romatschke and U. Romatschke, "Viscosity information from relativistic nuclear collisions: how perfect is the fluid observed at RHIC?," *Phys. Rev. Lett.* **99** (2007) 172301, arXiv:0706.1522 [nucl-th].

[23] P. Romatschke and U. Romatschke, *Relativistic Fluid Dynamics In and Out of Equilibrium*. Cambridge Monographs on Mathematical Physics. Cambridge University Press, 5, 2019.

[24] A. N. Kolmogorov, V. Levin, J. C. R. Hunt, O. M. Phillips and D. Williams, "The local structure of turbulence in incompressible viscous fluid for very large Reynolds numbers," *Proceedings of the Royal Society of London. Series A: Mathematical and Physical Sciences* **434** no. 1890 (1991) 9–13.

[25] A. N. Kolmogorov, V. Levin, J. C. R. Hunt, O. M. Phillips and D. Williams, "Dissipation of energy in the locally isotropic turbulence," *Proceedings of the Royal Society of London. Series A: Mathematical and Physical Sciences* **434** no. 1890 (1991) 15–17, https://royalsocietypublishing.org/doi/pdf/10.1098/rspa.1991.0076.

[26] U. Frisch and A. Kolmogorov, *Turbulence: The Legacy of A. N. Kolmogorov*. Cambridge University Press, 1995.

[27] G. 't Hooft, "A planar diagram theory for strong interactions," *Nucl. Phys. B* **72** (1974) 461.

[28] E. P. Wigner, *Characteristic Vectors of Bordered Matrices with Infinite Dimensions I*, pp. 524–540. Springer Berlin Heidelberg, 1993.

[29] D. J. Gross and E. Witten, "Possible third-order phase transition in the large-N lattice gauge theory," *Phys. Rev. D* **21** (Jan. 1980) 446–453. https://link.aps.org/doi/10.1103/PhysRevD.21.446.

[30] S. R. Wadia, "N = infinity phase transition in a class of exactly soluble model lattice gauge theories," *Phys. Lett. B* **93** (1980) 403–410.

[31] I. R. Klebanov, "String theory in two-dimensions," in *Spring School on String Theory and Quantum Gravity (to Be Followed by Workshop)*, pp. 30–101. Trieste Spring School, 1991.

[32] J. Bhattacharya, S. Bhattacharyya and S. Minwalla, "Dissipative superfluid dynamics from gravity," *JHEP* **04** (2011) 125, arXiv:1101.3332 [hep-th].

[33] J. Bhattacharya, S. Bhattacharyya, S. Minwalla and A. Yarom, "A theory of first order dissipative superfluid dynamics," *JHEP* **05** (2014) 147, arXiv:1105.3733 [hep-th].

[34] S. Putterman, *Superfluid Hydrodynamics*. Low Temperature Physics Series. North-Holland Publishing Company, 1974.

[35] S. S. Gubser, "Colorful horizons with charge in anti-de Sitter space," *Phys. Rev. Lett.* **101** (2008) 191601, arXiv:0803.3483 [hep-th].

[36] S. Sachdev, "What can gauge-gravity duality teach us about condensed matter physics?," *Ann. Rev. Condensed Matter Phys.* **3** (2012) 9–33, arXiv:1108.1197 [cond-mat.str-el].

19

Hermann Nicolai

Professor at the Max Planck Institute for Gravitational Physics and Director Emeritus

Date: 24 April 2013. Location: Potsdam. Last edit: 30 November 2020

In your opinion, what are the main puzzles in high-energy physics at the moment?

From the point of view of particle physics, the main open question is now what comes after the standard model (SM), or indeed, whether there is anything new at all, other than the Higgs boson. To be sure, there remain several important questions in particle physics, for instance, what is dark matter made of, or what is the mechanism driving baryogenesis, or cosmology-related issues, like what is the inflaton, and how precisely does it interact with the SM to produce SM matter particles upon reheating? Likewise, on the mathematical side there remain challenges like the Millenium problem of rigorously proving the existence of a mass gap in QCD. But overall the SM is in very good shape, and so is quantum field theory (QFT). In a larger perspective, however, there looms the unsolved problem of unifying quantum theory with general relativity, and this is undoubtedly the greatest challenge of all.

Why do we need a theory of quantum gravity?

In my view a chief argument for the necessity of a quantum theory of gravity is the incompleteness of the existing theories of physics, general relativity (GR) and QFT. On the GR side, the problem of spacetime singularities has been around for a long time; such singularities occur inside black holes, and at the beginning of the universe (big bang). At a singularity, space and time and, with it, the known laws of physics come to an end. On the QFT side there are the ultraviolet divergences that we have learned to deal with, though somewhat uneasily, in perturbation theory, and also the question of the existence of these theories as mathematically well-defined theories beyond perturbation theory. The status of QFT was greatly clarified in the early 1950s when Arthur Wightman formulated the axioms named after him [1, 2]. He quickly realised that free field theories satisfy these axioms (locality, relativistic invariance, positive energy spectrum), and so he thought it would only take a few weeks to produce non-trivial examples with interactions. However, to this date we still do not have a single example of a fully constructed four-dimensional quantum field theory which is non-trivial, that is, whose S-matrix is different from the identity. A case in point is ϕ^4 theory: we know it exists below four dimensions and we know it does not exist above four dimensions, but for the most interesting borderline case of four dimensions we have no fully rigorous proof, despite considerable effort. The folklore is that, in fact, all theories which are not asymptotically free do not exist in a rigorous sense. This means that perturbation theory, even though it works marvellously – for example, if you think of the LHC precision tests of the SM – will break down eventually.

So why is it giving the right results?

That is a big puzzle. Why do you get the correct prediction for the anomalous magnetic moment of the electron to 12 digits in precision from a theory that may not even exist in any mathematically rigorous sense? In my view, this indicates that at the Planck scale the QFT framework *must* give way to something else. This "something else" cannot be based on familiar notions of space and time, but must nevertheless cleverly allow standard QFT to emerge together with space and time.

So you are saying that the standard model can be continued with no problem all the way to the Planck scale?

This is certainly a possibility, and all indications from LHC so far point in this direction. The idea that the SM, possibly with some very minor modifications, may survive all the way to the Planck scale is not very popular with many of my colleagues, but maybe that is what nature has chosen. In my work with Krzysztof Meissner we try to explore this latter possibility, adopting an "agnostic" attitude with regard to what happens precisely at the Planck scale, and how precisely QFT merges into this unknown theory of quantum gravity. Remarkably, with this minimalistic approach it is possible to derive some restrictions from the assumption that (a minimalistic extension of) the SM works all the way up to the Planck scale [3].

What kind of restrictions?

If the Higgs boson had turned out to be much heavier than the observed value of 126 GeV, the theory would run into a Landau pole problem somewhere around 10 TeV, and then the SM would simply break down right there. Another restriction comes from requiring the effective potential to be bounded from below, as there may arise instabilities due to quantum corrections from the fermions which contribute with the opposite sign. Because this effect is proportional to the coupling (or the mass) of the fermion, and because the top quark is the heaviest fermion in the SM, it is the chief agent responsible for a potential instability. What is really remarkable is that the measured values of the Higgs mass and the top quark mass seem to indicate that the SM sits right on the verge of an instability that would set in around 10^{10} GeV. However, it is a small (and curable!) effect, so maybe nature is trying to tell us something here.

So your feeling is that no matter what they will do at the LHC or how powerful these accelerators will be the standard model will always be the correct description of the physics?

This is not a necessity, but definitely a possibility up to the Planck scale. And I am of course aware that many of my colleagues would have preferred a "supersymmetric" outcome of the LHC experiment. Maybe it sounds curious that I am not a believer in low-energy supersymmetry, having grown up as a student of Julius Wess in the 1970s. When the idea of low energy ($N = 1$) supersymmetry first came up [4] it was clear that this was a matter for experiment to decide. But then it did not show up at LEP, nor did it show up at the Tevatron, and after 10 years of running it has also failed to show up at LHC. Obviously, the idea will loose credibility the higher we push up the energy without seeing any traces of it. A more aesthetic objection (as opposed to experimental falsification) is that $N = 1$ supersymmetry (SUSY) models – apart from cancelling some quadratic divergences – do not exploit the one feature that in the early days was regarded as the main virtue of SUSY, namely that it can merge spacetime and internal symmetries, thus circumventing the Coleman-Mandula theorem which forbids such a "merger" within the framework of more standard kinds of symmetry [5]. But this unification/merging only takes place for extended SUSY theories (for which $N > 1$); it does not take place for $N = 1$. This is why you have to double up every boson by a fermion with the same internal quantum numbers and vice versa. This does not happen for extended SUSY, but there we don't know how to make fermions chiral.

So you say that you run the SM all the way to the Planck scale and then something is modified? Is that string theory?

For sure, the SM will be modified there for the reasons that I already mentioned earlier, but we still have no indication from experiment or observation that string theory is really the right theory. There has been a huge effort to derive the SM from string theory, especially with the heterotic string [6], widely considered to be the most promising pathway from quantum gravity to the SM. However, while there was clearly the hope in the original work of Candelas, Horowitz, Strominger and Witten that this should be a unique and beautiful path [7], this has not really panned out so far. Instead, we now get this plethora of string models, Calabi-Yau compactifications with and without fluxes, intersecting brane models, F-theory unification, and so on [8, 9]. Following this line of thought you are invariably led to introduce lots of new particles and couplings, no traces of which have been seen until now – in fact, to engineer away any immediate conflict with observation requires a great deal of ingenuity! This is even more so now that LHC has accumulated huge amounts of data not showing any signs of new physics. In other words, while you may get some features right, you never get just the standard SM *as is*. In this situation, many take recourse to the multiverse and anthropic reasoning [10], but I doubt that any useful prediction will ever come out of this. For instance, you can surely "explain" the motion of planets in terms of epicycles if you allow yourself to make the models arbitrarily complicated, with any number of free parameters, but I don't think anyone would find this acceptable. Better to stick with Occam's razor as a guiding principle!

What you are saying is that the standard model doesn't come naturally from string theory, right?

As I explained, this effort has been going on for almost 40 years, but we still have no compelling explanation why the SM is the way it is.

I think that the question is whether you believe in a framework where our universe is adjusted by measuring constants or you believe in an ultimate theory that predicts these values?

I am not sure about this ... there are many parameters that you cannot predict; they sort of accidentally have the values they have, like the distance between the Earth and the Moon, and no one would expect this to be a prediction of a fundamental theory. This statement may also apply to some "constants of nature." By contrast, if you think about the Einstein equation ... it gives rise to zillions of solutions, describing a huge variety of different physical situations, but the equation itself is a one-line statement. It is simply beautiful and unified. Analogously, I would prefer there to be a unique unified theory of quantum gravity which is fixed by similarly beautiful and simple principles like the ones that guided Albert Einstein to his theory, and that eventually would give rise to the myriads of physics phenomena that we see.

But wouldn't that equation have many solutions?

Of course the equation would have many solutions, but this is not like the multiverse idea. There isn't a beautiful equation there that everyone agrees on and that fits into one line. With the multiverse, you are not even sure what the precise theoretical framework is; it's more words than equations. And there appears to be no way to test or refute the idea, other than to say that the laws of nature must of course be compatible with our existence and with what we see.

So what is your expectation for the theory that describes physics at the Planck scale?

I wish I knew what it is, but I would think that there should be some beautiful principles describing it. Einstein was led to his equations by way of two conceptually very simple principles, general covariance and the equivalence principle. Unfortunately, at this time I cannot see any such conceptual simplicity in the different approaches to quantum gravity.

In your review article on quantum gravity you mention the two sides of Einstein's equations and that Einstein himself tried to give the right-hand side of the equation a geometric interpretation [11].

Turning timber into marble ...

The only approach I am aware of that tries to do something along these lines is the approach by Alain Connes in which geometry is described by some more abstract structures from which the standard model then comes out naturally [12]. Is this not along the lines that you agree with?

Well, Kaluza-Klein theories and supergravity also try to endow the right-hand side with a geometric meaning, by enlarging space and time with extra (bosonic or fermionic) dimensions! But yes, non-commutative geometry à la Chamseddine-Connes is certainly an attempt in this direction, and one that gives you the Higgs boson "for free." Nevertheless, I have some doubts regarding the claimed uniqueness of this proposal because this model of non-commutative geometry in some sense depends on what you put in: e.g., if you start with three-by-three matrices, you get $SU(3)$, and so on. So the question remains: *Why* is it $SU(3) \times SU(2) \times U(1)$? And why are there three generations? Here the non-commutative approach cannot offer an answer. But in all fairness, to get the answer "three" out of other approaches, you also have to bend and massage the theory in very elaborate ways; there is nothing here that distinguishes this number from many other possibilities.

People have been discussing $N=8$ maximal supergravity (SUGRA) and it seems to be finite [13–17].

It will be very hard to check if it is finite or not, but it is clear that this theory stands out among all quantum field theories, as it is the (almost) unique maximally supersymmetric extension of Einstein's theory in four dimensions. So I think the theory has some special role to play, but what exactly its role is in the scheme of things, I don't know. Regarding the finiteness, it was expected for 30 years that it would diverge at three loops, but then came this calculation proving three-loop and four-loop finiteness [17, 18], a truly amazing calculation, by the way. And more recently they have even been able to do five loops [19]! If you were to calculate this in terms of ordinary Feynman diagrams there would be more diagrams than cells in your brain (laughs) – just to give you a hint of the progress that has been achieved concerning the technology required! This calculation has now confirmed the expected finiteness of $N = 8$ supergravity at five loops, but still leaves the question of all-order ultraviolet (UV) finiteness up in the air. Current expectations are that a divergence might appear at seven loops or higher up [16], but even with this vastly improved technology it seems unlikely that it will be possible to verify the finiteness (or not) of the theory any time soon.

But suppose that it is finite to all orders ...

Then we have a QFT extension of Einstein's theory where magically all infinities cancel ... amazing! (laughs).

But does it have anything to do with the real world?

If you take it at face value, it would not seem to have any relation with the real world we live in. But then again, if you had proposed QCD as a theory of strong interactions in 1950, everyone would have said that you are *obviously* wrong, simply because no fractional charges have ever been seen in a particle detector.

So you think that if it is finite to all orders, it could be an extension of the standard model?

There is one curious fact first pointed out by Murray Gell-Mann, which has been on my mind for a long time, but I have no idea whether this is a mirage or a real hint. $N = 8$

SUGRA has 56 spin-$\frac{1}{2}$ fermions, and if you break all the supersymmetries (which you must), you remove eight fermions (Goldstinos) and you are left with 48, which is exactly the number of quarks and leptons in nature – three generations of 16 fermions, each including right-handed neutrinos [20]. If the theory is finite to all orders and if they do not find new fermionic spin-$\frac{1}{2}$ degrees of freedom at the LHC or at future collider experiments, one will have to reconsider this possibility. There are not so many theories that predict precisely the right number of fermions. Such theories do not grow on trees (laughs).

But isn't this against your idea that the standard model is not modified all the way to the Planck scale?

No, because with precisely 48 fermions there is no room for extras! Of course, the difficult task of embedding the SM into this Planck scale theory remains, which at the moment no one knows how to do. In any case, $N = 8$ SUGRA will not be the final answer, but perhaps part of it. As I said before, at the Planck scale I expect conventional spacetime notions to break down, while $N = 8$ supergravity is still within the framework of QFT. So if it is finite, and perhaps the unique finite extension of Einstein's theory, it just means that there is something right about it, but what replaces spacetime-based QFT at the Planck scale remains an open question.

Why have you tried to impose conformal symmetry in the standard model [3]?

Because the SM is classically conformally invariant, except for one term which is the term $m^2\phi^2$ in the potential. By following this idea, you start with a conformal theory and then you can try to explain the origin of masses from the conformal anomaly, the quantum mechanical breaking of the conformal symmetry. The nice thing about conformal symmetry is that you can have it classically, but it will almost always be broken by quantum effects. There are very few exceptions to this rule; for example, the famous $N = 4$ Yang–Mills theory is conformally invariant also quantum mechanically. Generically, when you start with a classical conformal quantum theory, it will break the symmetry at the quantum level. Curiously, you are here in a much better situation than with SUSY, because SUSY, once you have it, is very hard to get rid of. In 40 years no one has found a truly compelling mechanism to break SUSY. So in existing models it is simply broken by hand, which is not very appealing to me.

But can you predict anything from this? Is there is a unique way of breaking conformal invariance?

The first idea is to use the Coleman-Weinberg effective potential. This is a very old idea, though it is known not to work for the pure SM. In the wider context that we are exploring it is an open question whether it can be made to work, because radiative symmetry breaking is more difficult to implement technically. So we have tried to adapt it a little bit (as have other groups of researchers). For instance, you need to make reliable calculations for the effective potential with more than one scalar field. And eventually you would have to figure out how this ansatz fits with ideas about a ultraviolet complete theory of quantum gravity.

So, so far there is no prediction?

The prediction is that it is just the SM, and then maybe one or more extra scalar particle resonances. Logically this idea is as well motivated as many other ideas to go beyond the SM. And we think that it may actually help to solve some questions that are still open in the SM, such as the possible instability of the effective potential that I mentioned earlier; see, e.g., [21] and references therein.

Could you observe this resonance at the LHC?

For this you need to calculate decay widths, acceptance rates and so on. But even if it is there it will not be easy to see.

Why not?

First of all because it would couple only very weakly to SM matter. Second, if you had point particle collisions (like in an e^+e^- collider) you would be able to test the proposal exactly by fine-tuning the centre of mass energy. The problem with a hadron collider like LHC is that you're colliding composite particles, but the processes take place between constituents (partons, i.e., quarks and gluons). For instance, when you produce W-bosons, you have no complete control over how the energy is distributed between the partons, so it takes a special effort and great ingenuity to calibrate. CERN has been able to detect the Higgs by analysing zillions of collisions, where it happens only rarely that two partons have exactly the right energy and interact in the right way to produce the Higgs resonance.

Are there any prospects for building a linear collider?

The general expectation was that LHC should have seen lots of SUSY particles by now. If that had happened you could have argued, now that we have them we can build the linear collider and do precision measurements. But if there's nothing, the case for building a linear collider for that energy range is weakened, although you could still learn a lot from precision experiments. But let's wait and see, now with the upcoming LHC upgrade perhaps something will turn up at higher energy and luminosity.

I would like to shift the conversation to quantum gravity ... you mentioned also that $N = 8$ may be the unique extension of Einstein's theory. However, in three dimensions there are many ways of quantising gravity ...

That's because gravity does not have propagating degrees of freedom in three dimensions; it's just like quantising topology, and topology is usually characterised by a finite number of degrees of freedom.

I agree that that's why there are many different quantisations. Is there anything special about four dimensions that constrains the quantisation to some unique one or could there be many extensions?

Even in three dimensions, where you think you're dealing with a quantum mechanical system instead of a quantum field theory, we have not reached a complete consensus on how to properly do this. You might speculate that a given quantisation procedure could be ruled out by experiment, but we are not in three dimensions! In four dimensions we have a quantum field theory with propagating degrees of freedom, and we know that you can have inequivalent representations of the algebra of canonical commutation relations. Unlike in quantum mechanics, in QFT we thus have physically inequivalent realisations. It is one of the chief statements of loop quantum gravity (LQG) that their approach is inequivalent to the old Wheeler-DeWitt approach for precisely this reason. The Wheeler-DeWitt approach with metric variables has so far been a failure, but if you have an inequivalent representation, then maybe it can be made to work after all. And if you quantise covariantly, then there may be even more possibilities. So there are many prescriptions with different physics predictions, but none of them has led to a good answer, in the sense that this is now the answer that everybody agrees upon. By contrast, in string theory there is at least some consensus what the theory is, or should be.

Can you embed the SM into loop quantum gravity?

The SM can be embedded into LQG at a trivial kinematical level, by simply rewriting its Lagrangian in terms of Ashtekar (or rather, flux and holonomy) variables, but that does not mean very much. LQG does not make predictions about the matter content of the world, because you can do this with essentially any Lagrangian. In particular, LQG does not "see" the anomalies whose cancellation in the SM is one of the cornerstones of the theoretical consistency of the SM [22].

I understand, there is no real prediction ...

No, not in my opinion.

Can you apply LQG methods to string theory?

Yes, you can, but it does not give the same answers. String theory can be viewed as an example of a theory of quantum gravity in two dimensions with matter couplings. In the worldsheet description the string coordinates are like matter fields, and the Virasoro constraints are just the canonical constraints of quantum gravity (including the Wheeler-DeWitt equation). So right there you have a theory in front of your eyes where you can directly compare the different quantisation procedures. But LQG quantisation techniques will not give you the critical dimension [23].

But there has been a connection between topological strings and the topological sector of LQG [24].

Yeah, I heard about this some time ago ... there are some formal similarities, like the use of holonomy variables. But it looks like the excitement about the claimed connection was rather short-lived.

However, in your article, you say that there should be a convergence between the approaches [11].

If we were closing in on the right answer, agreement between the different approaches should definitely emerge. In 1924, Werner Heisenberg invented "matrix mechanics," while at about the same time, Erwin Schrödinger came up with his wave equation. At first people thought that these were two different theories, and Heisenberg argued publicly against Schrödinger. But within a year or so, it became clear that they were just different sides of the same coin.

But in this case, in quantum gravity (QG) there are all these approaches that start with completely different axioms.

Different axioms and assumptions, that may even be mutually inconsistent and contradictory.

So why there should there be a convergence of the approaches?

Well, you might argue that you have different theories here, with different physical predictions, and then you just let experiment decide. But that is very hard at the Planck scale! So I would prefer to see convergence, but we are probably still very far from the correct answer.

Why is this problem of quantising gravity so difficult?

It is difficult because we don't even know what the fundamental degrees of freedom are of quantum gravity and quantum spacetime. It is like trying to figure out the physics of atoms and molecules from the Navier-Stokes equation, the fundamental equation governing fluid mechanics. When Niels Bohr proposed his model of the hydrogen atom he was still thinking about it in terms of very classical concepts, like there's the electron which moves around the nucleus, and on those classical concepts he superimposed his quantisation conditions. In this way he could explain the hydrogen atom, but the model could not be generalised to other atoms. This came only with Heisenberg and Schrödinger and Paul Dirac. A decisive step was taken by Heisenberg, who postulated that one should only use observable concepts; so in setting up matrix mechanics, he judiciously avoided using concepts like the trajectory of an electron around the nucleus.

In quantum mechanics reality is not described by following the motion of a point particle in three-dimensional space but more abstractly, by describing the motion of a wave function in infinite-dimensional Hilbert space – a completely different conceptual setting. In QG we just have no idea what the right concepts are at the Planck scale. This is why there are these very different ideas about what quantum space time should "look like," such as spin networks and spin foams, or some variant of non-commutative space. String theory is also not very clear about what the fundamental Planck scale degrees of freedom are – D-branes, micro black holes, matrices? A fundamental string is often pictured as a Planck-size object that wiggles around in a smooth background, but in a true theory of quantum gravity there shouldn't be any spacetime background anymore for a Planck-size string to move in.

So there isn't really an answer from string theory for what quantum gravity should be?

There are more abstract ideas about what the configuration space of string theory could be, such as the abstract space of all conformal field theories out of which classical spacetime arises as a "condensate." Or a "supermembrane," or a collection of "D0-branes," both of which lead to a maximally supersymmetric matrix model [25, 26].

Couldn't this be answered in the context of AdS/CFT [27]?

The AdS/CFT proposal is the most prominent embodiment of the so-called holographic principle according to which the physics that "takes place" in some volume can be fully encoded in the surface bounding that volume, like for a hologram [28, 29]. Some of my colleagues think that this idea is *the* key to quantum gravity. In its original form, the AdS/CFT proposal posits a correspondence between string theory (and thus quantum gravity) in the "bulk" of $AdS_5 \times S^5$ and maximally supersymmetric Yang–Mills QFT on its boundary. One principal goal would thus be to describe what appears like a spacetime singularity in the bulk in terms of a perfectly non-singular and unitary QFT on the boundary. However, we are not there yet, as you can see for instance from the (to me rather confused, and confusing) "firewalls" debate that has been raging in the past years [30, 31] (and that now seems to have given way to the "quantum extremal surfaces" paradigm). Also, anti–de Sitter (AdS) represents a highly idealised environment. It is beautiful, but when you look outside your window it does not fit: the cosmological constant is positive, and no one has come up so far with a really good idea of how the scheme should work in those circumstances.

What about the asymptotic safety mechanism?

That is an altogether different approach.

In one of your papers you mention that nothing happens at the Planck scale in the asymptotic safety approach [11].

That is correct: while all other approaches, despite their differences, are united in the belief that something dramatic must happen to spacetime at the Planck scale, the asymptotic safety program is based on the assumption that QFT works all the way to the smallest distances and thus remains valid even beyond the Planck scale. The magic trick that is invoked to save the theory from failure in the ultraviolet is the assumed existence of a non-trivial fixed point of the renormalisation group evolution. Accordingly, a main feature of that approach is the hypothesis that spacetime remains a continuum also below the Planck scale.

And this is not your point of view?

I am skeptical, although there certainly has been much progress recently regarding the evidence for a fixed point [32]. However, what strikes me is that almost no matter what model they try, they invariably find evidence for a non-trivial UV fixed point. For instance, there appears to be nothing inconsistent about quantum gravity in, say, 47 spacetime dimensions and with almost any menu of matter fields. By contrast, I believe there must be a reason why we live in four dimensions, so I would rather like to see *inconsistencies*, that for instance would allow me to conclude that the spacetime dimension cannot be just any number. On the other hand, if the answer is (almost) *anything goes*, I would not find this very satisfying, like the multiverse. Furthermore, the formalism has severe difficulties dealing with Lorentzian signature.

For you, what is, if any, the most promising approach at the moment?

I am more on the supergravity/string side because I think in building a theory of quantum gravity one cannot disregard the fact that there is matter in the world, and not just gravity. The very existence of matter does impose restrictions, for instance on the fermion multiplets of the SM. One of the basic disagreements between LQG and string theory is that the LQG proponents claim that one can quantise Einstein theory *as is*, and that matter is not needed for consistency, at least not in a first approximation. But we know that this is not nature's way of arranging consistency. The cancellation of anomalies in the SM is a very important clue and a very non-trivial observational fact – it is one of the few true hints of trans-Planckian physics! So I think the actual existence of matter in the world, and the fact that it may be needed to cancel divergences, is significant.

So can you suggest at least an avenue?

If you want to know where I place my personal bets, then on something called E_{10} [33–37]. This is a huge symmetry with unsuspected links to maximal supersymmetry and supergravity, encompassing and unifying many duality symmetries of string theory. The history of physics has shown that symmetry is a very powerful organising principle of the laws of physics. Both GR and the SM owe their success in part to symmetry concepts, so why not go further in this direction? Never change a winning team, as they say …

Is E_{10} a symmetry of nature?

I wish I knew … But if the idea of exploiting symmetries as far as they can take us finally turned out not to work, then we would have to look for altogether different concepts.

But how did you come up with the idea of the E_{10}?

The appearance of "hidden" exceptional symmetries in maximal SUGRA theories has been known for 40 years. The E_{10} symmetry, first conjectured to appear by Bernard Julia in 1983 [38], is really a unique mathematical beast, and one that one knows almost nothing about. What got us (my colleagues Thibault Damour, Marc Henneaux and Axel Kleinschmidt, and myself) excited about this symmetry is that it does seem to play a key role in understanding the initial big bang singularity [33, 39–41]. This insight is based on the seminal work of Belinski, Khalatnikov and Lifshitz (or BKL, for short) on cosmological singularities [42], according to which the world becomes effectively one-dimensional near the singularity. This led us to the hypothesis that the full symmetry of whatever this theory is reveals itself only at the singularity. It is a bit like the high-energy limit of gauge theories, where you do not see the $SU(2) \times U(1)$ symmetry at low energies, but instead have to go to 100 GeV to see W and Z bosons and to reveal the underlying symmetry. Here it is even more unusual because we claim that the symmetry that governs this theory only appears right *at* the cosmological singularity, which is kind of weird because you think that at singularities everything breaks down. The idea is that when you reach the Planck regime and conventional notions of spacetime break down, you can still use the symmetry as a guiding principle towards "physics without space and time."

Is there any prediction or implication of this symmetry?

No, at the moment this is just an exploration of the mathematical possibilities and we hope that one day this fixes the theory, but then again who is able to make a valid prediction about Planck-scale physics (laughs)? For the moment we are content to know that we can take Einsteinian theory as far as this, and that this ansatz leads us to exciting new ideas about how to implement symmetries in physical theories. For example, one nice thing about the E_{10} symmetry is that we know that it cannot be realised as the symmetry of a spacetime-based field theory. Already right there you see that you must at some point give up notions of space and time if you want to understand how it acts and how it is realised.

I do not know everything about the E_{10} theory, but what is the theory just before you reach the singularity and you recover the E_{10} symmetry?

The idea is that when you reach the Planck scale you replace the spacetime field theory by an entirely group theoretical description based on E_{10}, such that the information about the spatial dependence gets somehow "spread out" over the Lie algebra. One possibility that suggests itself here is that as you try to move closer to the singularity there is actually some element of non-computability, in a mathematically precise sense. There may be an insurmountable barrier there.

But then the theory after that Planck time would flow into something else?

That "something else" would be an example of the physics without space and time, which I alluded to earlier. Nobody knows what E_{10} really is: as you move into this mathematical structure there is exponential growth and no matter where you stop, it keeps getting more and more complicated as you continue to move into it.

But what I meant was whether $N = 8$ is the theory that then leads to E_{10}?

The discovery of hidden symmetries of exceptional type in maximal supergravities was what gave rise to the E_{10} conjecture, although the setting is certainly more general than in supergravity.

But what happens as you progress away from the Planck scale?

$N = 8$ SUGRA would somehow sit in there. My point of view is that you take that theory as a starting point, and then move in the direction of even more symmetry, where there is no more spacetime but instead bigger symmetries which also mix spacetime and internal quantum numbers.

So is there a lot of investigation still to be done in this E_{10} symmetry direction?

I think we've just scratched the surface. For example, you can decompose the algebra in terms of sub-algebras that are easier to handle, and list all the Lie algebra elements up to a given "level." With my former PhD student Fischbacher we have listed all the subrepresentations up to level 28 where we already find 4.5 billion tensors [43] (laughs). You can try to compute the Lie algebra structure constants and you can even do it by hand at very low levels, and maybe a very sophisticated computer can go up to level 10, but that still falls short of 4.5 billion Lie algebra elements by far. And remember, it will get worse as we explore this structure further. A good way to picture the monstrous complexity of E_{10} is to think of it as a Lie algebra analogue of a Mandelbrot set!

But then do you hope to get any predictability from this programme?

I don't know but it seems to me that in order to understand what it really is, you need a global understanding of this Lie algebra. In mathematics, there has been very little progress on indefinite Kac-Moody algebras in the 50 years since their discovery. Quite a few years ago, a famous mathematician who wrote one of the seminal early papers on the subject told me that he had given up on it ... it's too difficult. But recently I heard that he wants to save this as the last problem to solve in his life (laughs).

So hopefully he'll figure it out. You mentioned that there's no approach that has given any prediction whatsoever but I hear that LQG does make a prediction about the quantisation of space and time.

LQG predicts a quantisation of areas and volumes at the Planck scale, but that will be difficult to check even indirectly. The other problem is that neither area nor volume operator as defined by LQG is a physical observable. For that they would have to commute with the Hamiltonian.

So why do they claim ... I don't understand ... so you cannot measure this volume in any way?

Defining physical observables in GR is extremely hard, and for pure gravity we do not know a single one. What can be done is to define a kinematical object which is given in terms of flux and holonomy variables, whose spectrum you can calculate – with the plausible outcome that these quantities become quantised because the argument utilises $SU(2)$ angular momentum-like variables, and we know that the spin is quantised. However, for a physical observable you need to have relational definitions, and in order to measure length you need matter to construct a measuring device. So what you would have to do is to couple gravity to matter and construct a "ruler."

What did you mean by "relational" definitions?

It is probably easiest to explain this notion in terms of a simpler example, following Einstein's explanation in his famous 1905 paper on special relativity. There he declares that every measurement of time is really a statement of simultaneity of two events: the arrival of a train in the station and the arrival of the pointer at a specific position on my watch are simultaneous events. Measurement of length is analogously based on coincidence, requiring a ruler which the object to be measured can be compared with. So both measurements are done by "relating" two different objects or events, and this process has to be phrased in the language of mathematics.

Why haven't LQG people considered the spacetime volume instead of three-dimensional volume?

It is basically a Hamiltonian approach, where you give up manifest spacetime covariance. In this approach there is always an issue with spacetime diffeomorphisms. So the hard part is *spacetime*, not space. Space is kind of kinematical.

I always hear the claim that they have a background-independent formulation of QG.

The canonical formulation of LQG is background-independent, but it is only independent of the *spatial* background. To get true *spacetime* background independence they would need to be able to fully control the quantum constraint algebra, which they have not accomplished so far. In full-fledged quantum gravity this is as difficult as solving the Wheeler-DeWitt equation. So in my opinion, LQG is not yet background-independent in the same sense that the Einstein field equations are background-independent.

Okay. Do you expect that the right theory of QG is background-independent in this sense or is it an emergent notion?

It is not entirely clear because to formulate gauge field theory entirely in terms of gauge-invariant objects is not easy, and it is even not clear whether it is meaningful. In particle physics, you always choose a gauge, and check only at the end of the calculation that the physical prediction does not depend on the chosen gauge. So I do wonder sometimes whether the notion of background independence as employed in LQG-like approaches is not too restrictive. A basic challenge for LQG approaches is to recover the smooth spacetime that we live in. You can trace this difficulty back to their absolute background independence. So it could well be that this postulate is realised in a more subtle way, not simply by starting with a formulation that outlaws background dependence from the outset.

And this new development of spin foam models is not rescuing this?

It is an attempt to lift the Hamiltonian formulation to a spacetime-covariant setting, but there as well recovering continuum spacetime is a major challenge.

So they don't recover it either?

It's not like in lattice gauge theory where you have a background and you approximate it by making the lattice finer and finer. Because it is background-independent you have some object but you have no measurement of length, so you have to introduce this notion somehow. This is difficult because it is also not so clear if it should be a spin foam or a sum of spin foams or if one should take a refinement limit. Even among the experts opinions are divided on how to do this right.

It seems that you are very agnostic about all approaches but I would like tell you that I've noticed that you've done quite some work in M-theory. Why are you still working on M-theory?

Because there I see beautiful mathematical structures, and these have always been for me a main attraction. For $N = 8$ SUGRA, there are still many open questions that you can address. Not so long ago I returned to the issue of consistent truncations in Kaluza-Klein theories, an old problem, and we have been able to find a complete solution [44, 45], which we had tried in vain for almost 30 years to find.

But your study of M-theory is purely aesthetic? Is it purely mathematics?

I hope not . . . of course, the ultimate motivation for our work is to link these ideas to physics, *real physics*, but I don't know whether I will live to see the success of these efforts. We need endurance and perseverance. It is not a matter of the next two weeks or even the next two years.

If a young person comes to you and asks which line of orientation should they follow . . .

I would try to find them exciting problems to work on, but then encourage them to follow their own ideas. The present situation is a little like physics just before 1900, when the basic laws seemed more or less under control, except for some small corners where a few things just did not want to fit. So I think the solution will come, if it will come, from thinking harder about the foundations of the theory, the basic concepts, and from adopting a new and different perspective. Eugene Wigner said of Einstein that he was not smart but perseverant. Einstein never accepted anything until he had found his own way of thinking about it, and that was one of the secrets of his success. When following up on some idea, it is always good to ask oneself, Why should it be true in the first place?

Don't you think that this happens because of sociological pressure?

There is a sociological pressure because there are more theoreticians than ever before and not so many tenured positions. And there is also a strong pressure to conform with what the majority is doing. If I give my students an interesting but non-mainstream problem then their citation rates and their chances of getting jobs or postdocs are significantly reduced.

But I'm convinced that true progress will come from someone who thinks orthogonally to what the majority thinks.

I understand but it seems that from the current state of affairs someone wouldn't have a chance . . .

Yes, it's difficult.

Why did you choose to do physics?

I was fascinated by all these questions we have been discussing. Other physicists prefer lab work, and they probably do more useful things than what I'm doing. But I have always been fascinated by the beauty of the laws of nature.

What do you think is the role of a scientist like you in society?

We are immensely privileged to live in a society and at a time which allows us to engage ourselves in esoteric activities that very few people really understand. It is not self-evident for society to pay theoretical physicists for doing things that from a practical point of view seem mostly useless. Less lucky people who grow up in places like North Korea or Somalia do not enjoy such privileges. For this I am grateful, and so we should share our insights with whomever takes an interest in them. Many are fascinated by the questions we worry about, so we should let them share the excitement.

References

[1] A. S. Wightman and L. Gårding, "Fields as operator-valued distributions in relativistic quantum theory," *Arkiv Fys.* **28** (1965).
[2] R. Streater and A. Wightman, *PCT, Spin and Statistics, and All That*. Landmarks in Physics. Princeton University Press, 2000.
[3] K. A. Meissner and H. Nicolai, "Conformal symmetry and the standard model," *Phys. Lett. B* **648** (2007) 312–317, arXiv:hep-th/0612165.
[4] J. Wess and B. Zumino, "Supergauge transformations in four-dimensions," *Nucl. Phys. B* **70** (1974) 39–50.
[5] S. Coleman and J. Mandula, "All possible symmetries of the *S* matrix," *Phys. Rev.* **159** (Jul. 1967) 1251–1256. https://link.aps.org/doi/10.1103/PhysRev.159.1251.
[6] D. J. Gross, J. A. Harvey, E. Martinec and R. Rohm, "Heterotic string," *Phys. Rev. Lett.* **54** (Feb. 1985) 502–505. https://link.aps.org/doi/10.1103/PhysRevLett.54.502.
[7] P. Candelas, G. T. Horowitz, A. Strominger and E. Witten, "Vacuum configurations for superstrings," *Nucl. Phys. B* **258** (1985) 46–74.
[8] A. Schellekens, "Big numbers in string theory," arXiv:1601.02462 [hep-th].
[9] W. Taylor and Y.-N. Wang, "Scanning the skeleton of the 4D F-theory landscape," *JHEP* **01** (2018) 111, arXiv:1710.11235 [hep-th].
[10] L. Susskind, "The anthropic landscape of string theory," arXiv:hep-th/0302219.
[11] H. Nicolai, "Quantum gravity: the view from particle physics," *Fundam. Theor. Phys.* **177** (2014) 369–387, arXiv:1301.5481 [gr-qc].

[12] A. H. Chamseddine and A. Connes, "Noncommutative geometry as a framework for unification of all fundamental interactions including gravity. Part I," *Fortsch. Phys.* **58** (2010) 553–600, arXiv:1004.0464 [hep-th].

[13] M. B. Green, J. G. Russo and P. Vanhove, "Ultraviolet properties of maximal supergravity," *Phys. Rev. Lett.* **98** (2007) 131602, arXiv:hep-th/0611273.

[14] R. Kallosh, "On a possibility of a UV finite N = 8 supergravity," arXiv:0808.2310 [hep-th].

[15] Z. Bern, J. Carrasco, L. J. Dixon, H. Johansson and R. Roiban, "Manifest ultraviolet behavior for the three-loop four-point amplitude of N = 8 supergravity," *Phys. Rev. D* **78** (2008) 105019, arXiv:0808.4112 [hep-th].

[16] M. B. Green, J. G. Russo and P. Vanhove, "String theory dualities and supergravity divergences," *JHEP* **06** (2010) 075, arXiv:1002.3805 [hep-th].

[17] Z. Bern, J. Carrasco, L. J. Dixon, H. Johansson and R. Roiban, "The complete four-loop four-point amplitude in N = 4 Super-Yang–Mills theory," *Phys. Rev. D* **82** (2010) 125040, arXiv:1008.3327 [hep-th].

[18] J. J. M. Carrasco, M. Chiodaroli, M. Günaydin and R. Roiban, "One-loop four-point amplitudes in pure and matter-coupled N <= 4 supergravity," *JHEP* **03** (2013) 056, arXiv:1212.1146 [hep-th].

[19] Z. Bern, J. J. Carrasco, W.-M. Chen, A. Edison, H. Johansson, J. Parra-Martinez, R. Roiban and M. Zeng, "Ultraviolet properties of $\mathcal{N} = 8$ supergravity at five loops," *Phys. Rev. D* **98** no. 8 (2018) 086021, arXiv:1804.09311 [hep-th].

[20] K. A. Meissner and H. Nicolai, "Standard model fermions and N = 8 supergravity," *Phys. Rev. D* **91** (2015) 065029, arXiv:1412.1715 [hep-th].

[21] A. Lewandowski, K. A. Meissner and H. Nicolai, "Conformal standard model, leptogenesis and dark matter," *Phys. Rev. D* **97** no. 3 (2018) 035024, arXiv:1710.06149 [hep-ph].

[22] H. Nicolai, K. Peeters and M. Zamaklar, "Loop quantum gravity: an outside view," *Class. Quant. Grav.* **22** (2005) R193, arXiv:hep-th/0501114.

[23] T. Thiemann, "The LQG string: loop quantum gravity quantization of string theory I: Flat target space," *Class. Quant. Grav.* **23** (2006) 1923–1970, arXiv:hep-th/0401172.

[24] R. Dijkgraaf, S. Gukov, A. Neitzke and C. Vafa, "Topological M-theory as unification of form theories of gravity," *Adv. Theor. Math. Phys.* **9** no. 4 (2005) 603–665, arXiv:hep-th/0411073.

[25] B. de Wit, J. Hoppe and H. Nicolai, "On the quantum mechanics of supermembranes," *Nucl. Phys. B* **305** (1988) 545.

[26] T. Banks, W. Fischler, S. H. Shenker and L. Susskind, "*M* theory as a matrix model: a conjecture," *Phys. Rev. D* **55** (Apr. 1997) 5112–5128. https://link.aps.org/doi/10.1103/PhysRevD.55.5112.

[27] J. M. Maldacena, "The large N limit of superconformal field theories and supergravity," *Int. J. Theor. Phys.* **38** (1999) 1113–1133, arXiv:hep-th/9711200.

[28] G. 't Hooft, "Dimensional reduction in quantum gravity," *Conf. Proc. C* **930308** (1993) 284–296, arXiv:gr-qc/9310026.

[29] L. Susskind, "The world as a hologram," *J. Math. Phys.* **36** (1995) 6377–6396, arXiv:hep-th/9409089.

[30] A. Almheiri, D. Marolf, J. Polchinski and J. Sully, "Black holes: complementarity or firewalls?," *JHEP* **02** (2013) 062, arXiv:1207.3123 [hep-th].

[31] A. Almheiri, D. Marolf, J. Polchinski, D. Stanford and J. Sully, "An apologia for firewalls," *JHEP* **2013** no. 9 (Sep. 2013). http://dx.doi.org/10.1007/JHEP09(2013)018.

[32] M. Reuter and F. Saueressig, *Quantum Gravity and the Functional Renormalization Group: The Road towards Asymptotic Safety*. Cambridge University Press, 1, 2019.

[33] T. Damour, A. Hanany, M. Henneaux, A. Kleinschmidt and H. Nicolai, "Curvature corrections and Kac-Moody compatibility conditions," *Gen. Rel. Grav.* **38** (2006) 1507–1528, arXiv:hep-th/0604143.

[34] T. Damour, A. Kleinschmidt and H. Nicolai, "K(E(10)), supergravity and fermions," *JHEP* **08** (2006) 046, arXiv:hep-th/0606105.

[35] A. Kleinschmidt and H. Nicolai, "Maximal supergravities and the E(10) coset model," *Int. J. Mod. Phys. D* **15** (2006) 1619–1642.

[36] T. Damour, A. Kleinschmidt and H. Nicolai, "Constraints and the E10 coset model," *Class. Quant. Grav.* **24** (2007) 6097–6120, arXiv:0709.2691 [hep-th].

[37] E. A. Bergshoeff, O. Hohm, A. Kleinschmidt, H. Nicolai, T. A. Nutma and J. Palmkvist, "E(10) and gauged maximal supergravity," *JHEP* **01** (2009) 020, arXiv: 0810.5767 [hep-th].

[38] B. Julia, "Kac-Moody symmetry of gravitation and supergravity theories," in *American Mathematical Society Summer Seminar on Application of Group Theory in Physics and Mathematical Physics*. 9, 1982.

[39] T. Damour, M. Henneaux and H. Nicolai, "Cosmological billiards," *Class. Quant. Grav.* **20** (2003) R145–R200, arXiv:hep-th/0212256.

[40] T. Damour, M. Henneaux and H. Nicolai, "E_{10} and a small tension expansion of M theory," *Phys. Rev. Lett.* **89** (Nov. 2002) 221601. https://link.aps.org/doi/10.1103/PhysRevLett.89.221601.

[41] A. Kleinschmidt and H. Nicolai, "E(10) cosmology," *JHEP* **01** (2006) 137, arXiv:hep-th/0511290.

[42] V. Belinskii, E. Lifshitz and I. Khalatnikov, "On a general cosmological solution of the Einstein equations with a time singularity," *Zh. Eksp. Teor. Fiz.* **62** (1972) 1606–1613.

[43] H. Nicolai and T. Fischbacher, "Low level representations for E(10) and E(11)," in *Ramanaujan International Symposium on Kac-Moody Lie Algebras and Applications (ISKMAA 2002)*, N. Sthanumoorthy and K. C. Misra, eds., pp. 191–227. AMS, 2003.

[44] B. de Wit and H. Nicolai, "Deformations of gauged SO(8) supergravity and supergravity in eleven dimensions," *JHEP* **05** (2013) 077, arXiv:1302.6219 [hep-th].

[45] H. Godazgar, M. Godazgar and H. Nicolai, "Testing the non-linear flux ansatz for maximal supergravity," *Phys. Rev. D* **87** (2013) 085038, arXiv:1303.1013 [hep-th].

20

Roger Penrose

Professor Emeritus at the Mathematical Institute at the University of Oxford

Date: 4 May 2011. Location: Leiden. Last edit: 18 March 2019

In your opinion, what are the main problems in theoretical physics at the moment?

At the moment quantum theory is really inconsistent as it contains two parts which don't actually fit together in a proper mathematical theory. So I think there has to be a revolution in quantum theory in order to solve the quantum measurement problem and I would say that this is the main problem confronting physics today.

Okay. Could you explain what the quantum measurement problem is all about?

The way that quantum mechanics is done in practice consists of two parts. One is the Schrödinger equation where you have unitary evolution; that is, you have a deterministic equation which does not involve probabilities. The quantum state just evolves and the equation tells you what the quantum state is at a later time. It describes the system in any given moment of time. The other part consists of performing experimental measurements and if you make a measurement, you don't measure what the quantum state is. Instead, you have to ask the system certain types of questions and the answers that you get depend probabilistically on that quantum state. At this point you infer that the system might do one thing or another thing with certain probabilities. But what are you actually doing when you make a measurement? Well, in order to make a measurement you have some piece of apparatus and that apparatus is made out of the same quantum constituents as everything else, thus should be treated as a quantum system. As such, the apparatus should be evolving according to the Schrödinger equation, which is a deterministic equation involving no probabilities. So this is the problem and people have endless different ways of coming to terms with this which, in my view, none of them really do.

You mean that there are many different interpretations of quantum mechanics, like the Copenhagen interpretation?

There are all sorts of interpretations. The Copenhagen interpretation is basically a pragmatic one [1], which is reasonable if you want to make quantum theory work and not worry too much about its foundations. According to this interpretation, you just say that the measuring apparatus is treated classically and it does one thing or it does another thing.

Thus, within the Copenhagen interpretation you never analyse the apparatus as a quantum system. There are also different versions of the Copenhagen interpretation. For instance, some people say that the state of the system is really a state of your knowledge but then the question is: Whose knowledge? I mean is it my knowledge or his or (laughs)? So the problem is not really resolved according to this interpretation either. There are also people who say that all things happen, that is, Schrödinger's cat is dead and alive at the same time and when you look at it you are in a superposition of seeing it dead and seeing it alive. Hence, for some reason which is not explained, your awareness splits somehow. None of these interpretations really makes sense to me.

Is the ultimate problem of quantum theory just a matter of interpretation?

I don't think it is. The theory really needs to be modified and the modification, in my particular opinion, comes in when gravitational effects start to become important. People tend to poo poo this idea because they think that since gravitation is so weak it cannot have any relevance. In my view, people are looking at the problem in the wrong way.

So what is the right way to look at it?

When people think that gravity is too weak, they think of it as a force like any other physical force. However, it is not a force according to Einstein's general theory of relativity. The arguments are rather technical but, basically, it's hard to make the principles of Einstein's theory and the principles of quantum mechanics come together. They have to change in some way. The main conflict that I see is in the principle of equivalence which is telling you that the gravitational field is equivalent to an acceleration. This principle was what led Albert Einstein into a curved space description of gravity and it is difficult to make it consistent with the superposition principle of quantum mechanics.

Why is this the case?

There are actually two problems but they're interconnected. One of them has to do with how you interpret time. In quantum mechanics you basically have an external time which just doesn't have anything to do with the physics. However, spacetime according to general relativity becomes altered and curved, and your notion of time is with respect to that spacetime. Now, if you have a displacement of a physical object, according to quantum mechanics two different possible locations of that object can simultaneously coexist. This means that you have to worry about two spacetimes being in a superposition and the fact that the notion of time in one of them can be different from the notion of time in the other one. Thus, the whole idea of what it means to have a stationary state becomes obscure. In quantum mechanics, if an object in one place is stationary (i.e., it would stay there forever if you left it there) and that same object in a different place is also stationary, you have to describe the object as being in a superposition of those two locations. However, within general relativity, what it means to say "stationary" depends on the spacetime that the object determines, which is different in both locations. According to the principles of Einstein's theory, you can't unambiguously form a superposition of the two spacetimes, unless you have a map that relates one into the other in a certain way. While this is possible to do if both spacetimes are flat, there is no determined way to map two curved spacetimes, which is problematic. To solve this problem I am not asking for a theory of quantum gravity.

Is that the case? I'm asking because you have mentioned incorporating gravity in quantum mechanics so I thought you were advocating for a theory of quantum gravity to solve the measurement problem.

No. People are usually confused about this when I discuss this subject and the reason is that they usually misunderstand my writings, which is probably my fault. When people speak of quantum gravity, they usually think of applications of the rules of quantum mechanics to gravitational theory, in particular to Einstein's general theory of relativity. This means that people have in mind the quantisation of gravity according to the standard principles of quantum mechanics. However, these principles don't really work when spacetime is curved because you don't know what time evolution means. Instead, you have two different notions of time evolution. So what I'm saying is that you should be looking at the problem the other way around, that is, not how quantum theory affects the structures of spacetime but how the principles of general relativity will affect quantum mechanics. Thus, quantum mechanics is not going to survive if it is left unchanged and we should be looking for a more even-handed marriage between the two, where the rules on both sides will be modified somehow. People usually accept that the way we describe the gravitational field will have to change because the other forces of nature were also required to change in accordance with quantum mechanics. Hence this is not an unreasonable expectation but what is not usually considered is that the rules of quantum mechanics may also have to change, which I think must be the case because of the measurement problem.

You mentioned several times that the usual principles of quantum mechanics don't really apply if spacetime is curved. So does that mean that you think that quantum field theory in curved space is ill-defined?

Yes, I think that it doesn't quite make sense but I also think that people accept this. Taking some classical theory, placing it in a curved background and quantising the theory has difficulties of its own. However, what I am thinking is not how to define quantum theory in a given curved spacetime but in a superposition of spacetimes where the notion of temporal evolution is different in one from in the other. There are different arguments for this; for instance, there is one argument which in some ways I like better which is directly appealing to the principle of equivalence, which in turn means that you need to define your quantum field theory in a superposition of two different vacua. Usual quantum field theory is not determined if you do not know what the vacuum state is, for which you can have different possibilities. But it's very difficult to define it in a superposition of different vacua without finding inconsistencies. When you deal with gravity, you need to deal with these problems.

Aren't the gravitational effects very small?

People often dismiss gravitational effects with this line of thought by stating that the gravitational effects are very tiny and that you shouldn't worry about it but this is because they are looking at it in the wrong way. They think of the reaction of quantum mechanics on spacetime as being very tiny, of the order of the Planck scale (10^{-33} centimetres), which is 20 orders of magnitude smaller than the scales of ordinary particles, and so is the Planck time (10^{-43} seconds), which is 20 orders of magnitude smaller than the fastest processes occurring in particle physics. However, what I am stating is that a displacement that involves a Planck mass difference will spontaneously reduce to one state or another in a Planck time. That's absurdly small, whereas in the case of Schrödinger's cat (being alive or dead), the amount of mass displacement of material from alive to dead will be such that the reduction time in this kind of scheme would be extremely tiny. So the cat would be alive or dead in a very tiny fraction of a second. If you want something where you have reasonable time scales (let's say, three seconds) then you need to superpose an object which you can just barely see.

Is this what you call gravitational OR?

That's right, gravitational OR, meaning *objective reduction* [2–6]. Diósi has a version of this scheme [7–10], which is basically the same as the one I'm proposing. He had already proposed this scheme before I did. I sort of came to it independently but with different motivations in mind. I was mostly motivated by general relativity, at least more strongly than he was. There are other schemes that have a similar character to his but this one is a specifically gravitational scheme. There are various proposals that people made for how to modify quantum mechanics to make the measurement process into an objective process, such as the Ghirardi-Rimini-Weber scheme [11, 12]. Some schemes have been refuted in the past but at the present moment there is no observation that tells you which is the correct scheme.

Do you think that there will be experiments which will be able to determine what is the correct scheme?

Earlier today there was a talk at the Lorentz Institute where Dirk Bouwmeester, who has been working on a scheme that I proposed about 10 years ago [13], spoke about possible experiments, though extremely difficult to carry out. These experiments are not going to allow you to detect some granularity of spacetime due to quantum gravity effects. That would be ridiculous as you would need experiments that would accelerate particles around the orbit of the earth or something like that, you see (laughs). Such things are off the wall. However, the experiments that Dirk Bouwmeester was talking about are experiments that try to probe the effects of gravity on quantum mechanics and not the other way around. So we might see some experimental results maybe within the next decade, which would be very exciting. On the theory side, I have no idea how much progress will be made in the next decade because to formulate a good theory for these effects will be a major challenge, most likely requiring a revolution of the same order of magnitude as general relativity.

Do you think that candidate theories of quantum gravity, such as string theory, solve these problems?

No, I don't think any of them does because of the simple reason that they use standard quantum field theory. They don't change the structure of quantum field theory. They try to change the physics by applying conventional quantum theory to strings or loop variables or whatever it is. In my opinion, the standard rules of quantum theory have to change if it's going to be successful.

Okay, but do you think that there's at least some hints coming out from these approaches about the nature of quantum gravity?

Maybe. I'm not complaining about people who are doing it. I think it's a good thing to explore and indeed you might see hints come out which could indicate what direction we should go. I just haven't seen any.

Okay. There is something that I read in your book which left me a bit in shock, namely, that you didn't really agree too much with the AdS/CFT conjecture in a mathematical sense [6].

Oh, I see. I had a problem with it at the level of degrees of freedom, which probably meant that what people are actually doing in AdS/CFT is not what they seem to be doing. I mean, at first sight, in the way it's often described, is that you have a quantum field theory on the boundary and this is supposed to be equivalent to string theory in the interior. I am worried about this because these spaces have different dimensions. Also, often what people mean when they do AdS/CFT is that they look at specific things on the boundary and things on the interior which are somehow related but it's not really a relation at the level of the full dynamics as far as I can see. I surely need to understand it better since people are doing something that makes sense and is true, even if the conjecture itself isn't true. There are enough aspects to it which do make sense and which can be applied in various areas as far as I can understand but I don't think it's really standard quantum field theory on the boundary. There are lots of difficulties with anti–de Sitter (AdS) spacetime anyway. It's a setup which has very badly posed problems, where you've got a timelike boundary and then you've got information coming in from infinity all the time. I've never gone into it deeply enough and what I need to do is to try to understand what they are doing, which at first sight seems odd.

In your book [6] you mentioned that dualities between theories seem mysterious and hint towards deeper mathematical structures. I was wondering if you thought that new mathematics needs to be developed in order to prove this conjecture?

I don't know. That may very well be. However, whether this conjecture is physics is another question. String theories, M-theory and all of that have some mathematical truth hiding behind them but it doesn't mean that it's physics. It may mean that some of the ideas will have relevance to physics but so far I don't see a physical theory there, at least not a believable one. Right at the beginning, some of the ideas of string theory were very attractive and maybe some of those ideas will return again as part of some coherent theory. My main problem with string theory is the fact that it has many extra dimensions, not with the strings. It's very hard to see how the information contained in those degrees of freedom in the spatial extra dimensions, which are rattling around all over the place, are suppressed. I can't see that happening.

In your book [6] you seemed to be more into loop quantum gravity, right?

In the book I was. I suppose I got a little bit discouraged about loop quantum gravity later. It kind of hit a brick wall. I don't think much has happened in loop quantum gravity since I wrote my book. They had this nice calculation that seemed to derive the entropy of black holes, which comes out right provided the so-called Barbero-Immirzi parameter takes a specific value [14]. Mathematically, it ought to take the value $i = \sqrt{-1}$ but in fact it has some real value which makes no sense to me. So there is something funny going on there. Additionally, loop quantum gravity doesn't have a proper Hamiltonian so it got a bit stuck. In my book, I wrote favourably about it because they were really taking the genuine issues of general relativity seriously, which the string theorists paid very little attention to, such as problems with general covariance and how to make a quantum field theory compatible with general covariance.

Could it be that approaches such as loop quantum gravity are wrong because they are quantising gravity directly instead of some other true microscopic degrees of freedom?

Well, there may be some interesting ideas along this line of thought. These ideas go back to Andrei Sakharov, who proposed that gravity was a kind of residual structure originating from deeper and more statistical processes taking place at a more microscopic level [15, 16]. In my opinion, this doesn't address the strangeness of gravity in general relativity. I mean, nothing else in physics affects the causal structure of spacetime; hence, to try and make it up out of some ingredients in that kind of way doesn't strike chords with me. I don't know exactly what people are doing along this direction so obviously I should shut up. There may be smaller ingredients of spacetime and there are people who discuss lattice structures, causal spaces, etc., some of which I myself played around with a long time ago (e.g., [17]). Even twistor theory [18–20], one might say, is aiming at something along those lines but I just don't think that gravity works like that.

Given that you mentioned the computation of black hole entropy within loop quantum gravity, I was wondering if you were not keen on the extremely accurate results for black hole entropy within string theory [21]?

There is a lot of hype connected with these things which I'm not too sure about. Quite honestly, I don't know what the current state is but if you look at the original arguments and context they are in, you notice that they are not even using the right number of dimensions (it was in a five-dimensional theory or something) and a lot of it involves Yang–Mills fields that don't exist in the physical world as far as we know. However, my main problem with it is that they are looking at black holes with no horizon; that is, the black holes are extremal which is not the case in a physical situation. The calculations are done in flat space; you count the number of string states and then you jack up Newton's constant, which is zero to begin with but which then takes higher values until you obtain the Bekenstein-Hawking formula that you sought after. There is very little physics in it that I can relate to, specially because, in my view, the whole point of computing black hole entropy is the existence of a horizon and these arguments don't even have a horizon. Thus, I don't see these calculations as proof of anything. The result might just be coincidental.

In 1799, Pierre-Simon Laplace, following Michell's ideas, obtained the right size for a black hole using Newtonian theory [22] but the argument was actually wrong. In Newtonian theory there are no horizons, no constant speed of light and no black objects. If you suppose that you had a large body which exceeded the Schwarzschild value you may calculate the escape velocity that is required for light to escape the surface of the body and you realise that it is greater than the speed of light moving freely in space. So light would fall back down onto the surface of the body and wouldn't be able to escape. However, if you look at the sky, light falls down from the stars onto the surface of this body and if you were to place a mirror on the surface of the body, light would be reflected with the same speed as it came in and it will escape. Why is that? Because by the time light reaches the surface, it's going faster than the speed of light travelling freely out there. Nevertheless, Laplace's calculation yielded the correct size for a black hole. All of this to say that maybe the Bekenstein-Hawking entropy argument derived in string theory is not so dissimilar. It's an argument that happens to give the right value.

Recently, you proposed a new model for cosmology called *conformal cyclic cosmology* [23–28]. What is this about?

This model was described in one of my books [25], which just come out, and is based on an idea that I had about five and a half years ago [23]. I used to give lectures on it as I thought it could be perceived as a completely crazy theory. I never believed that it was completely crazy but I liked to describe it in this way so as to disarm my critics. However, it has become serious now.

Does this theory explain the big bang?

Not exactly. In conventional cosmology, there is an initial big bang and eventually the universe begins to expand exponentially due to the cosmological constant. In the far future this universe would be dominated by photons, which are massless, so there is no way to measure scale. Thus the geometry of the universe can be treated as conformal: big and small become physically identical. In my scheme, this infinitely remote future transforms conformally into another big bang, which leads to another cycle (aeon) and so on and so forth. In the remote future of a given aeon, the claim is that you just have massless entities. In particular, black holes evaporate away.

What about the remaining mass in the universe?

Well, that is probably the weakest part of the argument; namely, all the rest mass has to decay away eventually, which does not mean that all particles decay into massless particles. Instead, the rest mass has to fade out somehow. The main reason for that is because of the existence of electrons. Electrons are the least massive charged particles so they can't decay into anything less massive. If the mass itself is not an absolute concept and fades out then you have charged asymptotically massless particles in the remote future. This is the most conjectural part of the theory, for which, at the moment, there's no convincing reason to believe in. However, in order to make the theory work properly, you must require it. Thus, in the remote future, you only have massless entities, which are not sensitive to scale but only sensitive to the conformal structure of the spacetime. If you would go back in time to the previous big bang, just earlier than the Higgs time, you find that again all particles were effectively massless and are not sensitive to the time scale. The big bang is conformally just a place like anywhere else but the claim is that the remote future of a previous aeon becomes the big bang of our own aeon. Surprisingly, the two aeons match smoothly onto each other due to the conformal geometry and the equations that govern the evolution of the universe are essentially deterministic. Furthermore, events like black hole collisions in the previous aeon can leave imprints on the microwave background of our own aeon. This claim that I am making is, as far as I can see, gaining support from the analysis of the WMAP observations [24, 26, 28].

So there are multiple big bangs, one after the other, but there is no need for any theory of quantum gravity?

That's the shock, yes. Naively, you expect that the best place to look for quantum gravity is at the big bang. This scheme, however, gives you classical equations. The reason is that, even though the curvatures are very large and the radius of curvature is of the order of the Planck size at the big bang, that does not mean that quantum gravity becomes important. It only becomes important when the Weyl curvature approaches the Planck scale. As you know, the curvature of spacetime is split into two parts; the gravitational degrees of freedom are described by the Weyl curvature, and the rest, which is the Ricci curvature, is directly determined by the matter. The Weyl curvature, which is the conformal part, takes the predominant role and the Ricci curvature has no role to play, given that scale is not important in this scheme. Thus, it doesn't matter what it is because the Weyl curvature goes to zero at the crossover between aeons. So the claim is that you don't actually need quantum gravity, which is a shock to me as much as to many of my friends but that seems to be the way it works.

Is there any prediction coming out of it?

Yes, one of them is the existence of circles that form patterns in the cosmic microwave background (CMB) [24, 26]. These circles, which are the consequence of black hole encounters in the previous aeon, are one of the biggest visible effects on our cosmic microwave background. I have published an article [24], together with my colleague Vahe Gurzadyan, which a lot of people complained about, in particular regarding the statistics we employed for analysing these circles. So we wrote another article in order to address some of these complains [29]. This other article [29] gives evidence for the existence of something out there in the sky that we hadn't noticed before and it consisted of two parts. In one part, we used the theoretical power spectrum (one of the triumphs of cosmology) as input in our theory and showed that there are no visible circles in the cosmic microwave background. On the other hand, if you input the observed power spectrum you do see them clearly. This tells us that the model does not predict circles in an arbitrary background but just only on a background that actually contains the circles, which we assume is the case in the observed power spectrum, and also that there are new features that this model predicts compared to the previously theoretically calculated power spectrum using the Lambda-CDM model [30].

The other thing we did was to plot on the sky all the instances of low variance circles and showed that these circles occur in concentric families. Some of these circles are quite big; certain areas of the sky are quite crowded, and they form a pattern. They are not random but if you do a simulation using the Lambda-CDM model then they're all over the place and they're not too many. In fact, they're far fewer in the simulations than in the observed sky and we have images on arXiv.org that show this [29]. So there is something there which is not the product of random data and is genuinely explained by this conformal cyclic cosmology scheme. As of March 2019, the experimental situation is dramatically changed [26–28].

Do you believe that it is possible to find the theory of everything that fixes all constants of nature?

We're not close to it yet. This raises the big question of whether the value of the constants of nature are calculable numbers or they are actually determined by some other principle like the anthropic principle. I hope that it is not the latter case. I would much prefer them to be mathematical numbers that could be calculated from some theory but that's a preference rather than an expectation.

You mentioned twistor theory earlier ... what's the story with twistor theory?

Since 2003 a lot of interesting things are happening with twistor theory. There's a certain irony in this story because I went to Princeton in 2003 to give a series of three lectures entitled "Fashion, Faith and Fantasy in the New Physics of the Universe" [31]. The fashion part of it had to do with string theory, the faith to do with quantum mechanics at all levels (laughs) and the fantasy to do with inflationary cosmology, which I'm not a fan of. I felt rather embarrassed going to Princeton to talk about these things, since some of the leading experts in at least some of these areas were there at Princeton. A few days after my first talk, I had a lunch appointment with Edward Witten. I was a little nervous because I thought he might be annoyed with my first talk, which was about string theory and why I was sceptical about some of the things in string theory. In fact, he wasn't at all interested in anything I might have said. I don't know if he even knew I was giving those talks. Instead, he said he wanted to describe something that might interest me. So he started explaining things about gluon scattering and so on. I looked at it and asked if he was talking about four dimensions. He said yes. So the first hurtle was gone and he continued talking about strong interaction processes in four dimensions using twistor theory. It was a piece of genuine physics and all these things I found interesting. What he was describing was the origin of the twistor theory approach to string particle theory. He told me that he was thinking of writing a short paper on it and asked me if I would be interested in looking at it. So I of course said yes and that he should send me a copy when it's ready. A couple of months later I got this hundred-page paper, which was about twistor-string ideas [32]. He had his own take on various things, which I was not totally happy with but the whole idea seemed very exciting and interesting.

Why did you begin developing twistor theory?

Actually, in case you're really interested in the origins of these ideas, the best place is to look for this article called "On the Origins of Twistor Theory" [33]. It was an article I wrote in honour of Ivor Robinson's birthday because he had various influences on some of its developments. Anyway, people ask me why was I working on twistor theory and it's a difficult question to answer. Usually, what drives people to progress is their worries about certain problems that they think need to be solved, but in the case of twistor theory, it was never quite like that. Twistor theory came about by trying to express in a nice geometrical form the notion of positive and negative frequencies in quantum field theory. In one dimension, people usually discuss a Fourier decomposition and describe positive or negative components (depending on conventions). However, this approach didn't quite satisfy me, partly because it's not manifestly conformally invariant. At that time, I had become interested in massless fields and the way you describe them in terms of spinors and conformal invariance, etc. I just found it all very beautiful, including the idea that somehow massless things are more primitive than massive ones. I wanted to understand how one could incorporate the positive frequency condition in the description of massless fields. So I used the concept of Riemann sphere in order to devise a complexified way of looking at spacetime, where the notion of positive frequency becomes very natural.

During these developments, I was somewhat influenced by Engelbert Schucking, who explained to me the importance of the positive frequency condition and was sympathetic towards the idea of introducing some kind of complex space. I was in Texas at the time, as well as Roy Kerr, who had an office on one side of mine and Ray Sachs further down. The motivations for following these ideas came from different places. Some came from looking at solutions to Einstein's field equations and seeing these complex structures appearing. These structures were hiding there somehow and I've always had a love affair with complex number analysis. All this was incredible and seemed to be very natural that it could be lying at the root of physics. I don't quite know why but it just seemed to me the kind of mathematics that has this magic and power to it.

How does the Riemann sphere help you express these twistor ideas?

In the Riemann sphere you have the real points which divide the space into two halves. I was looking for a higher-dimensional analogue of this that applied to spacetime as a whole. Eventually, the idea came to me that the analogue of the Riemann sphere's one-dimensional equator would be the five-dimensional space whose points represent entire light rays in spacetime. So the five-dimensional space light rays were like the Riemann sphere's equator, and the Riemann sphere itself was represented as a complex projective three-dimensional space called (projective) twistor space. A general point of twistor space stood for what is called a Robinson congruence, which is determined by a solution to Maxwell's equations, in which you displace light rays into the complex. This was the sort of thing I wanted to realise in the higher-dimensional Riemann sphere. What I did not know at that time was that this is actually something which comes up naturally when you look at the angular momentum of classical massless particles. For instance, when you look at the angular momentum structure for a massless particle with spin, you find very naturally that this complex structure comes up. So, it started at first as just a piece of mathematics with the aim of understanding how to express a space where the "real part" describes the light rays and where there is a more mysterious part which is the complex extension. It actually took me many years to realise that there was a physical point of view to this story and to write Maxwell's equations and the other massless fields in a way which fitted in with this twisted scheme. Then, it took me a lot longer than that to realise the role of cohomology in this mathematical construction, for which I needed the input from Michael Atyla, who explained to me about cohomology which I had forgotten about (laughs). So twistor theory divides the space into the top half and the bottom half of the higher-dimensional Riemann sphere, where the positive frequency goes one way and the negative frequency the other way and these give you the way in which you can express quantum field theory in twistor terms.

How do you exactly express quantum field theory in twistor terms?

I tried to develop a kind of quantum analogue of Feynman diagrams, called twistor diagrams, but I didn't really understand a lot of things about it. The person who really developed this was Andrew Hodges. He was a graduate student of mine and then he took some time off to write his famous biographical book on Alan Turing [34]. He later came back into twistor theory and almost single-handedly developed the subject of twistor diagrams [35–38]. For many years this field was developing in almost complete isolation, for which nobody took much of an interest, with only three or four graduate students of Andrew's working on this. So for a while no one picked up on this until Witten came along and did his twistor-string stuff [32]. Witten was interested in twistor diagrams too and he knew about twistors previously.

But Witten used it more like a tool to compute amplitudes, right?

I suppose that is what it turned out to be. It is a tool but in a way that's not so surprising. I think it's easy to get the wrong end of the stick about twistor theory. People tended to think that it's a revolutionary idea which was supposed to be quantum gravity in some crazy form whereas in a sense it's just a reformulation of standard physics. It's a reformulation that brings out different things. Some things are easier in twistor theory while other things that are easier in standard physics are very difficult in twistor theory. So it's a different perspective. However, when you are looking particularly at scatterings of massless entities, it's a very good tool. It's not such a surprise, in a sense it was expected I think.

But you think you can take twistor theory seriously and try to make a theory of quantum gravity?

I don't know. I would think that there is a chance of that in some sense. The main thing in that direction was the non-linear graviton construction [20], which shows you how to represent general complex anti-self dual, i.e., left-handed, solutions of Einstein's vacuum equations. I like to think of this construction as a non-linear wave function. Usually wave functions are meant to be linear and you can add them up according to the superposition principle of quantum mechanics. But if it is really a non-linear construction; maybe the superposition principle has to be modified in some way. However, the hanger has always been the googly problem [39], which is the fact that the left-handed graviton can be described very nicely in this scheme as a curve in twistor space but the right-handed graviton is quite problematic. On the other hand, at the linear level both left-handed and right-handed gravitons can be expressed in a straightforward way. However, when you attempt at making a non-linear version, the right-handed graviton is very obscure and we call this the googly problem. This is a term for which you have to be a member of the former British empire to understand because it is a cricket term (laughs). It means that the ball looks as though it is spinning in what is called "a leg break": you sort of throw the ball by twisting your hand in a certain way that leads to the ball bouncing off the ground in a manner that people don't expect (the ball spins in the right-handed direction instead of left-handed). All this to say that incorporating the non-linear degrees of freedom of the right-handed graviton in the left-handed graviton framework is difficult. Richard Ward in the meantime showed how to represent Yang–Mills theory in twistor terms [40, 41]. So it is not just in gravity but in all interactions, as far as we know, that we have problems incorporating right-handed degrees of freedom. This problem is maybe 40 years old.

You haven't pursued it further?

I have but the trouble is that I get too distracted. There is a program for how to do this, which might work out and involves relatively new ideas due to Andrew Hodges, which was stimulated by this new wave of interest in twistor theory [42]. So it may well be that using these ideas of Hodges, one can express the googly problem in a nice geometrical form which makes everything hang together. I do regard that as a key to a lot of things because once you can express gravity and Yang–Mills fields in this way, it will allow you to use twistor theory in ordinary particle physics in many different contexts. It would be a non-perturbative way of looking at it. There's also the way that David Skinner and Lionel

Mason [43], as part of this new a wave of activity, have expressed the general solutions of the Einstein equations. Basically they look at the left-handed solution, which one has complete knowledge of, and perturb away from that. This allows you to obtain the right-handed parts as a perturbation away from the general left-handed solutions. This is an interesting way of looking at it but it's not quite what I mean by a solution to the googly problem because you really want to have a non-perturbative approach.

If I ask you which direction should one pursue in order to find a theory of quantum gravity, what would you say?

Well I think that these twistor ideas are very important. I think that there is a theory of quantum gravity but it's not standard quantum field theory. Within this theory, one has to be able to see non-linearities appearing in the quantum formulation, which could be okay, but it necessarily means that Hilbert spaces are not the best concept to work with. Instead, one would need some other structure which would have curvature in it in some way, which some people have thought about. Then, in the weak field limit you would probably have an ordinary quantum particle that would behave like a graviton.

In your opinion, what has been the biggest breakthrough in theoretical physics in the past 30 years?

I think that the discovery of the positive cosmological constant, or dark energy as people call it, is one of the most important developments.

Does your model of cyclic cosmology explain anything about dark matter or dark energy?

"Explain" is an interesting word here. It requires a positive cosmological constant and without that it doesn't work. It also makes suggestions to where dark matter comes from because when you do the conformal rescaling across the cross-over, it's necessary to introduce a new scalar field and maybe that's the dark matter field. But it's still a little problematic because you've got to see what happens to the old dark matter and if it does somehow ultimately decay away and whether you need a lot of new dark matter every time there is a new eon. So there are issues there but it does make a suggestion for what field you need to have for the scheme to be consistent. This field has to start off as being massless but then it has to grow mass in not such a long time. This is perhaps the point where the Higgs mechanism plays a role and it would be interesting to tie that together.

Why did you choose to do maths and physics and not something else?

I was going to be on the medical side. I was one of three brothers. I had a sister too who at that stage was too young so we didn't know what she was going to do. But my parents were both medical and they concluded that the one to carry on the tradition was going to be me because my younger brother was only interested in chess and games and so on. My older brother was more interested in physics and mathematics and so I was the medical one. They had various reasons for choosing me. Secretly, I was going to be a brain surgeon and discover what made the mind work.

Well, you are not too far away from it, are you?

Well, yes not that far actually (laughs) [44]. So that was the idea, you open up people's heads and peep inside and see if you could see how it works. That was my ambition at that stage and so I was going to be a doctor and when I was entering the final two years at school, each one of us had to go and see the headmaster – the head teacher – one at a time and decide what subjects we would do in the following two years. So, he asked me what I wanted to do and I think I said (can't fully recall) that I wanted to do biology, chemistry and mathematics. He said "No, you can't do that combination." In those days they were very rigid and if you wanted to do biology then you had to do biology, chemistry and physics. If you did mathematics you were not allowed to do biology or something of that sort. So on the spur of the moment I said, "I'll do mathematics, physics and chemistry" because I didn't want to lose mathematics. When I got home my parents were rather annoyed with me because the one they thought was going to become medical was not on the right path anymore. Later they were all right because my little sister not only became medical but she also married another doctor and became a professor of genetics, which was my father's area. So they did all right eventually but they lost me to physics. In fact, physics only came to me in that way and my interest in it grew with time. I'm glad I didn't end up doing chemistry, goodness me (laughs).

I think everyone is glad for that. Final question: What do you think is the role of the theoretical physicist in modern society?

Oh gosh, that's a difficult question. Often a lot of technology originates from the physics of quantum mechanics and even from general relativity, such as the GPS system. In general, it takes a long time for these things to make their mark but it's clear that these things are important. However, I think it's terribly important just to understand what makes the world. I think that's one of the reasons that we are here for.

References

[1] J. Faye, "Copenhagen interpretation of quantum mechanics," in *The Stanford Encyclopedia of Philosophy*, E. N. Zalta, ed. Metaphysics Research Lab, Stanford University, winter 2019 ed., 2019.

[2] R. Penrose and M. Gardner, *The Emperor's New Mind: Concerning Computers, Minds, and the Laws of Physics*. Popular Science Series. Oxford University Press, 1999.

[3] R. Penrose, "On gravity's role in quantum state reduction," *General Relativity and Gravitation* **28** no. 5 (May 1996) 581–600. https://doi.org/10.1007/BF02105068.

[4] R. Penrose, "Quantum computation, entanglement and state reduction," *Philosophical Transactions of the Royal Society of London. Series A: Mathematical, Physical and Engineering Sciences* **356** (1998). https://doi.org/10.1007/BF02105068.

[5] R. Penrose, "On the gravitization of quantum mechanics 1: quantum state reduction," *Foundations of Physics* **44** no. 5 (May 2014) 557–575. https://doi.org/10.1007/s10701-013-9770-0.

[6] R. Penrose, *The Road to Reality: A Complete Guide to the Laws of the Universe*. Vintage Series. Vintage Books, 2007.

[7] L. Diósi, "Models for universal reduction of macroscopic quantum fluctuations," *Phys. Rev. A* **40** (Aug. 1989) 1165–1174. https://link.aps.org/doi/10.1103/PhysRevA.40.1165.

[8] L. Diósi, "Continuous quantum measurement and itô formalism," *Physics Letters A* **129** no. 8 (1988) 419–423. www.sciencedirect.com/science/article/pii/037596018890309X.

[9] L. Diósi, "A universal master equation for the gravitational violation of quantum mechanics," *Physics Letters A* **120** no. 8 (1987) 377 – 381. www.sciencedirect.com/science/article/pii/0375960187906815.

[10] L. Diósi, "Gravitation and quantummechanical localization of macroobjects," *Phys. Lett.* **A105** (1984) 199–202, arXiv:1412.0201 [quant-ph].

[11] G. C. Ghirardi, A. Rimini and T. Weber, "A model for a unified quantum description of macroscopic and microscopic systems," in *Quantum Probability and Applications II*, L. Accardi and W. von Waldenfels, eds., pp. 223–232. Springer Berlin Heidelberg, 1985.

[12] G. C. Ghirardi, A. Rimini and T. Weber, "Unified dynamics for microscopic and macroscopic systems," *Phys. Rev. D* **34** (Jul. 1986) 470–491. https://link.aps.org/doi/10.1103/PhysRevD.34.470.

[13] W. Marshall, C. Simon, R. Penrose and D. Bouwmeester, "Towards quantum superpositions of a mirror," *Phys. Rev. Lett.* **91** (Sep. 2003) 130401. https://link.aps.org/doi/10.1103/PhysRevLett.91.130401.

[14] A. Ashtekar, J. Baez, A. Corichi and K. Krasnov, "Quantum geometry and black hole entropy," *Phys. Rev. Lett.* **80** (1998) 904–907, arXiv:gr-qc/9710007.

[15] A. D. Sakharov, "Vacuum quantum fluctuations in curved space and the theory of gravitation," *Soviet Physics Uspekhi* **34** no. 5 (May 1991) 394–394. https://doi.org/10.1070%2Fpu1991v034n05abeh002498.

[16] H. Kleinert, "Gravity as a theory of defects in a crystal with only second gradient elasticity," *Annalen der Physik* **499** no. 2 (1987) 117–119. https://onlinelibrary.wiley.com/doi/pdf/10.1002/andp.19874990206.

[17] E. H. Kronheimer and R. Penrose, "On the structure of causal spaces," *Mathematical Proceedings of the Cambridge Philosophical Society* **63** no. 2 (1967) 481–501.

[18] R. Penrose, "Twistor algebra," *Journal of Mathematical Physics* **8** no. 2 (1967) 345–366. https://doi.org/10.1063/1.1705200. https://doi.org/10.1063/1.1705200.

[19] R. Penrose and M. MacCallum, "Twistor theory: an approach to the quantisation of fields and space-time," *Physics Reports* **6** no. (1973) 241–315. www.sciencedirect.com/science/article/pii/0370157373900082.

[20] R. Penrose, "Nonlinear gravitons and curved twistor theory," *General Relativity and Gravitation* **7** no. 1 (Jan. 1976) 31–52. https://doi.org/10.1007/BF00762011.

[21] A. Strominger and C. Vafa, "Microscopic origin of the Bekenstein-Hawking entropy," *Phys. Lett. B* **379** (1996) 99–104, arXiv:hep-th/9601029.

[22] C. Montgomery, W. Orchiston and I. Whittingham, "Michell, Laplace and the origin of the black hole concept," *Journal of Astronomical History and Heritage* **12** no. 2 (2009) 90–96.

[23] R. Penrose, "Before the big bang: an outrageous new perspective and its implications for particle physics," *Conf. Proc.* **C060626** (2006) 2759–2767.
[24] V. G. Gurzadyan and R. Penrose, "Concentric circles in WMAP data may provide evidence of violent pre-big-bang activity," arXiv:1011.3706 [astro-ph.CO].
[25] R. Penrose, *Cycles of Time: An Extraordinary New View of the Universe*. Knopf Doubleday Publishing Group, 2011.
[26] V. G. Gurzadyan and R. Penrose, "On CCC-predicted concentric low-variance circles in the CMB sky," *The European Physical Journal Plus* **128** no. 2 (Feb. 2013) 22. https://doi.org/10.1140/epjp/i2013-13022-4.
[27] V. G. Gurzadyan and R. Penrose, "CCC and the Fermi paradox," *The European Physical Journal Plus* **131** no. 1 (Jan. 2016) 11. https://doi.org/10.1140/epjp/i2016-16011-1.
[28] D. An, K. A. Meissner, P. Nurowski and R. Penrose, "Apparent evidence for Hawking points in the CMB Sky," arXiv:1808.01740 [astro-ph.CO].
[29] V. G. Gurzadyan and R. Penrose, "More on the low variance circles in CMB sky," *arXiv e-prints* (Dec. 2010). arXiv:1012.1486 [astro-ph.CO].
[30] A. Liddle, *An Introduction to Modern Cosmology*, 2nd ed. Wiley, 2003. https://cds.cern.ch/record/1010476.
[31] R. Penrose, *Fashion, Faith, and Fantasy in the New Physics of the Universe*. Princeton University Press, 2016.
[32] E. Witten, "Perturbative gauge theory as a string theory in twistor space," *Commun. Math. Phys.* **252** (2004) 189–258, arXiv:hep-th/0312171 [hep-th].
[33] R. Penrose, "On the origins of twistor theory," in *In Gravitation and Geometry: Volume in Honour of I. Robinson*, W. Rindler and A. Trautman, eds. Bibliopolis, 1987.
[34] A. Hodges, *Alan Turing: The Enigma*. Vintage, 1992.
[35] A. P. Hodges and S. Huggett, "Twistor diagrams," *Surveys High Energ. Phys.* **1** (1980) 333–353.
[36] A. P. Hodges, "Twistor diagrams and massless Moller scattering," *Proc. Roy. Soc. Lond. A* **385** (1983) 207–228.
[37] A. P. Hodges, "Twister diagrams and massless Compton scattering," *Proc. Roy. Soc. Lond. A* **386** (1983) 185–210.
[38] A. P. Hodges, "String amplitudes and twistor diagrams: an analogy," in Oxford 1988, *Proceedings, the Interface of Mathematics and Particle Physics*, D. Quillen, G. Segal, S. T. Tsou and G. B. Segal, eds., pp. 217–221. Oxford University Press, 1988.
[39] R. Penrose, "Palatial twistor theory and the twistor googly problem," *Philosophical Transactions of the Royal Society A: Mathematical, Physical and Engineering Sciences* **373** no. 2047 (2015) 20140237.
[40] R. Ward, "On self-dual gauge fields," *Physics Letters A* **61** no. 2 (1977) 81–82. www.sciencedirect.com/science/article/pii/0375960177908428.
[41] R. Ward and R. Wells, *Twistor Geometry and Field Theory*. Cambridge Monographs on Mathematical Physics. Cambridge University Press, 1991.
[42] A. P. Hodges, "Twistor diagrams for all tree amplitudes in gauge theory: a helicity-independent formalism," arXiv:hep-th/0512336 [hep-th].
[43] L. J. Mason and D. Skinner, "Gravity, twistors and the MHV formalism," *Commun. Math. Phys.* **294** (2010) 827–862, arXiv:0808.3907 [hep-th].
[44] S. Hameroff and R. Penrose, "Consciousness in the universe: a review of the Orch OR theory," *Physics of Life Reviews* **11** no. 1 (2014) 39–78. www.sciencedirect.com/science/article/pii/S1571064513001188.

21

Joseph Polchinski

Permanent member at KITP and Professor of Physics at University of California Santa Barbara

Date: 7 July 2011. Location: Benasque, Spain. Last edit: 3 March 2017

In your opinion, what are the main puzzles in theoretical physics at the moment?

Theoretical physics is a very broad subject so I can't answer this question for anybody else than myself. There are several puzzles that I am interested in or working on and, in some sense, they are all connected. The biggest puzzle, however, is still the same, namely, what is this theory? We have found many pieces of it and they all point towards some coherent whole. This has been the state of affairs since I first started working on string theory in the mid-1980s. There have been strong indications that this theory is a coherent and consistent theory and, with time, we learnt more and more about it but we still haven't found its complete form.

Now, having said that, there are many specific problems, in particular those related to cosmology such as understanding the initial conditions for the origin of the universe. There is a list of even more general questions that need be addressed such as, "What is the framework for describing the origin of the universe?", "What is time?" and "What are the observables?" These questions are all linked somehow and, presumably, understanding the mechanics of string theory – string theory defined broadly – should give us new ways of thinking about these questions.

Historically, there have been very difficult questions in quantum gravity like the cosmological constant problem and also questions related to black hole paradoxes. String theory, initially, when it first started out, was just a theory of strings breaking and joining together and it didn't seem to have much to say about these questions. As the theory was understood better, it gave new insights into some of these very difficult questions. To what concerns the really difficult questions related to cosmology, which I mentioned earlier, I'm still optimistic that it will also ultimately give new insights.

There's also more technical questions related to the subject of quantum field theory. Quantum field theory has been around since the 1900s and it was understood by Kenneth Wilson in the 1970s but we're still learning deeper things about quantum field theory. Due to the discovery of dualities [1], quantum field theory and string theory are now understood as two parts of the same structure and so understanding quantum field theory more deeply is another direction which is linked with all the previous ones that I mentioned. So, I have a sort of one big goal in mind which is to complete the theory but I expect that in the course of completing the theory we'll learn a lot of interesting things about many other parts of physics.

Do you see string theory as a theory of quantum gravity or more like a tool that you can use to address different problems of physics?

Nowadays there's a lot of people working on applications of string theory to condensed matter physics and heavy ion collisions and so on. Those are definitely really interesting directions to take but they should be considered only as string theory's hobby. Its real job is to be a theory of quantum gravity.

Do you think that string theory is the best candidate at the moment for a theory of quantum gravity?

Absolutely. Since it first became the most prominent candidate, the understanding of it has deepened in ways that were unexpected. It has answered questions, or suggested answers to questions, that weren't expected from the beginning. Developments like the AdS/CFT duality [2] have completely transformed it and to that transformation we also call "string theory." String theory, as originally thought to be a theory of strings, is sort of one of the limits of this broader theory. It's a very useful limit because it gives a relatively complete picture that you can use to extrapolate from. The truth, however, is that the history of discoveries within this field could have proceeded in a different order and then we would have given it a different name. Therefore, when I refer to "string theory" now, I mean the complex of ideas that includes the AdS/CFT duality, for example. I think that string theory is extraordinarily promising on its own right and it surely is the best candidate for a theory of quantum gravity as the remaining ones are not even close. It's just like that.

Okay, but do you think that it's important to follow other approaches to quantum gravity?

What we call string theory, in the broader sense that I just mentioned, has progressed by, in many cases, incorporating many ideas that came from outside, so to speak. Ideas that appeared in the context of cosmology, for example, such as inflation and eternal inflation, are now central to string theory. In the context of general relativity, people like Stephen Hawking, for example, encountered paradoxes while studying black holes and introduced the concept of black hole entropy. These and also ideas related to black hole complementarity and holography were hit upon from purely thinking about general relativity but they are now core parts of string theory. Furthermore, the large-N limit of gauge theory that was developed in the realm of quantum field theory is now a big part of string theory, though originally it wasn't [3]. There's a lot of good ideas that turned out to be part of this larger structuring that we now call string theory. So, definitely, one should be alert to good ideas coming from elsewhere.

Now, regarding the other direct approaches to quantising gravity, the question, concerning whether or not these have ingredients that are going to be incorporated into this larger picture, remains. There are two things that I look for in other theories. One of them is if these other theories have any holographic nature. I think that the holographic principle [4, 5] is a really deep thing that we've learned about the nature of quantum gravity and to the extent that I understand any of these other approaches, they are not holographic; instead, they're much more local and so they're distinctly different. The other thing is that I actually have a large problem with many of these ideas that seem to give up Lorentz invariance.

When you mentioned that you look for the holographic nature of a theory, what do you mean exactly?

Holographic in the sense that the fundamental variables of quantum gravity are not even approximately local. In order to describe gravity in an anti–de Sitter box, the fundamental variables that you use to describe it must live on the walls of the box. Consequently, if you're a physicist and you are sitting in the middle of the box, then you will see some kind of local geometry but this geometry is an emergent object. It originates from something that wasn't even approximately there at the starting point. Most attempts to quantise gravity, although they deviate from general relativity, start with general relativity but general relativity isn't holographic directly because it deals with a local metric. You can replace that local metric with more elaborate things but people who have tried to quantise gravity directly seem to retain this basic locality of general relativity. This is my understanding of these other directions.

I was just wondering, since you wrote in one of your papers that "all good ideas are part of string theory" [6], if you therefore think that these other approaches are not based on good ideas. Is that the case?

There are a lot of ideas that I think are not good ideas because a lot of them, though not all of them, give up Lorentz invariance. I actually put out a two-page paper expressing my concerns about some of these ideas [7]. There are very strong experimental tests on the breaking of Lorentz invariance, not just at high energies. Furthermore, a theory that breaks Lorentz invariance does not automatically pass all the standard tests. In fact, most don't. This fact has been largely ignored by people working in the subject, though not all of them, as many of them are aware of the problem, but many others ignore it or minimise it. This paper that I've put out [7] was just a reaction to a paper that came out the week before [8] stating that this is not a problem. I disagree completely. I even disagree with the title of the earlier paper [8]; I mean, I think they just have misstated what the issue is. So, I think that there are specific problems with many of these altered ideas. If we didn't have string theory, I certainly would be looking at each one of these more closely because the answer would have to be somewhere.

In any case, in the past there have been many examples of ideas that were incorporated into string theory. Another of such examples is supergravity, which was clearly a completely separate field and it wasn't clear at all that many of the results in supergravity were relevant to string theory. Now supergravity is an integral part of string theory. So, one does have to keep one's eyes open to other ideas.

David Gross' assessments of string theory at Strings conferences always state that the biggest problem with string theory is that people do not know what string theory is. This is also what you mentioned earlier to be the biggest problem. What do you think that string theory really is?

I have no idea. My way of proceeding has always been to ask simple and small physical questions and see what the answer to these questions implies. I do not really have a grand vision of what it is. It is hard to have such a vision ...

What are the fundamental degrees of freedom?

We sort of know the answer to this question in the case of anti–de Sitter space. It's the gauge theory variables, right? However, even in this specific case, the gauge theory variables describe in a nice way one set of observables, namely, the observables of someone who sits on the boundary of the space. The problem is, of course, that in quantum gravity, there aren't generally nice observables and all these observables are messy because you need to set up a system of clocks and rods. So, one of the things that I've been trying to do in recent years has been to try to understand observables in the interior of the spacetime and how they're represented in the gauge theory. I think it's much more messy because it's a property of gravity.

Now, when you try to understand cosmology it's even harder because things are changing in time and you don't have infinite time to make measurements in. That's why people, such as Maldacena [9], Leonard Susskind [10] and others, have been doing some quantitative things by dealing with really arbitrary precise measurements though at asymptotically late times. They're focusing on the kinds of observations that you could make at asymptotically late times where there might be some simple description like AdS/CFT. In that case, you could say that you would be describing the fundamental degrees of freedom only at late times. However, I'm not entirely sure that this point of view is really right.

In the past, this subject has taken directions that no one could anticipate. Actually, I shouldn't say "no one" because there are people like Lenny Susskind who have actually been really good at sort of taking a leap forward and discerning a general principle. In my case, I have been happy with just playing with something and then stumbling upon something important. Even when I discovered D-branes, it wasn't like, "I'm going to go out and discover D-branes and they're going to be the most important thing in theory." I was just interested in knowing what happens if you put a string in a very small box. It seemed like an interesting question and we knew the answer for closed strings so I wanted to know the answer for open strings. I mean, it was just a question that I wanted to ask and it was also a good question to give a graduate student [11, 12]. So a lot of good work comes out of trying to keep one's graduate student busy and so ...

Was it by means of this thought experiment that you discovered D-branes?

Yes. You put closed strings in a small box and then, because of T-duality, it starts behaving like a large box again. It's remarkable. It's an example of many things such as an example of emergent spacetime. No one at the time ... well, I shouldn't say "no one" because Petr Hořava at the same time [13] and also Michael Green [14], a little later, asked the same question but it wasn't something many people were thinking about. So, what happens when you put open strings in a small box? The answer was unexpected: these D-branes appeared. It took a few years before it was clear that these objects were really important to the theory. So, it wasn't really with any grand vision that I found D-branes.

Do you think that it's possible to formulate string theory in such a way that strings and D-branes are on equal footing?

If you look at the web of dualities and you consider the quantum theory as a whole, then there's a sense in which they are on equal footing. If you take a string and you turn up the coupling, it becomes a D-brane and if you turn up a different coupling, it becomes a soliton. So, physically, they are on equal footing. Similar behaviour occurs just in quantum field theory. In certain supersymmetric gauge theories, we know that there's a duality that relates electrically charged quanta and magnetic monopoles that look like solitons. However, to what concerns a single description, even in this context, we don't have a description that puts them together on the same footing. This is one of the reasons why I mentioned that the problem of understanding string theory is related to the problem of understanding quantum field theory. Both share some of the same problems.

Now, if you take the point of view that the gauge theory defines string theory in anti–de Sitter space, then there you don't start with either the strings or the D-branes. Instead, you just start with the quantum fields and hence the strings and the D-branes are just different kinds of states that you can make out of the basic quantum fields. In this situation, neither the strings nor the D-branes are fundamental objects.

So, then, that wouldn't be the way of making a non-perturbative formulation of string theory?

It would. A non-perturbative formulation of string theory doesn't have to have any strings in it. People suspected this even before the web of dualities was discovered. I gave a series of Les Houche lectures around 1994 in which I listed several situations where strings did not appear to be fundamental [15]. String perturbation theory, like quantum field theory perturbation theory, is an asymptotic expansion in the coupling, which means that it doesn't converge so it doesn't necessarily define the theory. So, when we refer to "string theory" now, we mean that it is some quantum theory, one of whose limits is string theory but that doesn't mean that the fundamental variables have anything to do with strings. It's hard for me to imagine what the fundamental variables would be.

Could it be that there are no fundamental variables and that you just have to change description depending on the situation?

It could be that it's a theory consisting of many different descriptions, none of which sort of captures the whole thing.

Do you like that?

I don't like that but you have to sort of really listen to what the theory is telling you. I'd be much happier if there was one description from which all of the other descriptions could be seen to emerge as different projections.

Do you have any idea of how that could work?

No. I want to think more about gauge theory. I spend a lot of time trying to slice up AdS/CFT in different ways because I think it's the deepest thing we know about the theory.

I remember that you expressed this statement, perhaps in a different form, in one of your papers, namely, that in a sense the AdS/CFT duality reduces the construction of quantum gravity to a solved problem [16].

Yes, but I believe that I also said that this was the case for the special situation of gravity with anti–de Sitter boundary conditions and for the observations made by an observer on the boundary of that space. That's still a lot. An observer sitting on the boundary of anti–de Sitter space can sort of throw things into the bulk of the space and see what comes out. It's similar to making experiments in particle physics. The observer can throw things in and make a black hole or make things collide at hyper-Planckian energies. This observer can even probe transitions in topology, that is, transitions from quantum states that have no geometry at all to states that describe, for example, Lin, Lunin and Maldacena geometries [17]. So, even though it's a limited construction of quantum gravity, it is a formulation that describes all these phenomena. The one thing it doesn't describe is cosmology because cosmology is not described by an anti–de Sitter space, which in some sense has some walls around it. Now, in physics, we often put a system in a box and then take the box away and that's how we define things. However, in a holographic theory, the system is the box itself and so it's harder to imagine how one can proceed in the exact same way.

String theory is criticised for not being a background-independent theory. Do you think AdS/CFT is an example of background-independent theory of quantum gravity?

In a sense, AdS/CFT is better than background-independent because it's holographic. One reason I don't worry so much about background independence is because background independence tells you what the theory doesn't have, meaning: it doesn't have a special background. However, we want to know what it does have; that is, we want to know what are the special principles or ingredients that we need to formulate quantum gravity. The holographic principle – that is, this idea that the fundamental variables are kind of projected on the boundary of the space – is something very novel. It's not something that you would arrive at just by saying that the theory doesn't have a background but, instead, it is something that has to be discovered.

Moreover, as I've mentioned, I'm a very big fan of thought experiments and thought experiments are always carried out in specific backgrounds. For example, Hawking's black hole evaporation process that led to the information paradox is taking place in a very specific background [18]. In that context, you have some asymptotically classical geometry and you've got a black hole. Of course, locally, as the black hole evaporates, the geometry is doing wild things but AdS/CFT captures this because, even though the geometry is classical at long distances, in the interior of the space the geometry is just as quantum as you want.

So, I'm a big believer in thought experiments. Thought experiments are actually carried out in backgrounds, so I'm happy to figure out what the theory is and what are the special properties it has that makes it work. I'm happy to think in a background and then let some deeper person come along and figure out what the background independent version is.

You mentioned that in the bulk of anti–de Sitter space you could have these wild spacetime excitations. Can one understand, using AdS/CFT, how spacetime looks like at the Planck scale?

Well, you have to ask a physical question.

Is spacetime discrete?

It's certainly not discrete in any ordinary sense. There's a certain kind of discreteness but it's not a local discreteness in space. In the gauge theory, you have some large value of N, where N is an integer. There is some form of discreteness because N is an integer and N is related to the Planck length but it's very far from any naive discreteness.

I'm asking this question about discreteness because, as you probably know, loop quantum gravity people have made a prediction for what the minimum volume of spacetime can be [19].

Now let me ask you a question about that. That's stated to be a prediction but have they ever explained how you measure this volume? They define it and the thing they define has a minimum but that's not physics. For it to be physics you have to specify some experimental protocol by which you would measure this number. Have they ever specified this protocol?

To my knowledge right now, I think that they at least have a thought experiment of how to do it.

What is that thought experiment, can you explain that?

Unfortunately, I don't know all the details.

I actually would like to know the precise statement and I would find a statement that predicts that there's this discreteness quite interesting if it involves more than a definition, that is, if there is also some way of measuring it.

However, I would certainly not claim that spacetime is discrete. In classical quantum mechanics there is a phase space and Planck's constant is kind of the minimum unit of this phase space. However, phase space isn't discrete; it's not a set of points. The way quantum mechanics works is much more subtle than phase space just being discrete and I think the way quantum gravity – I mean, certainly the way quantum gravity seems to work in AdS/CFT – is that it's much more subtle than just making spacetime discrete. Making spacetime discrete is kind of the first thing a person tries; that is, you build a lattice. However, the evidence from string theory is that the whole notion of geometry is an approximate notion and not a fundamental notion. Fundamentally, you have some quantum degrees of freedom. I don't know what they are but when you get enough of them together in a pretty typical state, they often organise themselves into a geometry.

But could this question of spacetime discreteness be addressed in the context of AdS/CFT?

When I think about the kind of experiments that you can do, not just in string theory but in general, I think about some kind of scattering experiment. Steven Giddings and others have emphasised that as you try to probe shorter distances by scattering things you start making black holes [20–22]. So, as you start to use shorter wavelengths and higher energies, you start making black holes. So, there's a kind of minimum length scale that you can probe anything. That length scale is the Planck length; it's the maximum resolution. So, talking about how spacetime looks at distances smaller than the Planck length does not make much sense since it's not something that you can measure.

It's possible that someone will come out with a cleverer experiment that gives a better answer. It's possible that we will end up understanding the full theory as just sequences of ones and zeros, which is pretty discrete. However, even if it's just ones and zeroes, it probably won't be ones and zeroes in space but ones and zeroes that have to be decoded in some way in order to find what the space is.

Is AdS/CFT a non-pertubative formulation of string theory?

It is because we know thanks to Ken Wilson what the quantum field theory is non-perturbatively [23–25]. So, yes, it is a non-perturbative formulation.

Some people criticise string theory because it seems that it has not been shown that it is a finite theory. Is this true?

So, in my book [26, 27], one of the things I did was to explain why for the bosonic string, which I really like much better because I like bosons (I like bosons because I have trouble keeping track of signs, which is embarrassing), it's just a very general property of the topology of the worldsheet that all divergences are infrared divergences. There are no ultraviolet divergences. So as a physical principle this makes me happy because it solves the problem that we're supposed to solve.

Now, the nice thing about the superstring is that, because of supersymmetry, those infrared divergences are supposed to cancel order by order. If I thought it was really important to prove that maybe I would try to do it, although again I'm not good with signs.

When you are first learning about the subject, there's a lot of focus on the order-by-order perturbation theory but the state that we live in is certainly non-perturbative and it's a very unsymmetric state. Whatever the beautiful symmetries of the underlying theory are, we are not in a state in which these are manifested, and what this means is that there's some kind of complicated dynamics going on. Certainly, non-perturbative effects are taking place and different orders of perturbation theory are competing with each other. So, I always felt that it was more important to understand how to think about the theory beyond perturbation theory instead of proving very rigid things order by order in perturbation theory.

Having said that, it would be satisfying if someone were to complete such proof. Proving things is not always that rewarding. For example, Gross, Wilczek and Politzer discovered asymptotic freedom [28, 29], got the Nobel Prize and are famous. However, no one knows who actually proved that asymptotic freedom is true non-perturbatively. It was a guy named Tadeusz Balaban who spent four years and 300 pages to prove this ([30] and earlier works). The physics required to prove it is something that we learn in our basic quantum mechanics and field theory classes but to actually prove that it is a non-perturbative statement is a tremendous technical exercise and it was not that rewarding. It's great that Balaban did this but these papers are not sort of useful in terms of figuring out the next mystery in physics. They're more important for sort of closing the book on an older chapter.

I've occasionally proven things; for example, I was very happy to finish a new proof of renormalisation when I was a postdoc [31] but I was happy not so much because I like proving things but because it was a proof based on new principles. I don't think that it is important to prove that this particular asymptotic expansion exists, since as I have mentioned, I have been of the opinion for several years that strings are fundamental. I'd much rather figure out what this asymptotic expansion is an approximation to.

On the other hand, it is a well-posed problem and I would be happy to see it proven. Nathan Berkovits and his collaborators [32] have been the ones who have done the most to kind of develop the formalism necessary to prove it. Actually, well . . . I'll say this because he did not tell me this in confidence. Edward Witten says that he has the proof but that he's never taken the time to write it down. So if I say this and you put it in your book maybe he'll read this and he'll take the time to write it down (laughs). However, I think that he probably feels the same way as I do, namely, that there are other more creative directions for his efforts. [After this interview, Witten did return to the problem, and complete the proof [33]].

People usually give the AdS/CFT correspondence as an example of emergent spacetime but is it emergent spacetime or just emergent space?

This is a really interesting question which I puzzle over a lot. There is certainly some notion of emergent time there because you also have a time direction on the boundary. Time is already present in the boundary theory but there are many time slices in the interior that kind of end on the same time at the boundary. There is no unique mapping between the boundary time and the interior time and so the time in the interior is emergent to some extent. In fact, I've been spending a lot of time thinking about this and trying to make it precise but it's illusive. The answer is somehow hiding within the gauge theory variables. However, we do not know exactly how to address it and so it's not clear how time actually emerges. The lesson that we've gotten so far from string theory, in particular T-duality and AdS/CFT, is that space is emergent.

Even though the theory is Lorentz invariance and hence to some extent space and time are on equal footing, there is also a sense in which they are not on equal footing. The problem of time in general relativity is reflected on AdS/CFT. Different time slicings of the spacetime are all on top of each other in the wave function of the universe, just as in AdS/CFT they are all on top of each other in the wave function of the boundary. So, you have to sort of discover time by thinking about how you do measurements. In the context of the wave function of the universe, the only way to talk about time is to consider a situation where there is a natural clock. I would like to hope that there is some concept out there that we are just missing and that would make us see this problem in a completely different way because we don't have the right language for addressing this question.

As I mentioned earlier, Juan Maldacena [9] and Leonard Susskind [10] tried to reconstruct cosmology from the observations carried out at late times. If that's the way one constructs cosmology, then time is emerging in precisely the same way that space emerges in AdS/CFT. The simple observables are only in the asymptotic future and observables at finite times are sort of reconstructed by looking backwards. I don't know if this is satisfying or not. The type of questions asked by Maldacena and Susskind are addressed at late times and involve a kind of global observer who is not limited by causality but who can look at correlations, for example between two photons, that no single physical observer can. So now you have to ask whether or not the fact that no physical observer can measure such correlations is an indication that you shouldn't be asking such questions at late times. On the other hand, the fact that the theory seems to describe these correlations in a simple way also suggests that perhaps we should listen to the theory. What we are hoping is that there will be at least one other revolution in our future which will be conceptually as important as AdS/CFT and will allow us to know what to do in these situations.

When people speak about emergence in this context they always think about the gauge theory, but if it is a duality, how do you know which description is fundamental and which is emergent? People seem to be taking the quantum field theory side and claiming that everything emerges on the bulk.

This is because in quantum field theory, again thanks to Wilson [23–25] and actually also thanks to Balaban, who proved some of the necessary theorems ([30] and earlier works), we can reduce any calculation to an algorithm. On the other hand, in the bulk we have many different approximations that hold in different kinds of regimes and so on, which complicates things. Whenever there is a duality, one can always hope that there is a single framework from which the two dual descriptions emerge.

We have a similar situation in string theory. There is one string theory but many different quantum field theories, each of them describing string theory with different boundary conditions. There should be one single theory that captures all the possible boundary conditions at once. At the moment, we don't even have the framework to address this question.

Are all gauge theories holographically dual to string theories?

If you start from one case where we know the answer like $N = 4$ super Yang–Mills theory, then you can break it down to smaller gauge theories; for example, for large N you have N D3-branes but you can separate them into smaller clumps and so you can in some sense derive dualities for any finite value of N. Then, you can perturb things in a way that breaks the supersymmetries. So there's a sense in which a very large set of quantum field theories, even with weak coupling and small N, describe what string theory does in certain special situations. That's almost a corollary that follows from Maldacena's discovery [2].

However, this kind of leads us back to the question of what's fundamental and what is not. Is the theory more than just a sum of all the quantum field theories? It seems like the theory must be more than just a sum of all the quantum field theories, especially because they only capture special boundary conditions. As I mentioned earlier, I sort of take small steps at a time and it's very hard to take a small step from anti–de Sitter space to something very different. You can try to, if you start perturbing the field theory in the ultraviolet by means of irrelevant operators. In this situation, you sort of start pushing the boundary outward. However, we don't understand quantum field theory well enough to give meaning to this.

Now that you mentioned "pushing the boundary," I remembered that you tried to push the boundary inwards with the purpose of developing a more local holographic principle [16].

There are holographic results that don't require an observer to be at some asymptotic boundary. These are results that kind of hold in the interior space. This is the case, for example, of the so-called entropy bounds, which are a very natural thing to think about from a purely gravitational point of view. We were therefore trying to push the boundary inwards in a very naive way, that is, in the obvious radial direction in order to describe local observers in the interior space holographically.

I have to say that a lot of what we have done is a repackaging of AdS/CFT. I mean, you can take AdS/CFT and you can repackage it and repackage it but at some point you hope to suddenly have a simpler formulation. This is a good thing to try to do but what hasn't happened yet is to encounter in the end a real feeling of rightness to it. However, it's useful as it allows you to understand better what AdS/CFT is and what it isn't.

Lately, I have been thinking a lot about the case where you have a black hole in the interior space and whether or not what is behind the horizon is visible from the boundary point of view. You can ask the question, "How about the observer behind the horizon and his Hilbert space?" Is that contained in the Hilbert space of the boundary theory? I'm not the first person to think about this and in fact many people over the years have said, "Yes it is." I've thought about it and I agree with them, basically for the same reasons; namely, an observer at the boundary can follow the black hole back in time until the time that it formed and that Hilbert space, before the black hole was formed, is definitely contained in the Hilbert space of the theory. You can then take those variables in the gauge theory and move them forward in time. It's not a very deep statement but it tells you two things: that the Hilbert space behind the horizon is in the gauge theory but also that it's in there in such a thermalised way that at the moment we don't know how to find it. We know it's in there but to find it requires some insight that we haven't had yet.

The firewall paradox, discovered shortly after this interview, shows that our understanding of the black hole interior is less complete than was thought [34, 35]. It is possible that the interior does not even exist. I think that no satisfactory solution has yet been given.

What is the firewall paradox and why does it suggest that there is no black hole interior?

AdS/CFT gave a partial resolution to the black hole information problem, giving convincing evidence that information is not lost. Gerard 't Hooft, Leonard Susskind and John Preskill suggested that the remaining paradox, an apparent duplication of information, was avoided because no single observer could see both copies. This was supported by various thought experiments [36, 37]. But our collaboration revealed that a more elaborate thought experiment still showed a contradiction between quantum mechanics and spacetime [34, 35]. If quantum mechanics is not modified, it seems that the black hole interior can't exist.

Why aren't you satisfied with the possible resolutions to this paradox? Isn't the Papadodimas-Raju [38, 39] proposal good enough?

There are a lot of ideas about what the firewall argument really means. All right now modify either spacetime or quantum mechanics. Papadodimas-Raju's argument [38, 39] is of the latter type. It is referred to as "background independent," but here this means a modification of quantum mechanics. It could be true, but would need a much more complete theory.

You mentioned earlier that one could not use AdS/CFT to address the problem of cosmology. However, people have thought about making a dS/CFT correspondence. If I'm not mistaken, you had a proposal of how to do it, right?

I don't think I did. I sometimes forget things I've done but I think I would remember this one.

It was related to this paper where you found conformal field theory (CFT) duals that were Argyres-Douglas-type field theories [40].

Was this one of my works with Eva Silverstein?

Yes.

Actually, I should say, she was the one who had the idea about dS/CFT and actually she is going to give a talk here and may talk about this.[1] In fact, she is the one who put those

[1] This interview took place during the workshop String Theory in Benasque.

words in the paper and that's why I forgot them (laughs). Also she is the one who has gone on to develop them [41]. Her idea has roots in her work with Shamit Kachru and others and she has pursued it over time. The core of the idea is that if you slice de Sitter in half down the middle, it's in some way like gluing together two copies of AdS/CFT.

One of the lessons that we learnt from string theory is that de Sitter (dS) is not eternal but can always decay [42]. We don't know if this holds true for every de Sitter construction but this statement seems pretty robust: de Sitter is unstable. This fact has led me to think that trying to formulate some holographic theory for de Sitter space is sort of going to lead nowhere. However, recently I have, sometimes, started thinking differently about this due to Maldacena, who says that there is a very tiny probability for de Sitter not decaying. There is a tiny piece of the wave function where de Sitter doesn't decay and we can still meaningfully talk about this piece of the wave function. We can calculate it and calculate the fluctuations around it. This is what dS/CFT does and it is a nice insight leading to the idea that there is something like a quantum field theory, but not exactly like a quantum field theory, that lives on the future boundary. What it calculates is the amplitude for the wave function of the universe to have a certain form at late times. On the one hand, this idea has a certain natural appeal that makes it worth exploring but, on the other hand, it could also just be the wrong extrapolation of what we already learnt. In order to give it some meat one needs to figure out what the quantum field theory is. If we had a specific example it could be possible to understand it better.

Is it a difficult problem to find such an example?

Yes. String theory has these funny solutions called the supercritical string solutions [43–45] which sort of have a rolling dilaton, which is a bit similar to de Sitter space though perhaps simpler to understand. So I keep thinking about this case and that it might lead to a simpler version of dS/CFT. All I can say is that I'm intrigued and that I have not contributed to this.

In this paper of yours with Eva Silverstein [40], it was written that the motivation was to formulate four-dimensional quantum gravity. I was just wondering if the fact that, in string theory, extra dimensions are usually there disturbs you?

So, we seem to live in a space where six dimensions are small and four are large but in all known examples of AdS/CFT the dimensions are similar in size. The main point of that paper was to explore the question, "What would you have to do to the field theory in order to get six small dimensions and four large?" So that was the goal and it's not surprising that it's hard to achieve because, from the point of view of just the balance of terms in the Einstein equations, it takes a lot of cancellations to get to a space like ours where some dimensions are very curved and some are not. So you need to get control over these big cancellations and we partly succeeded but we didn't really finish the job.

I'm not bothered by extra dimensions at all. The experimental value is at least three and until you've actually probed all the way to the Planck scale, all you can say is at least three because we don't know what happens at the Planck scale. Einstein had the vision that physics came from geometry and it works for gravity, so if it works for gravity then it should work for everything because physics is unified. However, if we are going to realise this idea then we need more geometry than what you naively expect. We live in this very rich universe and at every point there must sit some kind of dynamical structure that makes it possible. The idea that these extra dynamical structures involve extra dimensions is not so strange; it's just more copies of the structures we already know. We know that the dimensions that we see were vastly smaller in the past than they are today. There is no fundamental reason why the expansion of the universe was isotropic and why some expanded and some didn't. People used to get upset about the appearance of new particles. When the neutrino came along people were very upset with introducing a new particle so cavalierly. Now, we are very used to the idea that new particles can appear and there is going to be more of them. Dimensions are the same (laughs).

At some point you published a paper where you addressed whether wormholes existed in string theory or not [46]. Do they?

About 20 years ago, Sidney Coleman [47] and others [48] argued that there are instantons in quantum gravity and that they have the effect of making all the constants of nature random. Coleman, in particular, later argued that the constants of nature are variable in a way that allows you to predict their variations [49]. However, it was never clear what the rules were and the hope was that one would come back to this problem one day and understand what the rules were. So you have these Euclidean solutions which are saddle points in the path integral over metrics but no one tells you what saddle points you include and which you don't. Our logic was a little bit indirect but it's clear that by their nature a wormhole, where one end is here and the other end is in ancient Greece, is something very non-local. The remarkable thing is that Coleman explained that, although it looks non-local when you sum coherently, its effects are actually local but random. However, in AdS/CFT we certainly have locality in time on the boundary and there is a certain amount of indeterminacy of how you connect the bulk time to the boundary time, as we were discussing earlier, but it's not an infinite indeterminacy. There is a certain kind of periodicity there and so something which is as non-local in time as an Euclidean wormhole just doesn't smell like it can be captured by the gauge theory. This is kind of satisfying because, I have to say, having worked on that subject, I felt that it never had the ring of truth. I shouldn't praise Lenny Susskind too much but he was one of my collaborators at the time and at some point he basically said, "This is just so sickening, I don't believe that quantum gravity is really an integral over metrics and that these saddle points are really there." And I think he was right. The discussion here is for Euclidean wormholes. Recently, there has been some discussion of time-dependent wormholes, which are different [50].

Okay so you think the physics of one vacuum cannot affect the physics of the other vacuum?

Well, not through wormholes. After I wrote this paper on wormholes [46], I happened to be one day having dinner at a conference with Andrew Strominger, who will probably not want me to quote him on this. We both had too many to drink during that dinner and he told me that he thought that this paper of mine was a negative contribution to physics (laughs). I loved that. He feels embarrassed every time I remind him that he said this but what he meant was that he tends to believe that these solutions exist for a reason. Maybe they don't describe what we thought they described but they must have some interpretation that we haven't yet figured out. I'm happy to agree with him on that that. They are fairly robust and elegant solutions to the Euclidean Einstein equations and they surely describe some tunnelling process in some sense but not with the interpretation that we were giving it 20 years ago. So, by the way, you said that you would send me this to see if I said anything I shouldn't have said. So, I don't know, maybe I will wake up tomorrow and feel bad about having said this but I don't think so (laughs).

Good, because I think it's a million-dollar sentence.

Okay (laughs). Well, let me see how I feel tomorrow but I really I just loved that. I like Andy because he says what he thinks and that's refreshing.

Do you think that it's contained within string theory the mechanism that decides which vacuum we live in?

It would be nice. If you ask the cosmologists they say, "Well look, you've got inflation, you've got tunnelling, you are just going to explore all of this everywhere." Now, string theory seems to give the cosmologists what they said they needed. So, it's unfortunate; I mean, I wasn't happy with this development but I'm ready for things to turn out differently than what I expected. This seems to be the best way to interpret and understand the theory, that the vacuum we live in is just a random thing. So we got rid of the randomness of the wormholes and we got back the randomness from this other thing. I don't know if we're better off.

Have we found any compactification that can serve as a possible scenario for our current universe?

That's a good question. I don't work on this because a lot of other people are and so you should ask these other people. It depends on how you define it exactly, especially because with many of the vacua there are parameters that one doesn't know and things that one can't calculate. So you don't know if you get the right masses out of it and so on.

My question was meant to be a bit more specific. Has someone constructed a vacuum, which, for example, has the right value for the cosmological constant?

So the problem there, because the answer is 10^{-120} smaller than the input numbers, you would need to do the calculation up to 120 digits to know the answer and we don't have the calculational tools to do that. It's interesting because if the landscape is the right explanation for the cosmological constant, then there has to be enough vacua such that one of them is likely to be small enough. So, if somebody comes along and proves that string theory has only 10^{60} vacua and not 10^{120} vacua, it is not a crisis but it requires us to think since things don't work as we thought they were working. Now, 10^{120} turns out not to be such a large number. There's lots of places where numbers greater than that appear. For example, it was pointed out to me that if you simply take a rather large molecule and you count the number of spin states of its nuclei then you very quickly get to numbers as big as that. It follows an exponential scaling; that is, you take 2 to a big enough number. We already know that if you combine neutrons, protons and electrons you can get numbers like 10^{1000} very easily and so if you combine handles, branes and fluxes, it's not surprising that the power of exponentials generates many solutions. It is interesting to note that there seems to be some boundedness to the number. Unlike with classical groups, such as with $SU(N)$ where N can be as big as you want, no one has ever found an infinite number of Calabi-Yaus. They are all sporadic. It seems to be similar to what happens with exceptional Lie algebras, in which you have just a sporadic collection with some large but finite number.

String theory is often criticised for not being falsifiable, and many often say that it is not even a theory of nature. Do you agree with such criticisms?

The problem is nature, not string theory! What I mean is that nature (as reported by Max Planck) seems to predict that the fundamental length is far beyond experiment. So we are very dependent on theoretical evidence. And here I count six major successes [51, 52].

But does the string landscape really exist? Is the KKLT [53] mechanism viable?

The landscape is one of my six pieces of evidence. It (the multiverse) is the only theory able to explain the non-zero cosmological constant. I have never understood the arguments against the KKLT model. I recently looked more closely at it, having developed some new tools, and still could not understand it [54, 55].

You also worked on finding properties of condensed matter systems using holographic techniques [56–58]. Has holography brought any new insights into the physics of non-Fermi liquids?

AdS/condensed matter is very broad and one of the interesting things is in fact that people are looking at so many different things. Actually, when people began thinking about using AdS/CFT and applying it to QCD and heavy ion collisions, I really didn't take it seriously. I thought that it wasn't describing the real world but some hypothetical world that we didn't live in. However, Son, Starinets and Kvotun [59] and others came along and said, "Hey, actually there is one regime of behaviour in heavy ion physics where it's actually a good description, at least as good a description as any other that we have." I would never have thought of that and similarly with applications to condensed matter as well. I mean, I'm impressed by the diversity of things that people are thinking about.

About 20 years ago, I taught quantum mechanics and I decided that I would figure out what exactly this Fermi liquid theory was. I realised that it was really cool and then I heard about non-Fermi liquids and I thought that since I could understand Fermi liquids,

I would be able to figure out what non-Fermi liquids were but, of course, I failed. Now, 20 years later, everybody has failed. There are these phases of matter that have fairly robust properties and exist in various situations but there is no agreement on what they are and so it seems like a real target for someone who has new tools.

To study QCD using AdS/CFT seems reasonable because QCD is not that far from $N = 4$ super Yang–Mills, so it's not surprising that one could connect them. However, when you start talking about a lattice of electrons and nuclei, it becomes a very specific thing and so it's a little bit harder to say that $N = 4$ super Yang–Mills theory can capture the key properties of these systems. It is interesting to note that there is even a lack of framework for studying some of these condensed matter systems and so dozens of different groups seem to have a different answer and there is not agreement. A few years ago, a paper by Senthil Todadri appeared [60], where he tried to define a sort of scaling theory for non-Fermi liquids. This is interesting because it expresses the fact that the state of the art in this field is so rudimentary that even a scaling theory does not exist. It's not even a renormalisation group theory but just a scaling theory. So, I think that it's worth thinking about these new kinds of applications.

What do you think has been the biggest breakthrough in theoretical physics in the past 30 years?

So I think that we can start the list of breakthroughs in 1984 with the discovery of the heterotic string [61–63], the relevance of Calabi-Yau manifolds [64] and anomaly cancellation [65]. These are all cool. The relevance of Calabi-Yau manifolds was important because besides greatly impacting mathematics, it also had a big sociological impact. Later we found dualities [12–14, 66, 67], D-branes [11, 68, 69] and techniques for black hole entropy counting [70]. Certainly, dualities are amazing because we used to think that quantum field theory was just perturbation theory and a little more. However, I have to say, AdS/CFT [2] connects two things that no one had any clue were connected. If you count citations, you know, it has around 12,565 citations?[2] So that can be considered to be influential, to say the least. Now, if you had asked me about breakthroughs during the past 40 years, I would have probably said that it was asymptotic freedom [28, 29, 71, 72] but that's just a minus sign in some calculation (laughs).

Right. I think that someone would be a bit angry with that comment.

It's okay, it's okay. He can't fire me (laughs).

Why have you chosen to do physics? Why not something else?

Well, before I got to college, I was best at math but not the way a mathematician is, I mean, I was really interested in questions like, "What is gravity?" When I heard about Maxwell equations, although it wasn't very quantitative, I felt that it was really cool. I liked math but

[2] This was the citation count on 3 March 2017. On the 9 December it had 16,250 citations.

not so much abstract math; instead, I liked translating a physical situation into mathematics. I went to Caltech and, you know, as soon as you step into the campus, you're immersed in Richard Feynman lore. At that point, there was no doubt that I was a theoretical physicist. When I was there, we actually took our freshman physics from the Feynman lectures. In high school I was kind of ignorant. I didn't really know what science was but as soon as I found out what theoretical physics was, it was obvious that I was a theoretical physicist and that theoretical physics was what I had the best skills to do.

What do you think is the role of the theoretical physicist in modern society?

I cannot speak for all theoretical physicists who are working on many other things. I can only speak about what my role is in society. There's clearly a fascination with what we do. Those of us who got into this subject got into it because we wanted to understand something deep about nature. However, everybody else does too. They want to understand why things are as they are, what is a space, what is time and how things began. I think that it's good that there are people like us who are focused on these problems, trying to answer these questions and reporting back to the public on what we think we understand.

References

[1] J. Polchinski, "Dualities of fields and strings," *Stud. Hist. Phil. Sci. B* **59** (2017) 6–20, arXiv:1412.5704 [hep-th].

[2] J. M. Maldacena, "The large N limit of superconformal field theories and supergravity," *Int. J. Theor. Phys.* **38** (1999) 1113–1133, arXiv:hep-th/9711200.

[3] G. Hooft, "A planar diagram theory for strong interactions," *Nucl. Phys. B* **72** no. 3 (1974) 461–473. www.sciencedirect.com/science/article/pii/0550321374901540.

[4] G. 't Hooft, "Dimensional reduction in quantum gravity," *Conf. Proc. C* **930308** (1993) 284–296, arXiv:gr-qc/9310026.

[5] L. Susskind, "The world as a hologram," *J. Math. Phys.* **36** (1995) 6377–6396, arXiv:hep-th/9409089.

[6] J. Polchinski, "Quantum gravity at the Planck length," *eConf* **C9808031** (1998) 08, arXiv:hep-th/9812104 [hep-th].

[7] J. Polchinski, "Comment on [arXiv:1106.1417] 'Small Lorentz violations in quantum gravity: do they lead to unacceptably large effects?,' " *Class. Quant. Grav.* **29** (2012) 088001, arXiv:1106.6346 [gr-qc].

[8] R. Gambini, S. Rastgoo and J. Pullin, "Small Lorentz violations in quantum gravity: do they lead to unacceptably large effects?," *Class. Quant. Grav.* **28** (2011) 155005, arXiv:1106.1417 [gr-qc].

[9] J. M. Maldacena, "Non-Gaussian features of primordial fluctuations in single field inflationary models," *JHEP* **05** (2003) 013, arXiv:astro-ph/0210603 [astro-ph].

[10] W. Fischler and L. Susskind, "Holography and cosmology," arXiv:hep-th/9806039 [hep-th].

[11] J. Polchinski, "Dirichlet branes and Ramond-Ramond charges," *Phys. Rev. Lett.* **75** (1995) 4724–4727, arXiv:hep-th/9510017.

[12] J. Dai, R. G. Leigh and J. Polchinski, "New connections between string theories," *Mod. Phys. Lett. A* **4** (1989) 2073–2083.

[13] P. Horava, "Strings on world sheet orbifolds," *Nucl. Phys. B* **327** (1989) 461–484.

[14] M. B. Green, "Space-time duality and Dirichlet string theory," *Phys. Lett. B* **266** (1991) 325–336.

[15] J. Polchinski, "What is string theory?," in *NATO Advanced Study Institute: Les Houches Summer School, Session 62: Fluctuating Geometries in Statistical Mechanics and Field Theory Les Houches, France, August 2–September 9, 1994*, F. David and P. Ginsparg, eds. North-Holland, 1996. arXiv:hep-th/9411028 [hep-th].

[16] I. Heemskerk and J. Polchinski, "Holographic and Wilsonian renormalization groups," *JHEP* **06** (2011) 031, arXiv:1010.1264 [hep-th].

[17] H. Lin, O. Lunin and J. M. Maldacena, "Bubbling AdS space and 1/2 BPS geometries," *JHEP* **10** (2004) 025, arXiv:hep-th/0409174 [hep-th].

[18] S. W. Hawking, "Particle creation by black holes," *Communications in Mathematical Physics* **43** no. 3 (Aug. 1975) 199–220. https://doi.org/10.1007/BF02345020.

[19] C. Rovelli and L. Smolin, "Discreteness of area and volume in quantum gravity," *Nucl. Phys. B* **442** (1995) 593–622, arXiv:gr-qc/9411005. [Erratum: *Nucl. Phys. B* **456** (1995) 753–754.]

[20] S. B. Giddings and S. D. Thomas, "High-energy colliders as black hole factories: the end of short distance physics," *Phys. Rev. D* **65** (2002) 056010, arXiv:hep-ph/0106219.

[21] S. B. Giddings, "Black hole production in TeV scale gravity, and the future of high-energy physics," *eConf* **C010630** (2001) P328, arXiv:hep-ph/0110127.

[22] D. M. Eardley and S. B. Giddings, "Classical black hole production in high-energy collisions," *Phys. Rev. D* **66** (2002) 044011, arXiv:gr-qc/0201034.

[23] K. G. Wilson, "The renormalization group and strong interactions," *Phys. Rev. D* **3** (1971) 1818.

[24] K. G. Wilson, "Renormalization group and critical phenomena. 1. Renormalization group and the Kadanoff scaling picture," *Phys. Rev. B* **4** (1971) 3174–3183.

[25] K. G. Wilson and J. B. Kogut, "The renormalization group and the epsilon expansion," *Phys. Rept.* **12** (1974) 75–200.

[26] J. Polchinski, *String theory, Vol. 1: An Introduction to the Bosonic String*. Cambridge University Press, 2007.

[27] J. Polchinski, *String theory, Vol. 2: Superstring Theory and Beyond*. Cambridge University Press, 2007.

[28] D. J. Gross and F. Wilczek, "Ultraviolet behavior of nonabelian gauge theories," *Phys. Rev. Lett.* **30** (1973) 1343–1346.

[29] H. D. Politzer, "Reliable perturbative results for strong interactions?," *Phys. Rev. Lett.* **30** (1973) 1346–1349.

[30] T. Balaban, "Convergent renormalization expansions for lattice gauge theories," *Commun. Math. Phys.* **119** (1988) 243–285.

[31] J. Polchinski, "Renormalization and effective Lagrangians," *Nucl. Phys. B* **231** (1984) 269–295.

[32] N. Berkovits, "Finiteness and unitarity of Lorentz covariant Green-Schwarz superstring amplitudes," *Nucl. Phys. B* **408** (1993) 43–61, arXiv:hep-th/9303122 [hep-th].

[33] E. Witten, "Superstring perturbation theory revisited," arXiv:1209.5461 [hep-th].

[34] A. Almheiri, D. Marolf, J. Polchinski and J. Sully, "Black holes: complementarity or firewalls?," *JHEP* **02** (2013) 062, arXiv:1207.3123 [hep-th].

[35] D. Marolf and J. Polchinski, "Gauge/gravity duality and the black hole interior," *Phys. Rev. Lett.* **111** (2013) 171301, arXiv:1307.4706 [hep-th].

[36] C. R. Stephens, G. 't Hooft and B. F. Whiting, "Black hole evaporation without information loss," *Class. Quant. Grav.* **11** (1994) 621–648, arXiv:gr-qc/9310006 [gr-qc].

[37] L. Susskind and L. Thorlacius, "Gedanken experiments involving black holes," *Phys. Rev. D* **49** (1994) 966–974, arXiv:hep-th/9308100 [hep-th].

[38] K. Papadodimas and S. Raju, "An infalling observer in AdS/CFT," *JHEP* **10** (2013) 212, arXiv:1211.6767 [hep-th].

[39] K. Papadodimas and S. Raju, "Black hole interior in the holographic correspondence and the information paradox," *Phys. Rev. Lett.* **112** no. 5 (2014) 051301, arXiv:1310.6334 [hep-th].

[40] J. Polchinski and E. Silverstein, "Dual purpose landscaping tools: small extra dimensions in AdS/CFT," in *Strings, Gauge Fields, and the Geometry Behind: The Legacy of Maximilian Kreuzer*, A. Rebhan, L. Katzarkov, J. Knapp, R. Rashkov, and E. Scheidegger, eds., pp. 365–390. World Scientific, 2009.

[41] X. Dong, B. Horn, E. Silverstein and G. Torroba, "Micromanaging de Sitter holography," *Class. Quant. Grav.* **27** (2010) 245020, arXiv:1005.5403 [hep-th].

[42] S. B. Giddings, "The fate of four-dimensions," *Phys. Rev. D* **68** (2003) 026006, arXiv:hep-th/0303031 [hep-th].

[43] A. M. Polyakov, "Quantum geometry of bosonic strings," *Phys. Lett. B* **103** (1981) 207–210.

[44] R. C. Myers, "New dimensions for old strings," *Phys. Lett. B* **199** (1987) 371–376.

[45] A. H. Chamseddine, "A study of noncritical strings in arbitrary dimensions," *Nucl. Phys. B* **368** (1992) 98–120.

[46] N. Arkani-Hamed, J. Orgera and J. Polchinski, "Euclidean wormholes in string theory," *JHEP* **12** (2007) 018, arXiv:0705.2768 [hep-th].

[47] S. R. Coleman, "Black holes as red herrings: topological fluctuations and the loss of quantum coherence," *Nucl. Phys. B* **307** (1988) 867–882.

[48] S. B. Giddings and A. Strominger, "Loss of incoherence and determination of coupling constants in quantum gravity," *Nucl. Phys. B* **307** (1988) 854–866.

[49] S. R. Coleman, "Why there is nothing rather than something: a theory of the cosmological constant," *Nucl. Phys. B* **310** (1988) 643–668.

[50] P. Gao, D. L. Jafferis and A. Wall, "Traversable wormholes via a double trace deformation," arXiv:1608.05687 [hep-th].

[51] J. Polchinski, "String theory to the rescue," 2015. arXiv:1512.02477 [hep-th]. https://inspirehep.net/record/1408773/files/arXiv:1512.02477.pdf.

[52] J. Polchinski, "Why trust a theory? Some further remarks (part 1)," arXiv:1601.06145 [hep-th].

[53] S. Kachru, R. Kallosh, A. D. Linde and S. P. Trivedi, "De Sitter vacua in string theory," *Phys. Rev. D* **68** (2003) 046005, arXiv:hep-th/0301240.

[54] B. Michel, E. Mintun, J. Polchinski, A. Puhm and P. Saad, "Remarks on brane and antibrane dynamics," *JHEP* **09** (2015) 021, arXiv:1412.5702 [hep-th].

[55] J. Polchinski, "Brane/antibrane dynamics and KKLT stability," arXiv:1509.05710 [hep-th].

[56] S. A. Hartnoll, J. Polchinski, E. Silverstein and D. Tong, "Towards strange metallic holography," *JHEP* **04** (2010) 120, arXiv:0912.1061 [hep-th].

[57] T. Faulkner and J. Polchinski, "Semi-holographic Fermi liquids," *JHEP* **06** (2011) 012, arXiv:1001.5049 [hep-th].

[58] K. Jensen, S. Kachru, A. Karch, J. Polchinski and E. Silverstein, "Towards a holographic marginal Fermi liquid," *Phys. Rev. D* **84** (2011) 126002, arXiv:1105.1772 [hep-th].

[59] P. Kovtun, D. T. Son and A. O. Starinets, "Viscosity in strongly interacting quantum field theories from black hole physics," *Phys. Rev. Lett.* **94** (2005) 111601, arXiv:hep-th/0405231.

[60] T. Senthil, "Critical Fermi surfaces and non-Fermi liquid metals," *Phys. Rev. B* **78** (Jul. 2008) 035103. http://link.aps.org/doi/10.1103/PhysRevB.78.035103.

[61] D. J. Gross, J. A. Harvey, E. J. Martinec and R. Rohm, "The heterotic string," *Phys. Rev. Lett.* **54** (1985) 502–505.

[62] D. J. Gross, J. A. Harvey, E. J. Martinec and R. Rohm, "Heterotic string theory. 1. The free heterotic string," *Nucl. Phys. B* **256** (1985) 253.

[63] D. J. Gross, J. A. Harvey, E. J. Martinec and R. Rohm, "Heterotic string theory. 2. The interacting heterotic string," *Nucl. Phys.B* **267** (1986) 75–124.

[64] P. Candelas, G. T. Horowitz, A. Strominger and E. Witten, "Vacuum configurations for superstrings," *Nucl. Phys. B* **258** (1985) 46–74.

[65] M. B. Green and J. H. Schwarz, "Anomaly cancellation in supersymmetric D = 10 gauge theory and superstring theory," *Phys. Lett. B* **149** (1984) 117–122.

[66] A. Font, L. E. Ibanez, D. Lust and F. Quevedo, "Strong-weak coupling duality and nonperturbative effects in string theory," *Phys. Lett. B* **249** (1990) 35–43.

[67] E. Witten, "String theory dynamics in various dimensions," *Nucl. Phys. B* **443** (1995) 85–126, arXiv:hep-th/9503124.

[68] J. Polchinski, S. Chaudhuri and C. V. Johnson, "Notes on D-branes," arXiv:hep-th/9602052 [hep-th].

[69] J. Polchinski, "Tasi lectures on D-branes," in *Theoretical Advanced Study Institute in Elementary Particle Physics (TASI 96): Fields, Strings, and Duality*, C. Efthimiou and B. Greene, eds., pp. 293–356. World Scientific, 1997.

[70] A. Strominger and C. Vafa, "Microscopic origin of the Bekenstein-Hawking entropy," *Phys. Lett. B* **379** (1996) 99–104, arXiv:hep-th/9601029.

[71] D. J. Gross and F. Wilczek, "Asymptotically free gauge theories – I," *Phys. Rev. D* **8** (1973) 3633–3652.

[72] D. J. Gross and F. Wilczek, "Asymptotically free gauge theories. 2," *Phys. Rev. D* **9** (1974) 980–993.

22

Alexander Polyakov

Joseph Henry Professor of Physics at Princeton University

Date: 27 June 2014. Location: Princeton, NJ. Last edit: 2 December 2016

In your opinion, what are the main problems in theoretical physics at the moment?

There are several unsolved problems; for instance, understanding the theory of critical phenomena or conformal theories in three dimensions presents a great challenge. It's a complicated problem but solvable. The standard approach to critical phenomena via renormalisation group flow is insufficient to make progress in this and I think that we can do much better than that. My hope is that we can actually develop the theory in order to achieve the same level of understanding as that which has been achieved in two dimensions, namely, that we can more or less classify all possible conformal algebras and identify various fixed points quantitatively and so on. In my view, this will be a really important achievement that can be applied to a multitude of physical systems.

There are of course millions of unsolved problems but it's only worth discussing problems that I think can be solved at the moment. There are other problems which are maybe more interesting but I see no way to approach them right now. These problems are untimely and, as such, it's not the right time to tackle them.

I'm curious, which problems are you referring to?

Well ... for example, some problems in cosmology that people have been discussing such as the multiverse, etc. I think these questions are absolutely untimely, though of course fascinating. It is just like as if you didn't know about atoms but instead you knew about the phlogiston and you were asked to describe the behaviour of a superconductor. It would be a reasonably interesting problem but there would be no way in which you could approach it with phlogiston theory, meaning that it would be a complete waste of time.

Okay, so going back to the ones you think are solvable ...

Another problem that I think is close to a solution is the problem of turbulence in two and three dimensions. This is related to the first problem that I mentioned, namely, that of developing conformal field theories and I believe that there has been some recent progress in this field suggesting that it will be possible to say something concrete and non-trivial about this. Although, I must say, I have kept thinking, on and off, about turbulence for the past 40 years and I haven't found a solution so maybe I'm being over-optimistic as usual.

Do you have any idea of how to tackle this problem?

Yes, I think that I will try to use the operator product expansion of conformal field theory to fit the Navier-Stokes equations describing the turbulent flow. There's also another powerful tool – the theory of quantum anomalies – that has an analogous counterpart in the theory of turbulence. Actually, the Kolmogorov cascade [1] through the different scales resembles the axial anomaly in quantum electrodynamics. In the case of the axial anomaly, the symmetry is broken in the ultraviolet and then energy propagates through the scales to the infrared. In the case of Kolmogorov turbulence the same happens, in which case you pump energy at some very large scales and then there is a direct cascade, but together with an inverse cascade for which energy propagates from small scales to large scales, finally dissipating. I think that the apparatus of conformal field theory and the theory of anomalies in its present form is still not enough to find a solution to this problem but it could be extended to a certain extent in order to do so.

Recently, there has also been some quite tantalising progress in critical phenomena leading to the discovery of some rigorous inequalities using conformal bootstrap techniques [2–4]. These inequalities are almost saturated in nature. No one understands why it is the case that this saturation occurs but it is certainly a sign from heaven telling us that we should be working on that.

I see ... and for these problems that you mentioned, has string theory any role to play?

Oh yes, it's all tightly connected. In the case of critical phenomena, like the three-dimensional Ising model, you begin the analysis by presenting the partition function as a sum over random surfaces. This is, of course, non-critical string theory [5, 6] and moreover you find that this conjectured string theory must be fermionic; that is, it must be something like Ramond-Neveu-Schwarz string theory [7, 8] but modified. There is still a huge gap between this formal string theoretic representation and those inequalities that I mentioned but when the gap closes I think we'll have a solution. So I have very little doubt that ideas and methods of string theory will play an absolutely crucial role here.

So do you see string theory as a framework?

Yes, I always saw it as a framework for the past 40 years. This is not a recent viewpoint for me as I always saw it as a framework for solving various problems in physics.

What about the problem of quantum gravity; is string theory not meant for that?

Also the problem of quantum gravity and that's the most interesting thing about it, namely, that you can start by studying a boiling kettle and end up solving the problem of quantum gravity (laughs). That's fascinating.

Do you think that string theory is the only way to solve the problem of quantum gravity?

I have tried some other ways in the early 1970s. At that time, I thought that maybe we could describe the ultraviolet limit of quantum gravity by some conformal field theory. I tried to describe this but nothing concrete came out of it. String theory gives you a very concrete ultraviolet completion of quantum gravity, which is obviously consistent, highly nontrivial and very rich. This is not to say that I completely abandoned my earlier idea, as time to time I try to invent some conformal field theory that would fit quantum gravity, but so far it didn't work. The other approaches that people have followed outside string theory look rather poor. I have absolutely no prejudices but I love interesting formulas and there are plenty of interesting formulas in string theory while there are none in its competitors (laughs). If they arrive at something interesting I will be the first to look into it. Perhaps I will also arrive at something interesting and different than string theory but that has not happened yet.

Do you think it's good to keep an open mind and look for different ways of approaching this problem?

You should keep an open mind but if your mind is too open it's better to close it a little bit because otherwise you will be wasting your time. For me a good criterion, when I see some interesting formula, is that my heart starts beating faster. When I feel this, it's certainly worth thinking about it. Instead, if you see some banalities written on hundreds of papers with pompous titles then it is not worth your time. By the way, do you know what was the title of Max Planck's paper, published in 1901, that gave rise to quantum theory?

No, I don't know ...

The title was "On an improvement of Wien's equation for the spectrum" [9] (laughs).

You mentioned that sometimes you tried to think about quantum gravity as being described by a conformal field theory. Were you thinking about this in the context of holographic dualities?

Holography is a part of it but one should be careful when talking about holography because there's a lot of nonsense being said about it. The only good reasonable part of what is being said is certainly related to string theory. People usually refer to holography as gauge/gravity correspondence but that's actually misleading. Some of them, of course, mean the right thing but there's no gauge/gravity correspondence unless there's a gauge/string correspondence. Gravity cannot grasp all the degrees of freedom of the conformal field theory at the boundary. You must include all the massive string states in the bulk. Holography is interesting if you are referring to the right thing.

Okay, but was it in this sense that you meant that conformal field theories could describe quantum gravity?

No, not exactly in that sense. In the early 1970s I was reasoning as follows. We have gravity, which is a non-renormalisable theory, meaning that as the energy becomes much larger than the Planck energy we cannot get any information from the theory. We don't know what to do with it; however, there could be some regimes in which we could make progress. So let's assume that when the energies are much larger than the Planck scale there is no scale left at all and in this limit you have an ultraviolet fixed point; that is, you assume that there is a certain limit with conformal symmetry. The question, then, is how to find the operator product expansion consistent with diffeomorphism invariance. As I said, I have the impression that it's possible but there's no known practical way to proceed. It is possible to use some uncontrollable approximations but that doesn't interest me.

What has motivated you to develop string theory?

What motived me was precisely the attempt to solve the three-dimensional Ising model and QCD. In QCD it became clear at some point that I needed a completely different point of view. I was trying to solve the problem of confinement using instantons and it worked very in the Abelian theory – the $U(1)$ theory – where I think I found a very nice explanation of confinement [10–12], however ...

What is the explanation?

The idea is that you have instantons, which are random flashes of the field in Euclidean spacetime in four dimensions, and quarks get confused and can't propagate in this field. What happens then is similar to what happens in condensed matter systems in which there are too many impurities. In such situations, you end up having localisation and electrons cannot propagate. In the case of the Abelian theory, due to the fact that there are so many disturbances for the quark propagation, the quark stops and doesn't wander off to infinity; instead, it becomes localised. I think I was able to show quantitatively that this is indeed the case so it was a serious success [10]. I thought that only a few more months would be required to solve the non-Abelian case but it unfortunately turned out not to be the case. Instantons, themselves, get obliterated with perturbative fluctuations and the situation becomes more and more complicated so I was not able to have any success along these lines. Thus, I thought that maybe it would be necessary to have some totally different point of view in order to save the day and this different point of view basically comes from Faraday.

In Maxwell's description of the electromagnetic field, you describe it really as fields while Faraday was describing electromagnetism using lines of force. Faraday was, in fact, the first string theorist (laughs). Therefore, I conjectured that there should be, in the non-Abelian theory, an exact correspondence between the field description of the gauge theory and flux lines, which we now call strings. It is easy to understand that the description in terms of flux lines would be good for large distances while the field description would be good for small distances. In this way we would have a complete picture if indeed we would be able to find a string description that precisely corresponds to the field description. At the time, I was quite ignorant about the existing versions of string theory, which I think was only good because, besides the fact that the existing literature was complicated, there were claims in the literature that strings are consistent only at fixed dimensions. So I started to work out my own way of understanding strings, mostly because I couldn't understand what other people were doing, and the major surprise was the following: that in the case of gauge theory (and similarly, in the case of the three-dimensional Ising model), you want to describe the string propagating in four dimensions but it turns out that it in fact propagates as if it was a five-dimensional string. There was an extra dimension, which is what we now call the *holographic dimension*. I think that a good analogy to this would be a case in which you are shown some very complicated figure that moves in a strange way. You don't understand what it is as it's very hard to understand the laws of motion that govern it. However, some day you are taught that this is just a projection of the four-dimensional cube – the tesseract as it is called – and then everything becomes clear. That's the role of the extra dimension in string theory. It's just the natural habitat for strings. So that was my motivation in the late 1970s and early 1980s. Basically, it expresses my frustration with instantons (laughs).

And do you think that these ideas somehow can be applied to QCD? Can we understand confinement in QCD?

I'm pretty certain of that; moreover, I even have a more radical view on that. For me, it's not excluded that the fundamental definition of the standard model is not a field theory but a string theory instead. If we concentrate on a specific element of the standard model, namely, gluons, I can in principle imagine a string theory that in the low-energy limit will describe these gluons. However, if you try to understand it deeper you will find also some excited gluons, excited quarks and so on which would correspond to string modes. So, not only do I think that the QCD string should exist in an exact sense, but I also think that it may be more fundamental than QCD in its present formulation.

But then why is it so difficult to find the QCD string?

Oh well, that's a good question which I cannot answer. It's because physics is difficult until you find the right way (laughs). You see ... it's like trying to find the way out of labyrinth and you are not supposed to hit your head on the wall. We should try to find the right approach and the right principle and we haven't found it yet. It's hard and difficult.

Do you think that the string theory that you are after would be one of those that most people think about? Like 10-dimensional superstring theory with some supersymmetry-breaking mechanism that then leads to the standard model?

No, I don't think so but I could be wrong. I think it should be some non-critical string theory with or without supersymmetry, although it should contain some fermionic degrees of freedom. Many people study a lot of 10-dimensional compactifications, etc., but I wouldn't be looking in this direction.

At some point in one of your papers you defined string theory as a set of unifying concepts and methods in physics [13]. Do you still hold this opinion?

Yes, I think I still hold this opinion.

Okay, so how do you choose which concepts and methods should be included here?

I am not really choosing them on any philosophical ground. I'm just trying to solve problems of physics and write down some equations. I always try to remember all the examples I know of in order to see if there is some analogy that can help me. So it is just by trial and error.

What were the steps that led you to formulate what now is called the *Polyakov action*?

Actually, that action has the wrong name because the quadratic action that is called *Polyakov action* was very well known long before I started working in physics. Something like that appeared in the work of a mathematician whose name was Douglas in the 1930s, who solved the Plateau problem [14] – the problem of minimal area surfaces. Later, a supersymmetric version of this action was introduced in a nice paper by Brink, di Vecchia and Howe [15] and in another paper by Deser and Zumino [16]. My late friend Vladimir Arnold had a theory that "things are never called after the people who first invented them." So I really don't know why it's called the Polyakov action since I always referred to these original papers. However, there is another part of the same theory that I really invented and am proud of, which is now called the Liouville action and comes from the quantisation of the same theory [5]. This is really my invention but it is called after Liouville (laughs).

I did not know that the history of the Polyakov action had been that nonlinear. In one of your papers [17] you mentioned the relation between string theory and the theory of biological membranes. How does this come about?

When you deal with the statistical mechanics of any surface – the membrane being an example – you have technically the same problem as in string theory. This particular paper [17] dealt with what is now called the rigid string where you add some extrinsic curvature term to the string action. Actually, it's a very interesting and still largely unexplored possibility. One fascinating and not well-known fact is that this theory with the extrinsic curvature term in flat spacetime is equivalent to ordinary field theory in anti–de Sitter (AdS) spacetime. Let me explain this more precisely. Suppose that you are in AdS, for which you have a Liouville direction and a contour at the boundary of AdS. Then, you can calculate the minimal area of the surface whose end points lie in that contour, finding some functional of the contour, which is conformally invariant. The mathematical claim is that you will get the same functional in Euclidean spacetime using the action with the extrinsic curvature term. It is still unclear whether this will be the case at the quantum level so that's quite a fascinating area to research on because most things are still unknown.

Are you working along this direction?

From time to time I am. I don't devote myself completely to this but I'm hoping that some idea will come at some point.

Has someone else done work along this direction?

Yes. Actually, I collaborated with Juan Maldacena on this but we never published it since two mathematicians, Babich and Bobenko, made the same remark earlier on the classical level. The quantum theory is an open problem.

You mentioned earlier that string theory provides an ultraviolet finite theory. Has this been rigorously proven?

Who cares about rigour? I don't care if a statement is rigorous or not; I care whether it's correct or not. So, I don't know. There are people working on this and my guess is that it is finite. Whether it is proven or not . . . I think that sometimes I behave like a bad student: "I knew it before but forgot what the answer was" (laughs).

According to a string perspective of nature, what are we supposed to see at the Planck scale? Is spacetime supposed to be discrete or is there no spacetime at all?

My guess is that there is no spacetime at all. It is replaced by something totally different. The simplest example is found in the gauge/string correspondence in which case if you look at very short distances or at curvatures of the order of the Planck scale, then the theory is equivalent to free gauge theory formulated on some abstract space. In this case, there is no notion of geometry left so geometry doesn't become discrete or complicated or something. It could just disappear and what we observe at the end are ruins of geometry. No one knows for sure but I think that's the most likely possibility.

People sometimes criticise this particular aspect of string theory, because since it is not formulated in a background-independent way, how can it ever tell us something about geometry at the Planck scale?

The background is just a technical artefact. In principle, if you formulated it on a different background it would describe the same theory. A good analogy is, for example, if you think about the Higgs field and begin quantising it. It has some vacuum expectation value and you quantise it on a specific background. However, as you go to the ultraviolet regime it's totally unimportant if you change the parameters. In statistical mechanics, the analogue is that of an ordered phase versus a disordered phase. If you look at small distances you'll find precisely the same equations. At small distances the physics is totally insensitive to whether you quantise the Higgs field in one background or in another. So, I think that this criticism of background independence is ill founded. It just reflects a poor understanding of the concepts.

Okay. What about the problem of a non-perturbative formulation of string theory?

I would say that it would be interesting to have a non-perturbative formulation but let's again make an analogy with field theories. If you have a perturbative gauge theory you may also ask about its non-perturbative formulation. There are some non-perturbative effects in gauge theories like instantons but to formulate it non-perturbatively you need to discretise it and develop lattice gauge theory and so on. At this point, of course, one may worry whether this lattice formulation will have the correct continuum limit and things like that. Lattice gauge theories are beautiful, interesting constructions and it all makes sense; however, it has very little relation with the practical and useful things like instantons. For example, instantons break baryon conservation in the standard model and we can precisely study its effects. So practical and useful things have no relation to the theoretical question of how to define the theory non-perturbatively.

This may not be very clear so let me try to clarify. It would be very nice to have some analogue of this rigorous lattice formulation in the context of string theory but it is a secondary problem and it's not of any importance to me. The important thing is to learn how to make progress and how to solve concrete problems. For instance, how to calculate, in a given string theory, the generic features of random surfaces such as the fractal dimensions and study the singularities of the specific heat and so on. So the type of problems that interest me are related to calculating and studying these things. I don't give a damn about a rigorous non-perturbative formulation of string theory, which is not to say that I would not appreciate it if it appears, although I would not appreciate it if it appears in some mathematically complicated dirty way. If some beautiful idea appears, of course, I will look at it.

But I guess the question is if this is necessary to make progress or not?

Oh no, not at all. I'm pretty sure that it has nothing to do with progress.

What are the fundamental degrees of freedom and underlying principles of string theory?

This is related to your previous question of how you define string theory and so on. Again, you see ... theoretical physics is a piece-by-piece business. If you ask too general questions from the very beginning you will not make progress. The picture clarifies when you begin solving some small problem and then another problem and so on, gradually building intuition about what is right. So a priori I can't answer this question.

You seem to have a very different perspective from what I have encountered so far. If I understood you correctly, it could be that the string theory that will ultimately describe nature is not 10-dimensional but could be four-dimensional or something else. So, what is your opinion about the science communicators who tell the public that we now know that there are 10 dimensions and the world is supersymmetric and so forth?

I must admit that I don't know what they are saying and I avoid listening to them.

I see ... but when people ask you, for example, if you expect to see supersymmetry at the LHC, what would you say?

I think that any answer to this question is totally irrational because no one can have grounds for any intuition about whether supersymmetry will appear at the LHC or not. In other words, this question does not make any sense.

I understand that it is logically the right attitude to take but people do have very firm beliefs in string theory, right?

I'm open-minded to both possibilities. Any of them would be very interesting to me; after all, that's why they built the LHC, otherwise we would all be buying lottery tickets instead (laughs).

I wanted to ask you a question about the cosmological constant problem. Is this a problem that you would include in the category that you mentioned earlier of untimely problems or do you think there is a way to move forward right now?

First of all, I don't think that it's a very difficult problem. In fact, I think that it is the right time to attack this problem and we have enough understanding, tools and experience to do it. However, I have my own view on how to approach this problem and it's not the mainstream view.

Do you mean that it's not related to the multiverse or the anthropic principle?

Oh no . . . God forbid! I think that this problem will be clarified relatively soon. The problem is related to the fact that de Sitter spacetime cannot, necessarily, be in equilibrium, unlike Minkowski spacetime [18, 19]. To make progress here it's necessary to understand much better non-equilibrium quantum field theory and that's a technically difficult subject. I am still trying to accumulate some intuition in this subject. I have considerable intuition in equilibrium quantum field theories and conformal quantum field theories but the non-equilibrium case is something new and it takes some time before you acquire a critical amount of intuition about a given problem.

But why is non-equilibrium quantum field theory related to the problem of the cosmological constant?

It's basically due to group theoretical reasons. Minkowski spacetime is stable because particles are in representations of the Poincaré group and the Poincaré group has representations with positive energy – I am just saying trivial things now. This is why the Minkowski vacuum cannot decay as it cannot create particles and it cannot explode. It is a mathematical fact that in de Sitter spacetime there are no conserved quantities with definite sign. Any Hamiltonian that you could possibly write down for any field theory will be bottomless. So, the theory cannot stay as it is; it is not stable. If you assume that this is the case then that's the reason why we don't observe the cosmological constant: it went down the drain.

So the question which remains is how it went down the drain, what are the processes and the dominant time scales and so on. It's a complicated but doable problem.

Is there any research going on in this direction?

Mostly my students and I are doing this but people generally have had the tendency to ignore this problem.

In your opinion what has been the main breakthrough in theoretical physics in the past 30 years?

There have been many breakthroughs, for example, in both condensed matter and in cosmology. In cosmology, for instance, understanding the power spectrum of the CMB was a great achievement while in quantum field theory it was holography, even though I don't particularly like this name as it could very well be *photography* instead of *holography*.

Why "photography"?

There could be some correlation functions in the bulk that are not effectively described by the boundary. This is an open question and you don't want to go into this. In any case, I would rather refer to it as gauge/string correspondence. I think this is a very important concept for quantum field theory. Also, topological effects that were studied in field theory turned out to be very important in condensed matter physics. We now understand some other topological phases of matter such as topological insulators. So, there have been many breakthroughs in the past 30 years.

You said that I did not want to go into this but I cannot avoid asking: Are you a believer in photography or holography?

I have an open mind about it. So far I didn't arrive at any decisive arguments. I have a number of subtle arguments but when you have too many arguments it means that you don't really understand things properly.

Do you think that the most important breakthrough originating from string theory has been the gauge/string duality?

I think so but the meaning of the word "important" is difficult to quantify. To me, it's definitely the most interesting breakthrough coming from string theory.

Why did you choose to do physics?

Basically, when I was young I had my own radio transmitters and so on. I was interested in radios and then I realised that I was more successful in designing them than in building them with real hardware. So, I tried to read some books in order to calculate and design a receiver and transmitter. In order to do so you need some basic formulas for inductances, frequencies, etc. This led me to read some popular books without any success and, in fact, books in general physics didn't work well for me.

How old were you at this time?

I was about 13 or 14 years old. Later, in some second-hand bookstore, I found the book on mechanics by Landau and Lifshitz [20]. I experienced an epiphany when I started reading it. I loved it so much and the way of thinking expressed in the book impressed me. It was a very strong emotion. It was an emotional explosion when I saw that they started with the least action principle and developed things from that general principle all the way to the top. It's a wonderful book although my late friend Vladimir Arnold, who wrote a competing book on mechanics [21], has one course problem in his textbook consisting of finding the error on page 78 of the Landau-Lifshitz' book (laughs).

Is it a true problem?

I looked at page 78, of course. It was a little imprecision but for mathematicians it's a big deal.

What do you think is the role of a physicist like you in modern society?

That's a very hard question to answer. My observation is that the community of physicists is statistically slightly more decent than the average community. Of course, there is a lot of competition and lots of strong emotions between people and so on but generally my impression is that the community is morally healthier than, for example, the community of traders. The other thing is that physicists, on average, think more critically, which is not to say that some physicists sometimes have some very stupid opinions. However, statistically, I think it's slightly better than average and that's certainly positive. I think that that somehow is transmitted to society in one way or another.

References

[1] U. Frisch and A. Kolmogorov, *Turbulence: The Legacy of A. N. Kolmogorov*. Cambridge University Press, 1995.

[2] S. El-Showk, M. F. Paulos, D. Poland, S. Rychkov, D. Simmons-Duffin and A. Vichi, "Solving the 3D Ising model with the conformal bootstrap," *Phys. Rev. D* **86** (2012) 025022, arXiv:1203.6064 [hep-th].

[3] S. El-Showk, M. Paulos, D. Poland, S. Rychkov, D. Simmons-Duffin and A. Vichi, "Conformal field theories in fractional dimensions," *Phys. Rev. Lett.* **112** (2014) 141601, arXiv:1309.5089 [hep-th].

[4] S. El-Showk, M. F. Paulos, D. Poland, S. Rychkov, D. Simmons-Duffin and A. Vichi, "Solving the 3D Ising model with the conformal bootstrap II. c-Minimization and precise critical exponents," *J. Stat. Phys.* **157** (2014) 869, arXiv:1403.4545 [hep-th].

[5] A. M. Polyakov, "Quantum geometry of bosonic strings," *Phys. Lett. B* **103** (1981) 207–210.

[6] A. M. Polyakov, "Quantum geometry of fermionic strings," *Phys. Lett. B* **103** (1981) 211–213.

[7] P. Ramond, "Dual theory for free fermions," *Phys. Rev. D* **3** (1971) 2415–2418.

[8] A. Neveu and J. H. Schwarz, "Factorizable dual model of pions," *Nucl. Phys. B* **31** (1971) 86–112.

[9] M. Planck, "1. On an improvement of Wien's equation for the spectrum, *Verhandl. Dtsch. Phys. Ges.* **2** (1967) 202.

[10] A. M. Polyakov, "Quark confinement and topology of gauge groups," *Nucl. Phys. B* **120** (1977) 429–458.

[11] A. M. Polyakov, "Thermal properties of gauge fields and quark liberation," *Phys. Lett. B* **72** (1978) 477–480.

[12] A. M. Polyakov, "String theory and quark confinement," *Nucl. Phys. Proc. Suppl.* **68** (1998) 1–8, arXiv:hep-th/9711002 [hep-th].

[13] A. M. Polyakov, "A few projects in string theory," in *Gravitation and Quantizations. Proceedings, 57th Session of the Les Houches Summer School in Theoretical Physics, NATO Advanced Study Institute, Les Houches, France, July 5 – August 1, 1992*, pp. 783–804. 1993. arXiv:hep-th/9304146 [hep-th].

[14] J. Douglas, "Solution of the problem of Plateau," *Trans. Amer. Math. Soc.* **33** (1931) 263–321.

[15] L. Brink, P. Di Vecchia, and P. S. Howe, "A locally supersymmetric and reparametrization invariant action for the spinning string," *Phys. Lett. B* **65** (1976) 471–474.

[16] S. Deser and B. Zumino, "A complete action for the spinning string," *Phys. Lett. B* **65** (1976) 369–373.

[17] A. M. Polyakov, "Fine structure of strings," *Nucl. Phys. B* **268** (1986) 406–412.

[18] A. M. Polyakov, "Decay of vacuum energy," *Nucl. Phys. B* **834** (2010) 316–329, arXiv:0912.5503 [hep-th].

[19] A. M. Polyakov, "Infrared instability of the de Sitter space," arXiv:1209.4135 [hep-th].

[20] L. Landau and E. Lifshitz, *Mechanics*. Butterworth-Heinemann, 1976.

[21] K. Vogtmann, A. Weinstein and V. Arnol'd, *Mathematical Methods of Classical Mechanics*. Graduate Texts in Mathematics. Springer New York, 1997.

23

Martin Reuter

Professor at the Institute of Physics at the University of Mainz

Date: 25 June 2013. Location: Mainz. Last edit: 17 August 2020

In your opinion what are the main problems in theoretical physics at the moment?

I think this is an ill-defined question. If you ask this question to people who work on quantum gravity they will most likely consider it the most important issue. I don't. I find it interesting, but I understand that other physicists have different opinions. I believe that one of the most important issues, which so far has not been the target of my research on quantum gravity, is to understand dark matter.

Do we need a theory of quantum gravity to understand dark matter?

We cannot exclude it in our framework. There are hints that quantum gravity might have infrared properties that we don't know at all for the time being. Non-perturbative effects, analogous to those in the context of QCD, could be present and mimic something like dark matter. At the present moment it is very difficult to analyse this in analytical terms but it is completely plausible. Continuing our programme we will be able to better understand whether this is possible or not.

Is it important to find a theory of quantum gravity, though you might think this is not the most pressing problem in theoretical physics?

Yes, of course. I'm absolutely convinced that it is extremely important to search for it. If we can properly understand how to combine quantum mechanics and gravity, we might find, for instance, that there is no need to quantise gravity at all. This may not be what most people think but it is certainly a logical possibility. I do not personally think that this is the case given that the other three known fundamental interactions have to be dealt with as quantum theories. However, before we reach a conclusion there is a long homework that needs to be carried out.

If you find this theory, will you be able to explain things we observe at the moment or will you only be able to explain things that need very high energy scales?

It is my hope that if you really understand the theory, all of a sudden you discover structural explanations for things which are not "exotic" in any way, and which you never had associated with quantum gravity before. I could not yet tell you an example, but this is quite conceivable. In any case, I'm not convinced of the usual arguments that all predictive effects are suppressed by the Planck scale. This is typical of perturbation theory, and that's exactly what we want to overcome.

Do you think that it is possible to come up with several theories of quantum gravity that ultimately would not be the correct theory of nature?

That's quite possible of course. But this is sort of a very luxurious problem because for the time being we don't have any complete theory that works. So I would be very happy to discriminate between two that work, and ultimately experiment must tell us what is right and wrong. Before reaching that point we should really do our homework as theorists and find at least one consistent proposal. From the statistical physics and renormalisation group flow picture, I'm not worried that there will be many theories. When studying critical phenomena, we usually arrive at very few universality classes that keep things simple. In fact, I really hope that at some point I will find out that what other people are doing in another language, with other seemingly different variables, is in fact equivalent to what I am doing.

You just mentioned that there's many different approaches to quantum gravity that use different languages ... why is it so difficult to quantise gravity?

I think one of the reasons has its roots in the history of physics. Most of the time we have been very lucky to work with perturbation theory when dealing with particle interactions. There was not too much pressure to develop other mathematical tools besides perturbation theory. However, gravity is maybe the first theory where you really need to work non-perturbatively [1]. Most likely in 500 years, when you have all the mathematics in place, people will not call it particularly difficult. Looking back at the early days of quantum mechanics, people found it difficult because things as abstract as matrices were part of its formulation.

So you are saying that the biggest difficulty in quantising gravity is to figure out the mathematics of how to deal with non-perturbative effects?

Yes, that is correct. Most of the tools in the study of quantum field theory were developed in application to the three fundamental interactions in the standard model. The other methods we use in asymptotic safety are in spirit closer to statistical mechanics. Typically they come with fewer analytical tools so you couldn't do much with pencil and paper.

Do you think that there is no satisfactory theory of quantum gravity at the moment?

No final one. All approaches have specific advantages and specific disadvantages. So it is usually easy to explore something and more difficult to explore something else. Also, what is easy in one approach is typically hard in the other, and that's why we don't see the interrelations well. Some approaches have a very well-defined starting point, such as loop quantum gravity, while other approaches, you may sometimes complain, have a lack of mathematical rigour.

What approaches are you alluding to?

Well, for instance, in our approach we, in most studies, are forced to do truncations of the theory space that are not controlled (in the sense of error estimates) and there is no expansion parameter as in usual perturbation theory [2–11].

What does that mean exactly?

It means that it requires more work and new methods to control the errors you make because you have no small quantity to expand in. The way you judge whether a truncation in theory space is meaningful or not is much more sophisticated and indirect. This is something mathematical physicists sometimes are skeptical of. On the other hand, what you can do along the usual lines of perturbative expansions is simply too restrictive for our purpose in many cases, and therefore we need a different strategy for investigating things.

How do you choose the truncations of your theory?

Well to some extent they are inspired by classical considerations. You can perform classical or semiclassical expansions, then carry out a standard one-loop calculation, and look at what kinds of terms are generated beyond the classical ones. Then you include those new terms into the truncation and look at how they affect the flow.

Could you construct the full theory step by step?

In principle, yes, but in a realistic theory this has never been done, not even in cases where we believe that we know what is going on in physical terms. For instance, in three-dimensional statistical mechanics, nobody would deny that almost certainly there exists the famous Wilson-Fisher fixed point [12, 13]. However, the reason why we believe in it is not because of the existence of a rigorous proof. Instead, there are many different techniques, many different approximations and numerical simulations. You can compare paper and pencil methods to simulations, say. Additionally, there are multiple experiments within the universality classes described by that critical point. All these pieces of evidence hint at the existence of one and the same fixed point. This gives us the confidence that it is there. And this confidence does not only come from theoretical reasoning.

I think that we should also reach this kind of confidence in the context of gravity. It would be very important for different theoretical approaches and numerical simulations to converge. On the other hand, I will certainly during my lifetime never see a paper and pencil proof showing that this fixed point is there. I guess it is a harder problem than proving confinement in QCD; it's of a similar nature, however. There have been several Nobel Prizes for Yang–Mills theory or QCD but none of them was related to the infrared scales (to confinement or similar non-perturbative features). In gravity, nobody expects that the problem has an easy solution. I mean, a spin-off would be one million dollars from the Clay

Mathematics Institute for solving the problem for Yang–Mills (laughs), and gravity is even harder. My point being that the fact that there is no rigorous proof of the existence of a fixed point is not enough for not believing in the asymptotic safety approach as it currently stands.

So I guess you would make the same statement about Yang–Mills theory?

Yes, we all believe that Yang–Mills theory is the right framework for QCD. All this confidence, again, does not come from paper and pencil proofs that the infrared comes out like what we observe in nature. Instead, it comes from combining analytic calculations using perturbation theory at high energy scales, from chiral perturbation theory, from lattice simulations, from experiments and from phenomenological models [14–16]. All that fits together and provides a coherent picture which on the whole we can believe in. It is this picture that supports the idea that Yang–Mills theory has the correct behaviour in the infrared. I believe that we should try to reach the same kind of understanding for gravity as well.

Okay, but now taking the analogy with Yang–Mills theory a bit further ... we know how to write Yang–Mills theory even though we may not be able yet to prove its infrared behaviour. But how do we write the theory that you came up with?

That analogy breaks down because this tough non-perturbative sector in Yang–Mills theory is in the infrared while in gravity what I just considered analogous is the ultraviolet fixed point. So in gravity what is analogous to the Yang–Mills problem is finding the bare action. Of course the infrared sector of gravity can be very difficult as well, but already to write down the bare action might be horribly complicated.

However, even though it may not be nice and simple, the bare action is generated by a very simple fixed-point condition. In practice, to work it out is very hard and amounts to the first task of the asymptotic safety programme [8, 17–22]. The goal of this programme is to first find the bare action, then solve the flow, and then find the effective action.

The first step in the programme is already something much more ambitious than quantising a given theory because the predictive power would be much bigger than in ordinary quantum field theory. Usually, you specify a bare action by hand, while we don't have to do that. We basically compute it, with the only input being the fields and the symmetries, while the rest follows in principle. Of course, the program can be implemented only in approximations, but in principle the bare action is a prediction that can be computed with known methods.

But can't this problem be solved by just putting it into a computer?

Yes, there are efforts along these lines. To some extent the computer helps, for instance, in automatising simple classes of actions such as actions that depend on a function of the curvature scalar. However, this is still a very restricted class and in particular it's local, for which heat kernel techniques are still applicable. In principle, we should be more general and also take nonlocal actions.

Is the full theory supposed to be nonlocal?

Well, in every quantum field theory when you integrate out modes down to the effective level the action becomes nonlocal. But the question is, is the ultraviolet fixed point already nonlocal? There are a few possible answers one could give. The first one is that this is not a well-defined question; namely, you always must specify *local* in which fields. You could always introduce auxiliary fields and make nonlocal terms look local. This is in fact an interesting mechanism for how matter-like fields could emerge from pure gravity. Leaving that aside, let's just use the metric all the way from the ultraviolet to the infrared. Then the question is, is the fixed point a local functional of the metric? Our intuition tends to say yes but this is not necessarily true. Usually, we have a very clear notion of what is "infrared" and what is "ultraviolet." But now think of a big space of actions and the fixed point is just one point. We look at trajectories emanating from that point; they leave the point and flow towards the infrared. But there are also trajectories that run into the very same fixed point such that for these trajectories the fixed point is in their infrared. If you now apply your standard intuition you would say that the action is nonlocal for these trajectories. So this question is not settled yet. The answer is also likely to depend to some extent on the details of the formalism. However, in a way one should not be frightened of nonlocalities. If they emerge, it might just mean that you should use another field.

Could you back up a bit and explain what asymptotic safety is?

The main idea is due to Steven Weinberg [17–20] and it is essentially the following. In the Wilsonian view [23–25] of what it means to renormalise a quantum field theory you should find a trajectory in the space of all actions (what we call theory space) which extends both to infinitely large generalised momentum and infinitely small momentum. Both the limit of an ultraviolet cutoff going to infinity and an infrared cutoff going to zero must exist. Now, the way asymptotic safety (and that's the reason why it is called asymptotic safety) deals with the ultraviolet cutoff is to employ trajectories that run into a fixed point in the ultraviolet. If a trajectory does so then you are sure that it does not leave the space of well-behaved action functionals.

This reasoning often occurs in dynamical system theory. If you have a dynamical system, for instance describing the population of some species, and you want that the species lives forever, then (while it is not necessary) it is sufficient to find a fixed point into which the solution to the differential equation runs. Then it doesn't change anymore as you increase the time from any finite time to infinity. This is a simple way of controlling the infinite time behaviour.

Of course the crucial point is that you must show – and as far as we know this is indeed the case in gravity – that the fixed point is not only sufficient but also necessary to get a well-behaved theory. All the predictivity then comes from the fact that, typically, you find a family of trajectories with finitely many parameters only that emanate from this fixed point. They form the so-called *ultraviolet (UV)-critical hypersurface* of that fixed point.

Why is this fixed point referred to as a non-Gaussian fixed point contrary to a Gaussian fixed point?

Yes, it's a little bit of a misnomer which stems from scalar theories. According to the definition we adopted with Max Niedermaier in an early review [6], what it means to be a non-Gaussian fixed point is that the critical exponents are not the canonical ones. Basically, it is synonymous to saying that, in the language of lattice field theory, you take the continuum limit at a fixed point different from the one of perturbation theory [11].

Now going back to these flows from the fixed point, I guess you are looking at flows that go from pure classical Einstein gravity to something else in the ultraviolet?

We don't start in the infrared. I mean, logically, the first step is to find the fixed point, then to find its finite-dimensional UV-critical hypersurface, and only later look for trajectories going downward to classical Einstein gravity.

But how many fixed points are out there and how many different trajectories exist that flow to the same infrared physics?

The problem is that in the approximate calculations that we can do, we cannot yet fully explore the full size of this hypersurface, so …

Even in truncated theory spaces?

If I generalise the truncation then I make the full space bigger. But a priori, I don't know how many dimensions this adds to the embedded critical surface. For instance, in $f(R)$ truncations, what happens is that up to R^2 you get relevant directions [3–5, 9]. So the cosmological constant, R and R^2 give rise to parameters in the surface and so it's three-dimensional. Adding R^3 and so on, you only get irrelevant directions. In this class of truncations, you can add more and more powers of the Ricci scalar and you will not increase the number of parameters you have to determine experimentally. So, fixing three parameters experimentally, you can predict all the other coefficients of all R powers. This has been discovered by Roberto Percacci's and Frank Saueressig's groups.

What is the reason for that?

This is just the intrinsic three-dimensionality of this hypersurface. If you look at what is the unstable manifold of the fixed point, then you see that in three directions you're driven away when you lower the cutoff, while in all others you're attracted to it. But now you could ask whether, if I add something like the Riemann tensor squared, will it add a relevant direction or an irrelevant one? A priori, I do not know and I have to do the calculation. Whenever you generalise a truncation, you may add a relevant or an irrelevant direction. We have some intuition from Gaussian fixed points for which canonical dimensions are still a guide. So you can partly understand whether specific terms give rise to relevant or irrelevant dimensions. However, strictly speaking, you cannot determine the exact number of dimensions from any given truncation as it is a property of the full theory.

According to your intuition, you expect it to be only three-dimensional?

I think it's not many more, but it's certainly something one has to look into in more detail.

But let me go back to one of my previous questions. In the truncated versions of the theory, how many trajectories can you find from the ultraviolet to Einstein gravity in the infrared?

Yes, now there is a very important question: Are you happy with all end points that occur on this surface? In the truncations that we understand we can say that we are happy with all. Truncations of the type $f(R)$, or truncations that include some powers of R with additional terms involving the Ricci tensor squared and Riemann tensor squared (or something similar), all look like Einstein-Hilbert plus higher-derivative terms in the infrared in such a way that they're compatible with general relativity and current observations.

Okay, so you have all these different trajectories into the infrared that give you Einstein-Hilbert plus extra stuff?

Yes, but for the sake of the argument, assume that the UV-critical surface is three-dimensional. Then you have to measure Newton's constant, the cosmological constant and perhaps the R^2 coefficient in order to be able to determine the trajectory. The task is to pick one trajectory from that three-dimensional surface so you just have to provide three data points. In fact, if you take one of them as the absolute scale then it's basically two measurements that you have to perform in order to determine the trajectory. In QED we perform the measurements in the infrared regime; in QCD we do it in between the infrared and the ultraviolet. In gravity we would also measure the constants in between, at about one meter, say, or in the case of cosmology, at a few kiloparsecs (definitely not gigaparsecs).

So these three specific constants you can't predict?

Right; these you have to measure, but then all other constants are determined by these three parameters. This is the crucial difference when you compare with effective field theory. In effective field theory, you would have to measure all of them. This proliferation of undetermined couplings does not happen here. The principle that brings everything in order, so to speak, is that the points are known to lie in a finite-dimensional hypersurface.

And is there always only one fixed point?

In the examples we studied usually there's only one that is reliable. Sometimes there are mathematical artefacts that you can rule out as a serious fixed point. But for the time being, there is no choice to make in pure gravity.

If you had two that would be reliable I guess you would have some problems with predictivity?

Yes, but it has not happened yet. If we find two fixed points that are reliable and equally plausible, then they represent really two distinct universality classes. They will have their respective basins of attraction that can be realised by concrete physical systems. It should also be possible to "see" them in other theoretical frameworks: causal dynamical triangulations (CDT) or loop quantum gravity, say.

Is this because you think there should be some more fundamental approach for which your theory is an effective description or because loop quantum gravity or CDT gives macroscopics to your theory?

Not exactly. I was alluding to critical phenomena, where you don't care much about what is the microscopic dynamics. One thing one has to get used to here is that compared to the usual critical phenomena everything is upside down: what is the infrared world in critical phenomena is the ultraviolet world for asymptotic safety. This is a little bit like in turbulence, for instance, because the self-similar domain is in the deep ultraviolet, at short distances. So the same universality class means the same description of the deep ultraviolet in that case. Theories in a given class would provide the same microscopics, not macroscopics.

But can you say what the microscopic degrees of freedom are for quantum Einstein gravity?[1]

That's another good point. The answer is that they are also an outcome of the theory. The fields which we use as an input to the theory I refer to as the "carrier fields." For instance, we started by taking the metric as the carrier field; later we also took the spin connection and the vielbein as independent variables, or we required vanishing torsion and took the vielbein as the only carrier variable [26–31]. In any case, you have different theory spaces then, consisting of functionals of those fields.

And these would be what you would call the fundamental fields?

No! The carrier fields just "carry" the degrees of freedom. The notion of a degree of freedom is quite subtle. Let's define the number of degrees of freedom as the number of initial conditions you have to specify in order to have a well-defined Cauchy problem. But then for this definition to be applicable you must know what's the nature of the field equations or the action, in particular, how many derivatives are there? Because if you have for instance four derivatives and one field, it is like having two fields with two derivatives. So with four derivatives you have twice the number of degrees of freedom compared with two derivatives and one field. This shows that for a given set of fields, the number of degrees of freedom is determined by the structure of the bare action. But the bare action is what we compute, and this in turn means that how many degrees of freedom are there is an outcome of the theory in our approach.

This is conceptually different, for instance, compared to loop quantum gravity where they want to quantise a given theory which is spin two and has the Einstein-Hilbert action as the bare action. However, we leave all that open. We take the metric or the spin connection

[1] The asymptotic safety programme is usually referred to as *quantum Einstein gravity* (QEG).

and the vielbein as the symmetry-constrained carrier variables, and then we ask what action gives rise to a non-perturbatively renormalisable theory. The answer we can find by a clearly defined calculation, in principle at least. Once we have done that job, only then we learn which degrees of freedom we are truly quantising implicitly.

So how is your approach related to canonical quantisation?

It is in fact a very indirect relation; namely, once we have the fixed point for a given set of fields and symmetries, then that's an object that contains probably many complicated terms. If you want to phrase it in canonical terms, you need to take an additional step. You must analyse which Hamiltonian path integral after integrating out the momentum yields the Lagrangian path integral with the bare action we found. In this way you would reconstruct the Hamiltonian formalism underlying implicitly our theory. But this is a very indirect way of in retrospect discovering what we actually quantised. Strictly speaking this step is not necessary. For all predictions of the theory, you don't need that.

But has this procedure been carried out?

In toy models we have studied this "reconstruction problem" [32]. What one should do in principle is clear but in practice it is very involved. In addition, it has a sort of non-universality creeping in, namely, usually when studying the flow equations we don't need an explicit ultraviolet cutoff but if you want to define a functional integral, you need such a cutoff. So you have to make one more specification in order to arrive at a rigorous definition of this path integral, namely an ultraviolet cutoff scheme. To make contact with other approaches, this is certainly of interest and important, but it is not necessary for a matter of predictions.

So you said that in principle you could determine the number of degrees of freedom and their nature. In string theory we have some strings; in loop quantum gravity we have some spin network; what do we have in quantum Einstein gravity?

Well, it is something you can determine in any given truncation. If I take the Einstein-Hilbert truncation then the fixed point unavoidably is a functional with two time derivatives. It gives two helicity states to the graviton which are carried by the metric. But when I have four derivatives as in R^2 truncations, the counting seems to change, at least naively. People often are afraid of R^2 theories and their so-called ghosts. But this is unjustified because all the usual arguments against R^2 gravity don't apply here. Like in QCD, its vacuum is probably not the naive vacuum of perturbation theory. However, if the vacuum is something more complicated then, of course, it's not necessarily the case that the modes of the field you write down in the beginning are the propagating modes. In fact, in QCD the pions are not modes of a field you write down in the fundamental Lagrangian. This kind of question one has to understand much better to say something definite about the physical excitation spectrum.

When you said that if you take the Einstein-Hilbert truncation then it would be the graviton, does that mean that you can just take the Einstein-Hilbert truncation and find a quantum theory for it?

In a way, yes, but it is not closed under the renormalisation group flow; it's a truncation. All terms which the renormalisation group flow generates beyond R you must discard. So these truncations are approximations in the sense that the full critical surface is maybe three-dimensional and curved, and you approximate it by a two-dimensional plane, say.

I'm still confused. In many of your articles you claim that quantum Einstein gravity is not a quantisation of pure Einstein-Hilbert action [8].

No, Einstein-Hilbert is just an approximation. This statement means the following. If you forget about the necessity of doing approximations (imagine that you have a giant computer that can do all calculations to determine the fixed point) then most probably it's not pure Einstein-Hilbert that you will discover. At least, in all truncations we studied it was not pure Einstein-Hilbert.

But if this is the case, then your theory can never be the same as loop quantum gravity since you said yourself that they fix the action to be the Einstein-Hilbert action, right?

You are right. It looks so at first sight, and for several years I was very puzzled by that. But actually this is not so clear because of the reconstruction procedure that needs to be carried out in order to compare approaches. In particular, what loop quantum gravity does is to basically try to construct a path integral. In a way they define a measure and make a proposal for the action which is the Einstein-Hilbert action. What I call "defining the measure" is what they do in terms of holonomy and flux variables. But what we do is to compute something like an effective action at infinite cutoff scale. So there is a nontrivial step of connecting the fixed point's effective action to the bare action. There we have to make a free choice of an ultraviolet regulator. This then allows us to map the fixed point onto the bare action in a to-be-constructed functional integral. It might be that there exists a measure such that the connection process converts our complicated fixed point to the simple Einstein-Hilbert action. I believe this is a possibility which we must analyse further.

But then you wouldn't get any infrared corrections to the Einstein-Hilbert action?

That is not what I am saying. If we follow our flow and they evaluate their path integral to compute observables, and if the connection map is applied correctly then we must get the same answers in the infrared. Strictly speaking their Einstein-Hilbert action and our action, within the Einstein-Hilbert are not the same object. Ours is an effective action even at high scales. In standard matter field theories the ultraviolet limit of the effective action is essentially the same as the bare action. But in our formalism there are metric-dependent correction factors you have to apply. For gravity they become important and all the subtlety with the ultraviolet regulator of the path integral enters there.

So if you find the regulator that gives you the Einstein-Hilbert action then you would have proven the infrared limit of loop quantum gravity?

Yes.

Are there many choices you can make for the regulator or is there a natural one?

We have worked this out in detail for a definition of the measure which was convenient for computations by hand, so to speak [32]. In general there is no unique way to fix the regulator. I imagine that there are many ways of arriving at a satisfactory result. I cannot quantify it; in fact, attempts such as causal dynamical triangulations or maybe Euclidean triangulations or Regge calculus aim at defining a measure and they might all be different.

Can you be more specific about how you performed these comparisons?

The way we proceeded was by mapping our fixed-point action onto the bare action. The regularisation we employed aimed at making contact with simulations. Basically, it's a kind of one-loop determinant that relates the fixed point and the bare action. This is something that has to be computed in the regularisation scheme of the other people, so to speak, that is, those who use the functional integral.

Thus we proposed an object that you had to simulate on the computer if you wanted to make contact with our formalism [32]. This has not been done yet but it is clear how one could do it at least in principle. The problem is also that CDT is still far away from the fixed point and so it's too early to really make an attempt at comparing things [33, 34]. However, it should be possible to do it at some point.

In practice, the other approaches should compute with their measure what they expect to be our bare action (using the same regularisation procedure). So I would give them a horrible "observable" and they would compute its expectation value. The outcome is information which we can use to convert their results into a prediction for our fixed point. In particular, we would be able to extract properties of the fixed point like critical exponents and so on.

Is there a general argument that if you find one fixed point in the truncated version then that would be the fixed point of the full theory?

No. So, in principle, if you add extra terms, the fixed point could disappear. In fact in the first years each time we improved the calculation we expected that to happen. However, it never happened so far, because there are such a huge amount of "miracles" taking place (laughs). And indeed, if there does exist an exact fixed point in the full theory, all those miracles have a simple and natural explanation.

What happens then if you add different types of matter content?

That situation is again comparable to QCD, in which case, if you add too many quarks, you can destroy asymptotic freedom. In gravity, there is the pioneering work of Roberto

Percacci and his students, who added arbitrary multiplets of free fields [35–39]. Basically you can take any mixture of scalar, vector and Dirac fields, and very roughly you have something like a 50/50 chance that the fixed point remains there. You can destroy the fixed point by choosing particular matter contributions. Destroying the fixed point is usually not so much dependent on the number of fields but instead on the particular combination you take. In any case, there is no need of any fine tuning for the fixed point to be there, nothing as in supergravity, say, where you need a very fine compensation mechanism.

One of the biggest criticisms I heard about your approach was from one of Hermann Nicolai's past talks,[2] namely, that any type of matter can be coupled to quantum Einstein gravity and there are no restrictions at all.

Making such statements about any type of matter is an exaggeration. I mean, there is the possibility that the fixed point is destroyed by some types of matter and then you cannot even start flowing. There is also the possibility that even the exact trajectory could break down at some finite scale and then you also would say that specific matter could not be coupled.

I think one of the inconvenient truths about quantum gravity is the following. Many people have had the dream that solving the problem of quantum gravity would also explain very precisely why the standard model is what it is, why the specific parameters are what they are, why human beings are there, and why the climate change occurs (laughs). I think that's just not true. It would be nice if it happened to be like that, but I fear it is wishful thinking.

For instance, Yang–Mills theory is a very constrained setting in some sense. If you look at what kind of vertices with vector bosons you could write down with and without local gauge symmetry then it constrains a lot, but not so much that it would force on you a very specific kind of matter. You can couple matter so that everything changes its character, so there are some restrictions, but perhaps not as many as we would like.

One needs to understand the flows in the ultraviolet critical manifold better to say something more or better founded, but I don't see a reason why things should change dramatically. Perhaps we should better not hope to explain all mysteries of the matter sector, or at least not too many mysteries.

There are definitely some types of "mysteries" in the matter sector that can be explained. For instance, we found a fixed point in the coupled electron-photon plus graviton system [40]. This is actually an example where there are two fixed points and nature could have chosen any of the two. One of the two fixed points is more predictive and in our simple truncation it predicts the value of the fine structure constant. We can really compute the fine structure constant from the electron mass and the gravitational couplings. It is a comparatively simple calculation which provides a beautiful and particularly transparent proof of the principle. Numerically the analysis gives something like 1/100 with a certain error. The order of magnitude is right, but the precision is not enough to rule out the other fixed point.

[2] See www.youtube.com/watch?v=NXFcI7imgwE.

In a similar way the Wetterich-Shaposhnikov argument allows to state that asymptotic safety predicts a Higgs Mass of about 125 GeV [41].

How do they reach such a prediction?

Basically, you have to look at what you get from a suitable analytically accessible approximation to the exact coupled renormalisation group flow. If you follow the standard flow in perturbation theory, and you look at the Higgs' bare mass, and you require the trajectory to continue up to the Planck scale, then you are more or less unavoidably driven to that value of 125 GeV in the infrared.

Can you determine which is the true fixed point of QED by making some measurement in the infrared?

If you could do the calculation at an enormous precision and the measurements at the same enormous precision we can decide in principle which of the fixed points is the right one.

In principle, it is straightforward to re-do our matter calculation at a higher precision level, and do the gravitational side also in a better way, perhaps getting a bit closer to 1/137. However, I doubt you could do the theoretic homework at a level of precision so that, given the experimental precision, you could discriminate the fixed points. You would probably need to take the complete standard model into account. People are making progress with matter fields and computerising the calculations, so maybe in 10 or 20 years or so we could have an answer to these questions. It's definitely not impossible and not overly far-fetched.

How does quantum Einstein gravity deal with the basic divergence problem that people found when they tried to just quantise gravity perturbatively?

What is at stake is where the trajectories will go when you send the ultraviolet cutoff to infinity. Here is the big difference between perturbation theory and asymptotic safety. In the latter case they run into this fixed point. In the perturbative case, you perform a kind of massive continuum limit at the Gaussian fixed point. One has to be very careful here because it is not clear often what is the intended meaning of "perturbative." Perturbation theory has several different aspects, for instance, do you mean that you are expanding in a small parameter? This is not the issue. You can find the non-Gaussian fixed point even by expanding in small couplings. It's not this notion of being non-perturbative that is crucial; it is the non-Gaussian character of the fixed point that is important [1].

From a technical point of view, one of the essential differences with the typical perturbative calculations is that we can't take advantage of dimensional regularisation. This is something very dangerous when it ignores power divergences. There is an emphasis only on logarithmic divergences and so it is blind to an effect that is crucial for us, namely, that in the ultraviolet the dimensionfull Newton's constant runs to zero. This gives you effectively two powers of the momentum cutoff and makes diagrams by two orders in the ultraviolet cutoff more convergent.

For instance, the famous Goroff-Sagnotti term has Newton's constant G and it has a $1/\epsilon$ pole which is a logarithmic divergence [42]. But, in our picture, G times logarithm is logarithm divided by the ultraviolet cutoff squared. Therefore, in asymptotic safety the term is highly convergent.

Have you shown that this is actually the case?

There is something analogous in the simpler setting consisting of just an R^2 term with scalar matter which is also perturbatively non-renormalisable. There is work by Frank Saueressig and his students whereby they show that nothing special happens there [43], meaning that perturbative non-renormalisability doesn't do any harm to the asymptotic safety structure. Later another group around Frank Saueressig included the Goroff-Sagnotti term. By an enormous computation they demonstrated explicitly that indeed everything is as expected: asymptotic safety survives, and the Goroff-Sagnotti term does not seem to play any distinguished role [44].

You just mentioned that Newton's constant runs to zero. Obviously, you are assuming that it is not exactly constant but as far as we know right here, right now it is constant, so what are the implications of this?

The dimensionfull Newton's constant indeed flows to zero. It depends on the scale. Unfortunately, there are no experiments that could possibly observe this effect at present. For a long time we worked on phenomenological astrophysical applications [45–54] but the experiments are so "poor" that all bounds are really far away from constraining in any way what we expect for the running of Newton's constant. At the moment it appears that, according to all truncations that we considered, you would only start seeing effects close to the Planck scale. Considering the presence of the cosmological constant changes this slightly, but still you will need very high scales to observe a significant scale dependence of G.

You mentioned several times the R^2 coefficient. Does it not present any modifications to the physics we've observed?

If you modify classical general relativity by this R^2 term with a coefficient of the order unity, it is only at curvature scales of the order of the Planck scale that the R^2 term can compete with the other terms you have. It's remarkable how poor the bounds on such coefficients are. In particular, you can have a tremendously large coefficient in front of R^2 and nobody would ever have detected it. In fact, sometimes people wrongly describe QEG as if we take Einstein-Hilbert and evolve it upward. Well, this is certainly not what we do, nor do we expect that down here Einstein-Hilbert is the truth. There could be a lot more, and nobody would ever have observed it.

Okay, but, bottom line, at the moment you cannot come up with some experiment that would allow us to see these effects, right?

Not these ones which are related to local invariants. What could happen, however, is that there are effects in the far infrared, which are more difficult to treat analytically, but are analogous to the infrared effects in QCD that trigger confinement, chiral symmetry breaking and so on. In this case, there would be a chance to see them "in the sky."

The intriguing thing is that we can estimate where they would become visible; namely, they start to dominate where the local truncations break down. Now it is really very intriguing that they are predicted to break down close to the present Hubble scale or a bit earlier. So for a while we were exploring the idea that maybe the flat galaxy rotation curves have something to do with renormalisation effects in the far infrared [46, 49]. It is very hard to get an analytic handle on that; it's again this one-million-dollar problem but now in the infrared. So there are not many results on that. Anyhow, what I find remarkable here is that without any complicated calculations you can see that the breakdown of the Einstein-Hilbert truncation and similar ones happens when the dimensionless cosmological constant gets close to order unity. And we know that the dimensionfull running cosmological constant equals approximately the Hubble constant squared, at the Hubble scale. Thus at the Hubble scale, the dimensionless running cosmological constant is indeed of order unity in real nature.

What really goes on there I don't know; it might possibly be dark matter or modified gravity [54]. I find it a very attractive idea that ultimately something like MOND or similar could emerge. This ties nicely to one of your earlier questions of why we should look into quantum gravity.

Can the running of the cosmological constant explain inflation?

It can. We have examples where it happens. With Alfio Bonanno, we analysed possible cosmological implications by a "renormalisation group improvement" [55]. The results are quite inspiring, even though the method was less rigorous than the quantum field theory part of asymptotic safety because it deals with identifying scales. But at a qualitative level you can find cosmologies where you solve those problems that are usually solved by inflation using asymptotic safety. For instance, in the case of the horizon problem, we found that, so to speak, the light-cones open when you get very close to the big bang simply because the Friedman equations change by the running of G and Λ.

But those problems are solved by inflation, right, so what is the need for this?

Yes, but in our case there is no need of an ad hoc inflaton, and the problems are still solved. For instance, the generation of fluctuations is something we have a very natural mechanism for. The idea would be that the fixed point is something like a critical phenomenon in the deep ultraviolet [52] which gets blown up by the cosmological expansion. Fluctuations of the metric, by some sort of decoherence mechanism, would turn into the primordial density perturbations that we usually generate with an ad hoc scalar, say, during inflation.

In a first approximation, we have the prediction that the spectral index would be exactly one. This is fine at that zeroth order, but the clearly desirable more refined calculation has not yet been performed. It would be far more ambitious and there are still certain tools we are lacking to do it. Answering one of your questions about possible predictions, the result of this refined calculation would be one of them.

There are renormalisation group improved cosmologies where immediately after the big bang we can have easily 60 e-folds without any fine tuning. It's a type of inflation that is very beautiful because it does not require any inflaton as it is really just driven by the cosmological constant. When working with the usual framework for inflation there is always an issue regarding switching off inflation when you have a constant cosmological constant. In our approach, it is not constant. Only at low scales (sufficiently far below the Planck scale) it approaches a constant and so the universe ends in a standard FRW cosmology. The renormalisation group running of the cosmological constant has "switched off" the de Sitter phase.

Can you given any insight into what happened at the big bang with this approach?

In most of those effective cosmologies, the singularity is still there. However, I would be careful how to define what we call a singularity. If it is a divergence in a quantity that has no essential meaning, neither for the internal consistency of the theory nor for any observable, then maybe you can tolerate it. Or, if the mean field description we use breaks down, and the effective field equation yields a cosmological scale factor which goes to zero but has large fluctuations around zero, do you still call this a singularity? I often discuss this point with loop quantum cosmology people. It is not so easy to tell what it requires to resolve a singularity because in principle you should understand what is the set of observables. In any case, I have no strong opinion about this issue. For the time being it is only certain special solutions to the effective field equations that are at stake, not the asymptotic safety program as such.

However, I hope that the ideas on inflation driven by a varying cosmological constant that gets switched off automatically in the end and needs no inflaton, and the generation of density perturbations from metric fluctuations near the fixed point, will find their place in the final theory.

As far as I understand, quantum Einstein gravity has a completely different view on what happens to spacetime at the Planck scale. Compared with other theories, it states that it is still a continuum, right?

Yes and no. Perhaps I can make the following remark as an answer [8, 56]. In high school we are taught about quantum phase space and the Planck cells, with h as the unit, where you can put only one state in. There, you talk about lattices in phase space, and you have this picture of little boxes which partition phase space. This structure of a lattice in phase space can be described by the Moyal formalism in a completely continuum-based way. You can work with functions of commuting position and momentum p and q and the star product for those functions. The algebra of smooth functions equipped with the star product is powerful enough to describe this non-commutative geometry. So in a way, what I am trying to say is that there is no contradiction, in the sense that you can employ a continuum language and nevertheless describe very discrete and lattice-like physics.

So you are saying that this is what is happening within your theory?

I am not necessarily saying this is happening but I'm prepared to find something like that. There are indeed first hints pointing in this direction. For instance, in a quite subtle but well-defined manner the effective spacetimes of QEG are like "fuzzy" manifolds with a dynamically generated minimum length. There are limitations on the possibility of distinguishing spacetime points which go well beyond what we know from standard quantum field theories [57, 58].

From a different perspective, the vacuum of QEG shows features which are reminiscent of a fractal [59]. In particular, there are many instances of dimensional reduction which one also observes in various cases and in other theories. It means that in the ultraviolet of QEG there is a reduction of degrees of freedom compared with what is expected from standard perturbative quantum field theory in Minkowski space. This reduction is something you can very well exhibit in our formalism. We found also a kind of generalised C-function for four-dimensional quantum gravity which reflects this behaviour [60, 61]. But nevertheless, understanding the physical degrees of freedom, and closely related to that, the vacuum structure of QEG, is still on the agenda, and we also must develop new methods to explore them more directly. In particular it remains to formulate the outcome of the reconstruction process in a Hilbert space language.

By the way, there has been a semiclassical argument against the possible existence of a fundamental quantum field theory for gravity which was based upon a counting of states [10]. However, for a number of reasons the argument does not apply to QEG, and our reduction of degrees of freedom in the ultraviolet is one of them. The argument assumes that there are as many states in the ultraviolet as in an ordinary conformal field theory on a fixed, classical Minkowski space. But in QEG those states are not there as a manifestation of the theory's non-perturbative renormalisability and background independence.

Just to summarise your past few answers, you cannot conclude if spacetime is continuous or if it comes from some underlining discrete lattice?

What I tried to say is that the answer, neither from nature nor from QEG, can be a simple "yes" or "no."

If you provide me with a fully detailed description of an experiment which you plan to perform in order to settle this issue experimentally, then I can use QEG in order to work out an unambiguous prediction for the outcome of your experiment. Let's assume furthermore that QEG correctly describes not only your experiment, but also all similar (ideal) experiments geared towards answering the same question. Each experiment comes with a rule for projecting an extensive detector output (10,000 data points, say) on a single binary variable which assumes values "discrete" and "continuous."

My conjecture would be that even though all experimenters believe that they ask the same question of nature, by construction their devices agree on the discrimination of classical lattices and continua after all; they will not obtain a unique answer in the case of a quantum spacetime. Differences in the experimental details will cause changes in the 10,000 data points which in turn project on different binary answers then.

Stated differently, the structure of spacetime will reveal itself via observables related to concrete experiments, and not by the type of mathematical objects we use to formulate the theory. One formalism may use continuum notions while another formalism uses lattice methods and yet the same predictions are found for all possible experiments. So I don't see any contradiction for the time being. This is a little bit like early discussions of quantum mechanics. If you naively look at matrix mechanics and wave mechanics with the Schrodinger equation then, if you're not Paul Dirac (laughs), you also could think that discrete matrices and continuum differential equations cannot apply simultaneously. But we know that both are equivalent representations of the same thing.

I'm happy that you made this point clear because many people just say that in quantum Einstein gravity there's nothing unusual happening to spacetime in some sense, that it's still just a continuous description, so ...

It's a continuum description but that might mean close to nothing (laughs).

You mentioned dimensional reduction which leads us to the concept of effective dimensions. You also mentioned that other approaches observe similar behaviour. What did you mean by this?

There are several notions of dimension that you can try to compute, such as the spectral dimension or Hausdorff dimension. Their values don't have to agree even for classical fractals. These different types of dimension generalise different roles played by the classical dimension on smooth manifolds. It is possible to compare these dimensions in different approaches and we tried to do this to some extent [8, 56, 59]. In particular, we compared the spectral dimension with that of CDT. At present CDT is not yet close enough to the continuum limit to see the fixed point, but there is a regime, namely the semiclassical regime, where our predictions for the data are in good agreement with CDT. We find non-trivial deviations from the classical values, on which we agree. However, the simulations are not yet at the level that we can make a more detailed comparison.

I remember that in one of your papers you wrote that CDT is the counterpart of QEG [8]. Does this mean that you think they are the same theory?

Yes, I wanted to convey that possibility. Of course I don't know if they are really the same theory but if one is thinking or hoping that there are not too many different universality classes, the most minimal guess is that if the CDT programme and our programme ultimately work then they are the same theory. In the spirit of statistical field theory, they would perhaps amount to the continuum versus discrete approach to the same universality class. Admittedly, the evidence we have so far is not yet so strong. The comparison is done in the semiclassical regime which means in practical terms that it is very sensitive to the running of the cosmological constant. In pure gravity it is realised where the trajectory turns around this turning point, close to the Gaussian fixed point, which happens at the meV scale. This is the scale at which the k^4 running of the cosmological constant begins.

This very famous quartic divergence is observed in the simulations and we agree nicely on this part of the curve. However, I cannot judge how specific that is really. So I'm very much looking forward to and following closely new data from their simulations (laughs).

You probably have seen that some time ago it was shown that in two spacetime dimensions, quantising using CDT is equivalent to quantising Hořava-Lifshitz gravity [62]. Perhaps, though I don't know, the same could be true in four spacetime dimensions. Would you be happy with your theory also being some kind of quantisation of Hořava-Lifshitz?

I think this depends on which asymptotically safe universality class we pick. As I said, the input into our formalism is the field plus the symmetries. Up to now, we always require the symmetries to be four-dimensional diffeomorphisms in the case where the metric is the carrier field, plus local Lorentz invariance when vielbeins are the carrier fields. In the case of Hořava-Lifshitz perhaps the fields are the same but it has less symmetry because you introduce foliations which likely restrict to a smaller group of diffeomorphisms. If you do that, you get different flows which involve more invariants (given the fact that it has less symmetry). The first investigations of this type were performed by Frank Saueressig's group [63–65]. So also in this direction, work on other, that is, non-Einstein universality classes started. But nevertheless, in our present situation where no approach has led to a fully satisfactory theory yet, it is clearly premature to worry much about classifying all theories that can be constructed, and about their interrelations.

But CDT also introduces a foliation, right?

This is something that is not clear to me. I'm not sure what the CDT experts expect but the crucial question is if in the end foliation independence is restored. That I cannot judge. The hypothesis I put forward is the minimal one but I cannot exclude that there might exist another universality class. However, for a fact, if Hořava-Lifshitz exists as a quantum theory then that would amount to a fixed point in a theory space different from the one of QEG because it has different symmetries.

It appears to me that if the result in two dimensions that CDT is equivalent to Hořava-Lifshitz carries over for four dimensions then CDT will never be the discrete counterpart of QEG.

Another possibility, which might be unlikely though, is that for purely dynamic reasons the flow in the bigger space, where you require less symmetry (that is, not the full diffeomorphisms), or at least a certain subset of trajectories, which start from a fixed point and describe decent physics, accidentally respects the full diffeomorphism symmetry. This is conceivable if, so to speak, this remainder of the symmetry group is not essential. In other words, QEG is a consistent truncation of the bigger theory if the flow happens not to leave the theory space made of invariants which are invariant under the full diffeomorphism group. In principle this can happen but I don't know for certain what the outcome would be.

Some people have criticised QEG for having difficulties dealing with Lorentzian signature. Is this a valid criticism?

No, not at all. There are no indications that anything goes worse in Lorentzian signature; in the end, it might help even. Many people claimed such wrong things simply because they have problems imagining what a Lorentzian flow equation could look like. (Actually it would suffice to read Julian Schwinger's 1951 paper [66], for most purposes.)

Why did you choose to do physics?

Not because I wanted to understand quantum gravity at that time but just because I found it exciting, intriguing and fun.

What do you think is the role of a theoretical physicist in modern society?

If you had asked me this question 10 years ago I would have answered that we are trained to solve problems and just this ability to attack problems in a certain way is very much needed now in very diverse fields. However, today I would answer that there is something more basic that we can teach students and other people, namely, that it's sometimes worthwhile to sit down and work on something maybe for a week, for a month or for a year, be patient and not expect immediate outcomes. This is something that has become unfashionable. Our students who left for industry, banks or insurance companies often work with very similar formal tools and concepts as fundamental physics. But their answers they must come up with in very short periods of time. Patience is not so much needed anymore, while a lot of patience is needed for computing gravitational renormalisation group flows (laughs).

References

[1] O. Lauscher and M. Reuter, "Is quantum Einstein gravity nonperturbatively renormalizable?," *Class. Quant. Grav.* **19** (2002) 483–492, arXiv:hep-th/0110021.

[2] M. Reuter, "Nonperturbative evolution equation for quantum gravity," *Phys. Rev. D* **57** (1998) 971–985, arXiv:hep-th/9605030.

[3] M. Reuter and F. Saueressig, "Renormalization group flow of quantum gravity in the Einstein-Hilbert truncation," *Phys. Rev. D* **65** (2002) 065016, arXiv:hep-th/0110054.

[4] O. Lauscher and M. Reuter, "Flow equation of quantum Einstein gravity in a higher derivative truncation," *Phys. Rev. D* **66** (2002) 025026, arXiv:hep-th/0205062.

[5] M. Reuter and F. Saueressig, "A class of nonlocal truncations in quantum Einstein gravity and its renormalization group behavior," *Phys. Rev. D* **66** (2002) 125001, arXiv:hep-th/0206145.

[6] M. Niedermaier and M. Reuter, "The asymptotic safety scenario in quantum gravity," *Living Rev. Rel.* **9** (2006) 5–173.

[7] M. Reuter and F. Saueressig, "Functional renormalization group equations, asymptotic safety, and quantum Einstein gravity," in Proceedings, *Geometric and Topological Methods for Quantum Field Theory: Villa de Leyva, Colombia, July 2–20, 2007*, Hernan Ocampo, Eddy Pariguan and Sylvie Paycha, eds., 2010.

[8] M. Reuter and F. Saueressig, "Quantum Einstein gravity," *New J. Phys.* **14** (2012) 055022, arXiv:1202.2274 [hep-th].

[9] M. Reuter and F. Saueressig, *Quantum Gravity and the Functional Renormalization Group: The Road towards Asymptotic Safety*. Cambridge University Press, 1, 2019.

[10] A. Bonanno, A. Eichhorn, H. Gies, J. M. Pawlowski, R. Percacci, M. Reuter, F. Saueressig and G. P. Vacca, "Critical reflections on asymptotically safe gravity," arXiv:2004.06810 [gr-qc].

[11] R. Percacci, *An Introduction to Covariant Quantum Gravity and Asymptotic Safety*, vol. 3 of *100 Years of General Relativity*. World Scientific, 2017.

[12] K. G. Wilson and M. E. Fisher, "Critical exponents in 3.99 dimensions," *Phys. Rev. Lett.* **28** (Jan. 1972) 240–243. https://link.aps.org/doi/10.1103/PhysRevLett.28.240.

[13] J. Zinn-Justin, "Critical phenomena: field theoretical approach," *Scholarpedia* **5** no. 5 (2010) 8346. revision #148508.

[14] D. J. Gross, "Twenty five years of asymptotic freedom," *Nucl. Phys. B Proc. Suppl.* **74** (1999) 426–446, arXiv:hep-th/9809060.

[15] F. Wilczek, "Asymptotic freedom: from paradox to paradigm," *Proc. Nat. Acad. Sci.* **102** (2005) 8403–8413, arXiv:hep-ph/0502113.

[16] S. Bethke, "Experimental tests of asymptotic freedom," *Progress in Particle and Nuclear Physics* **58** no. 2 (2007) 351 – 386. www.sciencedirect.com/science/article/pii/S0146641006000615.

[17] S. Weinberg, *General Relativity: An Einstein Centenary Survey*. Cambridge University Press, 2010.

[18] S. Weinberg, "What is quantum field theory, and what did we think it is?," in *Conceptual Foundations of Quantum Field theory*, Tian Yu Cao, ed., pp. 241–251. Boston, 1996. arXiv:hep-th/9702027 [hep-th].

[19] S. Weinberg, "Living with infinities," 2009. arXiv:0903.0568 [hep-th]. https://inspirehep.net/record/814639/files/arXiv:0903.0568.pdf.

[20] S. Weinberg, "Effective field theory, past and future," *PoS* **CD09** (2009) 001, arXiv:0908.1964 [hep-th].

[21] R. Percacci, "Asymptotic safety," arXiv:0709.3851 [hep-th].

[22] A. Eichhorn, "An asymptotically safe guide to quantum gravity and matter," *Front. Astron. Space Sci.* **5** (2019) 47, arXiv:1810.07615 [hep-th].

[23] K. G. Wilson, "The renormalization group: critical phenomena and the Kondo problem," *Rev. Mod. Phys.* **47** (Oct. 1975) 773–840. https://link.aps.org/doi/10.1103/RevModPhys.47.773.

[24] K. G. Wilson, "Renormalization group and critical phenomena. I. Renormalization group and the Kadanoff scaling picture," *Phys. Rev. B* **4** (Nov. 1971) 3174–3183. https://link.aps.org/doi/10.1103/PhysRevB.4.3174.

[25] K. G. Wilson, "Renormalization group and critical phenomena. II. Phase-space cell analysis of critical behavior," *Phys. Rev. B* **4** (Nov. 1971) 3184–3205. https://link.aps.org/doi/10.1103/PhysRevB.4.3184.

[26] E. Manrique and M. Reuter, "Bimetric truncations for quantum Einstein gravity and asymptotic safety," *Annals Phys.* **325** (2010) 785–815, arXiv:0907.2617 [gr-qc].

[27] E. Manrique, M. Reuter and F. Saueressig, "Matter induced bimetric actions for gravity," *Annals Phys.* **326** (2011) 440–462, arXiv:1003.5129 [hep-th].

[28] E. Manrique, M. Reuter and F. Saueressig, "Bimetric renormalization group flows in quantum Einstein gravity," *Annals Phys.* **326** (2011) 463–485, arXiv:1006.0099 [hep-th].

[29] J.-E. Daum and M. Reuter, "Running Immirzi parameter and asymptotic safety," *PoS* **CNCFG2010** (2010) 003, arXiv:1111.1000 [hep-th].

[30] U. Harst and M. Reuter, "The 'Tetrad only' theory space: nonperturbative renormalization flow and asymptotic safety," *JHEP* **05** (2012) 005, arXiv:1203.2158 [hep-th].

[31] J. Daum and M. Reuter, "Einstein-Cartan gravity, asymptotic safety, and the running Immirzi parameter," *Annals Phys.* **334** (2013) 351–419, arXiv:1301.5135 [hep-th].

[32] E. Manrique and M. Reuter, "Bare action and regularized functional integral of asymptotically safe quantum gravity," *Phys. Rev. D* **79** (2009) 025008, arXiv:0811.3888 [hep-th].

[33] J. Ambjorn, A. Goerlich, J. Jurkiewicz and R. Loll, "Nonperturbative quantum gravity," *Phys. Rept.* **519** (2012) 127–210, arXiv:1203.3591 [hep-th].

[34] R. Loll, "Quantum gravity from causal dynamical triangulations: a review," *Class. Quant. Grav.* **37** no. 1 (2020) 013002, arXiv:1905.08669 [hep-th].

[35] D. Dou and R. Percacci, "The running gravitational couplings," *Class. Quant. Grav.* **15** (1998) 3449–3468, arXiv:hep-th/9707239.

[36] R. Percacci and D. Perini, "Constraints on matter from asymptotic safety," *Phys. Rev. D* **67** (2003) 081503, arXiv:hep-th/0207033.

[37] R. Percacci and D. Perini, "Asymptotic safety of gravity coupled to matter," *Phys. Rev. D* **68** (2003) 044018, arXiv:hep-th/0304222.

[38] P. Dona, A. Eichhorn and R. Percacci, "Matter matters in asymptotically safe quantum gravity," *Phys. Rev. D* **89** no. 8 (2014) 084035, arXiv:1311.2898 [hep-th].

[39] P. Dona, A. Eichhorn and R. Percacci, "Consistency of matter models with asymptotically safe quantum gravity," *Can. J. Phys.* **93** no. 9 (2015) 988–994, arXiv:1410.4411 [gr-qc].

[40] U. Harst and M. Reuter, "QED coupled to QEG," *JHEP* **05** (2011) 119, arXiv:1101.6007 [hep-th].

[41] M. Shaposhnikov and C. Wetterich, "Asymptotic safety of gravity and the Higgs boson mass," *Phys. Lett. B* **683** (2010) 196–200, arXiv:0912.0208 [hep-th].

[42] M. H. Goroff, A. Sagnotti and A. Sagnotti, "Quantum gravity at two loops," *Phys. Lett. B* **160** no. 1 (1985) 81–86. www.sciencedirect.com/science/article/pii/0370269385914704.

[43] D. Benedetti, P. F. Machado and F. Saueressig, "Taming perturbative divergences in asymptotically safe gravity," *Nucl. Phys. B* **824** (2010) 168–191, arXiv:0902.4630 [hep-th].

[44] H. Gies, B. Knorr, S. Lippoldt and F. Saueressig, "Gravitational two-loop counterterm is asymptotically safe," *Phys. Rev. Lett.* **116** (May 2016) 211302. https://link.aps.org/doi/10.1103/PhysRevLett.116.211302.

[45] M. Reuter and H. Weyer, "Renormalization group improved gravitational actions: a Brans-Dicke approach," *Phys. Rev. D* **69** (2004) 104022, arXiv:hep-th/0311196.

[46] M. Reuter and H. Weyer, "Running Newton constant, improved gravitational actions, and galaxy rotation curves," *Phys. Rev. D* **70** (2004) 124028, arXiv:hep-th/0410117.

[47] M. Reuter and H. Weyer, "Quantum gravity at astrophysical distances?," *JCAP* **12** (2004) 001, arXiv:hep-th/0410119.

[48] M. Reuter and F. Saueressig, "From big bang to asymptotic de Sitter: complete cosmologies in a quantum gravity framework," *JCAP* **09** (2005) 012, arXiv:hep-th/0507167.

[49] M. Reuter and H. Weyer, "Do we observe quantum gravity effects at galactic scales?," *EAS Publ. Ser.* **20** (2006) 251, arXiv:astro-ph/0509163.

[50] M. Reuter and H. Weyer, "On the possibility of quantum gravity effects at astrophysical scales," *Int. J. Mod. Phys. D* **15** (2006) 2011–2028, arXiv:hep-th/0702051.

[51] A. Bonanno and M. Reuter, "Entropy signature of the running cosmological constant," *JCAP* **08** (2007) 024, arXiv:0706.0174 [hep-th].

[52] A. Bonanno and M. Reuter, "Primordial entropy production and lambda-driven inflation from quantum Einstein gravity," *J. Phys. Conf. Ser.* **140** (2008) 012008, arXiv:0803.2546 [astro-ph].

[53] M. Reuter and E. Tuiran, "Quantum gravity effects in the Kerr spacetime," *Phys. Rev. D* **83** (2011) 044041, arXiv:1009.3528 [hep-th].

[54] D. Becker and M. Reuter, "Propagating gravitons vs. 'dark matter' in asymptotically safe quantum gravity," *JHEP* **12** (2014) 025, arXiv:1407.5848 [hep-th].

[55] A. Bonanno and M. Reuter, "Cosmological perturbations in renormalization group derived cosmologies," *Int. J. Mod. Phys. D* **13** (2004) 107–122, arXiv:astro-ph/0210472.

[56] M. Reuter and F. Saueressig, "Fractal space-times under the microscope: a renormalization group view on Monte Carlo data," *JHEP* **12** (2011) 012, arXiv:1110.5224 [hep-th].

[57] M. Reuter and J.-M. Schwindt, "A minimal length from the cutoff modes in asymptotically safe quantum gravity," *JHEP* **01** (2006) 070, arXiv:hep-th/0511021.

[58] M. Reuter and J.-M. Schwindt, "Scale-dependent metric and causal structures in quantum Einstein gravity," *JHEP* **01** (2007) 049, arXiv:hep-th/0611294.

[59] O. Lauscher and M. Reuter, "Fractal spacetime structure in asymptotically safe gravity," *JHEP* **10** (2005) 050, arXiv:hep-th/0508202.

[60] D. Becker and M. Reuter, "Towards a C-function in 4D quantum gravity," *JHEP* **03** (2015) 065, arXiv:1412.0468 [hep-th].

[61] D. Becker and M. Reuter, "En route to background independence: broken split-symmetry, and how to restore it with bi-metric average actions," *Annals Phys.* **350** (2014) 225–301, arXiv:1404.4537 [hep-th].

[62] J. Ambjørn, L. Glaser, Y. Sato and Y. Watabiki, "2D CDT is 2D Hořava–Lifshitz quantum gravity," *Phys. Lett. B* **722** (2013) 172–175, arXiv:1302.6359 [hep-th].

[63] A. Contillo, S. Rechenberger and F. Saueressig, "Renormalization group flow of Hořava-Lifshitz gravity at low energies," *JHEP* **12** (2013) 017, arXiv:1309.7273 [hep-th].

[64] G. D'Odorico, F. Saueressig and M. Schutten, "Asymptotic freedom in Hořava-Lifshitz gravity," *Phys. Rev. Lett.* **113** no. 17 (2014) 171101, arXiv:1406.4366 [gr-qc].

[65] G. D'Odorico, J.-W. Goossens and F. Saueressig, "Covariant computation of effective actions in Hořava-Lifshitz gravity," *JHEP* **10** (2015) 126, arXiv:1508.00590 [hep-th].

[66] J. Schwinger, "On gauge invariance and vacuum polarization," *Phys. Rev.* **82** (Jun. 1951) 664–679. https://link.aps.org/doi/10.1103/PhysRev.82.664.

24

Carlo Rovelli

Professeur de classe exceptionnelle and director of the quantum gravity group of the Centre de Physique Théorique at Aix-Marseille University

Date: 15 October 2020. Via Zoom. Last edit: 3 December 2020

What are the main problems in theoretical physics at the moment?

Two open problems stand out for their importance and because we may have a chance to solve them. They are of very different nature: one is quantum gravity, the other is dark matter. The problem of quantum gravity is that we do not know how to describe gravitational phenomena where quantum effects cannot be disregarded. Examples include the big bang, the centre of black holes and the end point of their Hawking evaporation. Finding a theory of quantum gravity capable of accounting for these situations is recognised as a major problem in theoretical physics. The problem of dark matter is of a very different sort: it originates from observations. There is compelling evidence that dark matter exists in the universe but it not formed by the common kind of matter we know.

Could these two problems be related?

They could. One possible scenario for dark matter, among the many existing ones, is that it could be made out of Planck-size black or white holes stabilised by quantum gravity effects. The evaporation process of primordial black holes might lead to white hole remnants [1]. White holes are classically unstable but small ones might be stabilised by quantum effects, as atoms are. This is an attractive scenario because it does not require any new ingredient in the world or any new dynamical law, besides what we already know to exist: gravity and quantum theory. It is a possibility I like, but for the moment it is just one hypothesis among many others. So, yes, there is a possibility for the two problems to be related, but we don't know.

Along with dark matter, dark energy is usually considered a major problem but you did not mention it.

The dark energy issue is over-rated. I do not think it is such a big mystery. It is the same problem as the polynomial divergence of the mass of the Higgs [2, 3].

Why is the problem of quantum gravity difficult?

General relativity has taught us that gravity is the dynamics of spacetime. Therefore to understand quantum gravity we have to understand quantum space and quantum time.

Hence we need a formulation of physics in which spacetime is only a classical approximation. This demands a conceptual rethinking of the structure of physical theories we are accustomed to. Conventional quantum field theory is formulated on a given spacetime. Therefore most of the tools developed in the context of quantum field theory are not directly appropriate for dealing with quantum gravity. No surprise that this takes time. I think that many theoreticians are still too attached to their conventional quantum field theory tools. They try to get back to those, because these are what they have been trained to use. This is why for instance my book *Quantum Gravity* spends many pages on basic conceptual issues [4]. The big revolutions in science have always taken long. But we are making progress, and we have some recent empirical input.

What kind of empirical input are you referring to?

We have gathered interesting empirical information about quantum gravity in the last decades. Here are some examples. A few years ago there was hope and enthusiasm for tentative quantum gravity theories that break Lorentz invariance at the Planck scale, such as Hořava-Lifshitz gravity [5]. This stimulated the search for astrophysical evidence of Lorentz invariance breaking. Today, we have pretty good evidence that in many astrophysical phenomena Lorentz invariance is not broken at the Planck scale. Hence these theories have lost appeal.

A second example is given by string theory, which has long been a leading candidate for a theory of quantum gravity. Most string theorists expected the theory to imply a negative cosmological constant. But the cosmological constant has been measured, and is positive. More recently, most string theorists expected low-energy supersymmetric particles to be detectable at LHC, and this has been excluded. There have been other phenomena suggested by string theory, such as production of black holes at CERN and modifications of gravity at the centimetre scale. All contradicted by experience.

To appreciate the relevance of these results, and why they have had the effect of diminishing the hope and the enthusiasm in Hořava-Lifshitz gravity or string theory, it is important to stress that none of these results falsifies these theories in the sense of Karl Popper. That is, Hořava-Lifshitz gravity or string theory (suitably adjusted) are still logical possibilities. But science rarely advances with single, decisive experiments. The common way science advances is via the accumulation of positive or negative incomplete bits of evidence. This slowly increases or decreases the credibility of the hypotheses considered. If the expectations of a research community are confirmed, they enhance (even if they do not prove) the ("Bayesian") credibility in a research direction. By the same token, the failure of predictions decreases (even if does not rigorously rule out) the likelihood that a theory is right.

Some years ago spin foam theories inspired Lorentz-breaking theories, such as those implementing relative locality [6]. Have these theories been ruled out and do you think that there are indications that loop quantum gravity (LQG) is not Lorentz-invariant?

LQG is Lorentz-invariant. Like classical general relativity, it is Lorentz-invariant locally: there is no notion of global Lorentz invariance in gravitational physics, in general. The

Lorentz invariance of LQG is particularly clear in the spin foam formalism. The spin foams' vertex amplitude can be written in terms of unitary representations of $SL(2, \mathbb{C})$ in a manifestly Lorentz-invariant way [7].

There is a persistent confusion among people who do not know the details of LQG: the idea that since LQG predicts a minimal area and a minimal volume [8], and since area and volume are not Lorentz-invariant, then LQG cannot be Lorentz-invariant. This is just a mistake [9]: to see why, recall that ordinary non-relativistic quantum mechanics predicts a minimum (non-vanishing) value for the L_z component of the angular momentum. L_z is not invariant under rotations, but its minimum eigenvalue does not prevent non-relativistic quantum mechanics to be invariant under rotations! The confusion is between a *classical* minimal quantity and a *quantum* minimal quantity. When changing frame, a minimum eigenvalue does not change; what changes continuously is the probability distribution of different eigenvalues. Rotating continuously a state with minimum value of L_z does not change the minimum eigenvalue of L_z; it changes the components of the state over different eigenstates. This is a good example of the conceptual difficulty that many physicists still have in thinking about quantum space.

You mentioned earlier the negative predictions of string theory but string theory is a framework in which one may develop different models. In particular, the string theory landscape includes an infinite number of metastable de Sitter vacua [10].

Whether you call it "theory," "framework" or something else, not much changes. A framework can be wrong for describing nature, as a theory can. Non-relativistic classical mechanics is a very flexible framework. It still fails to describes electromagnetism or gravity. If anything, adding flexibility adds vagueness. I do not think we need vague frameworks. We need a theory capable of computing precisely what happens at the end of the evaporation of a black hole, for instance.

Let me clarify that I don't want to seem too critical, as there is no point in that. All of us are trying to do our best in order to understand nature, right? We follow different research paths, and we are often unimpressed by the research programs of others. We may well be all wrong. I am giving voice here to a point of view that is skeptical about the likelihood for string theory to succeed as the fundamental theory of nature. I do so because you are asking, but I am expressing views that are not only those of researchers in my own field. They are increasingly widespread views in the community of physicists at large, which – let us not forget – is very much wider than those interested in quantum gravity.

String theory has been a spectacular attempt, but has failed the goals that it set itself and that many believed were close, 30 years ago: computing the parameters of the standard model, uniquely deriving $SU(3) \times SU(2) \times U(1)$, predicting the value of the cosmological constant, singling out a single unique theory, not a framework ... The recent negative empirical results have further depressed its plausibility, on top of these un-successes. My impression is that many still call themselves "string theorists" only because it remains a community of self-support, but few today are actually engaged in exploring the possibility that this is the fundamental theory of nature. AdS/CFT [11] is still fashionable. It started as a surprising result but it has become a sort of paradigm in terms of which to think; as such

I am afraid it is doing damage because it is trapping the way people think into a certain number of prejudicial assumptions. Our world is not described by anti–de Sitter spacetime. I think that what we need is to learn how to do quantum physics on a quantum spacetime, namely overcome conventional quantum field theory, and not to try to reduce the problem to a conventional (boundary) quantum field theory.

There are many approaches to quantum gravity but I guess the one you find most promising is LQG?

That is correct. I'm sure that LQG is a good theory of the world ... two days per week. The other five I have doubts (laughs). I am biased because I have been working all my life on this theory and because I live in a community in which we keep telling one another the same things. It's a bit like politics, isn't it? (laughs).

So, what are the arguments to favour loop quantum gravity? It is a rather well-defined theory of quantum gravity, although many aspects are still unclear. It's well-defined mathematically; there is a Hilbert space, operators, transition amplitudes that one can write explicitly order by order in a suitable expansion, theorems stating that there are no ultraviolet divergences (infrared divergences are still an open question). There are theorems connecting the theory to the right classical limit. In principle the theory can be used to compute. The main difficulties to do so are technical: amplitudes are expressed in terms of complicated integrals over non-compact groups. The theory gives a compelling picture of the discreteness of quantum spacetime at the Planck scale. It has developed significantly and continuously during the last 30 years. All the problems that were considered stumbling blocks at some point in time were overcome one after the other. When I hear criticisms from outside the community, usually these criticisms refer to the state of the theory a quarter of a century ago.

But the main reason for me to find LQG convincing has to do with some points that regard the philosophy of science. First, I think that the problem with quantum gravity is not related to unification. It's too soon to attempt having a theory of everything. We don't even know what dark matter is. Who knows what are all the ingredients of the universe? Too often physicists have mistakenly thought of being near the end of the road. Second, progress in physics has always been either from new empirical inputs or – as in the cases of Copernicus or Einstein – by taking seriously previous successful theories and the lessons they give us about the world and bringing the various pieces together. This is what LQG does. General relativity tells us that spacetime is a field; quantum theory tells us that classical fields are just approximations of some quanta. LQG takes these lessons all the way through. It addresses directly the key scientific problem: What is the quantum structure of spacetime in the bulk? The strength of LQG is that it builds upon two solid columns, just general relativity and quantum theory, without substantial additional hypotheses, and yet it yields a fantastically beautiful new picture of nature. It predicts discreteness of spacetime, a cosmological bounce, black holes evolving into white holes. It's just beautiful! (laughs).

Since you mentioned spacetime discreteness as a prediction, how would you measure it?

Area and volume operators in LQG have a discrete spectrum [8]. If we had sufficient precision, we could in principle measure any scattering amplitude (which has the dimensions of an area) and find that it cannot be less than $\sim \ell_P^2$ where ℓ_P is the Planck length. In practice, it is not impossible to think of delicate interference effects testing Planck time discreteness for instance [12], along the lines of some recently proposed tabletop quantum gravity experiments. But it is still a very long shot.

What is this tabletop experiment that you are referring to?

Only five years ago, I thought that talking about tabletop quantum gravity experiments was a crackpot thing to do (laughs). But there is a beautiful tabletop experiment that was proposed by Sougato Bose et al. [13] and by Marletto and Vedral [14]. It's a non-relativistic quantum gravity experiment, so it does not probe the Planck length or the Planck energy, but it nevertheless probes whether geometry can be in a quantum superposition or not (assuming general relativity). The experiment tests the effect of the gravitational interactions between two nanoparticles which are both quantum split. The interaction entangles the two particles and this can be ascertained by checking the Bell inequalities. If they are found entangled, the gravitational field that entangled them was necessarily in a quantum superposition. The experiment may be doable [15, 16]; if successful, it would rule out the possibility that the gravitational field is classical (not quantised), including Roger Penrose's idea that spacetime cannot be in a strong quantum superposition [17]. It would tell us that the spacetime geometry is quantised. A number of variants of the idea have appeared, for instance in [18]. However, since these are all non-relativistic experiments, they won't be able to distinguish between different approaches to quantum gravity, as these all agree that in the perturbative regime gravitons can be in a superposition.

To actually test relativistic gravity and Planck scale discreteness, sensibility must still be pushed a few orders of magnitude deeper. If that becomes possible, we might perhaps be able to test Planck time discreteness because the entangled spacetime would have two different notions of proper time, and the discreteness might perhaps have an effect in the difference between the two proper times [12].

Had you asked me 10 years ago if there was any chance of directly measuring quantum gravity effects, I would have replied , "Forget it," but things have changed. Earlier I mentioned that Lorentz-breaking theories at the Planck scale have been ruled out. So the Planck scale is not out of reach. After all, the $SU(5)$ gauge theory, that at some point was a favourite model for grand unification, is ruled out by proton decay experiments, which test a phenomenon taking place at 10^{16} GeV [19]. The Planck scale is just a few orders of magnitude higher. We just have to be smart when designing the experiments. Experimental fundamental physics is not just smashing up particles more and more brutally.

You also mentioned the cosmological bounce to be a prediction of LQG. Is this an established prediction?

I think that at this point there is evidence supporting the idea that the cosmological bounce is predicted by LQG [20]. By using LQG to backtrack the evolution of the universe until the big bang, one finds a bounce to a previous eon. Bounce theories have been around for a while, irrespective of LQG. LQG allows us to perform calculations of the fluctuations around the bounce and its effects in the CMB. There is a large body of literature trying to connect the cosmological anomalies that we observe today in the CMB with the models studied in LQG [20]. At the moment, we do not yet have solid predictions that are confirmed, because there are uncertainties both in the measurements and in the theoretical calculations, but this is a lively research direction that is moving fast towards predictivity. Another prediction of LQG is that at the end of black hole evaporation a white hole remnant forms.

My understanding is that loop quantum cosmology (LQC) is a truncation of LQG. So how can you be confident that the cosmological bounce is realised in the full theory?

To be confident is different from being certain. We are confident that QCD predicts confinement, even if this has not been proven. We do not control the full theory well enough to come up with a rigorous theorem proving the validity of the LQC approximation, but evidence is piling up, indicating that loop quantum cosmology is in fact equivalent to taking the full theory and studying it by means of a controlled approximation. There is a lot of ongoing work on this direction; for instance, intermediate steps between the full LQG and LQC are under investigation [21].

You mentioned white holes as a prediction for the end result of the Hawking evaporation process. How can you address black holes in LQG? Do you also work with a truncation of the theory as in loop quantum cosmology?

The transition from black holes to white holes [1] has been studied using the canonical formulation of LQG [22, 23] as well as the covariant spin foam formulation [24]. These two versions of the theory have the same kinematics, same Hilbert space, same operators, and same spin networks which are basis in the Hilbert space. But the first computes transition amplitudes using Hamiltonian methods while the second is a path integral–like approach. The transition from black holes to white holes has been studied in both. In the canonical approach the methods used are similar to those in loop quantum cosmology; that is, it amounts to reducing the degrees of freedom of the theory. By working with this reduced system, you can solve many things exactly.

Are these white hole remnants what you refer to as Planck stars in your papers [25]?

When matter, for instance, a star, collapses gravitationally, it falls towards the centre until its density reaches Planckian values. This is called the “Planck star” stage and is when quantum gravitational effects become strong. The main quantum effect is to bound the curvature from increasing further, because curvature is bound in LQG [26]. The result is a bounce. This is only one of the three distinct quantum phenomena affecting the black hole evolution [24]. The second phenomenon regards what happens when approaching the singularity where matter is not present. The third regards what happens when the Hawking evaporation reduces the area of the horizon to a Planckian scale. Recent LQG literature studies all these phenomena.

There is a popular belief that at the end of the evaporation a black hole would just pop out of existence: this is not sustained by any theory and is implausible. LQG appears to predict that the central singularity is replaced by a transition to an anti-trapped region and the horizon tunnels into a white hole horizon. The collapsing matter of the star that gave rise to the black hole reaches a minimal size stage, the Planck star, then bounces back, because quantum gravity prevents further collapse much like the fall of an electron on the nucleus is prevented by quantum theory. Ultimately, the entire black hole has tunnelled into a white hole.

You mentioned the spin foam approach and the canonical approach as two versions of the theory. Have they been shown to be equivalent?

Whether the spin foam approach is equivalent to the canonical approach of LQG is an open question. There are indications for this, but no solid proof.

Coming back to the issue of perturbative non-renormalisability and the issue of divergences, you mentioned that LQG does not have ultraviolet divergences. How does LQG solve the issue of perturbative non-renormalisability?

Overcoming the issues of divergences is one of the most beautiful aspects of LQG. It is something the community is proud of. The application of standard quantum field theoretical perturbation theory to gravity yields ultraviolet (or short-distance) divergences because the gravitational field is split into two components, one of which is used as a classical background. The infinities come from short distance, where the notion of distance is determined by this classical background. LQG, instead, works non-perturbatively. It does not split the gravitational field in two parts. Rather, it considers the full quantum state of the geometry. The key calculations show that area and volume operators have a discrete spectrum [8]. This means that if the full metric is taken as a quantum object, we see that there are no arbitrarily small scales. There is no infinite short distance at all in nature. Hence in the conventional approach there are ultraviolet divergences only because one erroneously assumes the gravitational field to be continuous and arbitrary small scales to exist.

By analogy, consider a harmonic oscillator. Energy is discrete. Imagine using a perturbative approximation around a background set of states where the energy is assumed to be continuous: in computing transition amplitudes one would include intermediate states with arbitrarily small energies, which do not correspond to anything physical. In LQG there are no trans-Planckian degrees of freedom. The theory cuts itself off using the standard discreteness of quantum mechanics. The spin foam formalism renders particularly clear that the intermediate states summed over do not include arbitrarily small Feynman, hence high-momentum, loops [27].

As a matter of fact, the mechanism underlying the finiteness of LQG is partly the same as that in string theory. Superficially, string theory includes ultra-Planckian degrees of freedom. But scattering computations, such as the old, beautiful results by Amati, Ciafaloni and Veneziano [28–32], have shown that if you increase the energy the string opens up in such a way that you can never probe arbitrarily small degrees of freedom. In this respect, string theory and LQG are similar, in contrast to asymptotic safety or other scenarios: there are no ultraviolet divergences and no infinite renormalisation. In both theories there is a fundamental scale (the Planck scale or the string scale) and the theory lives at that scale. Physics happens from that scale up, or as particle physicists like to say, from that scale down (laughs).

What about the issues with infrared divergences?

The best formulations of LQG are with a positive cosmological constant, and don't use $SO(3)$ or $SL(2,\mathbb{C})$ but instead use quantum groups, as for instance in [33]. But these are rarely used because they are complicated. In that formulation, there are theorems stating that there are no infrared divergences, because the quantum group acts as a long-distance regulator. The amplitudes are therefore truly finite at every order.

More work needs still to be done, however, because the infrared divergences are tamed by this setup, but their contribution is proportional to the inverse of the cosmological constant, which is a large number. Understanding the role of these large numbers is still unclear, as far as I understand.

So do I understand correctly that the finiteness of LQG does not rely on asymptotic safety [34, 35]?

Yes, this is my understanding. LQG, like string theory, does not rely on asymptotic safety. They are both genuinely finite. LQG in a sense is just like QCD on a finite lattice, a lattice of Planckian size. The difference is that the lattice is not fixed a priori: it is determined dynamically by the discreteness of the geometry, namely by the spectral analysis of the geometrical operators. Hence you don't need asymptotic safety. Asymptotic safety assumes that there are degrees of freedom at all scales and that the theory is local in a background spacetime [35]. From the asymptotic safety programme point of view, LQG, like a lattice theory, is a theory with an infinite number of local counter terms, not easy to tackle. In summary, it seems to me that asymptotic safety is a different philosophy in which you assume nature to work with infinitely small degrees of freedom.

Thomas Thiemann has recently developed Hamiltonian renormalisation techniques in order to try to fix ambiguities in the quantisation procedure of LQG [36], and there is some connection to asymptotic safety.

The ambiguities Thomas is concerned with characterise the Hamiltonian formulation of LQG. They are one of the reasons I prefer to work with the covariant spin foam approach.

In the covariant picture there is less freedom for ambiguities. This is similar to what takes place in QED. Richard Feynman had the great idea that there is just one single vertex connecting two leptons and one photon, with an associated amplitude. In the Hamiltonian language, instead, there are many corresponding terms to consider, one per each possible time orientation of the legs. The coefficients of each term need to be carefully adjusted for the result to be Lorentz-invariant. This kind of issue multiplies in quantum gravity. In the spin foam approach the demand that the vertex has all the symmetries of the theory restricts ambiguities considerably [37–39].

So should one be pursuing the spin foam approach instead of the canonical approach?

It's good that people are pursuing both. We are exploring a theory which we know little about. Ultimately I expect that there should be a version of the Hamiltonian theory that corresponds to the spin foam theory, and vice versa.

As far as I understand there are two versions of the canonical approach, one in which you quantise the theory and then try to solve the Hamiltonian constraint, and another in which you add matter and solve the constraint classically before quantising, allowing you to find all physical states [40]. Which version of the theory is the correct one?

Theories can have more than one version and all can be right. The important question is, What can you calculate with one version of the theory or the other? The idea of including matter and using it to gauge-fix is a good idea. It simplifies some issues, and has some limitations. The Polish group has obtained interesting results using variants of this idea [41]. At the end, it all boils down to different choices of gauge. If you add matter, you attach your coordinates to the matter fields which is a kind of gauge-fixing. If God is good, these two versions should be equivalent, although perhaps God is not always good (laughs).

Regarding these quantisation ambiguities, one of them is related to the Barbero-Immirzi parameter. How should we interpret this parameter?

The Barbero-Immirzi parameter is analogous to the θ parameter in QCD. It comes into LQG precisely in the same manner as the θ parameter enters the Yang–Mills action, that is, as the constant in front of a term with opposite parity than the usual term and which has no effect on the classical equations of motion. The parameter does not affect the classical theory but does affect the quantum theory. Taking it into account is required both in the canonical and as in the covariant pictures; it enters in the eigenvalues of the area and volume operators for instance. An open question is whether this parameter can only be determined empirically, via some measurement, or whether there is something fixing it in the theory. There is a school of people in LQG who think that black hole entropy calculations fix the Barbero-Immirzi parameter [42] but there are colleagues who say they can obtain the same result without fixing the parameter [43]. Whether the theory is consistent for specific values of the parameter and not consistent for other values is an open question, in my opinion.

Since you mentioned black hole entropy, some time ago Ashoke Sen published a paper stating that his entropy calculation did not agree with LQG calculations [44]. How should I interpret this result?

A theory of quantum gravity should be able to compute black hole entropy. We have calculations that reproduce the classical result $S = A/4$ where A is the area. These appeared in LQG [45, 46] and in string theory by Andrew Strominger and Cumrun Vafa [47] only a few weeks apart. The polemic you refer to was later, when people looked at logarithmic corrections to these results. Ashoke Sen had a paper on this stating that the method he used did not reproduce LQG results. He did not say that LQG was wrong! (laughs) [44]. This paper led to a discussion and a clarification. The point is that there is more than one kind of logarithmic correction. Those due to quantum field theory entanglement across the horizon are finite, independent from the cutoff and universal, but they are not the only ones. As the Euclidean path integral shows, there is also another contribution that depends on the statistical ensemble being fixed [48]. The LQG calculation fixes the horizon area, unlike Sen's calculation. Hence there is no contradiction.

So, do you think that the problem of black hole entropy is fully understood?

I do not. I think that neither in the LQG community nor in the string community it is clear how to think about black hole entropy. Within the AdS/CFT community, people are convinced that late Hawking quanta are correlated with early Hawking quanta (see, for instance, [49]). The argument is that since the black hole shrinks due to Hawking evaporation, the black hole area decreases and if the entropy depends on the number of states of the black hole, there are not enough states to purify the Hawking radiation, after the Page time. Thus Hawking radiation cannot be truly thermal. This scenario is forced by the AdS/CFT hypothesis, because this hypothesis *assumes* that all degrees of freedom are accessible from the boundary. However, this argument is disputed, because the Bekenstein-Hawking entropy may not be the full von Neumann entropy obtained by tracing all the degrees of freedom inside the black hole [50]. Instead, Bekenstein-Hawking entropy might just count horizon degrees of freedom, or short scale correlations across the horizon. But there are far more possible correlations with degrees of freedom deep inside the black hole that do not contribute to the Bekenstein-Hawking entropy. So, the exact relation between the Bekenstein-Hawking entropy and the full quantum theory of gravity is still open.

Can LQG find states that represent classical Minkowski spacetime?

This is a funny question because this was a problem in the theory 20 years ago.

(laughs) I am sorry to say but in the circles in which I move, this is still considered the main problem of LQG.

In those circles no one has even read a review paper about LQG for more than a decade, right (laughs)?

Hermann Nicolai et al. did write an evaluation of LQG in 2005 [51]; I understand this is 15 years ago ... by the way, what is your opinion about this review?

The objections that Nicolai et al. raised 15 years ago have been extensively addressed and answered. There have been detailed responses to that paper, clarifying the points misunderstood by those authors, for instance [52] and [53]. But LQG has much progressed since, and those discussions have only historical relevance today. Many of the crucial results in LQG came after that discussion.

Coming back to the issue of Minkowski spacetime, what is, then, the status?

There are coherent states that capture Minkowski spacetime and which can be used to calculate the graviton propagator [54] or n-point functions [55] in a perturbation theory around that Minkowski spacetime. One first has to build up Minkowski spacetime as a coherent state, as is usually done in quantum optics. In quantum optics, the state, say, of the electromagnetic field inside a capacitor can be written as a coherent state of (QED) photons. The mathematics of coherent states in LQG is very well developed [56–60]. Coherent states are not unique, which has led to long discussions in the community about who has the most useful ones (laughs).

Another important result about the classical limit of the theory is given by a family of theorems, the first one by Barrett et al. [61], showing that by taking a suitable large spin limit of the spin foam amplitude, that is, a limit in which you probe large distances compared to the Planck scale, the amplitude is related to the exponential of the classical Einstein-Hilbert action. One recovers the Regge action in the classical limit and then the Einstein-Hilbert action in the large-distance limit. Minkowski space is one of the solutions of the classical theory.

As I understand it, it is possible to construct many coherent states sharply peaked around Minkowski spacetime, so which one is the right one?

There is no single "right" one. There are different quantum gravity states that look the same at large scales. In standard quantum field theory the vacuum is unique. In fact, the uniqueness of the vacuum is one of the Arthur Wightman axioms. But in quantum gravity there is no reason for this to be the case. Using again the analogy with the electromagnetic field in a capacitor, you can build different coherent states formed by many photons that approximate the same classical configuration. The effects of this non-uniqueness of the Minkowski vacuum in LQG may play a role in the information loss issue [62]. It may be related to other quantum field theoretical phenomena that have recently raised attention [63]. The state best resembling a single and unique "vacuum" state in quantum gravity, on the other hand, is not Minkowski space; it is the state with no quanta of space: zero total volume and zero total area. This is a unique and well-defined state in the theory. All states can be built by creating quanta over it.

But in the canonical formulation of LQG, there is a Hamiltonian constraint that you have to solve. So has it been determined which of these coherent states are physical states?

The Hamiltonian constraint encodes the full general relativistic *dynamics*; to solve it exactly would amount to having the full solution of the Einstein equations, which is obviously out of the question. This is not the way physics works. What is to be done is to compute transition amplitudes between boundary states order by order in an appropriate expansion. This is equivalent to solving the Hamiltonian constraint approximately, order by order. Virtually everything we do in physics is approximations. The fact that we have a "non-perturbative" formulation of quantum gravity does not mean that we can compute everything exactly (laughs)! It only means we do not introduce an unphysical smooth background spacetime.

You mentioned a few times calculations of the propagator and scattering processes. Have these coherent states been used to calculate graviton scattering?

Yes. Coherent states are a key ingredient for the propagator and n-point functions. The reason is that the propagator describes at the first order the dynamics of perturbations over a spacetime; this spacetime must be specified in a non-perturbative theory, and its best description is via a coherent state [64]. The propagator has been computed to lowest order and it matches the one of conventional perturbative linearised quantum gravity [54]. The calculation was not easy, because in a sense LQG is naturally defined in the strong coupling limit, not in the weak coupling limit. But the calculation had an important historical role because it was first performed at an intermediate stage in the construction of the theory, when the dynamics was defined by the Barrett-Crane amplitude [65]. The propagator turned out to be wrong with that amplitude, and this led to the discovery of the amplitude that currently best defines the theory [37].

Regarding scattering, it should be said that we do not calculate the S-matrix seen by some distant observer at infinity. What we do is to consider a compact region of spacetime enclosed by a three-dimensional boundary. In particular, we are interested in a coherent state that represents flat space outside the region. We add excitations on the boundary and compute the associated transition amplitude.

What is the picture that LQG gives us of spacetime at the Planck scale? Is it continuous or is it discrete?

Both, as usual in quantum theory. Is light discrete or continuous? It is discrete because it is made up of photons: if we measure how many photons there are in a wavelength, we only get a *discrete* number, because the number operator has a discrete spectrum. But a generic quantum state of light is a *continuous* superposition of photon states. Hence expectation values can vary continuously. LQG is a conventional quantum theory, so the same happens. Light is the electromagnetic field, spacetime is the gravitational field. Geometric operators that express the geometry of spacetime have discrete spectra, like the number operator does; hence if you measure, say, the volume of something, the theory predicts that you can get only discrete values that are in the spectrum of the corresponding operator. The basis that diagonalises geometric operators is called the spin network basis. Basis states in this basis can be seen as discrete geometries, like n-photon states are discrete version of light. They are collections of discrete quanta of space, like photons are quanta of light. Unlike photons, however, these quanta do not move in space; they are space themselves. Their (discrete) quantum numbers do not represent position or momentum; they represent their size and adjacency relations. But generic states are *continuous* superpositions of these eigenstates. So expectation values can vary continuously.

This can be equally seen in the dynamics. Transition amplitudes are computed by summing over the intermediate structures called spin foams that can be interpreted as discrete four-dimensional geometries. But since the amplitude is obtained by summing over them, we cannot say that spacetime has a single discrete structure.

When describing spacetime at such small scales, one usually runs into the problem of time. Is this solved within LQG?

Yes, it is solved. A part of my book [4] is devoted to a careful discussion of all conceptual issues, including the issue of time. To a large extent, the problem has nothing to do with quantum gravity. It is an issue already in classical general relativity. This can be seen, for instance, by noticing that general relativity can be formulated in the Hamilton-Jacobi formalism, in a way that does not use any time variable at all. The reason is that in classical general relativity the gravitational field does not evolve in time. It's the evolution of the field itself which defines proper time. Coordinate time is an arbitrary quantity you attach to the evolution of the field; in general, there is no preferred time coordinate. This is confusing at first but it is a coherent way of describing the world, and in fact general relativity works very well. Correspondingly, in the quantum theory there is no Schrödinger-like equation, because the Schrödinger-like equation is an evolution equation in time. Instead, there are dynamical relations between phase space variables that define probability amplitudes. These are coded in the Wheeler-DeWitt equation, or in the spin foam transition amplitudes. The Wheeler-DeWitt equation is just the quantum version of the Hamilton-Jacobi equation of general relativity and the absence of a time variable in it is just the same as the absence of time in the classical theory. In the spin foam formalism, transition amplitudes between initial and final states are evaluated without mentioning time. The proper time lapsed between initial and final states is itself determined in terms of these amplitudes.

But is it clear that you have quantum diffeomorphism invariance in LQG?

Yes, it is clear. The formalism is in a diffeomorphism-invariant language to start with. The objects being talked about in the spin foam formalism are not (quantum versions of) metrics; they are (quantum versions of) geometries, in the language of general relativity. That is, equivalent classes of metrics invariant under diffeomorphisms.

What were the motivations behind your recent paper on gauge symmetry [66]?

This paper is not directly related to LQG. Gauge invariance has always hunted me. Why is the world described by gauge theories? We are taught in school that gauge symmetry is a redundancy in the mathematics of the theory and the world itself is gauge-invariant. But if so, why do we use this redundancy? Why don't we get rid of it? If we get rid of it, the theory becomes non-local and more complicated. So, why does the theory become local and simpler by adding redundancy? In the early days of gauge theory, these questions were discussed, then forgotten. But I always felt that there is something missing in this story.

I think that the idea that gauge symmetry is a purely mathematical redundancy is wrong. Things are trickier. We break the world into parts, for instance spacetime regions, subsystems, or distinct fields; in quantum mechanics we split the world into a quantum system and a measuring apparatus (or "the observer"). The key point is that physical degrees of freedom do not necessarily respect these separations. In general, physical degrees of freedom are not confined to subsystems; they are defined across subsystems. This is the reason underlying gauge. When we put together two systems, there are degrees of freedom that pertain to both and there are gauge-invariant couplings between non-gauge invariant variables in each. This means that the gauge non-invariant quantities of a system are not pure mathematics; they are the handle with which the system can couple to something else. For instance the gauge-invariant variables of electromagnetism are the electric and magnetic fields, but the way electromagnetism couples to fermions is via the Maxwell potential. Thus, quantities that are not gauge-invariant have a physical meaning once we look at how a system can interact with the rest of the world, and that's why we need them. If we think about gauge symmetry in these terms, we understand the physical meaning of gauge invariance and we are less confused about boundary charges and variables.

Recently, there has been renewed interest in BMS symmetry, boundary variables, boundary charges and similar notions, prompted by the work of Andrew Strominger and others [67]. I think all this becomes more comprehensible by looking at gauge symmetry in a more refined way than just saying it is mathematical redundancy. This also helps clarifying the discussion within LQG about the gauge invariance of area and volume operators. These are ways the gravitational field can couple to a measuring device.

What do you think has been the biggest breakthrough in theoretical physics the past 30 years?

Indeed, theoretical physics had a formidable stream of breakthroughs ... until 30 years ago (laughs)! I do not think that there is anything which we already recognise clearly as a major breakthrough in the past 30 years. However, there are theoretical results which may be considered as a major breakthrough in the future, *if* they turn out to be right. LQG's discreteness of space or AdS/CFT are examples: *if* any of them turns out to be related to the real world, they are major breakthroughs. For instance, an experiment, as discussed earlier, could probe the discreteness of geometry. Until we get more clarity we do not recognise the true theoretical breakthroughs. Until Hertz confirmed it, the great step taken by Maxwell was not really recognised. On the other hand, recent *observational* breakthroughs have been spectacular. It suffices to mention cosmological observations, gravitational waves and black holes, all recent results recognised with Nobel Prizes. We are in a golden age of gravitational physics.

Why did you choose to do physics?

I chose it late. I was not one of those kids who wanted to be a scientist since a young age. I was curious about what science is telling us about the world. Physics seemed to me to be at the core of science. While studying it at the university, however, I fell in love with quantum mechanics and general relativity. So here I am (laughs).

What do you think is the role of the theoretical physicist in modern society?

To push ahead our understanding of nature at a level which in some sense is fundamental (there are domains that are fundamental in other senses). I also write for the larger public, ask questions that can interest the philosophers and try to contribute to the wider cultural debate that shapes our worldview.

References

[1] E. Bianchi, M. Christodoulou, F. D'Ambrosio, H. M. Haggard and C. Rovelli, "White holes as remnants: a surprising scenario for the end of a black hole," *Class. Quant. Grav.* **35** no. 22 (2018) 225003, arXiv:1802.04264 [gr-qc].

[2] E. Bianchi, C. Rovelli and R. Kolb, "Is dark energy really a mystery?," *Nature* **466** (2010) 321–322.

[3] E. Bianchi and C. Rovelli, "Why all these prejudices against a constant?," arXiv:1002.3966 [astro-ph.CO].

[4] C. Rovelli, *Quantum Gravity*. Cambridge Monographs on Mathematical Physics. Cambridge University Press, 2004.

[5] P. Hořava, "Quantum gravity at a Lifshitz point," *Phys. Rev. D* **79** (Apr. 2009) 084008. https://link.aps.org/doi/10.1103/PhysRevD.79.084008.

[6] G. Amelino-Camelia, L. Freidel, J. Kowalski-Glikman and L. Smolin, "The principle of relative locality," *Phys. Rev. D* **84** (2011) 084010, arXiv:1101.0931 [hep-th].

[7] C. Rovelli and S. Speziale, "Lorentz covariance of loop quantum gravity," *Phys. Rev. D* **83** (2011) 104029, arXiv:1012.1739 [gr-qc].

[8] C. Rovelli and L. Smolin, "Discreteness of area and volume in quantum gravity," *Nucl. Phys. B* **442** (1995) 593–622, arXiv:gr-qc/9411005. [Erratum: *Nucl. Phys. B* **456** (1995) 753–754.]

[9] C. Rovelli and S. Speziale, "Reconcile Planck scale discreteness and the Lorentz-Fitzgerald contraction," *Phys. Rev. D* **67** (2003) 064019, arXiv:gr-qc/0205108.

[10] E. Silverstein, "The dangerous irrelevance of string theory," arXiv:1706.02790 [hep-th].

[11] J. M. Maldacena, "The large N limit of superconformal field theories and supergravity," *Int. J. Theor. Phys.* **38** (1999) 1113–1133, arXiv:hep-th/9711200.

[12] M. Christodoulou and C. Rovelli, "On the possibility of experimental detection of the discreteness of time," *Frontiers in Physics* **8** (2020) 207, arXiv:1812.01542 [gr-qc].

[13] S. Bose, A. Mazumdar, G. W. Morley, H. Ulbricht, M. Toroš, M. Paternostro, A. Geraci, P. Barker, M. Kim and G. Milburn, “Spin entanglement witness for quantum gravity,” *Phys. Rev. Lett.* **119** no. 24 (2017) 240401, arXiv:1707.06050 [quant-ph].

[14] C. Marletto and V. Vedral, “Gravitationally-induced entanglement between two massive particles is sufficient evidence of quantum effects in gravity,” *Phys. Rev. Lett.* **119** no. 24 (2017) 240402, arXiv:1707.06036 [quant-ph].

[15] R. J. Marshman, A. Mazumdar and S. Bose, “Locality and entanglement in table-top testing of the quantum nature of linearized gravity,” *Phys. Rev. A* **101** no. 5 (2020) 052110, arXiv:1907.01568 [quant-ph].

[16] T. W. van de Kamp, R. J. Marshman, S. Bose and A. Mazumdar, “Quantum gravity witness via entanglement of masses: Casimir screening,” arXiv:2006.06931 [quant-ph].

[17] R. Penrose, “On gravity’s role in quantum state reduction,” *Gen. Rel. Grav.* **28** (1996) 581–600.

[18] R. Howl, V. Vedral, M. Christodoulou, C. Rovelli, D. Naik and A. Iyer, “Testing quantum gravity with a single quantum system,” arXiv:2004.01189 [quant-ph].

[19] Super-Kamiokande Collaboration, “Search for proton decay via $p \to e^+\pi^0$ and $p \to \mu^+\pi^0$ in a large water Cherenkov detector,” *Phys. Rev. Lett.* **102** (Apr. 2009) 141801. https://link.aps.org/doi/10.1103/PhysRevLett.102.141801.

[20] I. Agullo and P. Singh, *Loop Quantum Cosmology: The First 30 Years*, Abhay Ashtekar and Jorge Pullin, eds., pp. 183–240. World Scientific, 2017.

[21] E. Alesci, T. Thiemann and A. Zipfel, “Linking covariant and canonical LQG: new solutions to the Euclidean scalar constraint,” *Phys. Rev. D* **86** (2012) 024017, arXiv:1109.1290 [gr-qc].

[22] R. Gambini and J. Pullin, “Loop quantization of the Schwarzschild black hole,” *Phys. Rev. Lett.* **110** no. 21 (2013) 211301, arXiv:1302.5265 [gr-qc].

[23] A. Ashtekar, J. Olmedo and P. Singh, “Quantum extension of the Kruskal spacetime,” *Phys. Rev. D* **98** no. 12 (2018) 126003, arXiv:1806.02406 [gr-qc].

[24] F. D’Ambrosio, M. Christodoulou, P. Martin-Dussaud, C. Rovelli and F. Soltani, “The end of a black hole’s evaporation – part I,” arXiv:2009.05016 [gr-qc].

[25] C. Rovelli and F. Vidotto, “Planck stars,” *Int. J. Mod. Phys. D* **23** no. 12 (2014) 1442026, arXiv:1401.6562 [gr-qc].

[26] C. Rovelli and F. Vidotto, “Evidence for maximal acceleration and singularity resolution in covariant loop quantum gravity,” *Phys. Rev. Lett.* **111** (2013) 091303, arXiv:1307.3228 [gr-qc].

[27] A. Perez, “The spin foam approach to quantum gravity,” *Living Rev. Rel.* **16** (2013) 3, arXiv:1205.2019 [gr-qc].

[28] D. Amati, M. Ciafaloni and G. Veneziano, “Superstring collisions at Planckian energies,” *Phys. Lett. B* **197** (1987) 81.

[29] D. Amati, M. Ciafaloni and G. Veneziano, “Classical and quantum gravity effects from Planckian energy superstring collisions,” *Int. J. Mod. Phys. A* **3** (1988) 1615–1661.

[30] D. Amati, M. Ciafaloni and G. Veneziano, “Can space-time be probed below the string size?,” *Phys. Lett. B* **216** (1989) 41–47.

[31] D. Amati, M. Ciafaloni and G. Veneziano, “Higher order gravitational deflection and soft Bremsstrahlung in planckian energy superstring collisions,” *Nucl. Phys. B* **347** (1990) 550–580.

[32] D. Amati, M. Ciafaloni and G. Veneziano, "Effective action and all order gravitational eikonal at Planckian energies," *Nucl. Phys. B* **403** (1993) 707–724.

[33] M. Han, "Cosmological constant in LQG vertex amplitude," *Phys. Rev. D* **84** (2011) 064010, arXiv:1105.2212 [gr-qc].

[34] S. Weinberg, "Ultraviolet divergences in quantum theories of gravitation," in *General Relativity: An Einstein Centenary Survey*, S. W. Hawking and W. Israel, eds., pp. 790–831. Cambridge University Press, 1979.

[35] M. Reuter and F. Saueressig, *Quantum Gravity and the Functional Renormalization Group: The Road towards Asymptotic Safety*. Cambridge Monographs on Mathematical Physics. Cambridge University Press, 2019.

[36] T. Lang, K. Liegener and T. Thiemann, "Hamiltonian renormalisation I: derivation from Osterwalder–Schrader reconstruction," *Class. Quant. Grav.* **35** no. 24 (2018) 245011, arXiv:1711.05685 [gr-qc].

[37] J. Engle, E. Livine, R. Pereira and C. Rovelli, "LQG vertex with finite Immirzi parameter," *Nucl. Phys. B* **799** (2008) 136–149, arXiv:0711.0146 [gr-qc].

[38] L. Freidel and K. Krasnov, "A new spin foam model for 4D gravity," *Class. Quant. Grav.* **25** (2008) 125018, arXiv:0708.1595 [gr-qc].

[39] W. Kaminski, M. Kisielowski and J. Lewandowski, "Spin-foams for all loop quantum gravity," *Class. Quant. Grav.* **27** (2010) 095006, arXiv:0909.0939 [gr-qc]. [Erratum: *Class. Quant. Grav.* **29** (2012) 049502.]

[40] K. Giesel and T. Thiemann, "Algebraic quantum gravity (AQG). IV. Reduced phase space quantisation of loop quantum gravity," *Class. Quant. Grav.* **27** (2010) 175009, arXiv:0711.0119 [gr-qc].

[41] M. Domagala, K. Giesel, W. Kaminski and J. Lewandowski, "Gravity quantized: loop quantum gravity with a scalar field," *Phys. Rev. D* **82** (2010) 104038, arXiv:1009.2445 [gr-qc].

[42] A. Ashtekar, J. Baez, A. Corichi and K. Krasnov, "Quantum geometry and black hole entropy," *Phys. Rev. Lett.* **80** (1998) 904–907, arXiv:gr-qc/9710007.

[43] J. Ben Achour, K. Noui and A. Perez, "Analytic continuation of the rotating black hole state counting," *JHEP* **08** (2016) 149, arXiv:1607.02380 [gr-qc].

[44] A. Sen, "Logarithmic corrections to Schwarzschild and other non-extremal black hole entropy in different dimensions," *JHEP* **04** (2013) 156, arXiv:1205.0971 [hep-th].

[45] C. Rovelli, "Black hole entropy from loop quantum gravity," *Phys. Rev. Lett.* **77** (1996) 3288–3291, arXiv:gr-qc/9603063.

[46] C. Rovelli, "Loop quantum gravity and black hole physics," *Helv. Phys. Acta* **69** (1996) 582–611, arXiv:gr-qc/9608032.

[47] A. Strominger and C. Vafa, "Microscopic origin of the Bekenstein-Hawking entropy," *Phys. Lett. B* **379** (1996) 99–104, arXiv:hep-th/9601029.

[48] D. N. Page, "Hawking radiation and black hole thermodynamics," *New J. Phys.* **7** (2005) 203, arXiv:hep-th/0409024.

[49] G. Penington, "Entanglement wedge reconstruction and the information paradox," *JHEP* **09** (2020) 002, arXiv:1905.08255 [hep-th].

[50] C. Rovelli, "The subtle unphysical hypothesis of the firewall theorem," *Entropy* **21** no. 9 (2019) 839, arXiv:1902.03631 [gr-qc].

[51] H. Nicolai, K. Peeters and M. Zamaklar, "Loop quantum gravity: an outside view," *Class. Quant. Grav.* **22** (2005) R193, arXiv:hep-th/0501114.

[52] T. Thiemann, "Loop quantum gravity: an inside view," *Lect. Notes Phys.* **721** (2007) 185–263, arXiv:hep-th/0608210.

[53] A. Ashtekar, "Loop quantum gravity: four recent advances and a dozen frequently asked questions," in *11th Marcel Grossmann Meeting on General Relativity*, Hagen Kleinert, Robert T. Jantzen and Remo Ruffini, eds., pp. 126–147. World Scientific, 2007. arXiv:0705.2222 [gr-qc].
[54] E. Bianchi and Y. Ding, "The time-oriented boundary states and the Lorentzian-spinfoam correlation functions," *J. Phys. Conf. Ser.* **360** (2012) 012044.
[55] C. Rovelli and M. Zhang, "Euclidean three-point function in loop and perturbative gravity," *Class. Quant. Grav.* **28** (2011) 175010, arXiv:1105.0566 [gr-qc].
[56] E. R. Livine and S. Speziale, "A new spinfoam vertex for quantum gravity," *Phys. Rev. D* **76** (2007) 084028, arXiv:0705.0674 [gr-qc].
[57] T. Thiemann, "Complexifier coherent states for quantum general relativity," *Class. Quant. Grav.* **23** (2006) 2063–2118, arXiv:gr-qc/0206037.
[58] E. Bianchi, E. Magliaro and C. Perini, "Coherent spin-networks," *Phys. Rev. D* **82** (2010) 024012, arXiv:0912.4054 [gr-qc].
[59] L. Freidel and S. Speziale, "Twisted geometries: a geometric parametrisation of $SU(2)$ phase space," *Phys. Rev. D* **82** (2010) 084040, arXiv:1001.2748 [gr-qc].
[60] A. Calcinari, L. Freidel, E. Livine and S. Speziale, "Twisted geometries coherent states for loop quantum gravity," arXiv:2009.01125 [gr-qc].
[61] J. W. Barrett, R. Dowdall, W. J. Fairbairn, F. Hellmann and R. Pereira, "Asymptotic analysis of Lorentzian spin foam models," *PoS* **QGQGS2011** (2011) 009.
[62] L. Amadei and A. Perez, "Hawking's information puzzle: a solution realized in loop quantum cosmology," arXiv:1911.00306 [gr-qc].
[63] A. Strominger, *Black Hole Information Revisited*, pp. 109–117. World Scientific, 2020. arXiv:1706.07143 [hep-th].
[64] C. Rovelli, "Graviton propagator from background-independent quantum gravity," *Phys. Rev. Lett.* **97** (2006) 151301, arXiv:gr-qc/0508124.
[65] J. W. Barrett and L. Crane, "Relativistic spin networks and quantum gravity," *J. Math. Phys.* **39** (1998) 3296–3302, arXiv:gr-qc/9709028.
[66] C. Rovelli, "Gauge is more than mathematical redundancy," *Fundam. Theor. Phys.* 199 (2020).
[67] T. He, V. Lysov, P. Mitra and A. Strominger, "BMS supertranslations and Weinberg's soft graviton theorem," *JHEP* **05** (2015) 151, arXiv:1401.7026 [hep-th].

25
Nathan Seiberg

Professor at the School of Natural Sciences, Institute for Advanced Study, Princeton NJ

Date: 27 June 2011. Location: Uppsala. Last edit: 19 September 2019

What are the main puzzles in theoretical physics at the moment?

There are many puzzles in the many different areas of theoretical physics. There are puzzles in condensed matter physics, the standard model of particle physics, puzzles in cosmology, puzzles in quantum gravity, string theory and the nature of spacetime. We can elaborate on any of these.

Let us start with the standard model, then. What are the questions/problems with the standard model?

There are questions regarding the origin of the quantum numbers and generations in the standard model and questions regarding the origin of the particular gauge group that the standard model exhibits. Additionally, we would like to understand where all the parameters in the standard model come from and why the spectrum of masses looks rather random. In general, it looks like the standard model is coming from some other model or theory and we already know that we need to extend it in order to account for neutrino masses. So that's already telling us for a fact that the need for extending the standard model is not just aesthetic. Besides that, the numbers that appear in the standard model, such as the values of the couplings, look rather mysterious. Where do these numbers come from? This question deserves an answer. Another issue with the standard model is the hierarchy problem. The mass of the Higgs particle is unstable, and understanding it is not just a mere technical exercise; it is likely to require fundamental progress. Usually, as you keep developing and understanding physics at smaller and smaller distances, or higher and higher energies, new things emerge. It would be very strange if we would all of a sudden find nothing interesting when going 15 orders of magnitude higher in energy.

It seems that you agree that the standard model has to be extended somehow, but what's the best way to do that?

You can wonder how to extend the standard model and people have suggested many different ways of doing that. It looks like nothing else rather than supersymmetry could work. Any model which tries to understand flavour, the associated quantum numbers and parameters usually immediately runs into problems of flavour changing neutral currents. Flavour-changing neutral currents point to an energy scale of about 10^{100} TeV, which is really beyond reach. Hence, it's hard to imagine that there is a model that explains these parameters and can be directly tested. With regard to the hierarchy problem, there are two classes of solutions, namely, those which are weakly coupled and those which are strongly coupled. However, the success of the precision measurements of the standard model really point to a solution which is weakly coupled. This fact has already narrowed the possibilities essentially to just supersymmetric extensions. Now, given that supersymmetry has not yet been discovered makes us nervous but it's not the end of the line yet because experiments have to scan the parameter space and see what's there.

The huge range of parameter space and the many superpartners that are supposed to be out there don't make you feel uncomfortable?

It is true that it is a huge parameter space but the essential features, I think, are independent of the majority of these parameters. Even in the most optimistic scenario that superpartners would be discovered in the hundreds of GeVs, the LHC will not be able to measure more than one or two dozen of numbers out of the 100 or so parameters. In some sense it's good news because it means that the search is not in the 100 or so parameters of the minimal supersymmetric standard model (MSSM) [1] but in a much narrower range.

Are you confident that people will find supersymmetry at the LHC?

No, but I'm confident that if supersymmetry is there, they will find it. It's really not trivial to find it because these experiments are very difficult.

If supersymmetry is in fact there, it has to be broken at some scale. You have proposed a mechanism that could break it assuming that we live in a metastable vacuum [2]. Can you explain this idea?

Well, the fact that our universe is not stable is not original. The idea that we live in a metastable state has been discussed in the 1970s and 1980s for various different reasons. There are cosmological and particle physics reasons for that discussion. The mechanism you mention originates from a slightly different point of view. If you look back at models in the context of what's known as gauge mediation, starting with papers in the mid-1990s by Dine, Nelson and their collaborators [3–5], in every single one of them it is assumed that we live in a metastable state. In some cases, people noticed that if they expanded around some ground state they would find nice properties and sometimes they noticed that if you go far enough in field space you'll find a much lower energy state. Sometimes they noticed it, sometimes they didn't and sometimes they even emphasised it. However, I think that the new understanding is that this is really inevitable and there's a good reason why it happens, which is basically due to a combination of facts. Your model has to satisfy points A, B, C [6, 7], and if that happens you have to be in a metastable state. Knowing all these constraints you know exactly that if you turn off one parameter or another, such that you don't satisfy one of the constraints, you will end up in a stable state.

Is there a physical motivation for this being the case or is it just a theoretical idea?

It's not an idea; you put in experimental constraints, which implies that you need to break supersymmetry and there are various requirements for supersymmetry-breaking. It has to be done in such a way that you need gaugino masses and don't want to have massless Goldstone bosons. So if you put all these things together, you show that you must be in a metastable state. The chain of reasoning is somewhat long but the input is very general. It uses things that we know and love. The output is that we're in a metastable state and indeed this is the case in all the examples you find in the literature. All I'm saying is that we shouldn't be surprised that it's true in all the examples because it's really inevitable. On the flip side, once you're not attempting to find a stable state, there are many more possibilities for model building. It's a lot easier to break supersymmetry dynamically if we live in a metastable state. So for model building, that's good news.

Does this mechanism for supersymmetry-breaking lead to falsifiable predictions?

Well, the predictions are falsifiable in principle. It's not clear that the LHC could reach the required energies. If the superpartners have masses of about 100 GeV then this mechanism includes messengers with energies of roughly the scale of symmetry-breaking, which is pushed to be 10 or 100 TeV. Thus, even if supersymmetry is discovered, and we find the superpartners, I cannot see direct implications of shorter-distance physics that determines phenomena at low energies. Having said that, the idea of gauge mediation does lead to direct predictions [6]. We can imagine measuring the spectrum of superpartners and use that to learn about whether symmetry-breaking is due to gauge mediation or not.

Could you explain what gauge mediation is exactly?

Okay. The idea is that there's a supersymmetric standard model, perhaps with a few other extensions, and that there's another almost decoupled sector which breaks supersymmetry in such a way that these two sectors talk to each other. That is, information about the supersymmetry-breaking sector is communicated to the observed sector, which is something like the MSSM. The question is, How is this communication taking place? One possibility is that there are ordinary gauge interactions between sectors as in the standard model. That's the first possibility and that goes by the name of *gauge mediation*. The alternative is that even if they don't talk to each other, there is some Planck-scale physics which allows for communication between them. In the context of gravity, there is always some Planck-scale physics which communicates the information of supersymmetry-breaking. This alternative possibility goes by the name of *gravity mediation*. In the context of gravity mediation, the scale of supersymmetry-breaking is a lot higher than within the context of gauge mediation.

Is that one of the reasons why people exclude the effects of gravity to try to make predictions at the LHC?

Right, so for the LHC you work only in the context of the MSSM and it doesn't matter where the numbers in the model come from. If you want to construct a more complete model in which supersymmetry is also broken, there are these two avenues that you have to choose from. Either everything happens at low energies and you go with gauge mediation or everything happens at higher energies and you go with gravity mediation.

Is there any reason to think that supersymmetry is broken at higher energies where gravity does play a role?

It's a lot easier to build models of gravity mediation, but there are some advantages in models of gauge mediation. At the moment I would keep an open mind for the two possibilities.

Is there any string theory realisation of these ideas?

People have discussed potential realisations in string theory. I think that some of these aspects associated with these realisations are more controversial than others, such as the existence of metastable states in string theory [8–13]. However, as far as I know, there isn't really a satisfactory model coming out of string theory that has all the properties we want. In fact, to be honest, even just within field theory every model has some parts that don't work but you can always imagine that it is possible to modify it somehow such that it would work.

I want to change a little bit the topic now. Do we need the theory of quantum gravity?

Absolutely yes.

Why? What phenomena can we not explain?

Well, there are two different questions: one is whether quantum gravity exists in nature and the second is whether there is a phenomenon we are aware of for which we need quantum gravity. Regarding the first one, I think that once particles are quantum, it doesn't make sense to separate gravity, in the sense that gravity should be treated as being quantum mechanical. Regarding the second question, you could have asked whether we see any evidence for quantum gravity. In fact, fluctuations in the microwave background are evidence for quantum gravity because the standard view is that they originated as quantum fluctuations in the early universe. That's why we really need quantum gravity.

Okay. But do you think there is not any other possible explanation for those quantum fluctuations in the early universe?

I think we need quantum gravity anyway because we have quantum mechanics in the world and we have a metric that describes spacetime, so why shouldn't the metric fluctuate?

You have proposed different kinds of properties that a theory of quantum gravity should have, such as the lack of global symmetry, etc. [14]. Are there good physical motivations for these properties?

Well, first of all, just to get the record straight, I shouldn't take credit for this proposal because these are old ideas which I just restated in a more modern language. Second, there is a lot of evidence for the lack of global symmetry. This evidence has many origins, ranging from the fact that the whole story of black hole evaporation would make no sense if there are global symmetries, to various examples of theories of quantum gravity such as weakly coupled string theories, strings in anti–de Sitter (AdS), etc. Our current understanding is that there cannot be any global symmetries in theories of gravity.

If a theory of quantum gravity exists, in your opinion, what properties should it have?

Well it should have a semiclassical limit where we obtain classical gravity; it should give rise to a macroscopic spacetime like the one we observe; it should be able to accommodate gauge interactions like those present in the standard model; and it should be self-consistent. I think it doesn't sound like a tall order, but it is (laughs).

Why do you think it's so difficult to find the theory of quantum gravity?

I think there are two reasons. One is sociological; namely, people tried and failed. The second is related to the idea that spacetime should be an emergent concept and that's very puzzling. We know examples where space is emergent, like in AdS/CFT [15] or various low-dimensional string models but time being emergent, that is more difficult to understand.

Why is time not emergent in AdS/CFT?

Because time exists along the timelike boundary where we have S^3 cross time. The direction which emerges, the fifth dimension, and other dimensions, are all spacelike. Our understanding of physics relies on the concept of time which is not emergent, so it is really puzzling how it can be emergent. For example, we make predictions about outcomes of experiments and we do the experiment and then we compare it with the predictions. So there's a natural order of the sequence of events – the idea of causality that the cause comes before the effect. For that you need to have some kind of a clock that tells you what happens before what. It's hard to see how such clock could be emergent. I wrote a little article called "Emergent spacetime" [16].

Yes, I know this article. You also mention in it the issue of locality. Do you think that the theory of quantum gravity has to be local?

In some sense, certainly, yes because the world around us looks local. However, whether we have locality at even shorter distances is more puzzling especially because the notion of distance should not make any sense at shorter distances. What I think happens is that the notion of space is emergent and the notion of locality must be replaced by something else which is more sophisticated, more fundamental. I have nothing concrete to propose and if I had something I would not be giving it to you in an interview. I would sit down and write it fast (laughs).

For a theory of quantum gravity should one demand nonlocality at short distances and locality at macroscopic distances?

At macroscopic distances the theory must definitely be local – there is no question about that. At shorter distances either it will be nonlocal or there will be something else which will replace the notion of locality while simultaneously ensuring all the nice things we know about locality. Locality means that information kind of propagates from point to nearby point in a smooth way but if the notion of space doesn't exist, what is locality? Again, I don't know what notion could replace locality. For me, trying to push it further is like being a classical physicist trying to force classical language on quantum phenomena before quantum mechanics was invented. The fact that momentum and position don't commute, the fact that phase space is not classical, that you need wave functions, probabilities, etc., is not something a classical physicist could understand. It will be a big step to come up with a notion that can replace locality.

Do you think string theory is the best candidate right now for a theory of quantum gravity?

It's the only candidate.

You don't think that approaches like loop quantum gravity or Hořava-Lifshitz gravity or . . .

Even putting them on equal footing is unfair. Loop quantum gravity is an interesting trick, for which there are a small number of papers that you can probably count with your own two hands. It's not a complete solid work. In string theory I would be surprised if there wasn't a body of work of the order of 100,000 papers. Additionally, the scope that they try to explain, the extent to which they explain it and the depth of their explanations is really beyond comparison. More specifically, there are problems with all these approaches.

But someone could argue that there are also problems with string theory, right?

I think that the problems in those approaches are much worse. For example, you asked me, "What do I want for a theory of quantum gravity?" and, at the very least, I would like it to have a semiclassical limit where we obtain classical gravity. Loop quantum gravity does not have that. They don't reproduce the $1/r^2$ law of Newton. So forget quantum mechanics and relativity, do they have a $1/r^2$ force (laughs)? They don't.

I understand, but they also argue, "Okay, we don't have a semiclassical limit but we have a background-independent theory."

I think this is wrong. String theory is background-independent in the sense that they refer to. This is a catchphrase that they use, which I don't think they really mean.

Okay, could you explain this point?

So, for example, in the context of AdS/CFT, there is background independence. We specify the boundary conditions, that is, the behaviour at infinity so the answer clearly depends on the boundary, but in order to understand what happens in the interior we have to sum over all possible geometries in that interior. This is more background-independent than any notion they refer to. They claim background independence in their theory but they don't have flat space as a background (laughs). I think that loop quantum gravity is a nice way of writing gravity using gauge-invariant variables but it's not a discipline; it's a small number of papers which didn't lead anywhere and some of them have technical problems.

I still get confused when speaking of background independence in string theory since the starting point is a fixed classical background.

Well it's classical plus small fluctuations. The important question is: What happens when the metric makes large field excursions? In particular, what happens near singularities? That's an interesting question and there's been a lot of progress but still many things are not understood. However, stating that string theory is not background-independent because of that is just not fair.

Another problem with string theory is that it appears to be very difficult to describe string theory in time-dependent backgrounds. Can we describe string theory in an expanding universe?

That's an interesting question. I think there has been progress but we are far from a complete answer. Additionally, it might be the case that de Sitter spacetime is not stable. If de Sitter spacetime is unstable then we'll have to learn how to formulate the theory in a metastable or unstable state but this is not to say that it is not possible. The question of how to formulate string theory in de Sitter spacetime might be an ill-formed question if strings cannot be in de Sitter spacetime to begin with. However, if strings could be in de Sitter spacetime for a very long time and eventually de Sitter spacetime decays then we might want to have a formalism to deal with that. In any case, first we need to understand whether de Sitter spacetime is stable or not, which is a question that can be answered independently of string theory. However, in the context of string theory, the question of stability is better posed because we know things make sense in flat space and we know how things make sense in AdS so we could try exploring de Sitter.

Since you worked so much on this [17], why is it so difficult to define string theory in time-dependent backgrounds?

Because many of the standard procedures we apply in time-independent backgrounds like just thinking about the Hamiltonian and diagonalising it are not really possible. It is important to stress that these issues of how to deal with time dependence have nothing to do with string theory. Whenever we have time dependence the questions are different. In particular, we should not diagonalise the Hamiltonian when we have time dependence. This is not to say that we can't solve it, as sometimes we can but the questions are different and in the context of string theory I think the main problem is that what we compute is an S-matrix element. We formulate something in the far past, something in the far future and we ask what's the amplitude that connects the two events. We should be able to obtain this amplitude if the time dependence is weak but in general it's more difficult, especially because there are issues of backreaction on the metric. There might be curvature and maybe even singularities in spacetime that one has to deal with. If the singularity is timelike, I think we have things under control but if the singularity is spacelike or null then the problem is much more subtle because it looks like space comes to an end; that is, space evolves and then *boom: it's the end.* We do not know how to think about these situations which is a problem independent of string theory. Hopefully, string theory will give us an answer to this. Indirectly we know it does through AdS/CFT, for example, but we don't know explicitly what an observer approaching the singularity feels. How will this observer formulate the laws of physics? What will this observer measure? We don't know the answers to any of these.

What about the problem of the singularity of the big bang? You proposed a model that could maybe address this question [18, 19], right?

Yeah, I don't think that model is correct. I proposed a model that one can pass through the singularity and I no longer think it's correct. My hunch is that as we approach the singularity, the notion of space and time, in particular time, will change. We've already talked about emerging space but I think that emerging time will be very important in order to understand such questions. So the idea of what one means by time would change and therefore questions like what happened before the big bang or what happens after that might just be the wrong questions to ask because perhaps the notion of time will not be what we think it is and something else will take its role.

Do you think string theory in its current state is not yet capable of addressing problems such as the big bang singularity?

It should be but we don't know how to do it and I'm sure what is missing is a new idea, a new insight. It's not that string theory is wrong. We don't have to throw everything away, start from scratch and come up with something totally different. I don't think this is what is needed. It's slow incremental progress and some young bright kid will come and explain it all at some point. Then, we will look back and see that it all makes perfect sense. This has happened many times in the past and it will happen again.

I want to shift a little bit the topic. You also did a lot of work in non-commutative gauge theories [20, 21]. What is the relevance of non-commutative gauge theories?

To the real world maybe not so much. In the context of condensed matter physics, as in the quantum Hall effect, there might be some relevance. In the context of particle physics probably not. However, non-commutative geometry is an interesting mathematical structure. String theory is a very rich field, which includes various subjects, such as non-commutative geometry and gauge theories in non-commutative space. Gauge theories in non-commutative space are interesting because they exhibit all sorts of surprising results that traditional field theories do not. So, for example, there is an issue of whether instantons can shrink or not, related to the connection between long and short distances, which works in non-commutative space but is totally different than what we are used to. So maybe one of these surprises which can exist in a non-commutative space and is incorporated and included in string theory can give us a clue about how to think about more general phenomena.

You published the most spiritual paper that I have ever seen, called OM theory [22].

Oh . . . was that spiritual?

Yeah! What is OM theory?

Well I don't think it's one of my main achievements . . . There were two Indian co-authors, Shiraz Minwalla and Rajesh Gopakumar, who thought of it. It's a theory that does not fit the standard rules of quantum field theory and in the end it's not as complicated as the full string theory. So it's kind of pushing the envelope as far as field theory is concerned, extending things that are not quite field theory, slightly nonlocal, not too nonlocal, and it's worth exploring.

There is another theory which is a lot more common which is known as little string theory [23–28]. Little string theory goes beyond the framework of field theory. It's clearly not a local quantum field theory. There is no good formulation of it. We know it exists as a limit of string theory. String theory is very rich, has many vacua, and we can take various limits which simplify the theory. Various field theories can be obtained as such limits of string theory by removing a lot of stuff. Non-commutative geometry is a limit we can take that removes many objects and we are left with something which is simpler than full string theory and is richer than ordinary local field theory. Little string theory is yet another such beast which is different than ordinary field theory.

What is minimal string theory [29–36] and have we learnt something from studying it?

Minimal string theory is a theory like string theory but a lot simpler. It has a lot of the complexity of string theory but it has far fewer states and the interactions are much more limited. I think we learned many things because a lot of problems within minimal string theory are exactly calculable. Some of the results are non-perturbative and some are true at all orders in perturbation theory. Additionally, some of the results hold for generic string theories. Obviously, minimal string theory describes a hypothetical reality. Our world has four dimensions and these models are formulated in one and two dimensions but the hope is that studying such toy models in fewer dimensions can lead us to extract general lessons that also hold in four dimensions. Indeed, there are some general lessons that can be extracted from lower-dimensional string theories, which materialise in higher-dimensional examples. This type of approach has been recurrent in physics over the years. Physicists take toy models that are simple but which have similar features to the real phenomena they wish to describe and they study them. Examples include the Ising model, as a prototype for critical phenomena, and the hydrogen atom, which teaches us a lot about quantum states and quantum transitions. Minimal string theory should be taken in this same spirit.

Do you think it's possible to have a consistent theory of string theory in four dimensions?

In a sense we do have it because we have compactifications to four macroscopic dimensions. The remaining dimensions are replaced by some abstract field theory and are effectively observed via the existence of other particles and fields.

What do you think has been the biggest breakthrough in theoretical physics in the past 30 years?

I think that the observation that the cosmological constant is non-zero is the biggest discovery of the past 30 years. Additionally, the discovery of the W and Z bosons in 1982 and 1983 was also very important. In more generality, I think that the existence of the cosmological constant and the whole story of the standard model of cosmology, which includes dark matter, dark energy, etc., and its concordance with WMAP experiments is now a beautiful chapter in physics, which used to be in complete chaos in the 1990s.

On the theoretical side, there's been a lot of developments in string theory, starting in the 1960s where the perturbative expansion was understood as well as strings in curved space and so forth. In the 1990s the first string duality was discovered as well as the relation between all sorts of string theories [37], culminating in the AdS/CFT correspondence [15]. More generally, I would say that this whole study of string theory over the last 25 years has seen enormous progress. Understanding the theory better and understanding its connections with mathematics and gauge theories has been impressive. I think this is an important part of human knowledge which will remain important in the future.

Why have you chosen to do physics; why not something else?

I always loved it. I actually did a lot of things when I was young so I experimented with many other things but maths and physics was always what I liked the most.

In your opinion, what is the role of the theoretical physicist in modern society?

First of all, theoretical physics as it stands is good: it is good for physics, it is good for technology, etc. Second, many theoretical physicists use their skills in other fields in order to find mathematical models for explaining different phenomena. Today, even in business people try to be more quantitative and make decisions based on quantitative tools rather than qualitative views. The way computers are searching, sorting, organising information, modelling various things ranging from designing drugs to building airplanes are types of activities that theoretical physicists have contributed to and can perform well. People in these businesses know this and they like to hire many physicists who got a PhD in theoretical physics. Theoretical physicists who for some reason didn't want to get a job or couldn't get a job in academia had no trouble finding a job in Wall Street, in industry or in the computer world, etc. So I think this is an important role in society.

References

[1] S. Dimopoulos and H. Georgi, "Softly broken supersymmetry and $SU(5)$," *Nuclear Physics B* **193** no. 1 (1981) 150–162. www.sciencedirect.com/science/article/pii/0550321381905228.

[2] K. A. Intriligator, N. Seiberg and D. Shih, "Dynamical SUSY breaking in meta-stable vacua," *JHEP* **04** (2006) 021, arXiv:hep-th/0602239 [hep-th].

[3] M. Dine and A. E. Nelson, "Dynamical supersymmetry breaking at low-energies," *Phys. Rev. D* **48** (1993) 1277–1287, arXiv:hep-ph/9303230 [hep-ph].

[4] M. Dine, A. E. Nelson and Y. Shirman, "Low-energy dynamical supersymmetry breaking simplified," *Phys. Rev. D* **51** (1995) 1362–1370, arXiv:hep-ph/9408384 [hep-ph].

[5] M. Dine, A. E. Nelson, Y. Nir and Y. Shirman, "New tools for low-energy dynamical supersymmetry breaking," *Phys. Rev. D* **53** (1996) 2658–2669, arXiv:hep-ph/9507378 [hep-ph].

[6] P. Meade, N. Seiberg and D. Shih, "General gauge mediation," *Prog. Theor. Phys. Suppl.* **177** (2009) 143–158, arXiv:0801.3278 [hep-ph].

[7] M. Buican, P. Meade, N. Seiberg and D. Shih, "Exploring general gauge mediation," *JHEP* **03** (2009) 016, arXiv:0812.3668 [hep-ph].

[8] I. Bena, M. Grana and N. Halmagyi, "On the existence of meta-stable vacua in Klebanov-Strassler," *JHEP* **09** (2010) 087, arXiv:0912.3519 [hep-th].

[9] I. Bena, G. Giecold, M. Grana, N. Halmagyi and S. Massai, "On metastable vacua and the warped deformed conifold: analytic results," *Class. Quant. Grav.* **30** (2013) 015003, arXiv:1102.2403 [hep-th].

[10] I. Bena, G. Giecold, M. Grana, N. Halmagyi and S. Massai, "The backreaction of anti-D3 branes on the Klebanov-Strassler geometry," *JHEP* **06** (2013) 060, arXiv:1106.6165 [hep-th].

[11] B. Michel, E. Mintun, J. Polchinski, A. Puhm and P. Saad, "Remarks on brane and antibrane dynamics," *JHEP* **09** (2015) 021, arXiv:1412.5702 [hep-th].

[12] D. Cohen-Maldonado, J. Diaz, T. van Riet and B. Vercnocke, "Observations on fluxes near anti-branes," *JHEP* **01** (2016) 126, arXiv:1507.01022 [hep-th].

[13] J. Armas, N. Nguyen, V. Niarchos, N. A. Obers and T. Van Riet, "Metastable nonextremal antibranes," *Phys. Rev. Lett.* **122** no. 18 (2019) 181601, arXiv:1812.01067 [hep-th].

[14] T. Banks and N. Seiberg, "Symmetries and strings in field theory and gravity," *Phys. Rev. D* **83** (2011) 084019, arXiv:1011.5120 [hep-th].

[15] J. M. Maldacena, "The large N limit of superconformal field theories and supergravity," *Int. J. Theor. Phys.* **38** (1999) 1113–1133, arXiv:hep-th/9711200.

[16] N. Seiberg, "Emergent spacetime," in *The Quantum Structure of Space and Time: Proceedings of the 23rd Solvay Conference on Physics*, David Gross, Marc Henneaux and Alexander Sevrin, eds., pp. 163–178. 2006. arXiv:hep-th/0601234 [hep-th].

[17] H. Liu, G. W. Moore and N. Seiberg, "Strings in a time dependent orbifold," *JHEP* **06** (2002) 045, arXiv:hep-th/0204168 [hep-th].
[18] J. Khoury, B. A. Ovrut, N. Seiberg, P. J. Steinhardt and N. Turok, "From big crunch to big bang," *Phys. Rev. D* **65** (2002) 086007, arXiv:hep-th/0108187 [hep-th].
[19] N. Seiberg, "From big crunch to big bang: is it possible?," in *Proceedings, Meeting on Strings and Gravity: Tying the Forces Together: 5th Francqui Colloquium: Brussels, Belgium, October, 19-21, 2001*, Marc Henneaux and Alexander Sevrin, eds., pp. 281–289. Franqui Scientific Library, 2003. arXiv:hep-th/0201039 [hep-th].
[20] N. Seiberg and E. Witten, "String theory and noncommutative geometry," *JHEP* **09** (1999) 032, arXiv:hep-th/9908142 [hep-th].
[21] S. Minwalla, M. Van Raamsdonk and N. Seiberg, "Noncommutative perturbative dynamics," *JHEP* **02** (2000) 020, arXiv:hep-th/9912072 [hep-th].
[22] R. Gopakumar, S. Minwalla, N. Seiberg and A. Strominger, "(OM) theory in diverse dimensions," *JHEP* **08** (2000) 008, arXiv:hep-th/0006062 [hep-th].
[23] A. Losev, G. W. Moore and S. L. Shatashvili, "M & M's," *Nucl. Phys. B* **522** (1998) 105–124, arXiv:hep-th/9707250 [hep-th].
[24] N. Seiberg, "New theories in six-dimensions and matrix description of M theory on T**5 and T**5 / Z(2)," *Phys. Lett. B* **408** (1997) 98–104, arXiv:hep-th/9705221 [hep-th].
[25] O. Aharony, M. Berkooz, D. Kutasov and N. Seiberg, "Linear dilatons, NS five-branes and holography," *JHEP* **10** (1998) 004, arXiv:hep-th/9808149 [hep-th].
[26] O. Aharony, "A Brief review of 'little string theories,' " *Class. Quant. Grav.* **17** (2000) 929–938, arXiv:hep-th/9911147 [hep-th].
[27] A. Giveon and D. Kutasov, "Little string theory in a double scaling limit," *JHEP* **10** (1999) 034, arXiv:hep-th/9909110 [hep-th].
[28] A. Giveon and D. Kutasov, "Comments on double scaled little string theory," *JHEP* **01** (2000) 023, arXiv:hep-th/9911039 [hep-th].
[29] E. J. Martinec, "The annular report on noncritical string theory," arXiv:hep-th/0305148 [hep-th].
[30] N. Seiberg and D. Shih, "Minimal string theory," *Comptes Rendus Physique* **6** (2005) 165–174, arXiv:hep-th/0409306 [hep-th].
[31] I. R. Klebanov, J. M. Maldacena and N. Seiberg, "D-brane decay in two-dimensional string theory," *JHEP* **07** (2003) 045, arXiv:hep-th/0305159 [hep-th].
[32] J. McGreevy, J. Teschner and H. L. Verlinde, "Classical and quantum D-branes in 2-D string theory," *JHEP* **01** (2004) 039, arXiv:hep-th/0305194 [hep-th].
[33] I. R. Klebanov, J. M. Maldacena and N. Seiberg, "Unitary and complex matrix models as 1-d type 0 strings," *Commun. Math. Phys.* **252** (2004) 275–323, arXiv:hep-th/0309168 [hep-th].
[34] N. Seiberg and D. Shih, "Branes, rings and matrix models in minimal (super)string theory," *JHEP* **02** (2004) 021, arXiv:hep-th/0312170 [hep-th].
[35] D. Kutasov, K. Okuyama, J.-w. Park, N. Seiberg and D. Shih, "Annulus amplitudes and ZZ branes in minimal string theory," *JHEP* **08** (2004) 026, arXiv:hep-th/0406030 [hep-th].
[36] J. Ambjorn, S. Arianos, J. A. Gesser and S. Kawamoto, "The geometry of ZZ-branes," *Phys. Lett. B* **599** (2004) 306–312, arXiv:hep-th/0406108 [hep-th].
[37] E. Witten, "String theory dynamics in various dimensions," *Nucl. Phys. B* **443** (1995) 85–126, arXiv:hep-th/9503124.

26

Ashoke Sen

Distinguished professor at the Harish-Chandra Research Institute, in Allahabad, India

Date: 17 February 2011. Location: Allahabad, India. Last edit: 7 December 2020

In your opinion, what are the main puzzles in theoretical physics at the moment?

Here I will take a narrower sense of theoretical physics and focus on that part which studies the fundamental constituents of matter and their interactions. From that viewpoint it seems to me that the main problem is trying to understand quantum gravity and finding a unified description of quantum gravity and everything else.

For many years now people have tried to quantise gravity and there are many theories in the market. Why is gravity so difficult to quantise?

There are some technical problems which are related to the ultraviolet divergence and renormalisation. Every field theory has certain divergences which come from very high-energy modes of the field but in conventional field theories one knows how to take into account those divergences and get finite results. In quantum gravity the problem becomes serious because high energy also means that gravity becomes strongly coupled since energy is directly the source of gravity. So these divergences are associated with the fact that high energy means strongly coupled gravity. That is one of the reasons why gravity cannot be quantised using the usual rules of treating these divergences in normal quantum field theories. This is a technical problem, but string theory can address this particular problem successfully. In string theory there are no ultraviolet divergences, which is due to the fact that the strings automatically imply a short distance cutoff due to their finite size. Besides this there are various conceptual problems in quantum gravity which string theory has only been able to address either partly or not at all. One problem is the problem of time: in gravity there is no distinction between space and time; that is, there is no preferred time, whereas when we do quantum mechanics we always use a preferred time. String theory still doesn't have a proper answer to this question. Then, there are problems with black holes which again string theory has only been able to address partially in special cases.

You mentioned that string theory can deal with the technical issues of divergences. Is superstring theory ultraviolet finite at every order in perturbation theory? Does superstring theory also solve the problem of infrared divergences?

Yes, superstring theory is ultraviolet finite and this has been known for a long time [1]. The scattering amplitudes of superstring theory are expressed as integrals over *moduli spaces* of Riemann surfaces – an abstract space that labels two-dimensional surfaces of different intrinsic shapes. This is a finite-dimensional integral, and the only source of divergence is from the blowing-up of the integrand. This happens near the boundary of the integration domain where the Riemann surfaces become singular. Such singularities have been long classified by the mathematicians and when one translates it back to physics, one finds that all such singularities arise from the region where the distance between two interaction points in a scattering process becomes infinitely large. This is long distance, i.e., infrared divergence, and not short distance (ultraviolet) divergence.

Now infrared divergences are quite common in quantum field theories, and they typically arise when we are doing something wrong. For example, if the effect of interaction makes a non-trivial change in the quantum ground state and we do not properly take this into account in our calculation, or if the interaction changes the masses of the elementary particles and we do not take this into account, we encounter infrared divergences. But quantum field theories have a well-defined procedure for making these corrections and removing the infrared divergences. The conventional formulation of string theory does not have this in-built mechanism for removing infrared divergences, but it is generally expected that using techniques borrowed from quantum field theories, one should be able to remove these divergences. Recently, progress has been made on this front by reformulating superstring theory as a quantum field theory. This quantum field theory has no ultraviolet divergences as in string theory, and has the usual in-built mechanism for removing infrared divergences [2–6].

You mentioned a few problems that string theory has not yet fully addressed such as the problem of time. Can you elaborate a bit more on why there is no clear concept of time in a theory of gravity but there is in a quantum theory?

The point is that in a theory of gravity because of general coordinate invariance you can always change your definition of time by mixing it with spatial coordinates: you can make a coordinate transformation to define a new set of time variables and in this new set of time variables the evolution will be quite different. Moreover, the definition of the vacuum, which is the state with lowest energy in a quantum field theory, is not clear since to define energy you need to have a concept of time because energy is the conserved charge associated with time translation symmetry. If there is no well-defined notion of time then we do not know what we mean by vacuum.

The other problems you mentioned concerned black holes. What kinds of problems do you have in mind?

Based on the classical understanding of black holes what we see is that black holes in some sense divide the spacetime into two parts: part of the spacetime inside the black hole cannot communicate with the part sitting outside the black hole. As a result, the usual rules of quantum mechanics, such as unitary evolution (that given the wave function at a given time you can follow its development all the way into the future) seem not to hold in the case of black holes. At least apparently, if we just look at the classical black hole solution, there is a loss of that unitary evolution because part of the spacetime is disconnected and you can never recover information from that part of spacetime. Now, it is generally believed that once you take into account the full quantum gravity this problem will cure itself.

Is this a statement of the black hole information paradox [7, 8]?

Yes, if we just naively follow the classical intuition of a black hole it looks like unitary evolution is lost but then it is also clear that you cannot use this semiclassical intuition in a full theory of quantum gravity. But how exactly quantum gravity repairs this is still an unsolved problem.

Okay, but what about the fuzzball proposal, which says that these parts are not really disconnected; instead, there is simply no space behind the black hole horizon [9]?

I think that right now this proposal is in a very rudimentary form and it is not clear that it can solve the problem. One of the drawbacks I see in the fuzzball proposal is the following. There is a geometric notion of the black hole entropy; that is, if we look at the classical black hole it produces a certain entropy. Now, the fuzzball proposal says that there are no real black holes, that you only have some classical solutions or classical configurations, and when you quantise them you recover the microstates of the black hole. However, this does not explain why the number of such classical configurations – more precisely, the number of states you get by quantising these configurations – is precisely the exponential of the area of the black hole horizon. I feel that any complete theory of quantum gravity must explain this: Why is it that this highly quantum notion of entropy (counting the number of quantum states and taking the logarithm of that number) is given by the classical result like the area of the event horizon? If you cannot explain this then you haven't really explained anything about black hole thermodynamics.

Okay, but before we go deeper into this problem, can you tell me what is your view on string theory, if you see it as a candidate for a theory of quantum gravity or more like a tool that can be used to solve other problems of physics?

Well, I certainly think of it as a candidate for a theory of quantum gravity. String theory was born for a different purpose but I think that one of the reasons why many people became interested in string theory since the mid-1980s is because it appeared as a candidate for a unified theory of both quantum gravity as well as all other fundamental interactions. In my mind that still remains the basic reason for doing string theory. It is true that string theory has also been used as a tool for addressing various other problems of theoretical physics but I think of this as a bonus. String theory has these other features, but for me that is not the real motivation for doing string theory. The motivation is really to understand a theory of quantum gravity and possibly of everything else.

I see. You mentioned that string theory wasn't meant for this; what was it meant for, then?

Well, string theory was originally devised for explaining strong interactions – not to explain gravity [10–12]. And the attempts to explain strong interactions using string theory failed. It never worked out the way people thought it would work. Now of course it has come back again but that is a different version [13, 14]. Initially, attempts to use string theory for strong interactions had various problems; it had a massless spin-two particle which was not part of the strong interaction spectrum. On the other hand, this was precisely why string theory become a candidate for a theory of quantum gravity. In a sense, string theory failed as a theory of strong interactions because string theory naturally incorporated gravity.

But why consider a string instead of a point particle in the first place? Why not a sphere or something else?

People have tried to formulate theories of extended objects. Naively, you would think that strings are successful because there is an intrinsic size and that makes the theory better.

Yes, but you could have taken any finite-size object ...

Yes, you could have taken anything else. So this is a very naive way of saying why string theory solves this problem. It is true that the finite size of the string helps but it requires a lot of mathematical steps from there to show that string theory is a fully consistent theory. This has not been achieved for other objects. Nobody has been able to take a theory of membranes, regard this as a fundamental object and build a fully consistent quantum theory [15, 16].

Well to me a sphere looks a bit more beautiful than a string, but maybe that is a harder problem?

(laughs) I think that people have made serious attempts to quantise what we call membranes and we also know now that membranes do appear as part of string theory, but to formulate a theory just based on membranes seems very hard [15, 16].

Do you think that it is important to try to establish other theories of quantum gravity or do you think that the whole community should work on string theory?

No, I certainly think that it is important to follow different approaches and if somebody comes out with a better theory of quantum gravity, with something that has a better chance of being the correct theory, then probably most people will start working on that.

But don't you think that what normally happens when someone tries to work on an idea outside the mainstream framework is that that someone is put aside by the community?

I don't think that they are put aside. Okay, maybe there is some kind of social phenomenon but it's more that from experience one finds that it is very hard to come up with a proposal which is fully consistent.

Right, but string theory has received many criticisms, the strongest of them being that it seems not be a verifiable theory.

I agree, but at least the internal consistency is something that string theory can rely on. In any attempt at finding a theory of quantum gravity, or in any attempt to find any theory at all, the first thing one must check is that the theory should be internally consistent. If there is something that we have learned from the whole history of modern science, it is that the language of mathematics is the right language to describe physics. So if something is not mathematically consistent it is hard to imagine how it could have anything to do with nature. Given that it is very hard to find a consistent theory of quantum gravity based on a conventional quantum field theoretic approach, and that string theory manages to do that, there is a strong indication that string theory is on the right track.

Okay, but what about theories such as loop quantum gravity, causal dynamical triangulations, non-commutative geometry, etc.?

Those approaches should be pursued. On the other hand, my feeling is that string theory is far ahead with respect to these other approaches, but maybe I am biased.

Sure, but couldn't you say that this is because there is 85% of the community working on string theory and 15% on the remaining ones?

True, but it is also true that before the mid-1980s very few people were doing string theory and the main reason why so many people started working on string theory was this powerful internal consistency of the theory. So, I think that the same thing will happen to these other theories if there is a major breakthrough, if there is major evidence that one of them is the right theory of quantum gravity, or has a better chance than string theory of being the right theory. If that happens, I am sure many people would be interested.

I see. So, you think that there is no other such theory out there that fulfils these requirements?

Yes, exactly. There hasn't been any such major breakthrough which indicates that one should approach gravity in a different way than the one set by string theory.

But has string theory made any prediction at all? Maybe something that we could observe, for example, at the LHC?

Well, the main problem of making predictions in string theory is not that in string theory you cannot calculate anything. The main problem seems to be that string theory has these many different phases or what we call vacua.

Just out of curiosity, how many vacua are we talking about?

Probably 10^{500}, more or less (laughs) [17, 18]. String theory is an internally consistent theory and because of that, in any one of these phases, you can in principle calculate everything. If you fix your attention on one particular phase you can predict particle masses, you can predict their interactions and everything. The problem is to find which phase of string theory describes our world. Now, string theory does have some phases in which there would be observable signatures at the LHC [19]. For example, there are a few phases that we can identify which have large enough extra dimensions [20, 21].

Okay, you mean that there is only a subset of all this vacua which could give some observable predictions at the LHC?

Yes, exactly, a very small subset, which you can identify by searching for vacua with large extra dimensions, for example. However, they are so unlikely that I find it very hard to believe that we actually live in such vacua. They are very unnatural from many perspectives. So, from that viewpoint I think it is very unlikely that we will see any signature of string theory at the LHC but for all I know we could be very lucky; we might live in such a vacuum and that it will show up in the LHC.

Right, but do you feel comfortable with the fact that string theory predicts all these extra dimensions; do you think that nature has all of those?

Well, I think that if string theory is the right theory then the world does have all these extra dimensions and it is surely consistent with what we know, because these extra dimensions have a small size. I think that the outstanding puzzle in string theory is not why we have small extra dimensions, but: Why are we living in such a big universe? Why does the standard model exist at an energy scale much smaller than the Planck scale? To explain why something is of the order of the Planck scale is very easy in string theory because naturally everything should be of the order of the Planck scale, which is the scale of quantum gravity. So, in this sense I don't find Planck-size extra dimensions unnatural because that is the natural scale.

Okay, a lot of the string community has been focused on the study of supersymmetry. What is supersymmetry exactly and will we be able to observe it in the near future?

First of all, supersymmetry is a hypothetical symmetry that relates fermions and bosons [22, 23]. If it is an exact symmetry then that will say that for every bosonic particle in the theory you should find a fermionic partner with the exact same mass. This, of course, is not observed in the universe, so supersymmetry has to be broken in some sense. There are various mechanisms for breaking supersymmetry [24–26], and it is still possible that we will find a signature of supersymmetry at the LHC [27]. In that case superpartners have a larger mass than what we see for the corresponding observable particles.

Now, in string theory supersymmetry is natural in the sense that a consistent string theory requires supersymmetry but typically it is also natural that supersymmetry is broken at the Planck scale (at the very high scale) ...

Sorry, why is it natural?

Because that is the natural scale for everything in string theory. So although supersymmetry is an intrinsic part of string theory, if it is broken the natural scale of breaking it is at the Planck scale. If supersymmetry is broken at an energy scale smaller that the Planck scale, you have to explain why this is so. In this sense I don't think that string theory predicts that you should see supersymmetry at the LHC. Now, there are other reasons why people expect that maybe we should observe supersymmetry at the LHC. These are related to the fine-tuning problem: that without supersymmetry it seems hard to explain why the masses of the standard model particles are so low but with supersymmetry there is a partial explanation (that if you somehow make them low in the classical limit then they will remain low). That's one of the reasons why people like supersymmetry. Another reason is the unification of the coupling constants: you find that the strong, weak and electromagnetic couplings get unified at high energy, which you don't see without supersymmetry. So I would say that if supersymmetry is found at the LHC it's good for particle physics; it's new physics after all, but by itself it will not say if string theory is right, and if we don't find it, it will not say that string theory is wrong.

So wouldn't you say that not finding supersymmetry would be a major blow for string theory?

I don't think so because the point is that supersymmetry at the LHC scale is not a definite signature of string theory nor is absence of supersymmetry a definite signature against the theory. However, I would say that supersymmetry is good for theoretical physics. Maybe I can put it in this way: supersymmetry is a symmetry that was predicted based on theoretical reasoning, and now if it is found through experiments it will be a major boost for theoretical physics or for this kind of mathematical reasoning, but not directly for string theory.

I see. Now, going back to this point of breaking supersymmetry ... when we formulate string theory without supersymmetry the theory predicts that we live in 26 dimensions but when you include supersymmetry you find 10 dimensions. In how many dimensions do theories that break supersymmetry live?

If you start from string theory and try to formulate a non-supersymmetric string theory, which is called a bosonic string theory, then that is naturally formulated in 26 dimensions. However, at least so far, no one has been able to make it into a consistent theory because it has a tachyonic mode and nobody knows what to do with it. This is a clear indication that at least the 26-dimensional vacuum is unstable and it is not clear if there is a stable vacuum in the theory at all. So in the conventional wisdom of string theory, that has not been considered a consistent theory. Only five consistent theories which are supersymmetric in 10 dimensions are regarded as genuine, consistent string theories.

Now when we talk about non-supersymmetric theories we mean the ones where you start with 10 dimensions and supersymmetry gets broken by the time you compactify down to four dimensions and here the natural scale of supersymmetry-breaking is again of the order of the Planck scale.

You mentioned five theories. What are the differences between them?

There are five different ways that strings can vibrate in 10 dimensions maintaining mathematical consistency. Initially, when string theory was formulated these were thought of as five different string theories. What was found later was that these five different theories are actually five different views of the same theory, and there is only one underlying theory [28–30]. You can think of this as a room with five windows. Suppose that you could look into this room through any of these five windows and from each window you have a good view of only a part of the room. In that case a person with normal vision may not realise that the different windows look into the same room, but a person with sophisticated technology may be able to discover this. In the same way, there are not five different theories but there are different ways of looking at it; it requires advanced mathematical techniques to discover this. These different viewpoints are related by some duality symmetries.

Now changing a bit the topic. Some of your recent research has been focused on the study of black holes. Why are these objects so interesting in the first place?

Black holes in essence provide a testing ground for any theory of quantum gravity. First of all, black holes are solutions in general relatively and they are unusual solutions in the sense that they separate our spacetime into two parts, where one part cannot communicate with the other. This makes it hard to apply the usual rules of quantum evolution. If we just think of a quantum theory on the background of black holes there are many puzzles and any theory of quantum gravity has to address these puzzles. This is the reason why understanding black holes in string theory is a good testing ground for showing whether string theory is really able to formulate a consistent theory of quantum gravity.

But has this study so far enhanced our understanding of quantum gravity somehow?

Yes, it certainly has to some extent, not to the extent that we would have liked, but the realisation that you can actually explain the entropy of a black hole in terms of the counting of states, which was done in the context of string theory starting with the work of Andrew Strominger and Cumrun Vafa [31], has certainly given a big boost [32]. This involves two completely different calculations: on one side you calculate the area of the event horizon, which is a purely geometric calculation, while on the other side you forget about the event horizon, you forget everything about geometry, and just try to identify what are the possible quantum states carrying the same charges that the black hole was carrying and you simply count their number. These two calculations agree: the logarithm of this number is precisely the area of the event horizon [33–42]. I think that this is certainly big evidence that string theory is on the right track but what we would like to do is to try to understand these quantum states directly in a geometric language. What has been done so far was to use supersymmetry as a tool to switch off the effects of quantum gravity and then do the counting of states. Because of supersymmetry you can argue that this effect of switching off quantum gravity has no effect on the counting and that is the reason why you can compare these two results. However, to truly understand quantum gravity from string theory what one needs to do is not to switch off quantum gravity but instead to just directly tackle the problem head-on: formulate string theory in the regime where quantum gravity is important, count the number of quantum states and then compare with the black hole entropy. That still has not been done and that I think would provide all the key insights into what black holes really are.

But this counting should be done with supersymmetry or without supersymmetry?

Even if you can do it with supersymmetry I think it would be a lot of progress because you have the full effective gravity in place. Supersymmetry may provide some calculational simplicity but if you can count the number the states with the full effective gravity in place I think that would show big progress.

Most people who work on these problems, as you said because it facilitates computations, always use supersymmetry. But wouldn't you say that it would be more realistic not to consider supersymmetry?

Yes, it is certainly more realistic to consider problems where supersymmetry is absent and I think that there has been a lot of progress in understanding non-supersymmetric black holes in the context of the AdS/CFT correspondence [13], where you can actually compare results which are beyond supersymmetry [32]. However, you cannot compare results very well in the sense that you always have some constant factors which you cannot calculate completely from first principles because you have to continue from a regime where gravity is strong to a regime where gravity is weak. But I think that qualitatively there has been a lot of progress in the understanding of the relation between classical black hole thermodynamics and the quantum description of gauge theories, even in the context of non-supersymmetric black holes.

But do you think that string theory will be able to give an answer to all these questions?

Yes, in principle, string theory has the tools for doing that. I mean, string theory is a complete theory so in this sense it should be able to provide answers to all these questions. What is happening is that right now it seems technically too complicated to try to address the full issue head-on. That is the reason why we proceed in small steps. We know that for supersymmetric situations we at least obtain the right answer when we switch off gravity and do the counting. Now, we should try to understand how to get the same result in the context of supersymmetric theories but with gravity in place. Once you understand how to do that calculation and when you know that the results agree then we can try to generalise it to the non-supersymmetric case [43]. If you want to solve the hardest problem in the first attempt you might be lucky and you may be able to solve the hardest problem, but it's always easier to proceed step by step, see what we have got and then try to make progress from there.

This comparison between the calculation on one side of the classical black hole entropy and on the other side of the counting of quantum states is based on the dual description of D-branes. But this is still a conjecture, right?

Yes, duality symmetries in string theory are conjectures. But I would say they are firm conjectures because there are so many ways that duality symmetries could have failed, and they have passed every single test. When duality symmetries came up they led to many predictions, things that you could calculate on both sides and compare. For example you compare huge integers which you can calculate completely independently from both sides, and they agree.

Okay, but don't you think that is just people trying to adjust their calculations so that they fit into the duality picture (laughs)?

No, it's not. You have no choice; you have some mathematical functions and you predict from completely different calculations what their Fourier expansion coefficients should be and you find that each of those Fourier expansion coefficients is exactly what you get from the other side. That still doesn't prove the duality because to prove it you have to show that every calculation you can do from the two descriptions matches exactly and that certainly has not been proven. The fact that it has worked in so many cases makes it a very strong case. It tells you with very high probability that duality is a symmetry of string theory. Now, we accept duality as a symmetry of string theory and simply use it to make further progress.

Would it be interesting to prove it exactly?

It would certainly be interesting to prove this but I don't really see it as a big obstacle to further progress. In any formulation of string theory one makes use of perturbation theory since one cannot solve the theory exactly. Even in quantum field theory which is a much simpler subject one does not know how to solve the theory exactly; you always use some approximation scheme and so does string theory. What this duality symmetry does is to give you new ways of approximating. An approximation scheme which would be valid in a certain domain but breaks down in some other domain can be replaced using duality symmetry, which tells you that there is another approximation scheme where you use a different string theory to study that regime. So it is more like an advantage, a tool that allows you to study string theory from another perspective and because this tool is available you can do a lot more things now than you could before, when you were limited to do perturbation theory around one particular point.

What do you think are the necessary steps to prove this duality symmetry?

I think that the first step missing in proving this duality symmetry is a full non-perturbative formulation of string theory: you first have to write down something like a path integral formulation for string theory just like you write down the path integral formulation for a quantum field theory. But that is only the first step. If we think of the $N = 4$ supersymmetric Yang–Mills theory, which is a conventional field theory, you can write down a path integral that describes the theory. This theory is supposed to have a self-duality (the strongly coupled theory and weakly coupled theory are the same) but even that has not been proven. So, to prove this duality symmetry you have to solve the theory exactly in some sense, and once you have solved the theory exactly then of course you can compare and study if the theory has this symmetry or not.

But do people believe that it is possible to solve this theory exactly?

I find it hard to believe that there would be an exact solution of string theory. Even exactly solved quantum field theories are very rare, so it is hard to see how string theory can be exactly solved. So, in this sense duality may not be proven exactly but nevertheless once you have sufficient amount of data you can use it more like a tool to make predictions.

Okay, now going back to the problem of counting the black hole microstates and comparing it with the classical result ... this computation has made large use of the quantum entropy function, a tool that you have developed [33–35]. What is exactly the purpose of this tool?

The problem is the following: when you compare the entropy of the black hole that you calculate geometrically and the microscopic entropy, what you do is that on the black hole side you calculate the area of the event horizon (and divide it by four times Newton's constant), which is the classical Bekenstein-Hawking entropy, and that you compare with the logarithm of a certain degeneracy of microscopic states that you calculate from a different perspective where you quantise a system of branes. Now, on the microscopic side this computation is completely well-defined; you can count the number of quantum states to the last digit – there isn't in principle any difficulty in the counting procedure once you fix the charges. The question now is: How well can we reproduce this result from the gravity side? To reproduce this result from the gravity side you have to go beyond the classical Bekenstein-Hawking formula because this formula is valid as long as gravity can be treated classically – that is, the quantum gravity effects are small – and the horizon is weakly curved. This happens when the charges that the black hole carries are large; that is, in the limit of large charges this formula becomes exact. So, if you want to focus just on the Bekenstein-Hawking entropy then you should compare with the corresponding result from the microscopic analysis in the limit of large charges, and the results agree. But the microscopic result is also available in the case of finite charges so then one would like to do better; one would like to ask if from the black hole side we can calculate these corrections for finite charge and see if the results agree.

This may look like a technical problem but there is an important underlying issue here which I'll now explain. Sometimes we can think of gravity as an emergent theory; in particular, because of the AdS/CFT correspondence and various other situations we see that there is a way of defining gravity just in terms of a dual description as a gauge theory. So, you think of gravity as an effective theory (perhaps not all string theorists share this view but many do); you don't think of gravity as a fundamental theory at all. On the other hand, we know that there is an underlying string theory in which the graviton emerges as one of the quantum states and as far as we know this string theory is good as a fundamental theory. Now you want to test whether that is really true or not, whether gravity is only an effective theory or if there is really an underlying fundamental theory behind gravity. In order to test it we want to see if gravity has the ability to reproduce the microscopic results exactly. If gravity was emergent in some limit then gravity would only be able to explain these microscopic results in that limit, where it emerges, in the limit of large charges where the classical gravity approximation is valid. For this you need to test if you can carry out this calculation in quantum gravity, get the corrections to the Bekenstein-Hawking entropy and compare the results with whatever you get from the microscopic side. So for this purpose it is important to understand what are the quantum corrections to the entropy, and when quantum gravity effects are large you should be able to correct the Bekenstein-Hawking formula.

Okay, but aren't these corrections supposed to be captured by Wald's entropy formula [44]?

Well, Wald's formula gives the classical corrections; it gives the corrections when the stringy effects are important but when quantum gravity effects are still ignored. In string theory there are two different kinds of effects. One effect comes from the fact that string theory is not exactly gravity; instead, it reduces to gravity at low energies. Even classical string theory has corrections to Einstein's equations and those corrections are captured by Wald's formula because Wald's formula is still classical; it takes into account higher-derivative corrections to the classical action, but it does not capture the quantum corrections.

But when do these quantum corrections become important?

The quantum corrections become important when the effective string coupling becomes significant at the horizon. There are two different expansions; both are determined in terms of the charges. How important the higher-derivative corrections are and how important the quantum corrections are are determined by different combinations of the charges. In some limit of charges you find that Wald's corrections are important but quantum corrections are negligible, and in some other limits both corrections are important. If you want to test whether gravity makes sense as a quantum theory you have to test the equality between the microscopic results and the gravity results in the domain in which quantum gravity corrections are important. In that domain it is important to know how to calculate quantum corrections to the entropy, not just Wald's corrections but also quantum effects. That is the goal of the quantum entropy function: it is a generalisation of Wald's formula and reduces to Wald's formula in the classical limit, when quantum corrections can be ignored, but it is a more general formula which tells you in principle how to also calculate quantum corrections [38].

You said "in principle"; has it been shown to be different?

Yes, it has been shown to be different, because it can predict, for example, some results about the microscopic spectrum which you couldn't get from Wald's formula. Normally in the microscopic theory you calculate the total number of states which is what you compare with the macroscopic entropy, but imagine that you wanted to calculate something different. Imagine that the microscopic theory has some discrete symmetry whose associated charge can take *n* different values and you want to count the total number of states weighted by the discrete symmetry charge. As an example, consider a situation where a given state can be either even or odd under a symmetry and you want to count the number of states which are even minus the number of states which are odd. If you want to calculate this number from the black hole side and try to apply Wald's formula you will get zero. However, the microscopic answer is non-zero. Now, using the quantum entropy function you can calculate what this answer should be and you find that the leading term that you get from the quantum entropy function matches with what the microscopic result gives. So, in this sense it is a test that gravity contains information beyond what the classical Wald's formula gives you.

You have also used the quantum entropy function to calculate quantum corrections to the entropy of Schwarzschild black holes [43]. How does this compare with calculations performed within the framework of loop quantum gravity (LQG)?

The quantum entropy function allows us to calculate the extremal black hole entropy beyond the Bekenstein-Hawking-Wald formula. In particular, one-loop quantum corrections involving massless fields give corrections proportional to the logarithm of the area. I used this method to calculate the logarithmic correction to the entropy of a variety of known supersymmetric black hole solutions in string theory. In many of these examples the exact microscopic result is also known, and by studying its behaviour for large charges carried by the black hole, one can extract the log corrections. In every case that was tested there was exact agreement between the gravity-side calculation based on quantum entropy function and the result from the microscopic side. This is just another test that string theory is an internally consistent theory.

One can generalise this analysis to also calculate log correction to the entropy of non-extremal black holes. I used this to calculate the log correction to the entropy of Schwarzschild black holes. Of course in string theory, there is no detailed microscopic understanding of Schwarzschild black hole entropy yet, and so this result cannot be tested against any microscopic results. There is a claim in the loop quantum gravity literature that they have a microscopic understanding of Schwarzschild black holes. There the leading result for the entropy is proportional to the area but the constant of proportionality is a free parameter of the theory and hence cannot be tested against the Bekenstein-Hawking result.

However the claim is that the coefficient of the logarithmic (area) term can be computed exactly. I compared my result for the logarithmic correction, which is based on the one-loop correction due to the graviton loop, with the answer obtained from the microscopic analysis of loop quantum gravity, and the results do not agree [43].

Do you agree with the claim made by Carlo Rovelli (page 463) that this tension between the two results has been resolved?

My understanding is that in LQG the leading-order entropy calculation does not give $A/4$, but the constant of proportionality depends on the Barbero-Immirzi parameter. In contrast, string theory gives $A/4$ whenever we can carry out the microscopic counting reliably.

Similarly, in all string theory examples, whenever one can calculate the logarithmic correction to the entropy from the microscopic and the macroscopic viewpoint, the results match perfectly. I have not seen any similar comparison in LQG. The results quoted in LQG just give one answer, and they do not agree with the semiclassical result for the logarithmic correction to the entropy. So I think the onus is on them to explain how the results they quote could be related to the semiclassical results. If the claim is that their result counts only the horizon degrees of freedom and not the contribution from outside the horizon, then my question would be how, in an interacting theory, can one systematically separate these two contributions to the entropy? And what computation in LQG will give the full result?

You mentioned earlier that with such types of calculation one could test whether gravity is a fundamental force or isn't but I'm still confused how that works exactly. Can you elaborate on this a bit more?

Well, it depends on what you mean by gravity as a fundamental force ...

I mean whether gravity emerges from a theory that does not contain gravity or not ...

Okay, so first let me explain this notion of duality. The whole idea of duality is that there are different formulations of the same physical theory. In one formulation, one set of degrees of freedom would appear as fundamental and everything else as emergent, but in the dual description, part of what was emergent in the previous description becomes fundamental and vice versa.

Okay, but then how do we know which picture is the true picture?

I don't think that there is a very nice answer to that question; it depends on which description you are using. So, when you ask if gravity is fundamental, the correct question to ask is not whether there is some description in which gravity is emergent. It is quite possible that there will be some description of the theory where gravity will be emergent. What you should really ask is: Is there any description where gravity is fundamental? Not if gravity is fundamental in every possible description. That's what I mean by asking whether gravity can be fundamental. We certainly know of examples where gravity is emergent. In the

AdS/CFT correspondence [13] one way of describing the theory is as a gauge theory so from that perspective gravity is certainly emergent. However, one believes that the same theory can be reformulated as a string theory in anti–de Sitter (AdS) space. If that makes sense, if you can define a perfect consistent theory of strings in AdS, then in that perspective gravity will be fundamental. Now, you can take two different viewpoints. You can say that the gauge theory is the only way of describing this theory while the dual string theory does not fully make sense and so gravity is necessarily emergent. However, if both descriptions are equally good in appropriate regimes then I would say that gravity is also fundamental. That is what we are trying to test here, whether or not there is one description of black holes in which gravity is regarded as a fundamental interaction (because you are using a theory of closed strings).

Does this make any change to our way of viewing space and time?

Not directly but I think that if you try to push this line of argument even further, that is, if you try to understand the full thermodynamics of black holes from the gravity side after reproducing some results like corrections to the entropy, it will ultimately require you to address all these fundamental questions such as how to define time.

So you think that it will be possible to define time precisely on the gravity side?

Yes, exactly. It is true that in the supersymmetric case there is a dual description of black holes in terms of quantum states of some supersymmetric system. These quantum states are different excitations of D-branes and various other objects that exist in string theory such as solitons and in some cases also fundamental strings. So the purpose of analysing these supersymmetric systems is that since you know the exact answer from one computation (which does not involve gravity), you can try to understand in detail how those results can be reproduced from the gravity side. We are trying to reproduce known results using a different technology where gravity is treated as fundamental and once we are able to do this, the next step is to generalise it to situations where the other method of computation cannot be used.

One of the things that makes people nervous is that, at least classically, black holes have a singularity, which seems to indicate that there is a problem with general relativity and that a consistent quantum gravity theory should somehow solve it. Does this dual description somehow get rid of the singularity?

No, it does not remove the singularity of the black hole, but in the extreme limits, which are the simplest to analyse, if we just focus on the horizon we do not see the singularity. So, there is the hope that in order to understand the entropy of the black hole we do not have to address the issue of the singularity.

Sure, but this does not answer the question of what happens to someone falling inside a black hole up there in the middle of the stars, does it?

True, it doesn't address directly the problem of the singularity, but what we are trying to do here is to understand one aspect of the black hole problem which is explaining the black hole entropy. It is not clear that these two problems are interconnected because the entropy seems to be a property of the horizon and not of whatever is inside. It is true that black holes have both of these unusual features (first there is an event horizon and then a singularity) but there are black holes which have an event horizon and an avoidable singularity (like the Reissner-Nordstrom black hole); that is, the singularity is timelike instead of being spacelike. This indicates that perhaps the existence of a singularity is not directly related to the existence of an event horizon and the associated problem of black hole entropy.

Okay, but do you think that string theory will be able to resolve the singularity problem?

Well . . . if string theory is the correct theory of quantum gravity then it better give a solution to all problems (laughs), no matter how easy or difficult the problem is. If we had an infinitely powerful understanding of how to compute things in string theory we would probably solve all problems at once but given our limited ability, we have to, basically, isolate specific problems and see how string theory addresses those problems.

I have already mentioned it before, but now it seems to be of more relevance since it could solve both problems at once: What about the fuzzball proposal? Couldn't we do these same exact computations considering fuzzballs instead?

My problem with the fuzzball proposal is the following: fuzzballs are classical solutions to general relativity coupled to other things, but so are black holes; so, I would think that if there are different classical solutions of the same theory all of them would contribute to the entropy. It is hard to see why some classical solutions would not be allowed and others would be; hence, my view of the fuzzball proposal is that if the fuzzballs are there then their contribution should be added to the entropy of the black hole.

But in these cases that you have studied haven't you seen that you didn't have to add their contribution to the entropy?

Let me be specific on this point. The context where fuzzballs have been studied in great detail is what we call two-charge black holes [45]. These are "black holes" because if we just think in terms of classical Einstein's gravity then these are singular solutions, but you can argue, by using Wald's formula, that once you consider stringy corrections you can get finite-area event horizons. However, this happens only in some descriptions, not in all possible descriptions. The reason for this is that Wald's analysis is classical but what is classical and what is quantum changes under duality transformations: some effects which are classical in one description become quantum in the dual description and vice versa. Now, what happens is that in some description there are fuzzball solutions (you can count how many of them are there) but in that description you don't find any black holes. In that description if you try to construct a black hole solution using the classical action you will find that the entropy is still zero. However, in the dual description, where Wald's formula makes the black hole into a finite area event horizon, you find that there are no fuzzball solutions; that is, there are no smooth classical solutions. So, for this particular system (the two-charge system), what seems to be happening is that in any given description of the system you either have fuzzballs or black holes but not both simultaneously. In this sense it is not contradictory with the idea that if there are both you have to add them to get the right entropy result.

Right, now changing a little bit the topic. In your opinion, what was the biggest breakthrough in theoretical physics in the past 30 years?

In the past 30 years it is hard to say. If you had asked for in the past 40 years it would be certainly the standard model. I think that in the past 30 years perhaps the breakthrough has been the understanding that we can address issues in quantum gravity. I'm not saying that there is a genuine breakthrough in the sense that we have solved the problem of finding a theory of quantum gravity but the realisation that string theory is a very natural candidate for unifying quantum gravity and the rest of the forces is a breakthrough. Maybe this is my narrow point of view; I'm thinking of theoretical physics only in this context, that you want to understand the basic constituents of matter. This realisation came before the past 30 years; it came in the late 1970s but I think that the major developments have all taken place during the last 30 years: that you have a candidate which has the chance of explaining, in principle, all that we know of.

Do you think that people will be using string theory 100 years from now?

(laughs) It is hard to predict what will happen 100 years from now but I strongly believe that string theory is on the right track and that it is the theory of quantum gravity. How exactly it will shape up in the next 100 years I have no idea, but the very fact that it has this internal mathematical consistency and the fact that gravity comes out so naturally makes it an ideal candidate for quantum gravity. If it is not the right theory then it is hard to see why string theory existed in the first place. Why is there a theory like string theory that seems to contain gravity so naturally and yet is not the correct theory of quantum gravity? So if you ask me to bet, I would say that string theory will still be there (laughs).

Why have you chosen to do physics? Why not fishing?

I think it was partly accidental. When I decided to take up physics in my undergraduate days I really had no idea of what physics was about. At least in Bengal, where I grew up, physics was considered one of the subjects that people should go into; most of the good students went into physics and that's how I started. Then, of course, I got interested in physics but if I had gone into some other field I would be doing something else.

I think that everyone would agree that it was a very good accident (laughs).

(laughs)

When physicists speak about their work many people criticise it, saying that it might be a bit useless for society. Do you think that this is reasonable criticism?

At present I think it is not very useful. It is hard to imagine how string theory will have any application in the next 50 years but if we think of the history of human development, every correct idea in theoretical physics which actually formed the foundations of our knowledge has found an application in some way or another. When quantum mechanics was being developed it was very hard to imagine what possible application it could ever have but now we know that almost all modern technology that we have around us is based on quantum mechanics. So, if you ask me now, I cannot imagine any way string theory will have an application in real life but probably the developers of quantum mechanics would have answered the same way about quantum mechanics if you had asked them at that time. My feeling is that everything that forms the foundations of our knowledge will eventually find some application. Let's forget about string theory and focus on the existence of W and Z bosons which is firmly established. If you ask: What use do they have for humankind in practical terms? At present no one can say that they have any use in our everyday life. But it is almost certain that in 100 years from now we will find some use for them because they also form the foundations of our understanding of nature. I cannot say the same about string theory with the same certainty because we still don't know if it is the right theory.

Okay, so maybe it would be a good idea to write a joint letter to the government saying that in 100 years' time this will be useful and please give us more money?

(laughs) Well ... we cannot say with certainty that string theory will be useful but we can say with certainty that the W and Z bosons will.

References

[1] A. Sen, "Ultraviolet and infrared divergences in superstring theory," arXiv:1512.00026 [hep-th].

[2] E. Witten, "Superstring perturbation theory revisited," arXiv:1209.5461 [hep-th].

[3] A. Sen, "Off-shell amplitudes in superstring theory," *Fortsch. Phys.* **63** (2015) 149–188, arXiv:1408.0571 [hep-th].

[4] A. Sen, "BV master action for heterotic and type II string field theories," *JHEP* **02** (2016) 087, arXiv:1508.05387 [hep-th].

[5] A. Sen, "Supersymmetry restoration in superstring perturbation theory," *JHEP* **12** (2015) 075, arXiv:1508.02481 [hep-th].

[6] A. Sen and E. Witten, "Filling the gaps with PCO's," *JHEP* **09** (2015) 004, arXiv:1504.00609 [hep-th].

[7] S. D. Mathur, "The Information paradox: a pedagogical introduction," *Class. Quant. Grav.* **26** (2009) 224001, arXiv:0909.1038 [hep-th].

[8] J. Polchinski, "The black hole information problem," in *Theoretical Advanced Study Institute in Elementary Particle Physics: New Frontiers in Fields and Strings*, Joseph Polchinski, Pedro Vieira and Oliver DeWolfe, eds., pp. 353–397. 2017. arXiv:1609.04036 [hep-th].

[9] S. D. Mathur, "The fuzzball proposal for black holes: an elementary review," *Fortsch. Phys.* **53** (2005) 793–827, arXiv:hep-th/0502050.

[10] Y. Nambu, "Quark model and the factorization of the Veneziano amplitude," in *International Conference on Symmetries and Quark Models, Wayne State U., Detroit*, pp. 269–278. 1997.

[11] H. B. Nielsen, "An almost physical interpretation of the N-point Veneziano model," submitted to Proc. of the XV Int. Conf. on High Energy Physics. (Kiev, 1970), unpublished.

[12] L. Susskind, "Harmonic-oscillator analogy for the veneziano model," *Phys. Rev. Lett.* **23** (1969) 545–547.

[13] J. M. Maldacena, "The large N limit of superconformal field theories and supergravity," *Int. J. Theor. Phys.* **38** (1999) 1113–1133, arXiv:hep-th/9711200.

[14] D. J. Gross and W. Taylor, "Two-dimensional QCD is a string theory," *Nucl. Phys. B* **400** (1993) 181–208, arXiv:hep-th/9301068.

[15] B. de Wit, J. Hoppe and H. Nicolai, "On the quantum mechanics of supermembranes," *Nucl. Phys. B* **305** (1988) 545.

[16] H. Nicolai and R. Helling, "Supermembranes and M(atrix) theory," in *Trieste 1988: Nonperturbative Aspects of Strings, Branes and Supersymmetry*, M. J. Duff, E. Sezgin, C. N. Pope, B. Greene, J. Louis, K. S. Narain, S. Randjbar-Daemi and G. Thompson, eds., pp. 29–74. 1998. arXiv:hep-th/9809103.

[17] S. Kachru, R. Kallosh, A. D. Linde and S. P. Trivedi, "De Sitter vacua in string theory," *Phys. Rev. D* **68** (2003) 046005, arXiv:hep-th/0301240.

[18] L. Susskind, "The anthropic landscape of string theory," arXiv:hep-th/0302219.

[19] N. Arkani-Hamed, S. Dimopoulos and G. Dvali, "The hierarchy problem and new dimensions at a millimeter," *Phys. Lett. B* **429** (1998) 263–272, arXiv:hep-ph/9803315.

[20] R. Bousso and J. Polchinski, "The string theory landscape," *Sci. Am.* **291** (2004) 60–69.

[21] M. R. Douglas, "The string theory landscape," *Universe* **5** no. 7 (2019) 176.

[22] J.-L. Gervais and B. Sakita, "Field theory interpretation of supergauges in dual models," *Nucl. Phys. B* **34** (1971) 632–639.

[23] P. Ramond, "Dual theory for free fermions," *Phys. Rev. D* **3** (May 1971) 2415–2418. https://link.aps.org/doi/10.1103/PhysRevD.3.2415.

[24] J. M. Maldacena and H. S. Nastase, "The supergravity dual of a theory with dynamical supersymmetry breaking," *JHEP* **09** (2001) 024, arXiv:hep-th/0105049.

[25] K. A. Intriligator and N. Seiberg, "Lectures on supersymmetry breaking," *Class. Quant. Grav.* **24** (2007) S741–S772, arXiv:hep-ph/0702069.

[26] S. Kachru, J. Pearson and H. L. Verlinde, "Brane/flux annihilation and the string dual of a nonsupersymmetric field theory," *JHEP* **06** (2002) 021, arXiv:hep-th/0112197.

[27] A. Canepa, "Searches for supersymmetry at the Large Hadron Collider," *Rev. Phys.* **4** (2019) 100033.

[28] A. Sen, "Dyon-monopole bound states, selfdual harmonic forms on the multi-monopole moduli space, and SL(2,Z) invariance in string theory," *Phys. Lett. B* **329** (1994) 217–221, arXiv:hep-th/9402032.

[29] A. Sen, "Strong-weak coupling duality in four-dimensional string theory," *Int. J. Mod. Phys. A* **9** (1994) 3707–3750, arXiv:hep-th/9402002.

[30] E. Witten, "String theory dynamics in various dimensions," *Nucl. Phys. B* **443** (1995) 85–126, arXiv:hep-th/9503124.

[31] A. Strominger and C. Vafa, "Microscopic origin of the Bekenstein-Hawking entropy," *Phys. Lett. B* **379** (1996) 99–104, arXiv:hep-th/9601029.

[32] A. Sen, "Microscopic and macroscopic entropy of extremal black holes in string theory," *Gen. Rel. Grav.* **46** (2014) 1711, arXiv:1402.0109 [hep-th].

[33] A. Sen, "Entropy function and AdS(2)/CFT(1) correspondence," *JHEP* **11** (2008) 075, arXiv:0805.0095 [hep-th].

[34] A. Sen, "Quantum entropy function from AdS(2)/CFT(1) correspondence," *Int. J. Mod. Phys. A* **24** (2009) 4225–4244, arXiv:0809.3304 [hep-th].

[35] A. Sen, "Arithmetic of quantum entropy function," *JHEP* **08** (2009) 068, arXiv:0903.1477 [hep-th].

[36] N. Banerjee, S. Banerjee, R. K. Gupta, I. Mandal and A. Sen, "Supersymmetry, localization and quantum entropy function," *JHEP* **02** (2010) 091, arXiv:0905.2686 [hep-th].

[37] S. Banerjee, R. K. Gupta and A. Sen, "Logarithmic corrections to extremal black hole entropy from quantum entropy function," *JHEP* **03** (2011) 147, arXiv:1005.3044 [hep-th].

[38] I. Mandal and A. Sen, "Black hole microstate counting and its macroscopic counterpart," *Nucl. Phys. B Proc. Suppl.* **216** (2011) 147–168, arXiv:1008.3801 [hep-th].

[39] A. Dabholkar, J. Gomes, S. Murthy and A. Sen, "Supersymmetric index from black hole entropy," *JHEP* **04** (2011) 034, arXiv:1009.3226 [hep-th].

[40] S. Banerjee, R. K. Gupta, I. Mandal and A. Sen, "Logarithmic corrections to N = 4 and N = 8 black hole entropy: a one loop test of quantum gravity," *JHEP* **11** (2011) 143, arXiv:1106.0080 [hep-th].

[41] A. Sen, "Logarithmic corrections to N = 2 black hole entropy: an infrared window into the microstates," *Gen. Rel. Grav.* **44** no. 5, (2012) 1207–1266, arXiv:1108.3842 [hep-th].

[42] A. Sen, "Logarithmic corrections to rotating extremal black hole entropy in four and five dimensions," *Gen. Rel. Grav.* **44** (2012) 1947–1991, arXiv:1109.3706 [hep-th].

[43] A. Sen, "Logarithmic corrections to Schwarzschild and other non-extremal black hole entropy in different dimensions," *JHEP* **04** (2013) 156, arXiv:1205.0971 [hep-th].

[44] R. M. Wald, "Black hole entropy is the Noether charge," *Phys. Rev. D* **48** no. 8 (1993) 3427–3431, arXiv:gr-qc/9307038.

[45] S. D. Mathur and D. Turton, "The fuzzball nature of two-charge black hole microstates," *Nucl. Phys. B* **945** (2019) 114684, arXiv:1811.09647 [hep-th].

27

Eva Silverstein

Professor of Physics at the Institute for Theoretical Physics at Stanford University

Date: 14 October 2020. Via Zoom. Last edit: 4 December 2020

What are the main problems in theoretical physics at the moment?

There are many. Classics include how to explain the phase diagram of high-T_c superconductors, how to derive the essential dynamics of various compact objects in the universe and how to formulate a quantum theory of gravity which is applicable to the observed universe (and to the physics beyond our horizon). Another which is dear to my heart is how to learn as much as we can about early universe physics from the available data combined with principled theory. Including biophysics, the list expands: How to develop more complete quantitative theories of evolution and of neuroscience? The list goes on; one doesn't get a chance to work on all of these (unless somebody finds a breakthrough in longevity).

Why do we need a theory of quantum gravity? What problems can't we solve without it?

In general, we know about both gravity and quantum mechanics, so we need a coherent framework in which they fit together, one without more outputs than inputs. A nuanced example of a problem for which we need quantum gravity arises in early universe cosmology (by which we mean the dynamics occurring nearly 14 billion years ago that leads to the origin of structure). This is sensitive to quantum gravity effects, according to a standard way of parameterising our ignorance of physics at energy scales we have not directly accessed known as Wilsonian effective field theory [1]. Large field inflation, where the inflaton field moves farther in its field strength (though not its energy) is sensitive to the symmetry structure of quantum gravity, and any inflationary model has some well-defined sensitivity to quantum gravity effects [2, 3]. This does not mean that we can turn the observations into an inference about an entire theory of quantum gravity, but it does mean that we need to understand some kinds of quantum gravity effects in order to fully understand inflationary cosmology. This sensitivity is particularly interesting because it ties to observations of the microwave background and to large-scale structure surveys.

Do you think string theory is the right framework to describe quantum gravity and the ultimate theory of nature?

I think this is very likely because string theory satisfies very nontrivial internal mathematical and physical consistency checks. One of them is how the theory fits together in terms of string dualities [4], that is, the many ways you can interpolate between weak and strong coupling or find specific topology-changing processes, even dimension-changing processes that connect different string formulations. That alone didn't need to be true but it makes it a more unified theory than it otherwise would be.

Another interesting example of physical consistency concerns the modelling of de Sitter spacetime and inflationary physics [2, 3]. In this setting there are some interesting structural constraints that the theory has. In particular, it doesn't have any hard cosmological constant in the dimension in which you define the theory. Instead, the potential energy of the scalar fields, in the Einstein frame in four dimensions, decays at weak coupling (or large radius) no matter what. So these types of models/solutions are metastable; that is, the potential energy rises, falls and rises again. The falling behaviour requires a negative contribution to the potential energy. In perturbative string theory there are objects with negative tension called orientifolds, which are part of the nuts and bolts of the theory, and which appear at intermediate stages in the expansion around weak coupling. These negative tension objects have a certain long-range effect on the geometry and do not cause any instability, as you might have naively thought would be the case when dealing with any other object with negative mass or tension, and are responsible for the falling behaviour.

String theory also reproduces predictions of what is called black hole thermodynamics. Here I am referring to black hole microstate counting that yields the classical Bekenstein-Hawking entropy as predicted in the 1970s for very special, but still very interesting, black holes [5, 6]. In fact, the negative tension objects, which cause the dip in the potential that I just mentioned, have the right properties such that black hole thermodynamics still hold. More precisely, if you consider some thought experiment in which you centre some black hole on a orientifold plane, you still recover the same black hole thermodynamics.

Do you think that there is no other theory that is also mathematically consistent and can reproduce black hole thermodynamics?

It's hard to exclude that; we don't know. However, it's been striking how in the past there seemed like there were different candidate theories with different starting points but then turned out to be unified with string theory. Thus, there is some indication that string theory covers the bases. In lower dimensions, there may of course be alternatives that are complete in and of themselves and don't need to have anything to do with string theory. But in a healthy number of dimensions (laughs), that is, in a theory that admits at least four large dimensions, string theory is the only theory we know well enough to pursue it as a concrete candidate.

One of the criticisms of string theory is that it cannot be falsified. Do you agree with such criticism?

I do not agree that this is a valid criticism, and I would like to stress that there is a very significant model-dependent intersection between string theory and empirical data [1, 3]. Let me answer this step by step, since it is an important question. Before answering directly, let's note that string theory could be falsified if one of its basic principles were violated (like unitarity of quantum mechanics or causality). I don't expect this, so let me address the question in the spirit I think it was intended.

A similar criticism would naively apply in a certain sense to quantum field theory. That is a framework which satisfies basic principles of unitarity and causality, but in order to use it to model physical systems one needs to specify a particular model within this general framework. At that point, one can test the model and constrain its parameters with experiments or observations.

String theory models can be tested to some extent in an analogous way, in particular, in regimes I mentioned before like early universe cosmology where the dynamics and predictions are sensitive to quantum gravity [7]. String theory reduces to a low-energy theory which has a large but highly structured space of solutions (more generally a large space of quantum mechanical trajectories). These are all connected as far as we know – and I want to stress that this includes different topologies and even dimensionalities (which we measure abstractly in terms of the density of excited states in the theory). When string dualities were discovered [4], and then generalised, the theory became more unified and less uniquely predictive in one fell swoop.

By working with this structured space of solutions, we discovered inflationary models and broadly defined inflationary mechanisms, some of which have distinctive observational signatures that enable us to constrain their parameter space. In this way, we could use CMB and LSS data to discover new parameters these scenarios predict, or alternatively falsify individual models, or a whole class of models, if the new effects are not seen at a certain threshold level. One can, and people did, extract from this essential lessons that were then incorporated into the very useful low-energy effective quantum field theory description of the observables. In this sense, string theory has already helped to generate a much more systematic understanding of empirical observables in early universe inflation.

What are these distinct signatures that you alluded to?

Primordial perturbations are visible in the CMB. One kind of observable that we have access to is related to correlation functions of perturbations of the inflaton field. This includes the two-point function – the *power spectrum* – as well as higher-point functions. These days we also study cases where we can access more about the shape of the probability distribution itself, including the tails of the distribution related to high-point correlators [8–10]. A second class of observables consists of the tensor perturbations, which get an imprint from primordial gravitational waves produced by inflation. Large field models make particular predictions for the observable that is often called the tensor to scalar ratio r.

This ratio is the most ultraviolet-sensitive but in these stringy models, it is grounded on symmetry principles. Another set of observables arises from relics like cosmic strings. The detailed predictions for the scalar and tensor perturbations and relics are in general very model-dependent.

String theory generates novel dynamics such as [8, 11–15] with signatures for these various observables [2, 3]. Some of them have significant interactions slowing the field, as in DBI and trapped inflation, leading to what we call equilateral non-Gaussianity generated by these inside-horizon interactions. Axions in string theory have potentials that are unwound via monodromy and yield r as well as oscillatory features. Multifield effects including non-adiabaticity leave certain imprints. A general lesson is what we called flattening [16]: the heavy fields coupled to the inflaton adjust in an energetically favourable way, causing the potential to flatten out at large field values. In specific examples of this we find interesting powers like a two-thirds power of the inflaton field, rather than, say, quadratic, for the inflaton potential. In general, the plateaus make sense from this point of view of heavy degrees of freedom from the UV completion adjusting energetically so as to flatten the potential.

Are you always on the lookout for data?

Yes, the times when a meaningful prediction arises in one's research are very special and important to capitalise on. I encountered in my own research several mechanisms for inflation from string theory, some of which output predictions (as in the examples we just discussed). As such, I sometimes get involved in actual collaborations on data analysis (laughs). In particular, I have works with Raphael Flauger and others, on certain aspects of oscillatory signatures in models of axion monodromy [8, 17–19].

But are people who are actually doing these experiments trying to constrain string theory models? What do they constrain exactly?

Absolutely, yes. This is clearly visible for instance from the Plank collaboration papers going back several years as well as the recent ones [7]. These analyses constrain both specific models, including the string-theoretic ones, as well as the parameter space of bottom-up field theoretic approaches to inflation. On the latter, one of my colleagues, Leonardo Senatore, has championed what is called the effective field theory of inflation – an effective theory of inflationary perturbations – which parameterises observational quantities in a more systematic way [20] than had been done before. However, what has happened repeatedly is that string theory models inform bottom-up approaches – filling in gaps in the latter by example. So it is fair to say that the interplay between bottom-up and top-down approaches has been concretely fruitful, in finding specific signatures and regions of parameter space that are important to constrain.

You also mentioned that people extracted essential lessons from string theory models that were incorporated into the low-energy effective field theory. What lessons do you have in mind exactly?

The structure of large field inflation appearing in string theory models, in particular axion models [14, 15, 21], is really interesting as the axion has a flattened large field potential, but one with an underlying periodicity. It contains elements of Andrei Linde's simplest model of inflation ($m^2\phi^2$) and of natural inflation, but incorporates the structure of string theory in a nontrivial way. The models have large field range and the effective potentials are not just parabolas of the form $M^2\phi^2$ but actually get flattened out due to relaxation effects caused by other fields. Moreover, it has residual oscillatory features as a result of the underlying periodicity [8, 18, 22]. Particular versions of the model are in fact falsifiable based on the tensor to scalar ratio r. These features of axion monodromy [14, 15], were quickly developed by Flauger et al. along with others from a bottom-up effective field theory point of view [23]. The predictions of the model involve a certain periodicity in the logarithm of the wave number k in cosmology. In summary, bottom-up approaches could have independently come up with these features by focusing on discrete shift symmetries of the kind that string theory models predict. However, they did not, as the focus was put on continuous symmetries. These lessons for low-energy effective field theory were drawn from string theory. An earlier example was equilateral non-Gaussianity, which was obtained perturbatively in higher-dimension interactions by Creminelli [24], but were naturally strong in DBI [25], which uses string theory to re-sum a series of such interactions that leads to novel dynamics slowing the field and generating nonlinear perturbations. Another example is heavy particle production, and multifield effects of the hyperbolic geometry of field spaces in string theory. Yet another early example is the scale of cosmic string tensions, taking into account gravitational redshift effects [11, 12].

Are there smoking guns specific to these models?

I think that would be a too strong statement to make but it depends on what you mean (laughs). The effective potential we derive from one example of these models goes like $\phi^{\frac{2}{3}}$, so if the tensor to scalar ratio r associated with these potentials would match experiment, and if you saw residual oscillations in the logarithm of k as well as corresponding effects in the large-scale structure, then I think people would tend to the idea that these sorts of string theory models are the right models (laughs). But even if this mechanism is correct, this depends on whether we can actually observe, for instance, large enough signatures of residual oscillations. That is very model-dependent and we don't anticipate it. But if all this evidence would pile up, then we would do everything humanly possible to calculate all the effects that arise from these models and see how well they fit with observations. This would be pretty compelling evidence, but would you be able to prove it? The answer is no, even in this most optimistic scenario (laughs).

You mentioned earlier analogies between string theory and quantum field theory. Do I understand correctly that you see string theory as a framework, as quantum field theory, where you have to build models and test against observations or do you think that string theory is able to predict every single constant of nature?

I do think as far as the connection to observations, it's more the former than the latter. However, string theory is unified and everything is connected dynamically so the analogy with quantum field theory is far from being complete. And one thing that I want to stress is that everyone, including myself (laughs), is well aware that quantum field theory has in fact been tested very extensively as a way to model physical systems while this interface between observables and string theory is still in its infancy. So, for practical purposes string theory is a framework in which you have a vast configuration space of solutions that you can use to construct explicit and well-defined classes of models for early universe physics and derive model-dependent predictions.

Right, but do you think that ultimately there will be some dynamical principle that will pick the right model or vacua from this vast space of solutions?

I think that's definitely possible. At this moment we don't know but we might get there. The way to understand whether that is possible or not is to study accurately and comprehensively the landscape of solutions as it's given rather than jumping to conclusions and making unfounded predictions (laughs). For instance, the number 10^{500} vacua is only a meaningful number in a very particular context, though highly quoted. Instead, what is the case is that there is an infinite number of possible sequences of solutions once you abandon the requirement of low-energy supersymmetry. The number 10^{500} is related to the idea that there may be a finite number of Calabi-Yau manifolds if one focuses on Ricci-flat manifolds for the internal dimensions [26]. However, most manifolds are not Ricci-flat; in fact, most of them are negatively curved. This is obvious if you think about the two-dimensional case where you have the sphere, the torus and an infinite sequence of higher-genus Riemann surfaces. This sequence proliferates even more in the case of higher-dimensions, which is important given that the dimensionality itself is actually a dynamical variable. This all sounds like you should just throw the whole thing away because it appears to be completely unpredictive (laughs). But making such statement would be a fallacious conclusion.

What I mean by that is that these infinite sequences may converge on particular predictions, rather than adding random possibilities to the plethora of string backgrounds. To make an analogy with quantum field theory, one can consider large-flavour models in which you have a gauge field coupled to matter and you proliferate the number of representations of the matter fields. What happens is that as you increase the number of representations the physics coalesces on a sort of unique set of predictions. There is analogous evidence that the landscape is also dominated by certain behaviour in specific limits in which the physics simplifies. For instance, as you increase the number of dimensions, the spectrum of particles/states related to the graviton field grows like D^2 where D is the dimensionality while the number of states related to axions grows like 2^D. Therefore such axion states are exponentially more common in comparison with those related to the graviton field. Thus, it may be that, when we complete a systematic study of the landscape, including the analysis of all kinds of limits, we will find that certain behaviour dominates. In general, the landscape predicts very characteristic physics such as the inflationary mechanisms we discussed before, in various corners of it that we have explored.

When discussing the landscape, people often refer to 10^{500} vacua and that anything can happen. I guess, given your answer, you disagree with such statements?

Neither statement is accurate. The number of possibilities is infinite: there are infinite sequences of possible topologies as well as dimensionalities to start from. However, those sequences do not introduce the possibility for anything to happen – the physics tends towards a particular behaviour in those limits, with corrections that become smaller and smaller as one goes further out in the sequence. This is analogous to the large-flavour number expansion that I mentioned in some ways. The landscape is vast and varied, but highly structured so it is not true that anything can happen. For example, there is not a stable cosmological constant – only metastable solutions. Large field inflation occurs – but with distinctive features derived from the landscape such as the underlying axion periodicity which is unwound by monodromy, with a flattened potential.

One of your articles has the daring title "The Dangerous Irrelevance of String Theory" [1]. What did you exactly mean by this?

This is a fun question; "dangerous irrelevance" is my favourite term in physics. The idea, which goes back to condensed matter theory, is that even if we do not directly excite physical degrees of freedom at a certain high-energy scale, nonetheless we can see concrete but more indirect effects of the underlying high-energy theory under certain circumstances. The circumstances involve long times, over which small effects can build up, or large ranges of field strengths, or large ranges of energy scale. The large field strength case of this occurs in inflationary cosmology as we discussed before.

To be a bit more specific in case it helps: in the Wilsonian parameterisation of our ignorance that I mentioned earlier, we write a series of corrections that might arise to the equations governing fields that we can directly access. These corrections are proportional to powers of the Newton constant, and behave like powers of energy or of field strength divided by the energy scale (the Planck scale) by which we know that gravitational interactions become strong as we go up in energy. If the field executes an excursion comparable to or greater than the Planck scale, then we had better know how those terms behave. These terms are called "irrelevant" because at small energies or field strengths, they do not matter. But as we just noted, they can matter – hence the term "dangerous irrelevance." A symmetry principle which prohibits a general series of such terms is one approach to theoretically controlling our calculations; this is an assumption about the symmetry structure of quantum gravity which we can check in string theory. In fact, in the axion models that we discussed earlier, discrete shift symmetry is one such example.

This concept is important because it explains in a principled way how there are some residual effects of quantum gravity on low-energy observables. I strongly prefer this term or ultraviolet sensitivity for this because it is the original term and more than adequate to describe the phenomenon (no re-branding with other terms is needed).

There have been many discussions about the existence of de Sitter (dS) vacua and inflationary solutions in string theory. Many doubt that they actually exist and there are conjectures that seem to prevent it [27]. What is your opinion about this?

My view is that there is overwhelming evidence for the existence of inflationary and de Sitter solutions. This includes classes of explicit models such as [28] and others, analysed at the appropriate level of approximation, and also various mechanisms that have been extensively analysed over the years and which hang together. These constructions contain a full set of compensating forces included to locally stabilise all the low-energy field configurations. Specifically, I have in mind supercritical models that I played with together with Alexander Maloney and Andrew Strominger [29], which follows an idea of a paper of mine [30]; the very well-studied KKLT mechanism [31] and variants [2]; models with stable power law directions, as opposed to exponential behaviour, which involve more general curvatures of the internal space [3, 32].

This could fail if there are (as yet unidentified) mistakes in the models, or if in every case corrections are larger than very standard estimates, and it is fair to say that their analysis is somewhat involved. But many physical systems are much more complicated than this but are quite successfully analysed and explained using the same, standard, approximation schemes. I would like to stress this point: the methodology here is widely applied in modelling physical systems, where there is never an exact treatment but there are well-defined approximation schemes that are tried and true in physics.

I say there is not an exact treatment of any real physics problem – including all of the standard, empirically tested physics that has ever been done – because that would require a complete, finished theory of quantum gravity (not to mention a mathematical level of rigour). But established physics results are based on overwhelming evidence, not an exact proof.

Another point to stress here is that my view – which in my experience is the typical view of those who engage most with the details – is the conservative point of view. Despite the absence of a general observational prediction, as we discussed above there is still a rich interface with observations at a meaningful but model-dependent level. A stronger claim of universal model-independent predictions of observationally accessible predictions would be an extraordinary claim, one contrary to the vast preponderance of evidence.

In fact there is a piece of evidence for these models connected to the more conceptual quantum gravity side of this subject. The metastability of the de Sitter solutions – something that is a structural constraint in string theory and an example of dangerous irrelevance – leads directly to a strong consistency test of the dS/dS approach to upgrading the famous AdS/CFT correspondence [33] to de Sitter. This dS/dS approach requires two sectors of strongly interacting degrees of freedom constrained by a residual lower-dimensional gravity, and the metastability property of uplifts of anti–de Sitter (AdS) to dS applied to the brane construction of AdS/CFT reproduces this [28].

There is a vocal set of researchers who indeed assert various conjectures [27], ones which are contrary to the evidence we just discussed from a more detailed and complete analysis of the extra dimensions of string theory. There are many papers on this lately, compared to the number of papers further developing the solutions. This is because it is relatively straightforward to assume these conjectures and derive consequences without delving into the details or methodology and logic of the existing constructions. Somehow these proposed statements have gone from being described as conjectures (which is already somewhat problematic) to being described and treated as criteria. This in my view is not a healthy situation, and in its extreme forms is rather counterproductive.

Incidentally, with De Luca and Torroba we have a new class of examples under development, which we are starting to write up, with a simpler set of stress energy ingredients. There is definitely more to learn from the structure of string theory and its relation to real and conceptual (thought-experimental) observables, and I remain enthusiastic about this enterprise.

Why are you working in this new case with De Luca and Torroba?

Despite the odd discourse currently, the subject is great and so, to me, it's important to really understand these more generic limits of the theory and that includes negative curvature internal spaces. The example that we are currently working on has a certain simplicity that we hadn't come across before. It is obtained from 11-dimensional supergravity by compactification on a hyperbolic geometry, taking into account a flux and the Casimir energy on that space. In terms of the ingredient list, it's as low as it has ever been (laughs). It is actually similar to the anti–de Sitter construction, obtained by compactification of M-theory in an S^7. In this case, instead, we consider a hyperbolic space in such a way that you obtain a strong Casimir effect that provides the crucial intermediate negative contribution for the uplift to de Sitter space.

As a side comment, I should mention that in the beginning of this subject, people focused on Calabi-Yau manifolds, which led to a beautiful interplay with mathematics. However, I think that this interplay also takes place in the case of hyperbolic geometry, an active area for which mathematicians have a great deal of understanding. We have only scratched the surface.

Several people have heavily debated the correctness of the KKLT mechanism. Do you think the KKLT mechanism makes sense?

It definitely makes sense. There have been a few papers making really unphysical, incorrect statements about it, such as that certain contributions to the potential are negative rather than positive. In my opinion, there is not much there (laughs). What was claimed in the original paper [31] was an outline for constructing the models, and many works followed up successfully on this approach. In that context, which is based on Calabi-Yau manifolds, it is not trivial to present a very specific example. However, on general grounds you have various ways of seeing that you clearly have the right interplay of forces needed to obtain consistent models. Detailed flux compactifications were constructed in this literature, and again I want to stress that the mechanism completely hangs together.

I do tend to question the enormous focus on low-energy supersymmetry (SUSY). The fact is that it has not been observed (as of now) and if you look at the theory at face value, generically it doesn't have it and it also doesn't need it for control. As with all physics, control can be achieved by using appropriate tools, including boring old perturbation theory. Other interesting models that are not discussed so often – and I guess that's why you never heard about them (laughs) – are those which break supersymmetry at a higher scale. As far as I can tell, a large part of the lack of attention to the less supersymmetric solution space is due to sociological reasons.

This last set of models you mention does not have low-energy SUSY?

No, it doesn't. Low-energy SUSY is a beautiful idea and I have worked on it for bottom-up reasons: it puts together the hierarchy problem and dark matter in an appealing way. However, the top-down theory generically doesn't have low-energy SUSY and doesn't need it. CERN has constrained the parameters of low-energy SUSY to a reasonable extent. So I would say that we don't need to fixate on this possibility.

But is it not the case that string theory requires SUSY, in particular, to cancel divergences and have control over the theory?

SUSY is only required at a very high energy scale – the string scale. Almost all backgrounds of the theory (solutions of the effective theory at or below the scale of the string tension) have very high scale supersymmetry breaking. This includes the generic supercritical limits of the theory and also negative curvature compactifications, which are much more common than Ricci-flat or positively curved ones. In essence, if we just count such parameters, string theory is at face value a high-dimensional theory of axions (the number of axions grows like 2^D, and at a fixed dimension they proliferate with topology).

Theoretical control is nonetheless available for practical purposes – in particular, SUSY below the Kaluza-Klein scale is not required for control of calculations. The well-controlled approximations we use to control calculations at weak coupling and mild curvature are the same as in standard physics, as we discussed before. In fact, SUSY makes de Sitter modelling harder in some ways – specifically in the sense of restricting contributions to the potential energy of scalar fields that need to be stabilised, although both cases work out.

What extended SUSY is best for is to interpolate between weak and strong coupling, which is very interesting as a theoretical tool. However, it is not relevant or needed for control of models of realistic physics.

How can there be such conjectures about the non-existence of de Sitter models [27] given all this overwhelming evidence?

The paper you refer to doesn't really engage with the details; they propose that there is some unidentified problem (laughs). What is more problematic is that the paper refers to incorrect analyses of some models, and it also completely ignores others. In its own study, the paper focuses on certain extreme limits of the theory, without generic sources of stress energy. These are situations that have been known for years not to lead to inflation or metastability. However, existing models of metastable de Sitter vacua are not that simple, as they include nontrival stress-energy sources. It is fair to say that de Sitter models are more complex than AdS ones, but again I would stress that much established physics is significantly more complicated, including what goes into observational cosmology over the known history of the universe. I do not find it productive to dwell on this.

Because of these de Sitter conjectures, some people started wondering if other mechanisms besides inflation could be found within string theory models, such as quintessence. And there were claims that quintessence could not be produced within string theory. Do you think this is correct?

Some people do make that strong a claim. I wouldn't. First, let me say that when it comes to stabilising the system, demanding accelerated expansion is a weaker condition and therefore easier to arrange than an actual de Sitter minimum. In the case of the axion models, axions appear in such a way that their contribution to the potential can play the role of a flux helping to stabilise moduli. These generalised fluxes are generic, and slowly vary with time because the axion rolls slowly. Some other landscape enthusiasts came to the conclusion that it's all about these metastable minima. I disagree with that. These types of slow roll could be common. However, I think everyone including myself would say that there is no a priori reason for that to be visible in some observations of the equation of state of the universe (laughs). Returning to your specific question about quintessence, Sandip Trivedi wrote a paper some time ago about quintessence using axion-type potentials and he thought it worked fine [34]. We find the general effect I mentioned of axions rolling in moduli stabilisation by generalised fluxes, e.g. in [21]. The landscape can support quintessence but we have no specific prediction that we should expect to see it observationally.

Is inflation the best solution to the cosmological problem?

My view is that inflation is very well tested as a paradigm – the measured power spectrum, which is a function of scale, could have been much different than measured in the CMB and LSS. It is very interesting to explore alternatives, even if (as seems to be the case so far) they do not work out, since it helps us understand better whether inflation could be inevitable. We cannot prove that (ultimately I don't think we prove theories in physics) but the paradigm fits the facts in great detail.

Can string theory cope with many types of inflation?

Yes, as we discussed earlier, string theory has several classes of inflationary models. Examples go by the names axion monodromy inflation, trapped inflation, brane inflation (slow roll or DBI), roulette inflation, fibre inflation, and others [2, 3]. These make various distinct predictions, and the data can discriminate among them. One general lesson is something we called "flattening" [16]: the additional heavy fields in string theory adjust in energetically favourable ways, giving large field inflation but not the simple parabolic potential proposed originally by Linde; instead, it gets flattened at large field range, reducing the prediction for the tensor to scalar ratio in axion monodromy. Such models remain viable currently as a result, but various single-field examples are falsifiable based on the forecasts for near-future CMB data. Another feature they have is an underlying periodicity: some aspects of the physics repeat during the process (like a wind-up toy).

My joke is that the timescale for these tests, albeit 5–10 years, counts as instant gratification for a string theorist.

Is the Lambda-CDM model the best we have for describing cosmological observations?

At the moment, no observations force us to additional parameters, but tests of early universe data or experiments in other regimes could measure non-trivial values for additional parameters.

Can we understand quantum gravity in the de Sitter models that you mentioned earlier? What are the basic issues that make this such a difficult problem?

We have sensitivity to some aspects of quantum gravity as we discussed before. The basic difficulties include the following. There is no extreme boundary where the system is pinned and non-gravitational, and there is no appreciable supersymmetry, so the weak curvature arises by some tuning which must also be reflected in the dual description.

You worked extensively in dS/dS correspondences and holography for de Sitter spacetime [13, 35]. Do we have a dual QFT description of de Sitter spacetime?

The dS/dS correspondence is obtained from the observation that a certain patch of de Sitter spacetime – which is greater than an observer-accessible patch – has two gravitationally redshifted regions which each reduce to those of AdS/CFT with the conformal field theory (CFT) on dS. This indicates a dual description in terms of two identical matter sectors with many degrees of freedom, interacting with a residual lower-dimensional gravity which does not decouple but which is more tractable than the original higher-dimensional gravity. This content – the two sectors – follows independently from the uplift to the de Sitter case of the "brane construction" which led to AdS/CFT. Recently, the nature of the two sectors has greatly clarified, as we identified them as deformations of CFTs by a generalised version of the $T\bar{T}$ deformation [35]. This is a solvable deformation whose energy levels match those of the appropriate finite patch of spacetime. Since then, we calculated entanglement entropies and other quantities in the limit of large numbers of degrees of freedom, finding precise agreement between the calculations done in the dual sets of variables. So at the moment, I think this is the most extensively developed approach to dS holography. It extends readily to a similar formulation of the late time physics of the metastable solutions, including the ultimate runaway to a more general Friedmann-Lemaitre-Robertson-Walker (FRW) spacetime.

What is the exact statement of the dS/dS correspondence? Can we formulate it similarly to AdS/CFT in terms of an equality between partition functions of IIB string theory and $N = 4$ super Yang–Mills theory?

It's not as strong but it's getting quite a bit closer, so let me explain. We can make the statement that we have a system which reproduces the geometry of de Sitter, and a generalisation tractable in the large-N limit produces the physics of matter fields on de Sitter. If you start from a pair of conformal field theories and for each one you deform it by the $T\bar{T}$ operator of Zamolodchikov [36], combined at each step of that sort of deformation with a cosmological constant, you get a theory whose energy levels are those of a patch of de Sitter spacetime that also reproduces certain entropy calculations [37]. To make it as explicit as AdS/CFT requires more work, but we see how to proceed to further close the gap in this approach.

But do you have a holographic understanding of quantum gravity in de Sitter spacetime?

It's not fully developed yet. At this point, if you're satisfied with pure gravity and provided you're happy with what we understand about the $T\bar{T}$ theory in itself which is still under development, then there is such a formulation, which is quite fun and encouraging. It is possible to move away from pure gravity and in fact there have been papers on this which involved additional contributions of matter fields at each step in the Zamolodchikov trajectory [38]. To describe de Sitter spacetime with matter holographically you would have to add operators that are not as easy to define as $T\bar{T}$. However, it is possible to make progress at large central charge c and to define such operators through large c factorisation in that limit. As it stands, we understand only this large c limit with bulk matter, and the task at hand is to move beyond that. At the moment, it is not as concrete as AdS/CFT.

Since you mentioned the concreteness of AdS/CFT, there is still some discussion about whether we understand the AdS/CFT dictionary to point that we can actually describe all local bulk physics from the boundary theory. Do you think there is still something to understand in this respect?

I think it's fair to say that we certainly haven't fully mapped that out. It would, however, be a completely different statement that it is impossible and more is needed than $N = 4$ super Yang–Mills theory. So is this a fundamental limitation or just our inability to calculate strong coupling physics in order to refine the dictionary? Steven Giddings expresses skepticism about this [39]. I do take seriously the old-fashioned notion that indeed the $N = 4$ super Yang–Mills theory is precisely dual to quantum gravity in asymptotically AdS. But yes, more work is needed to completely map out the dictionary.

Going back to the dS/dS correspondence, does it require large extra dimensions?

No. I can illustrate this with the latest de Sitter example in progress that I mentioned above. The idea is that you start with the S^7 compactification of M-theory to AdS_4 which is the M2-brane theory. Uplifting this S^7 through some topology change to a hyperbolic seven-manifold, you obtain a negative contribution to the Casimir energy. This gives you de Sitter spacetime that can have small extra dimensions. This is a dramatic uplift, in the sense that it changes the topology, but it can be stated very concretely.

An alternative possibility that we keep in mind is that the $T\bar{T} + \cdots$ formulation might not need to make use of string theory. At the level of pure gravity, the theory that reproduces de Sitter energetics and entropies only involves the $T\bar{T}$ deformation together with a cosmological constant. This connects well to your earlier question of whether there could be alternatives for quantum gravity. So, you know, we have our minds open to that (laughs).

But if you want to describe quantum gravity, not just classical gravity and matter fields, do you expect that you need to bring string theory back into the game?

I would tend to agree. I think that's the conservative view. But I should mention that it is not just classical gravity. The last step of the construction is to have two large c sectors, join the two partition functions and have quantum gravity in the lower dimensions such as three-dimensional quantum gravity, which is still challenging but less than in four. This is similar to the holographic interpretation of Randall-Sundrum theory [40, 41].

What about the dS/CFT correspondence [42]? Is that understandable in terms of string theory?

The dS/CFT proposal is an earlier approach, which is realised in a higher-spin theory and which also ultimately involves two identical CFT sectors. However, that one is more difficult to understand in the context of string theory, partly because of the metastability of the latter. Nonetheless, there may be truth to each approach – it is striking that they both involve two identical matter sectors.

Some people say that one of the hardest problems is to understand what are the observables of quantum gravity in cosmological scenarios. And before knowing that, we can't came up with a quantum theory of gravity. Do you have an answer for what these observables might be? Are they local observables?

They are not local observables, although in certain regimes of energy and at weak curvature, they can approximate local observables. We do not know the full set of observables, but calculations of transition amplitudes across the dS/dS patch as well as entanglement entropy and its match to the Gibbons-Hawking entropy provide some encouraging calculations so far.

It has recently been claimed that the black hole information paradox has been solved. Do you agree with this?

What has been claimed is that the entropy of the Hawking radiation as a function of time (or of the size of the evaporating black hole) can now be calculated, giving a unitary result where this entropy decreases to zero at late times rather than continuing to grow as Hawking's calculation by itself suggested. This is great progress, but not a full solution to the problem.

What is else is needed to solve the black hole information paradox?

What is needed is to understand the process by which information is transferred from the matter forming the black hole originally to the Hawking radiation. Euclidean quantum gravity lead to the calculations that brought us the Page curve. However, that calculation does not necessarily determine the physics or theoretical technology required to understand the information transfer. It has led to puzzles and questions of interpretation as well as technical questions.

What is the mechanism by which information is transferred from the inside to the outside of the black hole?

One effect which transfers significant information arises from the nonlocality of string theory (consistently with causality). As seen by a late infalling observer, the early matter that formed the black hole is highly boosted and has a large variance in its extent. This effect leads to long-range interactions that go beyond effective field theory, a necessary condition to explain information transfer. More calculations are required to establish quantitatively how much information this transfers, but we very recently found that there is nontrivial information transfer that goes beyond effective field theory.

What is the nature of this nonlocality of string theory that you mentioned?

Nonlocality in string theory consistent with causality has been around a long time. There is a particular aspect of it that I am focussed on which is something that Leonard Susskind introduced a long time ago and is called *string spreading* [43, 44]. It hadn't been clear whether this was a kind of an artefact of the light-cone gauge that he worked in but recently we made a point of checking for this effect in the string S-matrix. The analysis is quite subtle but we do find this signature. This effect arises because if you view a string with very high energy, thus with short time resolution, you see it as a very extended object, in particular, in the direction of relative motion, that is, the longitudinal direction. You see it as extended by an amount that goes like E/T where E is the energy and T the string tension. In a black hole, you have a lightlike separation between early and late time systems. Thus, the idea is to check how strong those interactions are and if they are enough. Recently, we addressed this question and found somewhat encouraging results [45, 46].

I would like to ask you two final general questions about string theory. Is string theory infrared and ultraviolet finite?

It seems to be. The infrared finiteness has to do with the same question in quantum field theory, where the wave function at late times in cosmology is well defined non-perturbatively although a naive perturbative treatment can break down.

Another issue that people often bring up is the fact the string theory seems to require extra dimensions. Do you agree with this statement or do you see extra dimensions as just a means to generate effective four-dimensional models?

It's an interesting point because just as we think of spacetime as emergent, so is its dimensionality. We define dimensionality abstractly in terms of the growth of the spectrum of states at high energies, and different dimensionalities are dynamically connected in the theory. This is indeed all consistent with reduction to only four large-radius dimensions.

Why did you choose to do physics?

I learned about special relativity at an impressionable age, and was hooked. I also had a very gifted high school physics teacher who conveyed the broader importance of things like Newton's equation $F = ma$. In addition to teaching how one can calculate trajectories using this, he stressed how amazing it was to determine a universal law of nature.

What do you think is the role of the theoretical physicist in modern society?

People are naturally curious, and having some members of society dedicated to physics research enables society as a whole to learn more about how the world works. Some of it is purely curiosity-driven and some more practical, and both are valuable.

References

[1] E. Silverstein, "The dangerous irrelevance of string theory," arXiv:1706.02790 [hep-th].

[2] D. Baumann and L. McAllister, *Inflation and String Theory*. Cambridge Monographs on Mathematical Physics. Cambridge University Press, 5, 2015.

[3] E. Silverstein, "TASI lectures on cosmological observables and string theory," in *Theoretical Advanced Study Institute in Elementary Particle Physics: New Frontiers in Fields and Strings*, Joseph Polchinski, Pedro Vieira and Oliver DeWolfe, eds., pp. 545–606. 2017.

[4] E. Witten, "String theory dynamics in various dimensions," *Nucl. Phys. B* **443** (1995) 85–126, arXiv:hep-th/9503124.

[5] A. Strominger and C. Vafa, "Microscopic origin of the Bekenstein-Hawking entropy," *Phys. Lett. B* **379** (1996) 99–104, arXiv:hep-th/9601029.

[6] A. Sen, "Microscopic and macroscopic entropy of extremal black holes in string theory," *Gen. Rel. Grav.* **46** (2014) 1711, arXiv:1402.0109 [hep-th].

[7] Planck Collaboration, Y. Akrami et al., "Planck 2018 results. X. Constraints on inflation," *Astron. Astrophys.* **641** (2020) A10, arXiv:1807.06211 [astro-ph.CO].

[8] R. Flauger, M. Mirbabayi, L. Senatore and E. Silverstein, "Productive interactions: heavy particles and non-Gaussianity," *JCAP* **10** (2017) 058, arXiv:1606.00513 [hep-th].

[9] M. Münchmeyer and K. M. Smith, "Higher N-point function data analysis techniques for heavy particle production and WMAP results," *Phys. Rev. D* **100** no. 12 (2019) 123511, arXiv:1910.00596 [astro-ph.CO].

[10] G. Panagopoulos and E. Silverstein, "Multipoint correlators in multifield cosmology," arXiv:2003.05883 [hep-th].

[11] S. Kachru, R. Kallosh, A. D. Linde, J. M. Maldacena, L. P. McAllister and S. P. Trivedi, "Towards inflation in string theory," *JCAP* **10** (2003) 013, arXiv:hep-th/0308055.

[12] E. J. Copeland, R. C. Myers and J. Polchinski, "Cosmic F and D strings," *JHEP* **06** (2004) 013, arXiv:hep-th/0312067.

[13] M. Alishahiha, A. Karch, E. Silverstein and D. Tong, "The dS/dS correspondence," *AIP Conf. Proc.* **743** no. 1 (2004) 393–409, arXiv:hep-th/0407125.

[14] E. Silverstein and A. Westphal, "Monodromy in the CMB: gravity waves and string inflation," *Phys. Rev. D* **78** (2008) 106003, arXiv:0803.3085 [hep-th].

[15] L. McAllister, E. Silverstein and A. Westphal, "Gravity waves and linear inflation from axion monodromy," *Phys. Rev. D* **82** (2010) 046003, arXiv:0808.0706 [hep-th].

[16] X. Dong, B. Horn, E. Silverstein and A. Westphal, "Simple exercises to flatten your potential," *Phys. Rev. D* **84** (2011) 026011, arXiv:1011.4521 [hep-th].

[17] CMBPol Study Team Collaboration, D. Baumann et al., "CMBPol mission concept study: probing inflation with CMB polarization," *AIP Conf. Proc.* **1141** no. 1 (2009) 10–120, arXiv:0811.3919 [astro-ph].

[18] R. Flauger, L. McAllister, E. Silverstein and A. Westphal, "Drifting oscillations in axion monodromy," *JCAP* **10** (2017) 055, arXiv:1412.1814 [hep-th].

[19] P. D. Meerburg et al., "Primordial non-Gaussianity," arXiv:1903.04409 [astro-ph.CO].

[20] C. Cheung, P. Creminelli, A. Fitzpatrick, J. Kaplan and L. Senatore, "The effective field theory of inflation," *JHEP* **03** (2008) 014, arXiv:0709.0293 [hep-th].

[21] L. McAllister, E. Silverstein, A. Westphal and T. Wrase, "The powers of monodromy," *JHEP* **09** (2014) 123, arXiv:1405.3652 [hep-th].

[22] R. Flauger, L. McAllister, E. Pajer, A. Westphal and G. Xu, "Oscillations in the CMB from axion monodromy inflation," *JCAP* **06** (2010) 009, arXiv:0907.2916 [hep-th].

[23] S. R. Behbahani, A. Dymarsky, M. Mirbabayi and L. Senatore, "(Small) Resonant non-Gaussianities: signatures of a discrete shift symmetry in the effective field theory of inflation," *JCAP* **12** (2012) 036, arXiv:1111.3373 [hep-th].

[24] P. Creminelli, "On non-Gaussianities in single-field inflation," *JCAP* **10** (2003) 003, arXiv:astro-ph/0306122.

[25] M. Alishahiha, E. Silverstein and D. Tong, "DBI in the sky," *Phys. Rev. D* **70** (2004) 123505, arXiv:hep-th/0404084.

[26] M. R. Douglas, "The statistics of string/M theory vacua," *JHEP* **05** (2003) 046, arXiv:hep-th/0303194.

[27] G. Obied, H. Ooguri, L. Spodyneiko and C. Vafa, "De Sitter space and the swampland," arXiv:1806.08362 [hep-th].

[28] X. Dong, B. Horn, E. Silverstein and G. Torroba, "Micromanaging de Sitter holography," *Class. Quant. Grav.* **27** (2010) 245020, arXiv:1005.5403 [hep-th].

[29] A. Maloney, E. Silverstein and A. Strominger, "De Sitter space in noncritical string theory," in *Workshop on Conference on the Future of Theoretical Physics and Cosmology in Honor of Steven Hawking's 60th Birthday*, G. W. Gibbons and E. P. S. Shellard, eds., pp. 570–591. Cambridge University Press, 2002. arXiv:hep-th/0205316.

[30] E. Silverstein, "(A)dS backgrounds from asymmetric orientifolds," *Clay Mat. Proc.* **1** (2002) 179, arXiv:hep-th/0106209.

[31] S. Kachru, R. Kallosh, A. D. Linde and S. P. Trivedi, "De Sitter vacua in string theory," *Phys. Rev. D* **68** (2003) 046005, arXiv:hep-th/0301240.

[32] A. Saltman and E. Silverstein, "A new handle on de Sitter compactifications," *JHEP* **01** (2006) 139, arXiv:hep-th/0411271.

[33] J. M. Maldacena, "The large N limit of superconformal field theories and supergravity," *Int. J. Theor. Phys.* **38** (1999) 1113–1133, arXiv:hep-th/9711200.

[34] S. Panda, Y. Sumitomo and S. P. Trivedi, "Axions as quintessence in string theory," *Phys. Rev. D* **83** (2011) 083506, arXiv:1011.5877 [hep-th].

[35] V. Gorbenko, E. Silverstein and G. Torroba, "dS/dS and $T\overline{T}$," *JHEP* **03** (2019) 085, arXiv:1811.07965 [hep-th].

[36] A. B. Zamolodchikov, "Expectation value of composite field T anti-T in two-dimensional quantum field theory," arXiv:hep-th/0401146.

[37] X. Dong, E. Silverstein and G. Torroba, "De Sitter holography and entanglement entropy," *JHEP* **07** (2018) 050, arXiv:1804.08623 [hep-th].

[38] T. Hartman, J. Kruthoff, E. Shaghoulian and A. Tajdini, "Holography at finite cutoff with a T^2 deformation," *JHEP* **03** (2019) 004, arXiv:1807.11401 [hep-th].

[39] S. B. Giddings, "Holography and unitarity," arXiv:2004.07843 [hep-th].

[40] L. Randall and R. Sundrum, "Large mass hierarchy from a small extra dimension," *Phys. Rev. Lett.* **83** (Oct. 1999) 3370–3373. https://link.aps.org/doi/10.1103/PhysRevLett.83.3370.

[41] L. Randall and R. Sundrum, "An alternative to compactification," *Phys. Rev. Lett.* **83** (Dec. 1999) 4690–4693. https://link.aps.org/doi/10.1103/PhysRevLett.83.4690.

[42] A. Strominger, "The dS/CFT correspondence," *JHEP* **10** (2001) 034, arXiv:hep-th/0106113.

[43] L. Susskind, "Strings, black holes and Lorentz contraction," *Phys. Rev. D* **49** (1994) 6606–6611, arXiv:hep-th/9308139.

[44] M. Karliner, I. R. Klebanov and L. Susskind, "Size and shape of strings," *Int. J. Mod. Phys. A* **3** (1988) 1981.

[45] M. Dodelson and E. Silverstein, "Long-range nonlocality in six-point string scattering: simulation of black hole infallers," *Phys. Rev. D* **96** no. 6 (2017) 066009, arXiv:1703.10147 [hep-th].

[46] A. Mousatov and E. Silverstein, "Recovering infalling information via string spreading," arXiv:2002.12377 [hep-th].

28

Lee Smolin

Faculty member at the Perimeter Institute for Theoretical Physics, Waterloo, Canada

Date: 15 July 2011. Location: Waterloo, Canada. Last edit: 16 October 2016

In your opinion, what are the main problems in theoretical physics at the moment?

Understanding the completion of quantum mechanics and the problem of finding a theory of quantum gravity are the key problems. Also, of special importance is the whole conundrum of questions concerning why this universe (rather than another possible world). Why do we see the laws of nature that we do and why these specific initial conditions that originated the universe that we are in? These are not just the problems of the moment; they are the key issues of the present period.

And while we are talking about the present moment, I should say that I am recently very interested in the role of time in fundamental physics and cosmology, and particularly in the hypothesis that the present moment is real, but the future is not, yet.

And why do we need a theory of quantum gravity?

We need it in order to have a unified picture of the world and therefore a unified understanding of the world. There are actual experiments in the domain where Planck's constant, Newton's constant and the speed of light are all relevant. At the present moment, we can really do experiments at the Planck scale and therefore we need to be able to predict what the outcomes will be. That is why we need a theory of quantum gravity.

I understand this as the necessary completion of the revolution begun by Albert Einstein in 1905, which overthrew Newtonian physics twice, resulting in quantum mechanics and general relativity. The revolution won't be over until we have a single theory rather than two unconnected incomplete theories. Our situation is analogous to that of Galileo and Kepler – each had a piece of the puzzle resulting from Copernicus's overthrow of Aristotelean physics, but the revolution would be incomplete until Newton's synthesis.

Now, whether the "quantum theory of gravity" is the quantisation of a classical theory of gravity or is arrived at through some process that is deeper, I don't know. I'm very open to the possibility that it is not the quantisation of a classical theory of general relativity.

What kind of experiments are you referring to?

Astrophysicists have been conducting several experiments that probe a domain of quantum gravity we call the relative locality limit. This is the domain in which $\hbar$ and G are taken to zero, while their ratio,

$$m_p^2 = \frac{\hbar c}{G}$$

is held fixed. Here c is also held fixed, so these are phenomena governed by two constants, c and m_p. Possible phenomena in this domain involve modifications of propagators or scattering amplitudes of order E/m_p. These might come from breaking or deformations of Lorentz or Poincaré symmetries.

For example, one can measure whether there is a time delay in the propagation of light, or, in other words (though not the most exact but good enough), one can measure if there is a dependence of the speed of light on the energy of incoming light rays due to quantum gravity effects. These effects would be of the order of energy over the Planck energy and hence are in this domain. Given a difference in speeds of

$$\frac{\Delta v}{c} \approx \epsilon \frac{\Delta E}{m_p},$$

a pair of photons emitted simultaneously with an energy difference of GeV between them accumulate around a second difference in arrival time after travelling a billion years. Such delays are in fact observable in observations of gamma ray bursts at the Fermi Gamma Ray Space Telescope. Interestingly enough, none has been seen leading to bounds on ϵ of order unity.

Also, the question of whether Lorentz invariance is broken or not is observable, in fact, at six or seven orders of magnitude past the Planck scale in experiments which test for effects of birefringence (see [1] for a review). So we are definitely in the era where quantum gravity is testable and we need to have theories that are able to make predictions. We have the opportunity of constraining ideas and theories of quantum gravity by experiment.

I should also mention that we call this domain the relative locality limit because it is best understood as a very particular weakening of the concept of locality. It is not really that photons of different energies travel different speeds; the deeper and more precise view is that there are modifications in the concept of locality that are energy-dependent and observer-dependent.

Why is it so difficult to quantise gravity?

It's not difficult – at least to get started. The situation we are in is that there is not one candidate but there are half a dozen instead. Some of them are more developed than others. The two that are more developed are loop quantum gravity (LQG) and string theory. LQG, certainly, is a consistent theory, which is the quantisation of general relativity. It has quantum dynamics, quantum kinematics, and we understand it plus we can do calculations with it. In string theory, there is, to a certain approximation, a unification of all forces. There are other approaches, which are not as well developed, that look promising such as causal dynamical triangulations (CDT) and asymptotic safety.

Each of these approaches achieves something non-trivial and surprising, which adherents point to to justify their interest in it. But by now each of these approaches has also hit at least one roadblock – a persistent issue that makes it unlikely to be the right theory. There are interesting lessons in both the successes and the failures, which can hopefully guide us in inventing an approach that goes all the way.

The difficult part is then not getting some results from combining general relativity and quantum mechanics. The real difficulty is in selecting the right way of doing it – that will go all the way and won't get stuck. The fact that it was not enough just to find a theory of quantum gravity is very interesting. One would not have thought that one could get halfway to quantum gravity in a half a dozen different ways.

Recently I have been reflecting on why we have failed to find an approach to quantum gravity that makes contact with observation and has the ring of truth. One question that I am convinced is paramount is the nature of time. For reasons I have argued in detail in my last two books I believe that we must embrace a temporal version of naturalism in which the present moment is real, and the future is not yet real and to an extent open. Related to this is the need, also argued for in those books, for a new framework for laws that can sensibly be applied to the universe as a whole, rather than subsystems of the universe.

I also think we fail to heed Einstein's distinction between constructive theories and principle theories. The former are descriptions of particular phenomena, such as Maxwell's equations. The latter are general principles that apply to all phenomena, such as the laws of thermodynamics or the equivalence principle.

Most of the well-studied approaches to quantum gravity, such as string theory, loop quantum gravity, CDT, causal sets, etc. are constructive theories. I believe they may contain parts of the story but to get the whole story we need to step back and invent novel principles. The holographic principle is one candidate for such, but I think it doesn't suffice – there is at least one missing principle. The reason is the holographic principle involves only the combination of constants $l_p^2 = \hbar G$. To get $\hbar$ and G separately into the story we need a second principle.

Why did you say that string theory is only valid up to a certain approximation?

Because it's only defined in perturbation theory and that perturbation theory is only demonstrated to be consistent or finite to genus two. Finiteness has not been proven to all orders.

Even more important is the problem that there is no non-perturbative and background-independent formulation of the theory in general. There is also the issue that the theory makes no predictions, because of both the landscape issue and the lack of a non-perturbative formulation.

There is the AdS/CFT correspondence [2] – and that is no doubt important. But the anti–de Sitter (AdS) background is a background; moreover, most of the calculations that verify the conjecture concern the planar $N \to \infty$ limit in which the bulk theory is a classical solution to general relativity or supergravity. The rest concern perturbation theory on a fixed background.

I am convinced that, at least at the level where the bulk theory is classical, there is an AdS/CFT correspondence, because the need for such a correspondence can be argued for on very general grounds without having to refer to supersymmetry or string theory. Several developments support this broader, but weaker, reading of the AdS/CFT correspondence. These include the holographic renormalisation group (RG) and related results from shape dynamics [3]. These do not rely on supersymmetry or strings and suggest a general perspective which is that the renormalisation group applied to a conformal field theory (CFT), or maybe even a more general quantum field theory, has a geometrical description given by general relativity with an added dimension corresponding to scale. Indeed the shape dynamics work explains exactly why the bulk theory generated by the RG of a CFT should have spacetime diffeomorphism invariance. There is recent work of Sung-Sik Lee and collaborators which adds evidence for this perspective [4]. There are also fascinating recent results suggesting a deep connection between geometry, renormalisation and entanglement entropy, which make no reference to supersymmetry or strings. From this perspective, the specific form of the AdS/CFT correspondence conjectured by Juan Maldacena could be a specific instance of this more general point of view.

You think you would be able to recover all AdS/CFT results with shape dynamics?

That's too strong of a claim but I suspect that many of the results will turn out to have nothing to do with supersymmetry or string theory. The correspondence is more general. At least in the limit where the bulk theory is classical, aspects of the AdS/CFT correspondence seem to me to be quite general and to hold independently of string theory and supersymmetry. However, to get stronger results, you need to impose more structure, such as supersymmetry.

Furthermore, whether shape dynamics give a correspondence between a quantum theory of gravity and a dual conformal field theory one dimension down, I don't know the answer. Part of it is that we don't yet know much about the quantisation of shape dynamics. Indeed, such a correspondence might even be a construction of a quantum shape dynamics.

What is the problem with string theory being only defined perturbatively? The standard model is only defined perturbatively and it is very useful ...

The fact that the standard model is only defined perturbatively tells us that it is only an effective field theory, and hence incomplete, requiring an ultraviolet completion. A part of this is the difficulty of defining chiral fermions – which are essential for the standard model – but cannot be included in a non-perturbative, lattice definition of the theory without violating locality. It is of course possible to regard string theory as just an effective theory, which would require a completion in a different framework. But, look, I don't want the focus of this interview to be a critique of string theory if you don't mind.

My view of string theory is very public and I have worded what I have to say very carefully. People can read what I wrote, which is in fact balanced between positive and negative remarks. Very quickly, my view of string theory is that there are some things that I'm very interested about in it and there are some things that I believe have not yet been demonstrated convincingly. I have put it all out on paper so we don't have to go through it again. Also, LQG has been developing faster in the last few years whereas not so much has changed since ten years ago with regards to my critique of string theory published in *The Trouble with Physics* [5].

This is not to say that some people who identify as "string theorists" have not done extremely interesting work. Two examples are all the work on amplitudes for supersymmetric gauge and gravity theories and the work on entanglement in AdS/CFT I mentioned above. But little of this work actually involves string theory (i.e., the hypothesis that the unified theory is constructed from the dynamics of two surfaces) and even less of it addresses the key long-standing obstacles to that hypothesis. I think that is very unfortunate, as I have for a long time believed that it would be important to give string theory a background-independent, non-perturbative formulation. I am puzzled that more people don't work on it.

Indeed, the heart of both string theory and LQG is a very compelling physical idea, which is that the physical excitations of a non-Abelian quantum gauge field are extended objects such as strings and branes. This can be expressed by the hypothesis that the vacua of these theories are dual superconductors, so that electric field flux is quantised. This picture gives rise to both string theory and LQG. When the dynamics of the quantised lines of electric flux are worked out in a fixed spacetime background one discovers there is an energy cost per length of the string – a string tension – and working this out you get string theory. Indeed, as Alexander Polyakov first argued, you get string theory one dimension up, so this is also the route to AdS/CFT. When the dynamics is instead worked out in a background-invariant manner, in which there is no fixed classical background metric, you get the Hilbert space of LQG. The dual superconducting vacuum can be defined in the absence of a background metric, and its excitations are spin networks. This vacuum is, in the absence of background fields, invariant under spatial diffeomorphisms and is called the Ashtekar-Lewandowski vacuum [6, 7].

Now the key to LQG is that general relativity can be expressed as a non-Abelian gauge theory. Indeed it's very close to a topological field theory – it is a constrained BF theory. So it can be quantised in a dual superconductor phase, and the result is that electric field flux is a measure of area. Hence area and volume are quantised. In fact, I first began studying the dynamics of flux quanta and Wilson loops in the Ashtekar formulation of general relativity as a warm-up to making a background-invariant formulation of string theory. So for me the two theories have always been closely related. Indeed, several times I have returned to the problem of making a background-independent formulation of string theory and it's a problem I continue to think about. Meanwhile, a lot of things have happened with LQG so I'd rather spend the rest of our time on that.

Sure, I just wanted to hear your opinion about it because there are also many criticisms of LQG so it would only be fair ...

Let's spend our time on LQG then. It is important to emphasise that LQG has gone through two stages. There was the early stage, of 1986–1995 when the focus was on the Hamiltonian theory, defined by imposing quantum constraints on a Hilbert space. Broadly speaking, there were successes and failures. The successes included the quantum kinematics, including the spatially diffeomorphism-invariant states and the boundary state spaces associated with horizons. The failures concerned the attempts to solve the quantum dynamics by defining regulated forms of the quantum Hamiltonian constraint acting on the kinematical state space. While this succeeded in defining such an operator and computing its kernel, my sense is that this part of the effort failed, because the resulting candidate for a physical Hilbert space did not have in it massless particles or long-range correlations. For this reason in the mid-1990s research turned to formulating the quantum dynamics in terms of path integrals. These are called spin foam models (or spin foams) and their study, beginning in the early 1990s, constitutes a second phase, which has been more successful at defining a sensible dynamics.

Many of the criticisms one hears, even now, about LQG concern the first phase, sometimes made by people who don't seem to have heard about all the work on spin foam models the last 20 years. This is like critiquing string theory pre-1995, before D-branes, Strominger-Vafa black hole entropy, M-theory, Maldacena and AdS/CFT.

Okay, let's go there then. What is loop quantum gravity?

There is a broader and a narrower LQG. The broader LQG is a method to quantise diffeomorphism-invariant gauge theories. It can be applied to any diffeomorphism-invariant gauge theory and has been applied to a wide variety of them including general relativity in $3 + 1$ dimensions, general relativity and supergravity in higher dimensions, various theories in $2 + 1$ dimensions and $1 + 1$ dimensions. Plus of course topological quantum field theories in three and four dimensions. So this is the broad sense: it's a precise method of quantisation; that is, you give me a diffeomorphism-invariant gauge theory in any dimension, with dimension two or greater, and we give you a Hilbert space of states, which is gauge-invariant and spatial diffeomorphism-invariant, and a procedure to construct the dynamics through a path integral.

You mentioned that LQG provides a method for quantising any diffeomorphism-invariant gauge theory. Thomas Thiemann applied LQG techniques to the quantisation of the string worldsheet and found completely different results from those obtained using the canonical methods in string theory [8]. How should this result be interpreted?

It is a commonplace that quantisation is ambiguous so a single classical system will have different inequivalent quantisations. This is the case here. In such cases there is no principle that determines which inequivalent quantum theory is correct; they are just mathematical models. At most one can describe nature. There is also no rule or principle that says that one quantisation procedure should be the right one for all systems. So all we learn is that there are inequivalent quantisations of the classical string, which should not be news to us.

You mentioned just before that you obtain a spatial diffeomorphism-invariant Hilbert space of Sates and not spacetime-invariant, right?

The kinematical theory has invariance under spatial diffeomorphisms. To get the full gauge invariance of spacetime diffeomorphisms you have to solve the dynamics. This is a key difference between general relativity (GR) and Yang–Mills theory. Since the gauge invariances of GR include local changes in the time coordinate, the Hamiltonian is a linear combination of constraints. So locally, finding the fully gauge-invariant states and solving the dynamics are intertwined. So, that's the broad meaning.

The specific meaning is a particular instance of this, in which it is applied to general relativity in $3 + 1$ dimensions, coupled to matter fields, coupled to gauge fields, chiral fermions or scalar fields, and that is quite a bit developed by now. The quantum kinematics has been understood rigorously for a long time. In 2002 there was a uniqueness proof [7, 9], namely, that the Hilbert space that we use for the quantum kinematics is unique when subjected to some natural assumptions; that is, there is a unique spatial diffeomorphism-invariant space of states in which theories can represent the Wilson loop operator and the area operator. In the last few years, the work of Bianca Dittrich and collaborators has shown that a technical loophole in those proofs can be exploited to construct new representations which have interesting properties, connected to topological quantum field theories [10].

There are also important recent results on black hole thermodynamics, which have resulted in derivations of the Bekenstein-Hawking entropy that get the 1/4 right for any value of the Barbero-Immirzi parameter. (The Barbero-Immirzi parameter is a kind of θ angle for quantum gravity. It also sets the quanta of area.)

Okay, but I thought that one of the big claims of LQG is that it can explain the Bekenstein-Hawking formula for black hole entropy under the condition that the Barbero-Immirzi parameter had to be fixed.

That is no longer the case. Those results have been superseded by the results I mentioned above that derive correct black hole thermodynamics for any value of the Barbero-Immirzi parameter. These results concern black holes modelled in spin foam models [11–14]. The key point is that the relevant ensemble is the one defined at fixed temperature. The earlier results concerned a different ensemble – one at fixed area – and that is where the Barbero-Immirzi parameter came in.

But most recent work has concerned the path integral approach to dynamics. In the last 10 years, there has been a convergence of results to a unique dynamics, which is called the new vertex [15, 16]. Before, there were other forms of the dynamics that were investigated and found fault with for different reasons. One good thing about the new vertex is that it leads to the emergence of the correct massless spin-two propagator. So one thing we know is that the theory contains the correct graviton degrees of freedom. An important insight is that the new vertex results from imposing on a topological quantum field theory a constraint that liberates precisely the massless spin-two degrees of freedom of gravitons. Moreover, that constraint is nothing but an expression of the first law of thermodynamics. As a result, one can directly derive the Einstein equations from a suitable semiclassical limit of the theory.

This is a beautiful story which was only realised recently [17]. The classical Einstein equations are coded into the measure of the full non-perturbative path integral, through constraints that break some gauge symmetries of a topological quantum field theory. The constraints, in turn, are equivalent to the first law of thermodynamics. In other words, the connection between the quantum, GR and thermodynamics, which we have been fascinated by ever since the work of Stephen Hawking, William Unruh, Paul Davies, etc., is coded into the measure of the spin foam path integral. This work was of course inspired by the beautiful result of Ted Jacobson, who showed that the Einstein equation can be understood to arise as an equation of state in a more fundamental theory [18]. This result complements earlier results which provide evidence for the hypothesis that spin foam models have both a semiclassical and a continuum limit, which is classical general relativity.

Related to this calculation of black hole entropy, Ashoke Sen published a paper [19] with corrections to black hole entropy using Euclidean gravity methods. These results seem to disagree with all previous results of corrections to black hole entropy in LQG. How should these new results be interpreted?

Sen's paper concerns the corrections to black hole entropy of the form of $c\ln(Area/\hbar G)$. These have been studied for a long time. They were in fact first studied in LQG [20, 21], and inspired Carlip to find such corrections to the Strominger-Vafa extremal black holes in string theory [22]. In LQG the result is $c = 3/2$ and there are arguments that value is universal. Sen finds a different number for pure general relativity. Is this a problem? I don't think so because the two calculations count different contributions to entropy.

The LQG calculation counts the log of the dimension of the Hilbert space of quantum degrees of freedom of the geometry of the horizon. This is a quantum gravity calculation; it requires that the horizon be a quantum geometry. Sen does instead a semiclassical calculation which counts the contribution to a partition function of linearised spin-two fluctuations in the exterior of the black hole horizon. The background spacetime is classical; what he counts is the semiclassical approximation to the entropy of the spin-two degrees of freedom in the spacetime to the exterior of the black hole. The degrees of freedom that are counted are different, so I see no reason to expect that the two calculations should agree.

Okay, you mentioned that there is a continuum and semiclassical limit that gives general relativity?

I am saying there are several different calculations which each provide evidence for the hypothesis that general relativity is the semiclassical and continuum limit of the spin foam models. Evidence is not rigorous proof – and we can explicate what is still needed, but it is progress.

We still need to define a version of coarse-graining and renormalisation for a spin foam model. Important steps to that have been taken by several people, including Bianca Dittrich and her group here at Perimeter Institute. Meanwhile, the thermodynamic arguments I alluded to above provide evidence for a related but different claim; that if any spacetime geometry emerges from a coarse-graining of a spin foam model, that geometry satisfies the Einstein equations. Another piece of evidence is the results on recovering the correct spin-two propagator [23]. Still more evidence is that the black hole thermodynamics comes out right.

Now let me describe some of the results regarding recovering general relativity. Some of these results concern an approximation to the spin foam amplitude in which all spins are taken to be large. These label the faces of the simplicial complex that defines a spin foam history. Large spin means large area in Planck units, so this is a semiclassical limit. What is found is that the spin foam action in this limit goes to the Regge action, which is the expression of the Einstein-Hilbert action evaluated on such a simplicial complex. Results on this were found by John Barett and collaborators, and by Laurent Freidel and Florian Conrady [24, 25].

But this last work by Freidel and Conrady is not a continuum limit, right?

Right, that's more like a kind of semiclassical limit. Evidence for the emergence of the Einstein equations in a different limit, related to the continuum limit, was also found by Elena Magliaro and Claudio Perini [26, 27]. This one is a continuum limit because they take a limit in which the Barbero-Immirzi parameter goes to zero, which means that the size of the energy quanta goes to zero, so that the spectrum of the area becomes continuous. Then, in that limit, they were able to use a semiclassical expansion in order to compute corrections in powers of the Barbero-Immirzi parameter, for small Barbero-Immirzi parameter.

So they were able to derive general relativity at the leading order in that expansion and corrections which include parity-breaking corrections, for example. It has been kind of conjectured for a long time that the Barbero-Immirzi parameter would be a measure of the quantum corrections that break parity and chirality [28]. For example, Claudio and Elena have a computation of the graviton propagator, which to leading order is what it should be with the right tensor structure, but there are corrections proportional to the square of the Barbero-Immirzi parameter which break parity.

I am slightly confused because I hear contradictory statements from different people and different papers which lead me to think that there is no such limit.

Let me just respond to that. First of all, we have been saying that. In LQG, we have, with a few exceptions, never exaggerated our claims. We have always been upfront with what the issues are, what the open problems are ...

I understand but I did notice in several papers many years ago that people were saying that there was a semiclassical limit if you introduce a cosmological constant [29, 30]. If I'm not mistaken, this is referred to as the Kodama state [31, 32], right?

The Kodama state came up early in the first phase of LQG, and was never rigorous. The results on the semiclassical and continuum limits of spin foam models are part of the second stage, and are far more detailed and sophisticated.The Kodama state is

$$\Psi(A) = e^{\imath \frac{k}{4\pi} S_{CS}(A)}$$

where $S_{CS}(A)$ is the Chern-Simons invariant of the Ashtekar connection and $k = 6\pi/\hbar G\Lambda$ is the level. There are two separate issues. First of all, is the Kodama state a semiclassical state for quantum general relativity with a cosmological constant Λ? Yes, at the level of a WKB wave function it's a semiclassical state associated with de Sitter spacetime. This reflects the fact – and it is a very interesting fact – that the Chern-Simons functional of the Ashtekar connection on a compact three-manifold solves exactly the Hamilton-Jacobi function for general relativity. This is closely related to the close connection between general relativity and BF theory – a topological field theory. It also is involved in the appearance of Chern-Simons theory as the boundary action on horizons. Moreover, it implies that the representation theory for spin networks must be quantum deformed in the presence of a cosmological constant, which serves as an infrared cutoff. There are a number of things that can be derived from that fact; for example, one can couple the theory to matter and derive quantum field theory for the matter in de Sitter space using the Wheeler-DeWitt equation, etc. [29]. It functions just fine as a semiclassical state. So at the semiclassical level the Kodama state has been useful.

We can go on to ask the second question: Is the Kodama state an exact state of full quantum gravity? But first, let me emphasise that there are lots of contexts in quantum many-body physics or quantum field theory where a semiclassical state is used to describe physics in a semiclassical approximation. In these cases one does not expect that the semi-classical state is an exact eigenstate of the quantum field theory or many-body Hamiltonian. They are not – but this does not detract from their usefulness in extracting the right physics. So if we can use the Coleman-DeLucia state to describe inflation, we can use the Kodama state to argue that de Sitter spacetime is the semiclassical limit to LQG with a positive cosmological constant, even if it is not an eigenstate of the full Hamiltonian. But let's go on and ask whether the Kodama state solves the full quantum constraints. The answer is that there is a particular ordering and regularisation of the Hamiltonian constraints which makes it so. There was understandably a lot of excitement about this a long time ago. But is it an honest physical state? That is, is it normalisable under the correct, physical, inner product? And are the choices of operator ordering and regularisation that you have to make in the Hamilton constraints to make it a physical state correct? The answers to these questions are not known, but I'm not optimistic.

Witten argued not, based on an analogy to Yang–Mills theory [33]. The analogy is incomplete, but there is still cause for concern. Freidel and I examined the linearisation of the Kodama state and found evidence for instabilities. This was re-examined by Magueijo and collaborators, who found that the linearised Kodama state is not the right vacuum for linearised gravitons on a de Sitter background. So while the idea of employing the Kodama state beyond the semiclassical approximation comes up from time to time, I am not optimistic.

It is interesting that what might be called the "ghost of the Kodama state" does persist in two aspects of the theory. First is the role of Chern-Simons theory as the dynamics on an horizon, which I introduced in 1995. Second is the fact that the cosmological constant has the effect of introducing a quantum deformation into the representation theory that defines the labelling on the spin networks and spin foams. I had introduced this in the 1990s, and apart from John Baez and Louis Crane, few took it seriously. However, recently this result was found in several different formulations of spin foam models and now is widely accepted.

The recent results using the new vertex are newer and mostly unrelated to the Kodama state. I can also say that I have been very impressed with this work. My work in the last years has not been focused on LQG but on a variety of things, such as quantum gravity phenomenology and some exploratory ideas like unification that we have just talked about. I have not been working in mainstream LQG which has consisted of the exploration and development of spin foam models. But I have been impressed by the recent results.

Okay. There is a claim in the literature that by using the spectral triple approach to LQG one can find a well-defined semiclassical limit [34]. Is this something useful?

I must say that I have not understood their work properly so I shouldn't comment, except to say it sounded promising. If you wanted an evaluation of that work you should ask some more mathematically minded people.

What are the fundamental degrees of freedom in LQG?

For pure gravity, the pure spin-two degrees of freedom. One can check this by linearising the theory or by counting the physical degrees of freedom, or else by computing the graviton propagator.

LQG is a very conservative research program. As I mentioned, it is a quantisation method for diffeomorphism-invariant gauge theories. So, the fundamental degrees of freedom are the degrees of freedom of whatever classical theory you are quantising. If you take classical general relativity, for example, its degrees of freedom are represented by the Ashtekar connection or some generalisation of it, or the canonical coordinates and the conjugate momentum related to the area of surfaces.

Is there a good reason for expecting that you can take general relativity, quantise it and obtain the real world at the Planck scale? Why not supergravity, for example?

Well, as I said earlier, you can apply the same method to supergravity and it has been done. I worked on supergravity in LQG very early on, with Ted Jacobson [35–37], then later with Yi Ling [38, 39]. We showed that $N = 1$ supergravity goes through, without significant changes to the results. We also set up and got a few results about 11-dimensional supergravity. For example, as Gambini first showed [40], the tensor gauge fields can be quantised using the methods of LQG. Higher N are complicated because you have BPS states, so there is an opportunity for someone to do a good piece of work understanding BPS states in LQG. There is also a recent research program by Thomas Thiemann and collaborators to apply LQG to 11-dimensional supergravity [41], which is at the heart of the whole string world, and give a quantisation of 11-dimensional supergravity. These succeed because, as I said, LQG is first of all a methodology.

However, one can be ambitious is two different directions. For instance, one can look for a way in which the fermions and gauge fields of the standard model might emerge from structures in quantum general relativity. There was a research program related to that in which braid excitations – excitations which look like braids – were found to propagate in a way that is characterised by some conserved quantum numbers and there were some speculations that one would be able to get matter out of quantum topology, which is an old idea. I was part of this research program together with Fotini Markopoulou and Sundance Bilson-Thompson [42].

Was this successful?

Not yet. We got a classification of states that agreed with the first generation of standard model fermions, but the third generation was not reproduced correctly.

I guess this would be some sort of unification, right?

Yes, if it would be successful but I don't expect it to be successful. In any case, I think it's worth investigating since there are these excitations which have braids and more complicated topologies and we want to understand their role in the theory. Another direction is that there is a natural way to unify general relativity and gauge fields in a non-trivial way by extending the dynamics in LQG either in the Ashtekar form of the constraints which was the program of Peter Peldan [43] or in the Plebanski action which is the more recent way that I developed in response to Garrett Lisi's work. Garrett Lisi had a conjecture related to using a certain gauge group E_8 to unify all the degrees of freedom [44] – an idea that he took from a very old paper by MacDowell and Mansouri [45]. I was then able to make a more elegant version of this idea, working together with him and Simon Speziale, not specifically for E_8 but for unifying gravity with any gauge group [46]. I think that it is an encouraging work and worth investigating.

Since you mentioned you worked on supergravity, do you expect that supersymmetry will be found at LHC?

I'm neutral about supersymmetry. I don't find it overwhelmingly beautiful, but it's an interesting structure which extends the symmetries of special relativity and I get that it might have helped resolve the hierarchy problem while giving us dark matter candidates. At the same time, the application of SUSY to beyond the standard model physics was considerably uglier than the basic idea, and introduced around one hundred new free parameters. So I never found low-energy supersymmetry compelling.

So to me it's a possible structure nature might employ, but it doesn't destroy anything about my worldview if supersymmetry is not there. I don't think that it's particularly elegant. If it had played a role in unification of particle physics and solved the hierarchy problem, I think it's fair to say that we would already have seen it experimentally. I'm not an expert but that's my understanding. It was a reasonable hypothesis as far as the hierarchy problem is concerned but I think that technicolour is a much more elegant hypothesis. I'm much more interested in the idea that the Higgs is a composite field, which is a result of some dynamical chiral symmetry breaking. However, if there is $N = 1$ supersymmetry in nature, it also won't disturb me. However, I think that nowadays there are even fewer reasons to believe that there is low-energy supersymmetry, as the constraints on super partners from the LHC have strengthened considerably.

In LQG is general covariance actually there at the quantum level? Is this clear?

By general covariance do you mean active spacetime diffeomorphisms? We do have a complete set of states that are in the kernel of the spatial diffeomorphism constraints – these make up the spatial diffeomorphism-invariant states. We can define a regulated version of the Hamiltonian constraint that acts on these states and discover an infinite-dimensional subspace of states that are annihilated by the Hamiltonian constraint in the limit in which the regulator is removed. So you might argue that we have the ingredients needed to construct a physical state space of spacetime diffeomorphism-invariant states. These were among the early claims we made with Jacobson [47] and Rovelli [48] and they are true. Our work was not mathematically rigorous, but these results were recovered within a rigorous framework by Thiemann, following work he did in collaboration with Ashtekar, Lewandowski and Corichi (see [7, 49]).

So, do we have a sector of physical, spacetime diffeomorphism-invariant states?

We do in the Hamiltonian approach, but I'm not convinced it is physically correct. First, to complete the construction of the physical Hilbert space we need to construct the physical inner product, which must exist on the kernel of the constraints and make the gauge-invariant states normalisable. It also must make the real-valued physical observables into Hermitian operators. One problem is that these must be operators that commute with the quantum constraints and it turns out to be not as easy to find these as it was to find states annihilated by the constraints. Furthermore, I do not believe that the regularisations of the Hamiltonian constraint that were studied by Thiemann and others give an operator that has the right dynamics coded into it. In particular, I don't believe that the theory has massless gravitons or long-range correlations. So it doesn't have the right low-energy limit.

My understanding is that after Thiemann made rigorous some ideas that many of us have been playing with, regarding the construction of the Hamiltonian constraint, and formulated the Thiemann constraint in closed form [49], it was realised pretty quickly that we did not have a low-energy approximation with massless gravitons. This was clear to me and pretty soon the interest in dynamics shifted a lot to spin foam models. There are some hand-waving arguments stating that because of the kind of discreteness that is present in LQG you shouldn't expect there to be an infinitesimal generator of time translations because there is not an infinitesimal generator of spatial diffeomorphisms. There are only finite unitary operators representing spatial diffeomorphism invariance. There are various versions of this argument; for instance, Fotini Markopoulou had some, Carlo Rovelli had some and I had some.

There are two other issues which the Hamiltonian approach to LQG faces. One is the question of whether there is an operator representing the ADM mass which is Hermitian and positive definite. This is unresolved. I showed in [50] that this would require certain conditions to be imposed on the physical inner product. The second is that Hamiltonian LQG has a fermion doubling problem. I suspected this for years, and we showed it with Jacob Barnett in [51]. (Although Gambini and Pullin find a way to fix the problem in a $1+1$-dimensional model [52].) This means that chiral fermions cannot be represented, but these are essential for the standard model. For these reasons many people gave up on the Hamiltonian approach to the dynamics of LQG back in the mid-1990s. Instead, they turned to develop path integral approaches to the dynamics – the spin foam models.

So, starting roughly in 1995, the interest in spin foam model began to develop. It took a few years to find out what is the right form of the spin foam amplitude so that new results could be derived from it. In order to demonstrate that there is really a strong form of spacetime diffeomorphisms invariance requires some results a lot stronger than the existence of a semiclassical limit where you find the semiclassical Einstein equations dominating the path integral. Issues of spacetime diffeormorphism and how they should be recovered or not at the quantum level is something that, for example, Bianca Dittrich has been very focussed on [53]. I don't know everyone you are talking to but if you talk to people in the LQG community you will find a range of people from highly enthusiastic and optimistic to concerned about technical issues. Probably Carlo Rovelli is on one extreme and if you listen to his talk at the Madrid meeting,[1] he says something like, "Here is the theory that we have; we don't know everything about it but it is well defined and has this and that property." I've tended to be on the more critical side. My role in the LQG community is not that different than my role with respect to the string theory community; it's just that LQG is a small community and hence it is less publicly manifest that I tended to be more critical and more conservative, especially in the last 15 years or so.

At the same time, there are aspects of LQG that seem to me so unexpected and beautiful that I am surprised they are not better known and appreciated outside the LQG community. Especially since one can imagine them being taken up and used by other approaches. I find it hard to believe they will not be part of the final story, whatever that is.

What results are underappreciated? What insights have been gained from LQG that you think should be part of the final story?

First of all, the fact that it is possible to make a genuinely background-independent quantisation of the gravitational field. There are many specific lessons we learned from this. The key idea of LQG is to write GR in first-order form as a theory of a connection. This immediately gives rise to a bundle of facts and insights connected with the close relation that exists between general relativity and topological field theory. This is that when written as a theory of a connection, the action of GR is that of a topological field theory–BF theory, on which some quadratic constraints have been imposed. This turns out to imply several fascinating insights:

[1] See www.cpt.univ-mrs.fr/~rovelli/LOOPS11/Rovelli-LOOPS11.html.

- The whole action is cubic in its variables. So the equations of motion of GR are quadratic equations. This is a drastic simplification from Einstein's action which is non-polynomial. Indeed the simplest form in which you could write a non-linear theory is one where the equations of motion are quadratic, so there can't be a simpler form.
- As a result one encounters only simple products of operators expressing the theory in operator form for the quantum theory – and the operator products can be defined in a way that respects spatial diffeomorphism invariance.
- If one imposes boundaries, the natural boundary action which arises is a Chern-Simons theory. This turns out to be the right way to describe horizons.
- Hence there is a natural realisation of holography, because the Hilbert space of degrees of freedom on the horizon has a dimension whose log is proportional to the area in Planck units.
- There is a key role played by the representation theory of quantum groups where the deformation parameter codes the cosmological constant.
- This results in a natural setting for extending the thermodynamics of quantum field theory on de Sitter spacetime to a thermal quantum gravity theory with positive cosmological constant [29, 54, 55].
- The connection with thermodynamics is coded into the simplicity constraints, as they are equivalent to the first law of thermodynamics.
- The theory makes contact with a body of results from topological quantum field theory; for example, the space of diffeomorphism-invariant states is classified by knot theory (because the states are in correspondence to the diffeomorphism classes of embeddings of certain labeled networks). So we have a natural setting for what mathematicians such as Louis Crane call the ladder of dimensions, linking CFT in $1 + 1$ to Chern-Simons theory in $2 + 1$ and up to BF theory in $3 + 1$.
- These structures play a role in quantum gravity in $2 + 11$ dimensions, where the theory is a topological quantum field theory related to Chern-Simons theory. Related to this is a close connection, in the limit where the cosmological constant is removed, to a deformation of Poincaré invariance, called κ-Poincaré. This makes $2 + 1$ quantum gravity, coupled to matter, an explicit realisation of the phenomenological frameworks of relative locality (or what we used to call deformed special relativity).
- The expression of the theory as a constrained topological field theory works not just for GR in $3 + 1$, but in higher dimensions, and it very elegantly can incorporate supergravity.

I feel very confident that these structures will be part of the eventual quantum theory of gravity.

One of the predictions of LQG is that the spacetime is quantised. There is an area operator and there is a minimum area. If this is a prediction, how can you probe it?

By measuring some really small areas (laughs), like black hole entropy, which is a consequence of the discreteness of the area. Corrections to the semiclassical Hawking radiation would be another way. One important question that we don't know the answer in 3 + 1 dimensions is whether, as a consequence of the discreteness of LQG, Lorentz invariance is either broken or deformed. I think that there is good theoretical evidence in the new spin foam models that there is a relativity of inertial frames and that there is no breaking of the relativity of inertial frames in the definition of the path integral. Whether, resulting from it, there is a phenomenology or not – that is, whether the excitations around the low-energy limit transform naively as in special relativity or there is a deformation of the Poincaré group – it's not settled. I hope for the second possibility and I've had some papers making very rough semiclassical arguments in this direction but it's far from being shown.

In 2 + 1 we know it's true that the low-energy description of 2 + 1 gravity coupled to matter is that matter moves in a non-commutative geometry, where the Poincaré algebra is quantum deformed. This has been understood precisely partly due to the work of Laurent and Etera Livine [56] and probably a bunch of other people. So this could potentially have phenomenological implications.

Do you find any evidence of a fractal structure at small scales?

I don't think so. In the mid-1980s Louis Crane and I published two papers speculating that the spectral dimension of excitations of quantum gravity would be dimensionally reduced at Planck scales to less than four [57, 58]. This resulted in a formulation in which the idea of asymptotic safety was realised, without the ghost that otherwise haunts renormalisable approaches to quantum gravity. This inspired me to play with networks of Wilson lines as a way of realising dimensional reduction, and this was part of the route to LQG.

In LQG there is a trivial sense of dimensional reduction in the sense that quantum geometry is based on networks of one-dimensional excitations. (This is discussed in [59, 60].) But there is no asymptotic scale invariance. There are simply no excitations below the Planck scale. This idea is still around; indeed, to my understanding it remains the case that dimensional reduction at short distances is necessary to realise asymptotic safety. And evidence for dimensional reduction has been found now in several approaches such as CDT [61] and asymptotic safety [62].

Is there any evidence for discreteness coming from other approaches to quantum gravity?

First of all, the most important question is: Do we have any evidence coming from experiment? And the answer is no. This is important because of experiments like those of the Fermi Gamma Ray Space Telescope in which one can look for Lorentz symmetry-breaking effects in the propagation of light. The effects might be of the order energy over Planck energy which is the most natural prediction if Lorentz invariance is broken [63]. There are strong constraints by now at the Planck scale, results which were published some years ago and which used data from gamma ray bursts [64, 65]. So we have good evidence that test discreteness, if it would show up as the breaking of Lorentz invariance. Similarly, there is good evidence that the GZK cutoff is really there in high-energy cosmic ray experiments, which was another prediction of the breaking of Lorentz invariance [1]. So, for me, experimental evidence is the most important thing.

What about theoretical evidence for spacetime discreteness from other approaches?

The causal set approach assumes this from the beginning, so they don't derive it. They claim that they have a form of discreteness which also preserves relativity of inertial frames [66].

And what about string theory?

The string vacua that people understand, say, around 10-dimensional spacetime, transform under the standard Poincaré algebra naively without any breaking of Lorentz invariance. There is theoretical evidence that in certain kinds of string scattering you can start to see that the longitudinal momentum is quantised. There are actually some papers of Leonard Susskind and others [67] which exhibit this beautifully.

In your book *Three Roads to Quantum Gravity* [68] you seemed pretty optimistic that all the different approaches could be combined and give one single theory. Do you still hold this opinion?

It would be great if that happened. I am disappointed that the two camps are still so disjoint that few think about this. First of all, there would have to be a background-independent formulation of string theory. I have long thought that the tools and mathematical structures in LQG could be used to formulate a background-independent form of string theory, and I have a few results which support this idea.

But how can this even be possible?

In the late 1990s to early 2000s I followed a research program to merge LQG and string theory, using a structure common to them both, which are matrix models [69–71]. This program asked two questions: (1) Are there background-independent matrix models? (2) Can the matrix models for string or M-theory be derived by expanding that theory around a classical solution? The results I found suggest that the answer to both questions is yes. I constructed background-independent matrix models based on matrix compactifications of Chern-Simons theory that were based on cubic actions. Expanding them around classical backgrounds I recovered the matrix versions of string and M-theory. One of the models I studied was specially elegant, as it incorporated the symmetries of string theory by the use of matrices whose elements are valued in the exceptional Jordan algebra.

There are some other directions in progress that aim to relate strings and loops. Thomas Thiemann and collaborators aim for a first principle quantisation of 11-dimensional supergravity using LQG methods [41]. Another approach is group field theory [72], which is a natural extension of matrix models, and also a generalisation of spin foam models.

Was there any explicit connection found? For example, Dijkgraaf, Gukov, Neitzke and Vafa found a connection some years ago between topological string theory and the topological sector of LQG [73]. Was anything as concrete as that found using these other ideas?

Perhaps I haven't been clear – the aim is not to show that string theory and LQG (in the narrow sense) are identical. They can't be. LQG is the quantisation of general relativity and as such has, so far as we know, only the massless spin-two degrees of freedom – two degrees of freedom per point. String theory has an infinite tower of degrees of freedom, for each of an infinite number of vacua. So they are different theories; string theory is very unlikely to be quantum general relativity. The aim was instead to use the techniques and lessons of LQG to effect a background-independent formulation of string theory. Here there are explicit, concrete results. First, the formulation of background-independent matrix models, expansions around which reproduce the matrix formulations of string and M-theory. Second, there are steps towards a background-independent quantisation of 11-dimensional supergravity, which it is conjectured to be a formulation of M-theory.

Something that I also would love to see, and I've said this for years and years, is LQG applied to systems with extended supersymmetries such that you could compare with string theory results about extremal or near-extremal black holes, which means supersymmetric black holes. There are results in $N = 1$ supergravity but they are not strong enough, so this is one reason why I'm very supportive of Thiemann's research direction [41].

Is the problem of making a background-independent theory dependent on boundary conditions?

No, because if there is a boundary there are fixed structures that must be set – boundary conditions, plus boundary terms in the action and Hamiltonian. The specification of these constitute fixing a background. So no theory defined on a spacetime with a boundary can be fully background-independent. Only theories defined on truly closed systems can be background-independent. If the theory is defined on a manifold, that manifold must be spatially compact. Hence, background independence means no boundaries and no asymptotic structure.

Furthermore, gravitational theories with boundaries such as asymptotically flat or asymptotically AdS are best understood as models of subsystems of the universe in which the complementary subsystem has been cut out and replaced by an approximate description in terms of boundary conditions. In other words, they are subsystems which are being idealised as isolated systems for the purpose of making an approximate description of the spacetime around a star or a galaxy. That is, spacetimes with boundaries are incomplete descriptions – they require a completion by a spacetime without boundaries.

Can you explain what you mean by this precisely?

We are trying to construct a theory of the whole universe. By definition, the universe is causally closed, so if it is described as a spacetime, that spacetime must be closed. Background independence goes back to the idea of Leibniz that everything is relational and that there are no absolute structures that you have to specify that give properties to degrees of freedom without themselves being dynamical. Anything that figures in the definition of a dynamical degree of freedom must itself be dynamical. That was the idea; that's the tradition of Leibniz, Mach, Einstein, etc.; and I think it's the most important idea in the search for quantum gravity.

So you think that the setup of AdS/CFT is not background independent?

Yes. Also, as I mentioned earlier, some of the results of AdS/CFT can be explained in a general setting that does not require supersymmetry, string theory or special dimensions. Nevertheless, I would also love to see the development of a real background-independent form of string theory. I'm very disappointed that as far as I know nobody is working on that problem anymore. I continue to think that string theory is an interesting hypothesis even though the landscape issue for me is a big issue. My problems with the landscape are related to the third type of question that I mentioned as the main questions that theoretical physics faces. These are the sort of cosmological questions of why these laws of nature and not others, what are the initial conditions, etc.

Is spacetime somehow emergent in LQG?

Yes.

Even time?

That's a very, very interesting issue. My own view of time has changed over the last years. Like many people working in quantum gravity, I used to believe that time was emergent from a timeless universe described by the Wheeler-DeWitt equation. I no longer believe this. Now I strongly believe that time is not emergent and that there is a fundamental time. Several reasons influenced my change of mind regarding time. The strongest has to do with the problem of understanding why we have these particular laws of nature rather than others.

My view is very close to that of Charles Sanders Peirce – the American pragmatist – who said in 1893 that physics would progress to the point that the question of "Why these laws?" would become important. He then said that the only possible explanation for why these laws, within science, would be that the laws are selected by a process of evolution. I have thought about this question a great deal and I also see no alternative to an evolutionary explanation – if we restrict ourselves to explanations that imply testable consequences. Peirce's demand that we must be able to explain the choice of laws basically restates Leibniz's principle of sufficient reason. It's not rational to simply describe the laws of nature. We must be able to justify why those are the laws and not others. Otherwise, there are important questions about nature that remain unanswered. Peirce said that the only possible rational or scientific – scientific is my interjection – way of accounting for laws is if they are the result of evolution. I take that to mean evolution in time. So, if laws result from an evolution in time, then time is older than law. Therefore, it can't be that time emerges from some fixed law.

I started to develop this idea in my work on cosmological natural selection, which is where I introduce the idea of the landscape, and recently I have been spending a lot of time on it. (By the way, it has happened to me more than once to have an idea and then to discover that Peirce had the same idea a century earlier. The guy must have been amazing.) I have written two books about the idea that time is fundamental. One of them goes much deeper than the *The Life of the Cosmos* [74] and it is about the reality of time [75]. It's basically the case for time not being emergent and time being real. The other is a book with philosopher Roberto Mangabeira Unger, which is a professional philosophy book about why laws have to evolve and how to deal with various issues that come up when you claim that the laws evolve [76].

But I always thought that all the results of LQG pointed towards the direction that there is no time and that time emerges at some point.

Yes, because there is this idea that the Wheeler-DeWitt equation is timeless and within some models you can realise that idea. Loop quantum cosmology provides us with some simple models within which that idea can be realised. However, I don't believe that that idea can be realised in a full quantum field theory with infinite number degrees of freedom. By the way, one of the reasons why I think that shape dynamics is important is that it shows that general relativity is equivalent to a theory with real time.

But is time emergent in LQG or not?

Time is emergent in the Hamiltonian approach to LQG which was based on constructing a physical Hilbert space made up of solutions of the quantum constraints. (Or rather would be, if that program could be carried all the way through.) This would be true in the spatial compact case. In the case with boundary or asymptotic conditions there is a non-vanishing Hamiltonian and it represents evolution according to a clock carried by an observer outside the boundary. So in that case time is not emergent. When the dynamics is addressed by path integrals, in spin foam models, it is not so clear that time is emergent. Some versions of spin foam histories have built-in causal structures, and in these models time is not emergent – even in the cosmological spatially compact case.

However, it should be stressed that the successful dynamical results of LQG or spin foam models, like the results about the graviton propagator and some of the results of the semiclassical limits, are best defined in the situation with a boundary. So they are not cosmological theories. So, it would not surprise me if the project of quantum cosmology consisting of deriving the world from the wave function of the universe fails at the full quantum field theory level, but that would not contradict any of the successful results about LQG so far.

Now, could you explain a bit your idea about cosmological natural selection (CNS)?

Let me emphasise that the landscape issue was on the table since a paper from Andrew Strominger in 1986 [77]. It didn't come about in 2003 with Leonard Susskind [78]. Andy Strominger got me worried because he found a vast realm of string vacua beyond Calabi-Yau manifolds. I began to worry about how the correct version of string theory would be picked out and I found some inspiration in population biology and how natural selection works. So, I took the idea that the space of parameters of theories in string theory was analogous to fitness landscapes in biology and then that's where the idea of cosmological natural selection came from.

I understood, because I understood how biology works, that if you want to get predictions or want to be able to test this idea with access to only one element of the population then that element has to be a typical member of the population and not an atypical member of the population. Therefore, the anthropic principle would never work. The anthropic principle was already on the table back then and it was obvious to me that it would never work. Eternal inflation was also on the table back then since Andrei Linde was already talking about the production of universes and it was also obvious that it wouldn't work. Nothing that has happened since has changed my mind about the fact that to get falsifiable predictions from a population of spacetime regions distributed on a landscape of possible low-energy laws, you need to assume that our universe is a typical member of that population. If our universe is an outlier, as assumed in eternal inflation and other anthropic theories, you can explain any fact that might be observed, so long as it is consistent with intelligent life. But if you could explain anything you cannot make any falsifiable predictions. This is why I get falsifiable predictions in CNS, because I assume our universe is a typical member of the ensemble on the landscape.

I have been making this simple point since my first paper on the subject in 1992, as well as in four books, and numerous papers and conference talks. To my knowledge, this straightforward objection has never been addressed by Linde, Guth, Vilenken or any of the other proponents of eternal inflation. So it is important to stress that the word *landscape* was taken from the biologists' fitness landscape. I chose it to emphasise the analogy.

The point that I was interested in was finding out what was necessary in order to have a cosmological scenario where the population of universes is somehow generated dynamically on a landscape that you could make predictions from. The two things that I understood right away from biology were that the distribution would have to be highly out of equilibrium as well as time-dependent and that our universe would have to be typical and not atypical such that there would have to be a mechanism that drove every universe to be bio-friendly (or all almost every universe to be bio-friendly). If you could set up a scenario which was like that then you could make predictions. The cosmological natural selection scenario [79] demonstrates that there are scenarios that follow those criteria which are predictive because they make real predictions.

What kind of predictions?

For example, that the upper mass of a neutron star has to be less than two solar masses. The reason comes from the work of Hans Bethe and Jerry Brown, who hypothesised that neutron star cores are dominated by kaon condensates (see [80] for a review). This softens the equation of state and lowers the upper mass limit over what it might have been. Recent work by nuclear astrophysicists finds that the heaviest neutron star consistent with a kaon condensate is two solar masses [80]. Given this, one notes that whether neutron stars are dominated by kaon condensates depends on the strange quark mass. If neutron stars are not kaon condensates then nature has missed a chance to make more black holes by adjusting the strong quark mass to below a threshold which – by condensing kaons – lowers the upper mass limit. So an observation of a stable neutron star above two solar masses falsifies the prediction of cosmological natural selection that the standard model parameters have been selected to optimise black hole production. The present observational evidence is consistent with this prediction. The heaviest well-measured neutron star mass is at 1.97 solar masses, while most cluster around 1.4 solar masses.

In this scenario, do the constants of nature change with time?

Yes. Not continuously, just at bounces which replace big bang singularities. At the time I came up with this idea, I started out working with the landscape of string theory and I got such flack when I would give talks and say that we are on the string theory landscape and are evolving. Also, I realised that, like in biology, there is a space of phenotypes and a space of genotypes. The space of phenotypes is analogous to the primary space of the standard model – it's the things that are actually measurable in an experiment. The string landscape is the space of genotypes and there is going to be some complicated relation between them. If I wanted to get real predictions, I would have to work directly on the space of phenotypes. So I would have to work like Mendel and Darwin and not like evolutionary geneticists do today.

But regardless of the model being right or not, do you expect that it's going to be possible to find the theory of everything that exactly predicts which vacuum we live in and which are the precise values or the constants of nature?

I hope so but only by taking very seriously what Charles Sanders Peirce wrote that I quoted, that is, the only way of explaining what the laws of nature are is if they are the result of evolution.

Do you consider the scenario of eternal inflation being one such scenario?

It doesn't satisfy the criterion that I mentioned, that our universe is a typical member of the population. Let me mention also another criterion, namely, that the dynamical mechanism that selects the population of universes has to be highly sensitive to low-energy physics. We have to be able to explain why the neutron and proton mass are what they are, why the Fermi constant is what it is, why the electron mass is what it is, etc. Therefore, the distribution of the population of universes has to be very sensitive towards low-energy parameters. It can't be, therefore, that the population is generated by very high-energy dynamics like what happens in eternal inflation. This is not new; it was obvious in 1986, 1987, and nothing has changed since then. There are no predictions coming from the dynamics of eternal inflation except for a very weak one, namely, that the spatial curvature must be negative, which is not very falsifiable because it could be arbitrary small. So, 25 years later there is nothing which is any different than it was in 1986 when I started looking at it. There are a lot more people interested in it and there is a lot more intelligence being thrown at what seems to me to be a non-solvable problem. I must also say that I don't believe that the cosmological natural selection scenario is the only cosmological scenario that satisfies all the criteria.

In your opinion, what has been the biggest breakthrough in theoretical physics in the past 30 years?

If you mean fundamental physics, a breakthrough would mean a revision of the laws of nature that has overwhelming experimental support. Very simply, there hasn't been one. The laws of nature are the same as they were in 1985. There have been three experimental discoveries: neutrino masses, dark energy and dark matter, but these were not the result of advances of theory.

One candidate is inflation – although that is slightly older than 30 years. But it's not yet been demonstrated convincingly. I'm convinced by Paul Steinhardt's arguments about the weakness of the claims about inflation having been already confirmed. I also have to confess that I hope it's not true because I would prefer that the explanation for the CMB sky comes from quantum gravity instead, which inflation, if true, would not let happen since it dilutes the quantum gravity effects. Nevertheless, it's a proper scientific idea; it was developed and it led to expectations (though not precise predictions) and there is evidence for those generic expectations. There is a huge interpretational issue which Paul made clear in his comments in the conference *Challenges for Early Universe Cosmology* at Perimeter Institute in 2011 but even if it's wrong it has been a very successful idea. Everybody involved in it should be given a lot of credit for it.

If you ask for something weaker than a breakthrough, which is progress in understanding various proposals for new laws of nature, there has been lot of progress over the last 30 years. I think that the idea that quantum gauge theories are to be understood through studying non-local excitations like loops or strings or branes, which is common to LQG and string theory, is the most influential idea in theoretical physics. It's also deeply influential in condensed matter physics; in fact, it came from condensed matter physics and it continues to be there, for example, in models and ideas about topological quantum computing.

Concretely about LQG and string theory or other ideas about quantum gravity it is too soon to make any statement. Both ideas are quite old: LQG is now 25 years old while string theory is 40 years old and both are worked on by relatively large communities of scientists. My sense is that things are kind of evening out; that is, the size of string meetings has fallen to something like 257 attendees and LQG meanings have had over 180 attendees in the last few years but the fact that there is a large number of people working on an idea doesn't make the idea right.

Do you think that your book *The Trouble with Physics* increased or decreased the distance between these two communities?

I'm not the right person to ask about that.

Why not?

Well, because I cannot be objective about it since I was the author of the book and, second, I have many doubts about whether or not I was the right person to be the author of that book. As a person I am very conflict-adverse and I hated the tone of the controversy. Also my own view of string theory is complicated, with strong pluses and strong minuses, and almost none of the reviews or commentary addressed the subtlety and complexity of the issues – and it was that tension that defined the book for me. I wouldn't have written the book if I didn't find a lot to be interested in in string theory and, at the same time, a lot to be disappointed by.

Also, very few got the fact that string theory was not the main focus of that book. The book was first of all an essay in the philosophy of science concerned with the question of how science works. It was chiefly addressed to the critique of science made by the philosopher Paul Feyerabend, which had had a huge influence on my view of science, but left me with some unanswered questions, which the book addressed. The main thesis of the book was an elaboration of Feyerabend's claim that controversy and disagreement within the scientific community play an important role in the progress of science. The key point for me was to propose an answer to Feyerabend's unanswered questions, which involved a characterisation of science as based in a community bound by a certain code of ethics.

String theory was in the book as a case study to illustrate these issues of scientific methodology. If you read the last part, and chapter 17 in particular, that should have been clear. In the early drafts, all the philosophy came first and string theory came later as one of several case studies. But I followed some good advice to drop the other case studies because they were beyond my expertise, and then flip the order to put the case study of string theory before the philosophy. That made it a less academic and more compelling read. But it left the book open to misinterpretation by people who didn't read it carefully or all the way through, and I'm sorry about that.

I did get a lot of positive feedback from physicists, from faculty members all over physics, saying that I nailed something. A few people in the string world misunderstood it and thought that it was a hostile attack. But many of those didn't read it. One way to put it is that the some of the more negative responses to the book verified some of its claims about the role of sociology and groupthink in science.

I was grateful for the review by Joe Polchinski, who read the book and explained what he thought I got wrong. Similarly, my debates with Brian Greene were respectful and constructive. I am sorry that the book angered several friends; this was not my intension and I was unhappy to have been the cause of their anger. Even if it was based on a misreading of the book, ultimately the author is responsible for leaving his text open to misinterpretations.

A few string theorists claimed that my book resulted in cuts to funding for research in string theory, presumedly because it was read by funding officers. I have no evidence that this ever happened; moreover, it remains the case that there are many more opportunities for funding, positions and prizes open to string theorists than to those who pursue other approaches to fundamental physics. And very little has changed to open up opportunities for those rare and brave young scientists who invent and develop their own approaches to getting physics out of its present crisis.

But this is by now a long time in the past. I think that in retrospect the picture of string theory given in the book was balanced and it has largely held up. I continue to be fascinated by what string theory accomplishes and continue to be distressed by its persistent unresolved issues. In addition to the others mentioned, I want to emphasise the lack of predictability because of the landscape issue. I have been worried about this since Andy Strominger explained to me in the late 1980s his discovery of vast numbers of string vacua. This led me to define and name the landscape problem and propose cosmological natural selection as a solution to it.

But the most interesting feedback I got about that book was not from physicists. I was very surprised by the feedback I got from people in other fields, such as economics, computer science, literary studies, and so forth. A lot of people from diverse fields got in touch with me to say that they have the same issues in their community, whether it's biology, neuroscience, artificial intelligence, computer science, linguistics, etc.

The book was principally about the role of controversy and disagreement in research, wherever it is found, and string theory was intended as a case study because it was the case that I knew. Outside of the physics world, I think the book was understood. Inside the physics world I don't want to say that it was completely misunderstood but a lot of people misunderstood it. I also have piles of email from physicists saying that I was so balanced and that I caught the pluses and minuses of different things right.

I have to say I was distressed by the controversy as it developed on blogs, especially the fact that people who hadn't read the book would nonetheless falsely characterise it in strong terms. It was not very constructive and it was not very reflective of the issues that I was interested in. I kind of jogged out of it; I mean, I haven't appeared much on blogs in the last several years. As far as my research goes, I've written several papers on string theory and since then I mostly worked on a range of topics from quantum gravity phenomenology to thinking through the consequences of the hypothesis that laws evolve on the concept of time in physics.

Since 1986, when we started doing LQG, I have strongly regretted that there is a separate string community and a separate LQG community. I never understood the reason for it because for me they were different ways of exploring the same idea. In my mind they were always the same thing. The idea of expressing gauge degrees of freedom of a gravity theory in terms of Wilson loops was something that I was already working on with Louis Crane before LQG, as a way to try to realise a background-independent version of string theory. Then I tried to apply to general relativity some of the technical things which I've understood with Louis about facing quantum gravity and diffeomorphism invariance using Wilson loops. So, for me it was the same thing and, furthermore, the ideas that went into LQG were the ideas that I brought in from the work of Polyakov, MacDowell and Wilson, which the high-energy community understood and the gravity community had to learn. I came from the high-energy community so the fact that LQG was categorised as an approach originating in the general relativity community and outside of the high-energy community, which is where Carlo and I were trained, always seemed to be bizarre to me.

Why have you chosen to do physics? Why not something else?

Well, there's a story but I've told that story in *The Life of the Cosmos* [74] so let me just refer to that. When I was 17, I was a high school dropout. I wanted to be an architect and I was studying differential geometry. I educated myself on differential geometry because I needed it to do structural calculations on ideas about structures in architecture I was playing with, which were generalised geodesic domes – extending geodesic domes to arbitrary curved surfaces. So I started reading about general relativity because it was in the differential geometry book that I was reading and then I read an essay by Einstein and his autobiographical notes, which appealed to me enormously as an adolescent. This idea that there is a true and beautiful, impersonal, eternal world of beauty and truth out there and that one can aspire to become part of it was very appealing to me. That romantic idea made me want to be a physicist. I don't believe that anymore. That's not the notion of truth I have and I'm not a Platonist anymore.

I'm very interested in physics and it's been kind of a mission for me to try to do my part to extend the knowledge of the world, but it's a wonderful community to be part of. It is sort of the opposite of what Einstein wrote about; that is, in my case, many of my close friends are in physics and I have the warmest and most meaningful interactions and personal relations inside the community of physics, so I love it and I feel very humble to be part of this community.

What do you think is the role of the theoretical physicist in modern society?

The role of a scientist is to make discoveries about nature. I think that human beings find themselves from the earliest times living in several different worlds. We live in a social world, we live in a world of imagination, we live in a mystical world and we live in a natural world. Science is the development of many centuries of disciplines about how to gain knowledge about the natural world. Politics is the development of many centuries of how to negotiate the social world, art is the development of how to negotiate the imaginative world, etc. So I think science is deeply embedded in human culture and our main job is to make discoveries about nature.

Beyond that, I believe there is a role for some articulate scientists to play in society as public intellectuals, especially in this period, because society has to make critical decisions on key threats to our general well-being that involve science, such as climate change.

References

[1] G. Amelino-Camelia and L. Smolin, "Prospects for constraining quantum gravity dispersion with near term observations," *Phys. Rev. D* **80** (2009) 084017, arXiv:0906.3731 [astro-ph.HE].

[2] J. M. Maldacena, "The large N limit of superconformal field theories and supergravity," *Int. J. Theor. Phys.* **38** (1999) 1113–1133, arXiv:hep-th/9711200.

[3] H. Gomes, S. Gryb, T. Koslowski, F. Mercati and L. Smolin, "A shape dynamical approach to holographic renormalization," *Eur. Phys. J. C* **75** (2015) 3, arXiv:1305.6315 [hep-th].

[4] S.-S. Lee, "Quantum renormalization group and holography," *JHEP* **01** (2014) 076, arXiv:1305.3908 [hep-th].

[5] L. Smolin, *The Trouble with Physics: The Rise of String Theory, the Fall of a Science, and What Comes Next*. Houghton Mifflin Harcourt, 2007.

[6] A. Ashtekar and J. Lewandowski, "Differential geometry on the space of connections via graphs and projective limits," *J. Geom. Phys.* **17** (1995) 191–230, arXiv:hep-th/9412073 [hep-th].

[7] A. Ashtekar and J. Lewandowski, "Background independent quantum gravity: a status report," *Class. Quant. Grav.* **21** (2004) R53, arXiv:gr-qc/0404018 [gr-qc].

[8] T. Thiemann, "The LQG string: loop quantum gravity quantization of string theory I: flat target space," *Class. Quant. Grav.* **23** (2006) 1923–1970, arXiv:hep-th/0401172.

[9] C. Fleischhack, "Proof of a conjecture by Lewandowski and Thiemann," *Commun. Math. Phys.* **249** (2004) 331–352, arXiv:math-ph/0304002 [math-ph].

[10] B. Bahr, B. Dittrich and M. Geiller, "A new realization of quantum geometry," arXiv:1506.08571 [gr-qc].

[11] E. Frodden, A. Ghosh and A. Perez, "Quasilocal first law for black hole thermodynamics," *Phys. Rev. D* **87** no. 12 (2013) 121503, arXiv:1110.4055 [gr-qc].

[12] A. Ghosh and A. Perez, "Black hole entropy and isolated horizons thermodynamics," *Phys. Rev. Lett.* **107** (2011) 241301, arXiv:1107.1320 [gr-qc]. [Erratum: *Phys. Rev. Lett.* **108** (2012) 169901.]

[13] E. Bianchi, "Entropy of non-extremal black holes from loop gravity," arXiv:1204.5122 [gr-qc].

[14] A. Ghosh, K. Noui and A. Perez, "Statistics, holography, and black hole entropy in loop quantum gravity," *Phys. Rev. D* **89** no. 8 (2014) 084069, arXiv:1309.4563 [gr-qc].

[15] L. Freidel and K. Krasnov, "A new spin foam model for 4D gravity," *Class. Quant. Grav.* **25** (2008) 125018, arXiv:0708.1595 [gr-qc].

[16] J. Engle, E. Livine, R. Pereira and C. Rovelli, "LQG vertex with finite Immirzi parameter," *Nucl. Phys. B* **799** (2008) 136–149, arXiv:0711.0146 [gr-qc].

[17] L. Smolin, "The thermodynamics of quantum spacetime histories," arXiv:1510.03858 [gr-qc].

[18] T. Jacobson, "Thermodynamics of space-time: the Einstein equation of state," *Phys. Rev. Lett.* **75** (1995) 1260–1263, arXiv:gr-qc/9504004 [gr-qc].

[19] A. Sen, "Logarithmic corrections to Schwarzschild and other non-extremal black hole entropy in different dimensions," *JHEP* **04** (2013) 156, arXiv:1205.0971 [hep-th].

[20] R. K. Kaul and P. Majumdar, "Logarithmic correction to the Bekenstein-Hawking entropy," *Phys. Rev. Lett.* **84** (Jun. 2000) 5255–5257. http://link.aps.org/doi/10.1103/PhysRevLett.84.5255.

[21] R. K. Kaul and P. Majumdar, "Quantum black hole entropy," *Phys. Lett. B* **439** (1998) 267–270, arXiv:gr-qc/9801080 [gr-qc].

[22] S. Carlip, "Logarithmic corrections to black hole entropy from the Cardy formula," *Class. Quant. Grav.* **17** (2000) 4175–4186, arXiv:gr-qc/0005017 [gr-qc].

[23] E. Bianchi and Y. Ding, "Lorentzian spinfoam propagator," *Phys. Rev. D* **86** (2012) 104040, arXiv:1109.6538 [gr-qc].

[24] F. Conrady and L. Freidel, "On the semiclassical limit of 4D spin foam models," *Phys. Rev. D* **78** (2008) 104023, arXiv:0809.2280 [gr-qc].

[25] F. Conrady and L. Freidel, "Quantum geometry from phase space reduction," *J. Math. Phys.* **50** (2009) 123510, arXiv:0902.0351 [gr-qc].

[26] E. Magliaro and C. Perini, “Regge gravity from spinfoams,” *Int. J. Mod. Phys. D* **22** (2013) 1–21, arXiv:1105.0216 [gr-qc].

[27] E. Magliaro and C. Perini, “Emergence of gravity from spinfoams,” *Europhys. Lett.* **95** (2011) 30007, arXiv:1108.2258 [gr-qc].

[28] C. R. Contaldi, J. Magueijo and L. Smolin, “Anomalous CMB polarization and gravitational chirality,” *Phys. Rev. Lett.* **101** (2008) 141101, arXiv:0806.3082 [astro-ph].

[29] L. Smolin, “Quantum gravity with a positive cosmological constant,” arXiv: hep-th/0209079 [hep-th].

[30] L. Smolin, “An invitation to loop quantum gravity,” in *Proceedings, 3rd International Symposium on Quantum Theory and Symmetries (QTS3)*, P. C. Argyres, J. J. Hodges, F. Mansouri, J. J. Scanio, P. Suranyi and L. C. R. Wijewardhana, eds., pp. 655–682. World Scientific, 2004. arXiv:hep-th/0408048 [hep-th].

[31] H. Kodama, “Specialization of Ashtekar’s formalism to Bianchi cosmology,” *Prog. Theor. Phys.* **80** (1988) 1024.

[32] H. Kodama, “Holomorphic wave function of the universe,” *Phys. Rev. D* **42** (1990) 2548–2565.

[33] E. Witten, “A note on the Chern-Simons and Kodama wave functions,” arXiv:gr-qc/0306083 [gr-qc].

[34] J. Aastrup and J. M. Grimstrup, “Lattice loop quantum gravity,” arXiv:0911.4141 [gr-qc].

[35] T. Jacobson, “New variables for canonical supergravity,” *Class. Quant. Grav.* **5** (1988) 923.

[36] T. Jacobson and L. Smolin, “Nonperturbative quantum geometries,” *Nucl. Phys. B* **299** (1988) 295–345.

[37] R. Capovilla, T. Jacobson, J. Dell and L. Mason, “Selfdual two forms and gravity,” *Class. Quant. Grav.* **8** (1991) 41–57.

[38] Y. Ling and L. Smolin, “Supersymmetric spin networks and quantum supergravity,” *Phys. Rev. D* **61** (2000) 044008, arXiv:hep-th/9904016.

[39] Y. Ling and L. Smolin, “Eleven-dimensional supergravity as a constrained topological field theory,” *Nucl. Phys. B* **601** (2001) 191–208, arXiv:hep-th/0003285.

[40] R. Gambini and J. Pullin, *A first course in loop quantum gravity*. Oxford University Press, 2011.

[41] N. Bodendorfer, T. Thiemann and A. Thurn, “New variables for classical and quantum gravity in all dimensions I. Hamiltonian analysis,” *Class. Quant. Grav.* **30** (2013) 045001, arXiv:1105.3703 [gr-qc].

[42] S. O. Bilson-Thompson, F. Markopoulou and L. Smolin, “Quantum gravity and the standard model,” *Class. Quant. Grav.* **24** (2007) 3975–3994, arXiv:hep-th/0603022 [hep-th].

[43] P. Peldan, “Ashtekar’s variables for arbitrary gauge group,” *Phys. Rev. D* **46** (1992) 2279–2282, arXiv:hep-th/9204069 [hep-th].

[44] A. G. Lisi, “An exceptionally simple theory of everything,” arXiv:0711.0770 [hep-th].

[45] S. W. MacDowell and F. Mansouri, “Unified geometric theory of gravity and supergravity,” *Phys. Rev. Lett.* **38** (1977) 739–742. http://link.aps.org/doi/10.1103/PhysRevLett.38.739.

[46] A. G. Lisi, L. Smolin and S. Speziale, “Unification of gravity, gauge fields, and Higgs bosons,” *J. Phys. A* **43** (2010) 445401, arXiv:1004.4866 [gr-qc].

[47] T. Jacobson and L. Smolin, “Covariant action for Ashtekar’s form of canonical gravity,” *Class. Quant. Grav.* **5** (1988) 583.

[48] C. Rovelli and L. Smolin, "The physical Hamiltonian in nonperturbative quantum gravity," *Phys. Rev. Lett.* **72** (1994) 446–449, arXiv:gr-qc/9308002.

[49] T. Thiemann, *Modern Canonical Quantum General Relativity*. Cambridge Monographs on Mathematical Physics. Cambridge University Press, 2007.

[50] L. Smolin, "Positive energy in quantum gravity," *Phys. Rev. D* **90** no. 4 (2014) 044034, arXiv:1406.2611 [gr-qc].

[51] J. Barnett and L. Smolin, "Fermion doubling in loop quantum gravity," arXiv: 1507.01232 [gr-qc].

[52] R. Gambini and J. Pullin, "No fermion doubling in quantum geometry," *Phys. Lett. B* **749** (2015) 374–375, arXiv:1506.08794 [gr-qc].

[53] B. Dittrich, "Diffeomorphism symmetry in quantum gravity models," *Adv. Sci. Lett.* **2** (Oct. 2008) 151, arXiv:0810.3594 [gr-qc].

[54] L. Smolin, "The ground state of quantum gravity with positive cosmological constant," *AIP Conference Proceedings* **646** no. 1 (2002) 59–76. http://scitation .aip.org/content/aip/proceeding/aipcp/10.1063/1.1524554.

[55] L. Smolin and C. Soo, "The Chern-Simons invariant as the natural time variable for classical and quantum cosmology," *Nucl. Phys. B* **449** (1995) 289–316, arXiv:gr-qc/9405015 [gr-qc].

[56] L. Freidel and E. R. Livine, "3D quantum gravity and effective noncommutative quantum field theory," *Bulg. J. Phys.* **33** no. s1 (2006) 111–127, arXiv:hep-th/0512113.

[57] L. Crane and L. Smolin, "Space-time foam as the universal regulator," *Gen. Rel. Grav.* **17** (1985) 1209.

[58] L. Crane and L. Smolin, "Renormalizability of general relativity on a background of space-time foam," *Nucl. Phys. B* **267** (1986) 714–757.

[59] F. Caravelli and L. Modesto, "Fractal dimension in 3D spin-foams," arXiv:0905.2170 [gr-qc].

[60] E. Magliaro, C. Perini and L. Modesto, "Fractal space-time from spin-foams," arXiv:0911.0437 [gr-qc].

[61] J. Ambjorn, J. Jurkiewicz and R. Loll, "Quantum gravity as sum over spacetimes," *Lect. Notes Phys.* **807** (2010) 59–124, arXiv:0906.3947 [gr-qc].

[62] O. Lauscher and M. Reuter, "Fractal spacetime structure in asymptotically safe gravity," *JHEP* **10** (2005) 050, arXiv:hep-th/0508202.

[63] L. Freidel and L. Smolin, "Gamma ray burst delay times probe the geometry of momentum space," arXiv:1103.5626 [hep-th].

[64] A. A. Abdo et al., "A limit on the variation of the speed of light arising from quantum gravity effects," *Nature* **462** no. 7271 (2009) 331–334.

[65] HESS Collaboration, A. Abramowski et al., "Search for Lorentz invariance breaking with a likelihood fit of the PKS 2155-304 flare data taken on MJD 53944," *Astropart. Phys.* **34** (2011) 738–747, arXiv:1101.3650 [astro-ph.HE].

[66] F. Dowker, J. Henson and R. D. Sorkin, "Quantum gravity phenomenology, Lorentz invariance and discreteness," *Mod. Phys. Lett. A* **19** (2004) 1829–1840, arXiv:gr-qc/0311055 [gr-qc].

[67] I. R. Klebanov and L. Susskind, "Continuum strings from discrete field theories," *Nucl. Phys. B* **309** (1988) 175–187.

[68] L. Smolin, *Three Roads to Quantum Gravity*. Science Masters series. Basic Books, 2002.

[69] L. Smolin, "Towards a background independent approach to M theory," *Chaos Solitons Fractals* **10** (1999) 555, arXiv:hep-th/9808192 [hep-th].

[70] L. Smolin, "A candidate for a background independent formulation of M theory," *Phys. Rev. D* **62** (2000) 086001, arXiv:hep-th/9903166 [hep-th].
[71] L. Smolin, "The cubic matrix model and a duality between strings and loops," arXiv:hep-th/0006137 [hep-th].
[72] D. Oriti, "Group field theory and loop quantum gravity," 2014. arXiv:1408.7112 [gr-qc]. https://inspirehep.net/record/1312968/files/arXiv:1408.7112.pdf.
[73] R. Dijkgraaf, S. Gukov, A. Neitzke and C. Vafa, "Topological M-theory as unification of form theories of gravity," *Adv. Theor. Math. Phys.* **9** no. 4 (2005) 603–665, arXiv:hep-th/0411073.
[74] L. Smolin, *The Life of the Cosmos*. Oxford paperbacks. Oxford University Press, 1999.
[75] L. Smolin, *Time Reborn: From the Crisis in Physics to the Future of the Universe*. Houghton Mifflin Harcourt, 2013.
[76] R. Unger and L. Smolin, *The Singular Universe and the Reality of Time*. Cambridge University Press, 2014.
[77] A. Strominger, "Superstrings with torsion," *Nucl. Phys. B* **274** (1986) 253.
[78] L. Susskind, "The anthropic landscape of string theory," arXiv:hep-th/0302219.
[79] L. Smolin, "Did the universe evolve?," *Class. Quant. Grav.* **9** (1992) 173–192.
[80] L. Smolin, "A perspective on the landscape problem," *Found. Phys.* **43** (2013) 21, arXiv:1202.3373 [physics.hist-ph].

29

Rafael Sorkin

Faculty member at the Perimeter Institute for Theoretical Physics, Waterloo, Canada

Date: 10 September 2014. Location: Copenhagen. Last edit: 18 October 2020

In your opinion, what are the main problems in theoretical physics at the moment?

Understanding quantum gravity should be first on the list. Then there is a large number of important questions such as understanding the nature of dark matter, quantum phase transitions (in particular phases that are essentially defined at absolute zero) or topological order in quantum systems, as it appears that we are lacking the analogue of an order parameter as used in the Landau theory of phase transitions. I also think that there are issues with the foundations of quantum mechanics which the many-worlds interpretation or the Copenhagen interpretation are inadequate to address. Another problem that I find important and which is close to some of my own interests is that of understanding the equation of state of nuclear matter, specifically at the kind of densities that you find in neutron stars. So far, people can predict the maximum mass of a neutron star within a factor of 10 or something like that. Understanding this better would be an important test of the ideas of QCD, nuclear physics and the Skryme model [1].

Can you explain what the Skyrme model is?

When the Skyrme model was first introduced, it was not clear what was the theory of strong interactions, so it could be viewed to some extent as an alternative. However, nowadays it is thought of as a low-energy approximation to quantum chromodynamics (QCD). The simplest Skryme model only includes up and down quarks, nothing else, so it describes a pion field. It is a non-linear sigma model in which the pion field is valued in SU(2), a topologically non-trivial space. Because of this there can be topologically distinct field configurations belonging to different homotopy classes. The vacuum corresponds to the trivial homotopy class. The non-trivial configurations have kinks, which are characterised by a winding number, in turn understood as baryon number. The interesting thing about the model is that a theory of nothing but pions can generate baryons such as protons. Within this context, the nucleon arises as a topological excitation, and it can acquire fermionic spin and statistics due to the existence of non-trivial loops in the configuration space of the Skyrme (pion) field [2, 3]. The model can thus describe nuclei, that is, nuclear matter, and also very dense nuclear matter. At the densities that one finds in neutron stars, one expects that approximating the system as a collection of individual nuclei will break down. Instead, one has to deal with the Skyrme field with large winding numbers.

How were you led to the Skyrme model?

It interests me because it offers a laboratory in which one can study certain questions of quantum gravity in a simplified setting. There is an analogy between baryons in the Skyrme model and topological geons in gravity. However, geons are more complicated, as one has to deal with the gauge/diffeomorphism invariance that is intrinsic to gravity. I worked on the gravitational case, together with John Friedman [4–6], and the motivation for my later work on the Skyrme model was to try to compute the ground state energy of different kinds of topological geons. We had found out that non-trivial topology of the gravitational field, i.e., of spacetime itself, could give rise to half-integer spin, and that opened up the door to the old dream of being able to get all known particles just from geometry. In fact, John Friedman and Atsushi Higuchi later wrote a paper showing that the main properties of the fields in the standard model can in principle arise just from topological effects in pure gravity [7]. Many of the quantum numbers that one finds in the standard model, including those of quarks, are already present in Kaluza-Klein theories thanks to the fact that gravity can exhibit spinorial/fermionic behaviour. Of course, the big open question is whether the mass of these putative objects can be anywhere as small as what it needs to be to match the known particles.

Within the Skyrme model, we know that skyrmions are unstable against collapse for the same dimensional reasons that gravitating matter is unstable against collapse. It is possible to stabilise skyrmions by introducing a so-called Skryme term that comes with a free parameter that can be used to tune the mass of the skyrmion to any value. In the context of gravity, the analogue of adding such term would be adding higher-derivative terms to the Lagrangian as is done in Lovelock gravity. However, in four dimensions we do not have such a thing so we need to find another stabilisation mechanism which can do the job that the uncertainty principle does in stabilising the hydrogen atom [8–10]. It's still unclear exactly how to do the calculation for gravity, but in the case of the skyrmion we could actually do it and we got quite a good fit to the masses of several low-lying baryonic states like those of the nucleon and the Δ.

So the main idea behind this is that topological excitations could in principle give rise to all particles?

Yes, that would be the ultimate dream [11, 12]. In fact, within the context of the causal set approach to quantum gravity [13–16] one may ask, Where does non-gravitational matter come from? There are different possible answers, but one of the possibilities is along these lines, which does not require the introduction of extra fields beyond those which can in principle emerge from the causal set itself as effective degrees of freedom.

I guess that this same idea could be used in other approaches to quantum gravity that start only from gravity?

It could, absolutely, as long as the resulting theory allows the spacetime topology to change. As far as I know, for instance, it is not yet clear if causal dynamical triangulation (CDT) allows for topology change [17–19].

Why do we need a theory of quantum gravity?

The principal reason is that physics, as it stands now, lacks a unified theoretical structure. This has been the reason why, ever since I was in graduate school, I was drawn to quantum gravity. At the present moment, physics is in need of a unified picture that moves beyond the models of forces and particles of the last century.

Besides the lack of a unified structure, there are many other problems, related to the big bang, dark matter, dark energy, as well as problems about understanding black holes, the singularities inside them, and the origin of their thermodynamic properties. We're now in a situation where we can hope for a lot of indirect and maybe even fairly direct evidence about how the universe looked around the Planck time. The indirect evidence will perhaps tell us how to extract the initial conditions at the big bang, and direct evidence from gravitational waves might reveal new physics at that time which is relevant to what's inside a black hole.

Do you think that understanding dark matter and dark energy requires a theory of quantum gravity?

I would be surprised if solving the riddle posed by the cosmological constant or "dark energy" would be possible without an understanding of quantum gravity. What people call the cosmological constant "problem" is about the nature of the quantum stress-energy tensor, which is at the heart of the theoretical issues that they are troubled by. By definition, understanding this at a deep level requires understanding the coupling of gravity to non-gravitational matter, and for that you will need quantum gravity. From another point of view, the problem is even more fundamental: Why is there a spacetime manifold at all and why is its radius of curvature so much smaller than the Planck value? It goes without saying that these questions belong to quantum gravity.

Regarding dark matter I am not sure but if you had to settle all the big questions at one stroke, it would perhaps be a more daunting task than necessary. For instance, it's not obvious that you have to understand all the details of the standard model and where it comes from in order to understand the cosmological constant. In the case of dark matter, it's very conceivable that it could be understood without quantum gravity, but it is equally conceivable that quantum gravity will be important. In fact, a student at Syracuse, Ajit Srivastava, once suggested that topological geons might be the source of dark matter [20]. Geons are purely gravitational objects which interact only gravitationally, and in a theory of gravity alone, one of them must be stable. Thus they could be a candidate for dark matter (although the scenario suggested by Ajit was not this simple).

There is also another possibility that I find interesting. There are reasons to believe that both discreteness and Lorentz invariance are manifested in quantum gravity. If that is the case, then there must be some fundamental nonlocality in nature [21]. Once you have nonlocality then at some scale you should encounter a nonlocal field theory. If that length-scale is large enough, it may have observational consequences either now or in the near future. It is very difficult to construct and study nonlocal field theories; in fact, no one knows how to do it precisely in a totally consistent manner. However, there is some recent work within causal set theory where it is suggested that a particular kind of nonlocality arises by replacing the D'Alembertian operator by a nonlocal integral kernel [22, 23]. Saravani and Aslanbeigi analysed the consequences of this, and they found that it gives rise to a continuum of what they called "off-shell particles" that interact very weakly with everything else and with themselves. Additionally, these particles could be created in the early universe. It's a proposal for dark matter that does not directly stem from quantum gravity but indirectly from the nonlocality inherent in a fundamental quantum discreteness.

Could it be the case that gravity should be left alone as a classical field and not be quantised?

I'm not sure how to interpret this question. If I interpret it as asking whether something like the semiclassical Einstein equation could be true, then the answer is no. In such a theory the metric still obeys deterministic equations of motion, but one sticks on the right-hand side of the Einstein equation the expectation value of the quantal stress-energy tensor. This is certainly inconsistent conceptually, due to the almost obvious reason that quantum events can give rise to many possible outcomes that are macroscopically utterly distinct (therefore with measurably distinct gravitational fields).

I wrote a paper with Daniel Sudarsky where we showed how the event of a large shell of matter collapsing and forming a black hole can be dependent on a microscopic quantum event [24]. On smaller scales such types of events are happening all the time, and you can go to the laboratory and test it. Even in our brains there are presumably quantum events taking place that lead us to take either one decision or the other. If you look at a spark chamber there are macroscopically distinct events taking place, and the gravitational fields associated with them aren't very large but they aren't zero either. Thus, fundamentally a theory based on the semiclassical Einstein equation is not a consistent idea, although in some approximation it can be useful for keeping track of the evaporation of a black hole and associated energy fluxes.

When people discuss the possibility of not quantising gravity, I do not think that they have in mind this naive theory that I just described. The origins of these ideas date back to Rosenfeld, who was a close collaborator of Bohr [25]. I don't know what Bohr's view was, but Rosenfeld's view was that gravity should not be quantum; it should remain classical. Without going deeply into his arguments, they had at their core the so-called Copenhagen interpretation of quantum mechanics in which there is an outside world that is classical and gravity is part of it. As such, gravity should also remain classical.

Regardless of the soundness of such arguments, one can actually write down field theories in which some fields are classical and some are quantum. Of course, I need to specify what I mean by a field being classical. In my point of view, and I think also that of many people, the characteristic difference between quantum dynamics and classical dynamics is quantum interference, as in the two-slit and three-slit experiments [26, 27]. So in any theory, if there is a field or particle that doesn't interfere with itself, I would call that a classical field or particle. (The statement that a certain field doesn't interfere with itself is a shorthand. What can interfere or not interfere are pairs of histories, and a full history includes the histories of all the fields that are present in the theory. When one says that the field F doesn't interfere with itself, one is really saying that two histories can only interfere with each other when F takes the same value in both.) Such hybrid formulations can be based on path integrals, and in fact many people have studied theories in which some fields interfere with themselves and some don't. Wave function collapse models [28] can be understood as theories of this sort, although they are not always described that way. Classical gravity coupled to quantum matter is not a logically inconsistent idea, though it may give rise to other problems.

The point of view that gravity remains classical is attractive in some ways, because if gravity does not have to be quantised, we avoid some very difficult conceptual problems. However, although there would still be gravitational waves in such a theory, as those are classical, there would be no gravitons. That would really put gravity in a funny situation vis-à-vis other fields. From an aesthetic perspective, it looks to me pretty artificial. Why should gravity not be subject to interference? Anyway, it's certainly conceivable, but it's not my point of view.

Some people think that gravity is not a fundamental force but some kind of emergent phenomenon. What is your opinion about this?

That's another hypothesis that could easily call into question the existence of gravitons. It is somewhat analogous to another question which I really don't know how to answer. Think about a fluid like water (not like superfluid helium), and ask yourself whether it is meaningful to speak about phonons in such a fluid. From a quantum point of view, if there is sound then there should be quanta of sound, phonons. However, you might argue that in drawing that conclusion, I am pushing the effective theory too far. I don't know the answer. I think some people would say that there are phonons and others wouldn't. I'd like to believe that there are but I'm not sure how one would test it.

But yes, it's certainly possible that gravity is an effective force. I believe that spacetime itself is an effective object, and if spacetime is emergent then gravity is too since the gravitational field is just the spacetime metric. But just because something is effective, I don't think it means that it can't have quantum properties. It depends on the regime in which it shows up. So, unless we believe that spacetime ceases to make sense at, say, TeV scales I think we have to believe that quantum interference will still be pertinent to the dynamcis of the metric.

As a matter of fact, one can give definite reasons why the characteristic scale on which spacetime breaks down could not be the TeV scale. I think this follows from the same argument that led to the heuristic predictions of the cosmological constant [29, 30]. Assuming that you are confident in the causal set account of the cosmological constant, if the fundamental scale were the TeV scale then the cosmological constant would be way too small, so small that it rules out the so-called scenarios of large extra dimensions at the LHC. There are also other ways in which you can argue against the breakdown of spacetime at large scales. If you believe in a fundamental discreteness and Lorentzian signature in the sense of having a notion of lightcone and ...

You mean that you think that causality is fundamental?

Causality in that sense is fundamental, in particular the causal structure of spacetime: its light-cones. For me this is one of the great acquisitions of twentieth-century physics. I personally cannot believe that we will go back to pre-relativistic notions of causality and spacetime where we deal with Galilean or Aristotelian invariance, distinguished rest frames, and instantaneous action at a distance. Those are logically possible but I can't believe it. So if you put discreteness and Lorentzian signature together you end up with causal sets [31].

But is it possible to combine causality with quantum fluctuations of geometry in a consistent way?

Well ... I certainly hope so otherwise causal sets is a non-starter (laughs).

But intuitively one thinks that if spacetime is fluctuating, it's very difficult to define anything like time or causality, right?

Intuitively that's right, but let us leave that question aside for a bit as I was trying to argue whether or not we should expect to see a breakdown of spacetime at the Planck scale. One of the strongest arguments for a discreteness scale comes from black hole entropy [32–37]. If you calculate the entanglement entropy between the inside and the outside of a black hole event horizon you will obtain a divergent answer unless you put in a cutoff [32, 37]. If that cutoff is of the order of the Planck scale, then you find a result that is comparable to the known black hole entropy. This points very strongly to the Planck scale, and not something much bigger, being the scale at which spacetime breaks down.

Okay, but now going back to causal sets ... you seem to value discreteness quite a bit but there are many approaches that have some discrete nature like loop quantum gravity (LQG), right?

LQG kind of has it. LQG is based on a continuum three-manifold that has some discrete string-like objects embedded in it. So LQG has some elements of discreteness but I don't think it's fundamentally discrete. Additionally, LQG is a species of canonical quantum gravity and I have strong reasons based on my own personal history to think that canonical quantum gravity is wrong. Canonical gravity will never be able to solve the difficulties which go by the name of *the problem of time*. So, even if such a theory would be fundamentally discrete, I wouldn't work on it. These difficulties have been there from the beginning and I believe they are still there.

So does causal set theory overcome the problem of time?

It does overcome it, yes, and we can talk about it, but in any case I don't think that there are other discrete approaches.

Isn't causal dynamical triangulations a discrete approach?

It could be, but its practitioners choose to view its discreteness as a cutoff analogous to the lattice cutoff in lattice QCD. At the end of the day they always want to take a continuum limit, and they view the discreteness as no more than a device to render the path integral convergent. However, it is true that they could take the opposite view of having a fundamental discreteness, but that would for me be like going back to an earlier stage of my thinking. I actually started my path in physics by working on Regge calculus [38–40], and CDTs are a particular form of Regge calculus [41–44]. I evolved away from Regge calculus toward topological posets [45] and from there to causal posets, which is what causal sets are [46]. One strong argument against CDTs, in case you do not take a continuum limit and assume the discreteness to be fundamental, is that it is not clear at all whether it is Lorentz-invariant. In fact, it is likely that it is not. If you take a particular triangulation, it will tend to pick out a preferred reference frame and distinguish one frame from another. Joe Henson has made these arguments clear in connection with spin foams [47] but it applies equally to CDTs without the continuum limit.

So you're saying that CDT is not locally Lorentz-invariant?

That's right, and causal sets are. However, it's true that my next favourite after causal sets is probably CDTs. I always have to state this at some point, namely, the fact that we know of analogue situations in which taking the continuum limit is just wrong. In other words, if there really is discreteness and you try to take a continuum limit, you tend to lose important physics. In the case of gravity, you can make a simple dimensional analysis to figure out what you will lose. If we set the speed of light to $c = 1$ then the Planck scale is of order $\sqrt{\hbar G}$ where G is Newton's constant. This is meant to be the discreteness scale. If you now want to take a continuum limit you have to make $\sqrt{\hbar G}$ go to zero, which you can do either by sending $\hbar$ to zero or G to zero. If you send $\hbar$ to zero and keep G finite you will be missing quantum effects. In this case you end up with classical relativity without quantum aspects. On the other hand, if you send G to zero you will be missing gravitational interactions and you'll end up with quantum field theory in a curved (and Ricci-flat) background, but without backreaction. Thus, if you have a fundamental discreteness and insist on taking a continuum limit the best you can get is one of those two possibilities but not both. Perhaps Jan Ambjørn or Renate Loll would dispute whether $\sqrt{\hbar G}$ actually is the discreteness scale but if we take that as a premise then what I am saying is true.

There is a more down-to-earth physical system which is analogous, namely one made of atoms (not atoms of spacetime but atoms of matter). There is the case, for instance, of the kinetic theory of gases. The counterpart of the Planck scale consists of two fundamental length and time scales, the mean free path ℓ_m and the mean free time t_m. These are two microscopic parameters that implicitly enter the macroscopic equations by influencing basic parameters like the viscosity. If you now want to take a continuum limit of kinetic theory, then you would have to take at least one of those two scales to zero. But what does that mean for the effective physics? The large-scale physics has two characteristic effects, one of which is sound/pressure waves and the other of which is diffusion or heat conduction. The diffusion constant is in order of magnitude ℓ_m^2/t_m, whereas the speed of sound is in order of magnitude ℓ_m/t_m. So if you take one of these parameters to zero (or both) you either lose sound (by getting an infinite propagation speed) or you lose diffusion. Either way, half of the physics is lost.

So, going back to the case of quantum gravity, if you insist on a fundamental discreteness and a Lorentzian structure, you end up with causal set theory, with the exception of hypothetical CDTs where you don't take a continuum limit.

As far as I know, the spin foam formalism incorporates discreteness [48], but is this not the kind of discreteness that you are talking about?

That's a good question, which you should put to someone who works on it actively. I don't want to speak for them but it is not entirely clear to me whether they intend to take a continuum limit or not. Also, as far as I am aware, they have not yet been able to construct a fully Lorentz-invariant model. Besides that, I think that the spin foam formalism introduces redundancies that actually lead to contradictions.

Why do you like causal sets?

What makes causal sets so attractive to me is that there are no extra structures. You can get spacetime just from causal relationships, because the discreteness allows you to count in such a way that it gives you information about the volume. Causal sets goes back to an observation by Bernhard Riemann. When I give an introductory talk on causal sets, depending on the audience, I show some quotes from Riemann's article, and then I show some quotes from a letter that Albert Einstein wrote (see [30] and references therein). If you put the two quotes together, you get a description of a causal set. Why is that? In the case of Einstein, despite the fact that he devoted most of his efforts to continuum theories, he argued for why physics is pointing towards spacetime discreteness. In this particular quote, he is stating that it is hard to take into account spacetime quasi-order, which I think is the German term for partial order or poset. Einstein is pointing at the causal structure of spacetime as key for a discrete theory. The particular quote from Riemann is actually even more clearly advocating for causal sets. In fact, that article by Riemann is one that everyone ought to read, as the origins of manifolds, differential manifolds and so on are explained. In that particular quote, however, he states that there are actually two kinds of manifolds; continuous manifolds and discrete manifolds. The discrete manifolds are the common ones that everyone understands, whereas the continuum manifolds are very confusing, hard to grasp, and we don't have many examples of them. Unfortunately, he didn't give examples of discrete manifolds because he said they're all over the place and everyone is familiar with it (laughs). But he did point out, and this is the crucial thing, that in a discrete manifold the metric relationships are inherent in the structure of the manifold itself, whereas in a continuum manifold, they are not inherent and need to be added in as extra structure.

For me, this is actually the strongest argument advocating for causal sets. They unify the spacetime metric with spacetime itself in a way that none of the other hypotheses does [49].

Do I understand correctly that a specific causal set structure gives rise to a unique metric?

The conception that lies behind causal set theory is that *order plus number equals geometry*. Actually, Riemann has another part of his article where he says that in a continuum manifold, if you consider a well-defined portion of the manifold, you can measure its size by bringing in something that is not part of manifold itself (e.g., by using a ruler), but if you're in a discrete manifold, you can measure the volume of a portion of it just by counting the number of elements that compose it. Funnily enough, he actually wrote that the individual portions of the manifold are called quanta (laughs), which by the way was completely unrelated to what we call quanta today.

Is the quanta here the units of volume?

Yes, in the case of causal sets, the quantum is the unit of four-volume. The hypothesis is that, as suggested by Riemann, the volume of a portion of spacetime is reflecting a certain number of elements of the causal set, namely the elements comprising that portion of spacetime. So that number gives you volume information and the causal information gives you the rest (i.e., the conformal metric). In the continuum, it's a rigorous theorem that if you know the causal relations among the points of the spacetime, and if the spacetime has no causal anomaly like closed timelike curves or closed causal curves (to be more specific, let's assume that it is a globally hyperbolic spacetime), then you can recover everything: topology, differentiable structure and the conformal metric (i.e., the metric up to a local scale factor).

So this doesn't hold for a Gödel universe?

That's right, a Gödel universe is a good example of a spacetime that could not emerge from a causal set. If you like, you can see the fact that we can't live in a Gödel universe as a prediction of causal set theory. And I think that's a well-verified prediction (laughs). Most people would not be astounded if I would tell them that there are no time machines or something like that.

Does this mean that Einstein equations do not come out of causal set theory?

No, it does not mean that; it means that some *solutions* of the Einstein equations are not physical. The Gödel universe is a quite non-trivial example compared with Minkowski space where you periodically identify time, which is still a solution of the Einstein equations but is excluded by causal sets.

Mathematically speaking, a causal set is a locally finite partially ordered set (poset) and it is embedded in its definition that it does not have any cycles. Therefore you cannot get closed causal curves from causal sets. It is possible to make a generalisation of causal sets that allows for causal cycles, but the resulting object does not yield anything that can be approximately described by a spacetime with closed causal curves. By the same token you also can't get spacetimes with curvatures greater than Planckian curvatures.

Interestingly, as the definition of causal set stands now, you also can't get anti–de Sitter (AdS) space. The reason is that defining a causal set as a locally finite partially ordered set tends to require (though its's not exactly equivalent to) global hyperbolicity, whereas AdS is not globally hyperbolic. To see more concretely what goes wrong, suppose you pick two points, one of which is in the past of the other. If the points are not too close to the boundary at infinity, the region of spacetime causally between them will be finite. But if you get too close to the boundary, the volume will be infinite. Infinite volume means an infinite number of elements, and that clashes with local finiteness.

So if you have a causal set, what dictates the dynamics or growth of this causal set?

Well, if I could fully explain that to you, we'd have the theory of quantum gravity, so I can't, but I can give you the status of things. There is a well-defined stochastic and classical growth process which can teach us a lot, but it's only a stepping-stone toward a quantum dynamics.

We know of two roads to a quantum dynamics for causal sets. One road goes via the path integral and the other goes via the dynamics of sequential growth, which is also in the end kind of like a path integral but in a less obvious way. I can explain later what growth dynamics is about but in order to carry out a path integral in practice, you seem to need an approach, similar to what the CDT people do, where you first write down an action for the causal set and then subject it to some kind of Monte Carlo simulation. You need to do an analytic continuation so that Monte Carlo becomes tractable, and then you wait and see if you get something like a spacetime at the end of the day. There is ongoing work in both directions I've just mentioned [16].

Going back to the statement that AdS cannot arise from causal sets, it means in relation to dynamics that AdS cannot arise from a growth process, or that the causal set action for AdS would not be well-defined (that is, it would contain divergent terms) and so you could not use Monte Carlo.

To discuss growth more in depth, it's necessary to talk about sprinkling first. Sprinkling is just a tool that helps us to describe a causal set, and it's not meant to have fundamental status within the theory [14]. The theory is built from causal sets and nothing else, except perhaps some extra matter fields, so let us focus on the case in which the causal set comprises all the underlying structure. Here there are also two different points of view: either reality is composed of a single causal set or it is composed of some kind of a superposition of causal sets. Most likely the latter is correct, but that brings in another level of complication, so let us just consider the case where a single causal set represents the spacetime we live in. How do you start from that causal set and derive the metric, the manifold, its dimension, notions of distance, time, etc.? It is in answering these questions that sprinkling plays a role. Sprinkling is used as a rule of correspondence between the discrete and its effective description as a continuum.

The idea of sprinkling is simple. You start with a globally hyperbolic spacetime and run a Poisson process. That is, you randomly throw darts at the spacetime with a certain density such that on average one dart lands within each Planck volume of spacetime. This is what is called sprinkling and it gives you a subset of the spacetime. Assuming the spacetime is compact – though it's also possible to consider infinitely extended spacetimes like Minkowski spacetime – there are general theorems that relate the volume of the spacetime to the number of darts you've thrown into it. The number of darts is not equal exactly to the volume but on average you find approximately that. If the region of spacetime that you are sprinkling is large then there will be very tiny fluctuations in that number, and in fact it's those fluctuations which are responsible for the prediction of a small cosmological constant [29, 30]. Anyway, once you have this collection of points, you can take a look at the causal relationships between the points, and since you started with a globally hyperbolic spacetime, the points together with their causal relationships form a causal set. So sprinkling can be used to obtain a causal set from a spacetime. In this manner, you acquire a concept of approximate equality between a given causal set and a given spacetime.

The idea, however, is that the causal set is more basic. So you would like to turn things around ... if I give you a causal set, is a specific spacetime a good approximation to it? The answer is: Could I have produced this causal set by some process of sprinkling that spacetime, or no way in hell could it have appeared by such process? This is a way to make precise the key inputs to the kinematics of causal sets; namely, the mathematical order-relation that defines a causal set corresponds to the light-cone structure in the spacetime, and the number of elements corresponds to the spacetime volume. Of course all the details rely on what I mean by *corresponds to*, and what I mean by that is that there exists a faithful embedding of that causal set into that spacetime. A faithful embedding is one that realises the order relations between the elements in the causal set and where the elements of the causal set are uniformly distributed throughout the spacetime. In fact, the only way we know how to distribute the points uniformly is via Poisson sprinkling.

Is the relation between spacetime and causal set unique?

This is what we call the Hauptvermutung of causal set theory [14]. The word comes from algebraic topology and means *main conjecture* in German. It's the name of a conjecture in algebraic topology which if you did not assume it to be true you would have had (historically) a lot of trouble with homology theory and things like that. Our Hauptvermutung is a conjecture which, if it were not true, would not easily allow us to relate causal sets to spacetimes.

Given a causal set, there is no reason why it should be faithfully embeddable in any spacetime. This is not surprising as quantum theories have this character in general. For instance, in CDT you can take an arbitrary triangulation, but there's no reason whatsoever that it will end up looking like a smooth spacetime. The same is true for causal sets. What the Hauptvermutung states is that the spacetime that a given causal set can be embedded in is unique, if it exists at all. There is a lot of evidence for this conjecture but it hasn't been proven yet. The idea is that the possibility of a faithful embedding characterises the spacetime uniquely except at very tiny scales. So the relationship between a causal set and a spacetime is unique going one way. Going the other way it is far from unique. If you fix the spacetime and sprinkle it several times you will almost never get the same causal set twice. This is because the "darts" will fall in different places, and so there will be some probability distribution over a large number of causal sets. How large? We don't have a good mathematical handle on this question, but it appears that the number of causal sets you could get by sprinkling a given spacetime grows exponentionally with the volume of the spacetime.

I imagine that you would like to start with some causal set and then somehow end up with something that approximates de Sitter spacetime . . . so what are the dynamics of causal sets?

As I said earlier, we do not have the full dynamics, but I can tell you a bit more about the two roads that are being pursued. One of those roads is conceptually similar to CDT, which first of all requires writing down an action. Within causal sets it's very hard to write a suitable action, because nonlocalities are an inevitable feature of the framework. Only in the past three to four years did we manage to find an action for causal sets which reproduces the Einstein-Hilbert action in some appropriate continuum limit [50–52]. What is this action? I will be oversimplifying it, but it looks a lot like Regge calculus or CDT. In CDT you count how many simplices there are of a specific sort and how many of another, then you sum over them and add in some coefficients. In causal sets it's exactly the same, except that instead of counting simplices (you do not have a simplicial complex) you are counting order intervals, an order interval being a pair of elements together with the elements causally between them. For each such interval you count the elements that comprise it. Then you separate the order intervals into kinds, add them up and introduce some coefficients to get an action.

This is one approach to dynamics, but it is very hard to simulate on a computer. The technique people usually use is Monte Carlo, which is always difficult but is even harder in this setting because Wick rotation is not available for causal sets. (Wick rotation gets rid of time by converting timelike directions into spacelike ones, producing a space in which causal relations are meaningless.) In CDT, they rely on Wick rotation, and they analytically continue to Euclidean signature. At the end of the day, they find a sphere, which is not de Sitter but an analytic continuation of it [53]. In contrast, the kind of analytic continuation we would use for the Benincasa-Dowker path integral could not produce a sphere or any manifold of Euclidean signature. It could conceivably produce de Sitter spacetime itself, but not de Sitter with time made imaginary by Wick rotation.

Are you able to show that your action is equivalent to the Einstein-Hilbert action somehow?

Yes, modulo boundary terms, and in an appropriate continuum approximation [50–52, 54]. The Einstein-Hilbert action emerges in the limit of low curvature compared with the Planck scale. A detail that hasn't been cleared up completely is a proper treatment near the light-cones in a curved spacetime where they develop caustics. Modulo this detail, we have indeed shown that one recovers the Einstein-Hilbert action.

Was this action constructed already with this limit in mind?

Actually, it was constructed by asking how one might attain two closely related objectives: first of all to obtain a manifold and, second, to obtain an effectively Lorentz-invariant and local action for it. Given these things, the Einstein equations would more or less follow automatically. Of course, and as with every approach, you'd have to worry about why the cosmological constant term in the effective action wouldn't dominate the scalar curvature term. Aside from that, you are almost guaranteed to get the Einstein equations if you can get a spacetime with an effectively local action.

I had been studying the simplest case of a scalar field, and asking in particular how to write down a local action for such a field on a fixed causal set, which is a good toy model for trying to understand how effectively local equations of motion can arise [55]. The expression for the action grew out of that study. We are now able to understand this simple example quite well, and there is a whole family of possible solutions (with one particular simplest solution in each dimension). Each possible solution gives you a sum over certain order intervals weighted by the values of the scalar field. In a specific continuum limit, or more precisely in the limit of infinite sprinkling density, this action evaluated in flat spacetime reproduces the free scalar field action, namely the integral of $\phi\Box\phi$. What happens in curved spacetime? There what comes out is $\phi\Box\phi - 1/2R\phi$ where R is the Ricci scalar. Thus, a curvature correction appears in the action. If you assume now that the field ϕ is constant then the entire dependence on it disappears and you are left with just the Ricci scalar. Voila.

So your hope is to use this action and Monte Carlo techniques to get de Sitter out?

Yes, but not particularly de Sitter, although that is a very simple example. The first thing we would like to see is whether we actually get four spacetime dimensions. It is not at all guaranteed which dimension it will produce if it produces a manifold at all. The notion of causal set itself doesn't prefer a particular dimension; in fact, it's compatible with any dimensionality or with the dimensionality changing from place to place. There is a particular dimensionality measure that you can attribute to causal sets which is called the Myrheim-Meyer dimension, and which is suitable for any small and approximately flat order interval. Whenever we are able to do Monte Carlo simulations, we will try to probe the simulations with different "measurements," and dimensionality would be the first variable we would like to measure. In fact, I should say that the action we wrote down, even though it's well-defined for any causal set, was actually written down with four dimensions in mind. If you evaluate that action in a causal set that turns out to correspond to a spacetime of some other dimension then you'll get something, but it might not be the Einstein-Hilbert action in that dimension at all. This is just to say that the type of action you start with for your Monte Carlo simulations might have an impact on the dimension that you observe at the end.

In two dimensions we actually have more control, and we understand in more detail how spacetimes can be reproduced from causal sets [56, 57]. The action appropriate to two-dimensional gravity is topological, so basically you are just doing an unweighted sum over causal sets [51]. It turns out that there's a well-defined family of causal sets which are two-dimensional and very easy to characterise mathematically. In higher dimensions, it's not so simple. If you ask me to produce a causal set which is going to have a Myrheim-Meyer dimension of four (and to do so without cheating, that is, without taking a known four-dimensional spacetime and sprinkling it), I wouldn't know how to do it. However, if you ask me to produce a causal set that has two dimensions, then there's a simple intrinsic condition that guarantees this. The condition is that the partial order be the intersection of two total orders, but it doesn't really matter what that means. If we impose this condition, then what comes out with extremely high probability is two-dimensional Minkowski space, or more precisely a so-called causal diamond within it. So you get a spacetime and you have automatically recovered its large-scale structure. I can't tell you that Minkowski spacetime solves the Einstein equations in two dimensions, because I don't know what the two-dimensional Einstein equations should be. In any case, I suppose the relative ease of dealing with two dimensions is analogous to CDTs where it's much easier to compute the two-dimensional sums [58] than higher-dimensional ones. In causal sets we don't need to evaluate the sums computationally, because we can appeal to precise mathematical theorems that tell you that you will get flat spacetime. With a little more work, I think you could also derive an effective action for the fluctuations around flatness.

Okay, so what is the other road to the dynamics?

The other road leads to a type of growth process, and I like to think that it is ultimately the right road to follow [59, 60]. However, at the moment, we are stuck halfway because currently we don't understand how to proceed from classical growth to quantum growth.

What I mean by classical growth is a certain family of birth processes which provide in particular an embryonic kind of causal set cosmology. The idea is that you start with the empty causal set, and then there is a birth. That produces the one-element causal set, which is not too interesting. Then there is another birth, after which the causal set has two elements. The second element has two choices: it could either lie to the future of the first element or be causally unrelated to it ("spacelike"). We posit that it cannot lie to the past of the first element, because that would make the time order of the growth process inconsistent with the intrinsic time of the causal set. In the end, we want the order relations defining the causal set to fully express physical temporality, and we don't want any external time that was introduced in order to describe the growth to retain physical meaning. Anyway, there are the two possibilities I mentioned, which are called the 2-chain and the 2-antichain. An antichain is a set of elements bearing no causal relation to one another, while a chain is the opposite, meaning that all its elements are related. (A chain looks like a sprinkling of $0+1$-dimensional Minkowski space.) Suppose the second birth produces a chain, and then a third element is born. That third element can augment the chain or it can be spacelike to it, and so on. As you can see, the number of possibilities increases as new elements are born, and in fact it explodes.

Classical sequential growth defined this way is mathematically a stochastic Markov process that is characterised by its transition probabilities. At any stage in the growth a large but finite number of causal relations will be possible. When a new element is about to be born it has different possibilities of relating to the other elements (choosing its ancestors, as one often says in analogy with an extended family tree). Those different possibilities occur subject to certain transition probabilities, that is, probabilities for the causal set to make a transition to another causal set with one additional element.

What sets the transition probabilities?

To answer this question let's go back to the foundations of general relativity. There are some basic principles which lead more or less uniquely to the Einstein equation. The most important are general covariance and locality. There has been a lot of debate about what general covariance actually means, starting with the Kretschmann-Einstein debate (see [61] and references therein). Kretschmann argued that general covariance was an empty condition because any theory could be written in a generally covariant way. (I remember hearing a nice talk about this by Karel Kuchar.) I actually think that the modern point of view aligns more with Kretschmann rather than Einstein. Here the example of topological three-dimensional gravity is instructive: people regard it as a generally covariant theory even though it is just flat space [62]. In fact, to make his point, Kretschmann wrote down the theory of a free scalar field in flat space, and then to make it into a generally covariant theory he wrote down another equation equating the Riemann tensor to zero (laughs). Anyway, let us define a generally covariant action functional to be one that depends only on the metric and no other fields or background structures other than the topological and differential structures of the spacetime manifold itself. Then the second principle, that of locality, tells us that you should be able to write your action as the integral of some local scalar. (A local scalar is by definition Lorentz-invariant, by which I just mean that its value at a point depends only on the metric in the neighbourhood of that point.) So if you put these two things together and throw in some hand-waving arguments for discarding higher curvature terms, you end up with the Einstein-Hilbert action.

To arrive at the classical sequential growth (CSG) dynamics for causal sets, we proceed in the same way, except that we can't invoke locality as a principle because we do not have it. So what do we do? We impose general covariance (which we do have) and then a causality condition to which we gave the name "Bell causality." Within the growth process, general covariance is equivalent to label invariance. Every time a birth takes place the new element acquires a label giving its order of birth. Labelling is the analogue of picking a particular coordinate system or a slicing of spacetime, which is exactly what we don't want to be significant. We want the resulting probabilities to depend on the causal set that is built up but not on the order in which the elements were born. This is what we call *discrete general covariance*. We also impose, as I mentioned earlier, that elements are not born to the past of elements born earlier.

Our second principle cannot be locality because that is a non-starter in the causal set context. Instead, we ask that the growth process comply with relativistic causality in a certain sense. In order to understand this, suppose that you have built a causal set up to some stage and now another element is to be born. It might be born over here so that it will group with this set of ancestors, or it might be born over there where it will acquire a different set of ancestors, or it might be that this new element does not relate to any of these elements, as it might be spacelike to all of them. The idea now is that a birth that takes place in a given region of the causal set cannot be affected by what's going on in regions spacelike to that region. The transition probabilities must reflect this. This is the condition we adopted, and we call it Bell causality, because when you think about it in the right way, it is closely analogous to the condition that goes into the derivation of the Bell inequalities in quantum mechanics.

It turns out that the two principles of discrete general covariance and Bell causality are very restrictive and lead to a unique set of rules for the transition probabilities – a unique class of dynamical laws governing classical sequential growth. The simplest (and in some sense the most generic) of the growth processes that one derives in this manner is called transitive percolation, a theory well studied by mathematicians and referred to in the combinatorics literature as the theory of random graph orders. (Interestingly enough, transitive percolation is time-symmetric in a certain sense and appears to be the only CSG model which is.)

I suppose that you have done several simulations with this growth process?

We have done some, but a lot is actually known analytically. The bad news is that it's pretty certain that no matter what choice of parameters you use, your CSG model will never produce a causal set that can faithfully embed in any spacetime (except for one-dimensional Minkowski). Nevertheless, you do produce some appealing toy models for cosmology, which are not too distant from a spacetime interpretation [63]. What I mean by this is that you can still define the causal set counterpart of a spatial slice in a cosmology, and thus you can meaningfully ask how the cosmological scale factor varies with cosmic time. You can then use this as a laboratory to get an idea of how quantum gravity might solve the most important puzzles of cosmology.

For one class of CSG models, even though the causal sets that they grow do not embed faithfully in any spacetime, they nevertheless exhibit something like spatial homogeneity and isotropy and exponential growth for some period of time. At sufficiently long times, however, the scale factor stops increasing and, modulo small fluctuations, sits at a fixed value for an extremely long time. But if you wait long enough, a giant fluctuation occurs and the scale factor goes down to zero, followed by a new epoch of expansion, that leads to a bigger cosmos than the previous one. This is what I have called a Tolman-Boltzmann universe [64].

The name of Boltzmann is attached to it because he raised the question of why we had such a low entropy in the past, which in the cosmological context, is like asking why we had such a small volume in the distant past. Boltzmann suggested that the low entropy was ultimately linked to a fluctuation. In these CSG models, the small volume actually does result from a fluctuation.

I am confused about how things can shrink since you can never remove an element from the causal set, right?

You don't remove any elements, and the causal set is always getting bigger. When I said that the scale factor grew or shrank, I meant that the volume of the spatial slices was growing or shrinking. The causal set counterpart of a spacelike slice is an antichain, because the points in an antichain are all causally unrelated. An antichain with a large number of elements is like a slice of a cosmos that has expanded to a large spatial volume. It is these antichains whose cardinalities grow and shrink. An element of the causal set that constitutes a big bang/big crunch is an element such that every other element of the causal set is either its ancestor or its descendant. When such an element comes into being, the scale factor shrinks to zero.

Why is the name of Tolman associated with this model?

Because Tolman's proposal for cosmology was a universe which shrinks, collapses, and then starts expanding again, via some process which he didn't know how to describe (see [65] and references therein). Although he didn't know how to describe it, he assumed that the second law of thermodynamics would continue to hold. In other words the entropy after the bounce would have to be at least as great as before. Starting the next cycle of expansion with a higher entropy accelerates the expansion and the universe grows to a bigger size, becoming more and more homogeneous in each cycle over larger and larger regions. This is exactly what we observe in these CSG models. If we could make a quantum version of these models it would stand as a potential alternative to the currently popular inflationary scenarios. That would be beautiful because for people working in quantum gravity, inflation is a kind of horror (laughs).

How do you actually define time in this growth process?

It turns out that temporal distance or timelike separation is much easier to define than spatial distance. When you want to know the temporal distance between two points in the continuum, you locate the geodesic between them and measure its length. But what is a geodesic? A geodesic is a longest curve, so we can say that you should seek out the longest curve between the two points and measure its length. We can take this as our measure of timelike separation in spacetime, and it works even when many geodesics join the same two points, as is often the case when gravitational fields are around. The counterpart of a longest timelike curve in a causal set is a longest chain, and all the properties you expect from relativity theory like the "twin paradox" work just the same. (This is another nice illustration of Riemann's point that in a discrete manifold you can deduce metric information just by counting.) How well does this counting distance agree with spacetime distance as defined by the metric tensor? Heuristically, it has to work in most situations, and it has been confirmed in simulations, but the only rigorous theorem I know of applies to Minkowski space and the causal sets that correspond to it. It tells you that the longest chain between two elements is a very good approximation to geodesic length when the length is large [66, 67].

So all that is for timelike distance. Spacelike distance is a whole different story, because there are no chains between causally unrelated elements. In fact, from a certain point of view, spacelike distance is not a very good notion in the continuum either. For instance, in de Sitter spacetime you can find pairs of spacelike-separated points which are joined by no geodesic at all. So what do you mean by the distance between them? Anyway, I don't want to go into it at length, but if you restrict yourself to your favourite antichain (an antichain being analogous to a spacelike hypersurface as I mentioned before), and you look only at pairs of elements within it, then it's not too difficult to come up with a concept of the distance between them.

Do you see the fact that you do not find a causal set that is faithfully embeddable in spacetime as a problem?

Well, I don't think it's a problem for the CSG dynamics. These models not being quantal, constructive interference is missing, and so I wouldn't have expected to recover spacetime from them. But it certainly means that no model of classical growth that we possess at this stage could be a theory of quantum gravity, and there's another, very simple reason for this: not only do the CSG models lack quantum interference, but the Bell causality condition that they incorporate is famously violated in the quantum world. Of course we are hoping to find a quantal generalisation of Bell causality that could lead to a theory of *quantum* sequential growth, but that's still in the future.

So are you saying that the process needs to be modified?

I think that the first question is, What can we do with it? I might have mentioned that the CSG model known as transitive percolation gives rise to something that resembles de Sitter, but I didn't explain in what sense. A de Sitter spacetime is maximally symmetric, which means that if you take any two points and consider the order interval between them, its volume can only depend on the proper distance between the points. Thus, there's a characteristic curve of de Sitter that gives this volume versus proper distance. You can ask a similar question in a causal set that's produced by transitive percolation, and for a fairly large range of distances you find that your curve falls almost right on top of the de Sitter curve [63]. This is a tantalising hint, but it's almost certainly not de Sitter in all respects.

You might have hoped, if you hadn't thought too much about the implications of Bell causality, that one of the CSG models would turn out to be quantum gravity. Lee Smolin, for instance, says that you really just have classical probabilities, but the inherent nonlocalities of your interactions conspire to mimic quantum effects. I agree with this in the sense that causal sets are inherently nonlocal, but at least in the context of CSG models, the type of nonlocality causal sets embody does not allow you to get quantum correlations. So I view classical growth models more as a conceptual laboratory that is generally covariant and fully background-independent, and in which you can study things like the problem of time and the kinds of observables that one can define. Within the context of these models we know what the good observables are, and I would say that this has allowed us to solve the problem of time. However, we are now stuck in the sense that we would like to write a quantum version of this dynamics and we do not exactly know how.

But what do you have in mind? Will these classical principles still be there in the quantum version?

The basic idea that I have in mind is that a dynamical law governing quantum sequential growth will be formulated as a certain kind of path integral, the kind that is called "decoherence functional" or "quantum measure." We would like to design such a path integral by generalising the transition probabilities that define classical sequential growth.

It looks like general covariance/label invariance can be carried over pretty straightforwardly to the quantum case. What's missing is a quantum version of Bell causality, or whatever would replace it, because Bell causality as we have it is an inherently classical notion that pertains to stochastic processes which are not quantum processes.

Understanding this largely boils down to answering another question: What is an appropriate criterion for relativistic causality from a path-integral point of view? If someone gives you an action or an amplitude, how will you know whether it incorporates relativistic causality, or even what that means precisely? No one knows how to answer this question, and part of the reason is that we lack a free-standing path-integral formulation of quantum mechanics itself [26].

What people usually do is ask the question: "Could Alice signal to Bob?" But who are Alice and Bob? They are external agents who stand outside the system itself, and who are supposed to tinker with something over here and then see whether any effect can be felt over there. But of course no agent can be external to the causal set, just as no agent can stand outside spacetime. Trying to incorporate homologues of external agents inside causal set theory would be a true nightmare. It would be like sealing Alice and Bob inside a box and taking them to be instruments that you have no idea how to build but which correspond to self-adjoint operators (laughs). Meanwhile, we don't have self-adjoint operators in the path integral ...

All this is to say that as far as I am aware, one doesn't know how to formulate relativistic causality within a path-integral framework. The moment we could do that, we could carry it over to causal sets. For the time being, I am a bit discouraged with this but there is a master's student, Jason Wien, whose project with me explores a certain consequence of Bell causality which could perhaps be turned into a dynamical ansatz for quantum sequential growth [68]. Maybe this will lead somewhere in the near future.

You've mentioned several times the intrinsic nonlocal nature of causal set theory. Where does this nonlocality come from?

It comes from the lack of local neighbourhoods. Think for a moment of the Maxwell equations, which were originally written in terms of fields which we now regard as effective variables that must be distinguished from the more fundamental microscopic fields of which they are in essence averages. If you open the book by Jackson [69], there is a whole chapter devoted to this problem. One proceeds by taking a spherical region that is neither too big nor too small: big enough to contain a lot of atoms but small enough that conditions are nearly the same everywhere within it. Then you average over the region to get an effective charge density, an effective polarisation tensor, an effective electric field, etc. These regions are local neighbourhoods, and the reason why you can pass so easily from the microscopic to a macroscopic description is because you have the notion of local neighbourhood at your disposal, over which you can average.

In the Lorentzian setting, you don't have local neighbourhoods. There is no invariant way to define them. That is the root of all evil but also the root of all the interesting consequences that arise when you combine Lorentz invariance with discreteness. In the Euclidean plane, the set of all points within a certain distance from a given point is an obvious local neighbourhood, a circular disk. What about in Minkowski spacetime? Well ... the whole light-cone is at zero distance from some given point, and light-cones extend all the way to infinity. So it's not a very local neighbourhood; in fact, it's not properly called a neighbourhood at all, as it stretches out to infinity. Even if we happened to live in a spatially closed cosmos, our past light-cone would reach all the way back to the early universe and would include regions that are now billions of light years away from us. Thus, you aren't able to average over local neighbourhoods in the Lorentzian setting. This is the source of the fundamental nonlocality in causal set theory. It is non-trivial to write down a local action for causal sets, because if you think to yourself, "Let me write down an action that just involves nearest neighbour couplings," all of a sudden you realise that you have an infinite number of nearest neighbours to deal with.

Is this one of the reasons why you think it's hard to quantise gravity?

It's an important reason, but there's also a larger context. I think everyone would agree that even though the electromagnetic field was part of its history from the very beginning, the quantum mechanics that developed out of atomic physics was not a field theory but a theory of action at a distance, primarily a theory of electrons and nuclei interacting via a Coulomb potential. Such a theory rests squarely on a notion of simultaneity and a distinguished time parameter. This is perhaps why canonical quantisation provides the most natural route to the Schrödinger equation.

But action at a distance conflicts with the Lorentzian nature of spacetime that was revealed by relativity theory. There's a tension between our theories of spacetime structure and our theories of quantum behaviour. The attempt to reconcile these two theoretical frameworks led to what we now call relativistic quantum field theory, which limits itself to ultra-local, point-like interactions and incorporates the idea of relativistic causality through the twin requirements of hyperbolicity and spacelike commutativity. Both of these rely heavily on a background causal structure and perhaps also on a background metric.

But a fixed causal structure is precisely what general relativity with its dynamical metric denies. If you want field operators to commute when they are spatially separated, you need to know how to distinguish a spacelike separation from a timelike one, and that's impossible a priori when the causal structure is variable. In other words, the need that even quantum field theory has for a rigid causal structure contradicts the dynamical nature of the causal structure which is built into our theories of gravity.

One popular attempt to resolve this tension employs canonical quantisation, and gives you what is called canonical quantum gravity. However, as I mentioned earlier I don't think that approach will ever solve the problem of time. So what are the alternatives? Well, you have what people sometimes call covariant approaches to quantising gravity, but in my opinion they are all too close to canonical formulations, and they also have other issues you should worry about.

The remaining alternative is the path integral, but as soon as you try to apply it to a dynamical causal structure, you will be led to the problem of how to elaborate a self-contained or "free standing" quantum theory based on the path integral. In other words, at some point you're going to end up asking the question: "Well, what does this path integral really mean?" Does it compute the transition amplitude between this state and that state? But what is a state? It's an element of the Hilbert space that you got from canonical quantisation, or if you didn't get your state from canonical quantisation, you still think of it as pertaining to a particular time. Thus you conclude that you now need a reference time, and the dreaded problem of time returns.

If someone could talk about quantum dynamics from beginning to end in terms of histories and amplitudes, never mentioning gadgets like self-adjoint operators, their eigenvalues, or measurements based on the von Neumann paradigm, we would have a formulation of quantum mechanics and of quantum field theory which would not have any obvious incompatibilities with gravity [70]. However, we don't have such a thing. This is the reason why I think that there is an intertwining between quantum gravity and quantum foundations [71]. I imagine that very few people working on quantum gravity would agree with me on this, but a lot of people working on quantum foundations would be very happy if what they are doing would actually be relevant for quantum gravity (laughs).

It's often said that causal set theory makes a good prediction for the smallness of the cosmological constant. Can you explain how this comes about?

I prefer to describe it as a heuristic prediction. In historical terms, it was undoubtedly a prediction, as it was put forward around 1990, but it was heuristic in the sense that it was deduced from an incomplete theory supplemented by some general facts and some guesses about how the more complete theory would look [29, 30, 72–74].

One of those guesses was that the full theory would be based on some sort of path integral or sum over causal sets, whose dynamics would most naturally be described as the kind of growth process you and I have been talking about. One ingredient in the argument was the fact that in quantum gravity the cosmological constant Λ is the conjugate variable to the spacetime volume V (as shown by the way it enters into the gravitational action). Another was the basic assumption, built into the rules of correspondence between causal sets and the continuum, that the equality between the spacetime volume and number of causal set elements is subject to Poisson fluctuations. Another was that the number of causal set elements functions as a time in the context of sequential growth.

Starting from such considerations, there were actually two slightly different predictions, the earlier one involving more hand-waving [29, 72] than the later one [30]. The later argument applied the uncertainty principle to gravity. It took V and Λ to be conjugate variables, and concluded therefore that the product of their fluctuations would be of order $\hbar$. Now, it is a feature of the Poisson process that if you hold the number of elements in the causal set fixed, the fluctuations of the volume V are of the order of $\sqrt{V}$ [30]. When V is very large the fluctuations of Λ are thus very small, and their magnitude is determined. In fact their predicted size placed them close to the threshold for being observable in the (then) coming years. As you can see, the prediction stemming from this train of thought concerned the magnitude of the fluctuations of the cosmological constant, but it did not answer the question of why the value about which Λ fluctuates is so close to zero.

As a side note, this analysis is connected closely with what people call (somewhat misleadingly) unimodular gravity [72, 75]. If in doing the gravitational path integral, you hold the spacetime volume fixed and don't allow it to fluctuate, you end up with unimodular gravity. In order to fix the volume you have to introduce a Lagrange multiplier, and that Lagrange multiplier combines with the cosmological constant, turning it into an undetermined constant of integration. In this unimodular setting, it's especially clear that there's no point in trying to calculate the value of the cosmological constant from the standard model, something which I think is often misunderstood by people.

Coming back to the heuristic prediction, the earlier argument not only yielded the characteristic size of the fluctuations in Λ, but it also implied that the mean value of Λ would be zero [29]. I would be the last to claim that the deduction involved in this earlier argument was airtight, but I can explain how one is led to it. If you look at the formula for the gravity action, you can immediately read off from it that one can think of Λ as the action per unit volume of spacetime as such (what my student Nosiphiwo Zwane called "the action of free spacetime"). But as we have discussed, spacetime volume equates to number of causal set elements. Therefore one can interpret the cosmological constant as an action per element of the causal set. What should that action be? We don't have any reason to think it's anything in particular, so suppose it were just random, suppose it were either $+1$ or -1, both being equally likely. Then if you want the total action, you just add up all these contributions of $+1$ or -1, and you get a mean of zero subject to fluctuations of the order of $\sqrt{N}$, where N is the number of causal-set elements. If you now want the action per element, you divide your result by N and you get Λ-fluctuations that go like $1/\sqrt{N}$. This argument yields the same small value for the fluctuations, but it tells us also that Λ will vanish on average. If we could take it seriously, we would have solved the riddle of the cosmological constant in *both* of its aspects.

Observations of the so-called dark energy are improving all the time and it could turn out that the cosmological constant actually is constant, in which case this model would be dead. There is a paper which, if it holds up, will change the game quite a bit [76]. It's a paper from the Baryon Oscillation Spectroscopic Survey that combined data from distant quasars with redshift-luminosity information from nearby galaxies, and their analysis concludes that the best-fit value of dark energy was a negative value of about -1.2 ± 0.8 at a redshift of 2.4, which, however, is still compatible with a zero or positive value. I think this indicates that there may already be some tension with the Λ being constant at earlier times, but we have to wait and see.

Why did you choose to do physics?

Well, I don't know why I chose to do physics. I think my parents, especially my mother, steered me in that direction, so it has always been in my mind that I would do physics. One thing I can recall clearly is that in college I was drawn to both mathematics and physics, and I had to choose between them. In one way, mathematics seemed more attractive because of the more inspirational way it was being taught at Harvard at that time. But despite this, I felt that mathematics was too remote from reality, and I wanted to be in touch with the world. However, looking at the world we live in now, I'm not sure how much I still want to be in touch with it.

What do you think is the role of the physicist in modern society?

I think that above all a physicist must not let himself or herself become a tool of the war-makers. This touches other disciplines besides physics, of course, but the historically sustained and intimate relationship between physics and the military marks out this particular duty of resistance as uniquely ours. As scientists, we also have a broader role to play. For example, we witness at present well-funded attempts to discredit the science behind truths of vital importance to humanity, truths like the reality of anthropogenic global warming, or the fact of our evolutionary kinship with animals and other forms of life. We need to resist such efforts as well.

References

[1] T. Skyrme, "A unified field theory of mesons and baryons," *Nuclear Physics* **31** (1962) 556–569. www.sciencedirect.com/science/article/pii/0029558262907757.

[2] R. D. Sorkin, "A general relation between kink exchange and kink rotation," *Commun. Math. Phys.* **115** (1988) 421.

[3] R. Sorkin, "Particle statistics in three-dimensions," *Phys. Rev. D* **27** (1983) 1787.

[4] J. L. Friedman and R. D. Sorkin, "Spin 1/2 from gravity," *Phys. Rev. Lett.* **44** (Apr. 1980) 1100–1103. https://link.aps.org/doi/10.1103/PhysRevLett.44.1100.

[5] J. Friedman and R. Sorkin, "Half integral spin from quantum gravity," *Gen. Rel. Grav.* **14** (1982) 615–620.

[6] R. D. Sorkin, "Introduction to topological geons," *NATO Sci. Ser. B* **138** (1986) 249–270.

[7] J. Friedman and A. Higuchi, "State vectors in higher dimensional gravity with kinematic quantum numbers of quarks and leptons," *Nucl. Phys. B* **339** (1990) 491–515.

[8] P. Jain, J. Schechter and R. Sorkin, "Quantum stabilization of the Skyrme soliton," *Phys. Rev. D* **39** (1989) 998.

[9] P. Jain, "Static properties of the nucleon as a quantum stabilized soliton," *Phys. Rev. D* **41** (1990) 3527–3530.

[10] P. Jain, J. Schechter and R. Sorkin, "Interpretation of the 'quantum stabilized Skyrmion'," *Phys. Rev. D* **41** (1990) 3855–3856.

[11] R. D. Sorkin, "Consequences of space-time topology," in *Proceedings of the Third Canadian Conference on General Relativity and Relativistic Astrophysics Held Victoria, Canada, May 1989*, A. Coley, F. Cooperstock and B. Tupper, eds., pp. 137–163. World Scientific, 1990.

[12] R. Sorkin, "On topology change and monopole creation," *Phys. Rev. D* **33** (1986) 978–982.

[13] L. Bombelli, J. Lee, D. Meyer and R. Sorkin, "Space-time as a causal set," *Phys. Rev. Lett.* **59** (1987) 521–524.

[14] R. D. Sorkin, "Causal sets: discrete gravity (notes for the Valdivia summer school)," in *Lectures on Quantum Gravity, Proceedings of the Valdivia Summer School, Valdivia, Chile, January 2002*, A. Gomberoff and D. Marolf, eds. Plenum, 2005.

[15] F. Dowker, N. Imambaccus, A. Owens, R. Sorkin and S. Zalel, "A manifestly covariant framework for causal set dynamics," *Class. Quant. Grav.* **37** no. 8 (2020) 085003, arXiv:1910.07292 [gr-qc].

[16] S. Surya, "The causal set approach to quantum gravity," *Living Rev. Rel.* **22** no. 1 (2019) 5, arXiv:1903.11544 [gr-qc].

[17] J. Ambjorn, J. Jurkiewicz and R. Loll, "Causal dynamical triangulations and the quest for quantum gravity," in *Foundations of Space and Time: Reflections on Quantum Gravity*, pp. 321–337. 4, 2010. arXiv:1004.0352 [hep-th].

[18] J. Ambjorn, A. Goerlich, J. Jurkiewicz, and R. Loll, "Nonperturbative quantum gravity," *Phys. Rept.* **519** (2012) 127–210, arXiv:1203.3591 [hep-th].

[19] R. Loll, "Quantum gravity from causal dynamical triangulations: a review," *Class. Quant. Grav.* **37** no. 1 (2020) 013002, arXiv:1905.08669 [hep-th].

[20] A. Srivastava, "Cosmological consequences of gravitationally Interacting Planck mass particles," *Phys. Rev. D* **36** (1987) 2368–2373.

[21] R. D. Sorkin, "Does locality fail at intermediate length-scales?" in *Approaches to Quantum Gravity – Towards a New Understanding of Space and Time*, D. Oriti, ed., pp. 26–43. Cambridge University Press, 2009. arXiv:gr-qc/0703099.

[22] M. Saravani and S. Aslanbeigi, "Dark matter from spacetime nonlocality," *Phys. Rev. D* **92** no. 10 (2015) 103504, arXiv:1502.01655 [hep-th].

[23] M. Saravani and N. Afshordi, "Off-shell dark matter: a cosmological relic of quantum gravity," *Phys. Rev. D* **95** no. 4 (2017) 043514, arXiv:1604.02448 [gr-qc].

[24] R. D. Sorkin and D. Sudarsky, "Large fluctuations in the horizon area and what they can tell us about entropy and quantum gravity," *Class. Quant. Grav.* **16** (1999) 3835–3857, arXiv:gr-qc/9902051.

[25] L. Rosenfeld, "On quantization of fields," *Nuclear Physics* **40** (1963) 353–356. www.sciencedirect.com/science/article/pii/0029558263902797.

[26] R. D. Sorkin, "Quantum mechanics as quantum measure theory," *Mod. Phys. Lett. A* **9** (1994) 3119–3128, gr-qc/9401003.

[27] U. Sinha, C. Couteau, T. Jennewein, R. Laflamme and G. Weihs, "Ruling out multi-order interference in quantum mechanics," *Science* **329** no. 5990 (Jul. 2010) 418–421. http://dx.doi.org/10.1126/science.1190545.

[28] A. Bassi, K. Lochan, S. Satin T. P. Singh, and H. Ulbricht, "Models of wave-function collapse, underlying theories, and experimental tests," *Rev. Mod. Phys.* **85** (2013) 471–527, arXiv:1204.4325 [quant-ph].

[29] R. D. Sorkin, "Spacetime and causal sets," in *Relativity and Gravitation: Classical and Quantum*, pp. 150–173. World Scientific, 1991.

[30] R. D. Sorkin, "Forks in the road, on the way to quantum gravity," *Int. J. Theor. Phys.* **36** (1997) 2759–2781, arXiv:gr-qc/9706002.

[31] R. D. Sorkin, "Does a discrete order underly spacetime and its metric?," in *Proceedings of the Third Canadian Conference on General Relativity and Relativistic Astrophysics Held Victoria, Canada, May 1989*, A. Coley, F. Cooperstock and B. Tupper, eds., pp. 82–86. World Scientific, 1990.

[32] R. D. Sorkin, "On the entropy of the vacuum outside a horizon," in *Tenth International Conference on General Relativity and Gravitation (held Padova, 4–9 July, 1983), Contributed Papers*, B. Bertotti, F. de Felice and A. Pascolini, eds., vol. 2, pp. 734–736. Consiglio Nazionale Delle Ricerche, 1983.

[33] L. Bombelli, R. K. Koul, J. Lee and R. D. Sorkin, "Quantum source of entropy for black holes," *Phys. Rev. D* **34** (Jul. 1986) 373–383. https://link.aps.org/doi/10.1103/PhysRevD.34.373.

[34] R. D. Sorkin, "The statistical mechanics of black hole thermodynamics," presented at *Symposium on Black Holes and Relativistic Stars (Dedicated to Memory of S. Chandrasekhar)*. 1997. arXiv:gr-qc/9705006.

[35] D. Dou and R. D. Sorkin, "Black hole entropy as causal links," *Found. Phys.* **33** (2003) 279–296, gr-qc/0302009.

[36] R. D. Sorkin, "Ten theses on black hole entropy," *Stud. Hist. Phil. Sci. B* **36** (2005) 291–301, arXiv:hep-th/0504037.

[37] R. D. Sorkin, "1983 paper on entanglement entropy: 'on the entropy of the vacuum outside a Horizon,' " in *10th International Conference on General Relativity and Gravitation*, vol. 2, pp. 734–736. 1984. arXiv:1402.3589 [gr-qc].

[38] R. Sorkin, "Time evolution problem in Regge calculus," *Phys. Rev. D* **12** (1975) 385–396. [Erratum: *Phys. Rev. D* **23** (1981) 565–565.]

[39] R. Sorkin, "The electromagnetic field on a simplicial net," *J. Math. Phys.* **16** (1975) 2432–2440. [Erratum: *J. Math. Phys.* **19** (1978) 1800.]

[40] J. Hartle and R. Sorkin, "Boundary terms in the action for the Regge calculus," *Gen. Rel. Grav.* **13** (1981) 541–549.

[41] T. Regge, "General relativity without coordinates," *Nuovo Cim.* **19** (1961) 558–571.

[42] J. W. Barrett, "The geometry of classical Regge calculus," *Classical and Quantum Gravity* **4** no. 6 (Nov. 1987) 1565–1576. https://doi.org/10.1088%2F0264-9381%2F4%2F6%2F015.

[43] R. Loll, "Discrete approaches to quantum gravity in four-dimensions," *Living Rev. Rel.* **1** (1998) 13, arXiv:gr-qc/9805049.

[44] A. P. Gentle, "Regge calculus: a unique tool for numerical relativity," *Gen. Rel. Grav.* **34** (2002) 1701–1718, arXiv:gr-qc/0408006.

[45] R. D. Sorkin, "A finitary substitute for continuous topology," *Int. J. Theor. Phys.* **30** (1991) 923–948.

[46] R. D. Sorkin, "A specimen of theory construction from quantum Gravity," in *The Creation of Ideas in Physics: Studies for a Methodology of Theory Construction, Proceedings of the Thirteenth Annual Symposium in Philosophy, Held Greensboro, North Carolina, March, 1989*, Jarrett Leplin, ed., pp. 167–179. Kluwer Academic Publishers, 1995. arXiv:gr-qc/9511063.

[47] J. Henson, "Macroscopic observables and Lorentz violation in discrete quantum gravity," arXiv:gr-qc/0604040.

[48] A. Perez, "The spin foam approach to quantum gravity," *Living Rev. Rel.* **16** (2013) 3, arXiv:1205.2019 [gr-qc].

[49] R. D. Sorkin, "Geometry from order: causal sets," *Einstein Online* no. Band 02 (2006) 02–1007.

[50] D. M. Benincasa and F. Dowker, "The scalar curvature of a causal set," *Phys. Rev. Lett.* **104** (2010) 181301, arXiv:1001.2725 [gr-qc].

[51] D. M. Benincasa, F. Dowker and B. Schmitzer, "The random discrete action for 2-dimensional spacetime," *Class. Quant. Grav.* **28** (2011) 105018, arXiv:1011 .5191 [gr-qc].

[52] F. Dowker and L. Glaser, "Causal set d'Alembertians for various dimensions," *Class. Quant. Grav.* **30** (2013) 195016, arXiv:1305.2588 [gr-qc].

[53] J. Ambjørn, A. Görlich, J. Jurkiewicz and R. Loll, "The nonperturbative quantum de Sitter universe," *Phys. Rev. D* **78** (2008) 063544, arXiv:0807.4481 [hep-th].

[54] M. Buck, F. Dowker, I. Jubb and S. Surya, "Boundary terms for causal sets," *Class. Quant. Grav.* **32** no. 20 (2015) 205004, arXiv:1502.05388 [gr-qc].

[55] R. D. Sorkin, "Scalar field theory on a causal set in histories form," *J. Phys. Conf. Ser.* **306** (2011) 012017, arXiv:1107.0698 [gr-qc].

[56] G. Brightwell, J. Henson and S. Surya, "A 2D model of causal set quantum gravity: the emergence of the continuum," *Classical and Quantum Gravity* **25** no. 10 (May 2008) 105025. https://doi.org/10.1088%2F0264-9381%2F25%2F10%2F105025.

[57] S. Surya, "Evidence for the continuum in 2D causal set quantum gravity," *Classical and Quantum Gravity* **29** no. 13 (Jun. 2012) 132001. https://doi.org/10.1088 %2F0264-9381%2F29%2F13%2F132001.

[58] J. Ambjørn, L. Glaser, Y. Sato and Y. Watabiki, "2D CDT is 2D Hořava–Lifshitz quantum gravity," *Phys. Lett. B* **722** (2013) 172–175, arXiv:1302.6359 [hep-th].

[59] D. Rideout and R. Sorkin, "A classical sequential growth dynamics for causal sets," *Phys. Rev. D* **61** (2000) 024002, arXiv:gr-qc/9904062.

[60] D. Rideout and R. Sorkin, "Evidence for a continuum limit in causal set dynamics," *Phys. Rev. D* **63** (2001) 104011, arXiv:gr-qc/0003117.

[61] R. Rynasiewicz, "Kretschmann's analysis of covariance and relativity principles," in *Expanding Worlds of General Relativity*, Einstein Studies, vol. 7, 1999, pp. 431–462.

[62] R. D. Sorkin, "An example relevant to the Kretschmann-Einstein debate," 2002. http:// philsci-archive.pitt.edu/371/.

[63] R. D. Sorkin, "Indications of causal set cosmology," *Int. J. Theor. Phys.* **39** (2000) 1731–1736, arXiv:gr-qc/0003043.

[64] X. Martin, D. O'Connor, D. P. Rideout and R. D. Sorkin, "On the 'renormalization' transformations induced by cycles of expansion and contraction in causal set cosmology," *Phys. Rev. D* **63** (2001) 084026, arXiv:gr-qc/0009063.

[65] H. Kragh, *Cyclic Models of the Relativistic Universe: The Early History*, vol. 14, pp. 183–204. Birkhäuser, 2018. arXiv:1308.0932 [physics.hist-ph].

[66] J. Myrheim, "Statistical geometry," report number CERN-TH-2538, 1 August 1978.
[67] G. Brightwell and R. Gregory, "The structure of random discrete space-time," *Phys. Rev. Lett.* **66** (1991) 260–263.
[68] J. Wien, unpublished manuscript.
[69] J. Jackson, *Classical Electrodynamics*. Wiley, 1975.
[70] R. D. Sorkin, "Quantum dynamics without the wave function," *J. Phys.* **A40** (2007) 3207–3222, arXiv:quant-ph/0610204.
[71] R. D. Sorkin, "Logic is to the quantum as geometry is to gravity," in *Foundations of Space and Time: Reflections on Quantum Gravity*, G. F. R. Ellis, J. Murugan and A. Weltman, eds., pp. 363–384. Cambridge University Press, 2010. arXiv:1004.1226 [quant-ph].
[72] R. D. Sorkin, "On the role of time in the sum-over-histories framework for gravity (paper presented to the conference on The History of Modern Gauge Theories, held Logan, Utah, July 1987)," *Int. J. Theor. Phys.* **33** (1994) 523–534.
[73] R. D. Sorkin, "A modified sum-over-histories for gravity," in *Highlights in Gravitation and Cosmology: Proceedings of the International Conference on Gravitation and Cosmology, Goa, India, 14–19 December, 1987*, B. R. Iyer et al., eds. Cambridge University Press, 1988.
[74] R. D. Sorkin, "Is the cosmological 'constant' a nonlocal quantum residue of discreteness of the causal set type?," *AIP Conf. Proc.* **957** no. 1 (2007) 142–153, arXiv:0710.1675 [gr-qc].
[75] W. Unruh, "A unimodular theory of canonical quantum gravity," *Phys. Rev. D* **40** (1989) 1048.
[76] K. S. Dawson, D. J. Schlegel, C. P. Ahn, S. F. Anderson, E. Aubourg, S. Bailey, R. H. Barkhouser, J. E. Bautista, A. Beifiori, A. A. Berlind et al., "The baryon oscillation spectroscopic survey of SDSS-III," *The Astronomical Journal* **145** no. 1 (Dec. 2012) 10. http://dx.doi.org/10.1088/0004-6256/145/1/10.

30

Andrew Strominger

Gwill E. York Professor of Physics at Harvard University

Date: July 2011. Location: Benasque. Last edit: 20 February 2016

In your opinion, what are the main puzzles in theoretical physics at the moment?

I wouldn't presume to have an opinion on what the main puzzles are, but I can tell you the ones that I think are the most interesting. The most interesting puzzles have to do with the fact that we don't have a consistent theory which incorporates both quantum mechanics and general relativity. There are many aspects of this puzzle, some of which are technical in the sense that we don't have a confirmed candidate for a mathematical structure that reduces to quantum mechanics in one limit and to general relativity in another limit. Other aspects are conceptual in nature, such as those related to black holes, for example, how information is stored in a black hole and what is the final state of its evolution.

These types of puzzles, which involves statistical mechanics in order to understand the structure of the information, general relativity and, in the context of black holes, also involve quantum mechanics, are the sort of puzzles that lie at the nexus of three main areas of modern physics. I think that it's impossible to resolve these puzzles without learning something really fundamentally new about the nature of our universe. Resolving these puzzles will not be possible without a real revolution in the way that we view the universe – a revolution on the same footing as the revolutions brought about by relativity and quantum mechanics. I don't know exactly how close we are to achieving this goal but I suspect that we are just at the beginning because we are able to ask the question but we don't know how far away the answer is.

I got puzzled because you mentioned that there were no consistent theories of quantum gravity . . .

No confirmed theories, I said. So there is a candidate, namely string theory, which we don't really have a proof that it's fully consistent but all the evidence points in that direction.

Do you see string theory as the best candidate in the market for a theory of quantum gravity?

Yes.

Do you think it's important to follow other approaches to quantum gravity such as loop quantum gravity [1, 2] or causal dynamical triangulations [3]?

Of course it's always important to follow other possible avenues. Indeed, string theory is at the end of several such avenues. In other words, there were several avenues in the past that looked promising and were pursued but turned out in the end to be the same approach. For example, 11-dimensional supergravity, the membrane theory and even non-Abelian gauge theories all led back to this one mathematical structure which we now call string theory. If you ask me about other specific proposals, such as loop gravity,[1] I think that you should only follow them insofar as they continue to look promising. If they don't pan out, don't lead to any interesting results or don't appear to be mathematically consistent, then you should turn to something else.

Yes, and do you think that loop quantum gravity ...

Loop gravity has yet to produce the one-loop correction to graviton scattering. That was a goal which the practitioners thought would be reached within a few months in 1986 and I asked them again several years ago when would they be able to do this and the answer was still a few months (laughs). The subject has been around for nearly 25–30 years and nothing has come out it.

One shouldn't pursue something just because it is an alternative. Specific approaches must have some intrinsic interest. I don't think there shouldn't be anyone working on loop gravity, since I think that each person should follow their own heart, insofar as what they find interesting, but I think that the prospects for loop gravity to metamorphosise into a mathematically self-consistent structure, let alone a candidate for a theory of nature, are exceedingly dim. I hold the same opinion about causal triangulations.

It could have happened that there was another inequivalent viable candidate for a mathematically consistent quantum theory of gravity and if that were the case I think that one should pursue it. However, I don't consider the two that you mentioned to be really viable candidates. I did consider them for the first 5–15 years of their existence but at some point you say, "Okay, put up or shut up!"

I was particular, keen on asking you this last question because I think that you are one of the very few string theorists who cited one loop quantum gravity paper, in particular, when you were discussing the successful computation of the entropy of black holes in a theory of quantum gravity.[2]

No, I didn't. I think that loop gravity failed in showing the right result. You are mistaken if you think I indicated that.

[1] The interviewee consistently referred to *loop quantum gravity* as *loop gravity* and to *causal dynamical triangulations* as *causal triangulations*.
[2] This refers to Ref. [5] of [4].

Okay, maybe I misunderstood it. When citing it, you did mention that there was a constant which was hard to obtain ...

Well ... they were off by some amount. Look, I don't want to provide you with a report of all the failures of loop gravity. That's not so interesting. In the early days of loop gravity, I spent a lot of time going through all the details with them and discussing it with many of them. So it's not something that I have ignored. I looked at it and I don't think that it is the right way to go. Many people have approached it seriously. Hermann Nicolai wrote a review detailing the shortcomings of loop gravity [1] and there was no coherent response from their community. I don't think that it's wrong that a few people continue to work on it; in fact, I think that's good since maybe something was previously overlooked. However, it's just not where I would put my money in.

In almost all David Gross's assessments of string theory, he says that the biggest problem in string theory is to know what string theory is. In your opinion, what is string theory? Is it a theory of strings, membranes?

I would say that the one thing we know about string theory is that it's not a theory of strings. In the early days we thought it was a theory of strings, i.e. that strings were the fundamental objects, but in the process of these many roads combining, the other roads did not involve strings. You can get to string theory from supersymmetric gauge theories or you can get to string theory from 11-dimensional supergravity. So there are many different ways of looking at it of which strings is only one. This indicates, I believe, that the notion of a string is not fundamental to string theory. As a field, we should have had another name.

What would that be M-theory?

I don't know, but not M-theory. M-theory became attached to something more restrictive. It has a meaning which is more restrictive than this whole mathematical structure that we call string theory. It is true that we began by looking at the dynamics of strings but that's not where we've ended up. The continuous use of that name may give the illusion that we have just looked at some little corner around the idea of a string as a fundamental object, whereas in reality we've been all over the place. We've connected many different ideas involving black hole physics and non-Abelian gauge theories, essentially all the ideas in modern theoretical physics beyond the standard model, quantum field theories and gauge theories. Everything that fits together under this kind of one big structure, for historical reasons, we continue to call string theory. I think that this confuses people.

String theory is very often criticised for not being background-independent. Do you think that this is valid criticism? Do you think that it's important to find a background-independent version of string theory?

Actually, I think that criticism is technically incorrect.

Why?

Because it's a statement about the classical theory, and the solution space of classical string theory is the space of conformal field theories, which is exactly as background-independent as the Einstein equations themselves. So, I think that people who state this are just confused. Classical string theory is exactly as background-independent as general relativity. What is not background-independent, for example, is string field theory [5]. The full quantum string theory has never been formulated in a background-independent way. However, I think that's far from being the most important criticism of string theory.

Okay, what do you think is the most important criticism? The fact that string theory is not falsifiable?

I wouldn't phrase it as not being falsifiable and I wouldn't even accept it as a criticism. It's just a statement about string theory that at the moment we cannot make. The current status of string theory is that we don't have a program for experimental verification. However, saying that you don't have a program for experimental verification is different from saying that it couldn't be confirmed. It could happen that when the LHC is fully ramped up, we see a Regge trajectory of resonances pointing to the graviton and that would be experimental evidence for string theory. That's not logically ruled out though I'm not holding my breath waiting for that to happen. On the other hand, the fact that it could happen demonstrates that string theory is a proposal for a theory of physics and a proposal for a theory of the world. It is just the case, at the moment, that we don't have an experiment which we can propose and execute that will tell us whether string theory is right or wrong. This is definitely an obstacle to making further progress.

Many people do not want to work on the subject because of this lack of connection with experiment and that's a reasonable point of view. On the other hand, we know for sure that we haven't figured out all the laws of physics and so some of us find it interesting to rattle around and do whatever we can to move forward. We don't have a 10-year detailed plan to write down all the laws of nature; instead, we have a set of ideas that lead to a set of interesting and interrelated connections for which there are many things to do and to understand better. One of the gratifying things about string theory is that we have had significant results that indicate that we might be on the right track. In the process of developing string theory, we have solved problems that we did not invent. We've solved problems or had something significant to say about problems that other people had posed.

Like which ones?

Certainly, one example that I have been involved in, is black hole entropy [6, 7]. The quantitatively conceptual insight, which we obtained, to the problem of understanding black hole entropy is not really limited to the string theory context, in which it was discovered. We began to understand how we can take the insights that we got by studying string theory and apply them to more general settings. Most of such examples are in pure mathematics. There have been hard problems in pure mathematics that were solved by methods that dazzled the mathematicians, using insights that we've got from string theory [8]. Certainly, this gives us the sense that we are onto something, even if we don't know exactly what it is.

I would like to quote Sheldon Lee Glashow at this point. You know that Glashow was one of the most outspoken critics of string theory and, when I explained to him our results on black hole entropy, he said to me, “Well . . . it’s clear that you are going somewhere, it’s not clear where you are going but you are going somewhere” (laughs). I think it’s one of the highest compliments which our field has ever received and I guess other examples of course are the understanding of the long-lived glueball states at RHIC [9] and the solution of Gerard ’t Hooft’s problem of the large-N limit of gauge theory [10]. Another example is the work of Belavin, Polyakov and Zamolodchikov in conformal theories, which were discovered from trying to understand string theory [11, 12]. Nowadays, conformal field theories are used every day in the study of condensed matter systems, and problems in the structure of two-dimensional conformal field theories were resolved by studying string theory. Also, it gave rise to many ideas for how to solve the hierarchy problem, which led to many of the proposals in modern phenomenology such as the Randall-Sundrum model [13, 14], the little Higgs model [15], etc.

So it’s really been a kind of flood of ideas which have come out of string theory. For example, my talk this morning was about how we can use ideas from string theory to address problems in fluid dynamics.[3] Nothing concrete has come out of it yet as the work is just a month old but there is a concrete connection between Einstein equations and the Navier-Stokes equation which can be used to understand better problems in fluid mechanics [16–19]. I even gave a talk last month at a fluid dynamics conference. So that’s a signpost that we’re not just making up our own problems, giving them to each other, and then patting each other on the back when we solve them. I think we’re doing much more than that.

You just mentioned fluid mechanics and your recent work on its relation to Einstein gravity [16–19]. Do you think that now it will be easier to tackle the problem of turbulence?

Well . . . first of all let me say that there has also been a huge body of work in which ideas in string theory have been used to solve high-temperature superconductivity and the dynamics of non-Fermi liquids [20]. The success of explaining the RHIC data [9] is in that general kind of territory but people are interested in more condensed matter applications and the jury is out on whether that’s going to be useful or not. On the other hand, the jury has not even convened on this particular subject and whether these ideas will help with turbulence or not. It’s a new concrete result but it’s a new one and it’s less than six months old, so it’s hard to say how it will flesh out.

You mentioned earlier the problem of black hole entropy. If we go through your record, we notice the many papers you have published on the subject of black holes. Why do you find them so interesting?

Some of my first works were on black holes and how one could use string theory as a tool to understand them [21–24]. I’m often referred to as a string theorist but I kind of think more

[3] This interview took place during the workshop String Theory in Benasque.

of string theory as a tool to be used to try to understand problems in physics rather than an end in itself. Black holes are interesting because they epitomise what we don't understand about the universe.

In a short online interview someone asked you the question, "What are black holes made of?" and you replied, "Well, that's a question I cannot answer." Can you answer now what a black hole is made of?

No, I can't answer it. We know something about what certain types of black holes in string theory are made of but we don't know have any confirmation that that's related to the real world. We have some very concrete and enticing aspects of black holes to think about, but do we know what they are made of? I'd say no. This is a question that we haven't answered.

What about these examples from string theory? Do we know what they're made of?

In string theory we know but we don't know if the real world is described by string theory. So when you asked this question, I presumed you were talking about the real world.

I understand, but in the string theory context, what are they made of then? It should hopefully give a hint of what reality is but maybe you disagree?

No, I think that the hints we get from string theory are probably putting us on the right track. However, I don't want to assert that the world is described by string theory. So what are they made of in the string theory context? First of all, in order to answer this, it is good to remember that the notions of space and time are not absolute in string theory; that is, they are only approximate concepts. It's clear, just from some very simple dimensional arguments that were made already by John Wheeler in the early 1960s, that the structure of space and time has to dissolve at short distances and be replaced by something else. Wheeler didn't really have anything to say about what that something else was and I think we don't really know even if string theory is exactly what is required to describe that something else. This is at the core of David Gross's question, "What is string theory?"

What is string theory is an important question, but the surprising thing about that question is that one can say an incredible amount of things about how string theory behaves without answering the question of what it is at a fundamental level. It's like a machine for which you press a button here and you see something coming out over there. You don't know what is going on inside the machine but you know what it does. So the question of what is a black hole made of, even in string theory, is difficult to answer at a fundamental level since it would be tantamount to answering what string theory actually is. Nevertheless, we can say something about it; namely, we can say that black holes behave like ordinary quantum systems [6, 7]. A black hole should be thought of as a quantum system but not a quantum system that lives in the volume occupied by the black hole; instead, it should be thought of as a quantum system that lives on the boundary of that volume. This is the so-called holographic principle [25, 26], which states that black holes are equivalent to quantum field theories that live on their boundaries. It's not quite a complete answer to the question of what is a black hole, but it's an answer to how it exactly behaves. Within this description, you can, in a very clever way, avoid conceptual paradoxes that Stephen Hawking brought up early in the 1970s [27].

So can you not say that black holes are just different excited states of this quantum field theory?

Yes, you can say that but I don't think that it fully answers the question of what is a black hole at a fundamental level because there's still something missing in that description. That description is already a long way ahead from where we were 30 years ago but it's not complete. It's not complete because we don't know if that description applies to every kind of black hole. We first understood that it applies to certain special black holes in string theory and we suspect, but we don't know for sure, that it applies to real-world black holes.

Real-world black holes have a universal property; namely, their entropy is just proportional to one quarter times its area and that we haven't explained from string theory. We hope that we could get the tiger by the tail but from just holding the tail you don't know what the tiger is. You just know that you've got this little tail and It's moving around. I think that's where we are now.

Related to this, you proposed a correspondence between real-world black holes and field theories, called the Kerr/CFT correspondence [28]. In this correspondence, are black holes also described by field theories that live on the boundary?

In the Kerr/CFT story the field theory lives on the boundary in precisely the same sense that it lived on the boundary in my work with Cumrun Vafa on supersymmetric black holes [6]. "Living on the boundary" is not a mathematically precise statement but you make it mathematically precise.

Have you been able to describe any new phenomena of Kerr black holes using this dual description? For example, could you describe emissions from the ISCO of the GRS 1905+105 black hole?

What we have discovered is that for rapidly rotating black holes, of which there have been a number observed up in the sky, if you get very near their horizon they exhibit a symmetry that no one noticed before. It's a very powerful and large symmetry group; in fact, it's an infinite number of symmetries. Systems with infinitely many symmetries, in fact the same infinite set, have been observed in the laboratory in condensed matter systems, such as edge states in the quantum Hall effect, which lead to all kinds of physical consequences.

These infinitely many symmetries of these extreme rotating black holes should have consequences for their behaviour. So can we see some of those? It's quite possible that we might be able to, eventually. One thing that I'm working on now, with some students, is not related to the GRS 1915+105 black hole but instead related to the MCG-6-30 nearby Seyfert galaxy which has a supermassive black hole in the middle and it is rotating at 99% of the speed of light so it's very close to extreme. There have been proposals that gravity wave detectors, not LIGO but LISA, could possibly see stellar mass objects falling into supermassive black holes at the centre of galaxies. I don't want to overstate this but we are looking at using this symmetry in order to compute the gravitational wave signal for a stellar mass object falling into the black hole. There is a possibility that the symmetry should be useful for understanding the data that comes out of this supermassive black hole [29].

There is kind of two sort of separate pieces of the Kerr/CFT correspondence. One is our discovery that there is this infinite-dimensional symmetry, which is this conformal group that acts on these extreme rotating black holes. I don't think that it's at the level of a rigorous mathematical theorem, but I wouldn't really say that it's a conjecture either. I think that it follows from the Einstein equations without further assumptions. The other piece is the conjuncture that this infinite-dimensional symmetry group turns out to be the symmetry group associated with a certain kind of two-dimensional field theory. The conjecture is therefore that there is some complete equivalence between these rotating black holes and these two-dimensional conformal field theories.

Do you know which two-dimensional conformation field theories we are talking about?

We don't know, we don't hope to know and it's not the goal of the programme to know that. Knowing which conformal field theory would be equivalent to knowing all the laws of physics up to the Planck scale and beyond. If you know which conformal field theory describes the system then you know everything. This work [28] is based on a bottom-up approach, not on a top-down approach. Measurements that we make about the laws of physics can be translated into properties of the conformal field theory. We can slowly understand the properties of that conformal field theory. It is a different way of describing the world, especially the world around a rapidly rotating black hole but it's a way that simplifies some of the problems. It also enables one to understand the origin of the mysterious Bekenstein-Hawking entropy law [30–32].

Is your point of view that these works [30–32] explained the origin of the Bekenstein-Hawking entropy law?

I would say that the observation that this infinite-dimensional symmetry group, which appears near the horizon of a black hole, can be heuristically argued to imply, but not prove, the statement that these black holes, just like black holes in string theory, have a dual description as a two-dimensional field theory. If you make that conjecture then you have an explanation of the Bekenstein-Hawking entropy law. That is one of a number of pieces of evidence that we have that justify taking that leap and making the full identification. It's incredibly tempting to do it and one should get experimental evidence for showing that it is correct to do so. But you know the saying: "If it talks like a duck and walks like a duck, then it is a duck." The black hole is definitely *talking like* a conformal field theory and *walking like* one, but we can't prove that it is a conformal field theory.

But you did embed one specific case in string theory, namely, the Kerr-Newman black hole [32], right?

In the context of string theory, it's true. But we don't know if string theory has anything to do with the real world so we are not allowed to assume that is the case in general. We can bang at this idea in many different ways and it seems to be holding up however we bang at it. It seems to be a very cohesive idea but whether we can look up at the sky and with confidence say that GRS 1915+105 is really a conformal field theory, I don't know. If I were to state that it is a conformal field theory then people would be critical and they would have the right to be critical.

Some people would say, "I want you to make a prediction that nobody else did." After all, when Albert Einstein discovered general relativity, it had this incredible sort of breath-taking mathematical coherence, and it postdicted fairly well the precession of Mercury's perihelion but nobody believed in it. It was too weird for the scientific community at the time. The knowledge of the anomaly in Mercury's perihelion precession was known for 40 years and everybody had their own pet theory, mostly based on different forms of dark matter which would explain it. So nobody really paid any attention to Einstein's theory, which was thought to be somewhat of an exotic theory.

It's always easy to make up a theory that explains something that has already been seen, so it wasn't until it made the prediction of the bending of light by the sun, which was experimentally measured, that it was formally established as a theory of physics. Whether we like it or not, we are going to be held to the same standard, which is the standard in physics. It's just how it is and until we make a prediction for which somebody goes out and tries to measure it, it shouldn't be thought of as established physics. As wonderful as all this stuff is, I think we have a very long way until we are able to do that and I don't know how to get there from where we are now but I'm still trying.

What do you think has been the biggest breakthrough in theoretical physics in the past 30 years?

I don't want to answer that question because I think that everything that has happened in the last 30 years is part of one synthetic whole. Everything that has been done is interconnected and different things have been explained in different ways. It's sort of one mass of theorists kind of moving together and I don't think it would be fair to kick out one thing and separate it from the rest.

Okay, why have you chosen to do physics? Why not something else?

It's fun and it's beautiful.

What do you think is the role of theoretical physics in modern society?

I think that most theoretical physicists pursue theoretical physics because they have a large amount of curiosity about the basic nature of the universe. I am referring to the type of questions you ask as a child, which I think everybody wants to know the answer to. I think that the world at large wants to move forward as a species and we move forward by trying to understand the universe around us. So I think that everybody wants to know that we are asking those questions and that we are trying to answer them. It shows that society, human spirit and human passion are alive. I doubt that people would like to hear otherwise.

References

[1] H. Nicolai, K. Peeters and M. Zamaklar, "Loop quantum gravity: an outside view," *Class. Quant. Grav.* **22** (2005) R193, arXiv:hep-th/0501114.

[2] T. Thiemann, "Loop quantum gravity: an inside view," *Lect. Notes Phys.* **721** (2007) 185–263, arXiv:hep-th/0608210.

[3] R. Loll, "Quantum gravity from causal dynamical triangulations: a review," *Class. Quant. Grav.* **37** no. 1 (2020) 013002, arXiv:1905.08669 [hep-th].

[4] A. Strominger, "Five problems in quantum gravity," *Nucl. Phys. B Proc. Suppl.* **192–193** (2009) 119–125, arXiv:0906.1313 [hep-th].

[5] A. Sen, "String field theory as world-sheet UV regulator," *JHEP* **10** (2019) 119, arXiv:1902.00263 [hep-th].

[6] A. Strominger and C. Vafa, "Microscopic origin of the Bekenstein-Hawking entropy," *Phys. Lett. B* **379** (1996) 99–104, arXiv:hep-th/9601029.

[7] G. T. Horowitz and A. Strominger, "Counting states of near extremal black holes," *Phys. Rev. Lett.* **77** (1996) 2368–2371, arXiv:hep-th/9602051.

[8] A. Strominger, S.-T. Yau and E. Zaslow, "Mirror symmetry is T duality," *Nucl. Phys. B* **479** (1996) 243–259, arXiv:hep-th/9606040.

[9] P. K. Kovtun, D. T. Son and A. O. Starinets, "Viscosity in strongly interacting quantum field theories from black hole physics," *Phys. Rev. Lett.* **94** (Mar, 2005) 111601. https://link.aps.org/doi/10.1103/PhysRevLett.94.111601.

[10] J. M. Maldacena, "The large N limit of superconformal field theories and supergravity," *Int. J. Theor. Phys.* **38** (1999) 1113–1133, arXiv:hep-th/9711200.

[11] A. M. Polyakov, A. Belavin and A. Zamolodchikov, "Infinite conformal symmetry of critical fluctuations in two-dimensions," *J. Statist. Phys.* **34** (1984) 763.

[12] A. Belavin, A. M. Polyakov and A. Zamolodchikov, "Infinite conformal symmetry in two-dimensional quantum field theory," *Nucl. Phys. B* **241** (1984) 333–380.

[13] L. Randall and R. Sundrum, "A large mass hierarchy from a small extra dimension," *Phys. Rev. Lett.* **83** (1999) 3370–3373, arXiv:hep-ph/9905221.

[14] L. Randall and R. Sundrum, "An alternative to compactification," *Phys. Rev. Lett.* **83** (1999) 4690–4693, arXiv:hep-th/9906064.

[15] M. Schmaltz and D. Tucker-Smith, "Little Higgs review," *Ann. Rev. Nucl. Part. Sci.* **55** (2005) 229–270, arXiv:hep-ph/0502182.

[16] I. Bredberg, C. Keeler, V. Lysov and A. Strominger, "Wilsonian approach to fluid/gravity duality," *JHEP* **03** (2011) 141, arXiv:1006.1902 [hep-th].

[17] I. Bredberg, C. Keeler, V. Lysov and A. Strominger, "From Navier-Stokes to Einstein," *JHEP* **07** (2012) 146, arXiv:1101.2451 [hep-th].

[18] V. Lysov and A. Strominger, "From Petrov-Einstein to Navier-Stokes," arXiv:1104.5502 [hep-th].

[19] I. Bredberg and A. Strominger, "Black holes as incompressible fluids on the sphere," *JHEP* **05** (2012) 043, arXiv:1106.3084 [hep-th].

[20] H. Liu, J. McGreevy and D. Vegh, "Non-Fermi liquids from holography," *Phys. Rev. D* **83** (2011) 065029, arXiv:0903.2477 [hep-th].

[21] M. J. Bowick, S. B. Giddings, J. A. Harvey, G. T. Horowitz and A. Strominger, "Axionic black holes and a Bohm-Aharonov effect for strings," *Phys. Rev. Lett.* **61** (1988) 2823.
[22] D. Garfinkle, G. T. Horowitz and A. Strominger, "Charged black holes in string theory," *Phys. Rev. D* **43** (1991) 3140. [Erratum: *Phys. Rev. D* **45** (1992) 3888.]
[23] G. T. Horowitz and A. Strominger, "Black strings and P-branes," *Nucl. Phys. B* **360** (1991) 197–209.
[24] J. Callan, Curtis G., S. B. Giddings, J. A. Harvey and A. Strominger, "Evanescent black holes," *Phys. Rev. D* **45** no. 4 (1992) 1005, arXiv:hep-th/9111056.
[25] G. 't Hooft, "Dimensional reduction in quantum gravity," *Conf. Proc. C* **930308** (1993) 284–296, arXiv:gr-qc/9310026.
[26] L. Susskind, "The world as a hologram," *J. Math. Phys.* **36** (1995) 6377–6396, arXiv:hep-th/9409089.
[27] S. Hawking, "Black hole explosions," *Nature* **248** (1974) 30–31.
[28] M. Guica, T. Hartman, W. Song and A. Strominger, "The Kerr/CFT correspondence," *Phys. Rev. D* **80** (2009) 124008, arXiv:0809.4266 [hep-th].
[29] A. P. Porfyriadis and A. Strominger, "Gravity waves from the Kerr/CFT correspondence," *Phys. Rev. D* **90** no. 4, (2014) 044038, arXiv:1401.3746 [hep-th].
[30] T. Hartman, K. Murata, T. Nishioka and A. Strominger, "CFT duals for extreme black holes," *JHEP* **04** (2009) 019, arXiv:0811.4393 [hep-th].
[31] A. Castro, A. Maloney and A. Strominger, "Hidden conformal symmetry of the Kerr black hole," *Phys. Rev. D* **82** (2010) 024008, arXiv:1004.0996 [hep-th].
[32] M. Guica and A. Strominger, "Microscopic realization of the Kerr/CFT correspondence," *JHEP* **02** (2011) 010, arXiv:1009.5039 [hep-th].

31

Leonard Susskind

Felix Bloch Professor in Physics at the Institute for Theoretical Physics at Stanford University

Date: 13 July 2011. Location: Waterloo, Canada. Last edit: 2 November 2016.

What are the main problems in theoretical physics at the moment?

(laughs) The problem of eternal inflation [1–4]. I think that's the biggest and hardest problem. I think it's the one that's going to determine how fundamental physics proceeds from here on. If eternal inflation is right we go one way, if eternal inflation is wrong we go another way and I can't think of any problem that I find bigger than that. Of course, the problem at the moment and maybe forever is a completely theoretical one and solving it means constructing a proper mathematical conceptual framework for it. I think we're at the moment fairly far from that.

Are you referring to the problem of quantum gravity?

Well, certainly we're talking about quantum gravity but we're talking about cosmology and we're talking about the question of whether a multiverse exists. Whether a multiverse does or doesn't exist, the outcome will radically determine the way we think about the laws of physics. If the multiverse exists and it's diverse and has lots of different kinds of places, we'll think about the laws of physics in one way. If the multiverse doesn't exist and the universe is all just one kind of thing, everywhere exactly the same, then we'll be thinking about the laws of physics in a totally different way. Where things go from here is going to be determined by our ability to come to a conclusion as to whether the universe has many different kinds of environments, many different kinds of laws of physics in different places or whether it's all the same. I can think of no problem bigger than that. Whether it will get solved, whether a consensus will be reached in the future, it's hard to tell. These things are always surprising.

Do we have any evidence for the multiverse?

We have three pieces of evidence and all of it is in a sense soft and squishy. First, as Steven Weinberg predicted, if there were a multiverse and if there were an anthropic explanation of the laws of physics, there would be a small cosmological constant [5]. There is, in fact, a small cosmological constant [6]. There is no other explanation for it except that the cosmological constant in a multiverse with lots and lots of different kinds of environments would take many different values. We would simply exist where the cosmological constant is small enough for us to exist, period. It would be like asking why the temperature of the Earth is what it is. If it were very different, we wouldn't be here. That logic may or may not be right but we'll come back to it in a minute.

The success of inflation is another piece of evidence. Inflation tells us that the universe is extremely big [7]. In fact, combining the fact that there's a cosmological constant, which tells us that there is an event horizon, with inflation, which tells us that the universe is much bigger than the event horizon, you conclude that there is a multiverse.

In which sense is there a multiverse?

In the sense that there is an event horizon, there are things behind it and we have to come to terms with what it means for a whole world to exist out there beyond the horizon. We have to understand what that means; we have to learn whether it means anything scientific and whether, for example, for statistical purposes, for counting purposes, for probabilistic purposes it makes sense to think about a world out there bigger than the region that we can ever see. So, in some sense, it's an experimental fact that, if not a multiverse, then at least something much bigger than the portion we'll ever be able to see is out there. Many of the questions that come up in relation to the existence of the multiverse, even if there isn't this vast landscape of different vacua, come up also when you try to make sense out of a universe where almost all of it is beyond our reach.

The third piece of evidence, which I'm not sure how seriously we should take it, is that string theory has given rise to a vast number of possibilities, which is now called the string landscape [8, 9]. The word was taken out of a context, the context being biology, where people speak about the landscape of biological designs (see, e.g., [10]). The landscape of biological designs is huge because there is a huge number of ways of rearranging the DNA. String theory provides a kind of framework in which there's a DNA, that is, the DNA is the nature of the compactifications of string theory. As far as it can be told, the number of possibilities grows exponentially the same way the number of possibilities for rearranging a DNA molecule does. Of course, this is not a hard and fast fact about string theory. It's something which seems likely to be the case if we ever make sense out of the mathematics behind it.

So we have a potential possibility for understanding something like a DNA in a landscape, the existence of a cosmological constant which has no other explanation and the fact that inflation makes the universe much bigger than anything we can ever see. Those three things combined together raise the question of whether a multiverse exists or not. It doesn't really matter whether you like this idea or you don't like this idea. The important question in physics today is whether it's correct or not. If you hate it or if you like it, you should be doing exactly the same thing, namely, trying to find out if it's correct. Kill it if you can or make it sensible if you can.

Regarding one of your assumptions, are you a firm believer in string theory?

Well, yes and no. I think there's a question of terminology in what you mean by string theory; like with any question of the same kind of nature, you have to set your terms of reference.

I would like to know what you think string theory is.

There's a very precise construction, a mathematical set of constructions, a whole vast collection of them. They consist of all the supersymmetric vacua that are interconnected and almost certainly form a mathematically well-defined structure. That's called string theory. People have won Fields Medal prizes for following this line of research[1] and there's very few doubts that it is a mathematically consistent framework that contains gravity and quantum mechanics. So in that sense, yes I believe in it; you cannot not believe in it; Do I believe that it's the theory of nature? Absolutely not. It is most certainly not the theory of nature but I've defined it in this narrow way. I've defined it as the set of supersymmetric vacua where the mathematics is well defined and we know the framework is consistent. However, we don't live in such a world.

But you assume that our world is in one of these vacua, right?.

No we don't. We certainly don't live in a supersymmetric world. We certainly do not. We do not live in a world which is asymptotically flat. We absolutely do not live in a world for which this narrow definition of what string theory is would apply. What the wider definition is, we don't know. We don't know how to define string theory as a larger framework. Most likely, string theory is a small part of a bigger set of things. Whether that bigger set of things can be explored by extrapolating the ideas from string theory or not is not known. There are ideas, people try to do it, it's very complicated, unreasonably complicated but my personal belief is that there is a wider structure that contains a lot more and that string theory is a part of it.

Is this wider structure M-theory?

No, no. M theory and string theory are the same thing. M theory is part of the supersymmetric space. We don't believe in M theory either. We don't live in 11 dimensions, we don't live in a supersymmetric vacuum. The world that we live in is not the world of string theory. So then comes the question: Is there a broader definition of string theory? Perhaps, my guess is yes but my guess is as good as yours. There most likely is a broader set of definitions for which string theory, as narrowly defined but defined well enough that you can be sure that it mathematically exists, would be part of. If it's part of a bigger framework, you might be able to discover that bigger thing by pushing against the boundaries of what we know. I don't think that, whatever the right theory is, it would be completely disconnected from string theory.

I don't think that there's more than one theory of quantum gravity. There may be more than one solution to its equations. String theory is a theory that has lots of different solutions obeying the basic rules of string theory. There are probably more extensive rules, or rules that are more general and have as their solution a broader class of things, which may include string theory. That's the way I'm inclined to think about it. But there's absolutely no doubt in my mind of two things: that string theory is a mathematically consistent

[1] In 1990, Edward Witten became the first physicist to be awarded the Fields Medal.

theory which contains gravity and quantum mechanics but that is narrow and supersymmetric. I have absolutely no doubt that the mathematical structure of string theory is consistent and I have absolutely no doubt that it's not our universe.

Okay, but so when you speak about the vast landscape in connection with the eternal inflation scenario, what kind of landscape are you talking about?

We are talking about something which goes beyond the narrow definition of string theory that I gave.

So you assume it exists even though you don't know that it exists?

That's right, we don't know that it exists. We don't have a mathematically precise formulation of string theory that includes things that potentially could be our world. We do not.

Most string theorists would argue – in fact, I think a wide range of string theorists would argue – that there's a good reason to think that the boundaries of the theory are larger than my definition. I would agree with that; I think they are. I think that maybe it's possible that the boundaries can be expanded by pushing string theory further. Maybe that's the only way we have in order to explore beyond the things we know. All of the constructions that are used for trying to study cosmology, for trying to study de Sitter space, for trying to study string theory with a cosmological constant, are beyond the rigorous framework which I'm certain exists.

But so, when people speak about the 10^{500} different vacua, are these all supersymmetric?

That number is a guess. Yes there are 10^{500} or more supersymmetric vacua [8]. Within the narrow framework of the mathematically precise version of the theory, they all have vanishing or negative cosmological constant. You can then use some approximate ideas, some incomplete guesses about the way the theory works and you can make some logical jumps. The logical jumps involve approximations which have worked in the past for other purposes. The logical jumps seem plausible but they are still logical jumps. Making these logical jumps you extrapolate from those 10^{500} supersymmetric vacua that there are 10^{500} versions of those vacua which have positive cosmological constant [11]. The arguments used in this extrapolation are, in my mind, good arguments. I don't intend to be negative about string theory; I intend to be precise about it.

Okay, so it's not clear that these vacua are there?

There are no known equations for which those vacua are solutions. You operate using a method of doing physics which has been highly successful, for example, in condensed matter physics. You operate as if your approximations were fundamental principles and then you use them, never really being quite sure what they are approximations to. You use approximations that have worked in the past but that you actually don't know what they are approximations to. A much better way of working is to have a well-defined set of equations and then approximate them. We don't have a well-defined set of equations so we're swimming in an ocean which is very big and we don't have a boat which has been constructed with the best precision that we would like it to have.

You mention that the biggest problem was to try to find a mathematically consistent formulation for eternal inflation.

I believe so and that, of course, would include, if string theory is the right way to think about it, generalising the ideas of string theory so that the equations would give rise to a bigger class of solutions. So these are not disconnected questions. In fact, they're probably the same question.

...of finding a full non-perturbative formulation?

Finding a full non-perturbative formulation of X-theory, whatever X-theory is, which allows de Sitter space vacua, will answer the question of whether there are de Sitter spaces. It would also answer the question about their stability. Are they metastable? Are they stable? If they're stable, they're not the ingredients for eternal inflation. They have to be able to create little bubbles of instability. If they are metastable, which string theory suggests [11], then there's no question that eternal inflation will result from these vacua.

So it's really the same question. To extend X-theory to include, with some precision, vacua which are not supersymmetric and in particular have positive cosmological constant is probably the same as finding a very precise and mathematically consistent framework for eternal inflation.

If you find X-theory, do you think that X-theory will tell you exactly which vacuum we live in?

No. It's like the theory of planets. The theory of planets is, of course, just quantum electrodynamics plus nuclear physics. Planets are made out of stuff, real stuff, atoms. The theory of planets does not tell you what kind of planet you live in. The theory of the multiverse will not tell you what part of it you live in. It will tell you what parts of it exist. It will tell you what parts of it can support life. It will tell you what the relative number of different possibilities are there, if we get it right, but it will no more tell us what kind of vacuum we live in than the theory of planets tells us what kind of planet we live in. It will not tell us what kind of patch of space we live in but it will enumerate the possibilities.

So the fact that we live in this vacuum is just random?

Not completely random. There are many that we couldn't live in. Just like we can't live on Pluto, we can't live on Jupiter and we can't live on the moon ... so it's not completely random.

So we need to invoke the anthropic principle?

Absolutely. It says that we live where we can live – that's called the anthropic principle [5, 9]. What determines, to some extent, the world we live in, is the requirement that it should be able to support life. Look, I talked about the biological designs. Now, as I'm sure you know, there's a huge landscape of possible DNA molecules. Almost all of them are unviable so if you would take a random DNA sequence, it most likely wouldn't produce living things that can do what we know is possible. The molecule might be able to exist but it wouldn't be able to think, it wouldn't be able to ask questions and it wouldn't be able to do anything. Is it an accident that our DNA happens to be of the kind which allows life or which allows complex behaviour? Of course not. All these issues of philosophical fine hair splitting, strong anthropic principle, the weak anthropic principle, etc., are silly. We might as well dismiss all the places where life isn't possible because we couldn't be there.

So if eternal inflation is correct, why doesn't all of a sudden the constants of nature change?

They do, due to bubble nucleation, in turn, due to the metastability of the vacuum. According to eternal inflation, our world in the past was different than it is now [3, 12, 13]. A bubble nucleated in that past world with different values of the constants and those constants shifted. They shifted sharply from one number to another. From one set of patterns to another set of patterns and now we're living in a world with a particular set of patterns but if we wait long enough something will happen. That's what eternal inflation is all about. It's about the metastability of the vacuum and the fact that things change.

So life just exists during short pockets of time?

Right. Pockets of time but in a big, infinite, inflating universe. I think that it is a serious question whether this can be made to have mathematical and conceptual sense and I don't think that this is a finished story by any means. What I do think is that this question, which is out there, which has been laid in our lap, has been put there due to the success of inflation, due to the existence of cosmological constant and due to some string theory ideas. It's been laid on our lap and now we have to answer the question: Is it right or not? That's what this conference is about.[2] I have to say that this story resembles the ... what is it? The blind men and the elephant?[3] Everybody sees a different story so at the moment it looks rather chaotic. I don't mean chaotic in the technical sense, I mean chaotic in the human sense. Everybody has a different idea, everybody has a different version of it and there's no coherent story there.

Does eternal inflation make any prediction that could somehow be observed?

There are ways to falsify it.

Wait long enough and see if it changes (laughs)?

No (laughs). There are ways to falsify it but you'll have to be very lucky. If the amount of ordinary inflation, not eternal inflation, in our patch was minimal, then we should be

[2] This interview took place during the conference *Challenges for Early Universe Cosmology* at the Perimeter Institute.
[3] See Wikipedia entry "Blind Men and an Elephant."

able to detect the curvature of space. Eternal inflation predicts that the curvature of space is negative, not positive. If positive curvature was detected, we go back to the drawing board. There are predictions that our sky could be full of bubble collisions and that could be observed. Our bubble could be colliding with another bubble [13, 14].

What would happen if this were the case?

We would just see patterns on the microwave sky. We would see round patterns of higher temperature as well as lower temperature and it would be a weak signal but it could be detectable with luck. This all hinges on not having too much ordinary inflation because ordinary inflation just stretches all signals and dilutes them to the point where you can't see [7]. So falsification is possible, confirmation is also possible, but I think that in either case the likelihood that the range of parameters will be such that we would be able to detect these things is not much. So that's why I said that it may never be possible to have absolutely convincing observational evidence about eternal inflation, in which case, theoretical consistency becomes all the more important because that's the only test that we really have. Hopefully not.

You mentioned before that you thought that there couldn't be many different theories of quantum gravity.

No. I said that I don't believe that there's more than one. There may be more than one way to look at that specific one. We have different ways to look at string theory, for example, using D-branes and so on. For all I know, although I doubt it, there may be a way to think about it where it looks like loop quantum gravity (LQG). I don't know.

Could that be the case?

I'm not going to say yes or no. There could be different ways of looking at it but my guess is that there is only one mathematical structure that contains gravity and quantum mechanics. The constraints that arise from combining the two are very tough. Now, of course, that could be wrong but I do think that there's only one theory with many solutions to its equations.

And that in one corner of the theory you could find LQG?

Perhaps, perhaps, perhaps.

So you don't have any problems with it?

I don't know anything about LQG but, you know, why not?

You worked a lot on the black hole information paradox. Could you explain what this paradox is really about?

I wrote a whole book about it [15]; do I have to explain it again?

Well, if you don't want to, you don't have to. Okay, let me ask another question, maybe simpler ... do you have any idea about what black holes are made out of?

(laughs) What they are made out of? I don't know what that means. All right, let me try to make a good question out of something that is not a bad question but which is imprecise. What is it that accounts for the entropy of a black hole? If you asked me: What is it that accounts for the entropy of a bathtub full of water? I would say that it's water molecules. So when you ask me what black holes are made out of, I would translate that into the question: What is it that accounts for the entropy of a black hole? Once you know that a bathtub full of water has entropy and the entropy is finite, then you're immediately led to ask: What are the microscopic degrees of freedom which account for that entropy? The answer is: atoms.

Once you know that a black hole has entropy then you're immediately led to the question: What is it that accounts for that entropy? What are the degrees of freedom which are being counted? I told you earlier that there's enough evidence that string theory is a mathematically consistent structure, whether or not it represents the real world. String theory contains gravity and so it must be able to answer the question of what black holes are made out of. It may not be the answer that is applicable to the real world but since it's a consistent theory it must have its answer for what black holes are made out of. In fact, it does: they're made out of the microscopic degrees of freedom of string theory.

What are these degrees of freedom?

Some black holes are made out of D-branes, some of them are made out of strings. There's a zillion different kinds of black holes with different charges but they are made of the basic elements and the basic structures of string theory.

So, according to you, are the microscopics of black holes well understood right now?

No, I don't think the microscopics of black holes are well understood. There is a way of counting black hole states for extremal black holes – charged, extremal, infinitely cold, black holes in particular – which is precise [16, 17]. It's rather remarkable as it gives the answer that Jacob Bekenstein [18] and Stephen Hawking [19, 20] found on totally different grounds. No such precise theory exists for Schwarzschild black holes at the moment. A less precise version of it exists, which was given by me [21, 22] and by Joseph Polchinski and Gary Horowitz [23], but just the fact that it's less precise and you can't pin down how precise it is exactly means that it's not completely understood.

On the other hand, things like the AdS/CFT correspondence [24] and matrix theory [25] tell us pretty clearly that string theory does not allow violations of the laws of information conservation [26–28]. For sure we're not making a mistake in thinking that information is conserved. That's one of the great things string theory has done, whether it's the right theory of nature or not. It removes the argument of [20] that says that whenever you combine gravity with quantum mechanics, you will inevitably violate the rules of information conservation. We know that it's not the case that those two things put together will always lead to loss of information. Once you know that, there's no longer a reason to believe that Hawking was right when he stated that information was lost in black holes [20]. In my mind, that is the major contribution of string theory, among many. It has told us that quantum mechanics and gravity can coexist. That's not a small thing.

So according to string theory, has the information paradox problem been solved?

Let's say that we know the outlines of an answer and we know what the answer has to be. There's always room for improvement, there's always room for understanding it better. I think that string theory has told us what the answer has to be. The answer is no, meaning that there will not be any loss of information. I think we can be absolutely confident about that but there's certainly room for enormous improvement of our understanding of how it works.

You mentioned AdS/CFT but do you think that string theory will be able to provide a dS/CFT correspondence?

A precise one? If you are thinking about an approximate one, which might be useful for ordinary inflation, not for eternal inflation, then I think Juan Maldacena has given a very convincing case for an approximate perturbative notion of dS/CFT which is useful for inflation [29]. My own feeling is that it can never be made exact, that it's part of perturbation theory, and that it's part of a bulk theory rather than a holographic theory. It's not a holographic theory because it doesn't have anything to do with information bounds and it doesn't have anything to do with entropy. So I think it's part of the approximate perturbative treatment of de Sitter space.

There may be some other entirely different construction that might have de Sitter space in it and that might in some ways have a conformal field theory (CFT) in it, in which case you might start calling it dS/CFT, even though it has nothing to do with what's presently called AdS/CT.

You have worked and written a lot about the black hole complementarity principle [30]. Recently, Gerard 't Hooft used this principle to construct a conformal invariant theory of gravity [31–33]. Do you think that this is a reasonable application of this principle?

It may be . . . look, you should never underestimate Professor 't Hooft. Never. It is always a mistake to dismiss anything he says. I don't know the details of what he's doing, I haven't studied it in detail. What I did have was an interesting discussion with him about a month ago and what I found interesting and encouraging is that through his own way of thinking about things, he was being pushed, against his will as he is not happy about it, towards the same kind of multiverse view of things, the same kind of semi-anthropic view of things. He is quite certain that his own equations, if they only have a small number of solutions, will yield a Planckian cosmological constant. The only hope for his equations is that they will turn out to have a huge number of solutions and so he said to me (and he's a very honest man), as much as he hates it, that he's being pushed in the same direction as string theorists. It's an honest view of things and I found it very encouraging that more than one point of view is pushing toward the same direction.

Another thing that I want to ask concerns the holographic principle [34, 35]. Do you think that the recent paper by Erik Verlinde [36] involving the notion of gravity as an entropic force and so on is a rightful application of this principle?

These ideas are very intriguing. I think that even Verlinde himself has backed down a little bit from them. There are contexts in which gravitational forces have an entropic character such as near the horizon of a black hole. There are also contexts where they're not entropic. Ordinary objects of zero temperature, far from each other and gravitating, are not entropic. In the context of string theory and in Verlinde's context they're called *Born-Oppenheimer forces*. They're obtained by integrating out high-frequency modes but they are not entropic. This comes out of Verlinde's description, which he is now trying to understand. So I think there is certainly some substance to what he's saying but I think it's still a work in progress. Erik is also one of these people who should never be underestimated.

What has been the biggest breakthrough in theoretical physics in the past 30 years?

The holographic principle [34, 35]. There's no doubt about it. Absolutely no doubt about it. It has been the big game changer in all ideas about how nature works. It's the only really big change since quantum mechanics. It's extremely radical, in the sense that it's extremely unintuitive, and it's not only a breakthrough conceptually that forces us to change our previous ideas but it also turned out to be the principal working tool of quantum field physicists now invading condensed matter physics, invading fluid dynamics and so on. So it's one of these things that not only was a radical conjecture about the way the world works in some very deep sense but it has had a remarkable history. It's not quite history yet but it's becoming a practical tool. In a very short period of time it went from being a wild-eyed conjecture of a couple of nut cases, meaning Gerard and myself, to being the standard working tool of the field. So I don't think that there has been anything as important as that in changing our view of physics.

Another person at the table:[4] Who formulated the first holographic principle? Was it Hawking and Bekenstein or someone after that?

Well, no, it was 't Hooft and myself [34, 35], who counted the maximum entropy in a space region leading to the conclusion that the world is made of two-dimensional stuff.

Why have you chosen to do physics? Why not something else?

I'm not sure there was any choice involved. I was very bad at everything else. I wasn't interested in anything else. Every time I had to do something else I just spent all my time

4 At this point several people had joined our table for lunch. James Hartle was the first to join but he left in the middle of this interview. There were also several other younger physicists at the table but I can't remember who they were.

dreaming about mathematics and physics. So, I don't think there was a choice. I never made a choice. Oh well, I did make the choice but I don't really think it was a choice that I had.

What do you think is the role of theoretical physicists in modern society?

(laughs twice) I don't know. If you want me to say something about the sociology of physics, I will tell you something which I'm finding really exciting, incredibly exciting. There has been a long period where physics has been dominated by people who are getting older and older and older. Even me, who happens to be a great-grandfather, still gets invited to conferences to give lectures. This is absurd. I've been very worried about it but in the very recent past I've begun to see a change. I come to a conference like this one and I see the best ideas coming out of the young people. I see young people not being afraid to push the old guys aside and this I find incredibly exciting. It looks to me like there is a turnover of generations taking place.

Now, that means that they're going to push me off the stage. Too bad but it means that there's a future to the subject. I see a couple of guys over there, who are in their early 20s and are two of the brightest people I've seen in physics.[5] They are not afraid to tell me and other people, "Crap! What you're saying is wrong." There's a guy over here (pointing towards someone at our table) . . . how old are you?

The person Leonard pointed at : I'm 36.

Well, 36 is still pretty young. He's[6] inventing paradoxes which are so absurd, so ridiculous, so peculiar that they deserve to be completely laughed at, except that they are going to be important. They are going to be important and he's doing it with Alan Guth but that's beside the point (laughter at the table). People like Freivogel, who can stand up there and express strong opinions about things, as well as Kleban . . .

Someone else at the table: That's dangerous, career-wise . . .

No, it's not dangerous. Albert Einstein got up and said that everyone was full of baloney, that they didn't know what they were talking about. My friend 't Hooft had never had any trouble telling people older than him that they were full of baloney. Can you imagine Richard Feynman being afraid of the older generation?

Someone else: Probably not . . .

Of course not!

I did forget to ask you one last question. Do you think that there are problems with quantum mechanics? Does it have to be modified? If not, what is your favourite interpretation?

[5] Pointing at two physicists who were in the Black Hole Bistro having their conference lunch. Again, I can't remember who they were.

[6] Referring to the 36-year-old.

I just wrote a paper with Raphael Bousso on the multiverse interpretation of quantum mechanics [37] (laughs). I guess I still think that's right. I don't think that quantum mechanics in the context of laboratory experiments has to be modified. In laboratory experiments, we know how to use quantum mechanics. It works. I don't see any need of changing it and, I guess, I think that to what concerns phenomena which take place on cosmological scales, it would be exactly correct. I'm guessing, of course, but I think it would be exactly correct. Nevertheless, I think that there are major puzzles about it and big confusions. These confusions may have to do with putting quantum mechanics into a cosmological context. It may be that the natural place to understand things like decoherence, probabilities and other subtle behaviour that puzzled people very much might be in the context of cosmology. So, does that require a modification of quantum mechanics? I'm not sure I would call it a modification of quantum mechanics because it wouldn't change quantum mechanics as we know it; instead, it might just modify and change the way we look at it and the way we think about it. On the scale of cosmological sizes and times, it may very well be that we have to . . .

[Someone at the table said, "Generalise it."]

Generalise it or have a different perspective about it.

[Someone else at the table said, "A small change like getting rid of locality might be enough . . ."]

That's not a small change. No, I think that we have to learn a new trick. We have to learn how to think about quantum mechanics from the inside. The rules of quantum mechanics always require an observer outside the system and a system. The observer is not part of the system. You can make the observer part of the system but then you have to put another observer outside it and so on. We're finally getting to the point now where we are saying, "Wait a minute. The universe as a whole doesn't have an outside." So we have to learn how to think about quantum mechanics from the inside. I guess that's going to require a change of perspective and a change in the way we're thinking about it.

But the idea behind your work with Bousso was that eternal inflation would give the foundations for a possible framework that could give meaning to the many-world interpretation of quantum mechanics, right?

Well, I would separate the two issues. First of all, I think that the foundations of quantum mechanics might have a big overlap with questions of cosmology. Then, I would separate that slightly from the issue of whether eternal inflation is the right theory and whether it can address these puzzles. I think it can but I would still separate the two questions. I feel more strongly that many of the puzzles of quantum mechanics are cosmological and feel fairly strongly that eternal inflation has something to offer in understanding it. We'll see. These are very interesting questions. How many years with these kind of conferences taking place will it take before there's some sort of consensus developing around a common set of ideas? There's some commonality. There is a strong sense that eternal inflation is the right problem to work on. The question of measures, which is a question of information, a question of probabilities, is at the locus but beyond that there's a lack of consensus about what's important, what's right and so on. That's healthy enough for me and it's fine.

[Someone at the table said, "There's lack of consensus because there's lack of experimental data ..."]

That's at least part of it but even in the early days of QCD and pre-QCD, people were wondering whether QCD or the Bag models or something else was the right theory. Then, somehow consensus forms. What precipitates the consensus? Sometimes it's a little bit illogical. Sometimes, of course, it's an experiment but sometimes it's a very complicated calculation that somebody does. It's so complicated that nobody can figure out how it works from the beginning and then the answer comes out just right. If you think about it, a priori, 't Hooft's calculation on renormalisation of the standard model [38] was not a good enough reason to explain the transition from not believing in the standard model to believing in it. However, in the end it was. Why? Because the calculation was so complicated, so difficult and it could have gone wrong in so many places that when it worked out people said, "Oh, that must be right." They should have realised that it was right long before that.

Another example is that people did not believe that black holes had an entropy and thermodynamic properties, as suggested by Bekenstein, until Hawking did his difficult calculation [20] and showed the existence of Hawking radiation. From the point of view of the present time, the calculation doesn't look that difficult but at that time it was a difficult calculation and it came out just right, agreeing with Bekenstein's idea that the entropy of the black hole is proportional to its area [18]. There were very good reasons for people to believe Bekenstein, in fact, deeper reasons than those that Hawking presented. Nevertheless, when the calculation by Hawking worked out just exactly right, it was the internal consistency of the calculation that convinced people and not some experimental phenomena. So that nailed it and everybody said, "Yes, black holes have entropy. Yes, black holes evaporate."

The same happened with the holographic principle. Everybody should have realised the holographic principle was right as soon as we said it, even though we didn't believe it at the time (laughs). However, when Maldacena nailed it with a very precise version of it, embodied in AdS/CFT [24], that was enough. Nobody thought otherwise after that. So I think that the hope, in the context of eternal inflation, would be that we achieve some sort of consensus in some similar way, that is, things coming together conceptually and mathematically in a surprising and coherent way. It hasn't happened, not yet. It could happen any day now or it could take 100 years. If people start talking the way they are talking here and once it becomes the centre of attention for lots and lots of people it could take two years, maybe three years, to have a more coherent idea of what this is all about. But it could also be that we are still far from that situation. It's hard to tell.

References

[1] P. J. Steinhardt, "Natural inflation," in *Nuffield Workshop on the Very Early Universe Cambridge, England, June 21–July 9, 1982*, G. W. Gibbons, S. W. Hawking and S. T. C. Siklos, eds., pp. 251–266. Cambridge University Press, 1982.

[2] A. Vilenkin, "Birth of inflationary universes," *Phys. Rev. D* **27** (Jun. 1983) 2848–2855. https://link.aps.org/doi/10.1103/PhysRevD.27.2848.

[3] B. Freivogel, Y. Sekino, L. Susskind and C.-P. Yeh, "A holographic framework for eternal inflation," *Phys. Rev. D* **74** (2006) 086003, arXiv:hep-th/0606204.

[4] A. H. Guth, "Eternal inflation and its implications," *J. Phys.* **A40** (2007) 6811–6826, arXiv:hep-th/0702178 [HEP-TH].

[5] S. Weinberg, "Anthropic bound on the cosmological constant," *Phys. Rev. Lett.* **59** (Nov. 1987) 2607–2610. https://link.aps.org/doi/10.1103/PhysRevLett.59.2607.

[6] Planck Collaboration, P. A. R. Ade et al., "Planck 2015 results. XIII. Cosmological parameters," *Astron. Astrophys.* **594** (2016) A13, arXiv:1502.01589 [astro-ph.CO].

[7] A. H. Guth, "Inflationary universe: a possible solution to the horizon and flatness problems," *Phys. Rev. D* **23** (1981) 347–356. http://link.aps.org/doi/10.1103/PhysRevD.23.347.

[8] M. R. Douglas, "The statistics of string/M theory vacua," *JHEP* **05** (2003) 046, arXiv:hep-th/0303194.

[9] L. Susskind, "The anthropic landscape of string theory," arXiv:hep-th/0302219.

[10] L. Smolin, *The Life of the Cosmos*. Oxford paperbacks. Oxford University Press, 1999.

[11] S. Kachru, R. Kallosh, A. D. Linde and S. P. Trivedi, "De Sitter vacua in string theory," *Phys. Rev. D* **68** (2003) 046005, arXiv:hep-th/0301240.

[12] B. Freivogel and L. Susskind, "A framework for the landscape," *Phys. Rev.* **D70** (2004) 126007, arXiv:hep-th/0408133 [hep-th].

[13] B. Freivogel, M. Kleban, M. Rodriguez Martinez and L. Susskind, "Observational consequences of a landscape," *JHEP* **03** (2006) 039, arXiv:hep-th/0505232 [hep-th].

[14] S. M. Feeney, M. C. Johnson, D. J. Mortlock and H. V. Peiris, "First observational tests of eternal inflation," *Phys. Rev. Lett.* **107** (2011) 071301. http://link.aps.org/doi/10.1103/PhysRevLett.107.071301.

[15] L. Susskind, *The Black Hole War: My Battle with Stephen Hawking to Make the World Safe for Quantum Mechanics*. Little, Brown, 2008.

[16] A. Sen, "Black hole entropy function, attractors and precision counting of microstates," *Gen. Rel. Grav.* **40** (2008) 2249–2431, arXiv:0708.1270 [hep-th].

[17] A. Dabholkar and S. Nampuri, "Quantum black holes," *Lect. Notes Phys.* **851** (2012) 165–232, arXiv:1208.4814 [hep-th].

[18] J. D. Bekenstein, "Black holes and entropy," *Phys. Rev. D* **7** (Apr. 1973) 2333–2346. https://link.aps.org/doi/10.1103/PhysRevD.7.2333.

[19] S. Hawking, "Black hole explosions," *Nature* **248** (1974) 30–31.

[20] S. W. Hawking, "Particle creation by black holes," *Communications in Mathematical Physics* **43** no. 3 (Aug. 1975) 199–220. https://doi.org/10.1007/BF02345020.

[21] E. Halyo, A. Rajaraman and L. Susskind, "Braneless black holes," *Phys. Lett.* **B392** (1997) 319–322, arXiv:hep-th/9605112 [hep-th].

[22] E. Halyo, B. Kol, A. Rajaraman and L. Susskind, "Counting Schwarzschild and charged black holes," *Phys. Lett.* **B401** (1997) 15–20, arXiv:hep-th/9609075 [hep-th].

[23] G. T. Horowitz and J. Polchinski, "A correspondence principle for black holes and strings," *Phys. Rev. D* **55** (1997) 6189–6197, arXiv:hep-th/9612146.

[24] J. M. Maldacena, "The large N limit of superconformal field theories and supergravity," *Int. J. Theor. Phys.* **38** (1999) 1113–1133, arXiv:hep-th/9711200.

[25] D. Bigatti and L. Susskind, "Review of matrix theory," *NATO Sci. Ser. C* 520 (1999). arXiv:hep-th/9712072 [hep-th].

[26] D. A. Lowe and L. Thorlacius, "AdS/CFT and the information paradox," *Phys. Rev. D* **60** (1999) 104012, arXiv:hep-th/9903237 [hep-th].

[27] N. Iizuka, T. Okuda and J. Polchinski, "Matrix models for the black hole information paradox," *JHEP* **02** (2010) 073, arXiv:0808.0530 [hep-th].

[28] K. Papadodimas and S. Raju, "Black hole Interior in the holographic correspondence and the information paradox," *Phys. Rev. Lett.* **112** no. 5 (2014) 051301, arXiv:1310.6334 [hep-th].

[29] J. M. Maldacena, "Non-Gaussian features of primordial fluctuations in single field inflationary models," *JHEP* **05** (2003) 013, arXiv:astro-ph/0210603 [astro-ph].

[30] D. A. Lowe, J. Polchinski, L. Susskind, L. Thorlacius and J. Uglum, "Black hole complementarity versus locality," *Phys. Rev. D* **52** (1995) 6997–7010, arXiv: hep-th/9506138 [hep-th].

[31] G. 't Hooft, "Quantum gravity without space-time singularities or horizons," *Subnucl. Ser.* **47** (2011) 251–265, arXiv:0909.3426 [gr-qc].

[32] G. 't Hooft, "Probing the small distance structure of canonical quantum gravity using the conformal group," arXiv:1009.0669 [gr-qc].

[33] G. 't Hooft, "The conformal constraint in canonical quantum gravity," arXiv: 1011.0061 [gr-qc].

[34] G. 't Hooft, "Dimensional reduction in quantum gravity," *Conf. Proc. C* 930308 (1993) 284–296.
arXiv:gr-qc/9310026 [gr-qc].

[35] L. Susskind, "The world as a hologram," *J. Math. Phys.* **36** (1995) 6377–6396, arXiv:hep-th/9409089.

[36] E. P. Verlinde, "On the origin of gravity and the laws of Newton," *JHEP* **04** (2011) 029, arXiv:1001.0785 [hep-th].

[37] R. Bousso and L. Susskind, "The multiverse interpretation of quantum mechanics," *Phys. Rev. D* **85** (2012) 045007, arXiv:1105.3796 [hep-th].

[38] G. 't Hooft and M. J. G. Veltman, "Regularization and renormalization of gauge fields," *Nucl. Phys. B* **44** (1972) 189–213.

32

Thomas Thiemann

Professor at the Department of Physics at the Friedrich-Alexander University Erlangen-Nürnberg

Date: 13 October 2020. Via Zoom. Last edit: 16 December 2020.

What are the main problems in theoretical physics at the moment?

I will give you a biased answer because it relies on my personal taste. I'm a physicist who likes to think about the most fundamental problems and not so much about using already known laws or equations and applying them to complex phenomena (e.g., superconductivity), as in condensed matter physics. In this setting in which you know what the Hamiltonian is, the problem is the difficulty in solving it. Having said this, for me the most interesting puzzles of contemporary physics are the problems of combining general relativity with quantum field theory, finding out what is the actual exact particle content of the world, and understanding how to make sense of interacting quantum field theory in four dimensions. To what concerns this last problem, I should mention that, to date, there is no model of an interacting quantum field theory in four-dimensional Minkowski spacetime where you can check that all Arthur Wightman axioms hold [1]. It is an unsolved problem and whoever solves it, specially in the case of Yang–Mills theory, will receive one of the prizes from the Clay Mathematics Institute.

Do you think that the two problems you mentioned, namely, that of combining general relativity with quantum field theory and the exact particle content of the world, are related?

If you assume that this unified theory that you are looking for admits a perturbative description then we know, for example, from the work of Martinus Veltman and Gerard 't Hooft that not everything is possible [2]. So it may be that only a specific mixture of particles can lead to a consistent theory of all interactions. As we may discuss later, within loop quantum gravity, we are faced with *the problem of time*, or the problem of gauge invariance, which simplifies considerably when you add matter to the theory [3–6]. So it is a possibility that the two problems are related but we don't know yet.

Why do we need a theory of quantum gravity?

First of all, it is already a very difficult problem to find one consistent candidate theory. In fact, people have been trying to understand quantum gravity since the 1930s starting with the works by many giants of physics such as Matvei Bronstein [7] and later Paul Dirac [8], John Wheeler [9], Bryce DeWitt [10] and Richard Feynman [11]. To date, this problem is still unsolved. Some people may consider this problem as just a problem of academic interest because, one might argue, the physical effects of combining gravity with quantum theory only become important at very high energies. However, the Nobel Prize this year has been awarded to Roger Penrose for, among other things, the discovery that classical general relativity is incomplete by itself. In particular, it predicts its own failure in the form of spacetime singularities, the most well known being the big bang singularity or the singularity at the centre of black holes where the spacetime curvature or matter energy density becomes infinite. It is believed that the occurrence of infinities implies that the current theory has been driven beyond its domain of validity. What is fascinating is that we can in principle "see" these singularities, at least indirectly: cosmological observations of the large-scale structure or of the CMB radiation might reveal the details of the initial singularity while at late stages of Hawking evaporation of black holes, which is a process that astrophysicists these days are interested in, you will have the chance to look into the black hole singularity. In both cases you learn something about quantum gravity. It follows that without the combination of general relativity and quantum theory, our understanding of nature remains incomplete.

Why is combining gravity and quantum theory a difficult problem?

It's difficult because general relativity and quantum field theory are based on different principles. If you open a book on quantum field theory, you will notice that you always need spacetime geometry to formulate quantum field theory, which is often taken to be Minkowski spacetime, and to add certain axioms. These axioms include, for instance, that of locality and causality, which states that observables located in spacelike separated regions can't communicate via propagating light. Various statements of this kind would not make sense if you did not know what the underlying spacetime geometry is. On the other hand, general relativity is all about solving the a priori unknown spacetime metric in terms of the available matter content. In this setting, there is a strong coupling between matter and geometry which in quantum field theory is ignored. It is difficult to remove this tension between the two theories.

So do you think that so far there is no successful approach to quantising gravity?

There are many approaches to quantisation and they all start from different corners where we think we know how to proceed. Historically, canonical quantum gravity was the first approach to quantum gravity that starts with the principles of general relativity and attempts to set up a non-perturbative approach. Loop quantum gravity is a modern incarnation of that programme. Quite the opposite are quantisation programmes such as string theory starting with the principles of quantum field theory except that particles are taken to be string-like, instead of point-like, solving some of the ultraviolet divergences. This type of programme is a perturbative approach to quantum gravity in which one perturbs around

Minkowski spacetime and proceeds as with any other quantum field theory. There are other non-perturbative starting points as well, such as the asymptotic safety programme [12] or causal dynamical triangulations [13], but nevertheless I think that everyone agrees that none of these programmes has been completed so far.

And do you think one of these programmes is more promising than the others?

Actually, I don't think so. I think that all programmes should be pursued simultaneously, and that there are ideas in these various branches that may actually be joined into something that is more complete than the individual programmes themselves.

But you do work on a specific approach, right?

Yes, I do. This choice arises from personal taste or maybe a personal belief of what kind of programme is more promising in making progress faster than the others. I believe more in the gravity-inspired branch of research, because, for instance, perturbative techniques are known to fail due to the so-called non-renormalisability of general relativity. This is why I think that non-perturbative approaches, including others such as asymptotic safety [14], are more promising. But I may actually be wrong; nobody knows.

I am sure you've heard that the AdS/CFT correspondence [15] is claimed to be a non-perturbative approach to quantum gravity. Do you agree with this claim?

Yes, I know about these claims, but there are so many conjectures involved that I cannot see through what is the actual precise statement or the currently secured status of the conjecture. I understand the original idea that there is a so-called holographic dictionary between the generating functional of Schwinger functions (with respect to some test function) that you can compute in supersymmetric Yang–Mills theory (which is a conformally invariant field theory (CFT)) on the boundary of an asymptotically anti– de Sitter (AdS) spacetime and the partition function of classical supergravity theory in the bulk of that spacetime (where the test function for the CFT is the boundary value of the supergravity field integrated over with those boundary values fixed) but there have been many improvements and modifications of this idea which I have lost track of. Currently, many string theorists seem to take the viewpoint that the super Yang–Mills theory *is* string theory and the task is to interpret the results of super Yang–Mills theory calculations in terms of quantum gravity.

Okay. So I understand that you prefer working on loop quantum gravity (LQG), as it is a gravity-inspired approach, but what can it accomplish better than other approaches?

What I like about LQG is that it's a rather conservative approach where you take classical relativity plus matter and you apply the principles of quantum mechanics in four spacetime dimensions. The matter content you consider is the matter content that has been observed in the standard model (Yang–Mills fields, fermions and the Higgs field) [3–5] or some well-motivated mild extensions thereof (such as an inflaton field). The formulation does not make any further assumptions such as the existence of extra dimensions, supersymmetry or departures from the point-like structure of quantum field theory. LQG is like a gauge field theory applied to the gravitational interaction in which you work with connections rather than metrics. In this sense, there is a kind of unification, as far as the language is concerned, with the other known forces of nature.

Since you mentioned the standard model, LQG is sometimes criticised for not "seeing" the usual anomaly cancellations in the standard model. Do you agree with such criticism?

I think that there is a misconception: LQG is based on the principle of a manifestly anomaly-free quantisation of all the gauge symmetries of the underlying theory. Thus anomaly cancellation is an integral building block in the very construction of LQG. As we can discuss later, there are still some unsolved problems precisely regarding this issue. What is maybe confusing is that people working on perturbative quantum field theory do not recognise this because the non-perturbative language that LQG is written in forces you to work with Hilbert space representations of the canonical (anti) commutation relations that are different from Fock representations.

Sometimes LQG is portrayed as being a constrained BF theory [16, 17]. Is this correct?

There is a formulation of general relativity that goes back to the works of Plebanski [18], which is the following. You start with a topological field theory, which, by definition, has a finite number of degrees of freedom. That theory is called BF theory [16, 17]. The gauge symmetries of BF theory are so numerous that the number of degrees of freedom is actually finite. One then imposes a "constraint" that destroys so many symmetries of the unconstrained theory that at the end of the day you end up with more degrees of freedom. It may appear counterintuitive at first but imposing those so-called simplicity constraints leads you to a formulation that is equivalent to classical general relativity, which has an infinite number of degrees of freedom. It's basically a mathematical trick in order to rephrase some of the questions of general relativity in terms of topological field theory where the actual dynamical content of the theory arises from these additional constraints. So, in this sense, those practitioners of LQG who follow the path-integral approach indeed quantise a constrained BF theory. The motivation for taking that detour is that topological field theories are rather easy to control from a mathematical point of view.

You mentioned that LQG is a lattice gauge theory approach to gravity. Usually, lattice approaches introduce artificial regulators that need to be removed at the end of the day. Is this also the case in LQG?

There are many similarities with lattice gauge theory but also many important differences. What is similar is that we work with gauge-invariant variables constructed from connections familiar from lattice QCD and known as Wilson loop variables which is also from where LQG derives its name. What is very different is that in lattice gauge theory one usually restricts to a limited class of lattices, the most common choice being rectangular lattices, while in LQG we consider all lattices of arbitrary topology at once, using all possible curves including knotted and braided ones meeting in vertices of arbitrarily high valence. What this means is that we actually work in the continuum from the get-go and do not need to take any scaling limit, contrary to usual lattice approaches. It is in fact quite non-trivial that this leads to a bona fide $*$-representation of the canonical commutation relations on which the Yang–Mills type gauge symmetries and certain kinematical diffeomorphisms act by unitary operators and that this representation is in fact uniquely selected if one imposes the physical requirement that those kinematical diffeomorphisms are represented unitarily [19–22].

There is one place at which a regulator is first used: when implementing the quantum Einstein equations (also known as the Wheeler-DeWitt constraint). However, exploiting the fact that the theory has kinematical diffeomorphism invariance allows you to remove its short-distance dependence. Still, the final operator depends on the diffeomorphism-invariant choices of the regularisation which we refer to as quantisation ambiguities.

You said that you do not need to take a scaling limit as in lattice approaches because in LQG you consider all lattices at once. Why is this approach not pursued in QCD, and wouldn't that differ from usual scaling limits of QCD?

There is a huge difference between QCD and quantum gravity: quantum gravity has the diffeomorphism group as a gauge group while QCD is a quantum field theory on a fixed Minkowski spacetime where diffeomorphisms play no role whatsoever. In QCD, lengths are measured with respect to the given Minkowski metric, while in quantum gravity the metric is not a classical background structure but rather a dynamical quantum operator (valued distribution). Therefore the very notion of length and scale in quantum gravity has no a priori meaning. What may be "long" in some coordinates is "short" in others; long and short are in some sense gauge-equivalent. This explains intuitively why scaling limits are taken profoundly differently in the two theories. In particular, as the physical states are supposed to be diffeomorphism-invariant, the scaling limit is *already taken* on the space of physical states (which also solve the Hamiltonian constraint). If one wants to speak about physical scales in quantum gravity, then such a notion of scale must be provided by the physical state itself. For instance, we could take a diffeomorphism-invariant operator measuring the physical length of a chunk of matter defined by the non-vanishing of some scalar defined by it and consider its expectation value with respect to a physical state. Then one could meaningfully say that that state describes a situation in which the corresponding observable is assigned that expectation value as length. But the operator itself has no a priori length scale at all. Such physical observables based on reference matter underlie the reduced phase space approach to LQG that we will talk about later. More information can be found in my textbook on LQG [23].

Okay, so in LQG you have a continuum lattice formulation in terms of connections and you apply the principles of quantum mechanics, as far as I know, via canonical quantisation. What else is missing?

Yes, we use canonical quantisation, which can be used in any given theory, for quantising this connection formulation of general relativity [24]. This is the first step and allows you to erect a platform from which you can start asking well-defined questions [25, 26]. The whole field really took off when people found the connection formulation [24, 27] because the degree of rigour that became available was lacking in the earlier formulation based on metrics. The hard task is to formulate and solve the dynamics of general relativity, which you may call the quantum Einstein equations or the Wheeler-DeWitt equations. These equations tell you how matter fields are related to the spacetime curvature, and how that system evolves quantum mechanically.

This second step consists of two parts. First one has to ensure that the quantisation you imposed on the connections is done in such a way that the quantum Einstein equations are mathematically well defined on the Hilbert space. It has in fact been shown that these equations can be given mathematical meaning. Mathematicians say that the constraint operators are densely defined on the Hilbert space which is a crucial step. The second part consists in solving these rather difficult equations, which is more difficult than solving the classical Einstein equations, as you might imagine. But before embarking on that and developing suitable approximation methods one should constrain and fix the quantisation ambiguities that I mentioned earlier.

Are you referring to the Barbero-Immirzi parameter?

The Barbero-Immirzi parameter is only a part of it. There are more severe quantisation ambiguities which need to be addressed before one can make reliable predictions. If you take the Klein-Gordon field in Minkowski spacetime, its action is quadratic, that is, polynomial in the Klein-Gordon field. If you take $\lambda\phi^4$ instead, the theory is more complicated but the action is still a fourth-order polynomial in the field ϕ. Gravity is much worse than these cases because the gravitational action is not even polynomial. Thus, you can imagine that in the case of gravity there are severe operator ordering problems that you encounter when quantising these composite field operators. These kind of operator ordering ambiguities are actually present in the quantisation of what we call the Wheeler-DeWitt operator. Such ambiguities cannot be avoided in an interacting quantum field theory and need to be addressed before solving the Wheeler-DeWitt constraint or making any predictions based on it.

How do you fix the Barbero-Immirzi parameter?

The Barbero-Immirzi parameter [28] is a parameter that you can introduce when performing canonical quantisation. It parametrises a family of canonical transformations and has no effect on the classical theory. However, this transformation may have a drastic effect quantum mechanically. A well-known example even for free fields where this is realised are

Bogoliubov transformations which classically are just canonical transformations but which in the quantum theory may map you out of the unitary equivalence class of representations of the canonical commutation relations. This is what happens also with respect to the Barbero-Immirzi parameter and it appears in the spectrum of what we refer to as the area operator [29]. If you could resolve distances of around the Planck length, you would be able to fix the Barbero-Immirzi parameter via an experiment.

And how do you fix the other ordering ambiguities that you mentioned?

These other ambiguities enter the dynamics of the theory and are not unfamiliar to people working on lattice QCD. As is usually the case in lattice QCD, the goal is to remove these ambiguities or to constrain them using renormalisation procedures [30]. We have recently embarked on that programme as well [31–35]. The idea is to apply these methods, which go back to Wilson and which have had quite a great success in other fundamental interactions, to the Wheeler-DeWitt operator (which comes with free parameters due to ordering ambiguities). It's usually the case that you cannot fix all the free parameters and so there will be some remaining parameters floating around once you use these methods. However, we call a theory renormalisable if the number of unfixable parameters is finite. These remaining finite number of parameters would also have to fixed by experiment.

Is there a relation between this and the asymptotic safety programme [36]?

Yes, there should be a rather precise relation between what people in asymptotic safety program are doing and the renormalisation program in the context of LQG in the covariant [37] and canonical [31–35] framework. However, at the present moment there are some differences in language that need to be translated in order to have a proper comparison. In particular, background-dependent techniques are used in asymptotic safety contrary to LQG, and in addition people work with Euclidean signature within the asymptotic safety programme instead of Lorentzian signature as in LQG. Once we understand these issues we will be able to make a precise statement regarding the relation between the two approaches.

How does LQG resolve the perturbative non-renormalisability encountered in perturbative quantum gravity? Does it rely on the asymptotic safety mechanism?

The whole starting point of LQG was that seminal result of Sagnotti et al. that perturbative quantum gravity (around Minkowski space) produces new counter terms starting from two loop calculations and presumably higher, and thus is believed to be a non-renormalisable theory [38]. Thus, we try to circumvent this no-go theorem by building a non-perturbative approach in which the interacting Hamiltonian (constraint) is well defined from the outset. The price you pay is that one has to depart from the language of perturbation theory; in particular, you have to change the Hilbert space representations on which the canonical commutation relations and $*$-relations are implemented in order such that the operators that you need to control in the theory, e.g., the Hamiltonian constraint, are densely defined. On the Fock spaces defined in perturbation theory, this operator is hopelessly ill defined.

So the simple answer is that perturbative non-renormalisability is avoided by using non-perturbative methods.

In that respect, we are indeed quite close to the idea of asymptotic safety that a perturbatively non-renormalisable theory may be non-perturbatively renormalisable [39, 40]. What this means is that in the Euclidian approach approach to quantum field theory (QFT) where you map the QFT to a statistical physics problem you look for a fixed point of the renormalisation group flow of the coupling constants defining the Euclidean action (which is the statistical physics Hamiltonian in one dimension higher). The renormalisation group flow drives a theory defined by a family of actions at finite resolutions to a consistent theory which has the property that integrating out the additional degrees of freedom of the action at finer resolution precisely gives the action at coarser resolution. The values of those coupling constants fixed this way define the counter terms mentioned before, but in contrast to the perturbative programme, at the fixed point, all but finitely many of them take prescribed values (this defines a theory to be non-perturbatively renormalisable). When translated back into the operator language you are dealing with a theory quantised on Hilbert spaces on which the free Hamiltonian is no longer well defined, only the interacting one is. Hence the name *non-perturbative fixed point* and now you see that we are quite close in language and methodology to the asymptotic safety programme. The recent renormalisation programme in LQG that I mentioned where we simply take the path integral renormalisation flow and translate it to the level of operator equations brings it even closer to the asymptotic safety idea.

Besides the problem of ordering ambiguities, you mentioned that it is also a difficult problem to solve the Wheeler-DeWitt equation. Is this what people refer to as solving the Hamiltonian constraint?

Yes, exactly. What one needs to do is to write down all the solutions and the physical Hilbert space in which these solutions are in fact normalisable.

But have people been able to classify these solutions or to find one physical solution in particular?

Actually, in one of my papers I have given an algorithm for how to do that in principle [41, 42]. This translates the problem into a computational one which could be the starting point for approximations.

I ask this because LQG is usually criticised for not being able to find a physical state that represents classical Minkowski spacetime. Is this because you cannot solve the Hamiltonian constraint?

There's several possible answers to this question. The most fundamental way of answering this question would be to find the complete solution space given by the Hamiltonian constraint and find the inner product such that all such solutions are normalisable. If you do not know the full solution space, you cannot find the inner product and without inner product you cannot define the S-matrix, etc. If you would find the solution space and the inner product then you still need to find a semiclassical state in that physical Hilbert space which represents Minkowski space. This has not been done.

What has been done is to find a state in the kinematical space of unphysical states, which is semiclassical in the sense that it is peaked like a coherent state on the initial data compatible with Minkowski spacetime. There are many of these kind of states because they are kinematical coherent states that are not fixed by some kind of dynamical principle [43–47].

But do these kinematical states solve the Hamiltonian constraint?

These semiclassical states solve the constraints in an approximate way. The reason is the following. Coherent states have the property that the expectation values of operators evaluated in those states are very close to their classical counterparts. This is the case for the basic operators but also for products of basic operators. It has been shown that this also extends to the kind of non-polynomial expressions that we encounter in the Hamiltonian constraint or the Wheeler-DeWitt equation [48]. Thus, to lowest order in $\hbar$, the expectation value of the operator is its classical counterpart. But the state itself is peaked on initial data which solves the classical constraint; therefore, by definition, to lowest order in $\hbar$ these expectation values solve the constraint. By the same method you can show that the fluctuations are small.

But are you guaranteed that such states would solve the constraint to higher orders in $\hbar$?

You can compute the corrections and such corrections are finite but of course non-zero as they measure the fluctuations. Clearly, these states are therefore only approximate solutions.

Okay, so summarising, if someone asks you whether there is a physical state representing Minkowski spacetime, your answer is yes?

To give you a precise answer to this question, I need to explain a few more details, in particular the fact that there are currently two different versions of the theory. In the vacuum case, with no extra matter, the only way to find physical states is the one I outlined just earlier; that is, you need to solve the constraint and find the inner product, which has not been done to date. To make progress people have focussed on semiclassical states in the kinematic Hilbert space that solve the constraints approximately in the way I just described. There are many classes of semiclassical states depending on the squeezing parameters you choose and they all solve the constraint to lowest order in $\hbar$ but their deviations at higher orders in $\hbar$ differ. So, as far as predictions are concerned, this does not provide a complete answer. In fact, this uncertainty is related to the ordering ambiguities that we discussed earlier.

Now, there is another version of the theory in which you bring matter into play and you perform what is called a reduced phase space quantisation [5, 6]. What this means is that you solve the constraints classically before quantising the theory. In this away, after quantisation you have the full solution space. This has been done using specific kinds of matter such that things are mathematically very convenient and serves as a proof of principle. In this case, of course, the dynamics of the theory depends on which material reference frame you are using and you have physical states that are candidates for Minkowski spacetime.

And in this version of the theory, do you know which of these candidate Minkowski states is the right one?

You could pose an analogous question to someone working on the $\lambda\phi^4$ theory: Do you know a semiclassical state whose trajectories in the Hilbert space stay close to a preferred classical solution of the $\lambda\phi^4$ theory for all times? Even more elementary, you could ask a similar question to someone working on quantum mechanics, in particular, to someone working with the anharmonic oscillator. This question is so complicated that nobody knows what are the good semiclassical states for the anharmonic oscillator. This is just to emphasise that this is a tough problem. One should certainly come up with an answer. Hence, the question is not answered yet, but at least in this version of the theory, whatever state you come up with will be a physical (gauge-invariant) state.

But is there any hope in solving this issue in LQG?

There are candidate coherent states for Minkowski spacetime within the reduced phase space approach peaked on Minkowski space initial data, so the hope is that one can show that they are stable for a long enough period of time. If that's the case then during that long period of time you can pose physical questions, say, scattering experiments, and obtain physical answers. Long time stability is hopefully enough to obtain the right physics.

These two versions of the theory, the one in which you quantise and then solve the constraint, and the one where you solve the constraint and then quantise, have been shown to be equivalent?

Already from analogous contexts of quantum mechanical models with only finitely many degrees of freedom, we know that these two procedures do not commute in general. So it is not known if they are the same theory but most likely not when you go beyond the semiclassical limit.

So these are two different theories and you need to find a way to figure out which one is the right one?

Well, to be honest, quantum gravity is so complicated that I would be the happiest person if someone would give me just one answer that works.

Okay, but what is missing in this second version of the theory in which you solve the constraints and then quantise?

We're not yet satisfied with it because of the non-polynomiality of the interactions which still leaves quantisation ambiguity imprints. Again, this is what we are trying to fix with the Hamiltonian renormalisation programme [31–35]. Also the mathematically convenient reference matter that is used is not experimentally confirmed, hence one should, and in fact can, use matter candidates that are discussed in certain extensions of the standard model of elementary particles. If that can be achieved then I would say that there's a consistent approach to quantum gravity that arrives at a certain answer which you can try to falsify.

Going back to the Hamiltonian constraint ... I have heard from people working on covariant approaches to LQG, i.e., spin foams, that they don't think that the Hamiltonian constraint gives the correct dynamics. Do you agree with this statement?

I'm actually not aware of such a strong version of the criticism as you phrased it. Classically, there's no doubt that the expression that we look at is the correct one. Then we apply standard techniques of quantisation, widely used in quantum field theory, like point-splitting regularisation, and find a precise answer. There are ordering ambiguities, as we discussed, but we only use standard recipes.

The quantisation is even consistent in the sense that the commutator between constraints is a linear combination of constraints. What is clear is that this linear combination is not the same linear combination as we have in the corresponding classical Poisson bracket calculation; that is, the "quantum structure constants" of the constraint algebra do not match the classical ones, although the kernel of the commutator is precisely the correct one. This is probably what you are referring to. It is my hope that this mismatch between the classical and the quantum theory can be removed again using renormalisation techniques. There is also very interesting work by Varadarajan along these lines [49, 50].

But have people shown that starting with spin foam models they can get the exact Hamiltonian and dynamics as in the operator approach to LQG? Are the two approaches equivalent?

When done properly, canonical and covariant approaches are equivalent. However, this equivalence between the two approaches to LQG has not been shown to date. So perhaps when people claim that the operator approach does not yield the correct constraint, it is because it doesn't arise from solving the equations of spin foam models. However, it is also not clear that the spin foam models that are currently used are correct. Spin foam models are still also under construction; in particular, they still lack a continuum limit; there are open issues with the correct quantum implementation of the aforementioned simplicity constraint, which in my view is the covariant face of the quantisation ambiguities; and there there are several versions of it, e.g., the BC model [51], the EPRL model [52], the FK model [53], the KKL model [54] or the group field theory formulation [55]. However, a spin foam path integral is supposed to be a projector on the kernel of the Wheeler-DeWitt constraints, so at the end of the day, both approaches must come to an exact match.

So, do I understand correctly that you are not a big fan of spin foam models?

Actually, I like very much this covariant way of thinking about the problem and if it would be easier then I would also work on it. However, at the moment this approach seems to actually be more difficult than the operator approach. For example, the spin foam sums are not manifestly convergent while the Hamiltonian constraint in the canonical theory is densely defined on the kinematical Hilbert space.

Is it clear that LQG is diffeomorphism-invariant at the quantum level?

In the operator approach to LQG, you have both spatial and temporal diffeomorphisms. Spatial diffeomorphism symmetry is built into the quantisation procedure and there is no doubt that it is implemented without anomalies. Temporal diffeomorphisms are generated by the Wheeler-DeWitt operator or the Hamiltonian constraint, and here the state of affairs is not yet satisfactory as I mentioned earlier.

What can be said in this case is that classically the Hamiltonian constraint implements spacetime diffeomorphism invariance. On the other hand, a kind of spacetime covariance has been proven in all quantisations of the Hamiltonian constraint. As I already mentioned, it has been shown that the generators of the Lie algebra close but the associated structure constants sometimes do not appear to be the correct ones, which makes people uneasy about whether we really have shown quantum spacetime covariance [56]. Again, this issue has roots in the quantisation ambiguities that need to be fixed first.

In the second approach in which you solve the constraint first and then quantise the theory, spacetime diffeomorphism invariance has been implemented classically and is not broken by quantisation. In this sense, it is clear that you have diffeomorphism invariance at the quantum level.

What are the main research directions that people within LQG work on?

Nowadays, the community is mostly pursuing spin foam models [57] and loop quantum cosmology [58–60]. Then there is a smaller group of people who work on quantum black holes and an even smaller group that is concerned with the foundations of the operator approach to LQG.

What is loop quantum cosmology (LQC)?

LQC was developed because the full exact theory is so difficult to solve. Thus, you would like to have a simplified version of the theory in which you can at least obtain definite answers to many physical questions. LQC employs a mini superspace formulation which only considers the cosmological sector of the theory. It then quantises that sector with the same methods as those used in the full LQG theory. In this case, you can quantise the constraint, find the physical Hilbert space and the full solution space as well as the Dirac observables. One of the interesting results that came out of LQC is that people found a bounce, instead of a singularity, which is quite inspiring.

But would features such as the bounce be there in the actual full theory?

That's the one-million-dollar question. We don't know. LQC needs to be further extended in that direction. In fact, people are now working on trying to extend their results to situations in which the solution space is non-homogeneous and non-isotropic in order to make contact with the full theory [61, 62].

What are the observables of LQG? Can you calculate, for instance, S-matrices within LQG?

S-matrices are often calculated in the context of the standard model and of course one should be able to calculate it also in quantum gravity. However, to be able to make contact with experiments at CERN, for instance, a lot of work needs to be done beforehand. First you need to specify in which spacetime you are going to perform the measurement. If you want to do it in Minkowski spacetime then you need to find a good candidate state for Minkowski spacetime that is stable for long enough times, etc. I can write down a list of things that one needs to do before attempting to find the S-matrix but this list will contain tough questions that go much beyond calculating Feynman diagrams.

People have done some S-matrix calculations (e.g., graviton propagators) using spin foam models and that is indeed very interesting [63]. However, these calculations should and can be improved significantly following a first principle derivation using gauge-invariant (so-called Dirac) observables, well-motivated physical states, etc.

Okay, so what observables can you define in LQG?

If you include matter in the theory you can define many observables. For instance, if you have four scalar fields you can define relative observables, that is, relative to that material reference frame [5, 6]. It is in fact an old idea that goes back to Bergmann and Komar (see, e.g., [56]). These are Dirac observables that can be constructed explicitly. In particular, you can define the area functional of a given surface that is fixed by the values of these four scalar fields. Taking the operator associated with that area functional, you can evaluate its spectrum, which would allow you to fix the Barbero-Immirzi parameter, for example. Furthermore, you can define the physical curvature scalar or even the Riemann curvature tensor.

But is there any experiment you can think of that would somehow measure the spectrum of this area operator?

No, these are of course gedanken experiments. You would need to have a resolution that we cannot access in a tabletop experiment. However, nowadays people are discussing the possibility of seeing such imprints in the cosmic microwave background or even in exploding primordial black holes as indirect probes of the spectrum of the area operator [64, 65]. These ideas are derived from LQC and, of course, what you want in the end is to derive such effects using the full theory.

What picture does LQG give about how spacetime looks at Planck scale? Is it continuous or discrete?

There's an exact answer to that. The spacetime manifold is not granular; it is a continuum. However, there is some inherent discreteness which expresses itself as discreteness in the spectra of certain operators that correspond to geometrical observables. This is similar to the case of electron orbitals. Nobody says that the electron is just sitting in a discrete set of circular orbits. The electron is actually sitting everywhere but its energy is quantised. This is similar to what happens in LQG.

But is there a notion of spacetime metric at such small scales?

Suppose that you are working with this version of the theory in which you solve the constraint classically and then quantise the theory. You can now construct the physical area operator and look for some kind of semiclassical states in which the fluctuations of this operator are quite small. Now you need to specify what you mean by the fluctuations being small. Small compared to what? In order to define it you can think about the expectation value of the operator in the semiclassical state and compare the magnitude of that expectation value with that of the corresponding fluctuations. However, you can make that expectation value smaller and smaller by looking at smaller and smaller areas. At some point, the magnitude of the fluctuations (these are non-vanishing because quantum geometry is non-commutative in LQG) will exceed the expectation value of this operator. When that takes place, you cannot associate to this geometry a smooth structure at that scale. In this sense, the geometry becomes fuzzy and how fuzzy it is depends on the actual state that you prepared.

If the geometry is fuzzy, how can you solve the problem of time?

The problem of time, at least in this approach in which you solve the constraints already at the classical level, is automatically solved because you have a physical Hamiltonian that drives the evolution of Dirac observables. In the other version of the theory, the problem of time is equivalent to the problem of temporal diffeomorphism invariance at the quantum level for which there is no definite answer at this point.

Some time ago, Hermann Nicolai published a paper with an outsider's view of LQG [66] and not much later you wrote a paper with an inside view [67]. Why did you feel the need to write this paper?

If you look at the acknowledgement section of that paper [66], you can see that the authors actually interviewed me quite extensively regarding many of the details of the theory. So, the technical statements made in that paper are more or less correct. However, to my taste, it was presented in a rather unfair way, which led me to write that reaction paper [67].

This unfairness concerns the following. When people looked at perturbative approaches to quantising gravity in the 1980s [38], they came to the conclusion that the theory was non-renormalisable. This means that you need an infinite number of counter terms each coming with a new undetermined parameter rendering the theory non-predictive. The authors of [66] concluded that nothing is gained from LQG regarding such issues and thus LQG is equally non-predictive. The message I wanted to convey in [67] is that they are in one sense right but in another incorrect. They are right in the sense that they understood the issues related to these quantisation ambiguities that we discussed. However, while we do have quantisation ambiguities, we don't have to sum a perturbative series which has probably zero radius of convergence. To fix the ambiguities we need to perform (Hamiltonian) renormalisation and find fixed points, which is not unusual in non-perturbative renormalisation programmes. However, the non-trivial achievement of LQG is its manifestly background-independent and non-perturbative formulation. That is: yes, we are not predictive yet; there are parameters to be fixed. But for each choice of parameters the answer we get is finite and can be written down in closed form! We do not need to perform higher loop calculations of indefinite order; we do not have to remove infinities order by order; we do not need to sum the orders and worry about convergence.

As you know, there have been claims in the past that you can calculate corrections to the Bekenstein-Hawking entropy using LQG [68–71]. However, some time ago Ashoke Sen published a paper where the logarithm corrections to the entropy that he calculated are not in agreement with LQG computations [72]. What is your opinion about this mismatch?

I have looked into this some time ago but not so much in detail. However, I came to the conclusion that the following is true. All these calculations of black hole entropy, both in string theory and LQG, assume some kind of semiclassical input and different inputs result in different answers. The resulting corrections are not just low-order corrections; they are significant corrections at the highest contributing order, specifically factors multiplying the area term in the entropy. So, I think that in order to make real progress we have to remove these semiclassical inputs and perform a full quantum computation.

So you are saying that in these calculations there is some kind of ambiguity?

The computations actually do not very much rely on the fine details of the quantum dynamics; that is, they are not proper quantum calculations. In the LQG side you define a classical black hole horizon and then count eigenvalues of the area operators. In the string side you do the counting for extremal charged black holes, which are not astrophysically interesting, and then use semiclassical arguments to extrapolate the counting to astrophysically relevant black holes. In both cases a pure quantum computation that allows you to prove the semiclassical correctness of the assumptions made would be far more convincing.

Related to string theory, several years ago, you applied LQG quantisation methods to the bosonic string worldsheet and you did not find a critical dimension [73]. Given that there are several quantisation methods in string theory and they all give the same critical dimension, how should I interpret the result you got?

This result arises due to differences in quantisation when looking at either the constraint algebra, which is usually the case in string theory calculations, or at the associated gauge group. In the latter case, you can impose certain additional requirements. For instance, you can demand that the representation of the gauge group (or diffeomorphism group) is either continuous or not. If it is continuous then looking at the constraint algebra or the gauge group does not make much of a difference. However, if you drop the requirement of the representation being continuous then the quantisations are different, in which case you do not find a critical dimension. Now you can ask: Why should I drop that requirement? If you don't then there is no contradiction in the methods. However, the point of my paper is that different quantisation requirements, which people may argue for or against, may provide different answers, even in the context of string theory.

Why didn't you apply this to the superstring?

At first I thought that people would be interested in this kind of approach since many have argued that we should try to combine LQG and string theory methods. However, the reaction from the string theory community was not very positive. So I was not very motivated to look at the superstring.

But some years ago you looked at quantising supergravity [74–76]. I guess your motivation was not to apply such methods to the quantisation of the superstring worldsheet?

Yes, the motivations were different. One of the motivations was the criticism that people working on supergravity and string theory usually give to LQG. In particular, it is often said that LQG can only be formulated in four spacetime dimensions. The other motivation was to make a small contribution to the AdS/CFT correspondence in the following sense. As I understand it, the AdS/CFT dictionary is better understood between supersymmetric Yang–Mills theory on the boundary of asymptotically anti–de Sitter (AdS) spacetime and classical supergravity in the bulk. Thus to translate supersymmetric Yang–Mills theory into string theory, it would be nice to have an intermediate regime in between string theory and classical supergravity, which should be quantum supergravity. So I thought it would be interesting to pursue this idea and understand what we can learn from it. The methods we used are the same as those in standard LQG, which are background-independent, which is important since in the bulk you have supergravity in all possible backgrounds but with asymptotically AdS boundary conditions. So these works also showed that LQG methods can be applied to supergravity in any dimension.

Were there any interesting results within AdS/CFT resulting from these works?

In these papers we just laid the foundations; in particular, we extended the methods to arbitrary spacetime dimensions and also formulated one of the more prominent supergravity theories, specifically 11-dimensional supergravity with corresponding additional supersymmetric matter, which was not considered before [74–76]. We also formulated the problem of counting the Bekenstein-Hawking entropy of higher-dimensional but not supersymmetric black holes [77]. I was not involved in later developments [78–80], which are being pursued by Norbert Bodendorfer – one of my former PhD students and who has looked at the connection between LQG and the AdS/CFT correspondence more closely.

Did you actually count the entropy of supersymmetric black holes and compared it with string theory results?

No, we did not. But I agree that it would be an obvious task. Technically, a missing step is to make the algebra of the quantum supersymmetry generators consistent (i.e., their anti-commutator should result in the Hamiltonian constraint) which has not yet been done and which again relates to the quantisation ambiguities that I have mentioned many times.

What do you think has been the biggest breakthrough in theoretical physics in the past 30 years?

This is, of course, a question that has an answer that depends very much on whom you ask. I am particularly interested in works which have scratched the very foundations of physics, and where you might not even know the exact equations or variables. Examples are the Nobel Prizes given to the fields of quantum field theory and general relativity that unravelled fundamental physics. You could list here the Nobel Prizes given to Hulse and Taylor for pulsars and gravitational waves; to 't Hooft and Veltman for renormalisation of Yang–Mills theory; to Gross, Pulitzer and Wilczek for asymptotic freedom; to Perlmutter, Schmidt and Riess for the accelerated expansion of the universe; and the Nobel Prizes for the discovery of neutrino oscillations, gravitational waves and the latest one to Penrose. I wouldn't call scientific breakthroughs any of the developments within quantum gravity as there hasn't been any verified prediction. After all, we want do physics, not pure mathematics.

Why did you choose to do physics and not something else?

When I was a teenager, I was interested in chemistry until I found out that chemistry is full of rules which I could not understand. Then I tried to find out where I could find the answers and that's how I came to physics.

What do you think is the role of the theoretical physicist in modern society?

The answer to this question has many facets. I think that the general public is not aware of the importance of theoretical physics in science because what people usually acknowledge or perceive are the applications that result from it after a long time of deepening new discoveries. Computers and everything that has to do with semiconductor physics are a good example as they are all based on quantum mechanics, which is now more than 100 years old, but the most interesting applications are maybe only 20 years old. The same is true concerning general relativity, which is more than 100 years old. In fact, the GPS system is perhaps the only real application of general relativity in everyday life and is only 20 years old. But I bet that more than 95% of the people who own a navigation system for their car do not know that its accuracy crucially depends on general relativity.

That is of course not their fault. Going back to your high school days, you probably remember that your physics course was not that thrilling for you, to say the least. Even the word *physics* to most people is frightening and it shouldn't be like that. Physics is fun but it is often presented as a very dry topic, inaccessible and very abstract, which is frustrating for most people. Things have improved in the recent past; in particular, outreach activities such as popular science TV programmes try to show that physics is fascinating and that when diving into the subject one can in fact understand it. It is nice to see that the number of female physics students is slowly increasing. But we should do more. Physics is the key science underlying major technological advances and people should be aware of that.

What would happen if people would come to the conclusion that physics can be spared? Politicians would direct taxpayers' money into different channels and a few decades later technological progress would be to a large extent only in the hands of companies that would of course invest mostly in activities that are granted to result in short-term profit. However, technological progress takes a quantum leap not from this kind of research but rather from the crazy ideas that are only possible in an environment for which profit plays absolutely no role.

References

[1] A. S. Wightman and L. Garding, "Fields as operator-valued distributions in relativistic quantum theory," *Arkiv f. Fysik, Kungl. Svenska Vetenskapsak.* **28** (1964) 129–189.

[2] G. 't Hooft and M. Veltman, "Regularization and renormalization of gauge fields," *Nucl. Phys. B* **44** (1972) 189–213.

[3] C. Rovelli, P. Landshoff, C. U. Press, D. Nelson, S. Weinberg and D. Sciama, *Quantum Gravity*. Cambridge Monographs on Mathematical Physics. Cambridge University Press, 2004.

[4] T. Thiemann, "Modern canonical quantum general relativity," arXiv:gr-qc/0110034.

[5] K. Giesel and T. Thiemann, "Algebraic quantum gravity (AQG). IV. Reduced phase space quantisation of loop quantum gravity," *Class. Quant. Grav.* **27** (2010) 175009, arXiv:0711.0119 [gr-qc].

[6] K. Giesel and T. Thiemann, "Scalar material reference systems and loop quantum gravity," *Class. Quant. Grav.* **32** (2015) 135015, arXiv:1206.3807 [gr-qc].

[7] M. Bronstein, "Quantum theory of weak gravitational fields," *Gen. Rel. Grav.* **44** (2012) 267–283.

[8] P. A. Dirac, "The theory of gravitation in Hamiltonian form," *Proc. Roy. Soc. Lond. A* **246** (1958) 333–343.

[9] J. Wheeler, "Superspace and the nature of quantum geometrodynamics," *Adv. Ser. Astrophys. Cosmol.* **3** (1987) 27–92.

[10] B. S. DeWitt, "Quantum theory of gravity. I. The canonical theory," *Phys. Rev.* **160** (Aug. 1967) 1113–1148. https://link.aps.org/doi/10.1103/PhysRev.160.1113.

[11] R. Feynman, "Quantum theory of gravitation," *Acta Phys. Polon.* **24** (1963) 697–722.

[12] A. Eichhorn, "An asymptotically safe guide to quantum gravity and matter," *Front. Astron. Space Sci.* **5** (2019) 47, arXiv:1810.07615 [hep-th].

[13] J. Ambjorn, A. Goerlich, J. Jurkiewicz and R. Loll, "Nonperturbative quantum gravity," *Phys. Rept.* **519** (2012) 127–210, arXiv:1203.3591 [hep-th].

[14] R. Percacci, "Asymptotic safety," arXiv:0709.3851 [hep-th].

[15] J. M. Maldacena, "The large N limit of superconformal field theories and supergravity," *Int. J. Theor. Phys.* **38** (1999) 1113–1133, arXiv:hep-th/9711200.

[16] G. T. Horowitz, "Exactly soluble diffeomorphism invariant theories," *Commun. Math. Phys.* **125** (1989) 417.

[17] L. Freidel, K. Krasnov and R. Puzio, "BF description of higher dimensional gravity theories," *Adv. Theor. Math. Phys.* **3** (1999) 1289–1324, arXiv:hep-th/9901069.

[18] J. F. Plebanski, "On the separation of Einsteinian substructures," *J. Math. Phys.* **18** (1977) 2511–2520.

[19] A. Ashtekar and J. Lewandowski, "Representation theory of analytic holonomy C* algebras," arXiv:gr-qc/9311010.

[20] A. Ashtekar, J. Lewandowski, D. Marolf, J. Mourao and T. Thiemann, "Quantization of diffeomorphism invariant theories of connections with local degrees of freedom," *J. Math. Phys.* **36** (1995) 6456–6493, arXiv:gr-qc/9504018.

[21] J. Lewandowski, A. Okolow, H. Sahlmann and T. Thiemann, "Uniqueness of diffeomorphism invariant states on holonomy-flux algebras," *Commun. Math. Phys.* **267** (2006) 703–733, arXiv:gr-qc/0504147.

[22] C. Fleischhack, "Representations of the Weyl algebra in quantum geometry," *Commun. Math. Phys.* **285** (2009) 67–140, arXiv:math-ph/0407006.

[23] T. Thiemann, *Modern Canonical Quantum General Relativity*. Cambridge Monographs on Mathematical Physics. Cambridge University Press, 2008.

[24] A. Ashtekar, "New variables for classical and quantum gravity," *Phys. Rev. Lett.* **57** (Nov. 1986) 2244–2247. https://link.aps.org/doi/10.1103/PhysRevLett.57.2244.

[25] A. Ashtekar and J. Lewandowski, "Quantum theory of geometry. 1: Area operators," *Class. Quant. Grav.* **14** (1997) A55–A82, arXiv:gr-qc/9602046.

[26] A. Ashtekar and J. Lewandowski, "Quantum theory of geometry. 2. Volume operators," *Adv. Theor. Math. Phys.* **1** (1998) 388–429, arXiv:gr-qc/9711031.

[27] J. Fernando Barbero, "Real Ashtekar variables for Lorentzian signature space times," *Phys. Rev. D* **51** (1995) 5507–5510, arXiv:gr-qc/9410014.

[28] J. Fernando Barbero, "A real polynomial formulation of general relativity in terms of connections," *Phys. Rev. D* **49** (1994) 6935–6938, arXiv:gr-qc/9311019.

[29] C. Rovelli and L. Smolin, "Discreteness of area and volume in quantum gravity," *Nucl. Phys. B* **442** (1995) 593–622, arXiv:gr-qc/9411005. [Erratum: *Nucl. Phys. B* **456** (1995) 753–754.]

[30] K. G. Wilson, "The renormalization group: critical phenomena and the Kondo problem," *Rev. Mod. Phys.* **47** (Oct. 1975) 773–840. https://link.aps.org/doi/10.1103/RevModPhys.47.773.

[31] T. Lang, K. Liegener and T. Thiemann, "Hamiltonian renormalisation I: derivation from Osterwalder–Schrader reconstruction," *Class. Quant. Grav.* **35** no. 24, (2018) 245011, arXiv:1711.05685 [gr-qc].

[32] T. Lang, K. Liegener and T. Thiemann, "Hamiltonian renormalisation II. Renormalisation flow of 1 + 1 dimensional free scalar fields: derivation," *Class. Quant. Grav.* **35** no. 24 (2018) 245012, arXiv:1711.06727 [gr-qc].

[33] T. Lang, K. Liegener and T. Thiemann, "Hamiltonian renormalization III. Renormalisation flow of 1 + 1 dimensional free scalar fields: properties," *Class. Quant. Grav.* **35** no. 24 (2018) 245013, arXiv:1711.05688 [gr-qc].

[34] T. Lang, K. Liegener and T. Thiemann, "Hamiltonian renormalisation IV. Renormalisation flow of D + 1 dimensional free scalar fields and rotation invariance," *Class. Quant. Grav.* **35** no. 24 (2018) 245014, arXiv:1711.05695 [gr-qc].

[35] K. Liegener and T. Thiemann, "Hamiltonian renormalisation V: Free vector bosons," arXiv:2003.13059 [gr-qc].

[36] M. Reuter and F. Saueressig, *Quantum Gravity and the Functional Renormalization Group: The Road towards Asymptotic Safety*. Cambridge Monographs on Mathematical Physics. Cambridge University Press, 2019.

[37] B. Bahr, B. Dittrich and S. Steinhaus, "Perfect discretization of reparametrization invariant path integrals," *Phys. Rev. D* **83** (2011) 105026, arXiv:1101.4775 [gr-qc].

[38] M. H. Goroff and A. Sagnotti, "Quantum gravity at two loops," *Phys. Lett. B* **160** (1985) 81–86.

[39] S. Weinberg, "Ultraviolet divergences in quantum gravity," in *General Relativity: An Einstein Centenary Survey*, S. W. Hawking and W. Israel, eds., pp. 790–831. Cambridge University Press, 1979.

[40] C. Wetterich, "Exact evolution equation for the effective potential," *Phys. Lett. B* **301** (1993) 90–94, arXiv:1710.05815 [hep-th].

[41] T. Thiemann, "Quantum spin dynamics (qsd). 2.," *Class. Quant. Grav.* **15** (1998) 875–905, arXiv:gr-qc/9606090.

[42] T. Thiemann, "Quantum spin dynamics. VIII. The master constraint," *Class. Quant. Grav.* **23** (2006) 2249–2266, arXiv:gr-qc/0510011.

[43] T. Thiemann, "Gauge field theory coherent states (GCS): 1. General properties," *Class. Quant. Grav.* **18** (2001) 2025–2064, arXiv:hep-th/0005233.

[44] T. Thiemann and O. Winkler, "Gauge field theory coherent states (GCS). 2. Peakedness properties," *Class. Quant. Grav.* **18** (2001) 2561–2636, arXiv:hep-th/0005237.

[45] T. Thiemann and O. Winkler, "Gauge field theory coherent states (GCS): 3. Ehrenfest theorems," *Class. Quant. Grav.* **18** (2001) 4629–4682, arXiv:hep-th/0005234.

[46] T. Thiemann and O. Winkler, "Gauge field theory coherent states (GCS) 4: Infinite tensor product and thermodynamical limit," *Class. Quant. Grav.* **18** (2001) 4997–5054, arXiv:hep-th/0005235.

[47] T. Thiemann, "Complexifier coherent states for quantum general relativity," *Class. Quant. Grav.* **23** (2006) 2063–2118, arXiv:gr-qc/0206037.

[48] K. Giesel and T. Thiemann, "Algebraic quantum gravity (AQG). III. Semiclassical perturbation theory," *Class. Quant. Grav.* **24** (2007) 2565–2588, arXiv:gr-qc/0607101.

[49] M. Varadarajan, "Quantum propagation in Smolin's weak coupling limit of 4D Euclidean gravity," *Phys. Rev. D* **100** no. 6, (2019) 066018, arXiv:1904.02247 [gr-qc].

[50] A. Laddha and M. Varadarajan, "Quantum dynamics," in *Loop Quantum Gravity: The First 30 Years*, Abhay Ashtekar and Jorge Pullin, eds., pp. 69–96. World Scientific, 2017.

[51] J. W. Barrett and L. Crane, "A Lorentzian signature model for quantum general relativity," *Class. Quant. Grav.* **17** (2000) 3101–3118, arXiv:gr-qc/9904025.

[52] J. Engle, E. Livine, R. Pereira and C. Rovelli, "LQG vertex with finite Immirzi parameter," *Nucl. Phys. B* **799** (2008) 136–149, arXiv:0711.0146 [gr-qc].

[53] L. Freidel and K. Krasnov, "A new spin foam model for 4D gravity," *Class. Quant. Grav.* **25** (2008) 125018, arXiv:0708.1595 [gr-qc].
[54] W. Kaminski, M. Kisielowski and J. Lewandowski, "Spin-foams for all loop quantum gravity," *Class. Quant. Grav.* **27** (2010) 095006, arXiv:0909.0939 [gr-qc]. [Erratum: *Class. Quant. Grav.* **29** (2012) 049502.]
[55] D. Oriti, "Group field theory and simplicial quantum gravity," *Class. Quant. Grav.* **27** (2010) 145017, arXiv:0902.3903 [gr-qc].
[56] T. Thiemann, "Canonical quantum gravity, constructive QFT and renormalisation," arXiv:2003.13622 [gr-qc].
[57] A. Perez, "The spin foam approach to quantum gravity," *Living Rev. Rel.* **16** (2013) 3, arXiv:1205.2019 [gr-qc].
[58] A. Ashtekar, T. Pawlowski, P. Singh and K. Vandersloot, "Loop quantum cosmology of k = 1 FRW models," *Phys. Rev. D* **75** (2007) 024035, arXiv:gr-qc/0612104.
[59] M. Bojowald, "Loop quantum cosmology," *Living Rev. Rel.* **8** (2005) 11, arXiv:gr-qc/0601085.
[60] I. Agullo and P. Singh, "Loop quantum cosmology," in *100 Years of General Relativity*, Abhay Ashtekar and Jorge Pullin, eds., pp. 183–240. World Scientific, 2017. arXiv:1612.01236 [gr-qc].
[61] L. Castelló Gomar, G. A. Mena Marugán, D. Martín De Blas and J. Olmedo, "Hybrid loop quantum cosmology and predictions for the cosmic microwave background," *Phys. Rev. D* **96** no. 10 (2017) 103528, arXiv:1702.06036 [gr-qc].
[62] B. Elizaga Navascués and G. A. M. Marugán, "Hybrid loop quantum cosmology: an overview," arXiv:2011.04559 [gr-qc].
[63] E. Alesci, E. Bianchi and C. Rovelli, "LQG propagator: III. The new vertex," *Class. Quant. Grav.* **26** (2009) 215001, arXiv:0812.5018 [gr-qc].
[64] I. Agullo, "Loop quantum cosmology, non-Gaussianity, and CMB power asymmetry," *Phys. Rev. D* **92** (2015) 064038, arXiv:1507.04703 [gr-qc].
[65] C. Rovelli and F. Vidotto, "Planck stars," *Int. J. Mod. Phys. D* **23** no. 12 (2014) 1442026, arXiv:1401.6562 [gr-qc].
[66] H. Nicolai, K. Peeters and M. Zamaklar, "Loop quantum gravity: an outside view," *Class. Quant. Grav.* **22** (2005) R193, arXiv:hep-th/0501114.
[67] T. Thiemann, "Loop quantum gravity: an inside view," *Lect. Notes Phys.* **721** (2007) 185–263, arXiv:hep-th/0608210.
[68] L. Smolin, "Linking topological quantum field theory and nonperturbative quantum gravity," *J. Math. Phys.* **36** (1995) 6417–6455, arXiv:gr-qc/9505028.
[69] K. V. Krasnov, "On quantum statistical mechanics of Schwarzschild black hole," *Gen. Rel. Grav.* **30** (1998) 53–68, arXiv:gr-qc/9605047.
[70] A. Ashtekar, J. Baez, A. Corichi and K. Krasnov, "Quantum geometry and black hole entropy," *Phys. Rev. Lett.* **80** (1998) 904–907, arXiv:gr-qc/9710007.
[71] A. Ashtekar, J. C. Baez and K. Krasnov, "Quantum geometry of isolated horizons and black hole entropy," *Adv. Theor. Math. Phys.* **4** (2000) 1–94, arXiv:gr-qc/0005126.
[72] A. Sen, "Logarithmic corrections to Schwarzschild and other non-extremal black hole entropy in different dimensions," *JHEP* **04** (2013) 156, arXiv:1205.0971 [hep-th].
[73] T. Thiemann, "The LQG string: loop quantum gravity quantization of string theory I: Flat target space," *Class. Quant. Grav.* **23** (2006) 1923–1970, arXiv:hep-th/0401172.
[74] N. Bodendorfer, T. Thiemann and A. Thurn, "Towards loop quantum supergravity (LQSG)," *Phys. Lett. B* **711** (2012) 205–211, arXiv:1106.1103 [gr-qc].

[75] N. Bodendorfer, T. Thiemann and A. Thurn, "Towards loop quantum supergravity (LQSG) II. p-Form sector," *Class. Quant. Grav.* **30** (2013) 045007, arXiv:1105.3710 [gr-qc].

[76] N. Bodendorfer, T. Thiemann and A. Thurn, "Towards loop quantum supergravity (LQSG) I. Rarita-Schwinger sector," *Class. Quant. Grav.* **30** (2013) 045006, arXiv:1105.3709 [gr-qc].

[77] N. Bodendorfer, T. Thiemann and A. Thurn, "New variables for classical and quantum gravity in all dimensions V. Isolated horizon boundary degrees of freedom," *Class. Quant. Grav.* **31** (2014) 055002, arXiv:1304.2679 [gr-qc].

[78] N. Bodendorfer, "A note on quantum supergravity and AdS/CFT," arXiv:1509.02036 [hep-th].

[79] N. Bodendorfer, A. Schäfer and J. Schliemann, "Holographic signatures of resolved cosmological singularities," *JHEP* **06** (2019) 043, arXiv:1612.06679 [hep-th].

[80] N. Bodendorfer, F. M. Mele and J. Münch, "Holographic signatures of resolved cosmological singularities II: numerical investigations," *Class. Quant. Grav.* **36** no. 24 (2019) 245013, arXiv:1804.01387 [hep-th].

33

Cumrun Vafa

Hollis Professor of Mathematics and Natural Philosophy at Harvard University

Date: 27 June 2011. Location: Uppsala. Last edit: 23 July 2015.

In your opinion, what are the main puzzles in theoretical physics at the moment?

The most important puzzle, for us string theorists, is that of having a definition of string theory. What is string theory? That is the main question. There are other more pragmatic problems; for example, the meaning of quantum mechanics remains unclear because quantum mechanics as a framework can be formulated in different ways.

Before string theory began to develop, we thought of the path integral as the potential definition of quantum theory or quantum field theory. After the development of string theory, we know that this definition may not be adequate because of the existence of dualities. Quantum field theory, when formulated in terms of the path integral, assumes certain fundamental degrees of freedom, but the existence of dualities suggests that this is not the right picture. Sometimes there are no fundamental degrees of freedom since their nature may change under dualities and it is not generally known what the right substitute for them is.

But is string theory the right framework to discuss what the fundamental degrees of freedom of quantum theory are?

I think that this question is unclear because what string theory is, itself, is unclear so I cannot answer it. However, history has shown that other consistent ideas and techniques that a priori appear not to be part of string theory, if shown to be consistent, will later become subsets of what we will call *string theory* in the future. In other words, string theory swallows other fields in some sense. My belief is that string theory is wide enough, or can be made general enough, to include everything that is mathematically consistent.

So you think that other approaches such as loop quantum gravity (LQG) [1, 2] . . .

To the extent that they are correct . . .

You don't think that LQG is correct?

I think that LQG is not on equal footing with string theory. The foundations of LQG cannot be compared to the foundations of string theory. Even though we don't have a full definition of string theory, it is far ahead in terms of development of the subject.

Is it then your opinion that if LQG and other approaches are shown to be consistent, they will be absorbed by string theory?

Absolutely.

Why is the problem of formulating a consistent theory of quantum gravity such a difficult problem?

It is not that difficult. The problem is that we don't know what the right theoretical framework for it should be. String theory has solved it in a very simple way; in fact, you find string theory by looking at a bunch of coupled harmonic oscillators. Systems of coupled harmonic oscillators are some of the most studied systems in physics and they are not complicated. It is, in fact, amazing that such systems describe quantum gravity. The question of why quantum gravity is described by string theory is perhaps the main question, but string theory itself doesn't look that complicated, at least not in the way we study the perturbation theory by using manifest dualities and so forth. Understanding what string theory is is complicated because of the existence of dualities [3]. For example, we don't have a formulation of string theory where the dualities have been a priori incorporated. This potential formulation would have a priori known that all the different string theories, related via dualities, are nothing but different coordinate patches of some bigger theory. This endeavour is of increased difficulty.

If a young student asks you what approach to quantum gravity they should work on, what would your answer be?

There is no question that string theory is the right framework to understand quantum gravity. By this I mean that it is closer to the truth than any other existent theory.

Is it worth exploring other approaches?

Well ... certainly being close-minded is not good. We should be open to other developments. But the fact that there exist other subjects does not justify exploring them if they are not on equal footing with string theory. I see each theory as being a tree with many branches that grow with time so there may be many trees with many branches but the roots of some trees are placed on a different level compared to the others.

You mentioned that string theory is a theory of quantum gravity, but one usually formulates string theory in a background that is treated classically, right?

String theory is for sure a theory of quantum gravity. String perturbation theory is formulated on a background which is treated classically and follows the same logic as any other type of theory, such as quantum field theory, when formulated as a perturbation theory. No matter what framework you choose, if you define it perturbatively, this is how it should be done. On the other hand, the question of how to formulate string theory in a background-independent way is, of course, a tough question and the fact that we still don't know the answer is related to the fact that we don't know what substitutes for the path-integral formulation of the theory, and hence to the fact that we don't have a good definition of string theory.

What is the idea behind what you have called the *swampland* [4]?

The swampland is an idea inspired by trying to understand what are the features characterising string theory. In a Wilsonian sense, there is a prescription for defining quantum field theories which includes defining the degrees of freedom and how the theory is renormalised, etc. In such cases there is a notion of what is a good renormalised quantum field theory. However, given this notion of consistent renormalised quantum field theories, we can now ask what is the analogous notion in the context of string theory and gravity. Theories which are not consistent – in the sense that they would be the analogue of non-renormalisable, or anomalous, etc. – are said, in the context of gravity, to be part of the swampland.

The swampland includes all theories which for some reason or another are not consistent. In the case of quantum field theory we know what these consistency conditions are, but in the context of string theory they are not all known and we would like to find such analogous conditions. We do have certain hints that indicate that certain theories cannot be realised, which we can argue based on anomalies and other properties. However, for certain theories we do not have good arguments to rule them out though we suspect that there exist good reasons for them to be discarded, which are yet to be discovered. These reasons would direct us towards the fundamental principles underlying quantum gravity. The hope is that the essential features characterising the swampland are also the essential features characterising quantum gravity.

Are some of these properties not physically motivated or not well understood?

"Physically motivated" is not quite the right statement. I would say that it is not well understood why certain theories are not realised and some are, but slowly people are finding the reasons why some theories are not appearing in the swampland. I suspect that black hole physics will play an important role in constraining the allowed number of theories since black holes can lead to several contradictions and paradoxes depending on which type of black holes may appear in a given theory. Understanding spacetime topology change will perhaps constrain the allowed black holes and I suspect that it will pinpoint the theories which are not consistent since it will tell you which manifolds and geometries are allowed. This has not been done but I think that it should be the case since gravity, after all, describes changes in geometry and topology and hence should encode which theories are not consistent.

As far as I understood, you have put forth a certain number of conjectures which would decide whether or not a theory would be in the swampland [4–6], right?

Yes, we have certain proposals for what they could be but none is complete nor proven. It is just a set of ideas.

But where did these ideas come from?

They are inspired by examples found in the context of string theory but they are not proven within string theory. We basically looked at a bunch of examples from string theory and observed that they all satisfy certain properties and then we take those properties and promote them to principles. For example, the volume of moduli space is compact. Is that a principle? Maybe. It didn't a priori have to be but what's the reason for it? That is the question.

Could you test these ideas in other theories of quantum gravity?

Well ... I don't know, maybe. You could investigate whether the volume of the scalar fields is finite or infinite in a given theory. So you could perhaps phrase it like in string theory.

One of these conjectures that you put forth was that gravity is the weakest force [5] – the weak gravity conjecture. Why would you think that this would lead to some restrictions on the theories?

Well ... it certainly leads to restrictions. For example, we saw that if you have a $U(1)$ gauge theory, the coupling constant cannot be too small and we can check in a concrete example how that works out. However, the motivation for this idea comes from string theory, since there we could not find counterexamples to this statement. We don't have a proof that this is true; however, we gave arguments inspired in string theory and also in effective field theory descriptions as well as in black hole physics as to why this yields a consistent picture. We don't have a proof but it seems to be reasonable.

Are there any phenomenological consequences of these restrictions?

Phenomenological consequences there are but interesting phenomenological consequences is another question. For example, the fact that there could be an extra $U(1)$ symmetry with such a small coupling is not a strong restriction but it is a restriction.

Is the standard model sitting inside the swampland or outside?

By definition it is not in the swampland. If it were in the swampland, we wouldn't be here (laughs). By "standard model" did you mean the model exactly as stated or the model that corresponds to our world? The answer depends on what you mean. If you just mean the Weinberg-Salam model [7, 8] without anything else then probably that is in the swampland; I don't know. But the correct model whose effective theory is the Weinberg-Salam model on our energy scale is presumably not in the swampland.

After having defined the swampland you say in one of your papers that the string theory vacua is a small island in the middle of the swampland [4]. But the string theory vacua is still pretty big, right?

Oh yes, absolutely. It is very big but of measure zero, I would say.

What do you mean by "measure zero"?

Let me give you a concrete example. We know that the ranks of the gauge groups in known string theory can in principle exceed a certain amount in any dimension. For example, if you have finite gravity in two dimensions and have some gauge group, what is the rank of that group? Is it 10 billion? Is it 20 billion? Maybe. However, we have never found such a large rank and therefore there is a bound, though we don't know exactly what it is. Now, you can easily imagine having a $U(N)$ gauge group and writing down an effective field theory. Then, if you take the rank to be 10 billion and divide it by infinity you find that the ratio is zero. In this sense, according to the known examples, it is of measure zero no matter what we do.

Do you think that a dynamical principle is contained within the current formulation of string theory that picks up the vacuum where we live in?

No, I don't think so; that would be very boring.

What do you mean by that?

I think it would be very boring if the physics was that rigid. I think that viewpoint, in which everything is fixed from the get-go, including the choice of the vacuum, is too restrictive. Nowhere in physics have we seen anything like that and it sounds very unphysical to me. It is a logician's paradise but I think it is a physicist's nightmare.

So is string theory more like a formalism?

It is similar to the way you think about Newtonian mechanics. Newtonian mechanics tells you what the possibilities are and then you have have to fix the constants that correspond to this universe that you are in. The question of how we ended up here is a question for cosmology. I don't think that there is a unique cosmology, neither do I think that the evolution which led to our world is unique. Why should there be a unique possibility?

Do you think that anthropic arguments are reasonable enough to fix these constants?

Anthropic arguments do not fix the constants but at least can point towards a physical situation in which there are no contradictions. I think that anthropic reasoning as a framework in which you try to fit certain ideas, or at least the bare bones of it, such as requiring the cosmological constant to be small in order not to contradict our existence, is fine. But trying to push it too much in order to argue that it can describe all the degrees of freedom of physics, such as giving you an estimate of the quark mass, is taking it too far. I think it depends on what aspects of anthropic reasoning you want to take into account. We should be worried about anthropic reasoning and be careful with it as it can dissuade us from finding the actual reasons for certain physical phenomena. Even though certain aspects of the world could be anthropic in principle, we should always be skeptical unless we don't have any other options.

Related to our cosmological evolution ... string theory predicts extra dimensions and some people often criticise this. Is there any good reason why we have four infinitely large dimensions and six very small ones?

There certainly could be a good reason for it. Together with Robert Brandenberger in the late 1980s we developed a model which could potentially explain why we have four large dimensions [9]. If that is the correct model or not is still unclear. On the other hand, the possibility of there being no explanation at all is not dismissed. It could be in the end that this just reflects the choice of background that we started with, which would be an input into the theory. Certainly, string theory does not necessarily have to have an explanation for it and to me the fact that we have all these possible backgrounds in string theory which are consistent with no cosmology at all, that is, time-independent (static) backgrounds with no evolution, seems to indicate that there is nothing in string theory that prohibits it from happening. So the question of why we have ended up with this particular vacuum or why it has evolved in this or that way is a question to which string theory may have nothing to say.

But the six extra dimensions which are now curled up, were they necessarily small in the beginning?

This kind of picture depends on the initial conditions. We don't know. The picture that I developed together with Brandenberger [9] is that everything is naturally Planck-scale size or string-scale size and therefore to explain that something is big we have to work hard. So we were trying to explain why something could be anomalously large and why we are left with only three large spatial dimensions in the end.

Was this programme successful?

It had some amount of success but I would say that the knowledge we have about string theory dynamics in the early universe and at very high energies is quite limited. We basically had to postulate our idea of what happens to string interactions at high energies without any proof.

Is there also a good reason why these dimensions all of sudden don't curl up again and the small ones don't unfold?

Well, the picture we had is that everything is curled up to begin with so it depends on what you assume. In string theory you could have had it either way.

One of the spinoffs from these ideas was the development of string gas cosmology [9–12]. What does this consist of?

The idea is that there are strings that form the background you started with and if any two strings meet they annihilate each other and space grows. And if there are too many strings moving in opposite directions and the spatial dimensions are large, it is likely that many strings will not meet each other and hence space will not grow. Our idea was that this mechanism of string annihilation would lead to an equilibrium.

Were there any observable consequences of this?

We were not at the point where we could give observational consequences. A lot of our results depend on how clearly we understand strings near the Hagedorn temperature and the corresponding interactions, which at this point is out of reach.

String theory is also criticised for not being able to work very well in time-dependent backgrounds.

I think that "not being able to work" is a strong statement. We don't know how it works and we don't know how to deal with it.

I would like to change a bit the topic ... what is F-theory [13]?

F-theory was formulated with the purpose of understanding certain backgrounds that can be constructed in type IIB string theory and are non-perturbative from the viewpoint of type IIB. According to F-theory, these backgrounds naturally arise from a geometry in two dimensions higher than type IIB, so it deals with elliptic fibred manifolds which describe the geometry associated with interesting configurations of 7-branes of type IIB.

My naive idea of F-theory is that you start with 12 dimensions, you compactify two and you get type IIB.

I think that's not necessarily the right picture because the geometry is certainly in two dimensions higher. Thus, if you wanted to have a geometry it would have to involve an extra 2-torus. You attach an extra 2-torus to the observable universe of type IIB to describe the manifold of F-theory. This you could say is an arbitrary construct but it can be brought to life, so to speak, by compactifying the dimensions of type IIB on a circle and then describing them via dualities from an M-theory perspective. In some sense, it suggests that there should be a physical 12-dimensional theory but it is not going to be Lorentz-invariant in 12 dimensions in the usual sense. So the question of what this 12-dimensional theory is is still open and I suspect that its signature would be (10,2) instead of (10,1). However, the details of exactly how to describe a consistent theory and how the non-covariance of it, namely the fact that it is not Poincare-invariant in 12 dimensions since there isn't enough supersymmetry in those dimensions, can be implemented is not known at the moment.

Why is it more powerful than type IIB?

Because it is non-perturbative and because the coupling constant of type IIB could be infinite at some loci on the manifold. We are varying the type IIB coupling constant and usually when the coupling constant is strong you go to a dual picture. Now the problem in F-theory backgrounds is that they have all these possible values of coupling constants and there is no picture where it is always weak. In some pictures, some region is weak, and in other pictures, it is strong. In summary, there is no overall picture. That is why we had to introduce a new notion that can allow it and it is not any known theory, so we called it F-theory.

But is this the ultimate theory?

No; in fact, in my opinion this is one of the things that is misunderstood in string theory, namely, that that there is no preferred perspective. For example, M-theory starts in 11 dimensions and unifies an interesting number of contexts when you go down to 10 and so forth. Some people interpret this as evidence for M-theory being the right description of the full theory. That is an incorrect picture and F-theory, in fact, provides counter-evidence. If you take the 11-dimensional theory, compactify it on a torus and shrink the torus you get a theory which has 10-dimensional Lorentz invariance. That doesn't make sense. From the viewpoint of M-theory, you should have only 9-dimensional Lorentz invariance. But F-theory says that you have 10-dimensional Lorentz invariance because we know that it captures type IIB. There is no preferred vantage point which gives you all the physics right. F-theory gets some of it right and M-theory gets some other part of it right.

Is there something else in more than 12 dimensions that is also interesting?

I don't think so. People have tried to increase the dimensions further and to describe other types of geometries but the main point is that each limit that you may take from these constructions may have its own duality description which is better than in any other picture. However, you do not know which description is the correct one. It's the same as asking which coordinates are good for a particular problem. It could be that for a specific problem you use a given set of coordinates, while for that other problem you use polar coordinates, Cartesian, etc. In Cartesian coordinates, everything is very beautiful and many problems can be solved but for other problems other coordinates may be more useful. I don't think we have a general overview, which would amount to a coordinate-independent way of thinking about this, and which would make M-theory and F-theory just coordinate patches.

So is it not possible to address the question of what are the fundamental degrees of freedom of the theory? Is it strings or membranes?

I think that's just the same thing as asking the question of which coordinate patch is better to work on.

Is the F-theory vacuum the only vacuum which can make any connection with the real world?

No, I wouldn't say so. First of all, I don't know if the F-theory corner is the right corner. Certainly, it seems very promising. We have explored it in the past few years, in particular with my colleague and former student Jonathan Heckman [14–18], and we are very happy with many of the results we have found but by no means would I say that this is the only way that F-theory can make contact with particle physics. We found it very surprising that it made a lot of contact with particle physics and made a lot of potential predictions with very little effort. This to me is a sign that there is something nice about it. But you know, it is just a picture and in string theory we can have other pictures.

Is it possible to know if the standard model is contained within this vacuum?

Standard model-like models are certainly contained in it but if our particular universe is contained in there I don't know. There are certain predictions that these models come up with. At least in simple classes of F-theory compactifications that we constructed, without being committed to any special type and trying to be as broad and universal as possible, assuming that gravity is decoupled and supersymmetry is maintained up to the weak scale, we developed a standard type of model where we could make phenomenological predictions [14–18]. For example, in the minimal version of it, we found a model in which if you assumed that supersymmetry is maintained up to the weak scale, which of course is a strong assumption, you could predict the existence of a charged $\tilde{\pi}$ which would be the $\tilde{\tau}$ with a lifetime of the order of a second to an hour. This range is very predicative, because if there is indeed a $\tilde{\tau}$ of the mass of the weak scale, that is, of half a TeV or so and with this lifetime, then we could observe it at the LHC [19–21]. Thus, if that would be the case we could potentially be in this class of vacua. However, if we don't find such particle, we cannot say that the theory does not allow for standard model-like models of our universe since we made certain assumptions in our analysis.

String theory has been always criticised for not being falsifiable. Could this be the first prediction of string theory?

Well, as I just said, this is not falsifiable either because if we don't find the $\tilde{\tau}$ then we may say that we have made a wrong assumption. But suppose that the LHC did find such a particle and a few more of these predictions would also be observed. Then, the model begins to be believable. It's like the rest of physics; that is, we don't prove a given theory but after a while we begin to believe in it because its predictions are verified by experiment. Such observations would constitute circumstantial evidence. However, if that does not happen then whatever LHC finds or the next step of physics determines we have to continue searching for the correct model within string theory.

So I cannot take this as the first prediction of string theory?

No. I would say that it is a prediction of a certain class of string models which are well motivated. But certainly not a proof. If they find it, it won't be a proof of string theory but instead circumstantial evidence for this class of models.

Do you think we'll find supersymmetry at the LHC?

It looks unlikely at this stage, unfortunately. If I were to bet the answer would be no. But if you told me that the LHC has already announced that the results of one of the theories that people have predicted (apart from the Higgs) is true then I would say that it must be supersymmetry. I don't believe that any other competing models have a higher chance to be correct than supersymmetric models but the likelihood of finding supersymmetry is low in my opinion. It is certainly true that supersymmetry is present at some scale, at least in the way we think about it in the context of string theory, but that scale could be the string scale or the Planck scale and hence useless for phenomenology.

Is there any mechanism contained in F-theory that would explain supersymmetry breaking?

Yes, we have proposed some in our papers [22] but there could be more. Supersymmetry breaking could be explained via situations in which D3-brane instantons wrapping around 4-cycles give rise to super potentials and hence would break supersymmetry. This is analogous to the mechanisms discussed in Polonyi-type models [23].

And this is possible to introduce at the right energy scale?

Yes, we give arguments for why supersymmetry breaking would be pushed from the grand unified theory scale all the way down because of the suppression factors of instantons.

Are there any F-theory implications for cosmology, for example, for dark energy or dark matter?

Well, we had certain predictions for dark matter but not for dark energy. In the context of the models we had, a 10–100 MeV gravitino was the major contribution to dark matter, along with other particles like the axion which also contributed to it [24]. Consequently, one of the predictions we made was that direct dark matter searchs will fail because these searches assume the existence of weakly interacting massive particles while gravitinos are too weak to be detected. In fact, so far these searches have failed.

Are you pretty confident about these models?

To be confident is a strong statement. I think I can say that we are happy with the models. Certainly, supersymmetry at low-energy scales is a strong claim. But one nice aspect of these models is that the $\tilde{\tau}$ that I mentioned earlier, which is charged and has a mass of about $1/2$ TeV or so, can get rid of some of the anomalies that have been observed in the abundance of light nuclei, such as lithium. These anomalies reflect a disagreement between the prediction of big bang nucleosynthesis models for the abundance of lithium and the observed abundance. The existence of a relatively long-lived charged particle like $\tilde{\tau}$ has been proposed in the literature as a potential way to solve these anomalies. The models we consider naturally incorporate this kind of picture. As such, they could potentially be the relevant models to look at.

Would it be possible to search for the $\tilde{\tau}$ at the LHC?

The class of models I have constructed with Heckman and collaborators [19–21] can be falsified and should be falsified or found to be correct at the LHC. Some of the predictions lie within the range of energies that the LHC will achieve. However, I think that it is likely that the models will be falsified at some point.

One of the most remarkable things that originated in string theory was the AdS/CFT correspondence [25], and you have done a lot of work to try to prove it [26–28]. Is it important to find such proof?

Proving it suggests an understanding of the correspondence. In that sense, I think it is important to prove it but by the term "proof" I do not mean a mathematical ϵ, δ proof but instead an understanding of it from a physicists' perspective. The correspondence is surprising and these works attempted at making it clearer as well as less surprising.

Is it okay to apply the correspondence to all different circumstances without proof?

Of course it is okay. We do not have a good understanding of most string dualities but we apply them and obtain new results. With time, more evidence for the existence of dualities piles up, to the point that they are by now believable. Applying the dualities doesn't shine light on the understanding of the dualities itself. As such, it would be nice to deepen our knowledge of dualities by exploring other directions, including finding such proofs.

Did you succeed in finding these proofs?

In the context of topological string theory we did find such proof [26] using the corresponding large-N theory, as in AdS/CFT, which I found together with Rajesh Gopakumar [29]. This correspondence involves D-branes in a model of topological strings with a large-N dual related to specific geometric transitions. In a paper with Hirosi Ooguri [26], we explained how this kind of duality could be potentially derived from a worldsheet perspective and gave an example as well as provided predictions that were verified. There should be a proof of AdS/CFT along these lines in the context of $N = 4$ super Yang–Mills theory.

Why is this a harder proof compared to topological strings?

Well ... the actual sigma model of AdS/CFT is more complicated, which means that we need to develop technical tools to deal with it. The proof that we put forth in the context of topological strings was based on the existence of a linear sigma model [26]. If we had an analogue linear sigma model for AdS/CFT we would be able to easily extend the proof. I suspect that it is just a hard, but important, technical problem which we might be able to solve in the future.

In the context of topological string theory you have proposed a kind of analogy between topological strings and the melting of crystals [30–32]. Are these realistic crystals?

No, they aren't realistic crystals.

How should we interpret this, then?

Well, it is just a mathematical physics model for the melting of crystals but not a realistic crystal model. In the context of this work, the melting of a corner of the crystal corresponds to string corrections to the geometry of classical particles. This is the essence of it.

But isn't it related to a more fundamental description of what spacetime should be at higher energies?

That's the kind of understanding we had. The picture of melting crystals is the analogue of a quantum gravitational foam which at large distances gives rise to string-scale physics. That is, at large distances it looks like classical molten crystals which correspond to classical geometry.

Is this the first time that string theory accurately probes Planck-scale physics?

This is the only case I know of in string theory where the notion of quantum gravitational foam has been made precise. In this context, the removal of each atom of the crystal corresponds to blowing up a point in spacetime. The ensemble of these blown-up points constitutes the moulting crystal, in turn determining the shape of the geometry. Thus, the average shape of these blown-up geometries is a complicated foam but looks smooth at large distances. It is a beautiful picture of how smooth geometry can arise from a foamy quantum description of gravity.

I think that everyone would agree that this is pretty amazing, but I suspect that the next step would be to understand how to do it for other string theories?

That's right, it's a good question. It should be possible but showing that smooth geometries emerge from a gravitational foam in topological string theory is simpler than in other string theories, though it can in principle be accomplished. I have tried to pursue this direction to some extent but it is not easy due to lack of analytic control.

Would the ultimate formulation of string theory be topological in some sense?

In an abstract sense it sounds plausible that there is a topological version of string theory but in a much more broader sense than what the term "topological" usually encompasses. Ultimately, I believe that the answer is going to be trivial once you find the right formulation. However, understanding in what sense it is topological (e.g., it should be background-independent) and what it means and so on is a hard task.

Is topological string theory background-independent?

Not even topological string theory is background-independent. The way Edward Witten, for example, has formulated it is in the context of a background with a manifold, where that manifold is itself part of the dynamics [33–39]. However, the zero modes and size of the manifold are frozen; that is, they are not part of the dynamics of topological strings.

So, in some sense, even in topological string theory the degree of background independence it is not quite satisfactory.

Do you have any idea of how one could proceed and find a background independent formulation of string theory?

No, not at this stage.

What do you to think has been the biggest breakthrough in theoretical physics in the past 30 years?

The discovery of dualities in all its different forms has, for sure, been the biggest breakthrough.

You have once made the statement that string theory *is definitely revealing the deepest understanding of the universe which we ever had* (see, e.g., [40]). Do you still hold this opinion?

Yes, absolutely.

Why have you chosen to do physics?

Because I enjoyed it (laughs).

What do you think is the role of theoretical physicists in modern society?

The main role of modern theoretical physicists is to educate the minds of people, both at universities as well as in public sphere, about what modern science is. You could have asked, "What should their role be?" but since you didn't . . .

Now I'm curious . . . . what should their role be?

The physicists' viewpoint is broad and applies to other areas beyond physics. This perspective has not been significantly conveyed to the general public. The way physicists model different systems and the problem-solving methods that were developed are not widely known beyond physics. We do convey many beautiful results, for instance, about black holes, but we could do a better job at communicating the methodology that physicists employ in order to solve difficult problems.

Do you think that such methodology would be useful in other areas?

Yes, for example, the concept of dualities could have an application in other areas that have nothing to do with physics. For instance, increasingly complex systems might admit simple descriptions. In a very rudimentary sense, this idea is the core notion of duality symmetries. Does this type of thinking have an application in society? I think so. We haven't found experimental evidence for string theory yet but we have found beautiful sets of ideas which underlie string theory and which should be conveyed clearly to the public.

References

[1] H. Nicolai, K. Peeters and M. Zamaklar, "Loop quantum gravity: an outside view," *Class. Quant. Grav.* **22** (2005) R193, arXiv:hep-th/0501114.

[2] T. Thiemann, "Loop quantum gravity: an inside view," *Lect. Notes Phys.* **721** (2007) 185–263, arXiv:hep-th/0608210.

[3] E. Witten, "String theory dynamics in various dimensions," *Nucl. Phys. B* **443** (1995) 85–126, arXiv:hep-th/9503124.

[4] C. Vafa, "The string landscape and the swampland," arXiv:hep-th/0509212.

[5] N. Arkani-Hamed, L. Motl, A. Nicolis and C. Vafa, "The string landscape, black holes and gravity as the weakest force," *JHEP* **06** (2007) 060, arXiv:hep-th/0601001.

[6] H. Ooguri and C. Vafa, "On the geometry of the string landscape and the swampland," *Nucl. Phys. B* **766** (2007) 21–33, arXiv:hep-th/0605264.

[7] S. Weinberg, "A model of leptons," *Phys. Rev. Lett.* **19** (Nov. 1967) 1264–1266. https://link.aps.org/doi/10.1103/PhysRevLett.19.1264.

[8] A. Salam, "Weak and electromagnetic interactions," *Conf. Proc. C* **680519** (1968) 367–377.

[9] R. H. Brandenberger and C. Vafa, "Superstrings in the early universe," *Nucl. Phys. B* **316** (1989) 391–410.

[10] J. Kripfganz and H. Perlt, "Cosmological impact of winding strings," *Class. Quant. Grav.* **5** (1988) 453.

[11] A. A. Tseytlin and C. Vafa, "Elements of string cosmology," *Nucl. Phys. B* **372** (1992) 443–466, arXiv:hep-th/9109048.

[12] S. Alexander, R. H. Brandenberger and D. Easson, "Brane gases in the early universe," *Phys. Rev. D* **62** (2000) 103509, arXiv:hep-th/0005212.

[13] C. Vafa, "Evidence for F theory," *Nucl. Phys. B* **469** (1996) 403–418, arXiv:hep-th/9602022.

[14] C. Beasley, J. J. Heckman and C. Vafa, "GUTs and exceptional branes in F-theory – I," *JHEP* **01** (2009) 058, arXiv:0802.3391 [hep-th].

[15] C. Beasley, J. J. Heckman and C. Vafa, "GUTs and exceptional branes in F-theory – II: Experimental predictions," *JHEP* **01** (2009) 059, arXiv:0806.0102 [hep-th].

[16] J. J. Heckman and C. Vafa, "F-theory, GUTs, and the weak scale," *JHEP* **09** (2009) 079, arXiv:0809.1098 [hep-th].

[17] J. J. Heckman and C. Vafa, "Flavor hierarchy from F-theory," *Nucl. Phys. B* **837** (2010) 137–151, arXiv:0811.2417 [hep-th].

[18] J. J. Heckman, A. Tavanfar and C. Vafa, "The point of E(8) in F-theory GUTs," *JHEP* **08** (2010) 040, arXiv:0906.0581 [hep-th].

[19] J. J. Heckman and C. Vafa, "From F-theory GUTs to the LHC," arXiv:0809.3452 [hep-ph].

[20] J. J. Heckman, G. L. Kane, J. Shao and C. Vafa, "The footprint of F-theory at the LHC," *JHEP* **10** (2009) 039, arXiv:0903.3609 [hep-ph].

[21] J. J. Heckman, J. Shao and C. Vafa, "F-theory and the LHC: Stau search," *JHEP* **09** (2010) 020, arXiv:1001.4084 [hep-ph].

[22] J. J. Heckman, J. Marsano, N. Saulina, S. Schafer-Nameki and C. Vafa, "Instantons and SUSY breaking in F-theory," arXiv:0808.1286 [hep-th].

[23] J. Polonyi, "Generalization of the massive scalar multiplet coupling to the supergravity," Hungary Central Inst Res - KFKI-77-93 (77,REC.JUL 78).

[24] J. J. Heckman, A. Tavanfar and C. Vafa, "Cosmology of F-theory GUTs," *JHEP* **04** (2010) 054, arXiv:0812.3155 [hep-th].

[25] J. M. Maldacena, "The large N limit of superconformal field theories and supergravity," *Int. J. Theor. Phys.* **38** (1999) 1113–1133, arXiv:hep-th/9711200.
[26] H. Ooguri and C. Vafa, "World sheet derivation of a large N duality," *Nucl. Phys. B* **641** (2002) 3–34, arXiv:hep-th/0205297.
[27] N. Berkovits, H. Ooguri and C. Vafa, "On the world sheet derivation of large N dualities for the superstring," *Commun. Math. Phys.* **252** (2004) 259–274, arXiv:hep-th/0310118.
[28] N. Berkovits and C. Vafa, "Towards a worldsheet derivation of the Maldacena conjecture," *AIP Conf. Proc.* **1031** (2008) 21–42, arXiv:0711.1799 [hep-th].
[29] R. Gopakumar and C. Vafa, "On the gauge theory/geometry correspondence," *AMS/IP Stud. Adv. Math.* **23** (2001) 45–63, arXiv:hep-th/9811131.
[30] A. Iqbal, N. Nekrasov, A. Okounkov and C. Vafa, "Quantum foam and topological strings," *JHEP* **04** (2008) 011, arXiv:hep-th/0312022.
[31] A. Okounkov, N. Reshetikhin and C. Vafa, "Quantum Calabi-Yau and classical crystals," *Prog. Math.* **244** (2006) 597, arXiv:hep-th/0309208.
[32] J. J. Heckman and C. Vafa, "Crystal melting and black holes," *JHEP* **09** (2007) 011, arXiv:hep-th/0610005.
[33] E. Witten, "Topological sigma models," *Commun. Math. Phys.* **118** (1988) 411.
[34] E. Witten, "Topological gravity," *Phys. Lett. B* **206** (1988) 601–606.
[35] E. Witten, "Topological quantum field theory," *Commun. Math. Phys.* **117** (1988) 353.
[36] E. Witten, "On the structure of the topological phase of two-dimensional gravity," *Nucl. Phys. B* **340** (1990) 281–332.
[37] E. Witten, "Introduction to cohomological field theories," *Int. J. Mod. Phys. A* **6** (1991) 2775–2792.
[38] E. Witten, "Mirror manifolds and topological field theory," *AMS/IP Stud. Adv. Math.* **9** (1998) 121–160, arXiv:hep-th/9112056.
[39] E. Witten, "Chern-Simons gauge theory as a string theory," *Prog. Math.* **133** (1995) 637–678, arXiv:hep-th/9207094.
[40] B. Greene, *The Elegant Universe: Superstrings, Hidden Dimensions, and the Quest for the Ultimate Theory*. W. W. Norton, 2010.

34
Erik Verlinde

Professor at the Institute of Physics at the University of Amsterdam

Date: 3 May 2011. Location: Amsterdam. Last edit: 3 November 2016

What are the main puzzles in theoretical physics at the moment?

I think that the main puzzles are related to cosmology and the description of the early universe. The equations we commonly use to describe the universe are still Einstein equations but there are many indications that these equations should have a different description. We do not fully understand the relation between these equations and quantum mechanics. I have the feeling that what we are doing now with string theory is bringing these different pieces together, which will ultimately lead to a change in the way we think about and describe the universe.

Do we need a new language for understanding physis in the early universe?

I think that the whole language that we usually use to describe the physics of the early universe will change. There are many problems regarding the cosmos such as those related to dark energy and dark matter. Only less than 1% of the energy in the universe is observed, 4% is sort of claimed to be understood and the remaining 96% people don't know what it is. I think that understanding this 96% is the biggest theoretical challenge in physics. Any serious theory of the universe must address this question.

Do you think that we need a theory of quantum gravity in order to address this question?

I think we need a very different way of looking at what gravity is and it should be combined with quantum mechanics, but that is not the same as saying that we need a theory of quantum gravity. Theories of quantum gravity assume that we can just quantise gravity directly, while what I just hinted at meant that there may exist some quantum theory from which we can understand gravity. This is a very different way of looking at the problem.

So where does string theory fit in all of this?

String theory is a physical framework like the framework of quantum field theory that we use for describing elementary particles. Frameworks are more like descriptions, and as with most frameworks, they may have hidden underlying principles. I'm convinced that what we do, as physicists and human beings, is to describe nature and, even though there is the possibility of there being some underlying fundamental principles in our descriptions, we may not always be smart enough to understand them.

So, I think that the most honest point of view is to consider any description/framework as an effective description and from there try to properly understand the frameworks and formulate them in terms of very general physical principles. This is the case for quantum field theory, for instance, in which case we understand how to formulate it from general principles. Similarly, in the case of thermodynamics, we know how to derive thermodynamic equations using very general starting points.

Bearing this in mind, I think that our laws of physics should be formulated in terms of very general starting points. The idea that, at this moment in time, we are able to guess what are the fundamental constituents of nature is unrealistic. There's almost zero chance that we are smart enough to figure that out and hence we should be happy with understanding a framework. String theory is one of such frameworks and there are ways in which we should be able to understand why string theory works so well based on general principles. It is a very different way of thinking when compared with starting with the assumption that we know what the fundamental constituents are and then trying to find a description.

Why is string theory working so well?

It is working very well in the sense that it has already established so much by connecting different fields of theoretical physics in such a way that makes it extremely unlikely that it will be falsified. In fact, it is a framework and therefore we cannot even speak of falsification.

But isn't the whole criticism of string theory the fact that it cannot be falsified?

No, the point is that it is a framework just like quantum field theory, which is a very efficient description of many physical phenomena. On the other hand, models (not frameworks) can be falsified. The idea that string theory should be able to predict its own vacua and thereby predict everything in the universe is simply wrong. This point of view is based on the idea that string theory is the ultimate theory, which is the wrong assumption. In quantum field theory, we are very happy with being able to write down the standard model and we didn't have to explore the whole landscape of other models that we could potentially write down. Analogously, that's not what we should do in the case of string theory either.

So do you think that the formalism is there, that it can give rise to many models, and what we have to do is to experimentally measure some parameters?

Yes. The idea that we have the final formulation of the universe is a wrong idea. Taking such idea as a premise, we are led to try to answer questions that we are simply unable to with the formulation at hand. However, if we are happy with the framework we have, we can try to understand it better.

But even as a framework, when people speak about the low-energy limit of string theory they claim that it should be possible to find a model that would be the correct extension of the standard model. As far as I am aware, the most conservative choice is the so-called minimal standard model [1] but this model has such a large number of parameters that should be tuned ...

Well, this model is based on the idea that supersymmetry is a symmetry of nature at low energies but I don't think that we need supersymmetry.

But then how do you satisfy the consistency requirements of string theory?

In some way there is always the need to require the existence of certain mechanisms for cancelling degrees of freedom in the ultraviolet (UV) regime, such as an asymptotic supersymmetry; otherwise, it would be problematic. However, I don't think that there's really any evidence for low-energy supersymmetry; neither is it a requirement of string theory. When you look at the standard model you start wondering about what happens between the low-energy scale that we observe now and the Planck scale, and you come up with the very traditional picture that everything should be unified at the Planck scale, in turn leading you to think of the existence of some kind of supersymmetry. This picture, in my opinion, is not really that developed yet and I think that there are other ways in which these problems can be solved.

If I understand you correctly, you think that you don't need supersymmetry at any scale. Is that it?

This is what I think of supersymmetry at the moment. It's a very beautiful idea but it also makes calculations a lot easier. Some of the consistent requirements that we demand from string theory are easy to get when using supersymmetry. In the 1980s people began finding finite models of field theories with supersymmetric properties. The first one was $N = 4$ super-Yang–Mills (SYM) theory [2, 3] and then $N = 4$ superconformal theories on the worldsheet were found [4]. Later, the $N = 2$ theory was proven to be finite [5, 6] and eventually people found the $N = 1$ theory [7, 8]. In the end, however, what we actually need in order to solve our problems and have consistent theories is conformal invariance and in fact there are models which are conformally invariant and do not have supersymmetry. With supersymmetry it's just easier to find and reach conformal fixed points and these models that I just mentioned are sort of the most symmetric models we can find. The idea that the world is described by the most symmetric model is an idea that I don't buy, as it just happens to be the class of models that are easier to construct and make calculations with. In essence, it's just a trick we use, like spherical symmetry, in order to simplify certain physical problems.

How should we tackle this problem of finding such theories without supersymmetry?

I think that we should look at wider-universality classes of models contained within string theory than those that we have been looking at so far. I mentioned the example of conformal field theories (CFTs), which is a much wider class than superconformal field theories that are widely used because supersymmetry simplifies the calculations. The class of conformal field theories is harder to work with but that doesn't mean that it's not describing the world. So I don't think that supersymmetry is an essential ingredient in string theory; it just happens to be the most symmetric case.

Do you have any idea of how to construct this wider class of conformal field theories?

I think that one needs, first of all, a sort of general existence theorem. We know that, quite generically, we can start with quantum field theories, integrate out degrees of freedom and look at how the theory behaves at low energies. CFTs can be obtained as a limit of this procedure and hence this is a general way from which we can understand these classes of theories. I'm almost sure that a similar understanding holds for string theory itself; that is, we should think of string theory as a limit or approximation (or a fixed point) of some underlying description.

You mentioned that supersymmetry was not an essential ingredient in string theory. Would 10 spacetime dimensions still be a requirement if you studied this wider class of theories?

No, but I do think that extra dimensions are possible. One thing that I really like about what we've learned in recent years is the idea that scale is an extra dimension and the way that this is manifest in AdS/CFT [9]. A scale is as much a dimension as any other dimension that we have. If I look left or if I look right, I have a finite range of what I can see but I can also look in a microscope and see smaller objects or I can look in a telescope *in the other direction*. This, for me, is the same as looking left or right but in a scaled sense and if I've placed myself in the world, I placed myself also somewhere in scale space. What I see goes from maybe the tiniest millimetre to maybe a few kilometres but that's localising myself in terms of scale just like you're localised right now. So, we just have to get used to the idea that scale is a dimension and that things don't need to look the same at every scale. Essentially, you can look at an object at various scales and you can say, "Well, I can use the language of cars to describe this object, I can use the language of atoms and then I can even go smaller" but it is just the same object described at different scales. It is the translation between those languages that you can sort of imagine happens in one direction. So I think scale is very likely to be eventually accepted as an extra dimension, not one that has translational invariance and so on but one that obeys certain rules such as scale transformations or specific ways of performing renormalisation. The reason why we require 10 dimensions in superstring theory is because of supersymmetry but if we accept that supersymmetry is not essential then you need one more dimension, which is the scale.

Only one more dimension?

Well, maybe you can imagine more but scale is certainly the one that plays a special role. But what is a dimension? In the context of string theory we have a very broad concept of dimension; that is, dimensions can be very small and have different properties than the dimensions we usually speak of, as sometimes we describe dimensions in terms of fermions. In the end, it doesn't even look like we are speaking of dimensions. So the definition of a dimension is not a very rigorous one, which has to do with the fact that space and time themselves already should be thought of as emergent, so any other dimension also. Thus, it's not very rigorous to ask how many dimensions we live in; it's just that the effective description of our world eventually needs a specific number of dimensions. Therefore if you ask me, "Is scale another dimension?" the answer is that it's only a dimension in the sense that I've just described earlier, as sort of an emergent dimension.

In the experiments at the LHC, for instance, people will be looking for signs of large extra dimensions or warped extra dimensions, which can play a role in these experiments. In fact, these large extra dimensions are the type of scale dimensions that I have been talking about and they were greatly inspired from the AdS/CFT correspondence [10]. So far, we've only seen evidence for the existence of those dimensions because we see that coupling constants do depend on the scale at which we do our experiments. As such, we already have been "moving" our objects in those directions and hence it's a dimension as any other. People usually have a very classical picture of what a dimension is and then try to think, "Well, maybe we can see particles moving in those directions" or something like that. However, this kind of reasoning, to me, is equivalent to not having enough imagination of what dimensionality really means. This kind of classical picture is inadequate to understand what happens at very tiny scales.

Several people have wondered about the dynamical principle that makes six extra dimensions curl up, given that the world, according to superstring theory, has 10 dimensions. Given your point of view of string theory as a framework, do you think that this type of question cannot be addressed?

Yes. I think this type of question assumes that we already found the ultimate theory of physics, which I think is too big of a claim.

People also say that string theory takes quantum mechanics for granted, ignoring its interpretational problems.

No, that's not entirely true.

But the way it works in string theory is that one ends up quantising things according to the usual rules of quantum mechanics, right?

Well ... string theory has not started from any formulation where you really know what the rules are. I think we're basically discovering a lot of things about string theory and I find it very intriguing that in string theory there are examples of certain quantum mechanical descriptions which emerge from something underlying it. For instance, there are objects for which we can compute partition functions like in statistical mechanics and then suddenly these partition functions start behaving like wave functions in a reduced system, where you only take the parameters on which those partition functions depend and you turn that into a quantum mechanical space. In other words, this is a way in which quantum mechanics emerges from an underlying description. I can even imagine, in some cases, that this underlying description may have different rules than the normal rules of quantum mechanics. So I think that quantum mechanics may even be derivable from underlying formulations, and we see examples of that in string theory.

What kind of examples do you have in mind?

The most well-known examples in fact predate string theory; for example, there is a relationship between partition functions of CFTs and wave functions in Chern-Simons theory [11]. Other examples are found in the context of topological string theories [12, 13]. People have seen that certain partition functions that they can calculate perturbatively, which also have some underlying description, eventually behave as wave functions on the moduli space of solutions. I think this can be made into a quite general statement, namely, that string theory can provide examples of quantum mechanical systems that don't exist microscopically but somehow can emerge at the microscopic scale. So I think that for every formulation that we so far have, like quantum mechanics, there may actually be something underlying it and that string theory can help us find specific examples of this.

One of the criticisms that people make of string theory is that it is not a background-independent formulation. Do you think this criticism points towards something unsatisfactory within string theory?

Yes, I agree that that is unsatisfactory and this is why I think string theory is also not finished or may even not really be a full description; that is, there may be something underlying it. If you want to claim that you have a complete, consistent theory which includes quantum mechanics and gravity, then there has to be a way in which you can understand how the background emerges from something else. Thus, assuming a classical background or some curved space from the very beginning can never be a full formulation of the theory. This is clear. So, I agree that if we claim that this is what string theory is, in the sense that we take a string and we quantise it or we take some other object like D-branes, it is not a satisfactory formulation of the theory. In that sense, I agree with that criticism but it's not the way that I've viewed string theory.

So, how have you viewed string theory?

Well . . . again, we have to understand what string theory is and, first of all, it is not complete. I really think that you can try to explain string theory as an emergent framework from something else underlying it.

That something else . . . is that M-theory [14]?

No. M-theory, in a sense, falls under the same category as string theory; that is, it assumes 11 dimensions, supersymmetry and has these objects that we know even less about.

M-theory claims that there's one object from which we can derive all string theories [14]. Well, what I think actually happens is there's a very large class of objects from which string theory can be obtained.

Some object in F-theory [15]?

No, no, no. I think that these are all frameworks, all descriptions. M-theory, F-theory and string theory are all the same thing and there are all kinds of relationships between them but none of them should be claimed to be the underlying description. We may not find it eventually and so I think we should probably accept the fact that what we have even in string theory is not a final theory but it's one that is derived from something underlying it. As I told you before, that gives you a very different perspective regarding the kind of questions that you should try to answer since you cannot with it explain everything about our world. For instance, you shouldn't try to think about the multiverse and all the possible realisations of this theory [16, 17] because this isn't what we've been doing with quantum field theory. You should be very happy with the fact that we have this very beautiful framework in which we can try to address some of the questions about how the world looks.

I have only one worry which is that string theory is too beautiful to be realistic. Let me give you an example. When I started doing physics, I learned quantum field theory which I thought was totally ugly because you had to do many things like throwing away infinities and so on. Then, I thought that this could never be the final way of thinking about it, which I still think is true, but we understand why it is a very good description of many things and that is because we know that there's some underlying formulation which makes everything eventually finite. It turns out that you don't have to know this formulation, as Kenneth Wilson taught us [18–20], but eventually we understand why it works. Later, I discovered CFTs (see, e.g., [21]) and they're mathematically very beautiful. They are so beautiful that if you start working on them you will almost believe that the world should be a CFT. Of course, it cannot be because it doesn't contain gravity and it's a Platonic limit of a quantum field theory with many beautiful things like dualities, which are kind of magical. String theory has this character also but we know that CFTs don't really describe physics; quantum field theories do. So, we have to find the analogue of what quantum field theory is to conformal field theory for the case of string theory. String theory is like a beautiful limit of this potential analogue and has such beautiful mathematics that people started believing that it is the real world, but it's not. We have to think in a broader context and I really believe that something more ugly is probably more close to nature.

So do you think that the fact that it's not background-independent doesn't make it fundamentally wrong?

No, I think it's actually a more consistent and more powerful framework than quantum field theory as it can address questions about gravity and about black holes which quantum field theory can never do. So it's much better and it's probably going to be the final theory, in the sense of being the limit of our capacity for understanding physics that we, as human beings, will construct but it may not be the final description of the world that we were looking for.

To which situations does the formalism apply? Does it only work in situations in which quantum excitations are small compared to the background spacetime?

For many practical calculations the formalism may be sufficient but if you really want to know what's going on underlying it, it may not be sufficient. So in that sense, you're right: only when you poke the system in some limited way do we have some understanding of how to describe it. There are examples for which we have done a lot more like counting microstates of black holes and so on but this is usually done in situations with a lot of supersymmetry and where there's more control [22]. Although, I think that we can eventually understand these things qualitatively even in general situations.

What about other theories of quantum gravity that are out there? Do you think we should try to do something along those directions or do you think that they are fundamentally wrong?

I think that they're fundamentally wrong. There are certain kinds of questions regarding the effective description of general relativity and its quantum properties that need to be understood. But it can only be done in an effective way so we have to introduce a cutoff and we have to ask questions that don't depend on that cutoff. This is the usual procedure in effective field theory and now we want to ask questions about gravity. Maybe such questions can be asked in lattice-like forms of gravity, but loop quantum gravity (LQG) [23] is a misconception. What people need to realise is that, using Wilson's way of thinking [18–20], we know that field theory has its limits. There's only effective quantum field theories in general and the only well-defined field theories that we know are CFTs. Some people think that there are asymptotically safe theories [24] that may be defined rigorously but I have some doubts about that, to be honest. I think that the only well-defined theories are CFTs and also perhaps integrable but renormalisable models where you have a running of coupling constants all the way up to the UV. This does not mean that those theories which are not well defined cannot be extremely valuable in certain regimes, as in the case of the standard model.

What are your issues with asymptotically safe theories or asymptotic safety?

These theories are logically and mathematically fine. My objection is partly a matter of taste and partly philosophical, but also pragmatic. I like to have a complete understanding of all phenomena, not just an effective description of some phenomena. The problems we have with black holes and quantum mechanics related to the Bekenstein-Hawking formula give an important hint that the number of degrees of freedom is finite. Asymptotically safe theories are continuum theories with an infinite number of degrees of freedom in the ultraviolet. These theories will never be able to explain the Bekenstein-Hawking entropy. All the hints we obtain from string theory and black hole physics tell us that gravity and also all other interactions are emergent from a deeper quantum mechanical description that is not a continuum quantum field theory (QFT). Otherwise, we cannot explain the value of Newton's constant as being related to entanglement entropy (e.g., the work of Ryu-Takayanagi [25]), because in asymptotically safe theories (and CFTs) the entanglement entropy is divergent and can only be computed by introducing a finite cutoff. The new language of quantum information is much more promising in my view, and this does not work very well with continuum theories with an infinite number of degrees of freedom.

Okay, going back to one of your earlier statements, I'm puzzled with your comment that LQG is a misconception. Is this related to your intuition that gravity is emergent and we are not supposed to quantise it?

As I said, quantum field theories have their limitations because they should be viewed as effective, and general relativity is known for not being quantisable without a cutoff. What LQG people do is to provide a certain cutoff but they don't realise that it is an arbitrary choice of which there are infinitely many. So, I think that they don't fully appreciate the whole Wilsonian framework, which tells us that any effective field theory where you provide a cutoff introduces in the UV infinitely many arbitrary parameters. The only way to get rid of them is to go to the infrared regime and don't ask questions about the UV. In any case, trying to quantise general relativity by doing something to the theory which makes it finite in the UV cannot be the final formulation.

Of course string theory has similar problems because it also provides many possible parameters. However, string theory at least makes it a lot better by reducing the very large number of parameters to a much smaller set. But this it does, I think, by being an effective description of what the UV looks like.

In general, I think that string theory is a cleverer idea and better motivated than any of these other ideas, which I think are just guesses. When I wrote my paper [26] people sent me emails saying that they knew what the microscopic model for what I was advocating was. I didn't believe it and I didn't even have to look at it because the possibility that we just know what it is like is really small. The probability is just one in a googolplex, so I don't think we have any chance of finding it. I think we can find general principles but not the microscopic model. Also, I think that LQG is too contrived and it doesn't explain anything.

Since, as you said, we don't know what the final theory is, isn't there room to follow approaches such as LQG?

No, the point is that's underestimating and actually underappreciating what string theory has already done. String theory has already connected so many things in theoretical physics that it's there to stay. If you now would throw away everything that we know about string theory, you would rediscover it afterwards. It's really there, in the sense that it's like quantum field theory. It's a very powerful framework. I think that people who follow other approaches should learn more about string theory as there is no other competing formulation. There's no comparison between string theory and other formulations.

But could you argue that these other approaches are not so developed simply because there isn't a sufficient number of people developing them?

No, no, no. There's something in string theory that cannot be ignored. As a theoretical physicist, I think you should take all theoretical knowledge and arguments, methods and concepts that have been discussed and put them in your toolbox for later use to address problems. You shouldn't ignore some things because you don't like them. So string theory, in that sense, may include any other method that people are developing. If there's some validity or something useful in LQG, it will be taken in. So I think that LQG people should do the same thing with string theory.

Okay, but do you think that string theorists do this with LQG?

There has been a short period in which there was some exchange of ideas. I went to some of the seminars and saw that they were making fundamental mistakes; otherwise, I would make an effort in trying to understand what might be good about it.

Are these mistakes the ones that you mentioned?

They're ignoring the fact that there are many infinitely choices and they don't really specify why they make the choice they make.

Yes, but couldn't one claim that they simply have a very broad formalism?

There is a conceptual reason which is related to what I think one should be looking at. It is the fact that gravity has many similarities with the laws of thermodynamics [27]. This should not be ignored; in fact, it should be taken very seriously and yet LQG doesn't. What they do is like taking hydrodynamics, quantising it and ignoring the fact that there's indications that there might be atoms. We wouldn't discover atoms by quantising hydrodynamics; instead, we derive hydrodynamics from an underlying atomic formulation. LQG, in turn, is taking hydrodynamics and quantising it and that cannot be right.

So you think that they don't have the right microscopic description?

No way, no way. If I would bet money on that it would be a safe bet.

But I was a bit puzzled with the fact that after you published your paper on the origin of the laws of Newton [26], Lee Smolin published a paper stating that applying arguments in your paper to LQG it was possible to derive Newton's law of gravity in an appropriate limit [28]. If they do not have the right microscopic formulation, how could Smolin get this out of LQG?

As I said, they have to do something to the equations to be able to quantise them; otherwise, there would be infinities everywhere but you can do this in infinitely many ways. Among those ways are those that lead to something that may look like what you want to get out of it and I think they have been massaging their equations so that they get something out that looks like what you want to get. However, I don't think it's based on general principles in any way. They just used the arbitrariness in their formulation to get the results that they wanted.

So you think that, in this case, the application of these principles that you suggested in your paper is wrong?

The principle works and I wrote that paper because I recognised a principle but I recognised it on the basis of how we work with string theory. Since I realised that maybe there's something underlying string theory, I wanted to know what are the general principles. I have been extending these ideas and I'm starting to get convinced that it's very closely related to string theory, in the sense that it is really very consistent with everything we know about string theory. I'm using string theory as a sort of inspiration and as a way of checking whatever things I develop. On the other hand, LQG is not an inspiration; it's not even an example of a theory that makes these ideas work. It could be true that there are more general formulations than string theory but I think that it's an infinite class of objects. Maybe what the LQG people are doing is trying to find one of those examples but when you realise that there is an infinite number of them, the chance of finding the right one is very small, so I stop thinking about finding the right one and instead prefer thinking about general principles. This is what I think people should be doing, and anyhow, I know that string theory is certainly a much more promising framework than any of these other ones.

As you mentioned that there may be something more general underlying string theory, it could be that in these other theories there are some good ideas that could be taken in, right?

People can have good ideas also in other fields and not only in string theory, but there are some fundamental flaws in the way that these things have been done. What I particularly dislike is that there is a very big difference between string theory and these other theories. In string theory you start from some very simple idea and you find that gravity is unavoidable; that is, you get gravity out of it [29]. In the other approaches people start from the Einstein equations and they don't even wonder where they come from. I think that is conceptually wrong. Thus, any approach that starts from the Einstein equations goes in the wastebasket for me. I think this is equivalent to ignoring the fact that it looks like hydrodynamics or thermodynamics. We are not quantising thermodynamics, we are not quantising hydrodynamics and therefore you should not quantise gravity.

Okay, now I want to go a bit deeper into your paper [26]. There you claim that gravity is an entropic force. What is an entropic force and in which sense is gravity an entropic force?

An entropic force is a force that is caused by some underlying microscopics and it is a macroscopic force that you only see when you look at large enough distances or low enough energies. The microscopic explanation of the force uses very different concepts than those used at the macroscopic scale. The way to understand the force, however, is not by thinking about what is the microscopic origin but by looking at the entropy (or what I call the amount of phase space of the underlying system) and how that is influenced by the macroscopic variables.

A very well understood example of an entropic force is the molecular force arising between two atoms. If, for example, you think about having two hydrogen atoms H_2 and take a look at their nuclei, you may wonder why they are bound together. Assuming you did not know much about electrons and atoms you could describe the attractive force between them as some kind of potential. But the actual reason is that the internal nuclei influence the phase space of the electrons. The electrons are like fast variables which you can integrate out, if you wish, and obtain this effective force. As such, the actual explanation of the force is that the amount of phase space (or information) associated with those electrons is being influenced by the positions of those nuclei. So what we do in physics all the time is to forget about many degrees of freedom.

Now, suppose that the phase space of those degrees of freedom is being influenced by parameters that are left over at the macroscopic scale. Then, there will be a backreaction force and that force is gravity. Gravity puts all the information that we forgot about the microscopic world in one force. Thus, gravity is a reaction force due to macroscopic objects influencing the amount of phase space of the underlying physics. This phase space is there in our world and there all kinds of evidence for it.

What kinds of evidence?

Dark energy and dark matter, for example. As I told you, understanding what dark matter and dark energy are precisely is one of the biggest problems in theoretical physics and the fact that we don't understand them is because we've been ignoring the possibility of that additional phase space.

But how do you explain what dark energy and dark matter are exactly?

So what I can do, and this is going to appear in a coming paper [30], is that I can calculate the volume of that phase space and see how much energy is contained in it. The amount of energy you get in that volume is the dark energy. So I'm sure that there is this phase space, which I refer to as "information" in that paper, because I don't have to specify what the microscopics are. Even if the phase space is the number of quantum states or something else I can always describe it in terms of bits. So the basic principle that I've used is the principle that if you influence that amount of information there's a force and I'm claiming that this force is gravity. The most basic form of entropy you can think of is information so this is an entropic force.

This for me is really an eye-opener in the sense of "Wow, this is the principle, this is really why gravity has something to do with thermodynamics." Why do the Einstein equations look like the equations of thermodynamics? Why is there entropy? It is there because it generates a force. You cannot ignore the microscopics. What makes gravity so special is that if you calculate the horizon area of a black hole, you will get the full microscopic phase space. Why would that be so? Gravity is a macroscopic force and it only exists at very macroscopic distances so how would it know about all these microscopics? There's an explanation for it, namely, because the microscopic phase space creates a force.

I'm turning the logic around and it suddenly becomes clear. That's the content of the paper [26]. This is why I wrote it in very simple equations because the concepts were important. I think that this point of view had been ignored because people wanted to have complicated equations but these types of concepts should have been studied much more than what people did. I think that there are lots of consequences when you take these principles seriously.

Okay, but something that I did not understand is how you distinguish between dark energy and dark matter. Do you know the answer?

I know some answers and I have done some calculations, although it's a subtle question, but let's go into the science. The point is that dark energy has to do with the leading answer, in the sense that you assume that you have full equilibrium at some given temperature and you look at what is the average energy everywhere. That average is dark energy. If you have a finite system and place it at some temperature there will also be fluctuations, so there are certain regions where it can be more dense and other regions where it can be less dense. If you accept the fact that the universe is described by a background which is de Sitter spacetime with some cosmological constant then what one has to do is to find some way of describing these fluctuations. Dark matter is a product of these fluctuations.

There's even a way in which ordinary matter can also be seen as stuff that's sort of out of equilibrium. First of all, a system like a black hole is also a system that's in equilibrium; that is, it has a certain temperature and behaves thermodynamically but when you start looking near the horizon you see that the temperature is causing fluctuations and even particles are coming out from it. The black hole is evaporating and this is the stuff that's coming out from the heat bath and going out of equilibrium. The force that tries to get it back is a force that's trying to bring the system back to equilibrium. Again, that force is gravity. It's the ordinary matter that we see and our laws of physics are describing ordinary matter but we're ignoring all of this phase space, which is actually the way we understand the counting of black hole microstates in string theory; that is, there is a very big phase space associated with black holes.

If you want to understand where the particles come from you have to reduce that phase space to a much smaller one where you think about individual particles moving on that space but there is a transition from one to the other. There is a region near the horizon where these thermal effects become important and where ordinary matter starts falling into the black hole. This is what gravitational collapse is about. It's basically the fact that these particles are becoming part of the heat bath that you started with. This creates some layer in which ordinary matter starts interacting with the heat bath. Our universe is filled with a heat bath; namely it's a de Sitter spacetime with a given temperature. This heat bath, as I mentioned, is dark energy. Ordinary matter is immersed in it but it's interacting with it and this intermediate layer is dark matter. Dark matter plays the role of what's in between ordinary matter and dark energy.

One thing that we don't understand about galaxies is how the rotation curves of stars can be explained using just Newtonian mechanics. Here, people have to assume the existence of dark matter in order to explain it, but if you accept the fact that the space outside the galaxy is not really empty, as it's where dark energy is living, then there may even be a phase space associated with that. So I think that dark energy, dark matter, ordinary matter and the way gravity emerges between them are all part of the same phase space. We know that 70% of the universe is roughly dark energy and that there is 4% of ordinary matter. One idea I have is that if I say that dark energy is the stuff that's in equilibrium and that ordinary matter is the stuff that is mostly out of equilibrium, then I can draw a Gaussian curve and with one or two sigma in deviation, I get the numbers for galaxy rotating curves approximately right. One needs to understand much better what is dark energy, dark matter and ordinary matter and to realise that our laws of physics are based only on the tail of this Gaussian distribution. Thus, it's no wonder that we don't fully understand the rest.

Okay. Since gravity emerges from microscopics, if I zoom to smaller scales, I do not see gravity anymore?

If you zoom in you begin explaining gravity in other ways. It would be an effective description of some other phenomenon that is going on. It doesn't mean that things don't fall anymore. Its analogous to the molecular force that I've explained earlier. It's the difference between describing the molecular force by it's own effective potential or by the way that you influence the electrons. The force is there; its. just a different way of describing the force. There is no fundamental force that binds two nuclei into a molecule but the force is there.

What are the general principles that you used to derive the laws of Newton in this entropic formulation?

I used this principle that there can be a force when there is a change in information or entropy. Then we also have learned that the amount of information that you can associate with a certain volume cannot be infinite; it must be finite. Furthermore, if you also accept that this information is not stored within the volume but it is actually in the boundary of the volume and that spacetime itself can be reconstructed from that information (that is, if you accept the holographic principle) and that information changes when you bring particles closer to each other, then you can explain gravity. Just to be clearer, gravity is a force but it's very closely related to what we call inertia, in the sense that it acts on the mass of an object and so there's no difference between the gravitational force and the force that you feel when you accelerate. Thus, the way to explain this from these principles is to note that the amount of information associated with a certain object, when you displace it, is proportional to its mass. If that's the case, then you can explain that every object falls in the same way because if you split two objects into two parts, each of them really carries the same fraction of the force, and they will fall with the same acceleration. If you understand that concepts such as spacetime that we have introduced to deal with macroscopic objects may not be inherent concepts microscopically but have instead been derived from microscopic physics, then you have to start deriving also the laws of Newton.

These laws must come from somewhere, even $F = m\vec{a}$. I think that the principles that I've just described allow you to understand this.

In string theory there are many indications that we should forget about the classical notion of spacetime. As I mentioned earlier, when we talked about background independence, if you want to understand gravity and background independence and all those things you have to start from a formulation in which spacetime itself is not the first thing you start from but it's something that you derive. The holographic idea that from things happening on the boundary one can reconstruct the bulk physics has certainly been incorporated in string theory and there we see already this other scale dimension emerging. In fact, I think that in a complete picture, all dimensions should emerge. Then you also can ask, "What causes the force in that direction?" because that direction doesn't exist per se, as it is emergent. However, if you look at the formulation where you have that direction there, you can create a force in that direction. But what will be the explanation of that force using a description in which that direction isn't there? What you find is that entropic considerations can explain the existence of forces in emergent directions. There's a certain way in which you can say that you are displacing an object in that direction, namely, that the amount of entropy or information is changing and that there's a force resulting from that change.

Okay, but I'm confused here. How does the holographic principle enter here precisely?

If you want to explain why gravity behaves as $1/r^2$ there is some way in which the holographic principle plays a role because in order to get Newton's law of gravity you must assume that the information is proportional to the area of a region of spacetime. Then, you can immediately get out the $1/r^2$ law, given that the area scales like r^2. In the end, Netwon's law comes out quite naturally.

What is the holographic principle, then?

The holographic principle is the idea that the amount of information that you can put inside a certain volume cannot exceed the area of the surrounding surface. If you think about the information in this room then eventually everything that happens inside this room can be captured by the bits that I put on the walls and not inside.

Is there any evidence for believing this?

Yes, gravity. I've turned around the logic but I think that the evidence is pilling up, in particular since it reproduces equations that we know about. I think that we have been making mistakes in the way we use quantum field theory or the standard model; that is, we believe that every particle is also associated with a field and that this field can vary arbitrarily throughout space. However, this makes you run into problems such as diverging results or the calculation of the vacuum energy that happens to yield a result which is 10^{20} times larger than the one being observed. So there's something wrong with the idea of having information spread over the entire volume. This point had been made already by Gerard 't Hooft almost 25 years ago [31]. So it's clear that our usual formulations of physics assume too many degrees of freedom and too much information. The amount of information must be smaller and gravity tells us how much information there is. So what I did in my paper [26] was to turn this around; that is, I assumed a given amount of information in accordance with the holographic principle and then I got gravity out of it. For me, this is some evidence that this principle works. This does not explain the principle itself but eventually we have to understand it because one of the consequences of what I just said is that if we understand the holographic principle then we also understand gravity.

If gravity is not fundamental, how would this be related to string theory? Aren't closed strings fundamental?

Closed strings are not fundamental. One of the properties of string theory, which is not sufficiently appreciated, is that closed strings moving throughout space (that is, the closed string propagator) can be viewed as a one-loop open string amplitude. This is because the same diagram (the cylinder) can be thought of as either an open-string one-loop amplitude or a closed-string propagator. So the classical effects, which is the closed-string propagator, are derived from quantum effects in the open string. Of course, you may say that you can look at the same diagram in one way or the other. For instance, if it was a very long tube then it must be a closed string and if it's a very short one, you can think about it as an open string. However, there's some point where you have to make the transition from one to the other but this is an arbitrary choice. I can choose to do it at a very long distance scale and maybe really insist that a closed string that goes from here to the moon is actually described by an open string which I integrate out. So, string theory has ways of describing gravity without using closed strings.

Is this clear or is just an idea?

It's clear in string theory because AdS/CFT and all those examples where we have dualities between gravitational theories and theories without gravity come about due to this open/closed string correspondence. So string theory has these ideas incorporated in it and hence the closed string is the result of integrating out certain degrees of freedom. This is where we see this entropic way of thinking at work; that is, closed strings incorporate the phase space of the open strings. So, actually, it should already be clear that a formulation which contains closed strings somehow must be an emergent formulation and cannot be the full fundamental description. Many people say that there is a duality between those formulations, as if they're totally equivalent, but we don't have a full definition of string theory that contains closed strings. In fact, we have never found such formulation.

What do you mean?

There's no formulation of string theory which is non-perturbative, complete and at the same time contains closed strings. We don't have that.

But is there a non-perturbative formulation just with open strings?

Yes, for certain situations, like for $N = 4$ SYM. It's a decoupling limit but it is one of the conformal field theories [9], which, as I told you earlier, are the only field theories that are well defined. They have a Hilbert space, they have operators and you can do everything you want and test them [32]. So when open strings are decoupled from closed strings, which means that there's no gravity in the formulation anymore, there may be complete formulations of the theory and AdS/CFT is an example. Then, you can relate this conformal field theory to a gravity theory.

Do you take AdS/CFT for granted? It has not been rigorously proven, has it?

I don't think that it makes sense to try to prove it beyond reasonable doubt. I think that it's already established.

But there's no mathematical rigorous proof, right?

No, but the point is that the principles from which it is derived are clear. What is not clear is what the gravity theory really is and how to understand it from the conformal field theory side. However, the fact that there is a link between conformal field theories and gravity in one extra dimension has been established beyond any doubt. I think that for a physicist it should be sufficient. I don't think that a mathematical proof of it is really important. When research in string theory began to be developed more seriously, people thought it was very important to prove that string perturbation theory was finite.

Yes and that has not been proven yet . . .

Well . . . Nathan Berkovits has been getting pretty close [33] and there have been earlier claims of it by Stanley Mandelstam (though unpublished) but people eventually lost their interest because, in a way, string theory had already gained enough credibility. I think that it's much more important to understand how to relate what we're doing with real physics; I mean, there are questions out there that we want to study instead of having to work out mathematical proofs of things that we already believe to be true anyway.

Recently, you have been awarded a big ERC grant. What are you going to do with it?

The number of things I can do with it is not infinite. I can hire graduate students, I can hire postdocs and maybe a little more. Of course, I'm going to continue with the research I started with. It's going to be concentrated on these ideas about gravity and cosmology and so on. So I have the possibility of attracting students and postdocs working on that for the next five years.

Life seems to be hard for a high-energy physicist as there are few jobs and it's very difficult to get them. Do you think there's something wrong with this academic system?

This has always been going up and down over the years. I think it's wrong when it happens that people who are very good can't easily see how their career will go on and have to move a lot between countries. But that there's competition is also good. I don't think we want to have too many scientists because I already feel that the field is producing lots of papers, many of which may be called irrelevant. So, I would not be in favour of too many jobs. The level and the quality of the science that is produced can even go up if there are fewer positions. By putting more money into a field you do not necessarily obtain more discoveries. Thus, young people have to be competitive, think that they are the best, do work and be good enough so that people will recognise their qualities and offer them jobs. This has always been the way. The question is whether four years of PhD is enough time to be prepared for it. I think it is if you are really focused and work hard. I think a young person has to realise that those are their most important years and it's like when you practice to become an athlete and want to go to the Olympics. These are some hard years of hard training and if you don't do it that way you won't qualify. It's just a top sport in some sense and it should be viewed and kept as such.

What has been the biggest breakthrough in theoretical physics in the past 30 years?

I think that the ideas that that came out of black hole physics like understanding the Bekenstein-Hawking entropy and how it appears in the context of string theory [34] is to me an amazing breakthrough. The fact that we first thought that quantum mechanics and gravity could not be combined and now we understand very deep links between them is one of the biggest achievements. I think that we live in very exciting times, almost as exciting as the times when people were developing quantum mechanics.

Is this string theory that you are referring to?

No, it's not. It's the link between gravity and quantum mechanics that came out of it. We will probably be discovering things which are eventually more general than what string theory is. AdS/CFT and the link between quantum mechanics and black holes was how these ideas appeared in string theory. However, I think these ideas are more general than string theory. From 1995 to 1997 we learnt so much about the relationship between gravity and quantum mechanics [9, 34–37]. These were such big discoveries, which I think we are still trying to understand what we really learnt from them. These papers paved the way toward some framework more general than AdS/CFT. One thing is certain: we've so far learnt quite a lot.

Why did you choose do physics and not something else?

I was intrigued by it and I think it is a wonderful science, in the sense that it has the beauty of mathematics and covers part of the reality that you see out there. I also thought about going in the direction of molecular biology and DNA physics. When I was young, I thought that those were the directions that one should go into. I also liked pure mathematics but physics to me was more playful and the challenges of thinking about what nature is and discovering something leads to some euphoria that I don't think you can get in other areas as much. There is something about understanding nature that really captures me. Plus, it's a fun science. Necessarily, the people in it are smart and you can travel and meet very interesting people. It's a nice profession.

What do you think is the role of the theoretical physicist in modern society?

I think that human beings since they started evolving questioned about the origin of the world. Actually, I like to use a cartoon by Sidney Harris precisely illustrating this when someones asks me this question.[1] Also, I think the general public likes this kind of knowledge. One of my students said that scientists should be in the entertainment industry, from an economic perspective. What he meant is that we should also bring our discoveries out and tell them to the general public. People will be entertained because they are interested in these kinds of questions. I think society as whole is eager to know about these discoveries. Also, our curiosity makes us discover things that lead to progress in our society in many directions and I think that without science we would not be here right now. So I think we have the duty to think about these things and I think that we should explain it to the general public. Spreading our knowledge in a wider context and making people aware of what is going on is one of our purposes. Many people are fascinated by these ideas even though they can't work them out themselves.

References

[1] S. Dimopoulos and H. Georgi, "Softly broken supersymmetry and SU(5)," *Nuclear Physics B* **193** no. 1 (1981) 150–162. www.sciencedirect.com/science/article/pii/0550321381905228.

[2] L. Brink, J. H. Schwarz and J. Scherk, "Supersymmetric Yang–Mills theories," *Nucl. Phys. B* **121** (1977) 77–92.

[3] L. Brink, O. Lindgren and B. E. W. Nilsson, "The ultraviolet finiteness of the N = 4 Yang–Mills theory," *Phys. Lett. B* **123** (1983) 323–328.

[4] L. Alvarez-Gaume and D. Z. Freedman, "Geometrical structure and ultraviolet finiteness in the supersymmetric sigma model," *Commun. Math. Phys.* **80** (1981) 443.

[5] P. Fayet, "Fermi-Bose hypersymmetry," *Nucl. Phys. B* **113** (1976) 135.

[6] N. Seiberg and E. Witten, "Monopoles, duality and chiral symmetry breaking in N = 2 supersymmetric QCD," *Nucl. Phys. B* **431** (1994) 484–550, arXiv: hep-th/9408099.

[7] S. Ferrara and B. Zumino, "Supergauge invariant Yang–Mills theories," *Nucl. Phys. B* **79** (1974) 413.

[8] R. G. Leigh and M. J. Strassler, "Exactly marginal operators and duality in four-dimensional N = 1 supersymmetric gauge theory," *Nucl. Phys. B* **447** (1995) 95–136, arXiv:hep-th/9503121 [hep-th].

[9] J. M. Maldacena, "The large N limit of superconformal field theories and supergravity," *Int. J. Theor. Phys.* **38** (1999) 1113–1133, arXiv:hep-th/9711200.

[10] N. Arkani-Hamed, S. Dimopoulos and G. Dvali, "The hierarchy problem and new dimensions at a millimeter," *Phys. Lett. B* **429** (1998) 263–272, arXiv:hep-ph/9803315.

[11] E. Witten, "Quantum field theory and the Jones polynomial," *Commun. Math. Phys.* **121** (1989) 351–399.

[12] M. Bershadsky, S. Cecotti, H. Ooguri and C. Vafa, "Holomorphic anomalies in topological field theories," *AMS/IP Stud. Adv. Math.* **1** (1996) 655–682, arXiv:hep-th/9302103.

[1] See carton in https://claesjohnsonmathscience.wordpress.com/2011/12/12/dark-age-of-physics/.

[13] H. Ooguri, A. Strominger and C. Vafa, "Black hole attractors and the topological string," *Phys. Rev. D* **70** (2004) 106007, arXiv:hep-th/0405146.

[14] E. Witten, "String theory dynamics in various dimensions," *Nucl. Phys. B* **443** (1995) 85–126, arXiv:hep-th/9503124.

[15] C. Vafa, "Evidence for F theory," *Nucl. Phys. B* **469** (1996) 403–418, arXiv:hep-th/9602022.

[16] M. R. Douglas, "The statistics of string/M theory vacua," *JHEP* **05** (2003) 046, arXiv:hep-th/0303194.

[17] L. Susskind, "The anthropic landscape of string theory," arXiv:hep-th/0302219.

[18] K. G. Wilson, "The renormalization group and strong interactions," *Phys. Rev. D* **3** (1971) 1818.

[19] K. G. Wilson, "Renormalization group and critical phenomena. 1. Renormalization group and the Kadanoff scaling picture," *Phys. Rev. B* **4** (1971) 3174–3183.

[20] K. G. Wilson and J. B. Kogut, "The renormalization group and the epsilon expansion," *Phys. Rept.* **12** (1974) 75–200.

[21] P. Francesco, P. Mathieu and D. Senechal, *Conformal Field Theory*. Springer, 1996.

[22] I. Mandal and A. Sen, "Black hole microstate counting and its macroscopic counterpart," *Nucl. Phys. B Proc. Suppl.* **216** (2011) 147–168, arXiv:1008.3801 [hep-th].

[23] A. Ashtekar and J. Lewandowski, "Background independent quantum gravity: a status report," *Class. Quant. Grav.* **21** (2004) R53, arXiv:gr-qc/0404018 [gr-qc].

[24] S. Weinberg, "Ultraviolet divergences in quantum gravity," in *General Relativity: An Einstein Centenary Survey*, S. W. Hawking and W. Israel, eds., pp. 790–831. Cambridge University Press, 1979.

[25] S. Ryu and T. Takayanagi, "Holographic derivation of entanglement entropy from AdS/CFT," *Phys. Rev. Lett.* **96** (2006) 181602, arXiv:hep-th/0603001.

[26] E. P. Verlinde, "On the origin of gravity and the laws of Newton," *JHEP* **04** (2011) 029, arXiv:1001.0785 [hep-th].

[27] T. Jacobson, "Thermodynamics of space-time: the Einstein equation of state," *Phys. Rev. Lett.* **75** (1995) 1260–1263, arXiv:gr-qc/9504004 [gr-qc].

[28] L. Smolin, "Newtonian gravity in loop quantum gravity," arXiv:1001.3668 [gr-qc].

[29] J. Scherk and J. H. Schwarz, "Dual models for nonhadrons," *Nucl. Phys. B* **81** (1974) 118–144.

[30] E. P. Verlinde, "Emergent gravity and the dark universe," *SciPost Phys.* **2** no. 3 (2017) 016, arXiv:1611.02269 [hep-th].

[31] G. 't Hooft, "On the quantum structure of a black hole," *Nucl. Phys. B* **256** (1985) 727–745.

[32] N. Beisert et al., "Review of AdS/CFT integrability: an overview," *Lett. Math. Phys.* **99** (2012) 3–32, arXiv:1012.3982 [hep-th].

[33] N. Berkovits, "Finiteness and unitarity of Lorentz covariant Green-Schwarz superstring amplitudes," *Nucl. Phys. B* **408** (1993) 43–61, arXiv:hep-th/9303122 [hep-th].

[34] A. Strominger and C. Vafa, "Microscopic origin of the Bekenstein-Hawking entropy," *Phys. Lett. B* **379** (1996) 99–104, arXiv:hep-th/9601029.

[35] J. Polchinski, "Dirichlet branes and Ramond-Ramond charges," *Phys. Rev. Lett.* **75** (1995) 4724–4727, arXiv:hep-th/9510017.

[36] T. Banks, W. Fischler, S. H. Shenker and L. Susskind, "M theory as a matrix model: a conjecture," *Phys. Rev. D* **55** (1997) 5112–5128, arXiv:hep-th/9610043.

[37] R. Dijkgraaf, E. P. Verlinde and H. L. Verlinde, "Matrix string theory," *Nucl. Phys. B* **500** (1997) 43–61, arXiv:hep-th/9703030.

35

Steven Weinberg

Jack S. Josey-Welch Foundation Chair in Science and Regental Professor, and Director of Theory Research Group at the University of Texas at Austin

Date: 22 July 2011. Location: Austin, TX. Last edit: 9 April 2017

What are the main puzzles in theoretical physics at the moment?

Well, you use the word "puzzle," which implies not just an open question but something we know but about which we feel frustrated because we don't understand it. There are certainly a number of such frustrations. One of the most obvious is something that we've looked at for around 30 or 40 years. We've been looking at the spectrum of the quark and lepton masses and they're really weird. The top quark is so much heavier than the up quark and the τ lepton is so much heavier than the electron. There seems to be some kind of pattern in the mixing angles but nobody can really tell what it is. Now, we have also been looking at neutrino masses and neutrino mixing angles, which are similarly puzzling. So it's not a new puzzle; it's just something that we keep looking at and not understanding.

A very big puzzle, even more puzzling in some sense, is the hierarchy problem, that is, the fact that we know that there are vastly different scales in physics, not just the difference between the electron and the τ lepton, which is just a factor of a thousand or so, but the difference between all of these masses and either the Planck scale or the scale at which the couplings come together according to renormalisation group theory. There are many suggestions about how to deal with this. Most of them involve some kind of slowly changing coupling which becomes strong at low energy and sets the scale of the low-energy masses. The classical example where this behaviour takes place is quantum chromodynamics. In this case, the fact that the mass of the proton is much smaller than the Planck mass is not surprising because the proton mass is set by the energy scale at which the strong coupling becomes strong and the coupling becomes strong very slowly as you go to lower energy. However, we don't know how to implement that in detail in other cases such as for the cases of the W and Z mass, the Higgs mass and so on. In fact, if the LHC discovers the Higgs boson, as most people expect, it will be a big mystery why its mass is only somewhere between 100 and 200 GeV and not 10^{18} GeV.[1] That's a huge problem.

[1] Indeed, the Higgs boson was discovered at LHC in 2012 [1] with a mass of approximately 125 GeV.

Do you think that gravity plays a role in solving any of these problems?

Well, it certainly plays a role in the hierarchy problem. I would imagine so. However, with the problem of the quark and lepton masses, I don't know.

So do you think it's necessary to have a consistent theory of quantum gravity to ...

Well (laughs), it's certainly desirable but we've made a lot of progress in physics without it. In a sense, we have a perfectly good quantum theory of gravity. It's just an effective theory with all possible generally covariant terms in the action that can be used to generate a perturbation theory in powers of the energy divided by the Planck mass. All infinities can be removed by renormalisation in that theory at every order in perturbation theory and only a finite number of renormalisations are necessary. The trouble with that theory is that it loses all its predictive power when you get to very high energies. It's not that it becomes inconsistent; it's just that you are expanding in powers of something larger than one, which is never a good idea. It's just like the fact that the theory of soft pions, which is also a non-renormalisable theory, loses its value when you try to calculate things at energies above a GeV. For quantum gravity, that energy is not of the order of magnitude of a GeV; instead, it's something like the Planck scale.

You mentioned that we had a satisfactory consistent theory that incorporates gravity. Which theory are you talking about exactly?

I'm not being specific. If, for instance, you want to write down a theory of the interaction of gravity with itself and the fields of the standard model, you just write down the most general Lagrangian you can imagine containing all possible terms consistent with the symmetries, including general covariance. That has an infinite number of terms. At any finite order in powers of energy divided by the Planck mass, there are only a finite number of terms. So, it's a good theory, which at lowest order is just the standard model. Gravity appears in it as a very weak coupling at low energy. So, the theory agrees with general relativity, in the context where general relativity works, that is, at astronomical scales. However, that theory becomes useless at very high energies. It's not a very specific theory but it's the most general possible theory. This is the way effective field theories work.

So what do you think of attempts like string theory?

Usually, you hope to find an underlying theory behind these effective field theories that works at all energies. For example, for the theory of soft pions, the theory underlying the effective field theory is quantum chromodynamics, which remains valid at all energies. It may not actually describe nature at the Planck scale because there are other things involved, but it remains mathematically consistent at all energies. We don't have a theory like that for gravity. String theory may be that theory and it is the best candidate right now for an underlying theory.

Okay. What about the proposal that you made some time ago that gravity could be an asymptotically safe theory [2–5]?

Well, I think it's a good idea to keep an open mind. It's certainly another possibility that, in fact, you don't need all these enormous complications of string theory and that quantum gravity, although not asymptotically free like quantum chromodynamics (i.e., the dimensionless couplings don't go to zero as you go to high energy), could be asymptotically safe (i.e., this infinite number of couplings all go to a fixed point). That's a possibility and we don't know whether nature is that way. There have been a number of studies since I made that original suggestion, mostly in Europe. So there are a number of calculations (see review [6]) that show that, in simple models, where you don't include all possible terms but only a few terms beyond the usual Einstein term in the Lagrangian, and maybe some matter fields, there is a fixed point. Not only that, but very importantly – and this is something I only learned in the last couple of years, which got me interested in the subject again – these calculations show that the surface of the trajectories that are attracted to these fixed points apparently has a dimensionality that doesn't change as you include more and more interactions in the model. It seems to be a three-dimensional surface in a space which may have eight or nine coupling constants. That is good evidence that it could be a useful theory because, obviously, as you include higher and higher interactions, if you needed more and more free parameters, it really wouldn't be very useful. However, if the trajectories that approach the fixed point form a finite-dimensional manifold in the infinite-dimensional space of all possible interactions, then you have a reason for taking these theories seriously. It's just like renormalisation theory in the conventional sense.

You might ask: Why is the Lagrangian of quantum electrodynamics so simple? Why don't you include, for instance, what's called the Pauli moment, that is, an arbitrary magnetic moment for the electron? The answer is that the theory wouldn't be renormalisable. Renormalisable theories pick out a finite-dimensional subspace of the space of all possible theories that are allowed by symmetries. That's a wonderful thing about renormalisation theory. I mean, that's the thing that originally, when I was a graduate student, got me excited about renormalisation theory, namely, that it was a rationale for simplicity.

Well, if you have a finite-dimensional surface of trajectories that are attracted to an ultraviolet fixed point, you have again a rationale for simplicity. It may not be simple, in the sense that maybe you cannot calculate things easily, but it's simple in the sense that there is only a small number of free parameters. So, I think that's the exciting thing about asymptotic safety and it has only emerged in the last few years through these calculations made in Europe. But even so, I find string theory very attractive and if I had to bet my life I would bet on string theory rather than on asymptotic safety. I think that both are directions that should be pursued and I don't have any good reason for choosing one over the other.

You mentioned that you could not include the Pauli moment in quantum electrodynamics because it would not be renormalisable. At the same time you have stated that renormalisable theories are as good as non-renormalisable theories [5]. What does this mean?

I intended it to be an aphorism. The narrow definition of a renormalisable theory is one in which all infinities can be eliminated by renormalising a finite number of couplings. Quantum electrodynamics and quantum chromodynamics are renormalisable in the narrow sense. A theory is renormalisable in the broad sense if all infinities can be eliminated by renormalising couplings but there can be an infinite number of couplings. Effective field theories, where you include all possible terms in the theory that are consistent with symmetries, are renormalisable in the broad sense, although they're not renormalisable in the narrow sense. As far as a physical requirement is concerned, that's just as good. However, as I said before, such effective field theories become useless at very high energy.

Regarding the asymptotic safety scenario ... what is the gravity theory exactly?

Well, there isn't any one theory. People have tried a number of different examples. Of course, these calculations require some truncation of the Lagrangian. So they may, for example, pick a theory of pure gravity in which you include not only the Einstein term $\sqrt{-g}R$ but also $\sqrt{-g}R^n$, where n goes up to, say, $n = 8$. That would be an example of a theory. Formally, it's not really a theory, because if you rigorously follow this philosophy you need to include all possible terms. However, it might be an approximation that could be of some value and you might also include matter fields, scalar fields, etc.

Can you include the standard model matter content?

You could but nobody has. It's too complicated but there's no reason not to do it. It's just a lot of work.

Would the theory be renormalisable in the broad sense?

In the broad sense, only if you would include all possible terms in the Hamiltonian. In practice, you have to keep only a finite number of terms and throw away infinities that would require renormalisation of higher terms. The consistency test is whether the results change radically when you include additional terms. The group in Europe has carried out these consistency tests. For example, they might include seven terms in the action and get certain numerical results and then they include eight terms and see if the numerical results change very much. Unfortunately, there are no theorems about the convergence of this procedure. Mathematically, it's completely murky. Nobody knows how these theories are supposed to converge.

Okay, but do you expect that, ultimately, this fixed point exists?

I wouldn't be surprised but I don't know. I don't have any arguments, except the extreme hand-waving argument based on dimensional continuation which is the argument I gave originally [2–5]. Namely, that in $2 + \epsilon$ dimensions you can do calculations in perturbation theory and show that there really is a fixed point where the couplings are of order ϵ and there is a finite number of trajectories that are attracted to it. Now, as you increase the dimensionality to four dimensions, you obviously expect things to change. The dimensionality of the surface of trajectories which are attracted to the fixed point changes because these critical exponents may change sign as you increase the dimensionality of spacetime, since trajectories that were repelled by the fixed point may now be attracted to it. However, you wouldn't expect that suddenly, as you increase the dimensionality of the spacetime from two to four dimensions, there is an infinite number of trajectories attracted to the fixed point. So, that's a hand-waving plausible argument, but beyond that I don't know.

And what would be the low energy limit of this theory?

The standard model plus Einstein gravity.

Just that?

Well, maybe the standard model with additional things in it. I don't know what else would be in it, maybe supersymmetry. I'm not a prophet (laughs), I don't know the answer to that.

Sure. So what do you expect to see at the LHC?

Hold on a second ... Jacques![2]

Jacques: Yes, sir.

While you were teaching quantum mechanics, did you ever think seriously about questions regarding the interpretation of quantum mechanics like Copenhagen versus many-worlds interpretation? Have you read the literature on that?

Jacques: I have read a little bit of the literature on it and have decided that, in my best interest, it would be best not to think about it (laughs).

Okay, then I won't bother you. Besides, I'm struggling with that and I finally wrote up my lecture notes on that but I think your attitude is very healthy on this particular matter (laughs). It has been my attitude throughout most of my life so I won't burden you with it.

Jacques: I can say one thing about the many-worlds interpretation. It seems to me that, though many things seem right about it, the many-worlds interpretation dodges the question when it replaces the phrase "reduction of the wave-packet" with "bifurcation of worlds," since it does not explain when that bifurcation happens. Okay?

No, it doesn't.

Jacques: The idea that there is some sharp place where that happens is exactly the problem one has with the other interpretation.

Exactly. For Niels Bohr, it is not much of an embarrassment because he accepts that the answer is not to be found in quantum mechanics itself whereas the many-worlds people think it is. Bohr can always say that "when the wave function collapses these are the probabilities, it's just an axiom" whereas the many-worlds people should find that as a result of the Schrödinger equation and there is no way to do that.

Jacques: That's right.

Well, that's fine. I think that answers the question very well (laughs).

[2] Speaking to Jacques Distler, who passed by Steven's office.

Jacques: That's it (laughs and leaves room).

Okay, so, I was asking what you thought was the most likely extension of the standard model to be seen at the LHC . . .

I don't know. I think that supersymmetry is very attractive but of course, there are problems with it. There are various versions of supersymmetry that mostly depend on how the breaking of supersymmetry is communicated to the low-energy particles that we observe in the laboratory. There's gravity-mediated supersymmetry and gauge-mediated supersymmetry, etc., and each one of these alternatives has drawbacks, such as things that seem very artificial or things that disagree with observation. So, there isn't any one supersymmetric theory that I would say is clearly the candidate that we want to test, which is a pity and makes you a little skeptical about supersymmetry, but I have an open mind.

Okay, so why do you like string theory so much?

It's very rich and very attractive mathematically. One of the things about string theory that Edward Witten has always emphasised is that it's not only a potential quantum theory of gravity but it's also a theory in which you cannot avoid the existence of gravity. To me, that's very attractive because in all other theories the following question arises: "Well, why should there be a massless spin-2 particle?" In the context of string theory, the answer is, "Because the string energy-momentum tensor is conserved." So, inevitably, there will be a massless spin-2 particle and I think that's very attractive.

So why haven't you ever worked on string theory?

I have worked on it. Back in the 1980s, I wrote several papers [7–10]. I have referred to them as papers of monumental unimportance. Well . . . many of the papers that I write have the purpose of teaching myself the subject and there are two or three such papers in string theory. I also gave a course on string theory one term. I don't work on it now probably because cosmology got so exciting that I couldn't do both and also, my mathematical training, because of my age, is not like the young theorists' training today. I didn't grow up with differential topology and so on. I tried to catch up on some of those things but for me it's much harder because I didn't learn those things when I was young. It's the same reason why I don't read Latin poetry. I didn't learn it when I was young but that doesn't mean that I don't admire the little bit that I know of it in both the case of superstring theory and the case of Latin poetry.

Since there is not really any way of figuring out which theory of quantum gravity is the right one, at least not at the moment, do you think that aesthetic reasons are a good guiding principle towards the right theory?

Aesthetic reasons are never arguments for the validity of the theory but they are arguments for taking a theory seriously enough to test its validity. I mean, you don't even bother worrying whether ugly theories are right or wrong. What do you get out of it (laughs)? On the other hand, you take theories that are beautiful very seriously and try to see whether they're right or wrong. It's possible that they're wrong, although, generally speaking, history has so far told us that nature doesn't waste beautiful theories.

Is string theory one of those theories?

Yes, I think so.

You mentioned that you now work on cosmology because it's so exciting. Could you explain why the cosmological constant problem is so difficult?

There are, in fact, two problems associated with the cosmological constant. One of the problems, which has been with us for a long time, is that theoretical physicists are aware that astronomical observations set an upper limit on the vacuum energy. That upper limit is very tiny compared to any estimate of what you would expect from fluctuations in quantum fields, in fact, smaller by many orders of magnitude. So much so that it really wasn't important whether the astronomers had gotten the number right to a factor of two or 10; I mean, the discrepancy was huge. That was the first cosmological constant problem but many people ignored it. I didn't. I gave a series of lectures at Harvard in 1988 exploring this problem [11].

Most of us thought that probably there was some fundamental principle of nature which we just hadn't discovered yet that stated that the vacuum energy is actually zero because if it's so small, why isn't it zero? We couldn't calculate it because it depends on a lot of things we don't know like the actual cosmological constant in the Einstein field equations, which could cancel any vacuum energy due to quantum fluctuations. So it's not something we could calculate but we thought that maybe there's some principle that states that it's zero. I could not find such a principle. I surveyed the literature and I concluded that nobody else had found such a principle. However, I did find one argument that said that there is one way of understanding not why it's zero but why it's small [12]. In fact, more or less as small as the astronomers tell us it is. That argument is that, in any kind of multiverse, astronomers will always find themselves in parts of the multiverse where the vacuum energy is small because otherwise galaxies couldn't form and so on [12]. So, I said, that would explain why the vacuum energy is small, though all it can do is explain why it's small enough not to interfere with galaxy formation. If it's much smaller than that, then this explanation is useless. I wrote a paper with two astrophysicists here at University of Texas, Martel and Shapiro, and we actually calculated what you would expect if that was the thing that made the vacuum energy small [13]. That was in 1998 and in that year the cosmological vacuum energy was discovered [14] and it turned out to be essentially just small enough not to interfere with the formation of galaxies. So that gave a certain encouragement to this point of view.

The second cosmological constant problem is: now that we know the value, or at least we think we know from these astronomical measurements starting in 1998, why is the cosmological constant comparable (though somewhat larger) to the mass-energy density at this moment in the history of the universe? In the distant past, it was much smaller than the mass-energy density, assuming that the mass-energy density is constant. In the future, it will be much larger than the mass-energy density because of the expansion of the universe. However, right now it's maybe three times larger than the mass-energy density. This anthropic argument does explain this because it says that the vacuum energy can't be much larger than the mass-energy density at the time galaxies formed and that the vacuum energy was larger than it is now because that was at the time when the universe was smaller. So actually, you would expect the vacuum energy to be maybe 10 times larger than the mass-energy density [12, 15]. It's actually only three times larger but, you know, what's a factor of three between friends?[3]

This anthropic reasoning has been criticised by many people. Do you think that these criticisms are well founded?

Well, it depends entirely on the existence of a multiverse. If there is a multiverse, in which constants like the vacuum energy vary from one part to another, then this reasoning is just common sense. If there isn't, then this reasoning is nonsense. We don't know.

Is there any evidence for the multiverse?

Well, the only evidence is that it appears in a natural way in some theories which are otherwise attractive theories; in particular, it appears in string theory and in theories of chaotic inflation [17].

Do you think that the required multiverse is there in string theory?

What we know in string theory is that there is a vast number of solutions and whether or not that represents actual physical parts of a multiverse, I don't know. There are a lot of papers about this. I'm not an expert on string theory so I can't judge it but some of the string theory experts think that string theory implies a multiverse and others probably don't; I don't know. Chaotic inflation theories, on the other hand, inevitably lead to a multiverse. It's attractive because it's a kind of theory in which you don't have to specify artificial initial conditions like a uniformly expanding universe.

Do you expect that it's possible to find an ultimate theory where the constants of nature are automatically determined or do you think that we will always have to use an anthropic reasoning?

The question is: Which constants? There are some constants that we know are determined anthropically like the distance between the earth and the sun. That distance used to be thought by Kepler to be a fundamental constant of nature and we now know that it's just a historical accident describing our particular solar system. This distance is anthropically

[3] This argument was further refined by Alexander Vilenkin, who succeed in predicting that it is three times larger than the mass-energy density [16].

determined within a certain range because if it were much larger or much smaller, we wouldn't have evolved to be asking about it. So now the question is: Are there other constants like that; for example, is the charge of the electron one such constant? We just don't know. I have an open mind. That is to say, my mind is open so that I don't know what the answer is but I am convinced that these are worthwhile questions.

Do you expect a final theory to be achieved by 2050?

(laughs) It could be tomorrow and it could be 2250. I don't know. Actually, from my point of view, there isn't much practical difference between 2050 and 2250.

What do you mean by that?

I won't be here in either case (laughs).

Do you think that, if this theory is found, theoretical physicists would lose their jobs?

No. Theoretical physics is potentially infinite. There are all kinds of fascinating questions that will not even be illuminated by finding a final theory. Questions about the nature of turbulence, about high-temperature superconductivity, about brain function, etc. I think it's an endless list of questions. However, there's a certain kind of physics that would come to an end, namely, the physics in a reductionist mode, which is what I and many elementary particle physicists have devoted time to. Luckily, this is not the only kind of interesting physics and it's not the only kind of mathematically interesting theoretical physics. There's lots to do.

Some people say that science cannot explain things but only describe them. If such a final theory would be found, would that be only a mere description?

I'd like to know what these people mean by "explain." I mean, if the newspaper is not delivered to your house in the morning and then you find out that the person who delivers it is ill, that explains why it wasn't delivered. You could say that it's only a description but to me, in the ordinary use of the word in English, it's an explanation and that's the way we use it in physics.

What do you think has been the biggest breakthrough in theoretical physics in the past 30 years?

So that takes me back to 1981. Well ... actually the standard model was already in pretty good shape by 1981. I think that the idea that string theory is a fundamental theory rather than just a theory of the strong interactions, if I remember correctly, was established in 1984 [18, 19]. So, is that the greatest breakthrough? Well, it is, in terms of creating the greatest amount of activity. Whether or not it has permanent value, we don't know because we don't know whether string theory describes the real world or not, but it's potentially the greatest breakthrough.

Why have you chosen to do physics?

When I was growing up, I was interested in chemistry and at a certain point, I learned that you can't understand why chemistry works the way it does without going into a deeper level, that is, the level of physics. At that deeper level, there was something called quantum mechanics that was very profound and to me it was like a young boy being invited to go to a school where you learn how to be a wizard (laughs).

What do you think is the role of the theoretical physicist in modern society?

I don't think that the discoveries of theoretical physics have much impact in society. I mean, I think that people who talk about the moral implications of the Heisenberg uncertainty principle or the theory of relativity just don't understand what they are talking about. However, I think that the fact that we can make progress in understanding nature at such a deep level is a wonderful example of what human beings can accomplish intellectually without relying on supernatural revelation.

References

[1] ATLAS Collaboration, G. Aad, T. Abajyan, B. Abbott, J. Abdallah et al., "Observation of a new particle in the search for the standard model Higgs boson with the ATLAS detector at the LHC," *Physics Letters B* **716** no. 1 (2012) 1–29. www.sciencedirect.com/science/article/pii/S037026931200857X.

[2] S. Weinberg, "Ultraviolet divergences in quantum gravity," in *General Relativity: An Einstein Centenary Survey*, S. W. Hawking and W. Israel, eds., pp. 790–831. Cambridge University Press, 1979.

[3] S. Weinberg, "What is quantum field theory, and what did we think it is?," in *Proceedings of the Conference on Historical Examination and Philosophical Reflections on the Foundations of Quantum Field Theory*, March 1–3, 1996, T. Y. Cao, ed., 241–251. Boston, 1996.

[4] S. Weinberg, "Living with infinities," 2009. arXiv:0903.0568 [hep-th]. https://inspirehep.net/record/814639/files/arXiv:0903.0568.pdf.

[5] S. Weinberg, "Effective field theory, past and future," *PoS CD* **09** (2009) 001, arXiv:0908.1964 [hep-th].

[6] M. Reuter and F. Saueressig, "Quantum Einstein gravity," *New J. Phys.* **14** (2012) 055022, arXiv:1202.2274 [hep-th].

[7] S. Weinberg, "Coupling constants and vertex functions in string theories," *Phys. Lett. B* **156** (1985) 309–314.

[8] S. Weinberg, "Particles, fields, and now strings," in *The Lesson of Quantum Theory*, edited by J. de Boer, E. Dal and O. Ulfbeck, eds. Elsevier Science Publishers B.V., 1986.

[9] S. Weinberg, "Covariant path integral approach to string theory," in *Jerusalem Winter School 1985*: 142.

[10] S. Weinberg, "Cancellation of one loop divergences in SO(8192) string theory," *Phys. Lett. B* **187** (1987) 278–282.

[11] S. Weinberg, "The cosmological constant problem," *Rev. Mod. Phys.* **61** (1989) 1–23.

[12] S. Weinberg, "Anthropic bound on the cosmological constant," *Phys. Rev. Lett.* **59** (1987) 2607.

[13] H. Martel, P. R. Shapiro and S. Weinberg, "Likely values of the cosmological constant," *Astrophys. J.* **492** (1998) 29, arXiv:astro-ph/9701099 [astro-ph].

[14] A. G. Riess, A. V. Filippenko, P. Challis, A. Clocchiatti, A. Diercks, P. M. Garnavich, R. L. Gilliland, C. J. Hogan, S. Jha, R. P. Kirshner, B. Leibundgut, M. M. Phillips, D. Reiss, B. P. Schmidt, R. A. Schommer, R. C. Smith, J. Spyromilio, C. Stubbs, N. B. Suntzeff and J. Tonry, "Observational evidence from supernovae for an accelerating universe and a cosmological constant," *The Astronomical Journal* **116** no. 3 (1998) 1009. http://stacks.iop.org/1538-3881/116/i=3/a=1009.

[15] S. Weinberg, *Dreams of a Final Theory: The Search for the Fundamental Laws of Nature*. Pantheon Books, 1992.

[16] A. Vilenkin, "Predictions from quantum cosmology," *Phys. Rev. Lett.* **74** (1995) 846–849. http://link.aps.org/doi/10.1103/PhysRevLett.74.846.

[17] A. Linde, "Eternally existing self-reproducing chaotic inflanationary universe," *Physics Letters B* **175** no. 4 (1986) 395–400. www.sciencedirect.com/science/article/pii/0370269386906118.

[18] L. Alvarez-Gaume and E. Witten, "Gravitational anomalies," *Nucl. Phys. B* **234** (1984) 269.

[19] M. B. Green and J. H. Schwarz, "Anomaly cancellation in supersymmetric D = 10 gauge theory and superstring theory," *Phys. Lett. B* **149** (1984) 117–122.

36

Frank Wilczek

Herman Feshbach Professor of Physics at the Massachusetts Institute of Technology

Date: 27 June 2011. Location: Uppsala. Last edit: 18 March 2019

In your opinion, what do you think are the main problems that theoretical physics is facing at the moment?

There are many. I think that we are still digesting the lessons of the standard model, which succeeded far beyond any reasonable expectation and is really profound. It is a very good representation of nature at a very basic level. And yet, just because it's so successful, we expect that it should be extremely beautiful and pleasing. However, if you look at it from this point of view there's clearly room for improvement. There are "coincidences" begging to be explained – for instance, the indications that the particles of the standard model can be unified into multiplets of a bigger symmetry, that gauge theories can be unified, and that the values of the couplings of the fundamental forces come together at a high-energy scale if there is low-energy supersymmetry. In addition, besides the unification of the couplings, you can also ask whether the masses get unified.

Why would the masses unify?

Well, in a unified gauge theory, in particular in the minimal $SU(5)$ model [1–3], the bottom quark and the τ lepton come together in one multiplet. So if you break the symmetry in a reasonably simple way, up at the unification scale they have the same mass. You can also be more ambitious and think about $SO(10)$ unification [4, 5], in which the top quark is also in the same multiplet. However, in this case and in the context of supersymmetric models, there is a further complication which is that the top quark and the bottom quark get masses from different Higgs doublets. So, the ratio of couplings gets mixed up with the ratio of vacuum expectation values. Those are some of the issues that we face today.

In addition we can ask whether there is a spectrum of supersymmetric particles and whether their masses also unify. Theoretically, that would enable a golden age for the program of unification and its contact with empirical work.

Haven't All these models such as $SU(5)$ and $SO(10)$ been ruled out because of issues related to proton decay [6, 7]?

No, not at all. The $SU(5)$ model, if implemented in a very minimalistic way, has been ruled out but not in general. In any case, though this point is not mentioned that often anymore, it would certainly be interesting to look harder. One thing that's encouraging in this regard is that $SO(10)$ naturally has right-handed neutrinos and the unification scale derived from other considerations also gives a pretty nice explanation of why neutrino masses are roughly what they are.

Is it not necessary to take gravity into account at the unification scales that you are talking about?

Well, it seems that calculations give a little gap of maybe a factor of 100 in scale [5] (laughs) and it would be very interesting to understand what is the explanation for that. The good news is that this gap insulates the unification from gravity. Thus, the usual assumption in these calculations is that gravity can be ignored and one hopes for the best (but if something went wrong then we'd reconsider). In the early days people laughed at these kinds of calculations because they thought it was ridiculous to extrapolate a result from 10^2 GeV to 10^{15} or 10^{16} GeV. But that's where the numbers take us, and to me it's another success of the program, and very suggestive, that the scale of unification is so close to the Planck scale – though still safely smaller, as I said.

Assuming the existence of supersymmetry?

Supersymmetry makes the unification work out quantitatively, significantly better than what you get without supersymmetry.

Do you think that supersymmetry is the most reasonable extension of standard model physics?

Yes, mainly for that reason (laughs).

But what about technicolour [8–11]? Isn't it helpful?

Well, technicolour could be helpful. It's conceptually simpler than supersymmetry since supersymmetry comes with a mathematical apparatus which is pretty cumbersome even nowadays, after years of streamlining. However, we've learned to love it, specially since field theories which have supersymmetry exhibit very special properties. Supersymmetry itself has had remarkable applications in pure mathematics and other areas. On the other hand, technicolour is just QCD all over again, changing the numbers a little bit (laughs).

The big problem with technicolour in its original version was that you have a very difficult time doing justice to the simplicity of the flavour sector, so to speak. The fermions certainly seem to behave as if they don't have the four-fermion interactions that you would expect to emerge from a strongly interacting sector responsible for electroweak symmetry breaking, i.e., technicolour. If you try to build an ambitious model of technicolour that doesn't have elementary scalars at all, then you have those problems, but if you include some elementary scalars it's a serious compromise, since the whole point was not to include them.

Nowadays, we have many precision tests of fundamental interactions. So far they've all come out dead on the standard model, without any modification. In supersymmetric models, if the particles are heavy you have some cancellations between fermions and bosons and the additional supersymmetric structure very much mirrors the structure you had before. So, the corrections aren't so big when dealing with supersymmetric models. On the other hand, if you have technicolour, unless you are very careful, your couplings may have nothing to do with the couplings we have known before and you might generate corrections that aren't there. Also, regarding unification of couplings, I think that no one would have been surprised in the early days if the calculation didn't work, since it involves a huge extrapolation. Who says that $SU(3) \times SU(2) \times SU(1)$ are the only gauge groups you have to worry about? But it works remarkably well. Technicolour complicates the task of unification, by adding a lot of new baggage.

So you think that it's not the best solution?

Yeah (laughs), to make a long story short. The only thing I can say in fairness is that I got so turned off by it after superficial study that I never made a really profound study. In addition, there are many other respectable physicists who devoted a lot of effort to technicolour or things that very much look like it, such as these large extra-dimension scenarios [12–15]. So, I wish them luck but that's not where my bets are. Actually, I don't wish them luck (laughs). I don't mind that they get tenured and so forth but I don't want nature to prove them right.

Would supersymmetry solve all the problems with the standard model?

No, definitely not. Everything I've said so far has limited scope. I've been stating that if you have supersymmetry, the unification scale should be roughly the electroweak scale, but not what the mechanism for supersymmetry breaking is, or what is heavier than what. Those details change the phenomenology drastically. For instance, supersymmetry might or might not produce a good dark matter candidate, which is an important difference (laughs). In the beginning of this conversation, you asked me about what the problems in theoretical physics were and a lot of them certainly have to do with the quest for unification.

Another important puzzle is dark matter. A lesson or hint that is coughed up by the standard model, which I think is very impressive, is the smallness of the QCD θ parameter [16].

What is the QCD θ parameter?

First of all, as background, the standard model is really impressive. If you take it at face value, you could say that it has a lot of parameters but every one of those parameters reflects something in nature. The standard model says that there should be an electron mass and there is an electron mass, and that there should be CP violation and there is CP violation. So you have this fantastic match between what the standard model says is possible and what actually happens. Some of the couplings in the standard model are bigger than others while some of them are disturbingly small, such as the Yukawa coupling (about 20^{-6}) that is responsible for the electron mass. That's a number very different from one so we have to explain it but the worst of all is that there's a possible parameter that the standard model allows involving the way gluons interact with each other in QCD [16]. That parameter is called the θ parameter. If you add this term to the standard model, you can put bounds on its value, as for the other coupling constants. Doing so, you discover that this parameter, which is an angle that a priori could be anything between 0 and 2π, is bounded by about 10^{-10}; that is, it must be really small. Presumably, that's not an accident.

Why is that bound there in the first place?

Because this new term induces time-reversal symmetry violation in the strong interaction, which would lead to things like the neutron having a significant electric dipole moment. If you introduce parameters in the standard model they are sometimes small like the electron mass but not as small as the θ parameter.

Why is the θ parameter small?

Various possibilities have been offered in order to explain why the θ parameter is so small [17] but the one that best stood the test of time and looks most attractive consists of introducing an addition to the standard model that allows you to implement a certain symmetry called Peccei-Quinn symmetry [16]. Once that is done, roughly speaking, what happens is that this θ parameter instead of being just a number that is inserted by hand, gets determined by dynamics. Then the dynamics chooses to make θ extremely small, which is very satisfactory.

Adding extra structure is adding extra quantum fields, which are associated with production or decay of new particles. It turns out that you get a lot of heavy junk but you also get a very light particle, which is the axion. It is a long but compelling story. The upshot is that axions are the potentially observable consequence of addressing this smallness of the θ parameter in a convincing way.

Do you expect to see the axion at the LHC?

No, because it interacts too weakly with matter and it's not significantly produced at the LHC. There are various constraints from astrophysics and cosmology on the properties of the axion. The net result of all the constraints is that, if the axion exists at all, it's got to be very light and very weakly interacting, which is a very unusual kind of pseudoscalar particle.

Is this why you think it could explain dark matter?

Yes. You can calculate how much axion matter gets produced in the big bang, and you find that you need to fine-tune the universe to avoid the axion being dark matter (laughs). If the axion exists, it could turn out to be the dominant component of dark matter.

Does the axion explain inflation?

It doesn't explain inflation. There have been models which attempt to use the axion itself or some of the structure that comes along with the scalar field in order to explain inflation. However, I think that these models consist of one speculation piled on another.

So when people speak of "axion cosmology" [18], what are they referring to?

They're talking about the axion as a dark matter candidate. It does get tied up with questions about inflation because how much mass density associated with axions you produce in the universe can be drastically affected by inflation. Inflation dilutes most kinds of matter to nothing but since the axion is a scalar field, its energy density doesn't get diluted by inflation, but inflation homogenises its value.

You've worked on axion cosmology quite a bit. What are the predictions that came out of these studies?

Yes, I'm obsessed with it [19–22]. The details are a little complicated, but the short version is that unfortunately axions are very hard to observe (laughs). It turns out that there are two interesting parameter ranges depending on whether inflation occurred before or after that scalar field acquired a vacuum expectation value. Hence, there are two different regions that need different kinds of experimental exploration. Currently, there is an experimental program that's going to cover one interesting parameter range and there are some recent ideas which I find quite fascinating that may enable us to explore in the foreseeable future the other interesting parameter range.

Why don't these models of axion cosmology need to include quantum gravity effects?

Very few things need quantum gravity effects (laughs). Nothing in fact (laughs). It is very hard to detect effects of quantum gravity. We don't know much about quantum gravity but it's conceivable that somehow it will affect axion production in the early universe. However, that's not part of the standard understanding.

So, in your opinion, we don't need quantum gravity at all?

We have quantum mechanics and we have gravity so of course we need quantum gravity but we don't need it for any immediate phenomenological purpose. On the contrary, no one's come up with a reasonable proposal for how you could observe any quantum gravity effect. There are some silly ideas but no serious idea so far. The hope is, because both gravity and quantum mechanics are profound theories, that by demanding that they be put together in a consistent framework you'll find that you are so constrained that you are led to ideas that have other important consequences. That's what string theory is really all about.

Do you think that string theory is the best candidate for a theory of quantum gravity available on the market?

Yes, but that's a low standard (laughs). String theory has so far held up as a consistent framework in which you can do both quantum mechanics and gravity and not run into any infinities or other obvious showstoppers. On the other hand, it's still vague. People don't know exactly what string theory is and certainly its predictions are extremely vague. I don't preclude its ultimate success, but I think it's plausible that there can be other developments that will be more illuminating about the questions that quantum gravity aspires to tackle, such as the state of the early universe or what happens deep inside black holes. This "wished-for theory" might even be an evolved form of string theory. It'd be a distilled form that uses fundamental principles that haven't been formulated within string theory but are somehow hidden there. I think that what Erik Verlinde is trying to do [23, 24] is something along these lines, that is, trying to understand quantum gravity in terms of different concepts that may be consistent with string theory but are more usable.

Do you think that the standard model is captured within the framework of string theory?

Well, you can accommodate it and there is no clear contradiction. However, it's certainly not captured in any useful detail (laughs). String theory has constructions that build many variants of the standard model with different numbers of flavour, different gauge groups, etc., and depending on how you count, there is some gigantic number of possible models (10^{500}) [25, 26], none of which is cleanly getting the standard model (laughs). They haven't all been examined by any means but it would be silly to say that string theory captures the standard model. It's more appropriate to say that string theory might accommodate it.

Has the standard model given us any hint on how to quantise gravity?

It's given us a very important indirect guidance. The fact that gravity seems to unify with the other interactions, together with the fact that we've learned quantum field theory works so well, is a clue that the graviton really is an elementary particle. Thus, general relativity, or some slight modification of it, provides the correct quantum theory of gravity up to very high scales. Those are tremendously important insights which weren't at all clear before. Earlier, it would have been very sensible to entertain the possibility that gravity is some kind of collective phenomenon or some form of composite gravitons.

Quantum field theory and Einstein gravity mesh very well together, noise to the contrary notwithstanding. You can write down the couplings between gravity and matter immediately, and in a quite restrictive manner. In this context, the minimal couplings work extremely well. That is very important information, I think, about the nature of gravity.

Thus, while attempts like that of Verlinde [23, 24] are bold and interesting, and I wish him luck, there's a very high bar they must clear before they stand comparison with the more straightforward approach of perturbative quantum gravity, which treats the graviton effectively as an elementary particle. To me, the gravitational field is just not so different from other quantum fields and we should be treating them all on the same footing.

People are now using a lot of AdS/CFT techniques to say something about QCD. Do you think this is reasonable?

It's a fine thing to try and it seems to be giving qualitative, maybe even some semi-quantitative, insight into what's happening in heavy ion collisions [27]. It's a fascinating area, still young, and it hasn't been properly digested. People are actively trying to take these lessons and bring them to condensed matter physics [28], etc. So, I'm very sympathetic to that exercise. It's definitely interesting, promising and important to pursue. One of the things that I really like is Ryu-Takayanagi's calculation of entanglement entropy [29, 30], which is really quite profound. I have spent a lot of time in my own career trying to calculate entanglement entropy and only got extremely limited results [31]. I'm very impressed that they have made a lot of progress. It's very pretty and, among other things, it gives you a vast generalisation of the black hole entropy formula. I really like that stuff (laughs).

Do you think people are losing their interest in string theory?

The fair thing to say is that the early expectation that string theory would lead to tremendous success along a broad front in a reasonably short period of time, similar to what took place in the early days of quantum mechanics or modern quantum electrodynamics or electroweak theory or QCD, pretty clearly hasn't happened and is not going to happen (laughs). Some good things have happened, but there's been no clear success in explaining any empirical fact at all, which is pretty disappointing (laughs). This is my third Strings conference. I was at the original one in Santa Barbara, which at the time wasn't even called "Strings," and I was at one in Amsterdam. The mood has changed (laughs). In the first conference there was euphoria, in the second still tremendous energy and optimism, now it's subdued.

Yes, I actually thought that this point of view was somehow conveyed by David Gross's speech in this conference.[1]

Yes, David gave a very subdued presentation (laughs). I've heard him talk on this subject over the years, many times, and to be fair, when he speaks to the public he emphasises excitement, but this was by the far the most downbeat speech I've heard from him (laughs). David is a very serious person and honest in his way, so I think it reflects reality.

Okay, let's not talk about sad things then ...

No, let's not (laughs) but just to complete the thought, I should repeat something he mentioned, and many others in the community have mentioned as well. This is that one main reason that it seems disappointing now is that expectations were so high in the early days. If you step away from those early expectations and look at what actually has been done, it's impressive.

Do other theories of quantum gravity like loop quantum gravity interest you?

I have looked at these alternatives as I've looked at technicolour. That is, they interested me enough that I took them seriously enough to look into them. After looking, I decided they

[1] This interview took placed during the conference Strings 2011.

weren't promising and I just didn't want to invest more effort into deeper study. My time is valuable enough to me that I'd rather spend it doing other things than pursuing theoretical ideas that seem to be problematic. If and when the technicolour people or the loop quantum gravity people can solve some of the obvious problems convincingly, then I'll re-examine my prejudices. Until then, life is too short (laughs).

Do you think that it's necessary to find a consistent theory of quantum gravity to explain quark confinement?

No. Quark confinement is perfectly well understood. Not only is the confinement of quarks understood [32] but also the hadron spectrum is calculated accurately in lattice gauge theory [33]. There are no mysteries there; it's just QCD. Additionally, confinement is manifest at strong coupling and there is no phase transition between strong coupling and weak coupling. I don't know what more you could ask for. Reasonable people may want to have more flexible tools that don't require massive calculations on a computer to see what's going on. From that point of view, it would be nice to have good models, based on QCD, for hadronic physics and nuclear physics. But I really think that the mysticism around confinement is totally misplaced, and that the problem of confinement is not only solved but also that we've gone way beyond proving confinement.

I always thought this wasn't well understood, given that QCD is not mathematically rigorously defined and that's why this is considered to be one of the Millennium problems?

Let me compare and contrast to another sacred theory, namely QED. QED is almost certainly not mathematically consistent. We know how to calculate quantities at lower orders in perturbation theory, but perturbation theory almost certainly does not converge [34]. Hence, you can't consistently remove the ultraviolet cutoff. So not only has QED not been proved mathematically to exist; it almost certainly doesn't exist and yet QED is very useful and by far the most accurately tested theory of physics [35, 36] (laughs). Mathematical rigour is the wrong criterion for a physical theory. QCD is much more satisfactory than QED, I would say, because of asymptotic freedom [37–40]. It's very plausible that if you compare results you get on a fine-grained lattice to the limiting case, the corrections will turn out to be quite small. In fact, there's overwhelming empirical evidence for this. Nowadays, lattice gauge theory people calculate hadron masses to great accuracy. Surely, it's not a coincidence that they get the right answer (laughs). So there is enormous numerical evidence that not only does the theory exist but that it correctly describes the strong interaction in quantitative detail. In my opinion, by the standards of theoretical physics, that's a proof and in fact it's probably the best controlled mathematical theory we have in all of theoretical physics (laughs).

What about the Higgs particle? Does it require new understanding?

Well, in the context of weakly coupled low-energy supersymmetry, the Higgs particle is just another particle (laughs). It is, however, one of the last pieces of the standard model that we want to observe, which has become such a tantalising thing because it has been the target for so long. In a real sense we have already discovered three Higgs particles, namely the longitudinal parts of the W and Z bosons (laughs). They're very well described by this mechanism of spontaneous breaking of the gauge symmetry, or more accurately, spontaneous breaking of a custodial $SU(2)$ that happens to be weakly gauged. So, this is just filling out the multiplet, which we've seen most of already (laughs).

It is often said that if the Higgs particle is found then the origin of mass is explained.

Yes, but that's ridiculous. The Higgs mechanism is the origin of the W boson mass and the Z boson mass but it's not the origin of most of the mass of you and me, which is almost all, say 95%, due to QCD. The quarks and gluons have a lot of energy and hence also the proton, whose mass is that energy divided by c^2. That's really the explanation for the mass of ordinary matter.

You have written a lot about the origin of mass, even in your book [41].

Well, that is because I think that it is a remarkable and underappreciated feature. The understanding of the origin of mass from no mass at all is historically important. Yet even professional high-energy physicists seem not to understand it (laughs). It's not controversial; it's just that people somehow haven't fully processed it. Physicists still mention the Higgs particle as the origin of mass, especially when speaking to the public. Some people know that they're oversimplifying, but other people don't seem to realise that there's a very different way of understanding the mass of ordinary matter, which is more beautiful first of all and, second, true. It's also profoundly important fundamentally because it gives a different perspective on why gravity is weak, which is one of the big qualitative mysteries about fundamental physics.

Does it give a different viewpoint on the strength of gravity or explain it?

When you say that "gravity is weak" you need to ask: Compared to what? You can compare the strength of gravity between protons to the other forces. By that measure, gravity is ridiculously weak, but its weakness is based on the mass of protons.

The mass of protons originates from QCD. There, we understand that we start with a small coupling, which has to renormalise up, but since the changes are logarithmic, that requires a big change of scale. Thus, we trace the weakness of gravity to the smallness of the proton mass, and then to asymptotic freedom. This does not give you a complete understanding, because it leaves open questions like why quarks don't have the Planck mass (or why electrons don't have the quark mass). This is related to the issue of why the electroweak scale is small (and why the Yukawa couplings are small). So understanding the feebleness of gravity is not completely settled but I would say that it is a very non-trivial part of the answer, maybe 50% of the answer.

Can you explain your idea that we live in a multilayered, multicoloured superconductor [41]?

Okay. The first thing that one should keep in mind is that the Higgs mechanism [42–44], which is used in the electroweak standard model, is really not just similar to but conceptually the same as what happens to photons inside a superconductor [45]. The equations for photons inside a superconductor look like the equations for massive vector bosons. This is because the Cooper pairs condense, acquire a vacuum expectation value and are charged. It's exactly the same mechanism, except that it is a $U(1)$ symmetry involved instead of $SU(2) \times U(1)$. Of course, the scale is very different and it happens in empty space, but superconductivity is profoundly similar to the Higgs mechanism conceptually and mathematically. Given that electroweak theory works so spectacularly well, I think that it is fair to say that we live within a kind of superconductor. If you extend that reasoning to the unified theories, you also need to have other very massive vector bosons, and it's hard to imagine that there's another essentially different mechanism that hasn't been discovered after all these years. Hence, it's probably the same mechanism, so we live in a multilayered superconductor.

Another topic that you worked on quite a bit is quantum statistics. What is an anyon?

For a long time it was thought that the only consistent possibilities for the behaviour of identical particles in quantum mechanics was that they can be either fermions or bosons. But I was never satisfied with the so-called proof of this. What I realised [46], which actually had already been realised by Leinaas and Myrheim a few years before in a widely ignored paper [47] (laughs), was that if you look at quantum mechanics in different numbers of dimensions you are led to different conclusions. Basically, in one space dimension the question is meaningless because when you try to exchange the positions of particles they have to move through each other and so you can't separate a purely intrinsic phase from an interaction. In one space dimension there exists these famous bosonisation techniques; for instance, in solid state physics you have Luttinger liquids [48, 49], which you can describe in terms of intermediate concepts. Anyway, in two spatial dimensions the careful analysis shows that there are many other possibilities for quantum statistics. In three dimensions (two spatial dimensions), you can think of the worldlines of the particles as forming a helix and you can count how many times one particle goes around another, which is a well-defined integer. In three or more spatial dimensions, you cannot define unambiguously how many times one particle has gone around another because you can always unwrap the worldlines. In these dimensions, all you can tell is whether the particles have been interchanged and that only produces a plus or a minus. However, in two space dimensions you can have a continuous parameter, which is what anyons are about.

Another way of thinking about this is in terms of vortices, which instead of line defects are like point-particles. If you have particles that are vortices and have electric charge then because of Aharonov-Bohm type phases, as one vortex goes around another you get extra phases and that's one way of realising anyons. Vector bosons could be massive so you don't get any other signature of interactions at long range except for these statistical factors and it can be generalised to non-Abelian phases, which is very interesting. Very shortly after I first thought of all this I wrote a couple of papers [50, 51] and it soon turned out that the quasiparticles in the quantum Hall effect are anyons theoretically [52].

Incidentally, there's a very thriving area consisting of something between mathematical physics and fantasy engineering, where people are talking about using non-Abelian anyons to do quantum computation because they have various advantages. For example, the interactions are long range and topological so they're not easy to disrupt and if you have non-Abelian things you can move around in a gigantic Hilbert space. So it's become a very popular topic in quantum computing and condensed matter physics, as the experimenters get serious about taking these ideas to practice.

Are anyons relevant for particle physics?

They are certainly not directly relevant to spacetime particles. In string theory, however, people are interested in things that are moving on sheets or branes of various dimensions. For instance, the mathematics of Chern-Simons theory is closely related to the description of anyons [50, 51].

Do you think that we'll be able to tell if anyons are out there?

Oh yes, they are out there, and decisive evidence could come any day now. There are powerful groups in Israel, MIT and Harvard, among others, working on it.

Will that give you another Nobel Prize?

Well, that's not for me to decide, but it wouldn't be completely inappropriate (laughs).

Have people tried to incorporate anyons in string formulations?

I don't know how hard it's been tried but it certainly hasn't had a big impact. It seems to me a natural thing to investigate, and people frequently ask me about it, but since I've never heard back, I assume that they didn't get anywhere.

You were one of the main developers of the membrane paradigm [53]. Why did this interest you so much?

I'm not sure. I've done a lot of work in my career on black holes [54–57] and I feel kind of guilty about it because it's uncomfortably removed from accessible reality. However, students really love it and time after time when I've tried to pull away my students have kind of dragged me back in (laughs). The questions are interesting, no doubt about it, and these bright young people come up with new ideas that are tantalising, which I can contribute to by bringing tools to bear, experience and expertise.

I'm not sure if you are aware of this proof of the membrane paradigm by Andrew Strominger and others [58] that appeared recently ... do you think this is a proper proof of your original ideas?

No, it's too early to tell how significant that work is. Ideas along those lines, although somewhat less precise, have been around for a very long time, in fact, since the early days of the membrane paradigm. The standard heuristic development of the membrane paradigm consists of taking the linearised Einstein equations and kind of extrapolating to a non-linear system that looks sensible. I don't really understand all the details of what they've done but what they seem to have proved is that if you make an appropriate definition which is not unnatural, you can make an exact classical mapping that's accurate in some approximation between the Navier-Stokes equation and some of the asymptotic fields on the gravity side. That's still several huge steps away from what I always envisioned in the membrane paradigm, which was some kind of effective field theory, ultimately a quantum theory, that reproduces the low-energy behaviour of black holes and the way it interacts with matter. I may be missing something but I think their work is several steps away from that goal.

In your opinion, what has been the biggest breakthrough in theoretical physics in the past 30 years?

Well, in the past 30 years it is not clear. I don't like that question very much because nothing has happened in the past 30 years which I consider to be at the same level of the commutation relations of quantum physics or the gauge principle [59]. If you would ask that same question in 1967, nobody would have said that the gauge principle in Yang–Mills theory was a breakthrough because it wasn't evident till much later. Maybe at that time the big breakthrough was thought to be the work of Gerard 't Hooft and Martinus Veltman showing that the quantum theory really existed [60]. If you're really serious about 30 years that also leaves out supersymmetry ... so I would say that it's really not clear at this point. However, if you really want to push me (laughs), though it is very early to tell, it may turn out that anomaly cancellation in string theory [61] or the AdS/CFT correspondence [62] will be considered to be major breakthroughs, but axions and anyons are not too shabby either.

Why have you chosen to do physics; why not something else?

I could easily have chosen to do something else. I think I was destined to do something involving mathematics of some kind but when I was an undergraduate I had also seriously considered trying to do some kind of neurobiology or artificial intelligence. I was also very interested in philosophy and mathematical logic. So, when I went to Princeton I started graduate school in mathematics not because I wanted to do pure mathematics since I was pretty sure I didn't but because I didn't know exactly what kind of applied mathematics I wanted to do. I was also very interested in physics and I read a lot of physics and really enjoyed the way group theory was used in physics. In particular, how directly group theory corresponded to physical reality was just like a religious experience to me. Anyway, the math building in Princeton was connected to the physics building, so it was very natural to wander over and go to seminars and colloquia in physics. It became clear to me at that time (in the early 1970s) that very exciting things were happening in physics that used the kind of mathematics that I liked. I found a very charismatic instructor, David Gross (laughs), and I just jumped in with both feet and never looked back.

In your opinion, what is the role of the theoretical physicist in modern society?

The theoretical physicist should do theoretical physics. However, at least some theoretical physicists should be concerned with sharing, both with the scientific public but also with the broader public, the marvellous things we've discovered. Theoretical physics nowadays is extremely esoteric and very few people can appreciate it without a lot of help. Also, the products of large parts of theoretical physics such as particle physics, string theory, etc., are very unlikely to have any economic benefit to mankind and to give rise to very direct technological applications in the foreseeable future. Hence, the products of these areas are really cultural – things like understanding the world better and expanding our minds. However, the value of a cultural product depends on it being part of the culture that is widely appreciated. So, I think that's an important role that some theoretical physicists should take on, and not to convey results but, even more importantly, to convey our message that honesty, cooperation and sharing of information leads to wonderful results.

References

[1] H. Georgi and S. L. Glashow, "Unity of all elementary-particle forces," *Phys. Rev. Lett.* **32** (Feb. 1974) 438–441. https://link.aps.org/doi/10.1103/PhysRevLett.32.438.

[2] S. Dimopoulos and H. Georgi, "Softly broken supersymmetry and SU(5)," *Nuclear Physics B* **193** no. 1 (1981) 150–162. www.sciencedirect.com/science/article/pii/0550321381905228.

[3] N. Sakai, "Naturalnes in supersymmetric GUTS," *Zeitschrift für Physik C Particles and Fields* **11** no. 2 (Jun. 1981) 153–157. https://doi.org/10.1007/BF01573998.

[4] H. Fritzsch and P. Minkowski, "Unified interactions of leptons and hadrons," *Annals of Physics* **93** no. 1 (1975) 193–266. www.sciencedirect.com/science/article/pii/0003491675902110.

[5] S. Dimopoulos, S. Raby and F. Wilczek, "Supersymmetry and the scale of unification," *Phys. Rev. D* **24** (1981) 1681–1683.

[6] R. Dermisek, A. Mafi and S. Raby, "SUSY GUTs under siege: proton decay," *Phys. Rev. D* **63** (2001) 035001, arXiv:hep-ph/0007213 [hep-ph].

[7] H. Murayama and A. Pierce, "Not even decoupling can save minimal supersymmetric SU(5)," *Phys. Rev. D* **65** (2002) 055009, arXiv:hep-ph/0108104 [hep-ph].

[8] S. Weinberg, "Implications of dynamical symmetry breaking," *Phys. Rev. D* **13** (Feb. 1976) 974–996. https://link.aps.org/doi/10.1103/PhysRevD.13.974.

[9] L. Susskind, "Dynamics of spontaneous symmetry breaking in the Weinberg-Salam theory," *Phys. Rev. D* **20** (Nov. 1979) 2619–2625. https://link.aps.org/doi/10.1103/PhysRevD.20.2619.

[10] C. T. Hill and E. H. Simmons, "Strong dynamics and electroweak symmetry breaking," *Phys. Rept.* **381** (2003) 235–402, arXiv:hep-ph/0203079. [Erratum: *Phys. Rept.* **390** (2004) 553–554.]

[11] K. Lane, "Two lectures on technicolor," arXiv:hep-ph/0202255 [hep-ph].

[12] M. Gogberashvili, "Hierarchy problem in the shell universe model," *Int. J. Mod. Phys. D* **11** (2002) 1635–1638, arXiv:hep-ph/9812296.

[13] L. Randall and R. Sundrum, "Large mass hierarchy from a small extra dimension," *Phys. Rev. Lett.* **83** (Oct. 1999) 3370–3373. https://link.aps.org/doi/10.1103/PhysRevLett.83.3370.

[14] L. Randall and R. Sundrum, "An alternative to compactification," *Phys. Rev. Lett.* **83** (Dec. 1999) 4690–4693. https://link.aps.org/doi/10.1103/PhysRevLett.83.4690.

[15] M. Gogberashvili, "Our world as an expanding shell," *Europhys. Lett.* **49** (2000) 396–399, arXiv:hep-ph/9812365 [hep-ph].

[16] R. D. Peccei and H. R. Quinn, "CP conservation in the presence of pseudoparticles," *Phys. Rev. Lett.* **38** (Jun. 1977) 1440–1443. https://link.aps.org/doi/10.1103/PhysRevLett.38.1440.

[17] M. Dine, "TASI lectures on the strong CP problem," in *Flavor physics for the millennium. Proceedings, Theoretical Advanced Study Institute in elementary particle physics, TASI 2000, Boulder, USA, June 4–30, 2000*, Jonathan L. Rosner, ed., pp. 349–369. World Scientific, 2000.
arXiv:hep-ph/0011376 [hep-ph].

[18] D. J. E. Marsh, "Axion cosmology," *Phys. Rept.* **643** (2016) 1–79, arXiv:1510.07633 [astro-ph.CO].

[19] M. S. Turner and F. Wilczek, "Inflationary axion cosmology," *Phys. Rev. Lett.* **66** (1991) 5–8.

[20] M. Tegmark, A. Aguirre, M. Rees and F. Wilczek, "Dimensionless constants, cosmology and other dark matters," *Phys. Rev. D* **73** (2006) 023505, arXiv:astro-ph/0511774 [astro-ph].

[21] F. Wilczek, "A model of anthropic reasoning, addressing the dark to ordinary matter coincidence," arXiv:hep-ph/0408167 [hep-ph].

[22] M. P. Hertzberg, M. Tegmark and F. Wilczek, "Axion cosmology and the energy scale of inflation," *Phys. Rev. D* **78** (2008) 083507, arXiv:0807.1726 [astro-ph].

[23] E. P. Verlinde, "On the origin of gravity and the laws of Newton," *JHEP* **04** (2011) 029, arXiv:1001.0785 [hep-th].

[24] E. P. Verlinde, "Emergent gravity and the dark universe," *SciPost Phys.* **2** no. 3 (2017) 016, arXiv:1611.02269 [hep-th].

[25] S. Kachru, R. Kallosh, A. D. Linde and S. P. Trivedi, "De Sitter vacua in string theory," *Phys. Rev. D* **68** (2003) 046005, arXiv:hep-th/0301240.

[26] L. Susskind, "The anthropic landscape of string theory," arXiv:hep-th/0302219.

[27] J. Casalderrey-Solana, H. Liu, D. Mateos, K. Rajagopal and U. A. Wiedemann, "Gauge/string duality, hot QCD and heavy ion collisions," arXiv:1101.0618 [hep-th].

[28] S. A. Hartnoll, "Lectures on holographic methods for condensed matter physics," *Class. Quant. Grav.* **26** (2009) 224002, arXiv:0903.3246 [hep-th].

[29] S. Ryu and T. Takayanagi, "Holographic derivation of entanglement entropy from AdS/CFT," *Phys. Rev. Lett.* **96** (2006) 181602, arXiv:hep-th/0603001.

[30] S. Ryu and T. Takayanagi, "Aspects of holographic entanglement entropy," *JHEP* **08** (2006) 045, arXiv:hep-th/0605073 [hep-th].

[31] M. P. Hertzberg and F. Wilczek, "Some calculable contributions to entanglement entropy," *Phys. Rev. Lett.* **106** (2011) 050404, arXiv:1007.0993 [hep-th].

[32] R. Alkofer and J. Greensite, "Quark confinement: the hard problem of hadron physics," *J. Phys. G* **34** (2007) S3, arXiv:hep-ph/0610365 [hep-ph].

[33] Z. Fodor and C. Hoelbling, "Light hadron masses from lattice QCD," *Rev. Mod. Phys.* **84** (2012) 449, arXiv:1203.4789 [hep-lat].

[34] F. J. Dyson, "Divergence of perturbation theory in quantum electrodynamics," *Phys. Rev.* **85** (Feb. 1952) 631–632. https://link.aps.org/doi/10.1103/PhysRev.85.631.

[35] S. G. Karshenboim, "Precision physics of simple atoms: QED tests, nuclear structure and fundamental constants," *Phys. Rept.* **422** (2005) 1–63, arXiv:hep-ph/0509010 [hep-ph].

[36] B. Odom, D. Hanneke, B. D'Urso and G. Gabrielse, "New measurement of the electron magnetic moment using a one-electron quantum cyclotron," *Phys. Rev. Lett.* **97** (Jul. 2006) 030801. https://link.aps.org/doi/10.1103/PhysRevLett.97.030801.

[37] D. J. Gross and F. Wilczek, "Ultraviolet behavior of nonabelian gauge theories," *Phys. Rev. Lett.* **30** (1973) 1343–1346.

[38] H. D. Politzer, "Reliable perturbative results for strong interactions?," *Phys. Rev. Lett.* **30** (1973) 1346–1349.

[39] D. J. Gross and F. Wilczek, "Asymptotically free gauge Theories – I," *Phys. Rev. D* **8** (1973) 3633–3652.

[40] D. J. Gross and F. Wilczek, "Asymptotically free gauge theories. 2," *Phys. Rev. D* **9** (1974) 980–993.

[41] F. Wilczek, *The Lightness of Being: Mass, Ether, and the Unification of Forces*. Basic Books, 2008.

[42] P. W. Higgs, "Broken symmetries and the masses of gauge bosons," *Phys. Rev. Lett.* **13** (1964) 508–509.

[43] T. W. B. Kibble, "Symmetry breaking in non-abelian gauge theories," *Phys. Rev.* **155** (Mar. 1967) 1554–1561. https://link.aps.org/doi/10.1103/PhysRev.155.1554.

[44] F. Englert and R. Brout, "Broken symmetry and the mass of gauge vector mesons," *Phys. Rev. Lett.* **13** (Aug. 1964) 321–323. https://link.aps.org/doi/10.1103/PhysRevLett.13.321.

[45] P. W. Anderson, "Plasmons, gauge invariance, and mass," *Phys. Rev.* **130** (1963) 439–442.

[46] F. Wilczek, "Magnetic flux, angular momentum, and statistics," *Phys. Rev. Lett.* **48** (Apr. 1982) 1144–1146. https://link.aps.org/doi/10.1103/PhysRevLett.48.1144.

[47] J. M. Leinaas and J. Myrheim, "On the theory of identical particles," *Nuovo Cim. B* **37** (1977) 1–23.

[48] S.-i. Tomonaga, "Remarks on Bloch's method of sound waves applied to many-Fermion problems," *Progress of Theoretical Physics* **5** no. 4 (07, 1950) 544–569, http://oup.prod.sis.lan/ptp/article-pdf/5/4/544/5430161/5-4-544.pdf. https://dx.doi.org/10.1143/ptp/5.4.544.

[49] J. M. Luttinger, "An exactly soluble model of a many-Fermion system," *Journal of Mathematical Physics* **4** no. 9 (1963) 1154–1162, https://doi.org/10.1063/1.1704046. https://doi.org/10.1063/1.1704046.

[50] F. Wilczek and A. Zee, "Linking numbers, spin, and statistics of solitons," *Phys. Rev. Lett.* **51** (Dec. 1983) 2250–2252. https://link.aps.org/doi/10.1103/PhysRevLett.51.2250.

[51] D. P. Arovas, J. R. Schrieffer, F. Wilczek and A. Zee, "Statistical mechanics of anyons," *Nucl. Phys. B* **251** (1985) 117–126.

[52] D. Arovas, J. R. Schrieffer and F. Wilczek, "Fractional statistics and the quantum Hall effect," *Phys. Rev. Lett.* **53** (Aug. 1984) 722–723. https://link.aps.org/doi/10.1103/PhysRevLett.53.722.

[53] M. Parikh and F. Wilczek, "An action for black hole membranes," *Phys. Rev. D* **58** (1998) 064011, arXiv:gr-qc/9712077 [gr-qc].

[54] S. R. Coleman, J. Preskill and F. Wilczek, "Growing hair on black holes," *Phys. Rev. Lett.* **67** (1991) 1975–1978.

[55] S. R. Coleman, J. Preskill and F. Wilczek, "Quantum hair on black holes," *Nucl. Phys. B* **378** (1992) 175–246, arXiv:hep-th/9201059 [hep-th].

[56] C. G. Callan Jr. and F. Wilczek, "On geometric entropy," *Phys. Lett. B* **333** (1994) 55–61, arXiv:hep-th/9401072 [hep-th].

[57] M. K. Parikh and F. Wilczek, "Hawking radiation as tunneling," *Phys. Rev. Lett.* **85** (2000) 5042–5045, arXiv:hep-th/9907001 [hep-th].

[58] I. Bredberg, C. Keeler, V. Lysov and A. Strominger, "Wilsonian approach to fluid/gravity duality," *JHEP* **03** (2011) 141, arXiv:1006.1902 [hep-th].

[59] C. N. Yang and R. L. Mills, "Conservation of isotopic spin and isotopic gauge invariance," *Phys. Rev.* **96** (Oct. 1954) 191–195. https://link.aps.org/doi/10.1103/PhysRev.96.191.

[60] G. 't Hooft and M. Veltman, "Regularization and renormalization of gauge fields," *Nucl. Phys. B* **44** (1972) 189–213.

[61] M. B. Green and J. H. Schwarz, "Anomaly cancellation in supersymmetric D = 10 gauge theory and superstring theory," *Phys. Lett. B* **149** (1984) 117–122.

[62] J. M. Maldacena, "The large N limit of superconformal field theories and supergravity," *Int. J. Theor. Phys.* **38** (1999) 1113–1133, arXiv:hep-th/9711200.

37

Edward Witten

Professor at the School of Natural Sciences, The Institute for Advanced Study, Princeton, NJ

Date: 22 May 2014. Via email. Last edit: 22 May 2014

In your opinion, what are the main challenges/problems in theoretical physics at the moment?

Personally, I would like to understand what string theory really is. I think it is our main real idea to go beyond the established framework in physics and it is a real challenge that we don't understand it well.

Is it important to construct a theory of quantum gravity? Why do we need it and can it solve any of these main problems?

It might be that trying to make a theory of quantum gravity will help us understand the standard model better, or will give us a better understanding of cosmology or even of quantum mechanics. But at any rate quantum mechanics and gravity are there and we are bound to be curious about how they can work together.

Why is it hard to combine general relativity with quantum mechanics?

I would say it is hard because quantum field theory is a very tight and rigid framework. That is one of the reasons that physics made so much progress in the twentieth century – based on only fragmentary observations about (say) the weak interactions, because the framework of quantum field theory is so tight, a few clues could go a long way. Quantum field theory is so tight that one can say (with very high but not 100% confidence) that gravity doesn't fit in, and it is so tight that it is hard to change it in any way while maintaining its consistency. Indeed, string theory is the one real idea about how to do so.

Due to the lack of experimental data, there exist a plethora of different approaches to quantising gravity. Which of this approaches, in your opinion, is closer to a true description of nature and why?

I would say your premise is a little misleading. String theory is the only idea about quantum gravity with any substance. One sign is that where critics have had interesting ideas (non-commutative geometry, black hole entropy, twistor theory) they have tended to be absorbed as part of string theory. Another sign is the way string theory has been successful in generating new insights about standard quantum field theory and even about geometry.

Acknowledgements

This book required a tremendous effort to complete and it wouldn't have been possible without the support of many friends, family and colleagues over the years, whom I feel compelled to briefly acknowledge, even though there are not enough words to describe the value of their support.

In all honesty, I suspect this book would have been completed much later, if ever, if not for my friend Rasmus Grønfeldt Winther. I lost count of how many times I discussed this book with him. He gave me so many comments, suggestions, feedback and motivational speeches. He suggested that a third party would transcribe the interviews and he facilitated the contact with Cambridge University Press. He also introduced me to Lucas McGranahan, an editor who significantly improved parts of this book. Rasmus has been a constant in my life for many years and I just hope to be able to repay him one day, despite my innate Azorean pessimism.

Over the years, I received an enormous amount of help and comments from my friend Konstantinos Zoubos. He suggested multiple questions in several of the interviews and he revised some of the interviews, in particular those which the interviewees did not think upheld the required quality standards at first. I hold him dear and I also hope to help him one day as much as he has helped me in all sorts of facets of life.

From the very beginning, Niels Obers, my PhD supervisor who later became my friend, encouraged me to pursue this project. He gave me several suggestions of potential questions to ask and helped me making contact with some of the interviewees. I've also had, and continue to have, so many discussions about physics and academic life with him that I am certain that I inherited from him many of the ways I behave within the scientific community today. I'm grateful for this legacy, which I hope to be able to uphold and perhaps one day also transmit.

My multiple discussions about aspects of quantum gravity with my PhD co-supervisor and friend Troels Harmark, who was also encouraging from the start, have contributed for the betterment of this long lasting project. I used the many discussions about the interviews I had with him to elaborate further questions, which aimed at an improved understanding of several topics. I hope he will enjoy reading this book.

During the first few years of the process of completing this book, I was friends with Jesper Grimstrup and had many discussions with him about quantum gravity and his own approach to this problem, which is mentioned in some of the interviews. He suggested questions that I should ask certain interviewees and also pointed me towards relevant literature, in particular that in which string theory is criticised. I certainly benefited from this interaction.

I also acknowledge and am grateful for all the effort and time that all interviewees dedicated to this book. Overall, the interviewees were explicitly supportive and often offered useful comments and pointed me towards relevant literature. Of all the interviewees, I want to thank in particular Jan Ambjørn and Jan de Boer for multiple discussions about different approaches to quantum gravity and the sociology of the scientific community. I regard their opinions highly, though I do not always agree with them.

All my collaborators had a significant role to play in my current understanding of physics which is indirectly reflected in this book. I benefited greatly from my interactions with Niels Obers, Troels Harmark, Pavel Caputa, Joan Camps, Marta Orselli, Andreas Vigand Pedersen, Jakob Gath, Matthias Blau, Jyotirmoy Bhattacharya, Nilay Kundu, Marco Sanchioni, Vasilis Niarchos, Akash Jain, Javier Tarrío, Nam Nguyen, Thomas Van Riet, Enrico Parisini, Jelle Hartong, Emil Have, Bjarke Frost Nielsen, Yangyang Cai, Geoffrey Compére, David Garfinkle, Samuel Gralla, Jan de Boer, Richard Green and Luca Giomi.

In addition, I received help in various ways from other colleagues, either via suggestions, facilitating contact with some interviewees, comments or the reviewing of some interviews. In particular, Alejandra Castro, Adolfo Guarino, Riccardo Argurio, Andrés Collinucci, Gaston Giribet, Pierre Vanhove, Stefan Vandoren, Shiraz Butt, Yuki Sato, Daniel Elander, Lisa Glaser, Irene Amado, Jan Zaanen, Blaise Goutéraux, Lárus Thorlacius, Sameer Murthy, Simon Ross, Ricardo Monteiro, Óscar Dias, Jorge Santos, Vítor Cardoso, Harvey Reall, Shinji Hirano and João Gomes. I also acknowledge the important role taken by my colleagues at the University of Amsterdam, in particular Diego Hofman, Alejandra Castro, Erik Verlinde, Jan Pieter van der Schaar, Miranda Cheng, Jan de Boer, Marcel Vonk, Ben Freivogel and Daniel Baumann in making the string theory group lively and exciting, and how that has had a positive impact on my scientific development.

I wouldn't have been able to pursue this project if not for Mathias Blau, Geoffrey Compére, Jan de Boer and the board of the Dutch Institute for Emergent Phenomena (DIEP), who gave me a salary and were generally supportive of this endeavour. I also acknowledge the funding bodies that made it indirectly possible, namely, Fundação para a Ciência e a Tecnologia de Portugal; the Innovations- und Kooperationsprojekt C-13 of the Schweizerische Universitätskonferenz (SUK/CRUS); the ERC Starting Grant 335146 HoloBHC; the Netherlands Organisation for Scientific Research (NWO) through the NWA Startimpuls funding scheme and the DIEP cluster at the University of Amsterdam.

My travels over the years in order to conduct the interviews were possible due to the generosity of some of my friends and couchsurfers, in particular Ying Que, Luisa Hlawatsch, Rodrigo Carvalho, Brittany Huerta, Melissa Glidewell, Christina Woolner, Mohamed Bille, Chris Ferone, Brian Dickens, Scott Beibin and Elizabeth Jane Cole.

I value all scientists, artists, musicians and volunteers that participated at Science & Cocktails over the years. Interactions with all of them have brought a lot of joy to my life and shaped the way I view science and research as a whole.

Several of my friends have given me a lot of support and helped me going through difficult periods during this long period of time. Pedro Lucas has been a steady friend in my life since I was 14 and has encouraged me in many ways, including in pursuing this book. I have been very lucky to have him around for so long.

Christina Okai, Carmen Espinosa, Lars Erik Schmidt, Christo Buizert, My Larsdotter have over the years made my life joyful and some of them have seriously helped me sometimes, which I am eternally grateful for.

Many friends have inspired me, supported me or been present during these years, even if some of them just during limited periods of time: Bruno Gomes, Soraia Rodrigues, Johan Paulsson, Luís Bicudo, Bruno Lacerda, Thomas Greg Corcoran, Maria Shatokhina, Aurora Ribeiro, Rimantas Vančys, Sara Soares, Maaike Happel, Stéphane Detournay, Wout Merbis, Shyam Gopalakrishnan, Michael Rexen, Henrik Schütze, Thomas Bolander, Olga Ordeig, Pedro Solá, Rui Barreto, César Alvernaz, Tiago Bom, Ana Bela Monteiro, Nikolaj Høi, Rui Costa, Rui Paperini, Pedro Almeida, Lene Harbo, Gitte Harbo, Benjamin Glorieux, Sofie Carstensen, João Almeida, Albino, João Porto, Saba Abdul Hussein, Seraina Klopfstein, Christof Schüepp, Theo Jenk, Tanina Jenk, Ben Gräve, Carla Minnema, Gunjan Lakhlani and Ken Mimasu.

For most of the duration of this book, I was lucky to have spent a lot of time in Fabrikken, Christiania, which has been a socially supportive environment. I built a home there and gained an extended family that has very often helped me unconditionally. I am really grateful for all the time I've spent so far with Thomas Jørgensen, Peter Plett, Nuka Forchhammer, Sara O'Connor Aastrup, Rob Collings, Joana Lima Ramos, Maxime Danos Albertsen, Linnea Kemppainen, Hjalte Bested Møller, Sofie Hjorth, Remi, Pipaluk Supernova, Ulla Morgenthaler, Ella Forchhammer, and many others that stopped by even if just for a short period of time. Thomas, who I view as an older brother, has been a pilar in my life from whom I have learnt a lot from and who has motivated me to pursue many projects, including this one. Rob and Joana have saved my life once, which is hard to ever repay. My heart still aches, and probably always will, for having moved away from them recently.

Being away from my family most of the time makes it hard but I was very lucky to continuously receive warm and kind support from my family-in-law who were quick to accept me. I am quite grateful for all the help of, and time I spent with, Hanne, Marianne, Thomas, Erik, Christa, Ida, Clara, Dagmar and Lars Schmidt as well as Bodil Thomsen, Jannet Jacobsen, Torsten and Philip Hesselbjerg, Michael and Birgitte Hansen.

The strongest pillar throughout my life and during this period of time has been my family, who have shaped and supported me unconditionally. I cannot express all that I have received from Conceição Nascimento, Tomás Azevedo, Maria Isabel Armas, Zoraida, João, Fernando, João Fernando and Aquilina, Fernando e Lila Nascimento, Luís and Maria Manuela, Alda Maria and Renato, and Carlos Manuel Saldanha, without whom all would have been lost, as well as from José, Jácome, José and Isabel, Margarida,

Maria and Francisco Armas, Isabel Armas and Victor Medina, Simão, Frederico, Teresa, David, Beatriz, Guilherme, Maria and Manuela Armas. In addition, I thank Ana, Catarina, Rúben, Ana, Marta and Rita Azevedo, Juliana, Paulo, Simone and Miguel Nóbrega. I also acknowledge my Decq Mota, Salema, Saldanha, Stattmiller, Bettencourt, Menezes, Morais, Azevedo, Campos, Greaves and Peixoto "cousins" for the role they have played in my life. I specially thank my unofficial godmother Rosa Greaves for the support over the years.

Most importantly, I would have not managed to pursue and complete this project, and if I had it would have had no meaning, if not for Signe Hansen, whom I was lucky to meet and cannot live without, and our two beautiful beings Bertrand and Marguerite Armas.

Index